7th International Conference on Theoretical and Applied Physics (ICTAP 2017)

Journal of Physics: Conference Series Volume 1011

Yogyakarta, Indonesia
6 – 8 September 2017

ISBN: 978-1-5108-6288-3
ISSN: 1742-6588

Printed from e-media with permission by:

Curran Associates, Inc.
57 Morehouse Lane
Red Hook, NY 12571

CURRAN ASSOCIATES INC. proceedings.com

Some format issues inherent in the e-media version may also appear in this print version.

Printed by Curran Associates, Inc. (2018)

For permission requests, please contact the Institute of Physics
at the address below.

Institute of Physics
Dirac House, Temple Back
Bristol BS1 6BE UK

Phone: 44 1 17 929 7481
Fax: 44 1 17 920 0979

techtracking@iop.org

Additional copies of this publication are available from:

Curran Associates, Inc.
57 Morehouse Lane
Red Hook, NY 12571 USA
Phone: 845-758-0400
Fax: 845-758-2633
Email: curran@proceedings.com
Web: www.proceedings.com

Table of contents

Volume 1011

The International Conference on Theoretical and Applied Physics

6–8 September 2017, Yogyakarta, Indonesia

Accepted papers received: 16 April 2018
Published online: 9 May 2018

Preface

Papers

Advanced Materials

Biophysics and Medical Physics

Instrumental Physics

Interdisciplinary Physics

Laser and Optoelectronics

Nanoscience and Nanotechnology

Theoretical and Computational Physics

IOP Conf. Series: Journal of Physics: Conf. Series **1011** (2018) 011001 doi:10.1088/1742-6596/1011/1/011001

Preface

The 2017 International Conference on Theoretical and Applied Physics (ICTAP 2017) held on September, 6-8, 2017 is an event where prominent researchers and scientists from around the world are joining together to share their latest research results and exchange their ideas. In addition, direct contacts among the researchers and scientists will therefore promote not only national but also international research networking as well as collaboration in the future.

The seminar, which is being organized by Universitas Gadjah Mada and Physical Society of Indonesia, has accepted 219 abstract, around 100 papers are going to be presented in this seminar. Finally, under a tight peer-review process by at least two expert referees for each paper, selected papers from ICTAP 2017 will be published in Journal of Physics: Conference Series (JPCS) published by IOP - Institute of Physics United Kingdom, which is a periodical indexed by Scopus.

Finally, we are grateful to Faculty of Mathematics and Natural Sciences Universitas Gadjah Mada, Indonesian Physical Society (HFI) and all sponsorships for providing technical supports to this seminar.

Last but not least, we also wish to thank reviewers for invaluable comments and suggestions and the organizing committee for arranging all needed facilities to the success of this seminar.

Chairman of ICTAP 2017,

Dr. Ahmad Kusumaatmaja

IOP Conf. Series: Journal of Physics: Conf. Series **1011** (2018) 011002 doi:10.1088/1742-6596/1011/1/011002

List of Committee

International Advisory Board :

Prof. Dr. Kusminarto (UGM, Indonesia)

Prof. Dr. Jakrapong Kaewkao (NRPU, Thailand)

Prof. Dr. Jasper Knoester (University of Groningen, the Netherlands)

Prof. Dr. Hideyo Kunieda (Nagoya University, Japan)

Dr. François Beauducel (Insitut de Physique du Globe de Paris, France)

Prof. Dr.Ing. Mitra Djamal (ITB, Indonesia)

Dr. Bambang Widyatmoko (LIPI, Indonesia)

Dr. Pramudita Anggraita (BATAN, Indonesia)

Prof. Dr. Suparmi (UNS, Indonesia)

Prof. Dr. Khairurrijal (ITB, Indonesia)

Chairman :

Dr. Ahmad Kusumaatmaja

Co-Chairman :

Dr. Restu Widiatmono

Scientific Committee:

Dr. Iman Santosa

Dr. Kuwat Triyana

Dr. Fahrudin Nugroho

Dr. Moh. Adhib Ulil Abshor

Dr. Sholihun

Dr.rer.nat. Sintia Windhi Niasari

Dr. Ade Anggraini

Dr. rer.nat. Wiwit Suryanto

Dr. Edi Suharyadi

Treasurer :

Dr. Chotimah (UGM, INA)

Dr. Juliasih Partini (UGM, INA)

Support:

Shidiq Nur Hidayat, S.Si, M.Sc

List of Plenary Speaker :

Prof. Dr. Muhammad Nurhuda (Universitas Brawijaya, Indonesia)

Prof. Dr. Jasper Knoester (University of Groningen, the Netherlands)

Prof. Dr. Hideyo Kunieda (Nagoya University, Japan)

Dr. François Beauducel (Insitut de Physique du Globe de Paris, France)

IOP Conf. Series: Journal of Physics: Conf. Series **1011** (2018) 011003 doi:10.1088/1742-6596/1011/1/011003

Sponsor Acknowledgments

The International Conference on Theoretical and Applied Physics 2017 has been organized by Physical Society of Indonesia (HFI) with the support from Universitas Gadjah Mada, KGC Scientific, Gestrindo, PT. Era Mitra Perdana, Thermoscientific, and GeneCraft Labs.

Opening ceremony by Chairman of Physical Society of Indonesia, Prof. Dr. Mitra Jamal

IOP Conf. Series: Journal of Physics: Conf. Series **1011** (2018) 011004 doi:10.1088/1742-6596/1011/1/011004

Plenary Session by Dr. François Beauducel

IOP Conf. Series: Journal of Physics: Conf. Series **1011** (2018) 011004 doi:10.1088/1742-6596/1011/1/011004

Parallel session

Group Photo

IOP Conf. Series: Journal of Physics: Conf. Series **1011** (2018) 011004 doi:10.1088/1742-6596/1011/1/011004

Registration Desk

IOP Conf. Series: Journal of Physics: Conf. Series **1011** (2018) 011005 doi:10.1088/1742-6596/1011/1/011005

Peer review statement

All papers published in this volume of *Journal of Physics: Conference Series* have been peer reviewed through processes administered by the proceedings Editors. Reviews were conducted by expert referees to the professional and scientific standards expected of a proceedings journal published by IOP Publishing.

Mineral identification of black-jade gemstone from Aceh Indonesia

Ismail, Akmal Nizar, Mursal

Department of Physics, Faculty of Mathematics and Natural Science, Syiah Kuala University, Banda Aceh 23111, Indonesia

Corresponding author's e-mail address: ismailab@unsyiah.ac.id

Abstract. One of the gemstones in Aceh Indonesia is called black-jade where the name of black-jade is a local name. Unfortunately, detail information about this gemstone is still limited. No one knows whether this gemstone can be categorized as jade or not until this study is presented. We have utilized X-Ray Fluorescent (XRF) to study the black-jade gemstone from Aceh Tengah (Takengon) and Nagan Raya regions in Indonesia. Our results show that the black-jade gemstone from Aceh Tengah contains 39.6% of SiO_2, 35% of Fe_2O_3, 17% of MgO, 3% of CaO, and 2% of NiO. While, the black-jade gemstone from Nagan Raya contains a little bit less SiO_2 but more Fe_2O_3 than that of black-jade from Aceh Tengah: 38.4% of SiO_2, 39% of Fe_2O_3, 17% of MgO, 0.5% of CaO, and 2.6% of NiO. By comparing the results to the available mineral data (jadeite, nephrite-actinolite, nephrite-tremolite, serpentine-clinochrysotile, serpentine-antigorite, and vesuvianite), we found that oxide compounds contained in the black-jade gemstone from Aceh are found in the serpentine-antigorite, except H_2O. The total difference between the oxide compositions in black-jade and serpentine-antigorite is 43% with its average difference of 11%. This means that the oxide composition in black-jade is almost the same as in the serpentine-antigorite. Accordingly, the black-jade gemstone from Aceh Indonesia is a type of serpentine-antigorite-jade.

1. Introduction

Jade is one of the popular gemstones in the worldwide. People use jade for medicinal purposes, jewelry, ornaments, and religious artifacts. In Chinese, jade symbolizes nobility, perfection, constancy, and immortality. The gemstone is known for many fine nuances of green, but also in shades of white, black, gray, yellow, orange, and violet. Jade is a natural gemstone which is rare and highly valued [1]. Jade is found in many places in the world, such as in Burma, Myanmar, United States, Russia, British, Canada, Turkey, and Indonesia. In Indonesia, jade is found in several areas such as West Sumatra, North Sumatra, and Aceh.

Some studies on jade have been done previously. Jade is considered as the minerals composed of either nephrite or jadeite [2, 4]. However, their chemical compositions are different [4]. Nephrite is categorized into group tremolite or antigorite composed by Ca_2 $(Mg,Fe)_5$ Si_8O_{22} $(OH)_2$ Si_2O_6. While jadeite is group pyroxene composed of Na(Al,Fe) Si_2O_6. Serpentine and vesuvianite can also be categorized as jade which is called as serpentine-jade and vesuvianite-jade. A Recent study shows that Turkish purple jade contains 63.54% of SiO_2, 20.17% of Al_2O_3, 6.71% of Na_2O, 1.78% of Fe_2O_3, and 2.7% of CaO [5]. This Turkish purple jade can be categorized as jadeite-jade.

Recently, gemstones from Aceh Indonesia are quite popular. These gemstones were ranked into the first rank in the Indonesian Gemstone Stone Contest in Jakarta in 2014. Nonetheless, the research on

IOP Conf. Series: Journal of Physics: Conf. Series **1011** (2018) 012001 doi:10.1088/1742-6596/1011/1/012001

precious stones in Aceh is still limited. Nurul Aflah *et al.* researched jade from the mountain Beutong Aceh. They found that the various types of jade from the area mentioned above are the result of the existence of a steady path in the area [6]. Akmal *et al.* have identified the mineral content in the bio-solar gemstone from Aceh. They found that bio-solar gemstone contains minerals of CaO (59.8%), SiO_2 (19.7%), Fe_2O_3 (11.1%), Al_2O_3 (7.5%), and NiO (1.3%) [7]. By comparing their results with the international mineral database, they found that the bio-solar gemstone from Aceh is a type of vesuvianite [7]. Another Aceh's famous gemstone is called "black-jade." The name of black-jade is a local name. However, what kind of minerals contained in this precious stone is still unknown. It is unknown whether this gemstone can be categorized as a type of jade or not.

We have identified the mineral contents in black-jade gemstone from Aceh Indonesia by using X-Ray Fluorescent (XRF). The results are compared with the previously published mineral data and reported in this paper.

2. Experimental Details

The samples of black-jade gemstone used in this study were obtained from Nagan Raya and Aceh Tengah districts in Aceh, Indonesia. The color of black-jade gemstone from Aceh is dark green, and it is beautiful as shown in figure 1.

Figure 1. Black-Jade Gemstone from Aceh.

The sample of black-jade gemstone was crushed into powder. Identification of mineral content from black-jade gemstone sample from Aceh was conducted by using X-Ray Fluorescent (XRF), Brand of PANalytical MiniPal Type 4. The experiment was performed in the Laboratorium Sentral Mineral FMIPA Universitas Negeri Malang, Indonesia. Data were collected at room temperature.

3. Results and Discussions

By using XRF, we found that the black-jade gemstone from Aceh Tengah Indonesia contains 39.6% of SiO_2, 35% of Fe_2O_3, 17% of MgO, 3% of CaO, and 2% of NiO. While, the black-jade gemstone from Nagan Raya contains a little bit less SiO_2 but more Fe_2O_3 than that of black-jade from Aceh Tengah: 38.4% of SiO_2, 39% of Fe_2O_3, 17% of MgO, 0.5% of CaO, and 2.6% of NiO. In general, the compounds contained in the black-jade from Aceh Tengah are the same as that from Nagan Raya. Detail compounds or minerals contained in black-jade gemstone are listed in table 1.

The next step is to compare the composition of the black-jade gemstone above with some previously published jade data as shown in table 2. In this case, we use the data of the black-jade gemstone from Aceh Tengah (Takengon). The comparison is made only for the composition greater than or equal to 5%, while data content of compound below 5% is ignored. The oxide compounds in black-jade from Aceh (result from this work) are compared with jadeite, nephrite-actinolite, nephrite-tremolite, serpentine-clinochrysotile, serpentine-antigorite, and vesuvianite.

IOP Conf. Series: Journal of Physics: Conf. Series **1011** (2018) 012001 doi:10.1088/1742-6596/1011/1/012001

Table 1. The content of oxide compounds in black-jade gemstone by using XRF.

Type of oxides	Composition (%) Black-jade Nagan Raya	Composition (%) Black-jade Takengon
Fe_2O_3	38.9	35.1
SiO_2	38.4	39.6
MgO	17.0	17.0
NiO	2.62	2.13
Cr_2O_3	1.35	1.37
P_2O_5	0.66	-
MnO	0.44	0.79
CaO	0.42	3.28
Re_2O_7	0.10	-
ZnO	0.04	0.79
La_2O_3	0.07	0.05

Table 2. Comparison of the composition of oxide compounds contained in black-jade with the oxide content in other gemstones

Type of oxides	Black-jade (%)	Jadeite (%)	Nephrite Actinolite (%)	Nephrite Tremolite (%)	Serpentine Clinochrysotile (%)	Serpentine Antigoride (%)	Vesuvianite (%)
SiO_2	39.60	58.61	54.86	59.20	43.36	39.90	38.03
Al_2O_3	0	22.38	0	0	0	0	14.34
Na_2O	0	15.11	0	0	0	0	0
Fe_2O_3	35.10	3.89	0.47	0	0	17.92	0
MgO	17.00	0	16.11	24.80	43.63	30.15	5.67
CaO	0	0	12.03	13.80	0	0	39.43
FeO	0	0	10.61	0	0	0	0
H_2O	0	0	0	0	13.00	11.98	0

First, let's compare black-jade with jadeite. From the table 2, we find that the main oxide contained in jadeite is SiO_2 which is 58.61%. Black-jade also contains SiO_2, but it is only 39.60%. The difference between SiO_2 composition in black-jade and SiO_2 composition in jadeite is about 19%. Black-jade does not contain Al_2O_3 and Na_2O. However, jadeite includes 22.38% of Al_2O_3 and 15.11% of Na_2O. Black-jade contains 35.1% of Fe_2O_3 while it is only 3.89% in jadeite. Its difference is about 31%. Furthermore, black-jade includes 17% of MgO while jadeite does not contain MgO at all. The total difference between oxide compositions in black-jade and jadeite is 105%, and its average is 21%. The summary of comparison between the oxide composition in black-jade and jadeite is listed in table 3.

Table 3. The composition of oxide compounds in black-jade with jadeite.

Type of oxides	Jadeite (%)	Black-jade (%)	Difference (%)
SiO_2	58.61	39.60	19.01
Al_2O_3	22.38	0	22.38
Na_2O	15.11	0	15.11
Fe_2O_3	3.89	35.10	31.21
MgO	0	17.00	17.00
Total Difference			105
Average Difference			21

Similarly, we compare the oxide composition in black-jade with nephrite-actinolite. Table 2 shows that nephrite-actinolite contains 54.86% of SiO_2. But black-jade contains only 39.60% of SiO_2. Its difference is about 15%. The largest difference between oxide compositions in black-jade and nephrite-actinolite is Fe_2O_3 that is 34.6%. The total difference between oxide compositions in black-jade and nephrite-actinolite is 70%, and its average is 14%. Next, let's compare black-jade with nephrite-tremolite. As shown in table 2 nephrite-tremolite contains 59.20% of SiO_2. But black-jade contains only 39.60% of SiO_2. Its difference is about 19.6%. Nephrite-tremolite contains 24.80% of MgO and 13.80% of CaO which are similar to black-jade. The largest difference between oxide compositions in black-jade and nephrite-tremolite is Fe_2O_3 that is 35.1%. The total difference between oxide compositions in black-jade and nephrite-tremolite is 73%, and its average is 18%. Subsequently, let's compare black-jade with serpentine-clinochrysotile. From table 2, we can see that the composition of SiO_2 in black-jade is almost the same as in serpentine-clinochrysotile. However, it is quite different for the composition of Fe_2O_3 which is 0% in serpentine-clinochrysotile, but it is 35.15% in black-jade. On the other hand, serpentine-contains 13% of H_2O which is 0% in black-jade. Serpentine-clinochrysotile contains 43.63% of MgO which is much less in black-jade, it is only 17%. The total difference between oxide compositions in black-jade and serpentine-clinochrysotile is 79%, and its average is 20%. Next, let's compare black-jade with serpentine-antigoride. From the table 2, we find that the main oxide contained in jadeite is SiO_2 which is 39.90% which is about the same as in black-jade. Serpentine-antigoride contains 17.92% of Fe_2O_3 which is smaller than that in black-jade. Serpentine-antigoride contains 30.15% of MgO which is larger than that in black-jade. Serpentine-antigoride also contains 11.98% of H_2O, but it is 0% in black-jade. The total difference between oxide compositions in black-jade and serpentine-antigoride is 43%, and its average is 11%. The last one, let's compare black-jade with vesuvianite. From the table 2, we find that the main oxide contained in vesuvianite is CaO which is 39.43%. But CaO compound is about 0% in black-jade. Vesuvianite contains 38.03% of SiO_2 which is about the same as in black-jade. Vesuvianite also contains 14.34% of Al_2O_3. However, black-jade does not contain this compound. Vesuvianite contains 5.67% of MgO which is smaller than that in black-jade. The total difference between oxide compositions in black-jade and vesuvianite is 98%, and its average is 20%. The summary of comparison between black-jade gemstone and other six kinds of gemstones is listed in table 4.

Table 4. Total and average differences oxide compositions in black-jade and other gemstones.

Kind of Gemstone	Total Difference (%)	Average Difference (%)
Jadeite	105	21
Nephrite-Actinolite	70	14
Nephrite-Tremolite	73	18
Serpentine-Clinochrysotile	79	20
Serpentine-Antigoride	43	11
Vesuvianite	98	20

When we compare black-jade to jadeite, we found that its total difference is 105% and its average difference is 21%. These mean that black-jade is not jadeite because the composition of oxides in black-jade is so different than that in jadeite. Table 4 shows that the smallest total difference of oxide compositions in black-jade with other six gemstones is serpentine-antigorite that is 43% with its average difference of 11%. These mean that the oxide compositions in black-jade are almost the same as in serpentine-antigorite. As shown in table 2, all minerals contained in the gemstone black-jade are found in the serpentine-antigorite, except H_2O. Therefore, the black-jade is a type of serpentine-antigorite.

4. Conclusions

The gemstone black-jade from Aceh Tengah contains 39.6% of SiO_2, 35% of Fe_2O_3, 17% of MgO, 3% of CaO, and 2% of NiO. The oxide compounds contained in black-jade gemstone from Aceh Tengah and Nagan Raya Aceh are about the same. It is found that all minerals contained in the gemstone black-jade from Aceh are found in the serpentine-antigorite, except H_2O. The total difference between oxide compositions in black-jade and serpentine-antigorite is 43% with its average difference of 11%. Thus, we conclude that the black-jade gemstone from Aceh Indonesia is a type of serpentine-antigorite-jade. This finding is essential for gemstone society in Indonesia. All gemstones from Indonesia should be studied to obtain their specifications. Thus, it is interesting to identify minerals contained in other gemstones from Indonesia which are not identified yet.

References

[1] Harlow GE and Sorensen SS 2005 Jade (nephrite and jadeite) and serpentine: metasomatic connections *International Geological Review* **47** pp 113–146
[2] Prewitt CT and Burnham CW 1966 The crystal structure of jadeite, $NaAlSi_2O_6$ *American Mineralogist* **51** pp 956–975
[3] Htein W and Naing AM 1994 Mineral and chemical compositions of jadeite jade of Myanmar *Journal of Gemology* **24** pp 269–276
[4] Harder H 1995 Trace elements as colouring agents in jadeites *The Journal of Gemology* **24** pp 508–511
[5] Murat H, Basevirgen Y, and Chamberlain SC 2012 Gem Turkish purple jade: Geological and mineralogical characteristics *Journal of African Earth Science* Vol **63** pp 48 – 61
[6] Aflah N, Muchlis, Halimi K, Nufus H, Maysura Z, Simatupang MZ 2014 Classification of Jades (Giok) Beutong Aceh based on Mineral Composition. *Journal Natural* Vol.**14** No.2 pp 19-22
[7] Akmal N, and Mursal I 2016 Minerals Identification of Bio-Solar Gemstone from Aceh *Journal of Aceh Physics Society* Vol. **5** No. 2 pp 22-26

IOP Conf. Series: Journal of Physics: Conf. Series **1011** (2018) 012002 doi:10.1088/1742-6596/1011/1/012002

The structural and electrical properties of polycrystalline $La_{0.8}Ca_{0.17}Ag_{0.03}MnO_3$ manganites

F Ruli[1], B Kurniawan[1*] and A Imaduddin[2]

[1]Department of Physics, FMIPA Universitas Indonesia, Kampus UI Depok, Depok 16424, Indonesia
[2]Metallurgical Research Center, Indonesian Institute of Sciences, Puspiptek-Serpong 15314, Indonesia

*Corresponding author: bkuru@fisika.ui.ac.id

Abstract. In this paper, the authors report the electrical properties of polycrystalline $La_{0.8}Ca_{0.17}Ag_{0.03}MnO_3$ manganites synthesized using sol-gel method. The X-ray diffraction (XRD) patterns of polycrystalline $La_{0.8}Ca_{0.17}Ag_{0.03}MnO_3$ samples reveal an orthorhombic perovskite structure with *Pnma* space group. Analysis using energy dispersive X-ray (EDX) confirms that the sample contains all expected chemical elements without any additional impurity. The measurement of resistivity versus temperature using cryogenic magnetometer was performed to investigate the electrical properties. The results show that the electrical resistivity of polycrystalline $La_{0.8}Ca_{0.17}Ag_{0.03}MnO_3$ exhibits metalic behavior below 244 K. The temperature dependence of electrical resistivity dominantly emanates from electron-electron scattering and the grain/domain boundary play a important role in conduction mechanism in polycrystalline $La_{0.8}Ca_{0.17}Ag_{0.03}MnO_3$.

Introduction

Perovskite manganites with general formula $La_{1-x}A_xMnO_3$ (A = divalent ions) have been a popular research subject due to its high magnetoresistance (MR) and magnetocaloric effect [1–5]. A theoretical explanation of relation among the level of A doping, electrical conductivity and magnetic properties was initially explained by Zener double exchange interaction [6]. It is a transfer of e_g electron from Mn^{3+} ions to Mn^{4+} ions through O^{2-} ions [6,7]. According to this model, the electrical conductivity is proportional to the level of A doping [6].

A perovskite material $La_{0.8}Ca_{0.2}MnO_3$ has a high MR and magnetocaloric effect but its Curie temperature (T_c) and metal-insulator transition temperatur (T_{M-I}) below room temperature [8–10]. Monovalen ions oxidizes Mn^{3+} ions to Mn^{4+} ions twice as much as divalent ions [11]. For low doping monovalent Ag, Pi et al reported that doping Ag in La site of lanthanum manganite induced an increase of MR, Curie temperature and metal-insulator transition temperature [12]. In this present work, the authors partially subtitute divalent ion Ca in $La_{0.8}Ca_{0.2}MnO_3$ with monovalent ion Ag to form a perovskite manganite $La_{0.8}Ca_{0.17}Ag_{0.03}MnO_3$. We study the structural, and electrical properties of that perovskite mananites.

Experimental

The perovskite manganite material $La_{0.8}Ca_{0.17}Ag_{0.03}MnO_3$ was synthesized by sol-gel method. Stoichiometric amounts of La_2O_3, $Ca(NO_3)_2.4H_2O$, $AgNO_3$, $Mn(NO_3)_2.4H_2O$ were used as precursor materials. La_2O_3 was dissolved in nitric acid solution. Then, lanthanum nitrate solution was mixed

IOP Conf. Series: Journal of Physics: Conf. Series **1011** (2018) 012002 doi:10.1088/1742-6596/1011/1/012002

with other metal nitrate solution. Citric acid was added to the solution with molar ratio of 1.2 between citric acid and metal nitrate. the pH value of solution was adjusted to 7 by adding amonia solution while stirring the solution at temperature of 80 °C until the gel was formed. The gel was heated at 120 °C for dehidration. It was burned at 600 °C for 5 h to remove produced organic materials. The materials obtained from this process were ground by using an agate mortar. The powders were pressed into pellets and then heated at 900 °C for 24 h.

Identification of the phase and structural analysis of the sample was determined by X-ray diffractometer (XRD) using monochromatic CuKα radiation. The composition of elements was analyzed using energy dispersive X-ray spectroscopy (EDX). The electrical properties of samples was determined by cryogenic magnetometer at temperature 12-255 K.

Results and discussion

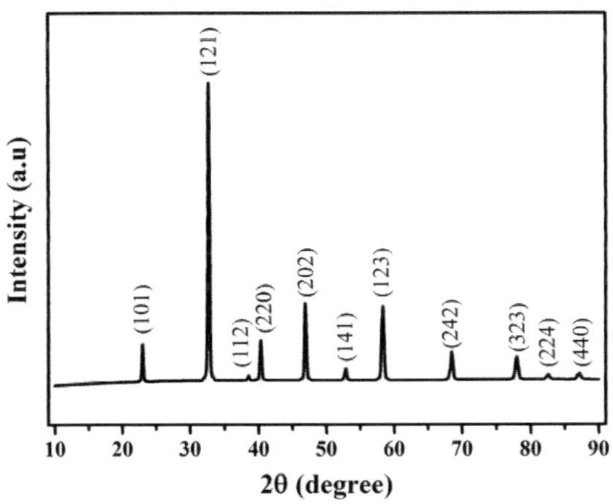

Figure 1. XRD patterns of $La_{0.8}Ca_{0.17}Ag_{0.03}MnO_3$

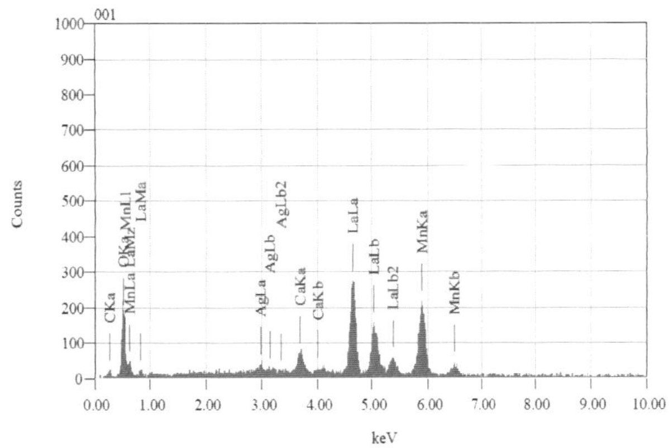

Figure 2. The EDX results for $La_{0.8}Ca_{0.17}Ag_{0.03}MnO_3$

The XRD pattern of polycrystalline $La_{0.8}Ca_{0.17}Ag_{0.03}MnO_3$ is shown figure 1. The analysis results using Rietveld refinement method confirm that the sample is a single phase since no peaks belongin to the other phase detected. The diffraction peaks can be indexed in an orthorombic perovskite structure with space group *Pnma*. The refinement results are shown in table 1. The quality of the refinement of the XRD data is evaluated by the goodness of the fit indicator χ^2. The Energy Dispersive X-ray (EDX) analysis shown in figure 2 reveals the presence of all expected elements. It shows that there is no quantifiable loss of each elements.

Table 1. Lattice parameters, unit cell volume and googness of fit (χ^2)

Parameters	$La_{0.8}Ca_{0.1}Ag_{0.1}MnO_3$
a (Å)	5.472
b (Å)	7.739
c (Å)	5.499
Volume (Å^3)	232.89
Goodness of fit	1.055

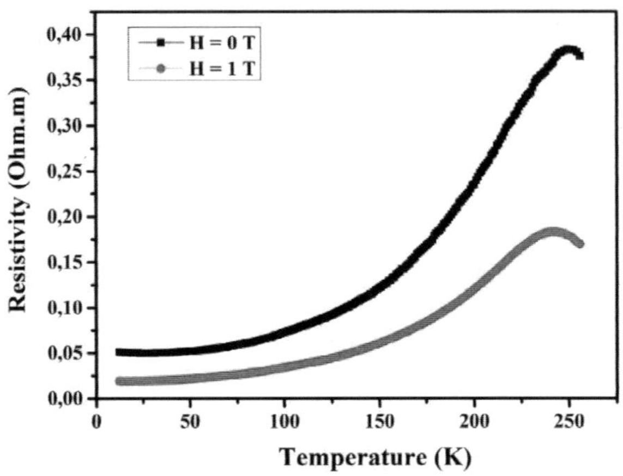

Figure 3. The temperature dependence of electrical resistivity of polycrystalline $La_{0.8}Ca_{0.17}Ag_{0.03}MnO_3$

The temperature dependence of electrical resistivity of polycrystalline $La_{0.8}Ca_{0.17}Ag_{0.03}MnO_3$ at zero applied magnetic field and 1 T is shown in figure 3. The electrical resistivity decreases when magnetic field 1 T is applied. The polycrystalline $La_{0.8}Ca_{0.17}Ag_{0.03}MnO_3$ shows a metal-insulator transition temperature ($T_{M\text{-}I}$) at 244 K. Below 244 K, The electrical resistivity of policrystalline $La_{0.8}Ca_{0.17}Ag_{0.03}MnO_3$ exhibits metalic behavior which electrical resistivity increases with increment of temperature. To analyze the conduction mechanism at metalic region, the experimental data of policrystalline $La_{0.8}Ca_{0.17}Ag_{0.03}MnO_3$ at zero applied magnetic field and 1 T are fitted using equation [3,13,14]:

$$\rho(T) = \rho_o + \rho_2 T^2 + \rho_{4.5} T^{4.5} \quad (1)$$

Where ρ_o, ρ_2 and $\rho_{4.5}$ are resistivities due to grain/domain boundary scattering, electron-electron scattering, and combinations of electron-electron, electron-magnon and electron-phonon scattering processes, respectively [3,13,14]. The fitting results are shown in figure 4 and the fitting parameters are presented in table 2. The quality of fitting is evaluated by the square of linear correlation coefficients R^2 [3]. The values of ρ_o, ρ_2 and $\rho_{4.5}$ decrease when a magnet field 1 T is applied. The value of ρ_2 is higher than that of $\rho_{4.5}$ which indicates that the temperature dependence of electrical resistivity dominantly emanates from electron-electron scattering. The value of ρ_0 is the highest indicates that grain/domain boundary play a important role in conduction mechanism in polycrystalline $La_{0.8}Ca_{0.17}Ag_{0.03}MnO_3$.

Table 2. Fitting paramater of electrical resistivity data of polycrystalline $La_{0.8}Ca_{0.17}Ag_{0.03}MnO_3$

	ρ_o (Ohm.m)	ρ_2 (10^{-3} Ohm.m/K^2)	$\rho_{4.5}$ (10^{-6} Ohm.m/$K^{9/2}$	Adj. R-Square
H=0	0.047	0.223	2.116	99,95%
H=1	0,019	0,054	1,252	99,99 %

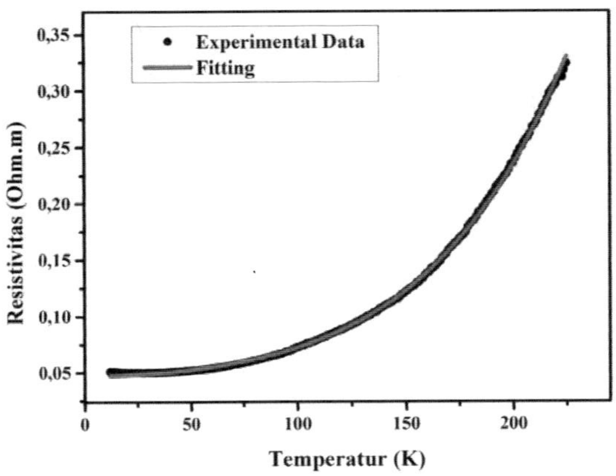

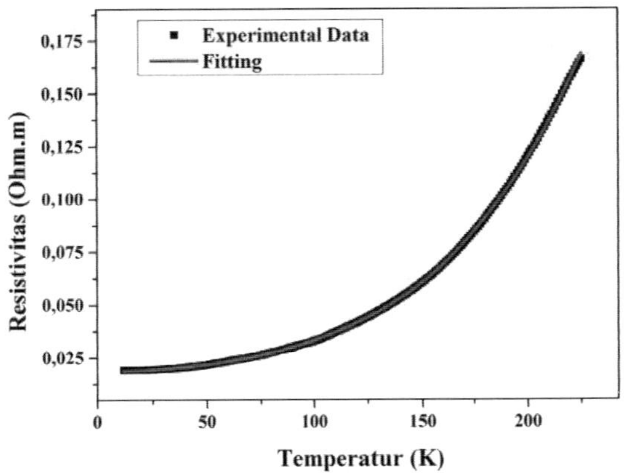

Figure 4. Theoretical fitting of resistivity data in metallic region at zero magnetic field (left) and applied magnetic field 1 T (right)

Magnetoresistance (MR) as a function of temperature measured at a temperature of 12-255 K is shown in figure 5. The magnetoresistance (MR) is calculated using equation [15]:

$$MR(\%) = \frac{\rho_0 - \rho_H}{\rho_0} \times 100\% \tag{2}$$

Where ρ_0 and ρ_H are the resistivities in the zero magnetic field and the presence of magnetic field, repectively. The polycrystalline $La_{0.8}Ca_{0.17}Ag_{0.03}MnO_3$ has the highest value of MR at lowest temperature and decrease with temperature increment. At temperature near $T_{I\text{-}M}$, the value of MR increase with increment of temperature. Magnetoresistance (MR) effect in polycrystaline lanthanum manganite emanate to two contribution. First, the intrinsic MR originates from suppression of spin fluctuation when the magnetic field is applied [16]. The intrinsic MR reaches maximum around the paramagnetic-ferromagnetic transition temperature [17]. The second contribution emanates from spin polarized tunneling process at grain boundary which is dominant at low temperature [16,17]. The spin at the grain boundary is highly disorder [18]. That spin aligns to the applied magnetic field direction which reduces electron scattering by the grain boundary and increase electron tunneling [9,16]. The magnetoresistance of polycrystalline $La_{0.8}Ca_{0.17}Ag_{0.03}MnO_3$ reveals an extrinsic MR at lower temperature and an intrinsic MR around $T_{I\text{-}M}$.

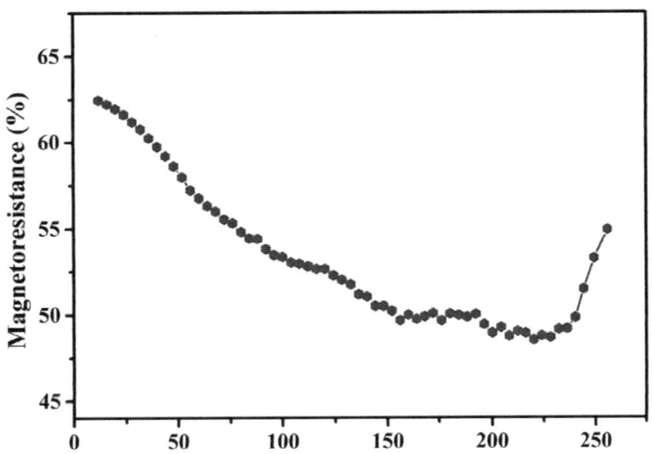

Figure 5. Magnetoresistance as a function of temperature

Summary
The structural and electrical properties of polyicrystalline $La_{0.8}Ca_{0.17}Ag_{0.03}MnO_3$ samples (x = 0, 0.03) prepared by sol-gel method are investigated. The X-ray diffraction (XRD) patterns of polyicrystalline $La_{0.8}Ca_{0.17}Ag_{0.03}MnO_3$ samples reveale an orthorhombic perovskite structure with *Pnma* space group. Analysis using energy dispersive X-ray (EDX) confirm that the sample contains all expected chemical elements without any additional impurity. Below 244 K, The electrical resistivity of polyicrystalline $La_{0.8}Ca_{0.17}Ag_{0.03}MnO_3$ exhibits metalic behavior. The temperature dependence of electrical resistivity dominantly emanates from electron-electron scattering and the grain/domain boundary plays a important role in conduction mechanism in polyicrystalline $La_{0.8}Ca_{0.17}Ag_{0.03}MnO_3$. The magnetoresistance of polyicrystalline $La_{0.8}Ca_{0.17}Ag_{0.03}MnO_3$ reveals an extrinsic MR at lower temperature and an intrinsic MR around T_{I-M}.

Acknowledgements
This work was supported in part by research grant of Ministry of Research, Technology and Higher Education (contract No. 2699/UN.2.R3.1/HKP.05.00/2017) and Indonesia Endowment Fund for Education (LPDP).

References
[1] M. Phan and S. Yu 2007 *IEEE Trans. Magn.* **308** 325
[2] S. Xi, W. Lu and Y. Sun 2012 *J. Appl. Phys.* **111** 63922.
[3] S. O. Manjunatha, A. Rao and G. S. Okram 2015 *J. Alloys Compd.* **640** 154 .
[4] J. C. Debnath and A. M. Strydom 2015 *J. Alloys Compd.* **621** 7.
[5] A. G. Gamzatov and A. B. Batdalov 2012 *Phys. Solid State* **54** 70 (2012).
[6] C. Zener 1951 *Phys. Rev.* **82** 440.
[7] P. W. Anderson and H. Hasegawa 1955 *Phys. Rev.* **100**.
[8] R. Skini, A. Omri, M. Khlifi, E. Dhahri and E. K. Hlil 2014 *J. Magn. Magn. Mater.* **364** 5.
[9] R. Skini, M. Khlifi, M. Wali, E. Dhahri, E. K. Hlil and P. Lachkar 2014 *J. Magn. Magn. Mater.* **363** 217.
[10] Z. B. Guo, Y. W. Du, J. S. Zhu, H. Huang, W. P. Ding and D. Feng 1997 *Phys. Rev. Lett.* **78** 1142.
[11] S. Ravi and M. Kar 2003 *Phys. B* **348** 169.
[12] L. Pi, M. Hervieu, A. Maignan, C. Martin and B. Raveau 2003 *Solid State Commun.* **126** 229.
[13] S. Manjunatha, A. Rao, V. P. S. Awana and G. S. Okram 2015 *J. Magn. Magn. Mater.* **394** 130.
[14] S. Hcini, S. Zemni, A. Triki, H. Rahmouni and M. Boudard 2011 *J. Alloys Compd.* **509** 1394.
[15] P. Siwach, U. K. Goutam, P. Srisvastava, H. K. Singh, R. S. Tiwari and O. N. Srivastava 2006 *J. Phys. D: Appl. Phys.* **39** 14.
[16] H. Y. Hwang, S. Cheong, N. P. Ong and B. Batlogg 1996 *Phys. Rev. Lett.* **77** 2041.
[17] S. Das and T. K. Dey 2005 *Solid State Commun.* **134** 837.
[18] S. P. Isaac, N. D. Mathur, J. E. Evetts and M. G. Blamire 1998 *Appl. Phys. Lett.* **72** 2038.

IOP Conf. Series: Journal of Physics: Conf. Series **1011** (2018) 012003 doi:10.1088/1742-6596/1011/1/012003

A comparative study between titania and zirconia as material for scattering layer in dye-sensitized solar cells

N M Nursam, J Hidayat, Shobih, E S Rosa L M Pranoto

Research Centre for Electronics and Telecommunication – Indonesian Institute of Science
Kampus LIPI Building 20 level 4 Jl. Sangkuriang, Cisitu, Bandung 40135, Indonesia

Email: natalita.maulani.nursam@lipi.go.id

Abstract. The photoanode of dye-sensitized solar cells (DSSC) is typically composed of nanocrystalline titania (TiO_2) layer that has been sensitized with light-absorbing dye molecules. Large portion of the light, however, could not be efficiently absorbed due to some physical reasons, such as TiO_2 crystal size (typically 10-25 nm) that makes the photoanode remains partially transparent to the visible region in the solar spectrum. One of the ways to improve the light harvesting efficiency in DSSC could be achieved by employing an additional scattering layer over the TiO_2 electron transport material. In this contribution, we evaluate the effect of light scattering properties on the performance of DSSC. Specifically, the light scattering properties provided from two different scattering materials, i.e. additional TiO_2 scattering layer and zirconia (ZrO_2) scattering layer, were compared. Both layers were deposited using screen printing technique under the same condition on top of 8 μm thick TiO_2 photoanode layer. All samples subsequently received the same thermal annealing treatment at 500 °C and sensitized with ruthenium-based synthetic dyes. Our results revealed that the thickness of the scattering layer for both TiO_2 and ZrO_2 had a significant effect on the solar cell performance. The best photoconversion efficiency was achieved by samples that were coated with one screen-printing cycle, giving an overall efficiency of 3.50 % and 4.02% for TiO_2 and ZrO_2, respectively.

Keywords: efficiency; hydrothermal; photoelectrode; solar cell; TiO_2

1. Introduction

Dye-sensitized solar cell (DSSC) based on porous nanocrystalline films offers low fabrication cost yet with high attainable efficiency. As one of the entities of third generation photovoltaic technology, DSSC was made to achieve high photoconversion efficiency beyond the Shockley-Quisser limit of 31% [1]. The performance of DSSC has been progressing considerably since its breakthrough in 1991 [2].

There have been many efforts reported to increase the performance of DSSC. The DSSC photoelectrode is typically formed by highly porous titanium dioxide with a particle size in the range of 10-50 nm [3, 4]. This specification provides a high surface area that allows an effective monolayer adsorption of the dye molecules at the surface of TiO_2. However, according to Rayleigh and Mie scattering theory [5], particle size above 200 nm is typically required to be able to scatter light in the visible light range between 500-700 nm. Consequently, the TiO_2 particles are transparent to most of the incident light during the solar cell operation. Hence, additional layer with large particle size could be employed to reflect the light back into the TiO_2 layer, thus increasing the amount of light captured by the cell and minimizing the loss of radiation. Optical enhancement of the porous film has also been shown to improve the performance of TiO_2 photocatalyst [6]. Light scattering effect could be achieved by employing different structures or materials, for example using large TiO_2 beads [7], rutile TiO_2 [8], TiO_2 with core-shell [9], or hollow sphere structure [10]. Among the approaches above, the use of large TiO_2 and ZrO_2 particles have been some of the most frequently implemented techniques. However, the comparison between the effectiveness of those two materials in improving the DSSC performance has not yet been explored. In this contribution, we report the enhancement in photoelectric conversion efficiency of TiO_2-based DSSC by applying two different materials the scattering layer, namely anatase TiO_2 and monoclinic ZrO_2. These two materials were particularly selected as they are abundant, versatile, stable, and can be readily synthesized using relatively simple techniques such as via wet chemistry method. The extent of their roles as scattering layer will be investigated thoroughly by analyzing the optical properties as well as measuring the electrical parameters of the cells.

2. Experimental Section

2.1. Cells preparation

DSSC photoanodes with an active area with a size of 1×1 cm^2 was prepared by depositing TiO_2 paste (Dyesol, 18NR-AO opaque paste) on top of fluorine-doped tin oxide substrate (Dyesol, 15 Ω/sq, 3.2 mm thick) using the screen-printing technique. The TiO_2 layer was printed in two cycles for each sample, giving a total thickness of approximately 8-10 μm. After the deposition, the TiO_2 films were subsequently annealed at 500 °C for 15 min (RTC L4-310 conveyer belt furnace) to strengthen the bonding between the TiO_2 particles and the substrate.

Before proceeding to the sensitization step, all samples received an additional coating of either TiO_2 (Dyesol, TiO_2-WER) or ZrO_2 (Solaronix) reflective layer. To analyze the effect of thickness, the deposition of each scattering layer was varied to 1, 2 and 4 cycles. The thickness of the scattering layer was approximately 2, 4 and 6 μm after 1, 2 and 4 cycle of printing, respectively. The samples were labeled as TiO_2-x or ZrO_2-y, wherein x and y represent the number of coating cycle for TiO_2 and ZrO_2, respectively. Meanwhile, a sample without any scattering layer was referred as a control. Another annealing at 500 °C for 15 min was subsequently conducted on all samples. The sensitization process was carried out by soaking all samples in ruthenium-based dye $[RuL_2(NCS)_2]$:2 tetra-nbutylammonium, where L=2,2'-bipyridil-4,4'-dicarboxylic acid (Dyesol, Z907) with a concentration of 20 mg/L in ethanol. Meanwhile, the counter electrodes were prepared by sputtering Pt target using DC sputtering (ARC-12 M) under argon gas with a pressure of 4mTorr for approximately 20 min. After both electrodes were attached together using thermoplastic sealant, an acetonitrile-based liquid electrolyte containing I^-/I_3^- redox mediators (Dyesol, EL-HPE) was then injected through the spacer using a syringe. The electrolyte filling holes were then sealed using the two-part hermetic sealing compound to protect the cells from the external environment.

2.2. Characterizations

The morphology of the final films was studied using field emission scanning electron microscopy (FE-SEM, JEOL JB-4610F) operated at 15 kV. All samples were sputter coated with carbon before the analysis. The crystal phase and structure of the films were characterized using X-ray diffraction (XRD) analysis (Shimadzu Maxima XRD-7000 X-ray diffractometer) with Cu Kα source at $\lambda=1.54056$ Å. After the construction of the cells, the electrical parameters of the finished samples were immediately measured using current-voltage (I-V) measurement system (National Instrument) under sun simulator with AM1.5 filter and an intensity of 500 W/m^2 (Oriel, 91192).

3. Results and Discussion

Figure 1 shows the surface morphology of the TiO_2 photoanodes without and with scattering layers. The control sample (Figure 1a) essentially represented the photoactive layer, showing mesoporous TiO_2 with average particle size of approximately 30-50 nm. It can be seen that the particle size of the scattering materials (Figure 1b and c) were in the order of hundreds of micrometers, which are much larger compared to the particle size of the TiO_2 photoactive layer. It is clear that both TiO_2 and ZrO_2 reflective layers existed as aggregates. However, it can be qualitatively seen that ZrO_2 formed larger agglomerations compared to TiO_2. The particle size of TiO_2 is estimated at ~200 nm, while the particle size of ZrO_2 was larger than 400 nm. In addition, it can be seen that the TiO_2 reflective layer was separated by large voids between the particles.

Figure 1. SEM images of TiO_2 photoanodes (a) without scattering layer, (b) with TiO_2 reflective layer, and (c) with ZrO_2 reflective layer.

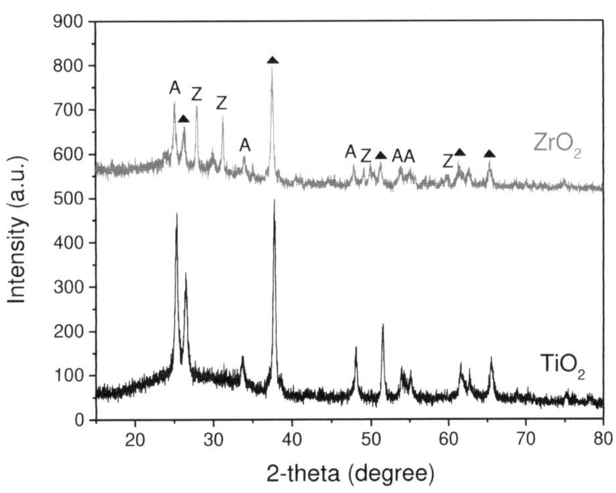

Figure 2. XRD patterns of photoanodes with TiO_2 and ZrO_2 scattering layer (both samples were deposited in 1 cycle). The black triangle (▲) symbols represent the crystal planes of the FTO glass, while the A and Z represent the crystal planes of anatase TiO_2 and monoclinic ZrO_2, respectively. Note that the patterns were shifted along the *y*-axis for clarity.

Figure 2 shows the XRD spectra obtained on the photoanodes with reflective layers. The wide-angle XRD was used to determine the crystal phase of the reflective layers. The XRD pattern shows that the TiO_2 reflective material consisted of anatase phase based on JCPDS 21-1272. This was confirmed by the presence of strong peaks located at 2θ~25.4° that can be assigned to (101) plane of anatase titania. Meanwhile, from JC-PDS 37-1484, the XRD pattern of the ZrO_2 reflective layer only showed monoclinic phase, as indicated by the strong peak at 2θ~28.1° and 31.3°. These results indicated that both TiO_2 and ZrO_2 only existed as single phase crystal.

The effect of scattering layer was subsequently characterized by the photovoltaic performance of the constructed cells. Figure 3 shows the current-voltage (*I-V*) curves for DSSC with and without scattering layer, while the electrical parameters of the cell are summarized in Table 1. For both TiO_2 and ZrO_2 scattering layer, 1 cycle scattering layer deposition gave the best photoconversion efficiency among all variations, showing almost ten folds enhancement in the efficiency as compared to that of the control sample. The improvement was mainly attributed by the significant rise in the photocurrent density, as indicated by the I_{SC} that increased from 0.817 mA in the control sample to 7.860 mA and 7.879 mA when TiO_2 and ZrO_2 scattering layer was respectively added. The increase in the I_{SC} suggested that the additional scattering layer may also serve as the anchoring spots for the dye molecules, thus not only giving enhancement from light scattering but also increasing light absorption through better dye loading. However, the I_{SC} was found to drop when the thickness of the scattering layer was increased further, suggesting a decrease in charge collection efficiency as previously reported by Son *et al.* [11].

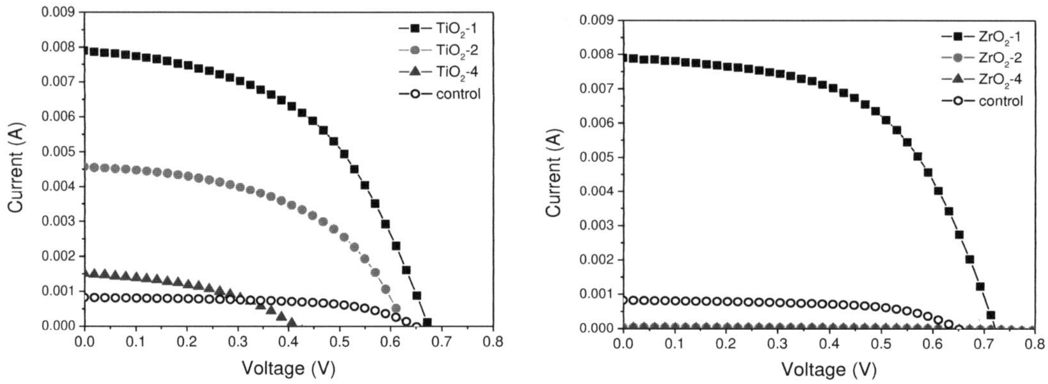

(a) (b)

Figure 3. *I-V* curves of DSSC made with (a) TiO_2 and (b) ZrO_2 scattering layer with various deposition cycles, compared to the cell without any scattering layers (control).

Table 1. Photovoltaic properties of the DSSC prepared without and with scattering layer with different scattering material and thickness.

	Control	TiO_2-1	TiO_2-2	TiO_2-4	ZrO_2-1	ZrO_2-2	ZrO_2-4
P_{max} **(mW)**	0.312	2.635	1.423	0.262	3.099	0.008	0.003
V_{max} **(V)**	0.488	0.447	0.427	0.264	0.488	0.366	0.305
I_{max} **(mA)**	0.640	5.887	3.332	0.995	6.348	0.022	0.010
I_{sc} **(mA)**	0.817	7.860	4.547	1.478	7.879	0.038	0.018
V_{oc} **(V)**	0.631	0.672	0.611	0.407	0.713	0.611	0.488
FF	0.606	0.499	0.512	0.436	0.551	0.349	0.353
Efficiency (%)	0.42	3.50	1.88	0.35	4.02	0.07	0.03

Table 1 shows that the best performing sample in this work was achieved by sample ZrO_2-1 with an efficiency of 4.02%. The reason for this could be related to the particle size of the reflective layer. As previously confirmed by the SEM analysis, ZrO_2 had a bigger particle size (i.e., >400 nm) than TiO_2, which consequently reflected light in the visible light wavelength (400 nm<λ<700 nm) more efficiently according to the Mie wavelength dependent scattering theory. Overall, the deterioration in the performance of the cell with the increase in the scattering layer thickness could be attributed to two plausible factors. First, as previously mentioned, the decrease in the efficiency was associated with the decrease in I_{SC}. This trend was especially profound for the ZrO_2 samples. It was suspected that the denser nature and larger aggregates in the ZrO_2 layer might have affected the infiltration of dye (during the impregnation process) and hindered the electrolyte contact with the photoactive layer. These possibilities were supported by the significant decrease in I_{SC} for ZrO_2-2 and ZrO_2-4. Secondly, the fill factor (*FF*) also decreased with the addition of scattering layer. Thicker films typically lead to higher series resistance in the cell [12]. Thus it was fairly expected that the increase in the scattering layer thickness would result in lower *FF*, although such decrease was not significantly observed in this work.

4. Conclusions

In this study, we have presented a comparison of the material characteristics and photovoltaic performance between TiO_2 and ZrO_2 as a scattering layer in DSSC. Both materials were shown to exist as large aggregates, which was an order of magnitude larger than the particle size of the mesoporous TiO_2 photoactive layer. Likewise, both materials showed similar characteristics in term of the crystal phase, i.e., both had only single crystal phase that is anatase TiO_2 and monoclinic ZrO_2. The addition of TiO_2 and ZrO_2 scattering layer significantly enhanced the cell performance to some extent. The highest improvement was obtained when ZrO_2 deposited through 1 printing process was applied. However, the increase in the scattering layer thickness has a negative influence on the cell performance due to incomplete dye and electrolyte penetration as well as the increase in the internal resistance. The results of this study could be used as standard guidelines for applying TiO_2 and ZrO_2 as a reflective layer in TiO_2-based DSSC.

5. Acknowledgement

The authors gratefully acknowledge research fundings from Kegiatan Unggulan LIPI and DIPA Pusat Penelitian Elektronika dan Telekomunikasi (P2ET) LIPI for supporting parts of this research. Research Center for Physics (P2F) and Research Center for Geotechnology of LIPI are thanked for providing SEM and XRD access, respectively. Travel funding was provided by the INSINAS Research Grant Program 2017 from the Ministry of Research, Technology and Higher Education of the Republic of Indonesia with a title "Design, Fabrication, and Up-scaling of Monolithic Dye-sensitized Solar Modules for Indoor Applications" (contract no. 04/INS-2/PPK/E/E4/2017).

6. References

[1] Hagfeldt A, Boschloo G, Sun L, Kloo L and Pettersson H 2010 *Chem. Rev.* **110** 6595.
[2] O'Regan B and Gratzel M 1991 *Nature* **353** 737.
[3] Kroon J M, Bakker N J, Smit H J P, Liska P, Thampi K R, Wang P, Zakeeruddin S M, Grätzel M, Hinsch A, Hore S, Würfel U, Sastrawan R, Durrant J R, Palomares E, Pettersson H, Gruszecki T, Walter J, Skupien K and Tulloch G E 2007 *Prog. Photovoltaics Res. Appl.* **15** 1.
[4] Hu B, Tang Q, He B, Lin L and Chen H 2014 *J. Power Sources* **267** 445.
[5] Deepak T G, Anjusree G S, Thomas S, Arun T A, Nair S V and Sreekumaran Nair A 2014 *RSC Adv.* **4** 17615.
[6] Nursam N M, Wang X and Caruso R A 2015 *J. Mater. Chem. A* **3** 24557.
[7] Huang F Z, Chen D H, Zhang X L, Caruso R A and Cheng Y B 2010 *Adv. Funct. Mater.* **20** 1301.
[8] Sun X, Zhou X, Xu Y, Sun P, Huang N and Sun Y 2015 *Appl. Surf. Sci.* **337** 188.
[9] Tétreault N, Arsenault É, Heiniger L-P, Soheilnia N, Brillet J, Moehl T, Zakeeruddin S, Ozin G A and Grätzel M 2011 *Nano Lett.* **11** 4579.
[10] Hore S, Nitz P, Vetter C, Prahl C, Niggemann M and Kern R 2005 *Chem. Commun.* 2011.

[11] Son M-K, Seo H, Kim S-K, Hong N-Y, Kim B-M, Park S, Prabakar K and Kim H-J 2012 *Int. J. Photoenergy* **2012** 1.

[12] Han L, Koide N, Chiba Y, Islam A, Komiya R, Fuke N, Fukui A and Yamanaka R 2005 *Appl. Phys. Lett.* **86** 213501.

IOP Conf. Series: Journal of Physics: Conf. Series **1011** (2018) 012004 doi:10.1088/1742-6596/1011/1/012004

Effect of Low Doped Calcium on the Magnetic Properties of $La_{0.7}Ba_{0.297}Ca_{0.003}MnO_3$ for Magnetocaloric Application

B Kurniawan[a], S Winarsih, and T Komala
Department of Physics, Faculty of Mathematics and Natural Sciences (FMIPA)
Universitas Indonesia, Depok 16424, Indonesia

E-mail: bkuru@fisika.ui.ac.id or bkuru07@gmail.com

Abstract. Effect of low doped calcium concentration to the magnetic properties of $La_{0.7}Ba_{0.3}MnO_3$ material has been reported in this paper. This compound was chosen because it has high potential to be applied in electronic devices. The most intensive research in the last decade is about magnetocaloric effect (MCE). Magnetic refrigeration will replace Freon R12 that is used in conventional refrigeration with magnetic material which has MCE properties. It is due to conventional refrigeration produce chlorofluorocarbons (CFC) and hydrochlorofluorocarbons (HCFCs) gases which causes ozone depletion. The magnetic measurement was performed by using Vibrating Sample Magnetometer in the range of temperature between 300-400 K under external magnetic field until 20 kOe. By using derivative of magnetization as a function of temperature of the material's result, Curie temperature (T_c) of the material is 332.7 K. By using equation (5) for calculating magnetic field dependence of magnetization at various temperature result, magnetic entropy change value (ΔS_M) of the material can be obtained. The ΔS_M of the material is -2.928 $Jkg^{-1}K^{-1}$. The material has highest magnetic entropy change value around its T_c. It is due to there is strong lattice spin coupling around its T_c and would be weak at below and above its T_c.

1. Introduction

Lanthanum manganites $La_{1-x}A_xMnO_3$ with A is divalent ions doped have been intensively explored in the last two decades. It is because they have multi physical properties such as magnetoresistance, magnetocaloric effect (MCE), and microwave absorbance so that these materials can be applied to various electronic devices [1–4]. One of those electronic devices is magnetic refrigeration which use MCE principal [5,6]. Increment of conventional refrigeration usage causes ozone depletion on our earth because this refrigeration produces chlorofluorocarbons (CFC) and hydrochlorofluorocarbons (HCFCs) gases [7]. Magnetic refrigeration is expected to decrease that effect. In magnetic refrigeration, magnetic material which has MCE properties will be used to replace Freon R12 that is used in conventional refrigeration [7]. MCE properties value and external magnetic field strength play an important role to the magnetic refrigeration system.

There are two conditions in MCE, adiabatic and isothermal. In adiabatic condition, when external magnetic field is applied to the system, its magnetic entropy will be reduced. In the same time, lattice entropy and electronic entropy of the system will be increased because of spin lattice connections and vibrations. It is happened to compensate the loss energy of the system. Thus, there is a decrement of system temperature which depends on the magnetic field strength [7]. The magnetic entropy will back to be normal by capturing energy from the lattice when external magnetic field is removed from system. The consequence system temperature returns to its initial temperature [7]. Meanwhile, in isothermal condition, the magnetic spins of the material tend to polarize in the same direction with an external magnetic field. It is caused magnetic entropy of the system reduce so that the system temperature also decrease [7]. When external magnetic field is removed from the material, magnetic spin back to randomly oriented. Hence the system temperature will be increase [7].

Technically, isothermal condition is easier to measure MCE properties of the material than adiabatic condition [5,6].The ΔS_M as a function of external magnetic field from 0 to H_0 is given by Eq. (1) and Eq. (2).

$$\Delta S_M(T,H_0) = S_M(T,H_0) - S_M(T,0) \tag{1}$$

$$\Delta S_M(T,H_0) = \int_0^{H_0} \left(\frac{\partial S(T,H)}{\partial H}\right)_T dH \tag{2}$$

Thermodynamics Maxwell relation is used to relate the magnetization data and magnetic entropy change as described in Eq. (3) [5,8]:

$$\left(\frac{\partial S}{\partial H}\right)_T = \left(\frac{\partial M}{\partial T}\right)_H \tag{3}$$

Then, we can calculate the isothermal magnetic entropy:

$$\Delta S_M(T,H) = S_M(T,H) - S_M(T,0) = \int_0^H \left(\frac{\partial M}{\partial T}\right)_H dH \tag{4}$$

At a discrete field and different temperature, the magnetic entropy change can be approximated as described in Eq. (5):

$$|\Delta S_M| = \sum \frac{(M_n - M_{n+1})_H}{T_{n+1} - T_n} \Delta H_0 \tag{5}$$

Where T is temperature, H_0 is variation of external magnetic field, S_M is magnetic entropy, M is magnetization, M_n and M_{n+1} are the magnetization values when measured at temperature T_n and T_{n+1} [5,8].

Previous research proved that $La_{0.7}Ba_{0.3}MnO_3$ has Curie temperature (T_c) of 336 K with magnetic entropy change (ΔS_M) of -1.67 $Jkg^{-1}K^{-1}$ while $La_{0.7}Ca_{0.3}MnO_3$ has T_c = 267 K with ΔS_M = -6.40 $Jkg^{-1}K^{-1}$ [5,6,8]. Curie temperature is transition temperature from metallic-ferromagnetic state to the insulator-paramagnetic state. At this temperature, the material has highest magnetic entropy change value [5,6,8]. Hence, doped calcium (Ca) to the $La_{0.7}Ba_{0.3}MnO_3$ is aimed to decrease its T_c near room temperature and increase its magnetic entropy change.

In this work, we reported magnetic properties of $La_{0.7}Ba_{0.297}Ca_{0.003}MnO_3$ polycrystalline material to observe effect of low doped calcium on the $La_{0.7}Ba_{0.3}MnO_3$. This work will be completed our study about low doping calcium concentration to the physical properties of $La_{0.7}Ba_{0.3}MnO_3$ where our previous publication explained about structure, resistivity, and magnetoresistance properties of this material [9]. Low doped concentration study is important to be observe in order to understand the characteristic change gradually due to calcium concentration. Sol-gel method was chosen to obtain the material because by using this method, homogenization of particle distribution of the material will be achieved [10]. So, it will influence its physical properties.

2. Experiment Method

The bulk sample material of $La_{0.7}Ba_{0.297}Ca_{0.003}MnO_3$ was obtained by using a sol-gel method. La_2O_3, $BaCO_3$, $CaCO_3$, and $Mn(NO_3)_2.4H_2O$ with high purity from Merck were used as starting precursors. Firstly, stoichiometric amounts of all precursors were weighed. Each of carbonate and oxide precursors were mixed with nitric acid until transparent solution were formed. Meanwhile, nitrate precursor was mixed with aquabidestilate solution. After that, all of those precursors were mixed become one solution. Secondly, citric acid was added to the solution with citric acid/total metal ions molar ratio is 1:1. Thirdly, ammonia solution was used for making its pH's solution become 7. Forthly, stirred that solution at 353-363 K until gel was formed and then heated at 393 K for 3-4 hours to evaporate the excess solvent. After dried gel was obtained by that process, it was heated at 873 K for 8 hours for decomposing the organic precursors. Finally, it was heated at 1123 K for 10 hours until powder was formed. The obtained powder was compacted and sintered at 1573 K for 30 hours to crystal growth process.

Vibrating Sample Magnetometer (VSM) was used to measure magnetic properties of the material in the range of temperature 300-400 K under external magnetic field until 20 kOe. By using derivative of magnetization as a function of temperature of the material's result, we can analyze the T_c (Curie temperature) of the material. After that, magnetic field dependence of magnetization at various temperature around its T_c was measured. These data were used for calculating magnetic entropy change (ΔS_M) as described in Equation 5 [5,6].

3. Results and Discussion

The bulk polycrystalline $La_{0.7}Ba_{0.297}Ca_{0.003}MnO_3$ was successfully obtained by sol-gel method. Our previous publication showed that the material is single phased with rhombohedral structure and R-3c space group, it also does not has additional elements or in the other words, compositional purity of the material already confirmed in here [11].

Temperature dependence of magnetization at 100 Oe under Zero Field Cooling (ZFC) is depicted in **Figure 1**. The T_c of the material is determined by minimum value of derivative of this graph (see inset Figure 1). From this measurement, it is obtained that T_c of the material is 332.7 K. This result proved that Ca substitution to the $La_{0.7}Ba_{0.3}MnO_3$ material has lowered its T_c although only in low doped concentration.

Figure 2 demonstrate the magnetization dependent magnetic field at various temperature of the material. There is a saturation condition by increasing magnetic field strength at T = 297 K which is this temperature is below Curie temperature. It is inferred that the material has ferromagnetic behaviour which has ordered spin at below Curie temperature. On the other hand, it has linear graph with raising external magnetic field strength at above Curie temperature. Hence, the material is in paramagnetic condition which has disordered spin. It is clearly seen that there is magnetic transition from ferromagnetic become paramagnetic with increasing temperature more than T_c. This kind of behaviour plays an important role to the MCE properties as shown in Eq. (5).

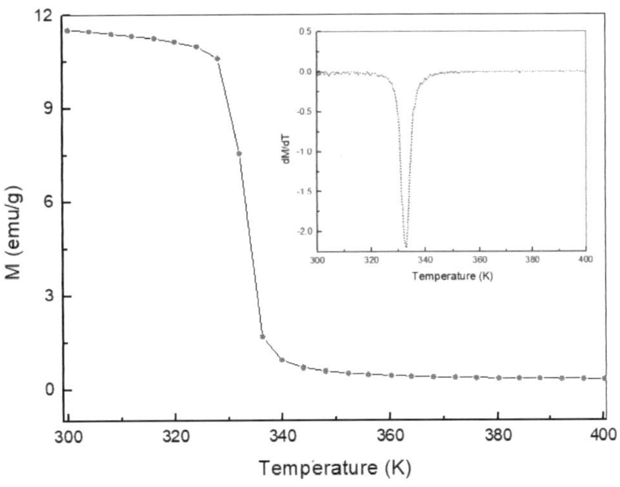

Figure 1. Temperature dependence of magnetization at 100 Oe of polycrystalline $La_{0.7}Ba_{0.297}Ca_{0.003}MnO_3$. In set figure is derivative of magnetization vs temperature data.

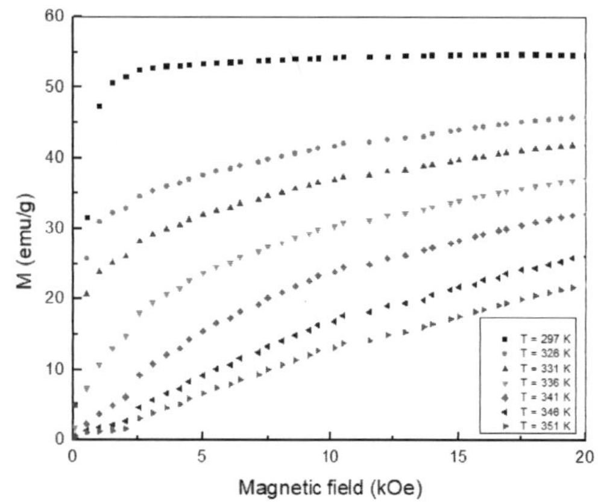

Figure 2. Magnetization dependent magnetic field at various temperature of polycrystalline $La_{0.7}Ba_{0.297}Ca_{0.003}MnO_3$.

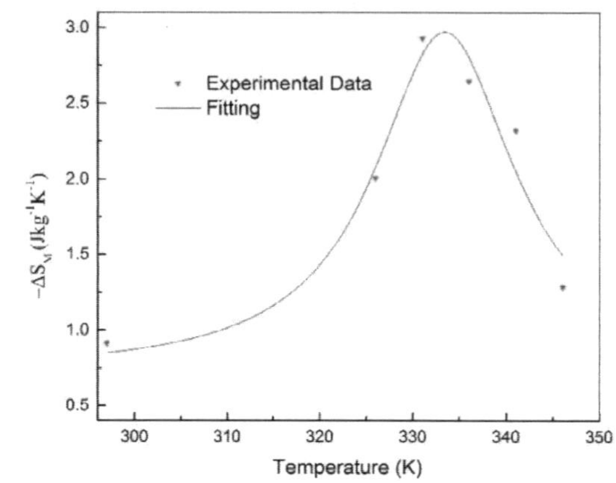

Figure 3. Magnetic entropy change (ΔS_M) as a function of temperature of polycrystalline $La_{0.7}Ba_{0.297}Ca_{0.003}MnO_3$.

The data from figure 2 was used to calculate the magnetic entropy change of the material using Equation (5). Magnetic entropy change as a function of temperature is shown in **Figure 3**. From this measurement, it is obtained that polycrystalline $La_{0.7}Ba_{0.297}Ca_{0.003}MnO_3$ material has ΔS_M = -2.928 $Jkg^{-1}K^{-1}$. This value is proved that magnetic entropy value of $La_{0.7}Ba_{0.3}MnO_3$ has been increased by calcium low doped in La-site. It might occur because double exchange (DE) and super exchange (SE) interactions is influenced by divalent ions (in this case the divalent ions are barium and calcium) substitution in La site [12,13]. DE interaction is simultaneous hopping electron from Mn^{3+} ion to Mn^{4+} ion via O^{2-} ion which leads metallic ferromagnetic state in lanthanum manganites materials [14]. On the other hand, SE interaction is coulomb interaction between Mn^{4+} ion and Mn^{4+} ion. SE interaction causes electron could not travel easily so that causes the material has insulator-paramagnetic behaviour [13,14]. These interactions can artificially control by tuning doping concentration and doping ion in La and Mn sites [12,13].

Around Curie temperature of $La_{0.7}Ba_{0.297}Ca_{0.003}MnO_3$, magnetic entropy change of the material reaches maximum value. It is due to there is an additional of magnetic entropy value of the material around its transition temperature [3,8]. Strong spin lattice coupling at that point is the main factor of this phenomenon and it would be weak at below and above Curie temperature of material [3,8].

4. Summary

The effect of low doped calcium on the magnetic properties of $La_{0.7}Ba_{0.297}Ca_{0.003}MnO_3$ for magnetocaloric application has been succesfully studied. The material has Curie temperature (T_c) = 332.7 K with magnetic entropy change (ΔS_M) = -2.928 $Jkg^{-1}K^{-1}$. This value is proved that magnetic entropy value of $La_{0.7}Ba_{0.3}MnO_3$ has been increased by calcium low doped in La-site. It might occur because double exchange (DE) and super exchange (SE) interactions is influenced by divalent ions (in this case the divalent ions are barium and calcium) substitution in La site. The material has highest magnetic entropy change value around its T_c due to spin lattice coupling at this point.

5. Acknowledgment

This work is financially supported by Universitas Indonesia under research grant “Hibah PITTA (Hibah Publikasi Internasional Terindeks untuk Tugas Akhir Mahasiswa Universitas Indonesia” with grant contract number is 634/UN2.R3.1/HKP.05.00/2017. The authors are also grateful to Department of Physics, Universitas Indonesia for research facilities.

6. References

[1] Manjunatha S O, Rao A, Lin T-Y, Chang C-M and Kuo Y-K 2015 *J. Alloys Compd.* **619** 303–10.
[2] Manjunatha S O, Rao A, Subhashini and Okram G S 2015 *J. Alloys Compd.* **640** 154–61.
[3] Ayadi F, Regaieg Y, Cheikhrouhou-koubaa W, Koubaa M and Cheikhrouhou A 2015 *J. Magn. Magn. Mater.* **381** 215–9.
[4] Islam M S, Hanh D T, Khan F A, Hakim M A, Minh D L, Hoang N N, Hai N H and Chau N 2009 *Phys. B* **404** 2495–8.
[5] Phan M, Tian S, Yu S and Ulyanov A N 2003 *J. Magn. Magn. Mater.* **256** 306–10.
[6] Phan M and Yu S 2007 *J. Magn. Magn. Mater.* **308** 325–40.
[7] Gomez J R, Garcia R F, Catoira A D M and Gomez M R 2013 *Renew. Sustain. Energy Rev.* **17** 74–82.
[8] Phan M, Yu S, Hur N H and Jeong Y 2004 *J. Appl. Phys.* **96** 1154–8.
[9] Winarsih S, Kurniawan B, Imaduddin A and Manaf A 2017 *AIP Conf. Proc.* **1862** 1–5.
[10] Gaur A and Varma G . D 2006 *J. Phys. Condens. Matter* **18** 8837–46.
[11] Winarsih S, Kurniawan B, Manaf A, Saptari S A and Nanto D 2016 *J. Phys. Conf. Ser.* **776** 1–5.
[12] Coey J M D and Molnar S Von 1999 *Adv. Phys.* **48** 167–293.
[13] Salamon M B and Jaime M 2001 *Rev. Mod. Phys.* **73** 583–628.
[14] Viret M, Ranno L and Coey J M D 1997 *Phys. Rev. B* **55** 8067–70.

High purity Fe_3O_4 from Local Iron Sand Extraction

Y E Gunanto[1], M P Izaak[1], E Jobiliong[2], L Cahyadi[2], W A Adi[3]
[1] Dept. of Physics Education, Faculty of Education, University of Pelita Harapan, Karawaci, Tangerang 15811, Indonesia
[2]Faculty of Science and Technology, University of Pelita Harapan, Karawaci, Tangerang 15811, Indonesia
[3]Center for Science and Technology of Advanced Materials, BATAN, Tangerang Selatan 15314, Indonesia

E-mail: yohanes.gunanto@uph.edu, maya.izaak@uph.edu, eric.jobiliong@uph.edu, lina.cahyadi@uph.edu, dwisnuaa@batan.go.id

Abstract. Indonesia has a long coastline and is rich with iron sand. The iron sand is generally rich in various elements such as iron and titanium. One of the products processing of the iron sand mineral is iron (II) (III) oxide (magnetite Fe_3O_4). The stages of purification process to extracting magnetite phase and discarding the other phases has been performed. Magnetite phase analysis of ironsand extraction retrieved from Indonesia have been investigated.The result of analysis element of iron sand shows that it consists of majority Fe around 65 wt%. However, there are still 17 impurities such as Ti, Al, Ce, Co, Cr, Eu, La, Mg, Mn, Na, Sc, Sm, Th, V, Yb, and Zn. After extraction process, Fe element content increases up to 94%. The iron sand powder after milling for 10 hours and separating using a magnetic separator, the iron sand powders are dissolved in acid chloride solution to form a solution of iron chloride, and this solution is sprinkled with sodium hydroxide to obtain fine powders of Fe_3O_4. The fine powders which formed were washed with de-mineralization water. The X-ray diffraction pattern shows that the fine powders have a single phase of Fe_3O_4. The analysis result shows that the sample has the chemical formula: $Fe_3O_4$with a cubic crystal system, space group: Fd-3m and lattice parameters: $a = b = c = 8.3681$ (1) Å, $\alpha = \beta = \gamma = 90°$. The microstructure analysis shows that the particle of Fe_3O_4 homogeneously shaped like spherical. The magnetic properties using vibrating sample magnetometer shows that Fe_3O_4 obtained have ferromagnetic behavior with soft magnetic characteristics. We concluded that this purification of iron sand had been successfully performed to obtain fine powders of Fe_3O_4 with high purity.

Keywords: Iron sand, local resources, purifying, magnetic, Fe_3O_4, and high purity

1. Introduction

In Indonesia, approximately 2 billion tons natural iron sand resources are spread along the southern coast of Java, Sumatra, West Nusa Tenggara, and so on with the main content of iron oxide, titania, silica and alumina [1]. In Indonesia, iron sand is the source of raw material of iron/steel industry [1] and cement [2]. Also, the product of iron sand processing can be Fe_3O_4. By processing the natural resources of iron sand into Fe_3O_4, the added value can be increased several number fold and can be used for import substitute products to enhance the competitiveness of the national industry.

Magnetite (Fe_3O_4) is an interesting material to develop because it is very applicable so that many researchers are interested in developing it [3-9]. Some of the treatment methods used are milling and precipitation [3], coprecipitation [4,7], precipitation and sol-gel [5], mechanical vibration [6], coprecipitation with polyethylene [8], mechanical alloying [9]. Through the Fe_3O_4 calcination process, barium hexaferrite may be produced as a microwave absorber [3], for biomedical [10], and others.

In this paper, we will report Fe_3O_4 synthesis results from the natural iron sand of the titanomagnetite type.The iron sand powder after milling for 10 hours and separating using a magnetic separator, the iron sand powders are dissolved in acid chloride solution to form a solution of iron chloride, and this solution is sprinkled with sodium hydroxide to obtain fine powders of Fe_3O_4. Therefore this study aims to obtain nanoparticle of Fe_3O_4 through the iron sand extraction.

2. Experiment

The iron sand of titanomagnetite is retrieved from Lampung, Indonesia. The purifying process of iron sand into iron oxide product in the form of high purity Fe_3O_4 is shown in Figure 1.

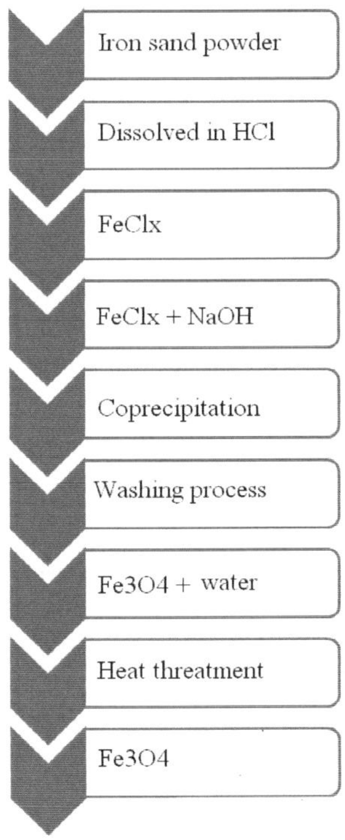

Figure 1. The purifying process of iron sand into iron oxide product

The iron sand of titanomagnetite was milled by using high energy milling (HEM), then dissolved in chloride acid (HCl) technical quality with the ratio of 250 g/L HCl to solve iron chloride. 5 M sodium hydroxide technical quality was added to a solution of iron chloride at a temperature of 80 °C for 1 hour. The precipitate was washed with demineralization water until pH of 7 with the purpose of disappearing the residual salts, then the precipitate was separated from the solvent using a permanent magnet and after that dried in an oven at a temperature of 100 °C. The dry powder sample was re-milled and then heated at a temperature of 400 °C in a furnace for 5 hours, so that obtained a Fe_3O_4 powder.

The Fe_3O_4 powder was analyzed by X-ray diffractometer (XRD) of Pan Analytical product with CuKα radiation (λ = 1.5406 Å). The qualitative and quantitative analysis of phase was carried out by GSAS program. The particle morphology of the Fe_3O_4 powders was observed by the scanning electron microscope (SEM) using JED 2300 JEOL. The elemental composition was analyzed by using energy dispersive spectroscopy (EDS). Finally, the magnetic properties were measured by using vibrating sample magnetometer (VSM) of Oxford product.

3. Result and discussion

Figure 2 shows the result of microstructure observation on iron sands sample of titanomagnetite type extracted by using scanning electron microscope.

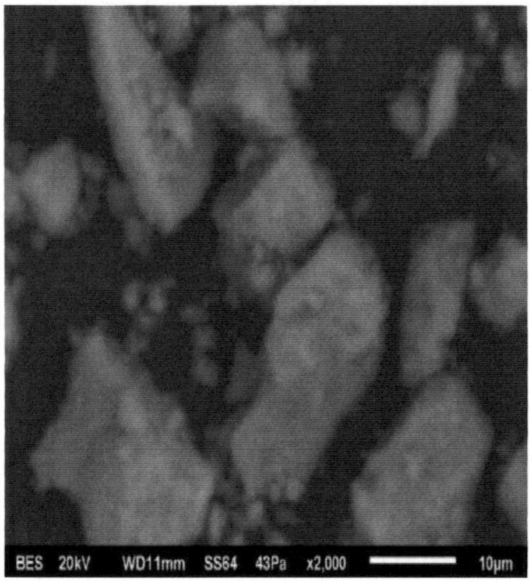

Figure 2. The surface morphology of iron sand samples of titanomagnetite type results from the extraction process

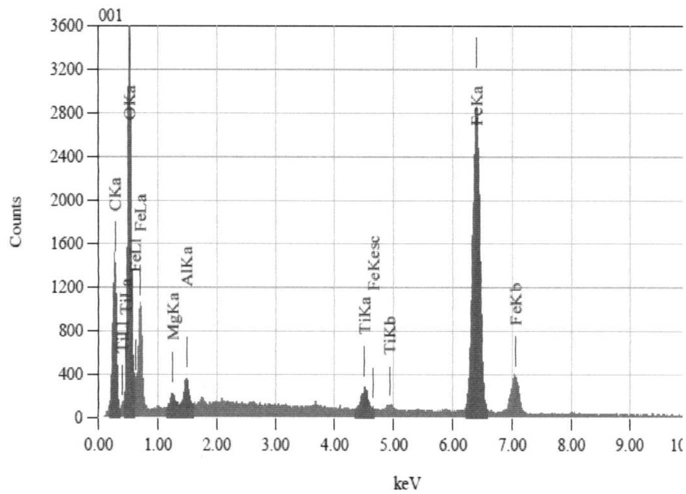

Figure 3. The elementary analysis using energy dispersive spectroscopy.

In Figure 2 it appears that the surface morphological observations show that the samples have the same form and color, although there is still a wide particle of about 1.8-10 μm. It is assumed that the sample has a homogeneous phase. However, photographs of these surface morphologies also cannot determine the phases contained in the titanomagnetite iron sand samples of these extractions. For the analysis of the existing phases in the iron sand, x-ray diffraction is used, while for the analysis of the constituent elements, energy dispersive spectroscopy is used.

Figure 3 shows the result of elementary analysis using energy dispersive spectroscopy on titanomagnetite iron sand sample. The resulting energy spectrum as shown in Fig. 3 shows that the most dominant elements are oxygen (O) and iron (Fe) which are consecutively at 0.525 keV and 6.398 keV energy with K-wavelength. This means that the extraction of titanomagnetite iron sand sample has successfully separated the iron (Fe) element from other impurities. In detail, the element content present in the iron sands of this type of titanomagnetite is shown in Table 1.

In Table 1 it appears that the Fe content of total iron sand samples of titanomagnetite type of this extraction is 79.12%. These results indicate that the iron content in the titanomagnetite iron sand sample has been successfully performed.

IOP Conf. Series: Journal of Physics: Conf. Series **1011** (2018) 012005 doi:10.1088/1742-6596/1011/1/012005

Tabel 1. The result of elementary analysis using energy dispersive spectroscopy

No.	Element	Content (% weight)
1.	Iron(Fe)	79.12 ± 0.21
2.	Titanium (Ti)	1.32 ± 0.11
3.	Magnesium (Mg)	0.66 ± 0.11
4.	Aluminum (Al)	1.10 ± 0.09
5.	Carbon (C)	1.62 ± 0.06
6.	Oxygen (O)	16.18 ± 0.11

Figure 4 shows the results of the measurement and structural analysis that can identify the phase of the x-ray diffraction pattern of titanomagnetite iron sand samples extracted.

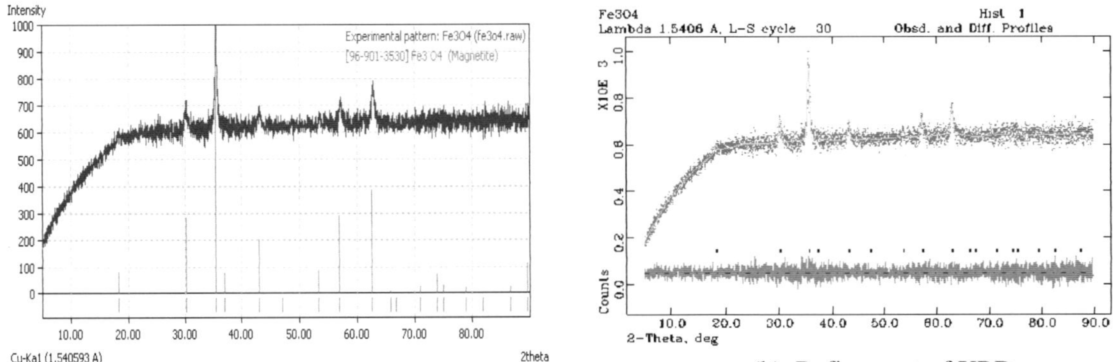

(a) XRD identification (b) Refinement of XRD

Figure 4. The X-ray diffraction pattern shows that the fine powders have a single phase of Fe_3O_4.

Figure 4 shows that there have been formed Bragg diffraction peaks which are suspected to have been single phase. The phase identification refers to JCPDS number 19-629, i.e. the magnetite phase with the cubic crystal structure and lattice parameters: $a = b = c = 8.3681$ (1) Å. Similar results were obtained by Sunaryono et al. [4] and also Rusianto et al. [6]. Thus, based on the results of the analysis of the x-ray diffraction pattern, the amount of phase content contained in the iron titanomagnetite iron extraction is shown in Table 2.

Table 2. The phase mass fraction contained in the iron sands sample of the titanomagnetite type of extraction

No.	Mineral	Empirical formula	Mass fraction (%)
1.	Magnetite	Fe_3O_4	100

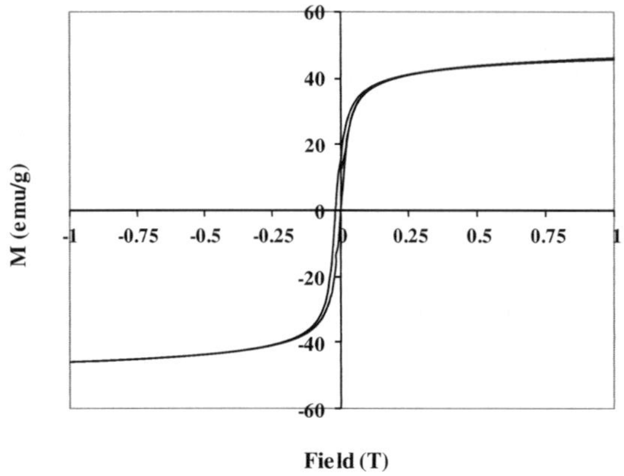

Figure 5. Magnetic properties of iron sand samples of titanomagnetite type of extraction.

Figure 5 shows the magnetic properties of iron sands samples of titanomagnetite extracted species. The relationship between *M* and *H* is not linear. This indicates that the sample of titanomagnetite iron sand of this extract has ferromagnetic properties.When the magnetic field intensity of *H* is increased to reach *H* = 1 Tesla, the value of *M* has a saturation point called saturation magnetization *Ms* around 43 emu/g; then when the magnetic field *H* is lowered, the curve path does not return through the original curve. At the price of *H* = 0, the magnetization of *M* has 18.5 emu/g. This condition is called remanent magnetization *Mr*.Furthermore, the intensity value of the magnetic field *H* is lowered (negative value), the *M* curve will intersect the axis of 226 Oe magnetic field denoted as *Hc*. This Hc intensity is what is required to make the flux density *B* = 0 or eliminate the flux in the material. The intensity of the *Hc* magnet is called the coercivity of the material or the coercivity field. Based on this hysteresis curve, the titanomagnetite iron sand sample of this extraction is a magnetic material.

4. Conclusion

The samples Fe_3O_4 have been successfully extracted by coprecipitation method from local iron sand. The sample has a homogeneous phase. However, photographs of these surface morphologies also can not determine the phases contained in the titanomagnetite iron sand samples of these extractions. The resulting EDS shows that the most dominant elements are oxygen (O) and iron (Fe). The XRD data show that the sample has a single phase with the structure of spinel ferrite particles. The functional group's analysis by using FTIR shows that transmittance peaks of the iron sand extraction appeared at wave numbers around 600 cm^{-1} that indicated the presence of the Fe–O bond vibrations. The hysteresis loop shows that all of the samples have the ferromagnetic behavior with a relatively small of coercivity value.

Acknowledgments

This work was supported by Superior Research of Higher Education 2017 No. DIPA: SP DIPA-042.06.1.401516/2017, Development of Iron Oxide Nanoparticles from Iron Sand Processing in Support of National Industry.

References

[1] Ministry of Industry Republic of Indonesia 2014 Profile of Steel Industry

[2] Rochani S, Pramusanto, Sariman, and Anugrah R I 2008 *Indonesian Mining Journal* Vol **11** No 11 pp 1-17

[3] Rusianto T, Wildan M W, Abraha K, and Kusmono 2015 *International Journal of Technology* Vol.**6** pp1017-1024

[4] Sunaryono, Taufiq A, Mashuri, Pratapa S, Zainuri M, Triwikantoro, and Darminto 2015 *Materials Science Forum* Vol **827** (Switzerland Trans Tech Publication) pp 229-234

[5] Yulianto A, Sulhadi, Azis A L I, and Dayati E 2013 *Malaysian Journal of Fundamental and Applied Science* Vol **9** No 4 pp221-215

[6] Rusianto T, Wildan M W, Abraha K, and Kusmono 2015 *Indian Journal of Engineering & Materials Science* Vol **22** pp 175-180

[7] Mufti N, Atma T, Fuad A, and Sutadji E 2014 *Proc. Int. Conf. on Theoretical and Applies Physics* **1617** (*AIP Conference Proceedings*)pp 165-169

[8] Bukit N, Frida E, Simamora P, and Sinaga T 2015 *Journal of Chemistry and Material Research* Vol **7** No 7 pp 110-115

[9] Jalil Z, Rahwanto A, Mustanir, Akhyar, and Handoko E *2016 Proc. Int. Symposium on Current Progress in Mathematics and Science* **1862** (*AIP Conference Proceedings*)pp 030023-1-030023-4

[10] Zélis P M, Muraca D, Gonzales J S, Pasquevich G A, Alvarez V A, Pirota K R, and Sánchez F H 2013 *Journal Nanoparticle Res.* **15** 1613-1616

IOP Conf. Series: Journal of Physics: Conf. Series **1011** (2018) 012006 doi:10.1088/1742-6596/1011/1/012006

Experimental Study of Acid Treatment Toward Characterization of Structural, Optical, and Morphological Properties of TiO_2-SnO_2 Composite Thin Film

M N Fajar[1], R Hidayat[2], Triwikantoro[1] and Endarko[*1]

[1]Physics Department, Institut Teknologi Sepuluh Nopember, Sukolilo, Surabaya, 60111, East Java, Indonesia

[2]Physics of Magnetism and Photonics Research Division, Faculty of Mathematics and Natural Sciences, Institut Teknologi Bandung, Jl. Ganesha 10, 40132, West Java, Indonesia

[*]Corresponding email address: endarko@physics.its.ac.id

Abstract, The TiO_2-SnO_2 thin film with single and double-layer structure has successfully synthesized on FTO (Fluorine-doped Tin Oxide) substrate using the screen printing technique. The structural, optical, and morphological properties of the film were investigated by XRD, UV-Vis, and SEM, respectively. The results showed that the single and double-layer structure of TiO_2-SnO_2 thin film has mixed phase with a strong formation of casseritte phase. The acid treatment effect on TiO_2-SnO_2 thin film decreases the peak intensity of anatase phase formation and thin film's absorbance values. The morphological study is also revealed that the single layer TiO_2-SnO_2 thin film had a more porous nature and decreased particle size distribution after acid treatment, while the double-layer TiO_2-SnO_2 thin film Eroded due to acid treatment.

1. Introduction

Recently, DSSC has gained extensive attention due to their environment-friendliness, simple fabricating procedure and low-cost alternative to conventional Si solar cells [1-2]. A typical DSSC is composed of a nanoporous TiO_2 photoanode and a Pt counter electrode separated by an iodide/triiodide electrolyte [3]. The photoanode is usually prepared from TiO_2 nanoparticles on a transparent conducting oxide (TCO), while the counter electrode is a thin layer of Pt deposited on another TCO substrate [2].

Titanium dioxide (TiO_2) is one of the most studied extensively as photoanodes materials in DSSCs specially anatase phase due to its excellent stability and photoactivity. As the main material of the photoanode, characteristic of TiO_2 such as structure and morphology will undoubtedly affect the performance of DSSC [4]. Photoanode in DSSCs must have some characteristic of the porous structure, a large surface area which can increased molecules dye-loading capacity in photoanode. Some strategy to got the criteria are TiO_2 composite with SnO_2 and acid treatment on its surface.

Tin(IV) oxide, SnO_2, has seen considerable usage in composites with TiO_2 for photocatalysis, and battery applications. (Composite TiO_2 nanomaterial). Tin oxide has some advantages over TiO_2

because of higher diffusion rates of electrons in bulk SnO_2 [5]. Some research obtains that the acid treatment of photoanode surface can be enhanced the porosity characteristic and the performance of DSSC [6-7]. However, Hao [8] reported the decreased performance of acid-treated DSSCs, so this inconsistent result needs further clarification. To better understand those effect, acid treatment on surface photoanode should be investigated.

In this article, composite film TiO_2/SnO_2 was synthesized by double and single-layer screen printing method on FTO glass with TiO_2 and SnO_2. The influence of the acid treatment on Structural, Optical, and Morphological photoanode is investigated by X-ray Diffraction (XRD), UV-Vis, and Scanning Electron Microscopy (SEM). It is found that results of acid treatment was different for each type of composite structure (single and double layer composite) and also can change the optical, and morphological thin film.

2. Experimental

2.1. Deposition of Blocking Layer

Blocking layer TiO_2 was made by depositing a solution of TiO_2 blocking layer in FTO glass using spin coating method. Spin coating speed is 5000 rpm for 30 s. Finally, blocking layer with substrate was heated periodically at 100 °C for 15 min, 300 °C for 15 min, and 450 °C for 30 min.

2.2. Preparation of SnO_2 paste

The preparation of SnO_2 paste begins by dissolving 0.13 g of ethylcellulose was added into 3 mL of isopropanol solution and stirring for 1.5 h. Further 0.25 g of SnO_2 powder was added to the solution and continued stirring for 2 h.

2.3. Preparation of composite TiO_2/SnO_2 paste

0.05 g of ethylcellulose was added to mixing solution of 0.25 mL terpineol, and 0.25 mL of ethanol then stirred with a temperature of 80 °C for 10 min. 0.09 g of TiO_2 paste and 0.09 g of SnO_2 paste dissolved into ethylcellulose solution then stirred at 60 °C for 30 min.

2.4. Preparation of thin film

Each paste was deposited on the FTO glass using screen printing method. An active area of 0.5 × 0.5 cm^2 was formed in the FTO glass conductive part. Four types of thin film samples are TiO_2, SnO_2, single layer TiO_2 / SnO_2 composite, and a double-layer TiO_2 / SnO_2 composite. For a double-layer composite TiO_2 / SnO_2, SnO_2 and TiO_2 paste was deposited with TiO_2 above SnO_2. After the deposition process, the samples were gradually heated from 100 °C for 15 minutes, 300 °C for 15 min and 450 °C for 30 min.

2.5. Acid treatment

All thin film samples were immersed in the 2M HNO_3 solution for 20 min. All samples were cleaned with distilled water to remove the remaining HNO_3 from the immersion process. The use of nitric acid is expected to improve the photoanode performance for DSSC as some previous studies [6,12]

3. Result and Discussion

3.1. Thin film XRD analysis

The diffraction pattern in the form of an amorphous solid is obtained from the glass substrate diffraction pattern. With qualitative analysis using software match! It is shown in Figure 1 that the TiO_2 has anatase phase. The crystal plane 101 at ° 2Theta around 25.3 ° is the dominant crystal plane with the highest intensity in the TiO_2 thin film. While on other semiconductor materials, it was found that SnO_2 has cassiterite phase with the highest intensity in the (110), (101) and (211) planes.

The diffraction patterns formed from CDL (thin film double-layer TiO_2/SnO_2 composite) and CSL (thin film single-layer TiO_2/SnO_2 composite) are a combination of the peaks from TiO_2 and SnO_2 in a particular crystal plane. The CDL diffraction pattern is dominated by the crystalline planes of the SnO_2

cassiterite phase while the CSL diffraction pattern shows the crystal plane 101 derived from the anatase phase TiO_2 and the crystal plane (110) from the SnO_2 cassiterite phase. Li et[9] also obtained the same diffraction pattern on the TiO_2 / SnO_2 composite film. The addition of SnO_2 to the composite was also able to reduce the peak (101) of TiO_2, a similar result obtained by Arunachalam [10] in his study of the effect of the addition of Sn for the fabrication of composite TiO_2/SnO_2 thin films.

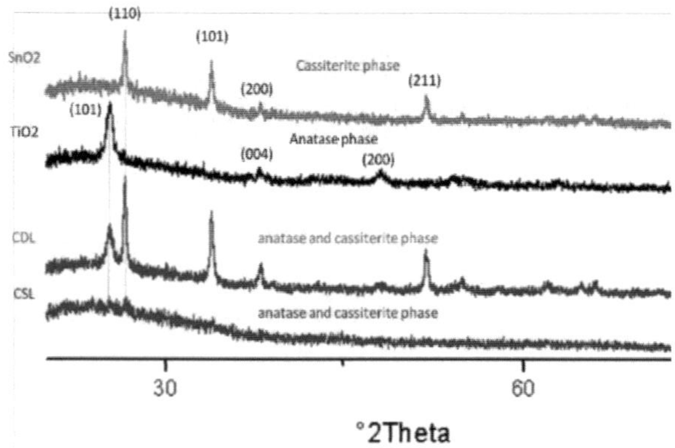

Figure 1. XRD shows diffraction pattern of SnO_2, TiO_2, CDL(Composite Double Layer), and CSL(Composite Single layer) thin film

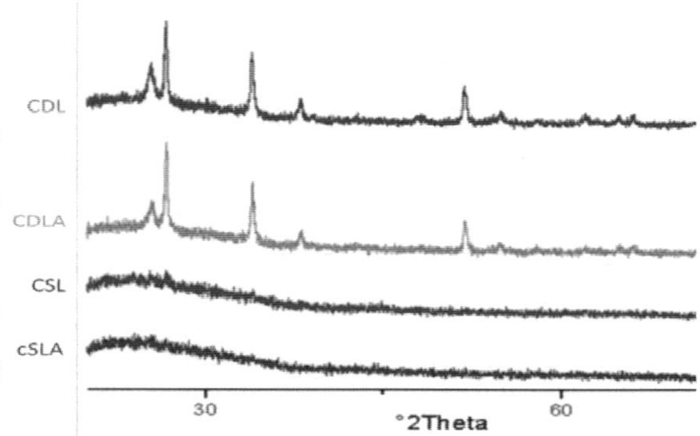

Figure 2. XRD shows diffraction pattern of CDL and CSL thin film after the immersed process in HNO_3(CDLA and CSLA)

Figure 2 shows the diffraction pattern of CDL and CSL thin film after the immersed process in HNO_3 solution which called CDLA and CSLA. It can be seen that there is an intensity reduction at anatase phase in all the diffraction peaks in CDL and CSL thin film due to the reduced number of particles in the thin film. This effect also indicates a decrease in the degree of crystallinity in the thin film [11-12].

3.2. Morphological study

The morphology of the thin layer of CDL (Figure 3a) shows more pore than the thin layer of CSL(Figure 3e) on its surface. After immersion with an HNO_3 solution in thin film, it was found that there was an erosion of the coating and the reduction of TiO_2 particles in CDLA thin film(Figure 3d) and more pore formation on CSLA thin film (Figure 3c). This is due to the difference of thin film structure at the time of immersion with an HNO_3 solution.

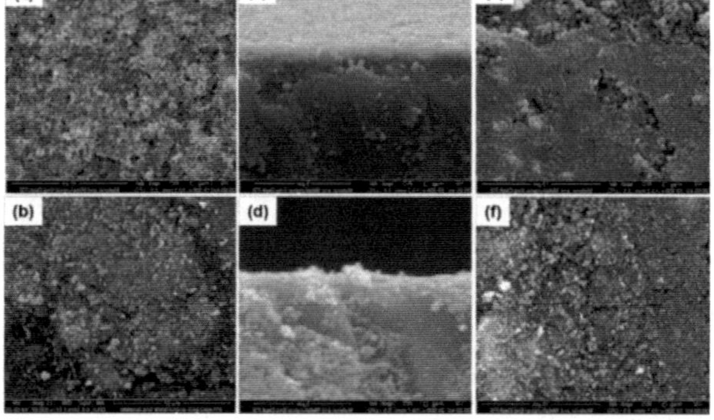

Figure 3. The SEM micrograph shows the surface morphology of the thin film (a) CDL, (b) CDLA, (c) cross section of CDL, (d) cross section of CDLA, (e) CSL, and (f) CSLA

The SEM analysis also provides particle size distribution information from the thin film. Figure 4 shows the average particle size of the CDL thin film is between 0-150 nm with the dominance of the most particle size being approximately 40 nm. In the average CSL shows the particle size is in the

IOP Conf. Series: Journal of Physics: Conf. Series **1011** (2018) 012006 doi:10.1088/1742-6596/1011/1/012006

range 0- 250 nm with an almost equal distribution of particle size with the dominance of the most particle size being approximately 50 nm. The acid treatment also gives effect to the particle size distribution. In CDLA thin film, the particle size becomes larger while decreased particle size of CSLA.

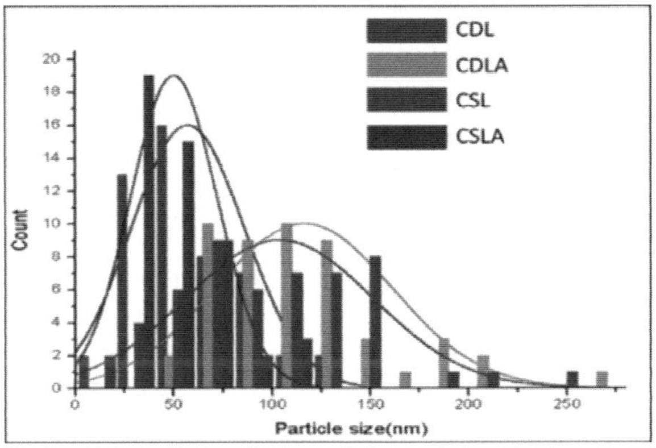

Figure 4. The average particle size of CDL, CDLA, CSL, and CSLA thin film obtained from SEM micrograph.

3.3. Optical characteristic study

We obtained results in the graph of absorbance which illustrates how the optical characteristics of the thin film using UV-Vis. Figure 5 shows that the absorbance wavelength of the TiO_2 thin film in the 300-600 nm while SnO_2 is in the 300-870 nm wavelength (Figure 5a). CDL thin film has a wavelength range of 300-870 nm and the highest absorbance value when compared with the TiO_2 and SnO_2 thin films. This is due to CDL thin film is consist of two layers TiO_2 layer which has a high absorbance value and a SnO_2 layer which has a wavelength range 300-870 nm. Figure 5b shows absorbance values over a specified wavelength range of the TiO_2, SnO_2, and CSL thin films where the CSL thin film absorbance is lower than TiO_2 thin film in the range wavelength 300-600 nm but higher than SnO_2 thin film in the range wavelength 300-870. This is due to CSL thin film was a composite layer of TiO_2 / SnO_2 so that the absorbance is lower than the TiO_2 thin film but has the same wavelength range as the SnO_2 thin film. Batmunkh et al. also obtained similar results [13].

Acid treatment gives affect decreased absorbance for each thin film(figure 5 c and d). Decreased in absorbance of CDLA shows indicates the erosion of TiO_2 layer after immersed in the HNO_3 solution, this result also obtained by Wen et al.[14] where the decreased due to the increased molar acid solution in thin film immersion. Decreased in absorbance of CSLA indicates of increased porosity on TiO_2 thin film after immersed in the HNO_3 solution, this result is also obtained by Watson et al.[6]. Both the erosion and increased porosity of TiO_2 give effect to the decreased in the absorbance due to the amount of light passed through the thin film.

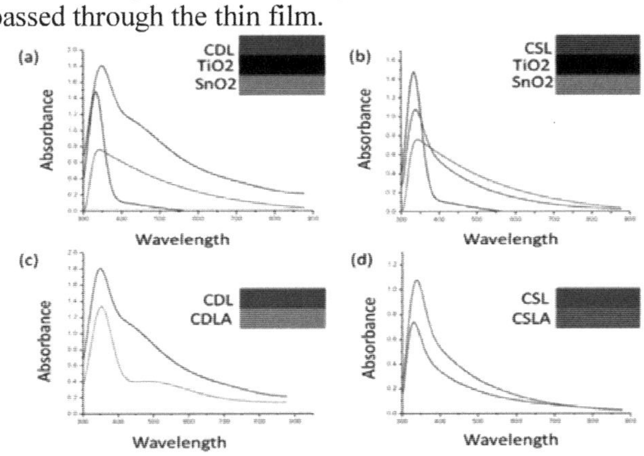

Figure 5. UV-Vis shows the absorbance graphs of (a) TiO2, SnO2, and CDL; (b) TiO2. SnO2, and CSL; (c) CSL and CSLA; (d) CDL and CDLA

4. Conclusion

In summary, the single and double-layer structure of TiO_2-SnO_2 thin film has mixed phase with a strong formation of casseritte phase. The acid treatment effect on TiO_2-SnO_2 thin film decreases the peak intensity of anatase phase formation and thin film's absorbance. The decreased absorbance on thin film due to morphological study which also revealed that the single layer TiO_2-SnO_2 thin film had a more porous nature and decreased particle size distribution after acid treatment, while the double-layer TiO_2-SnO_2 thin film Eroded due to acid treatment.

Acknowledgments

All authors would like to thank the Ministry of Research, Technology and Higher Education of the Republic of Indonesia, and LPPM-Institut Teknologi Sepuluh Nopember for providing Laboratory Grant 2017 with contract number: 697/PKS/ITS/2017 and the Physics of Magnetism and Photonics Research Division, FMIPA-Institut Teknologi Bandung over all their facilities during this study.

References

[1] Wu M, Lin X, Wang T, Qiu J, and Ma T 2011 Low-cost dye-sensitized solar cell based on nine kinds of carbon counter electrodes *Energy Environ. Sci.*vol.**4** no.6 pp 2308–2315
[2] Suhaimi S, Shahimin M M, Alahmed Z A, Chyský J, and Reshak A H 2015 Materials for enhanced dye-sensitized solar cell performance: Electrochemical application *Int J Electrochem Sci* vol. **10** no.4 pp 2859–2871
[3] Ye M *et al* 2015 Recent advances in dye-sensitized solar cells: from photoanodes, sensitizers and electrolytes to counter electrodes *Mater. Today* vol. **18**, no. 3 pp 155–162
[4] Adawiyah S R and Endarko 2017 Structural and morphological characterization of TiO_2-SnO_2 thin film prepared by combining doctor-blade and sol-gel techniques *IOP Conference Series: Materials Science and Engineering* vol. **188** p 012062
[5] Ranasinghe C S K *et al* 2015 Tin oxide based dye-sensitized solid-state solar cells: surface passivation for suppression of recombination *Mater. Sci. Semicond. Process.*, vol. **40**, pp. 890–895
[6] Watson T, Charbonneau C, Bryant D, and Worsley D 2012 Acid treatment of titania pastes to create scattering layers in dye-sensitized solar cells *Int. J. Photoenergy*, vol. **2012**
[7] Siti Aida I and Srimala S 2010 Effect of pH on TiO_2 nanoparticles via sol gel method
[8] Hao S, Wu J, Fan L, Huang Y, Lin J, and Wei Y 2004 The influence of acid treatment of TiO_2 porous film electrode on photoelectric performance of dye-sensitized solar cell *Sol. Energy*, vol. **76** no. **6** pp 745–750
[9] Li H, Xie C, Liao Y, Liu Y, Zou Z, and Wu J 2013 Characterization of Incidental Photon-to-electron Conversion Efficiency (IPCE) of porous TiO $_2$/SnO $_2$ composite film *J. Alloys Compd.* vol. **569** pp 88–94
[10] Arunachalam A, Dhanapandian S, and Manoharan C 2016 Effect of Sn doping on the structural, optical and electrical properties of TiO_2 films prepared by spray pyrolysis *J. Mater. Sci. Mater. Electron.* vol. **27** no. 1 pp 659–676
[11] Yun H -G, Bae B -S, Jun Y, and Kang M G 2011 Effective Methods for the High Efficiency Dye-Sensitized Solar Cells Based on the Metal Substrates in *Solar Cells-Dye-Sensitized Devices*, InTech
[12] Park D -W *et al* 2011 Nanocrystalline TiO_2 films treated with acid and base catalysts for dye-sensitized solar cells *Adv. Powder Technol.* vol. **22** no. 6 pp 771–776
[13] Batmunkh M, Dadkhah M, Shearer C J, Biggs M J, and Shapter J G 2016 Tin Oxide Light-Scattering Layer for Titania Photoanodes in Dye-Sensitized Solar Cells *Energy Technol.*, vol. **4**, no.8 pp 959–966.
[14] Wen Q *et al* 2016 Hydrothermal treatment of a TiO_2 film by hydrochloric acid for efficient dye-sensitized solar cells *New J. Chem.* vol. **40** no. 4 pp 3233–3237

The Optical Properties of Thin Film Reduced Graphene Oxide/Poly (3,4 Ethylenedioxtriophene):Poly (Styrene Sulfonate)(PEDOT:PSS) Fabricated by Spin Coating

Arinta W Rokmana[1], A Asriani[1], H Suhendar[1], K Triyana[1,2], A Kusumaatmaja[1,2], and I Santoso[1,a*]
[1]Departemen Fisika, Universitas Gadjah Mada, Sekip Utara BLS 21 Yogyakarta, INDONESIA
[2]Interdiciplinary Nanomaterial Research Group, Universitas Gadjah Mada, Jl. Kaliurang Km. 4 Sekip Utara Yogyakarta, INDONESIA

*Corresponding Author: iman.santoso@ugm.ac.id

Abstract. Reduced Graphene Oxide (rGO) has been successfully synthesized from Graphite powder through chemical process using modified Hummers method by removing $NaNO_3$ from reaction formula. Hydrazine hydrate 80 wt% has been chosen as reductor to eliminate the epoxy group in GO. FTIR and Uv-Vis spectroscopy result showed that Graphene Oxide (GO) and rGO were formed. Our produced rGO then used to fabricated the composite thin film rGO/PEDOT:PSS by spin coating at room temperature. The optical constant of thin film rGO/PEDOT:PSS were calculated from the absorbance spectrum of Uv-Visible spectra. The result showed that the value of coefficient absorbance of rGO dropped from 4.7×10^6 m^{-1} to 1.3×10^6 m^{-1} after doped with 0.02 mL PEDOT:PSS, then increase with the addition volume concentration of PEDOT:PSS. The value of extinction coefficient decrease from 0.31 to 0.08 after rGO doped with 0.02 ml PEDOT:PSS and then increase with the addition concentration of PEDOT:PSS. Our result show that thin film rGO/PEDOT:PSS was more transparent than that of thin film rGO.

Keywords: Thin film, Graphene Oxide, reduced Graphene Oxide, PEDOT:PSS, Optical properties.

1. Introduction

Graphene, a novel and an exciting material composed of sp^2 bonded carbon atoms densely packed in a honeycomb crystal lattice have received significant attention in recent years, due to its excellent properties like high electron mobility, high termal conductivity, extraordinary elasticity and stiffnes. The electron mobility graphene will be up to 150,000 cm^2/Vs at a temperature of 300 K and 6,000 cm^2/Vs at a temperature of 4K [1]. Universal conductance of graphene is about $6.08\times10^{-5}\ \Omega^{-1}$ at 0.1

eV to 0.6 eV [2]. Near the Fermi energy, graphene has linear band energy [3,4,5] which make graphene has semimetal and very high conducting and mobility properties.

Optical properties of graphene also have unique properties with transparence up to 97.7% [5] in the visible range. Universal of optical conductivity graphene ($\sigma_0=\pi e^2/2h$) at infrared to visible range [6,7,8], also calculation of graphene absorbance in infrared to visible range spectrum showed that graphene absorbance is constant ini $\pi\alpha$=2.93% [9]. These features have made graphene has wide range of application in various industries such as photovoltaic cells. Due to its high electron mobility and transparence properties have made graphene potentially to be applied as an alternative transparent electrode in photovoltaic device.

Besides its application as transparent electrode, graphene can be applied as donor/acceptor material in hetero junction solar cell. However, to be applied as donor/acceptor material graphene should have doped with conductive polymer. Doping can be a feasible method to open band gap in graphene also in controlling electron or hole. Zahibi et al. [10] report that thin film graphene doped with P3HT can be applied as donor material in solar cell and it has a power conversion efficiency up to 5.17%.

On the other hand, PEDOT:PSS is conducting transparent co-polymer widely used in photovoltaic device as the hole transporting layer and even as a transparent electrode [11, 12]. Blending graphene with PEDOT:PSS can be controlling its hole transport due to hole majority charge at conductive polimer PEDOT:PSS. Also doped graphene with PEDOT:PSS could be improved electrical and mechanical properties as reported by Chen et al. [13]. Chen reported that optical thin film graphene doped with PEDOT:PSS up to 720 S/cm.

The investigation of optical properties of rGO doped with PEDOT:PSS is very important to utilize graphene as transparent electronic devices. In this work, we use reduced graphene oxide (rGO) which obtained by chemical reduction of GO. GO was synthesized using modified Hummer's method [14]. We were focused on the optical properties of thin film rGO/PEDOT:PSS.

2. Experimental

2.1. Synthesis of graphene oxide

Graphite oxide was synthesized using a modification Hummer's method by removing $NaNO_3$ from reaction formula. Fristly, 3 mg of graphite powder was added to 70 ml of concentrate H_2SO_4 under stirring for 1 h in ice bath. Then 9 gr of solid $KMnO_4$ were subsequently added to the mixture and stirred for 3 h while the temperature was kept below 20 $^\circ$C. Successively, the mixture was vigorously stirred for about 20 h while the temperature was kept at 40 $^\circ$C using oil bath. Then, 150 ml of distilled water was added to mixture and stirred for 15 min at 95 $^\circ$C. After 15 min, the heat was turning off and additional 500 ml distilled water was added then followed by a slow addition of 15 ml H_2O_2 (30%) until the color of mixture changed from dark brown to yellow. To remove ions, the mixture was washed with HCl (10%). Next the mixture was washed with distilled water and then centrifuged at 4000 rpm for 30 min to remove the acid. At last, the resulting filter cake was dried in oven (100 $^\circ$C) during 4 h.

2.2. Reduction of graphene oxide by hydrazine

rGO was obtained by reducing GO with hydrazine hydrate as reducing agent. A homogeneous GO aqueous suspension was achieved by dispersing 300 mg of graphite oxide in 100 ml of distilled water using ultra sonication. Then 100 μL of hydrazine hydrate was added and the mixture stirred for 3 h in oil bath to kept the temperature at 70 $^\circ$C. Finally, a black rGO dispersion was obtained. The resulting filter cake was washed using distilled water and then dried in oven (80 $^\circ$C) during 4 h.

2.3. Preparation of thin film rGO/PEDOT:PSS

PEDOT:PSS solution was mixed with rGO dispersion (5 mg mL^{-1}) in various volume concentration (0.02 mL; 0.04 mL; 0.06 mL; 0.08 mL) and sonicated for 30 min to form a stable mixture. The mixture was then spin coated using vacuum spin coater VTC-100 on clean quartz substrate at 1000 rpm for 60 s. Then the resulted film then dried at 100 °C for 5 min.

2.4. Characterization

Uv-Visible (Uv-Vis) absorption spectra of powder of GO and rGO also thin film GO, rGO, and rGO/PEDOT:PSS were obtained on Shimadzu UV-1700 spectrophotometer in the spectrum range of 200-800 nm in ambient condition and at room temperature. FTIR ABB MB was used for Fourier Transform Infra Red (FTIR) characterization. GO and rGO powders were pulverized and pressed into pellets. Their spectra were measured in the range 1000-4000 cm^{-1}.

2.5. Optical Constant Calculation

By using Uv-Vis absorbances data, we could calculate imaginary part of refractive index of rGO and rGO/PEDOT:PSS by using following equation:

$$\alpha = \frac{absorbance\ value \times 2{,}3026}{d} \quad (1)$$

with α is coefficient of absorbance and d is a thin film thickness. From equation above we could get imaginary part of refractive index (k) as write by following equation:

$$k = \frac{\lambda\alpha}{4\pi} \quad (2)$$

3. Results and discussion

3.1. The characterization of GO and rGO

GO was synthesized from natural graphite using the modified Hummers method. The resulting product of GO powder is light brown in color and water dispersible. The GO was reduced by hydrazine hydrate. When hydrazine hydrate was added to the GO suspension, the color of GO suspension gradually changed from light brown to black, this change indicate that the GO was reduced by hydrazine hydrate. The reduction of synthesis Graphene Oxide (GO) and reduced Graphene Oxide (rGO) was characterized with Uv-Visible and FTIR spectroscopy. Absorbance spectrum of GO and rGO result by using Uv-Vis spectrophotometer was shown in Fig. 1 The absorbance result for GO showed two peaks at 230 and 300 nm that can be correlated to the electronic transition from $\boldsymbol{\pi} \rightarrow \boldsymbol{\pi}^*$ and $\boldsymbol{n} \rightarrow \boldsymbol{\pi}^*$ orbitals, respectively. These two peaks are characteristics in GO absorbance. UV-Vis spectroscopy at Fig. 1also showed the absorbance spectrum of rGO, which only one peak at a wavelength of 268 nm appears. This peak can be correlated to the $\boldsymbol{\pi} \rightarrow \boldsymbol{\pi}^*$ orbital transition. This peak is the characteristic for rGO absorbance spectra. At longer wavelength, the absorbance has constant value that similar to pristine monolayer graphene.

Fig. 2 shows the FTIR spectroscopy result of GO and rGO. The GO spectra shows wide peak at 3430 cm^{-1} (O-H stretching) which shows that GO has hydrophilic properties. It can be mentioned that the reducing FTIR spectra at 1000 cm^{-1} and 1700 cm^{-1} as GO is reduced to rGO inferring that C=O and C-O bounding is decreased. As it seen from FTR spectra, there were reducing of C=C bounding which shows at 1600 cm^{-1}. It can be happened due to the presence of energetic Hydrazine.

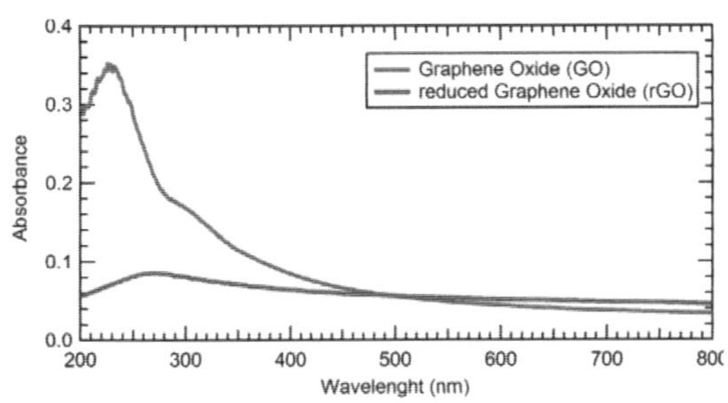

Figure 1. The absorbance of GO and rGO

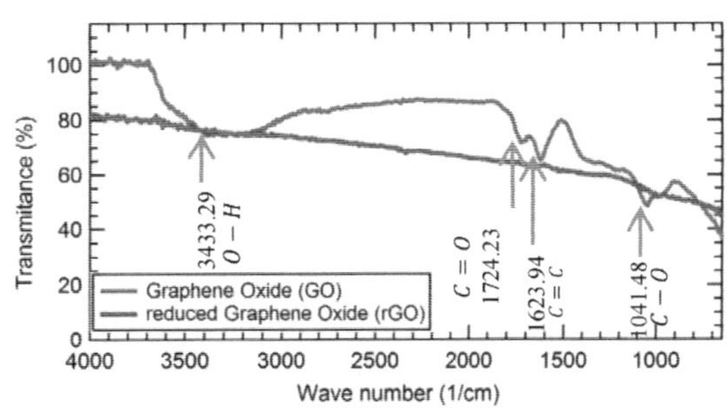

Figure 2. FTIR spectrum of GO and rGO

3.2. The optical constant of thin film rGO/PEDOT:PSS

Fig. 3 and 4 is show absorbance and coefficient absorbance of thin film rGO and rGO/PEDOT:PSS, respectively. Its shows that the coefficient absorbance value decreased from dropped from 4.7×10^6 m^{-1} to 1.3×10^6 m^{-1} after doped with 0.02 mL PEDOT:PSS. The addition concentration of PEDOT:PSS (0.4 mL, 0.6 mL, 0.8 mL) increase the coefficient absorbance value.

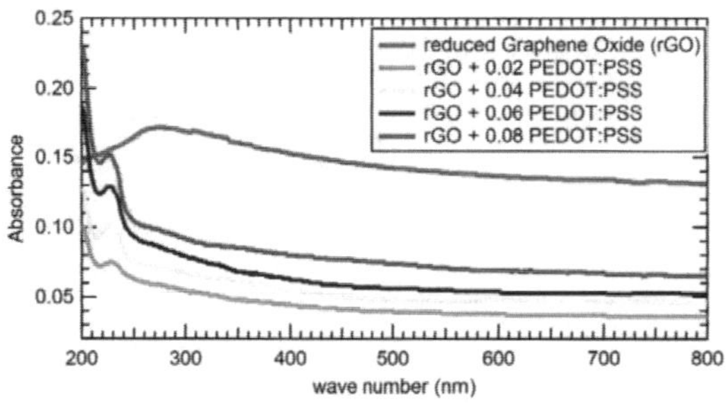

Figure 3. The absorbance of thin film rGO and rGO/PEDOT:PSS in various concentration of PEDOT:PSS

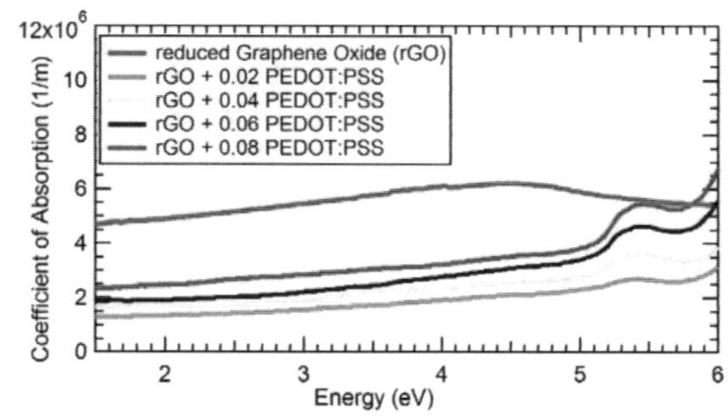

Figure 4. Coefficient of absorbance of thin film rGO and rGO/PEDOT:PSS in various concentration of PEDOT:PSS

Fig. 5 shows coefficient extinction (k) of thin film rGO and rGO/PEDOT:PSS that shows the decreased k value of rGO after doped with 0.02 mL PEDOT:PSS from 0.31 to 0.08, as volume concentration of PEDOT:PSS increased the value of k also increased. It could be mention that transparency of rGO increased after doped with 0.02 mL PEDOT:PSS, and after increasing the volume concentration into 0.4 mL, 0.6 mL, and 0.8 mL the transparency of thin film rGO/PEDOT:PSS was decreased. But the k value of rGO/PEDOT:PSS was lower than the k value of rGO, it indicate that thin film rGO/PEDOT:PSS more transparent than thin film rGO. The dropped k value of thin film rGO after doped with 0.02, 0.04, 0.06 and 0.08 mL PEDOT:PSS shows that thin film rGO/PEDOT:PSS was more transparent from thin film rGO. At low energy as indicated by increasing of k values, the carrier is increased as PEDOT:PSS added, inferring that more carriers responds the photon at lower photon energy.

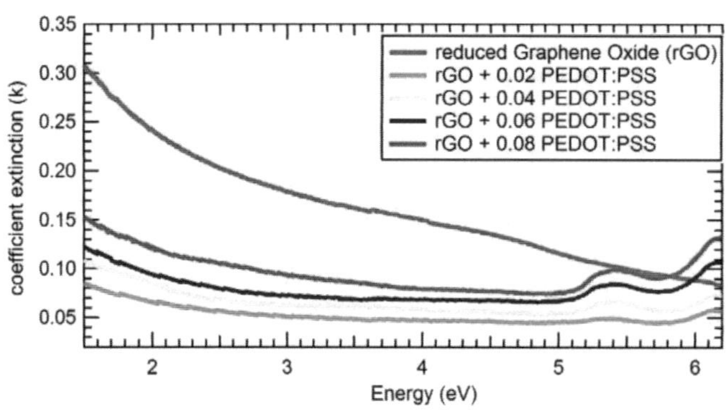

Figure 5. Coefficient extinction of thin film rGO and rGO/PEDOT:PSS in various concentration of PEDOT:PSS

4. Conclusion

From *k* value data, we got to know that thin film rGO/PEDOT:PSS was more transparent from thin film rGO. So its shows that thin film rGO/PEDOT can be applied as transparent electrode in photovoltaic device.

References

[1] Kuzumenko A B, Van H E, Carbone F and Van D M D 2008 *Phys. Rev. Lett.* **100** 117401
[2] Novoselov K S, Geim A K, Morosov S V, Jiang D, Zhang Y, Dubonos S V, Grigorieva I V and Firsov A A 2004 *Science* **306** 666
[3] Geim A K and Novoselov K S 2007 *Nat. Mater.* **6** 183-191
[4] Bernard J Z, Zulicke U and Ziegler 2010 *Physica E* **42** 755-758
[5] Nair R R, Blake P, Grigorenko A N, Novoselov K S, Booth T J, Stauber T, Peres N M R and Geim A K 2009 *Science* **320** 1308
[6] Peres N M R 2010 *Rev. Mod. Phys* **82** 2673
[7] Santoso I, Gogoi P K, Su H B, Huang H, Lu Y, Qi D, Chen W, Majidi M A, Feng Y P, Wee A T S, Loh K P, Venkatesan T, Saichu R P and Goos A 2011 *Phy. Rev. B* **84** 081403R
[8] Gogoi P K, Santoso I, Saha S, Wang S, Castro N A H, Loh K P, Venkatesan T and Rusydi A 2012 *Europhys. Lett.* **99** 67009
[9] Ando T, Zheng Y and Suzuura H 2002 *J. Phys. Soc. Jpn.* **71** 13118
[10] Zabihi F, Chen Q, Xie Y and Eslamian M 2016 *Superlattice and Microstructure* **100** 1177-1192
[11] Razza S, Giacomo D F, Matteocci F, Cina L, Palma L A, Casaluci S, Cameron P, D'Epifanio A, Licoccia S, Reale A, Brown M T and Carlo A D 2015 *J. Power Sorce* **277** 268-291
[12] Hau S K, Yip H L, Zou J and Jen A K 2009 *Organic Electronic* **10** 1401-1407
[13] Chen Q, Zabihi F and Eslamian M 2016 *Synthetic Metals* **222** 309-317
[14] Chen J, Yao B, Li C and Shi G 2013 *Carbon* **64** 225-229

IOP Conf. Series: Journal of Physics: Conf. Series **1011** (2018) 012008 doi:10.1088/1742-6596/1011/1/012008

Microwave Absorption Properties of $La_{0.8}Ca_{0.2-x}Ag_xMnO_3$ (x=0.05; x=0.15) Synthesized by Sol-Gel Method

B Kurniawan[1], **W Laksmi**[1], **N A Sahara**[1]

[1]Department of Physics, FMIPA Universitas Indonesia, Kampus UI Depok, Depok 16424, Indonesia

Email: bkuru@fisika.ui.ac.id

Abstract. Microwave absorption properties of $La_{0.8}Ca_{0.2-x}Ag_xMnO_3$ (x= 0.05; 0.15) is reported in this paper. Lanthanum manganite materials was reported as a potential absorber material [1][2][3]. In this paper, the material was synthesized by sol-gel method, calcined at 550°C, and sintered at 900°C. The material was characterized by X-Ray Diffractometer (XRD), and we found that the materials were single phased. Through SEM-EDS characterization it is found that the materials have compositional purity. The resistivity of the materials is obtained by four point probe method, and it is shown that Ag doped decreases the resistivity of the materials. Reflection loss of $La_{0.8}Ca_{0.15}Ag_{0.05}MnO_3$ reaches -4.470 dB and $La_{0.8}Ca_{0.05}Ag_{0.15}MnO_3$ reaches - 7.953 dB.

1. Introduction

Research on microwave absorber material has been conducted by a number of researchers. Microwave Interference phenomenon is the reason why the material with a good microwave absorption is absolutely necessary. Perovskite manganite material with a formula ABO_3 is one of common material that examined in terms of microwave absorption. Group of lanthanum manganite material with A is Lanthanum, B is manganese, and O is oxygen, have been examined in terms of the microwave absorption, magnetoresistance properties and magnetocaloric.

Modification by dope site-A with another element done to see the impact on three properties that has been mentioned before. For example the addition of Silver on site-A [2]. The addition of silver on site-A makes this material has wide microwave absorption from 2 GHz to 18 GHz. The wide range of microwave absorption is the profit for this material to be used in wider frequency range. Other studies have found that Lanthanum Manganite doped by Calcium on Lanthanum site causes the reflection loss at microwave absorption increases to -31 dB [3]. But it has a very thin absorption region, about 11 GHz up to 13 GHz. This study will report the results of Lanthanum Manganite material doped with Silver and Calcium at once. The composition is $La_{0.8}Ca_{0.2-x}Ag_xMnO_3$ with x = 0.05 and 0.15. The method that used in this research is sol-gel method. This method is relatively easy and have a lower sintering temperature compared to solid state methods [4] and so on.

2. Experiment

All the raw materials that used during the synthesized are nitrate compound. This could simplify the process of sol-gel, because nitrate can be soluble in aquabidest. The entire raw materials blended on the magnetic stirrer. The temperature kept at 80°C for 3 hours, with a pH of 7 until the entire solution turned into a gel and then hardened. Afterwards the samples entered the dehydration process for 3 hours. Then calcination process at 550 ° C over 5 hours, this process eliminated organic compounds or nitrates which were still contained in the material. The last is the process of sintering at a temperature of 900 ° C for 24 hours. A number of tests conducted after passing through this process.

3. Results and discussions

3.1. X-Ray Diffractometer

The samples were characterized by X-Ray Diffractometer. In the Figure 1, we can see that the additional of Ag composition, with x=0.15 changes the crystal structure of material. This marked by HKL value that changes, detected by the HKL of the highest peaks on the pattern. Based on data processing the crystal structure of $La_{0.8}Ca_{0.15}Ag_{0.05}MnO_3$ material is orthorhombic. Meanwhile crystal structure of $La_{0.8}Ca_{0.05}Ag_{0.15}MnO_3$ material is rhombohedral. Previous research reported that Lanthanum Manganites with Ca and Ag doped changed it crystal structure [5]. Based on scherrer's formula, the crystallite size of material by x=0.15 must be smaller than x=0.05. This is confirmed by SEM results at the next chapter.

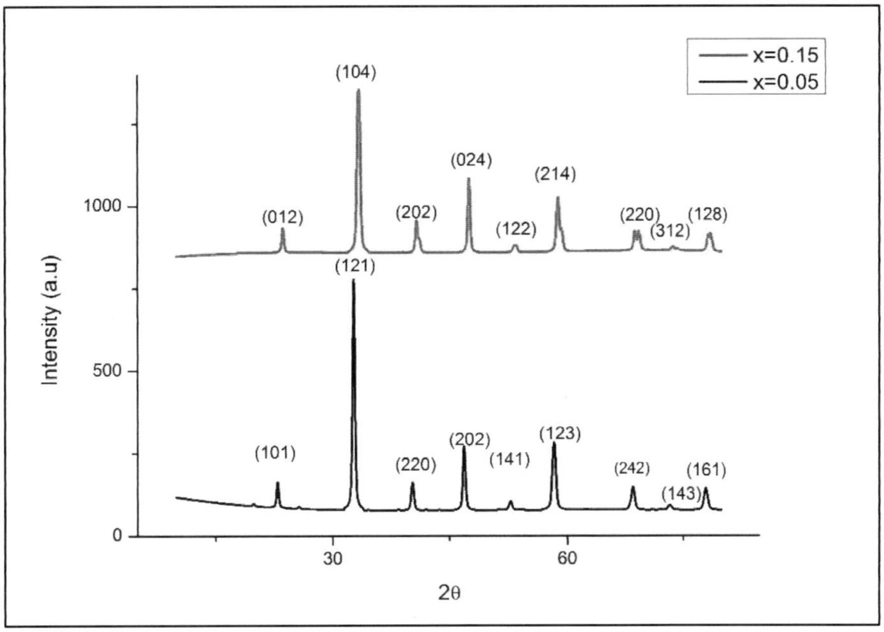

Figure 1. XRD Patterns of $La_{0.8}Ca_{0.15}Ag_{0.05}MnO_3$ and $La_{0.8}Ca_{0.05}Ag_{0.15}MnO_3$

Goldschmidt tolerance factor calculation verified the changing of crystal structure material. The Goldschmidt equation (τ) shows below:

$$\tau = \frac{r_A + r_O}{\sqrt{2}(r_B + r_O)} \qquad (1)$$

With r_A is the ionic radii of La^{3+}, Ca^{2+}, and Ag^{+}. Meanwhile r_O and r_B are the ionic radii of O^{2-} and $Mn^{3+/4+}$. Perovskite manganite structure distorted into rhombohedral when $0{,}96 \geq \tau \geq 1$. But when $\tau < 0{,}96$ the crystal structure turn into orthorhombic. The calculation using ionic radii reference shows that $La_{0.8}Ca_{0.05}Ag_{0.15}MnO_3$ has τ=0.95 and $La_{0.8}Ca_{0.05}Ag_{0.15}MnO_3$ τ=0.96 [6]. This calculation verified that the crystal structure of material was changing.

Table 1. Lattice parameter of $La_{0.8}Ca_{0.15}Ag_{0.05}MnO_3$ and $La_{0.8}Ca_{0.05}Ag_{0.15}MnO_3$

Material	a (Å)	b (Å)	c (Å)	V (Å^3)
$La_{0,8}Ca_{0,15}Ag_{0,05}MnO_3$	5.500	7.740	5.467	232.476
$La_{0,8}Ca_{0,05}Ag_{0,15}MnO_3$	5.514	5.514	13.366	352.011

3.2. SEM-EDS results

SEM and EDS results of $La_{0.8}Ca_{0.05}Ag_{0.15}MnO_3$ and $La_{0.8}Ca_{0.15}Ag_{0.05}MnO_3$ are reported. Figure 2 and figure 3 show the morphology of materials. By these pictures we can approximate the size of particles. For of $La_{0.8}Ca_{0.15}Ag_{0.05}MnO_3$ and $La_{0.8}Ca_{0.05}Ag_{0.15}MnO_3$ the size of materials are 203 nm and 136 nm respectively. These results verified the previous approximation with Scherrer's formula that obtained the crystallite size with more silver composition is smaller then the other one.

Figure 2. The morphology of $La_{0.8}Ca_{0.15}Ag_{0.05}MnO_3$

Figure 3. The morphology of $La_{0.8}Ca_{0.05}Ag_{0.15}MnO_3$

To verified that the materials are compositionally pure, the EDS characterization are reported. The composition of materials shown in table 2.

Table 2. The results of EDS characterization

	At% (Calculation)					At% (Observation)				
X	La	Ca	Ag	Mn	O	La	Ca	Ag	Mn	O
0.05	16	3	1	20	60	15.77	2.96	0.95	17.54	62.7
0.15	16	1	3	20	60	15.64	2.05	1.98	17.71	62.62

The additional of silver shows by the increasing from silver percentage at the materials. However this method is semi-quantitative calculation. The composition of the elements not 100% accurate, according to the formula $La_{0.8}Ca_{0.2-x}Ag_xMnO_3$. But as shown in table 2, the additional of x relates with the increasing of Ag in atomic percent. Also there is no element that is shown in the table. It is caused by the purity of the samples that is not contained any elements besides La, Ca, Ag, Mn, and O.

3.3. Resistivity by four point probe methods

The resistivity of materials decreases as the composition of Ag increases. $La_{0.8}Ca_{0.15}Ag_{0.05}MnO_3$ material resistivity is 0.036 Ω.m, and $La_{0.8}Ca_{0.05}Ag_{0.15}MnO_3$ material resistivity is 0.013 Ω.m. The resistivity decreases may caused by the electrical properties of Ag as a good electrical conductor that has very small resistivity. Another conjecture this is caused by crystal structure of material that changed.

The microwave absorption material itself, represented by reflection loss. This value shows how much material absorb the microwave that given. The reflection loss calculated by the equation below [7].

$$RL=20\log\left|\frac{z_{in}-z_0}{z_{in}+z_0}\right| \qquad (2)$$

Reflection loss relates with impedance in (z_{in}) and impedance 0 (z_0). The impedance itself relates with permittivity (electrical property,ε), permeability (magnetic property, μ), and also material thickness. These three factors may influence the ability of materials to absorb the microwave. The research before claimed that higher resistivity at semiconductor resistivity range, could posses more reflection loss. Although our result shows that the material resistivity decreases after doped by more Ag, these value still in the semiconductor resistivity range. Further discussion is reported in the next part.

3.4. Vector Network Analyzer Characterization

Reflection loss of materials are investigated by vector network analyzer. This characterization investigated the reflection loss of materials at X-Band frequency range, 8-12 GHz. The maximum reflection loss of $La_{0.8}Ca_{0.15}Ag_{0.05}MnO_3$ reaches -4.470 dB at frequency 19.24 GHz, and $La_{0.8}Ca_{0.05}Ag_{0.15}MnO_3$ reaches -7.953 dB at frequency 14.68 GHz. In the X-Band frequency range (8-12 GHz) the maximum reflection loss of $La_{0.8}Ca_{0.15}Ag_{0.05}MnO_3$ and $La_{0.8}Ca_{0.05}Ag_{0.15}MnO_3$ only reach -4.310 dB and -4.674 dB respectively. Although the resistivity of $La_{0.8}Ca_{0.05}Ag_{0.15}MnO_3$ is lower than $La_{0.8}Ca_{0.15}Ag_{0.05}MnO_3$ the reflection loss shows the opposite. With x = 0.15, the material has deeper reflection loss.

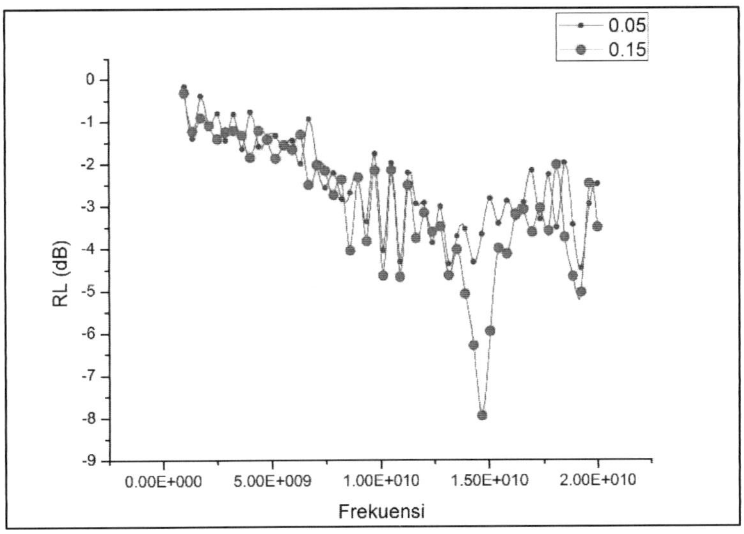

Figure 4. The reflection loss of $La_{0.8}Ca_{0.15}Ag_{0.05}MnO_3$ and $La_{0.8}Ca_{0.05}Ag_{0.15}MnO_3$

As we can see at the equation (2) the reflection loss depends on impedance. Impedance itself determined by the equation below [8].

$$z_{in}=\left(\frac{\mu}{\varepsilon}\right)^{\frac{1}{2}}\tanh\left(j2\pi fd(\mu\varepsilon)^{\frac{1}{2}}\right) \qquad (3)$$

$$z_0=\left(\frac{\mu_0}{\varepsilon_0}\right)^{\frac{1}{2}} \qquad (4)$$

Where f is electromagnetic frequency, d is material thickness, μ permeability, ε is permittivity. Equation 3 and 4 show that is not only dielectric property that can determine the value of impedance in, but also magnetic properties, the dimension of samples, and also the frequency that apply on the materials. The VNA characterization shows that imaginary permeability of $La_{0.8}Ca_{0.05}Ag_{0.15}MnO_3$ higher than $La_{0.8}Ca_{0.15}Ag_{0.05}MnO_3$. The magnetic loss of material represented by imaginary permeability [8]. It shows that $La_{0.8}Ca_{0.15}Ag_{0.05}MnO_3$ has higher reflection loss caused by its magnetic loss.

4. Conclusion

XRD pattern shows that all the samples were single phase. Lattice parameters show that the crystal structure is orthorhombic for x=0.05 and rhombohedral for x=0.15. The resistivity for x=0.05 and x=0.15 respectively 0.036 Ω.m and 0.013 Ω.m. VNA characterization shows that the Reflection loss of $La_{0.8}Ca_{0.15}Ag_{0.05}MnO_3$ reaches -4.470 dB and $La_{0.8}Ca_{0.05}Ag_{0.15}MnO_3$ reaches - 7.953 dB.

Acknowledgement

This research fully sponsored by PITTA 2017. With grant contract number: 633/UN2.R3.1/HKP.05.00/2017 "Rekayasa material berbasis lanthanum".

References

[1] Y. L. Cheng, J. M. Dai, D. J. Wu, and Y. P. Sun, "Electromagnetic and microwave absorption properties of carbonyl iron /La0.6Sr0.4MnO3," *J. Magn. Magn. Mater.*, vol. 322, no. 1, pp. 97–101, 2010.

[2] S. Huang *et al.*, "Effect of Ag substitution on the electromagnetic property and microwave absorption of LaMnO 3," *J. Magn. Magn. Mater.*, vol. 324, no. 19, pp. 3149–3153, 2012.

[3] F. Jiang, J. Zheng, L. Liang, M. Zhang, and Y. Wang, "Microwave absorbing properties of La0.1Ca0.9MnO3 porous microsphere synthesized by method of precipitation," *J. Mater. Sci.*, no. 3, pp. 2243–2247, 2015.

[4] R. Shreekala, M. Rajeswari, S. P. Pai, S. E. Lofland, and V. Smolyaninova, "Improved properties of La2 / 3Ca1 / 3MnO3 thin films by addition of silver," vol. 2857, pp. 2–5, 1999.

[5] M. Koubaa, W. Cheikhrouhou-koubaa, and A. Cheikhrouhou, "Effect of Ag doping on structural , magnetic and magnetocaloric properties of La0.65Ca0.35-xAgxMnO3 Manganites," vol. 473, pp. 5–10, 2009.

[6] B. Y. R. D. Shannon, M. H, N. H. Baur, O. H. Gibbs, M. Eu, and V. Cu, "Revised Effective Ionic Radii and Systematic Studies of Interatomie Distances in Halides and Chaleogenides," 1976.

[7] Y. L. Cheng, J. M. Dai, X. B. Zhu, D. J. Wu, Z. R. Yang, and Y. P. Sun, "Enhanced Microwave Absorption Properties of Intrinsically Core / shell Structured La 0 . 6 Sr 0 . 4 MnO 3 Nanoparticles," *nanoscale Res. Lett.*, pp. 1153–1158, 2009.

[8] S. Zhang and Q. Cao, "Electromagnetic and microwave absorption performance of some transition metal doped La 0 . 7 Sr 0 . 3 Mn 1 − x TM x O 3 ± 1 (TM = Fe , Co or Ni)," *Mater. Sci. Eng. B*, vol. 177, no. 9, pp. 678–684, 2012.

Electrical, Magnetic and Microwave Absorption Properties of M-type Barium Hexaferrites ($BaFe_{12-2x}Co_xNi_xO_{19}$)

Susilawati[1], A Doyan[1], H Khair[1], M Taufik[2], Wahyudi[2]

[1]Magister Science, Post Graduate of Mataram University, Mataram, Indonesia
[2]Faculty of Education and Teacher Traning of Mataram University, Mataram, Indonesia

Email: susilawatihambali@yahoo.co.id

Abstract. M-type barium hexaferrites synthesis with Co-Ni doping ion ($BaFe_{12-2x}Co_xNi_xO_{19}$) based on natural iron sand of Loang Balok beach, Lombok, Indonesia, to be applied as a microwave absorbent material using co-precipitation method. The materials used in the synthesis process are magnetite minerals (Fe_2O_3 and Fe_3O_4), 12M HCl, NH_4OH 37%, $CoCl_2.6H_2O$ and $NiCl_2.6H_2O$. This research to investigate the effect of doping ion concentration variation (x = 0.0, 0.6 and 1.0) and calcination temperature (T = 80, 600, and 800°C) on electrical and magnetic properties and microwave absorption as well. The samples were characterized using Vibrating Sample Magnetometer (VSM) and Network Vector Analyzer (VNA). The result from VSM showed that the coercivity value decreased when doping ion concentration and calcination temperature increased (0.151 Tesla at 600°C for x = 0.0 and 0.044 Tesla at 800°C for x = 1.0. The value of magnetic saturation and the magnetic remanence increased with increasing ion concentration (Ms = 0.327 emu/g at x = 0.0 increased to 35.4 emu/g at x = 1.0) and Mr = 0.148 emu/g for x = 0.0 increased to 15.6 emu/g at x=1.0, this indicates that the sample has been soft magnetic. The result from VNA showed that the electrical conductivity values measured in the range 8.0-15.0 GHz indicate that the sample is a semiconductor (6.149×10^{-6} -5.975×10^{-4} S/cm). It also showed that the microwave absorption properties increased at higher concentration of doping ions and the calcination temperature would increase the value of Reflection Loss (RL). The maximum RL value of the sample is -14.47 dB at 12.38 GHz, and the absorption coefficient of 96.43%. These results indicate that the $BaFe_{12-2x}Co_xNi_xO_{19}$ sample can be applied as a microwave absorbent material on X-band to Ku-band frequency.

Keywords: Barium M-Hexaferrites, electrical and magnetic properties, natural iron sand, microwave absorber

1. Introduction

The rapid technological developments have an impact on increasing pollution of electromagnetic interferences (EMI) in the environment that can affect life, so research on microwave absorbent is an important topic. One of the materials that can be utilized as a microwave absorbent is Barium M-Hexaferrites (BaM). The ferrites (Fe) is basic material of synthesizing permanent magnet BaM, in which the iron element is obtained by separating the magnetite material from the natural iron sand. The utilization of natural iron sand is currently less optimal because it is only used as a mixture of cement. In fact, the natural iron sand contains iron oxide in the form of magnetite (Fe_3O_4), maghemit (Y-Fe_2O_3) and hematite (α- Fe_2O_3) which has the potential to be processed into various products with higher economic value. The natural iron sand used in this study came from the Loang Balok beach in Sekarbela Sub-district, Mataram City, Lombok, Indonesia. The BaM has a hexagonal molecular structure ($BaFe_{12}O_{19}$), high saturation magnetization (78 emu/g), large coercivity field (6700 Oe), high Curie temperature (450°C), chemical stability and good corrosion resistance, with magnetic and electric properties adjustable according to the required application [1]. However, the high coercivity field causes weak absorption properties so it is less effective to use as microwaves absorbent material. To overcome this problem, we need substitution of a Fe^{3+} ion with other metal cations of similar sizes, such as: Al^{3+}, Ga^{3+}, and Cr^{3+} in order to decrease the magnetism of BaM from hard to soft magnetic [2]. A number of studies have been conducted to modify the magnetic properties of BaM such as the use of Co-Zn, Co- Mn, and Mn -Ni ions [3,4], Mn-Ti, Co-Ti, Ni-Ti, and Zn-Ti [5], As well as Co-Zn and Ni-Zn [6]. The substitution carried out in this study would use a combination of transition metal elements Co-Ni, because these two elements have radii and ionic configuration similar to Fe^{3+} (ionic radii Fe = 0.077 nm; Co = 0.072; and Ni = 0.069) was then analyzed for its effect on the electrical and magnetic properties of $BaFe_{12-2x}Co_xNi_xO_{19}$ based on natural iron sand produced to be applied as a microwave absorbent material.

2. Materials and Methods

The material used in sample synthesis was 368.937 grams of natural iron sand in magnetite compound (Fe_3O_4), $BaCO_3$ Merck KGaA 64271 Darmstadt. The solvent utilizes hydrochloric acid (HCl) 37% 12.063 M, NH_4OH solution 6.5 M and distilled water. The materials for doping Cobalt (II) Chloride Hexahydrate ($CoCl_2.6H_2O$) PA and Nickel (II) Chloride Hexahydrate ($NiCl_2.6H_2O$) PA. The synthesis process was carried out as follows. The separation of natural iron sand by using a permanent magnet (3.6 T) to separate the magnetite material from the sand. The Ferrite (Fe) content testing in a magnetic material with Atomic Absorption Microscopy (AAS) type Perkin Elmer Analyst 400 and X-Rays Fluorescent (XRF) Rigaku Supermini 200 type. The separation of Fe_3O_4 as the basic ingredient of synthesis. The sample synthesis using co-precipitation method [3]. The Vibrating Sample Magnetometer (VSM) Oxford type for its magnetic properties, the Network Vector Analyzer (VNA) Advantest R3770 in range frequency 300 kHz - 20 GHz at 220 volt and 1 ampere to measure the electrical properties and the microwave absorption.

3. Results and Discussion

The magnetic material obtained from 50.033 g of sand of Loang Balok Beach after separation using permanent magnet is 46.775 g. The analyzed using AAS was obtained iron 12.816 mg/g, the result from XRF show that iron element is 62.1% weight in the Fe_2O_3 compound of 51.5% weight. These results were reinforced by XRD show that the dominant phase generated is hematite (Fe_2O_3) and magnetite (Fe_3O_4) phases corresponding to the average crystal size of 574.5967Å.

The magnetic properties of samples described in Table 1 and Figure 1. Theoretically BAM synthesized based $FeCl_3$ from the manufacturer has a value of Ms of 72 emu/ g and Hc value of 6.7 T. When compared with the results obtained have shown impaired Hc Which is very significant to 0.151 T for sample $BaFe_{12}O_{19}$ (x = 0,0); 0.060 T for samples $BaFe_{10,8}$ $Co_{0,6}$-$Ni_{0,6}$ O_{19} (x = 0,6) and 0,044 T for sample $BaFe_{10}$ Co-NiO_{19} (x = 1,0). It can be concluded that the samples $BaFe_{12-x}$ $Co_xNi_xO_{19}$ made from iron sand nature doped by metal Co-Ni could reduce the magnetic properties of samples from hard into a soft magnetic that can be applied as a mineral absorber of microwaves. The other studies have shown that mineral absorbing radar waves most
Preferably at low coercive value of 0.0506 Tesla at x = 0.3 [7].

Table 1. The Magnetic properties of BaM with different concentration of ion doping Co-Ni.

x	Sample	Hc (T)	Mr (emu/g)	Ms (emu/g)
0	$BaFe_{12}O_{19}$	0,151	0,148	0,327
0,6	$BaFe_{10,8}(Co\text{-}Ni)_{0,6}O_{19}$	0,060	2,84	6,79
1,0	$BaFe_{10}(Co\text{-}Ni)_{1,0}O_{19}$	0,044	15,6	35,4

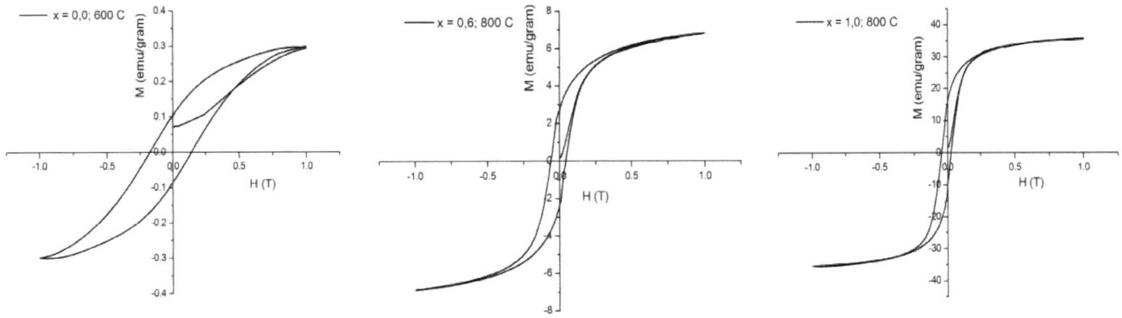

(a) (b) (c)

Fig. 1. The Hysteresis loops of a). $BaFe_{12}O_{19}$ b). $BaFe_{10.8}Co\text{-}Ni_{0.6}O_{19}$ c). $BaFe_{10}Co\text{-}NiO_{19}$

The electrical properties of the sample were characterized by using VNA, the conductivity value was in the range of $2.892E^{-05}$ - $8.059\ E^{-04}$ S/cm for sample $BaFe_{12}O_{19}$ (x = 0.0); $7.634E^{-05}$ - $8.078\ E^{-04}$ S/cm for sample $BaFe_{10.8}\ Co_{0.6}\text{-}Ni_{0.6}\ O_{19}$ (x = 0.6); and $7.318E^{-05}$ - $6.086\ E^{-04}$ S/cm for sample $BaFe_{10}Co\text{-}NiO_{19}$ (x = 1.0). These results indicate that the conductivity value is in the range of semiconductor properties since they are in the range of 10^{-7} - 10^{3} S/cm [8]. As a microwave absorbent, the mineral must be semiconductor because it can convert microwave into heat energy, form an electric field on an absorbent surface. The current will flow as the surface current, when the current flows into the absorber, the microwave energy will be converted in the form of heat energy [9].

Figure 2(a) showed that the permittivity (ε) and permeability (μ) values of the samples in complex numbers. The real value indicates the magnitude of the microwave energy stored in the material and the imaginary one shows the amount of microwave energy which is dissipated in the form of energy loss [10]. This combination of real and imaginary values are the determinant of the frequency of microwave absorbent material. The real permittivity value (ε') for all three samples tended to be constant at frequency (10.0-11.8) GHz and to be increased at frequency (12.0-15.0) GHz. The value of ε' increases with the increase of the doping mole fraction and the calcination temperature (from 17.8 at x = 0.0 to 20.09 at x = 1.0), but does not apply to doping x = 0.6 with the decreasing ε' value. The increase in the value of ε' is due to the transformation of Fe^{3+} ions into Fe^{2+} ions due to doping causing polarization on the sample surface [11]. A decrease in the value of ε' occurring in the sample by doping x = 0.6 because the on covering leads to the occurrence of lattice distortion thereby increasing electron scattering and resistivity [12].

Figure (2b) showed that a graph of the relationship between the imaginary permittivity (ε") of the sample at the measurement frequency range (10.0-15.0) GHz. Compared with the real permittivity value of the sample, the resulting imaginary permittivity value is lower. This means that the ability to form an electric field in the sample is higher when compared to the ability of samples in releasing electrical energy. The resulting permeability (μ) value (Figure 2b) decreases as the doping mole fraction and the calcination temperature increased. For the sample x = 0.6 it yields a real part of permeability (μ') lower than the sample x = 0.1 and results for imaginary part of permeability (μ") higher than the sample x = 0.1 compared to x = 0.6.

IOP Conf. Series: Journal of Physics: Conf. Series **1011** (2018) 012009 doi:10.1088/1742-6596/1011/1/012009

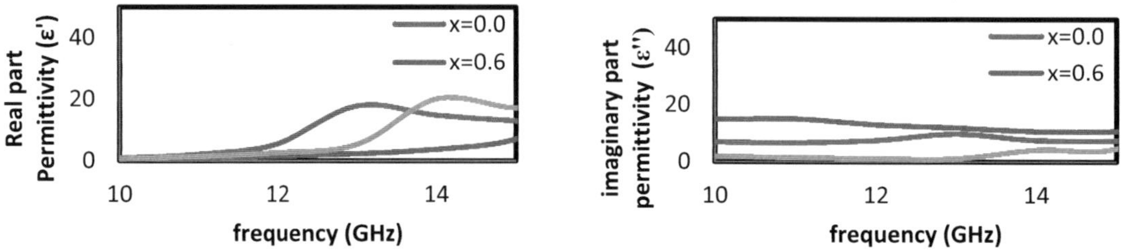

Fig.2(a). The Value of Sample Permittivity

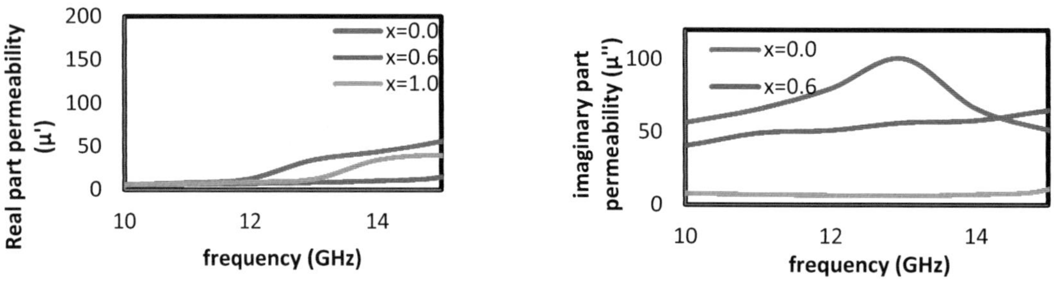

Fig. 2(b). The value of Permeability

The identification of microwave absorption properties was done through characterization using VNA resulting in data s parameters consisting of values S_{11} and S_{21} in the form of real and imaginary numbers used to determine the Reflection Loss (RL) value of the sample. The real and imaginary part of permittivity (ε) dan permeability (μ) was using equation (1) and (2)

$$\varepsilon = \frac{\lambda_0^2}{\mu}\left\{\frac{1}{\lambda_c^2} - \left[\frac{1}{2\pi L}\ln\left(\frac{1}{T}\right)\right]^2\right\} \tag{1}$$

$$\mu = \frac{1-\Gamma}{\Lambda(1-\Gamma)\sqrt{\frac{1}{\lambda_0^2}-\frac{1}{\lambda_c^2}}} \tag{2}$$

$$Z_{in} = Z_0\sqrt{\frac{\mu_r}{\varepsilon_r}}\tanh\left(i\left[\frac{2\pi L}{\lambda_0}\right]\sqrt{\mu_r\varepsilon_r}\right) \tag{3}$$

Where Z_{in} is impedance through to the sample and Z_0 is impedance free space ~ 377 Ω. The value of reflection loss (RL) was calculated using equation (4).

$$RL = 20\log\left|\frac{(Z_{in} - Z_0)}{(Z_{in} + Z_0)}\right| \tag{4}$$

Figure 3 shows the absorption intensity value (RL) in the frequency range (8.0-15.0 GHz) for each sample. The maximum RL value for the $BaFe_{10}Co\text{-}NiO_{19}$ sample at a frequency of 12.38 GHz was -14.47 dB with an absorbency coefficient of 96.43%. The RL value and absorption coefficient indicate the ability of microwave absorption. The greater the negative RL value caused the greater the absorption capacity of the material to absorbent microwave [13]. Based on the data it is found that the higher concentration of Co-Ni doping fraction and calcination temperature cause the higher RL value and the resulting absorption coefficient.

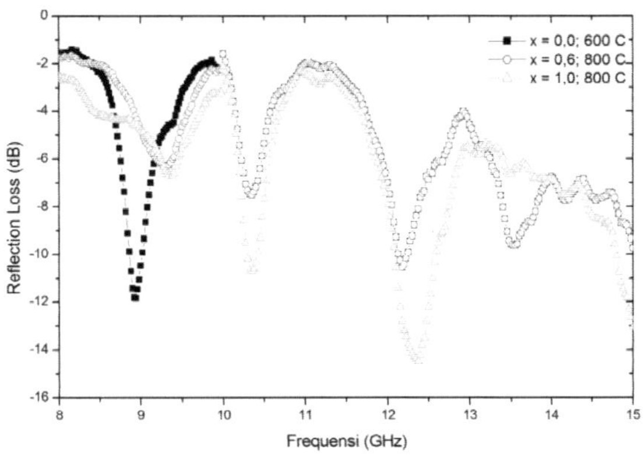

Fig. 3. The Value of Reflection Loss (RL) for Different Frequency and Concentration of Ion Doping

4. Conclusion

The $BaFe_{12-2x}Co_xNi_xO_{19}$ sample of the resulting natural iron sand has the dominant phase of Barium Iron Oxide and Hematite. The electrical properties of the sample show that the conductivity value in the range of 6.149 x 10^{-6} - 5.975 x 10^{-4} S/cm as a semiconductor. The magnetic properties of the samples indicate that soft magnetic, since the sample coercivity value has decreased significantly from 0.151 to 0.044 Tesla with an increase in the concentration of ion doping and temperature. The nature of microwave absorption yields a maximum RL value of -14.47 dB at 12.38 GHz frequency, and absorption coefficient of 96.43%. These results indicate that the $BaFe_{12-x}Co_xNi_xO_{19}$-based sample of natural iron sand can be used as a microwave absorbent material on x-band frequencies up to Ku-band.

Acknowledgement

We acknowledge the support from the analytical chemistry laboratory of Mataram University and the member of material physics research groups from the Physics Department of Bandung Institute of Technology.

Funding Information

We thank the education department of research and technology for providing the research grant.

References

[1] Saidah I N, and Zainuri M, 2012 Effect of pH Variation of HCl Solvent in Barium Synthesis M-Hexaferrit With Doping Zn ($BaFe_{11}$, 4Zn0, $6O_{19}$) Using Co-precipitation Method *Journal of Science and Art ITS* **(1)**. B41-B46

[2] Vinnik D A, Zherebtsov D A, Mashkovtseva L S, Nemrava S, Bischoff M, Perov N S, and Niewa R, 2014 Growth, Structural and Magnetic Characterization of Al-substituted Barium Hexaferrite Single Crystals *Journal of Alloys and Compounds* **615** pp 1043-1046

[3] Susilawati, Doyan A, and Sahlam, 2017 Synthesis and Characterization Materials M-Barium Hexaferrite Doping Ions Co-Mn Nano Particle. In IOP Conference Series: Materials Science and Engineering *IOP Publishing* Vol.**196** No. 1 p 012016

[4] Susilawati, Doyan A, and Munib M, 2015 Synthesis by Co-precipitation Method and Characterization of Nickel-doped Barium M-Hexaferrite ($BaFe_{12}O_{19}$) *Proceeding ICMSE* ISBN 9786021570425
[5] Sözeri H, Deligöz H, Kavas H, and Baykal A, 2014 Magnetic, Dielectric and Microwave Properties of M-Ti Substituted Barium Hexaferrites (M = Mn^{2+}, Co^{2+}, Cu^{2+}, Ni^{2+}, Zn^{2+}), *Ceramics International* **40** (6) 8645-8657
[6] Agustianto R, and Widyastuti W, 2014 Mol Fraction and Ph Variation against Magnetic Properties and Micro Barium Hexaferrite With A Sol-Gel Auto Combustion Method *Journal of Engineering ITS* **3** (1) F108-F112
[7] Rahmawati, Kusumawati D T, Rohmawati L, 2015 Variations of Fe_3O_4 Addition in Pani/Fe_3O_4 Alloys As Micro Wave Absorbents *Journal of Physics* vol**. 04** Number 01 pp 1-5.
[8] Said L M, 2013 Analysis of the Conductivity Properties of Electrics in Several Types of Materials with Remote Potential Methods *Journal of Teknosains* vol.**7** p. 66-67.
[9] Mahmood S H, Aloqaily A N, Maswadeh Y, Awadallah A, Bsoul I, Awawdeh M, and. Juwhari H 2015 Effects of Heat Treatment on the Phase Evolution, Structural, and Magnetic Properties of Mo-Zn Doped M-Type Hexaferrites In Solid State Phenomena *Trans Tech Publications* Vol. **232** pp 65-92
[10] Wagner T R, 1998 Preparation and Crystal Structure Analysis of Magnetoplumbitetype $BaGa_{12}O_{19}$ *J. Solid State Chem.* **136** pp 120-124
[11] C Li, Wang J, Wang J N, 2012 Magnetic and Microwave Absorbing Properties of Electro spun Ba(1-x) $La_xFe_{12}O_{19}$ Nanofibers *Journal of Magnetism and Magnetic Materials* **324** (7) 1305-1311.
[12] Zhang W, Bai Y, Han X, Qiao L, 2013 Magnetic Properties of Co-Ti Substituted Barium Hexaferrite. *Journal of Alloys and Compounds* **546** pp 234-238.
[13] Nature R S, Moradi M, Rostami M, Nikmanesh H, Moayedi R, Bai Y, 2015 Structural, Magnetic and Microwave Absorption Properties of Doped Ba-hexaferrite Nanoparticles Synthesized by Co-Precipitation Method *Journal of Magnetism and Magnetic Materials* **381** 1-9

IOP Conf. Series: Journal of Physics: Conf. Series **1011** (2018) 012010 doi:10.1088/1742-6596/1011/1/012010

EXAFS Study on $LiFePO_4$ Powders Produced From Two Sol-Gel Routes

V S I Negara[1], C Latif[1], W Wongtepa[2] and S Pratapa[1]

[1] Department of Physics, Institute Technology Sepuluh Nopember (ITS), Surabaya, Indonesia

[2] Synchrotron Light Research Institute (SLRI), Thailand

*Corresponding author: suminar_pratapa@physics.its.ac.id

Abstract. The local structure of $LiFePO_4$ powders has been investigated using Fe K-edge Extended X-Ray Absorption Fine Structure (EXAFS) Spectroscopy data. The synthesis of LFP powders was carried out using two different sol-gel methods. The raw materials for Fe source were ironstone and commercial precursor of $FeCl_2{\cdot}4H_2O$. Synthesis using natural materials produced two phases, namely $LiFePO_4$ olivine and $Li_3Fe_2(PO_4)_3$ nasicon, whereas that using a commercial product produced a single phase of $LiFePO_4$ olivine. The EXAFS data for both samples were collected at Synchrotron Light Research Institute (SLRI), Thailand. Fitting of the model on the experimental curve provided parameters that can be interpreted as the distance between Fe as the absorber and the nearest atoms on the LFP materials. The EXAFS data analysis has shown that synthesis of LFPs using different Fe sources gives slightly different nearest-neighbor distances, namely Fe-O of 0.21% - 0.23%, Fe-P of 0.14% - 0.16%, Fe-Fe of 0.12% for both samples, respectively.

1. Introduction

Lithium batteries have the required criteria as energy storage media for portable electronic devices since they exhibit large capacity and energy density, high electrochemical potential, durable and rechargeable [1], [2]. The lithium battery components are cathode, anode, electrolyte, and separator. The common cathode materials include $LiMn_2O_4$ [3], $LiCoO_2$ [4] and $LiFePO_4$ [5]. Among these cathode materials, $LiFePO_4$ is very promising because it has a theoretical capacity of about 170 mAh/g, a specific energy of 0.59 Wh/g, the density of 3.60 g/cm^3, the potential of 3.5 V, non-toxic and environmental friendly [1], [6]. However, it has a low electrical conductivity of approximately 10^{-9}–10^{-10} S/cm and slow diffusion rate about 10^{-12}-10^{-14} cm^2/s [7]. These characteristics are attributed to the general structural nature of the material.

The local structure is the distribution of atoms around the absorbing atom. The local structure examines the distance between the absorber atom and the nearest neighbor atoms, while the crystal structure studies about the distance between the diffraction planes in a crystal building. The local structure can be determined through Extended X-ray Absorption Fine Structure (EXAFS) data analysis. The EXAFS spectrum is affected by photoelectron scattering in the absorption range from ~30 to ~1000 eV above absorption edge and provides information about the local structure, the number of coordination, interatomic distance, Debye-Waller disorder parameter [8], [9].

Methods that can be used in $LiFePO_4$ synthesis included solid-state reaction, sol-gel, and hydrothermal.The sol-gel method, however, has advantages on controlling the morphology and particle size of the product, resulting in homogeneous particles and high phase purity. In this research, the basic materials used are a natural material, namely ironstone, and commercial product, i.e. $FeCl_2{\cdot}4H_2O$. The effect of the Fe source on the local structure of LFP was investigated using EXAFS. Differences in local structure may affect the physical properties, such as the electrical conductivity. The distance between the absorber atom and the nearest neighbor atoms should be larger than the ionic radius of Li^+ as the charge carrier, in order to make the ionic Li^+ move around the neighboring atoms.

In this study, we compare the results of LFP characterization produced by two different sol-gel methods. The studied issues are the formed phases and the local structure of LFP.

2. Experimental Section

2.1 Synthesis using natural materials

The ironstone as the Fe source was dissolved in HCl 37% at 100 °C. Then filtering was done to separate the solution with the nonmetallic elements. Li_2CO_3 (lithium carbonate, Merck) and $NH_4H_2PO_4$ (ammonium dihydrogen phosphate, Merck) were dissolved in HCl 37% and deionized water. Each solution was mixed dropwise with continuous stirring at temperature 100 °C for 1 hour. After that NH_4OH (ammonium hydroxide) was added to the solution until pH 4 was achieved while stirring was continued until a gel has formed. The gel was dried at temperature 300 °C. The precursors were milled for 6 hours using planetary ball milling and alcohol as mixing medium. The resultant was dried and calcined at 600 °C for 10 hours under a flow N_2 atmosphere at a heating rate of 5 °C min^{-1}. The product was a powder and denoted as sample A.

2.2 Synthesis using commercial materials

From resulting of stoichiometry, $FeCl_2{\cdot}4H_2O$ and $NH_4H_2PO_4$ were dissolved in deionized water, and Li_2CO_3 was dissolved in HCl 37%. The solutions were mixed dropwise with continuous stirring. The next steps are similar to those in the use of natural stone except that (1) the pH was 7, (2) gel formation temperature of 80 °C, (3) drying at 120 °C and (4) calcination at 700 °C for 10 hours under a flow Ar atmosphere with the heating rate of the furnace was 5 °C min^{-1}. The powder product was designated as sample B.

2.3 Analysis of EXAFS data

EXAFS measurement was carried out at beamline 8 at the Synchrotron Light Research Institute (SLRI), Thailand. Analysis of EXAFS data was processed using *Athena* and *Artemis* program using FEFF8 program-based theoretical standard calculation [10]. The files that can be fed on FEFF8 are "feff.inp" using the *Atom* program. For crystalline materials, the Stand Alone *Atom* program can also directly translate crystallographic data (cif) into the form of "feff.inp"[11]. The sequence of EXAFS data processing is first, normalized then interpolated to k-space. Information which can be obtained from an EXAFS data analysis includes the distance between atoms, the number of coordination and Debye-Waller factors (σ^2) [11].

The identification phase was verified by X-ray diffraction (XRD) using Match!2 program. The morphology of sample was observed by Scanning Electron Microscope (SEM). The local structure was obtained from *Extended X-ray Absorption Fine Structure* (EXAFS) measurement.

3. Result and Discussion

3.1 Identification phase of $LiFePO_4$

Figure 1 shows the X-ray diffraction pattern of Samples A and B. Sample A contains two phases, i.e. olivine $LiFePO_4$ (PDF 083-2092) and nasicon $Li_3Fe_2(PO_4)_3$ (PDF 078-1465), whereas Sample B contains only olivine. The presence of two phases in Sample A is due to the ironstone used as the Fe source has two Fe ions, i.e., Fe^{2+} and Fe^{3+}. LFP olivine has a Fe^{2+} oxidation state and LFP nasicon has a Fe^{3+} oxidation state [12]. Meanwhile, synthesis using $FeCl_2{\cdot}4H_2O$ as the Fe source gives single phase olivine. The reason is that the base material has only one type of Fe ion in the formation of LFP, that is Fe^{2+}.

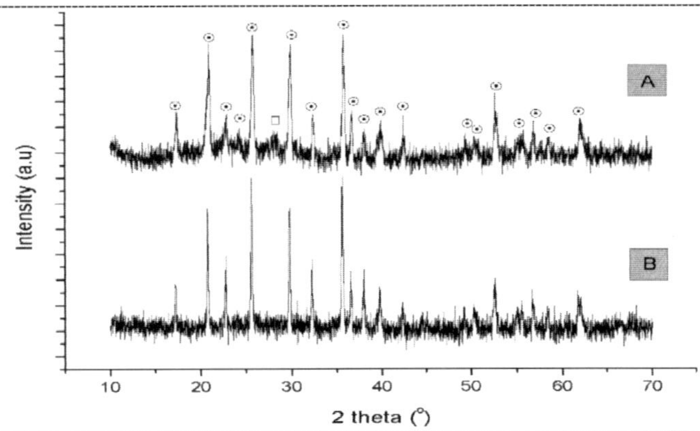

Figure 1. X-ray diffraction patterns of LFP synthesized using natural material (A) and commercial material (B). (⊙= $LiFePO_4$; □= $Li_3Fe_2(PO_4)_3$)

3.2 Local structure of $LiFePO_4$

The normalized background-subtract k^3-weighted $[k^3\chi(k)]$ Fe EXAFS pattern for LFP which the synthesis from Samples A and B are shown in Figure 2. It can be seen that in principle they have a similar peak shape.

Data $\chi(k)$ describes the oscillation of backscattering waves in the area around ~10Å, which provides information around the central atom (Fe absorber) [13]. The k-weight spectrum in k-space defines the area of the spectrum to be analyzed and depends on the element used. The $\chi(k)\cdot k^3$ patterns in k-space oscillate periodically indicating that $\chi(k)$ as a wave function derives from regularly arranged atoms. This ordered arrangement is then extracted in the form of atomic positions using Fourier transforms by changing EXAFS oscillations χ from k-space to r-space.

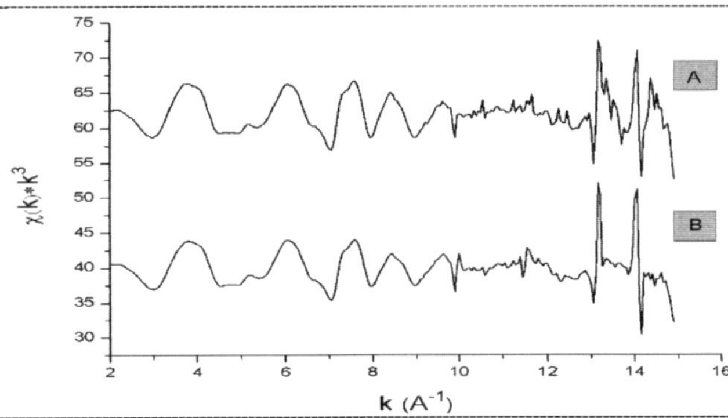

Figure 2. Normalized background-subtract k^3-weighted $[k^3\chi(k)]$ Fe EXAFS for LFP for samples A and B.

The results of the Fourier transform of the background-subtract k^3-weighted $(k^3\chi$ (k)] was used to determine the radial structure function of EXAFS Fe K-edge of LFP - see Figure 3. The function of the radial structure of the result of a match between the experimental results and the model obtained from the Fourier transform of the EXAFS Fe K-edge spectrum. The first high peak can be associated with the scattered electron that is released at the nearest atom of the oxygen-containing absorber. The second and third peaks appear as the effect of the scattered electron that is respectively released by phosphor and iron. Both samples have the same shape and intensity of Fe-O, Fe-P and Fe-Fe peaks that imply similar local structures [13].

The fitting of the Fe-edge EXAFS patterns provides an output of parameters that interprets the distance between Fe absorbing atom to the nearest atoms in the LFP materials as summarized in Table 1. The distance between atoms and coordination numbers for LFP samples A and B are not much different which indicates that their local structures are similar [13]. The natural LFP (A) has a slightly shorter interatomic distance as compared to the commercial LFP (B). The difference between interatomic distances for both samples is 0.12% for Fe-Fe, 0.14% - 0.16% for Fe-P and 0.21% - 0.23% for Fe-O. The difference in the distance between the atoms for each sample or the shift of atoms from its equilibrium position can be due to a charge difference of the structure that can generate an internal electric field. This internal electric field may affect the movement of charged particles in the structure. Charged particles with free movement would result in a greater electrical conductivity so that it affects the electrical properties of the LFP. It is interesting to note that Li^+ ion has an ionic radius of 0.9 Å [16], while the distance between Fe-O atoms is about 1.93140 Å and 1.93576 Å for LFP A and B respectively. Thus Li^+ ions in both LFP samples may perform intercalation equally well.

Table 1. Structural parameters for LFP samples A and B from Fe K-edge EXAFS data fitting.

	Sphere	Z_a-Z_b [a]	*CN* [b]	*r*(Å) [c]	σ^2 (Å²) [d]	*R-factor* (%)
	1	Fe-O	2	1.93140	0.02677	
	1	Fe-O	1	1.97600	0.05139	
	1	Fe-O	1	2.07410	0.16056	
Sample A	2	Fe-P	1	2.70140	0.00156	1.88
	2	Fe-P	1	3.07830	0.51775	
	2	Fe-P	3	3.11730	0.05089	
	3	Fe-Fe	4	3.71310	0.27080	
	1	Fe-O	2	1.93576	0.02740	
	1	Fe-O	1	1.98036	0.06359	
	1	Fe-O	1	2.07846	0.13774	
Sample B	2	Fe-P	1	2.70576	0.00209	1.85
	2	Fe-P	1	3.08266	0.58178	
	2	Fe-P	3	3.12166	0.09857	
	3	Fe-Fe	4	3.71746	0.16697	

[a] Central absorber and scattering atom correlation
[b] Coordination number
[c] Interatomic distance
[d] Debye-Waller factor

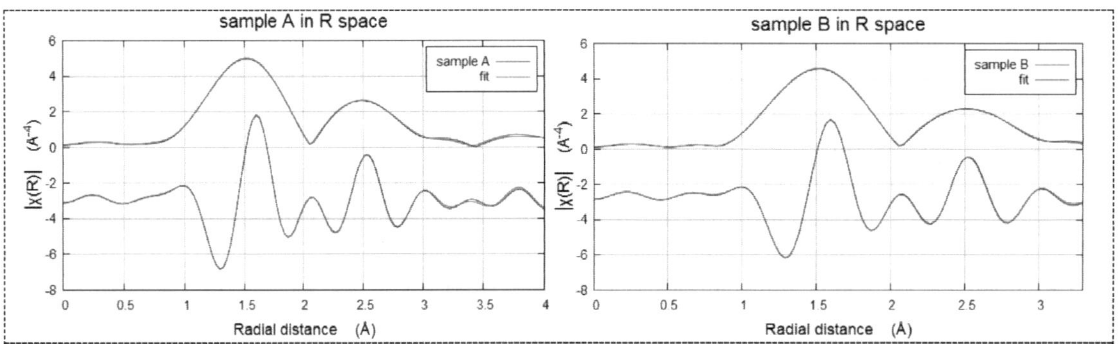

Figure 3. The result of fitting experimental data and EXAFS Fe K-edge model data for LFP

samples A and B.

4. Conclusions

The result of LFP synthesis using natural materials has two phases namely $LiFePO_4$ olivine and $Li_3Fe_2(PO_4)_3$ nasicon and for LFP synthesis using commercial materials have a single phase that is $LiFePO_4$ olivine. Analysis of their local structures using EXAFS data resulted in acceptable fittings with an R-factor value of less than 0.02. Both materials have almost the same peak shape and intensity which implies that their local structures are similar.

5. Acknowledgement

We thank the Synchrotron Light Research Institute (SLRI), Thailand for providing the beam-time through Beam Time Project Number 2618 (2017) and the BL8 scientists for their engaged support during the measurements. This research was also supported by PMDSU research (Pendidikan Magister Menuju Doktor untuk Sarjana Unggul), under contract 490/PKS/ITS/2017 awarded to CL.

6. References

[1] Wu B, Li N and Ren Y 2011 $LiFePO_4$ Cathode Material INTECH Open Access Publisher.
[2] Oswal M et al. 2010 A Comparative Study of Lithium-Ion Batteries Univ. South. Calif.
[3] Chew S Y et al. 2009 Thin Nanostructured $LiMn_2O_4$ Films by Flame Spray Deposition and In Situ Annealing Method *J. Power Sources* vol **189** no 1 pp 449–453.
[4] Mizushima K et al. 1980 Li_xCoO_2 (0<x<-1): A New Cathode Material for Batteries of High Energy Density *Mater. Res. Bull* vol **15** no 6 pp 783–789.
[5] Yamada A et al. 2001 Optimized $LiFePO_4$ for Lithium Battery Cathodes *J. Electrochem. Soc* vol **148** no 3 pp A224–A229.
[6] Padhi A K et al. 1997 Phospho-Olivines as Positive-Electrode Materials for Rechargeable Lithium Batteries *J. Electrochem. Soc* vol **144** no 4 pp 1188–1194.
[7] Satyavani T V S L et al. 2016 Methods of Synthesis and Performance Improvement of Lithium Iron Phosphate for High Rate Li-Ion Batteries: A Review *Eng. Sci. Technol. Int. J* vol **19** no 1 pp 178–188.
[8] Koningsberger D C and Prins R 1988 X-ray Absorption: Principles, Applications, Techniques of EXAFS, SEXAFS, and XANES Wiley, New York.
[9] Rehr J J and Albers R C 2000 Theoretical Approaches to X-ray Absorption Fine Structure *Rev. Mod. Phys* vol **72** no 3 pp 621–654.
[10] Ravel B and Newville M 2005 ATHENA, ARTEMIS, HEPHAESTUS: Data Analysis for X-ray Absorption Spectroscopy Using IFEFFIT *J. Synchrotron Radiat* vol **12** no 4 pp 537–541.
[11] Muchlis K 2017 Study on Crystal Structure and Local Structure Using Wide Angle X-ray Scattering (WAXS) and X-ray Absorption Spectroscopy (XAS) Synchrotron Radiation Data Institut Teknologi Sepuluh Nopember, Surabaya.
[12] Efhana D P et al. 2014 The Influence of Holding Time Variation of Calcination Process Cathode Material of Lithium Ferrophosphate *J. Tek. POMITS* vol **1** pp 1–5.
[13] Hsu K F et al. 2009 Formation Mechanism of $LiFePO_4$/C Composite Powders Investigated by X-ray Absorption Spectroscopy *J. Power Sources* vol **192** no 2 pp 660–667.
[14] Fornasini P and Grisenti R 2015 On Exafs Debye-Waller Factor and Recent Advances *J. Synchrotron Radiat* vol **22** pp 1242-1257.
[15] Rehr J J et al. 2017 XAFS Techniques for Catalysts, Nanomaterials, and Surfaces. USA: Springer International Publishing Switzerland.
[16] Shannon R D 1976 Revised Effective Ionic Radii and Systematic Studies of Interatomic Distances in Halides and Chalcogenides *Acta Cryst* **A32** pp 751-767.

Growth of $CH_3NH_3PbI_3$ Perovskite on Stainless Steel Substrate Layered by ZnO Nanoparticles Using One-Step Spin Coating Route

A Fuad[1,2*], A A Fibriyanti[1], N Mufti[1,2], A Taufiq[1,2], S Maryam[1], and N Hidayat[1,2]

[1]Department of Physics, Faculty of Mathematics and Natural Sciences, Universitas Negeri Malang (State University of Malang), Jl. Semarang No. 5, Malang 65145, Indonesia

[2]Central Laboratory for Mineral and Advance Materials, Faculty of Mathematics and Natural Sciences, Universitas Negeri Malang (State University of Malang), Jl. Semarang No. 5, Malang 65145, Indonesia

*Corresponding Author: abdulloh.fuad.fmipa@um.ac.id

Abstract: In this work, we report the preparation of $CH_3NH_3PbI_3$ perovskite using one-step spin coating route for solar cell application. CH_3NH_3I• PbI_2•DMF•DMSO complexes were coated on stainless steel as a subtrate layered by ZnO nanoparticles as an electron transport layer. To obtain samples with a special performance, we annealed the samples at a temperature of 100, 120, and 140 ^{0}C for 10 minutes. The samples were then characterized by means of XRD, SEM/EDX, and Spectroscopic Ellipsometry. The analysis of XRD data presented that the $CH_3NH_3PbI_3$ perovskites were successfully prepared and crystallized in tetragonal structure confirming from crystalline planes (110) and (220). Meanwhile, the particle size of the samples prepared at a temperature of 100, 120, and 140 ^{0}C presented 42.96, 54.73, and 55.19 nm, respectively with coincide with the SEM images. The results indicated that the increase in temperature during synthesis influenced the particle growth. Furthermore, the characterization using Spectroscopic Ellipsometry exhibited that the $CH_3NH_3PbI_3$ successfully layered on the substrate sizing nano metric scale that open high opportunity to be applied to solar cells with high performance.
Keywords: $CH_3NH_3PbI_3$, perovskite, nanoparticle, one-step spin coating, solar cell

1. Introduction

Organometal halide $CH_3NH_3PbX_3$ ($CH_3NH_3^+$ = MA, X=I, Cl or Br) based Perovskite Solar Cells (PSC) become the main promising candidate of photovoltaic tools in the recent years since they have some excellences such as band gap engineering capabilities (~1.5 eV) [1], the high mobility in bringing the charge and the low recombination speed as well as the big absorption power [2,3]. The characteristic absorption and perovskite light sensitivity are at the wave length of 300-800 nm, especially around 500 nm [4]. In 2009, Power Conversion Efficiency (PCE) of PSC increased from 3.8 % [5] to approximately 21% in 2016 [6]. PSC with PCE which is more than 20% consists of n-i-p structure, in which the perovskite functions as intrinsic semiconductor, TiO_2 as an electron transport (n-type layer) and Spiro-OMeTAD as a hole transport (p-type layer) [7]. The increase in the efficiency of perovskite solar cells is influenced very much by the layer structure [8], chemical composition of perovskite and its fabrication method [9].

In general, there are two methods to make perovskite solution of solar cell namely one step and two steps [10]. Yazdi et al. reported that the perovskite film producing by using one-step method was easier in controlling solution. Furthermore, by giving anti solvent, the perovskite had a good crystal structure and presenting a higher efficiency than that of two steps method [11]. However, PSC efficiency through one step spin coating method is influenced by solvent control, preheating condition, substrate surface condition and annealing temperature [12]. In 2016, Zhu et al. reported that PSC fabrication used one step spin coating method with the variation of the preheating temperature of 40 ^{0}C, and 60 ^{0}C and the efficiency was 18.64% with the good stability after it was stored for 20 days at the preheating temperature of 40 ^{0}C [10]. Although PSC research develops fast, the toxicity, stability, the selection of layer structure type and the size of crystal grain are rarely learned. Therefore, the further study of such topics is needed in PSC development.

In this work, PSC was grown up by using one step spin coating consisting of Stainless steel/ZnO/$CH_3NH_3PbI_3$ layer. Stainless steel functions as a substrate layer since it is good and resistant to corrosion, heat, and wear out compared to plastic and polymer [13]. For electron transport layer, ZnO was used as the change of TiO_2, since the ZnO conductivity is higher than TiO_2[14]. $CH_3NH_3PbI_3$ was used as perovskite material since its fabrication process is relatively easy, its hole mobility is high and its recombination speed is low [15]. Meanwhile, this work did not use the spiro-OmeTAD material because in original form, spiro-OmeTAD has low conductivity so that it still needs dopant and its price is relatively expensive [16]. In this work, solar cell perovskite was fabricated by using one step spin coating method with the variation of annealing temperature to get the proper film texture for PSC by knowing the structure, morphology, and thickness of the $CH_3NH_3PbI_3$ film.

2. Materials and Methods

2.1 Growth of ZnO Nanoparticles

ZnO Nanoparticles were synthesized by using sol gel spin coating method on a stainless steel substrate. Before using the stainless steel, it was washed with ethanol, acetone, and deionized water in an ultrasonic cleaner. Zinc Acetate Dihydrate (ZnAc) was diluted in ethanol and it was stirred at a temperature of 70 ^{0}C. Subsequently, it was added with Monoethanolamine (MEA) into the solution for two hours. The solution was spun on a substrate with the speed of 3000 rpm. The preheating and annealing process was then done at a temperature of 150 ^{0}C for 10 minutes and 450 ^{0}C for 2 hours.

2.2 Growth of Perovskite Solar Cells

The $CH_3NH_3PbI_3$ as the PSC film layer was prepared by using one step coating method and the result of the solution was layered on ZnO film. First was making methylammonium iodide (CH_3NH_3I) powder by mixing hydroiodic acid (HI) with methylamine (CH_3NH_3). It was continued by mixing the CH_3NH_3I powder that had been prepared with Dimethylsulfoxide (DMSO) solution and N,N-dimethylformamide (DMF) that was stirred for 25 minutes. The solution resulted in the spin coating on the layer surface of ZnO nanoparticles/stainless steel. The coating process was done through two ways; they were by adding and without adding the anti solvent of anti-solvent diethyl ether. After that, the sample of $CH_3NH_3PbI_3$ of PSC film was preheated with the face-up position for 30 seconds. Last, the $CH_3NH_3PbI_3$ PSC film was annealed with the temperature variations of 100, 120 and 140 ^{0}C.

2.3 Characterization

The crystal structure of $CH_3NH_3PbI_3$PSC film was characterized by using X-ray Diffraction (XRD). The morphology of sample surface was characterized by using Scanning Electron Microscopy (SEM), and Ellipsometry Micropack SpecEl 2000 was used to know the thickness of ZnO nanoparticle film layer and $CH_3NH_3PbI_3$ perovskite in one system.

IOP Conf. Series: Journal of Physics: Conf. Series **1011** (2018) 012011 doi:10.1088/1742-6596/1011/1/012011

3. Result and Discussion

The diffraction patterns of $CH_3NH_3PbI_3$ PSC are presented in Figure 1.

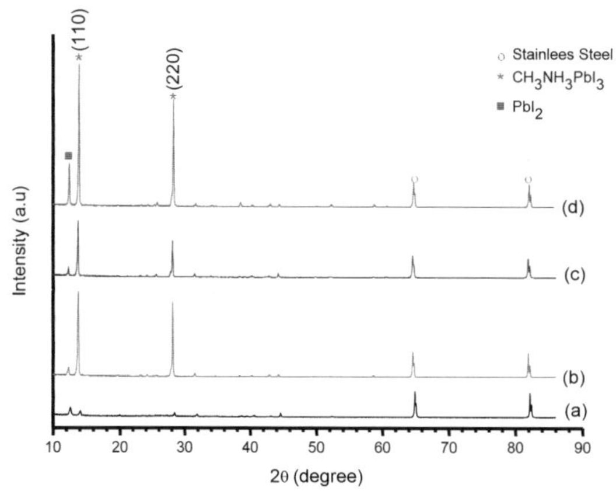

Figure 1. Diffraction Pattern of $CH_3NH_3PbI_3$ PSC with the variation of anneal temperature; a) Without anti-solvent at 100 ^{0}C, and with anti solvent b) 100 ^{0}C, c) 120 ^{0}C, d) 140 ^{0}C

Figure 1a. Shows the diffraction pattern of $CH_3NH_3PbI_3$ without using anti-solvent indicating that PbI_2 phase was more dominant the original $CH_3NH_3PbI_3$ phase. Actually, the phase that should be formed more dominantly was $CH_3NH_3PbI_3$. Meanwhile, the sample added with anti-solvent shown in Figure 2b, 2c, and 2d representing that the diffraction peak of $CH_3NH_3PbI_3$ PSC film with the annealing temperature variation had the strongest diffraction peak at 13.95^0 and 28.23^0; this phenomenon was in line with the study conducted by Zhu et al. (2016) indicating that the original compound of $CH_3NH_3PbI_3$ had been formed with tetragonal crystal structure and group of 14/mcm [17]. There were three other peaks appearing at the angle of 2θ 12.44^0, 64.69^0, dan 81.95^0. The peak of 12.44^0 angles was indexed as the PbI_2 residue [3], and the other peaks were included in the diffraction peak of stainless steel substrate [18]. Meanwhile, we can see at the diffraction pattern of $CH_3NH_3PbI_3$ at a temperature 140 ^{0}C that there was the impurity of PbI_2 with the relatively high intensity peak compared to the intensity of the simple at a temperature of 100 and 120 ^{0}C. This phenomenon was because of the higher the temperature, the anti-solvent would more easily steam so that it changed the crystal phase.

To know further the film crystallinity, the result of XRD diffraction pattern was also analyzed by Scherrer equation to state the size of the crystal grain. The results of the calculation are presented in Table 1.

Table 1. Particle Size of $CH_3NH_3PbI_3$ Perovskite

Annealing Temperature (^{0}C)	FWHM (rad)	Particle Size (nm)
100	0.00354	42.96
120	0.00278	54.73
140	0.00275	55.20

In Table 1, the particle size increased by the decrease in FWHM value. The change of FWHM and the size of crystal grain was due to the growth period as the result of temperature change. The particle size of $CH_3NH_3PbI_3$ perovskite with the annealing temperature of 100, 120, and 140 ^{0}C was 42.96, 54.73, and 55.20 nm, respectively. This phenomenon was in line with the study conducted by Zhu et al. (2016) stating that the film crystallinity increased by the reduce of impurity interstitial of $CH_3NH_3PbI_3$ and the increase in crystal size [17]

The morphology $CH_3NH_3PbI_3$ PSC film surface sample was characterized by using SEM. The results of SEM with the variation of annealing temperature are shown in Figure 2.

Figure 2. The SEM result of $CH_3NH_3PbI_3$ PSC with the variation of annealing temperature: a) 100 ^{0}C, b) 120 ^{0}C, c) 140 ^{0}C

Figure 2 indicates that $CH_3NH_3PbI_3$ film was growth inhomogeneously over the substrate surface. However, by the increasing temperature, the morphology was more randomized with the high density. The high density of crystal can cause the quality of electrical and optical characteristics of film decrease so that it automatically causes the PCE of solar cell reduces [19].

The thickness of film becomes an important parameter because it can influence in improving light-harvesting properties [17]. The thickness of layer structure of the $CH_3NH_3PbI_3$ film was characterized by using Ellipsometry Spectroscopy. The characterization results are presented in Table 2.

Table 2. Thickness of $CH_3NH_3PbI_3$ PSC Film

Annealing Temperature (^{0}C)	Thickness of ZnO (nm)	Thickness of $CH_3NH_3PbI_3$ (nm)
100	154.1	79
120	155.1	79
140	155.5	79

Table 2 shows that the thickness of each sample at an annealing temperature of 100, 120, and 140 ^{0}C was similar that was 79 nm, with the thickness of ZnO layer functioning as an electron transport layer was 154.1, 155.1, 155.5 nm, respectively. This case indicates that the change of annealing temperature did not influence the film layer thickness.

4. Conclusion

The $CH_3NH_3PbI_3$ PSC film that was grown through one step spin coating route had a tetragonal crystal structure. The highest crystal orientation of all samples with the variation of annealing temperature directed to the peak of (110). The crystal grains of $CH_3NH_3PbI_3$ were in nano size with the good crystallinity and it grew up on the substrate surface of ZnO nanoparticles/stainless steel holistically that had the thickness of nanometer scale. The high annealing temperature caused the crystal grains were formed randomly with the high density. In addition, it also caused the appearance of PbI_2 residue and increased the crystal density that bécame one of the causes of the decrease in solar cell efficiency. Thereby, in this work, it can be concluded that the proper annealing temperature to produce the solar cell with high performance is 100 ^{0}C.

Acknowledgements
The authors would like to thank the *Kemenristekdikti* for providing funding of Hibah Penelitian PPT for AF 2017.

References

[1] Löper P, Stuckelberger M, Niesen B, Werner J, Filipič M, Moon S-J, Yum J-H, Topič M, De Wolf S and Ballif C 2014 Complex Refractive Index Spectra of CH 3 NH 3 PbI 3 Perovskite Thin Films Determined by Spectroscopic Ellipsometry and Spectrophotometry *J. Phys. Chem. Lett.* 141210002314002

[2] Song T B, Chen Q, Zhou H P, Jiang C Y, Wang H H, Yang Y, Liu Y S, You J B and Yang Y 2015 Perovskite solar cells: film formation and properties *J. Mater. Chem. A* **3** 9032–50

[3] Yu J, Chen X, Wang Y, Zhou H, Xue M, Xu Y, Li Z, Ye C, Zhang J, van Aken P A, Lund P D and Wang H 2016 A high-performance self-powered broadband photodetector based on a $CH_3NH_3PbI_3$ perovskite/ZnO nanorod array heterostructure *J. Mater. Chem. C* **4** 7302–8

[4] Gao X, Li J, Baker J, Hou Y, Guan D, Chen J and Yuan C 2014 Enhanced photovoltaic performance of perovskite CH_3 NH_3 PbI_3 solar cells with freestanding TiO_2 nanotube array films. *Chem. Commun. (Camb).* **50** 6368–71

[5] Kojima A, Teshima K, Shirai Y and Miyasaka T 2009 Organometal halide perovskites as visible-light sensitizers for photovoltaic cells *J. Am. Chem. Soc.* **131** 6050–1

[6] Bi D, Yi C, Luo J, Décoppet J-D, Zhang F, Zakeeruddin S M, Li X, Hagfeldt A and Grätzel M 2016 Polymer-templated nucleation and crystal growth of perovskite films for solar cells with efficiency greater than 21% *Nat. Energy* **1** 16142

[7] Burschka J, Pellet N, Moon S-J, Humphry-Baker R, Gao P, Nazeeruddin M K and Grätzel M 2013 Sequential deposition as a route to high-performance perovskite-sensitized solar cells. *Nature* **499** 316–20

[8] Kim H S, Im S H and Park N 2014 Organolead halide perovskite: new horizons in solar cell research *J. Phys. Chem. C* 140127150857007

[9] Jeon N J, Noh J H, Kim Y C, Yang W S, Ryu S and Seok S Il 2014 Solvent engineering for high-performance inorganic–organic hybrid perovskite solar cells *Nat Mater* **13** 897–903

[10] Zhu W, Kang L, Yu T, Lv B, Wang Y, Chen X, Wang X, Zhou Y and Zou Z 2017 Facile Face-Down Annealing Triggered Remarkable Texture Development in CH3NH3PbI3 Films for High-Performance Perovskite Solar Cells *ACS Appl. Mater. Interfaces* **9** 6104–13

[11] Ahmadian-Yazdi M R, Zabihi F, Habibi M and Eslamian M 2016 Effects of Process Parameters on the Characteristics of Mixed-Halide Perovskite Solar Cells Fabricated by One-Step and Two-Step Sequential Coating. *Nanoscale Res. Lett.* **11** 408

[12] Kang R, Yeo J S, Lee H J, Lee S, Kang M, Myoung N S, Yim S Y, Oh S H and Kim D Y 2016 Exploration of fabrication methods for planar CH3NH3PbI3 perovskite solar cells *Nano Energy* **27** 175–84

[13] Zhu S, Zhao J, Zhang X, Cui H and Zhou Z 2016 Controllable Synthesis of ZnO Nanorod-nanosheets Hierarchical Nanostructure on a Stainless Steel Mesh Via A Facile Hydrothermal Growth **2007**

[14] Djuriˇ A 2012 Improving the performance of ZnO-based perovskite solar cells 7–9

[15] Ahn N, Son D Y, Jang I H, Kang S M, Choi M and Park N G 2015 Highly Reproducible Perovskite Solar Cells with Average Efficiency of 18.3% and Best Efficiency of 19.7% Fabricated via Lewis Base Adduct of Lead(II) Iodide *J. Am. Chem. Soc.* **137** 8696–9

[16] Paek S, Qin P, Lee Y, Cho K T, Gao P, Grancini G, Oveisi E, Gratia P, Rakstys K, Al-Muhtaseb S A, Ludwig C, Ko J and Nazeeruddin M K 2017 Dopant-Free Hole-Transporting Materials for Stable and Efficient Perovskite Solar Cells *Adv. Mater.* 1606555

[17] Zhu W, Bao C, Wang Y, Li F, Zhou X, Yang J, Lv B, Wang X Y, Yu T and zou zhigang 2016 Coarsening of one-step deposited organolead triiodide perovskite films via Ostwald ripening for high efficiency planar-heterojunction solar cells *Dalt. Trans.* **78**56–65

[18] Olaseinde O A, Van Der Merwe J and Cornish L 2014 Characterization and Corrosion Behaviour of Selected Duplex Stainless Steels in Acidic and Acidic-Chloride Solution *Adv. Chem. Eng. Sci.* **4** 89–93

[19] Kim H Do, Ohkita H, Benten H and Ito S 2016 Photovoltaic Performance of Perovskite Solar Cells with Different Grain Sizes 917–22

IOP Conf. Series: Journal of Physics: Conf. Series **1011** (2018) 012012 doi:10.1088/1742-6596/1011/1/012012

Microwave Hydrothermal Synthesis of Reduced Graphene Oxide: Effects of Microwave Power and Irradiation Time

La Agusu[1,*)], La Ode Ahmad[2], Desna Anggara[1], Alimin[2], Seitaro Mitsudo[3], Yutaka Fujii[3], and Hiromitsu Kikuchi[3]

[1]Department of Physics, Halu Oleo University, Indonesia
[2]Department of Chemistry, Halu Oleo University, Indonesia
[3]Research Center for Development of Far-Infrared Region, University of Fukui, Japan

*) Corresponding author: la_agusu@yahoo.com

Abstract. Reduced graphene oxide has been synthesihzed by one-pot microwave assisted hydrothermal method. Effects of microwave power and irradiation time to its crystal structure and electrical conductivity were investigated. Here, graphene oxide, firstly, were synthesized by modified hummers method and subsequently mixed with Zn as a reducing agent. Then it was transferred to modified domestic microwave oven (800 watts) with glass distiller equipment for completely reduction process. Three different power levels (240, 400, 630 watts) and two cases of irradiation times (20 and 40 minutes) were treated. XRD study shows that irradiation time variation is more effective than the variation of power level. Power level of 270 watts and for 40 minutes microwave irradiation are enough for producing estimated bilayer rGO with graphene interlayer of ~0.4 nm. Bilayer graphene and water molecule (~0.3 nm) may vibrate the same manner and perhaps they are accepting the same temperature. Graphene seems to be re-arranged in unspecified way among the thermal pressure, temperature gradient and/or water surface tension between graphene and water induced by microwave, in order to achieve thermal equilibrium through out the system The electrical conductivity rGO/PVA (60/40 %w) paper are ranging from 15.6 to 43.4 mS/cm.

1. Introduction

Graphene, two dimensional single layer graphite carbon lattice, has evoked much attention due to its unusual electronic, thermal and mechanical properties [1-6]. During the last decade, many approaches have been formulated for the preparation of graphene, including micromechanical methods [7], epitaxial growth [8], the ultrasonic exfoliation of graphite [9], chemical vapor deposition (CVD) [10], and the chemical reduction of graphene oxide (GO). Among them, the chemical reduction of graphene oxide was considered as a promising method for large-scale production of graphene. It involves conversion of the resulting graphite oxide to graphene oxide (GO), exfoliation process, and the subsequent reduction of GO.

Graphene can be reduced by different ways such as chemical, thermal, electro-chemical, hydrothermal, microwave, gamma ray and photo reductions. As for chemical reduction, different kinds of chemicals have been utilized to reduce GO. In many cases, strong reducing agents such as hydrazine monohydrate, hydroquinone, sodium borohydride, and hydrohalic acid have been used as the reducing agents [11-14]. However, these chemicals may harm the environment or be cost-prohibitive when used in the large-scale industry. Moreover, graphene layers tend to agglomerate and form graphitic structures when strong chemicals are used. Recently, a significant interest has been directed to use metals (Fe, Al, Zn) as reduction agent of GO [15-17]. Nevertheless, metal particles may remain as impurities and need more treatment to remove it completely.

While conventional thermal treatment has been well-recognized method for reduction of graphene by removing oxygen functionalities from the surface of GO [18], the unconvetional thermal method using a microwave as thermal source can rapidly reduce graphene oxide. Microwave irradiation to dry powder of GO was recently found to effectively reduce GO in several seconds [19]. Microwave heating tend to be more effective to reduce GO in lower temperature and shorter period compare to reduction process by the conventional heating. In addition, microwave hydrothermal method was also provide excellent property compare to conventional hydrothermal technique since the microwave device can deliver electromagnetic wave energy homogeneously entire the wet chemicals [20].

In this study, we investigate effect of microwave power and interaction time to the performance of rGO using a modified domestic 2.4 GHz microwave oven as the hydrothermal reactor. Graphene oxide was firstly synthesized using, a wet chemical, hummers method. Then, it was rigorously mixed with Zn for 30 minutes and transferred to glass container that was connected water cooled condenser for microwave irradiation.

2. Experimental sections

2.1 Synthesis of GO

GO was prepared by the oxidation of graphite powder (Sigma Aldrich), without pre-treatment, according to modified hummers method re-described in [21]. Typically, a 2 g of graphite powder was added, under stirring, to 98 mL of concentrated H_2SO_4 in ice bath. Under robust agitation, KMnO4 (8.0 g) and 4 gram NaNO3 were added slowly, colors of suspension changed to dark green at the end. Stirring process was continued for additional 24 h where temperature was raised to ~35° C, color changed from green purple to be light brown. Then, aquades (200 mL) was added slowly to the suspension for about 1 h. Here, colors changed to be dark brown. Additional 15 mL H_2O_2 was added slowly to the suspension, turning the color of the solution to yellow. After that, the mixture was centrifuged, filtered, washed with 10 ml HCl 37% and several times with aquades to remove the remaining metal ions. The resulting solid, graphite oxide was dried in air at 105° C for 12 h. The resultant graphite oxide was then diluted to 200 mL of aquades, stirred overnight and then it was sonicated for 120 min at 53 kHz.

2.2 Microwave hydrothermal reduction of GO

Reduction process of GO involved addition of Zn and irradiation of microwave energy. A 0.6 g of Zn was mixed in water based GO dispersion, and subsequently added by, under stirring, 50 mL of hydrochloric acid for another 30 minutes. Then, it was exposed to electromagnetic wave radiation in the domestic microwave oven at different powers (240, 400, and 560 watt) and duration (20 and 40 minutes). Sample names and treatment condition were summarized in Table 1. Here, sample names A, B, C, D, E, and F were corresponding for condition of microwave power-irradiation time: 240 W-20 minutes, 400 W-20 minutes, 560 W-20 minutes, 240 W-40 minutes, 400 W-40 minutes, and 560 W-40 minutes, respectively. At the end, to remove the remaining Zn, solution was finally washed several times with HCl and aquades until pH = ~7. The reduced graphene oxide (rGO) was collected by centrifugation to remove the unexfoliated graphite.

2.3 Characterization

Crystalline structure and phase purity of the as-prepared rGO were examined using powder X-ray diffraction (XRD). XRD pattern was recorded with graphite-monochromated Cu-Kα (γ = 1.54056 Å) radiation on Rigaku RINT UltimaIII ranging from 10° to 70° (2θ) with a scanning speed of 2° per minute. The surface morphology of the resulting products were investigated by a scanning electron microscopy (SEM) TESCAN. Fourier transforms infrared spectroscopy (FTIR) spectra were recorded on a Fourier transform infrared spectrometer (Bruker Vertex V70). Electrical conductivity was studied using a four point probe measurement where the electrode spacing was 1.0 mm.

3. Results and Discussions

Figure 1 shows XRD spectrum of as-prepared rGO at different conditions (powers and irradiation times). In the case of microwave irradiation time is 20 minutes, peaks of 002 crystal plane of rGO are 19.45°, 19.13°, and 20.28°, corresponding to power of 240 W, 400 W, and 560 W, respectively. Peak tends to shift to the left by increasing microwave power, indicating the decrease of calculated inter-layer distance from 0.456 nm to 0.438 nm (see Figure 2a). By contrast, peak is drifted to the bigger diffraction angle (Figure 1b), namely layer-by-layer stacking distance is changing from 0,41 nm to 0.43 nm, if the power is adjusted from 240 W to 560 W and the interaction time is 40 minutes. The inter-layer or layer-by-layer stacking distance (d) was calculated by using the well-known Bragg's law $n\lambda = 2d\sin\theta$. Here, the wavelength, λ is 0.15406 nm.

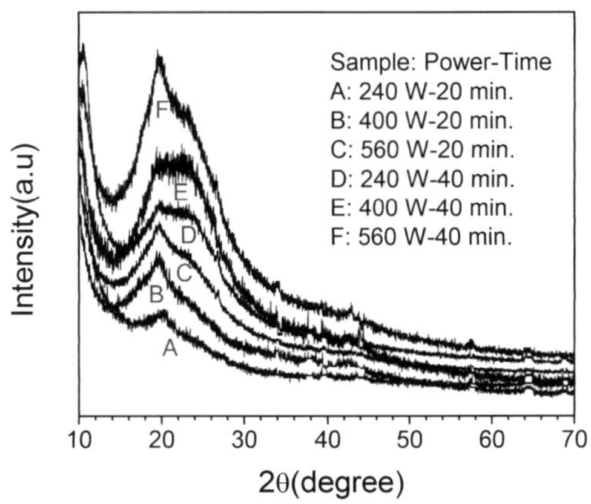

Figure 1. XRD diffraction spectrum of the as-prepared rGO

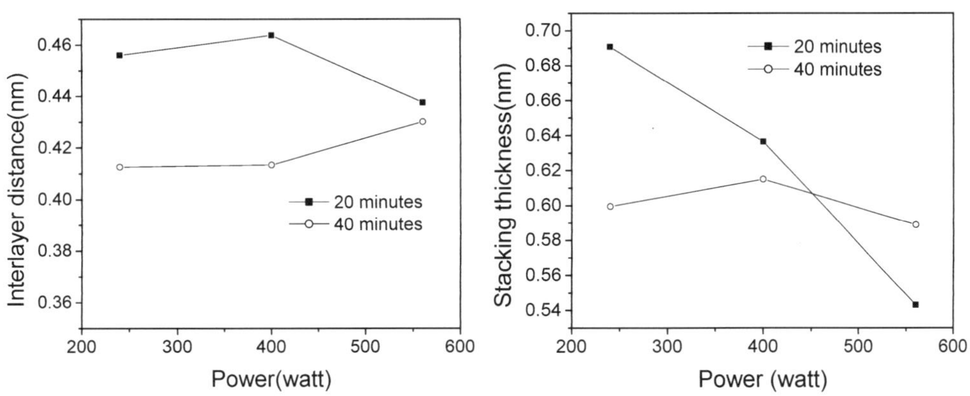

Figure 2. Inter layer distance and stacking thickness

The stacking thickness of graphene layers can be estimated from (002) diffraction peak of XRD data [23, 24] using the Scherrer formula

$$D = k\lambda / \beta \cos\theta \qquad (1)$$

where k is a constant value ($0< k <1$) and β is the full-width of half maximum. Here, $k = 0.89$ was used in all calculation. Number of layers was estimated from $N = (D + d)/d$.

For irradiation of only 20 minutes, a rapidly decreasing of stacking thickness of layered rGO from 0.69 nm to 0.54 nm as the microwave power raising has been observed from XRD data. However, when irradiation period is about 40 minutes the stacking thickness is very slightly altered by changing the microwave power. An interesting phenomenon found in estimation of number of layered graphene, in all cases a bilayer graphene was found, except in case of power level of 240 W and irradiation of 20 minutes, the three layers graphene is exist. It seems that, in this case, packets of bilayer graphene are favorably formed in the water under irradiation of microwave power. The fact that bilayer graphene with interlayer distance of 0.4 nm is very similar to the size of water molecule (~0.3 nm) in the 'eyes' of a very long wavelength of microwave, bilayer graphene and water molecule are vibrating in the same manner and perhaps they are accepting the same temperature, even though the graphene and water molecule have different dielectric constant. Graphene may be re-arranged in 'unspecified' way among the thermal pressure, temperature gradient and/or water surface tension between graphene and water induced by microwave, in order to achieve thermal equilibrium through out the system [21]. Although, high densification, smaller size, and less pores have been specified as the microwave effect or non-thermal effect on the sintering process of materials compared to the conventional heating at the same temperature, the mechanism responsible for such effect is still unclear. The physics of the microwave processing of materials had remained unexplored and hence these complex phenomena are 'less understood'.

Water dispersion test in Figure 3 shows that dispersion can be prevented almost in the same period for all samples. It confirms existing of the same layers of graphene in all samples processed in different powers and irradiation times.

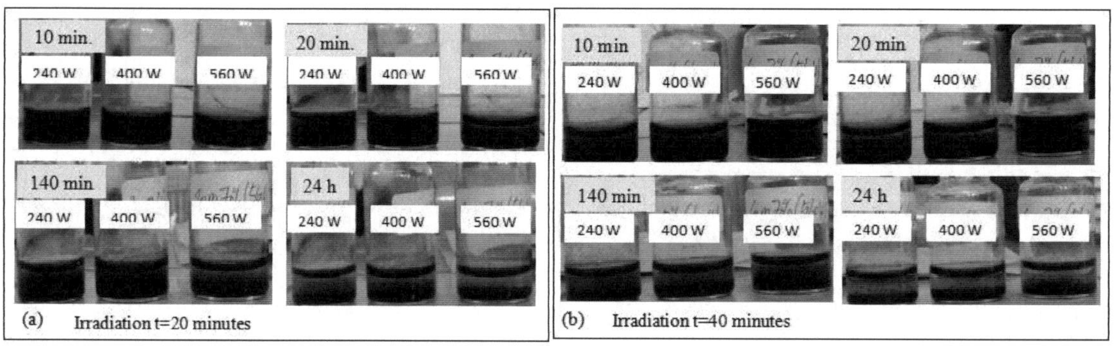

Figure 3. Dispersion test of graphene in water where graphene are proceeded by irradiation of (a) 20 minutes and (b) 40 minutes. In all cases, microwave power 240, 400, and 560 W are applied.

Table 1. Electrical conductivity of PVA/rGO (40:60 %w)

Power (W)	Irradiation time (minutes)	Conductivity (S/cm)
240	20	0.026
400	20	0.041
560	20	0.033
240	40	0.036
400	40	0.043
560	40	0.016

Electrical conductivity of PVA/rGO (40:60 %w) paper are shown in Table 1. Conductivity is ranging from 0.016 S/cm (560 W, 40 minutes) to 0.043 S/cm (400 W, 40 minutes). In electrical conductivity consideration, processing of graphene using microwave, the power should be no more than 400 W. Morphology of PVA/rGO paper using rGO processed by microwave irradiation at 400 W and 40 minutes are shown in Figure 4. It has micrometer scale of pores (Figure 4a) and morphology of clumped graphene layers are shown in Figure 4b.

FTIR spectra confirm reduction of graphene seen in Figure 5. In the spectrum, the peaks located at 3424 and 1087 cm^{-1} are attributed to the stretching vibration of hydroxyl groups. A very weak interaction peak at 1720 which corresponds to stretching vibration of C=O, can be assigned to a complete reduction of graphene oxide. More importantly, the shifting peak at around 1573 cm^{-1} (in-plane vibration of C=C) by changing processing condition by microwave are the strong indication of interaction between graphene layer and microwave radiation.

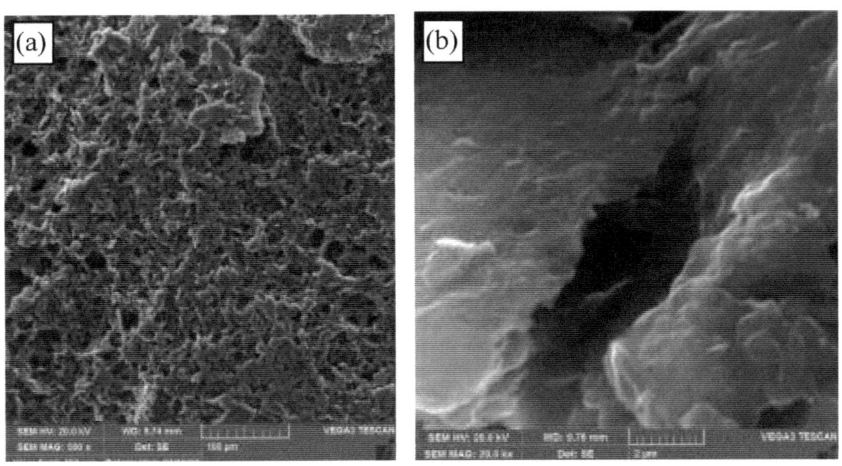

Figure 4 SEM image of PVA/rGO (40:60 %w) paper, (a) low magnification and (b) high magnification.

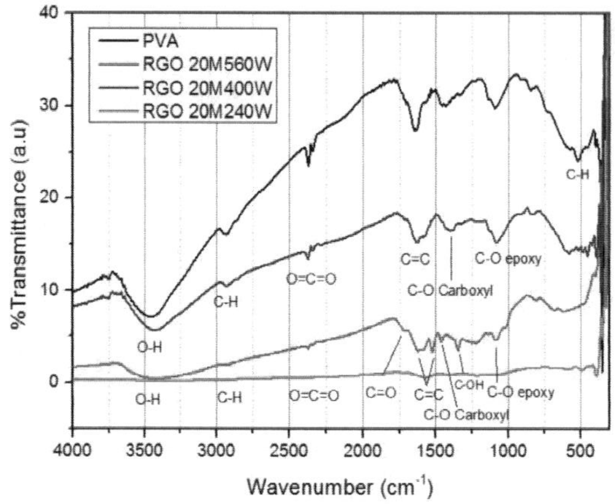

Figure 5. FTIR spectrum of pVA/rGO paper processed at different microwave power and irradiation time of 20 minutes

4. Conclusion

Hydrothermal synthesis of reduced graphene oxide by utilization of microwave irradiation has been demonstrated where effect of microwave power and irradiation time to the crystal structure of graphene are investigated. XRD study shows that irradiation time variation is more effective than varying the power level. A 270 watts and 40 minutes of microwave irradiation are enough for producing the bilayer graphene with graphene interlayer of ~0.4 nm. The electrical conductivity rGO/PVA (60/40 %w) layers are ranging from 15.6 to 43.4 mS/cm. Processing graphene should use power not more than 400 W in order to reach high electrical conductivity.

Acknowledgment

This work was partially supported by Fundamental Research Grant 2017 from the Ministry of Research and Higher Education.

References

[1] Sun Y, Wu Q, and Shi G 2011 *Energy Environ. Sci.*, **4** 1113-1132.
[2] Geim A K 2009 *Science* **324** 1530-1534.
[3] Zhang L, Wang H, Li X C, Xia F L, Liu Y, Xu X Y, Gao J and Xing F B 2015 *Electrochim. Acta* **172** 42-51.
[4] Lu Y H, Zhou M, Zhang C and Feng Y P 2009 *J. Phys. Chem. C* **113** 20156-20160
[5] Tang M, Wu T, Xu X, Zhang L and Wu F 2014 *Mater. Res. Bul.* **60** 118-129
[6] Li X, Xu X, Xia F, Bu L, Qiu H, Chen M, Zhang L and Gao J. P. 2014 *Electrochim. Acta* **130** 305-313
[7] Novoselov K S, Geim A K, Morozov S V, Jiang D, Zhang Y, Dubonos S V, Grigorieva I V, Firsov A A 2004 *Science* **306** 666-669.
[8] Yang W, Chen G, Shi Z, Liu C C, Zhang L, Xie G, Cheng M, Wang D, Yang R, Shi D, Watanabe K, Taniguchi T, Yao Y, Zhang Y and Zhang G 2013 *Nature Materials* **12** 792–797
[9] Chen G, Wen W G and Wu D J 2004 *Carbon* **42** 753-759
[10] Zhou H, Yu W J, Liu L, Cheng , Chen Y, Huang X, Liu Y, Wang Y, Huang Y and Duan X 2013 *Nature Communications* **4** 2096

[11] Stankovich S, Dikin D A, Piner R D, Kohlhaas K A, Kleinhammes A, Jia Y, Wu Y, Nguyen S T and Ruoff R S 2007 *Carbon* **45** 1558–1565
[12] Wang G, Yang J, Park J, Gou X, Wang B, Liu H and Yao J 2008 *J. Phys. Chem. C* **112** 8192–8195
[13] Muszynski R, Seger B and Kamat P V, 2008 *J. Phys. Chem. C* **112** 5263–5266
[14] Moon I K, Lee J, Ruoff R S and Lee H, 2010 *Nat. Commun.* **1** 73
[15] Fan Z J, Kai W, Yan J, Wei T, Zhi L J, Feng F, Ren Y, Song L P and We F 2011 *ACS Nano* **5** 191–198.
[16] Fan Z, Wang K, Wei T, Yan J, Song L and Shao B 2010 *Carbon* **48**, 1686–1689
[17]Yang S, Yue W, Huang D, Chen C, Lina H and Yang X 2012 *RSC Adv.* **2**, 8827–8832
[18] Schniepp H C, Li J L, McAllister M J, Sai H, Alonso M H, Adamson D H, Homme R K P, Car R, Saville D A and Aksay I A 2006 *J. Phys. Chem. B* **110** 8535-8539

[18] Liu S, Sun H, Suvorova A, Wang S 2013 *Chemical Engineering Journal* **229** 533–539
[19] Voiry D, Yang J, Kupferberg J, Fullon R, Lee L, Jeong H Y, Shin H S and Chhowalla M 2016 *Sciences* **353**(6306) 1413-1416
[20] Zhou W, Zhang F, Liu S, Wang J, Du X, Yina D and Wang L 2014 *RSC Adv.* **4** 51362-51365
[21] Parmar H, Asada M, Kanazawa Y, Asakuma Y, Phan C M, Pareek V and Evans G M 2014 *Langmuir* **30** (33) 9875-9879
[22] Agusu L, Mitsudo S, Ahmad L O, Herdianto, Fujii Y, Ishikawa Y, Furuya T and Ahmad L O N 2017 *AIP Conference Proceedings* **1801** 050006.
[23] Saikia B Y, Boruah R K, and Gogoi P K 2009 *J. Chem. Sci.* 121(1) 103-106
[24] Manoj B and Kunjomana A G 2012 *Int. J. Electrochem. Sci.* **7** 3127 - 3134

Electrospinning β-SiC fibers from SiC nanoparticles dispersed in various polymer solutions as the electrospinning agents

A Fuad[1,2*], N Fatriani[1], C I Yogihati[1], A Taufiq[1,2] and E Latifah[1]

[1] Department of Physics, Faculty of Mathematics and Natural Sciences, Universitas Negeri Malang (State University of Malang), Jl. Semarang No. 5, Malang 65145, Indonesia
[2] Central Laboratory for Mineral and Advanced Materials, Faculty of Mathematics and Natural Sciences, Universitas Negeri Malang (State University of Malang), Jl. Semarang No. 5, Malang 65145, Indonesia
*Corresponding Author: abdulloh.fuad.fmipa@um.ac.id

Abstract. Silicon carbide (SiC) fibers were synthesized by electrospinning method from SiC nanoparticles dispersed in polymer solutions, i.e., polyethylene glycol (PEG) and polyvinyl alcohol (PVA). The SiC nanoparticle used in this research was synthesized from sucrose and natural silica via a sonochemical method. The natural silica was extracted from local pyrophyllite by a sol-gel method. The characterization was performed via x-ray fluorescence (XRF), X-ray diffraction (XRD), scanning electron microscopy (SEM). The XRD characterization results showed that the sample possessed a β-SiC phase and formed a cubic-structured crystal with a lattice parameter of $a = b = c = 4.3448$ Å. The use of PEG and PVA in the electrospinning process resulted in fractal and fiber structured SiC, respectively.

1. Introduction

Various non-oxide ceramics are used for many industrial applications. Silicon carbide (SiC) serves as an exclusive example of such non-oxide ceramics due to its superior characteristics such as good mechanical property, high level of thermal and electrical conductivity, excellent chemical stability, high radiation resistance, and so forth [1]. Such distinctive properties have turned SiC to be a promising material to be applied as a nuclear, turbine, and furnace reactor, and a jet engine, or even for the needs of space exploration [2–4]. Moreover, fiber structured SiC can also be applied toa solar panel or as a photocatalyst in a hydrogen production via water splitting process [5,6].

Commonly, the synthesis of SiC fibers is conducted by chemical vapor deposition (CVD), template, arc-discharge, carbothermal reduction, and molten salt synthesis methods [7–9]. Several obstacles are found in such methods, which include the needs of a very high synthesis temperature, an expensive and rare precursor, and high and complex technology. Therefore, it is necessary to find an alternative that allows a low price and simple synthesis of SiC fibers.

This research aimed to synthesize silicon carbide fibers through a straightforward and inexpensive method. In order to achieve such purpose, the synthesis was performed using sucrose and silica extracted from pyrophyllite as precursors. Besides reducing the production cost, the utilization of silica extracted from pyrophyllite as a primary precursor is also improving the functional value of local resources in Indonesia.

2. Experimental

2.1. Silica extraction from pyrophyllite
The silica was extracted from pyrophyllite via a sol-gel method. The content of Si and other elements in the pyrophyllite are presented in table 1. Furthermore, 20 grams of pyrophyllite powders were put into a beaker, and then added with 500 ml 5M sodium hydroxide solution and stirred at 120 °C. Then, the mixture was filtered. The filtrate in the form of a sodium silicate was then titrated using a 5M hydrochloric acid solution until it reached pH 7. The silica gel formed in the titration process was rinsed using distilled water to be then drained.

Table 1. Element contents in pyrophyllite.

No.	Element	Content of element (%)
1.	Si	64.70
2.	Al	16.00
3.	Fe	8.95
4.	Ti	7.17
5.	Ca	1.32
6.	Other elements	1.86

2.2. Synthesis of SiC nanoparticles
The SiC nanoparticles were synthesized using a sonochemical method. 3 grams silica and 9 grams sucrose were added to a 240 ml distilled water and ethanol mixture. Next, the mixture was stirred and sonicated at 70 °C. The result was then put into a furnace for the following preheating process for 1 hour at 180 °C. After that, the process continued by annealing at 950 °C in Argon ambiance until SiC nanoparticles were produced.

2.3. Synthesis of SiC fibers
0.1 grams SiC nanoparticles was added in to a 2 ml PEG solution (Mw ~6,000 Da) and a PVA solution (Mw ~186,000 Da). In order to ensure the even dispersion of the SiC nanoparticles, each of the solutions was stirred and sonicated for 1 hour. The sonicated solution was then put into a syringe with 25 gauge needle. After that, the electrospinning process was performed by using the parameters presented in table 2.

Table 2. Electrospinning process parameters.

No.	Parameter	Value
1.	Tip to collector distance (TCD)	10 cm
2.	Pump speed	0.005 ml/min
3.	Electrospinning period	1 hour
4.	Applied voltage	20 kV

2.4. Characterization
The percentages of the elements contained in pyrophyllite and natural silica were identified by an x-ray fluorescence (XRF, PANalytical) characterization. The SiC nanoparticles formed a crystalline phase that was identified via x-ray diffraction (XRD, Phillips X'Pert Pro) characterization. Morphology of the SiC fibers resulted from the electrospinning process was observed via a scanning electron microscopy (SEM, FEI INSPECT-S50) characterization.

3. Results and Discussion

3.1. Element contents in natural silica

Elements contained in natural silica extracted from pyrophyllite were identified via XRF. The XRF characterization results are presented in table 3. According to the XRF results, it is known that the extracted silica contains 95.4 % Si element, which indicated an improvement on the silica purity after the extraction process compared with the initial pyrophyllite that only contained 64.7 % Si element (table 3).

Table 3. Element contents in natural silica.

No.	Element	Content of element (%)
1.	Si	95.40
2.	Ca	1.30
3.	Ti	0.12
4.	Fe	0.43
5.	Other elements	2.75

3.2. Crystalline phase, size, and structure of SiC nanoparticles

The diffraction pattern of the Sic nanoparticles is displayed in figure 1. All of the diffraction peaks formed were in good agreement with β-SiC diffraction peaks based on ICSD No. 01-072-1708, without any other phase. The results of the Rietveld analysis identified that the SiC nanoparticles are cubic-structured with a lattice parameter of a = b = c= 4.3448 Å. Mostly, the SiC nanoparticles have a size of 18.1 nm as measured via a Scherrer equation.

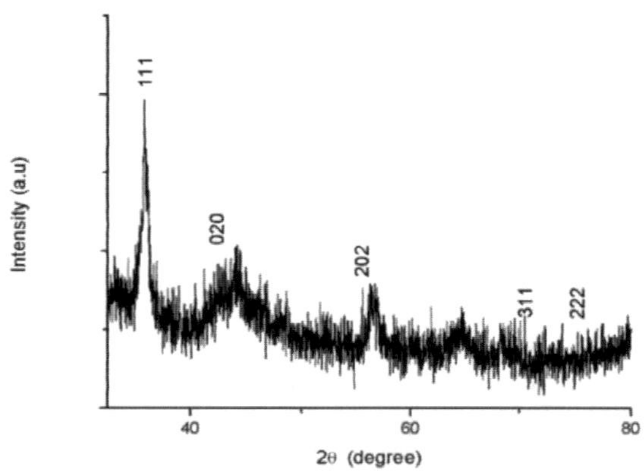

Figure 1. The diffraction pattern of SiC nanoparticles synthesized by a sonochemical method.

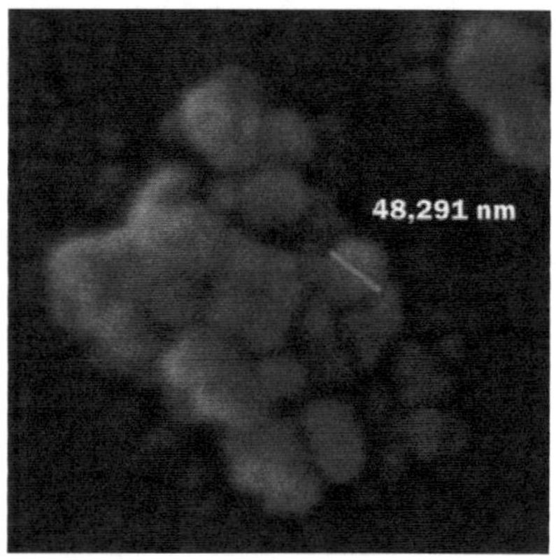

Figure 2. The morphology of SiC nanoparticles

3.3. Morphology of SiC nanoparticles

The SiC nanoparticle morphology is shown in figure 2. It can be seen that the SiC nanoparticles tend to agglomerate and form a bigger sized aggregate. It indicates an occurrence of agglomeration in the sample. The average size of the SiC nanoparticles was 48.291 nm.

3.4. Morphology of SiC fibers

Figure 3 displays the morphology of the SiC synthesized via electrospinning method using the PEG solution. The use of PEG produced fractal or irregular-structured SiC. It was due to the overly low molecular weight (Mw ~6.000 Da) of the PEG used in the process. The molecular weight of a polymer represents the length of such polymer chain. A low molecular weight of a polymer will have a short chain that is breakable during the electrospinning process [10]. The synthesis of the SiC fibers by using PEG with a higher molecular weight (Mw ~300.000 Da) is proven to be able to produce SiC fibers in diameters ranging from 200-800 nm [11].

Figure 3. Morphology of SiC fibers synthesized by electrospinning method using a PEG solution

Figure 4 shows the SEM characterization results of the SiC synthesized via electrospinning method using a PVA solution. The utilization of PVA in the electrospinning process produced beaded SiC fibers, which indicates that the PVA concentration was still too low. A polymer solution with a low concentration will be mostly formed by the solvent molecule with less polymer molecule. A high ratio of the solvent molecule to polymer molecule causes the solvent surface area to dominate the polymer solution during the electrospinning process. The high surface tension of the solvent will cause it to maintain its spherical form along the polymer chain leading to the production of the beaded fibers [10].

Different with PEG, the use of PVA in an electrospinning process can result in fiber-structured SiC (figure 4). It is because PVA has a higher molecular weight than PEG. The high molecular weight of PVA indicates that it has a longer polymer chain, which tends to form an entanglement. In an electrospinning process, when the solution was injected from the needle to the collector, the polymer chain of the solution would stretch because of an electric field. An entanglement of a polymer chain would prevent it to break during such stretching process so that it could result in a continuous production of fibers [10].

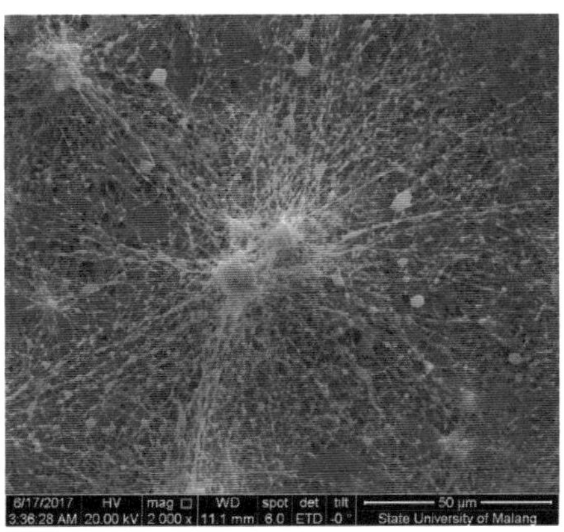

Figure 4. Morphology of SiC fibers synthesized by electrospinning method using a PVA solution of 10% w/v.

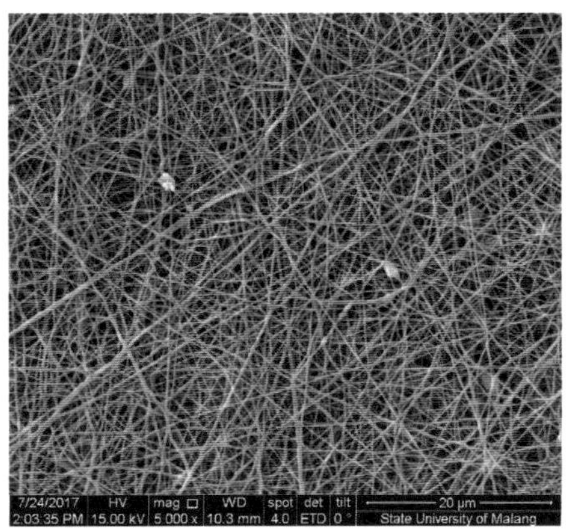

Figure 5. Morphology of SiC fibers synthesized by electrospinning method using a PVA solution of 12.5% w/v.

The growth of the beaded fibers can be avoided by increasing the molecular number of PVA such as by increasing the concentration of the PVA. Figure 5 shows the morphology of the β-SiC that produced by electrospinning technique for the concentration of PVA 12.5% w/v. The mean diameter of the β-SiC fibers is 228 nm. The balance comparison of PVA with solvent yang tended to reduce the surface tension of solvent and the solvent distributed homogeneously along with the fibers [10]. Therefore, it produced the fibers with a homogeneous diameter as presented in figure 5.

4. Conclusion

In this research, β-SiC fibers have been successfully synthesized using an electrospinning method from SiC nanoparticles dispersed in a polymer solution. The produced SiC is in the form of a cubic-structured β-SiC phase with a crystal size of 18.1 nm. The use of PEG in an electrospinning process led to the production of fractal-structured β-SiC while the utilization of PVA resulted in β-SiC fibers. Hence, the β-SiC fibers were well-synthesized by an electrospinning method using PVA as the electrospinning agent.

References

[1] Ding D, Zhou W, Zhang B, Luo F, Zhu D 2001 Complex permittivity and microwave absorbing properties of SiC fiber woven fabrics *J. Mater. Sci.* **46** 2709
[2] Mu Y,Zhou W, Luo F, Zhu D, Qing Y, Huang Z 2014 Electromagnetic interference shielding effectiveness of SiCf/SiC composites with PIP–SiC interphase after thermal oxidation in air *J. Mater. Sci.* **49** 1527
[3] Qin X H, Xiao B L, Dong S M, Jiang D L 2007 SiCf/SiC composites reinforced by randomly oriented chopped fibers prepared by semi-solid mechanical stirring method and hot pressing *J. Mater. Sci.* **42** 3488
[4] L Xiong, Y Xu, Y Li, X Xia 2008 Synthesis of SiC ceramic fibers from nuclear reactor irradiated polycarbosilane ceramic precursor fibers *J. Mater. Sci.* **43** 4849
[5] Ceballos-Mendivil L G, Cabanillas-López R E, Tánori-Córdova J C, Murrieta-Yescas R, Zavala-Rivera P, González J C 2014 Synthesis and characterization of silicon carbide in the application of high temperature solar surface receptors *Energy Procedia.* **57** 533
[6] Wang B, Wang Y, Lei Y, Wu N, Gou Y, Han C, Xie S, Fang D 2016 Mesoporous silicon carbide nanofibers with in situ embedded carbon for co-catalyst free photocatalytic hydrogen production *Nano Res.* **9** 886
[7] Li Y B, Xie S S, Zou X P, Tang D S, Liu Z Q, Zhou W Y, Wang G 2001 Large-scale synthesis of β-SiC nanorods in the arc-discharge *J. Cryst. Growth.* **223** 125
[8] Li Z, Gao W, Meng A, Geng Z, Gao L 2008 Large-scale synthesis and Raman and photoluminescence properties of single crystalline β-SiC nanowires periodically wrapped by amorphous SiO2 nanospheres 2 *J. Phys. Chem. C.* **113** 91
[9] Zhao B, Zhang H, Tao H, Tan Z, Jiao Z, Wu M 2011 Low temperature synthesis of mesoporous silicon carbide via magnesiothermic reduction *Mater. Lett.* **65** 1552
[10] Ramakrishna S 2005 *An introduction to electrospinning and nanofibers*, World Scientific
[11] Khan S, Khan A, Kordesch M E 2007 Fabrication and Optical Characterization of Silicon Carbide Nanofibers *MRS Online Proc. Libr. Arch.* p 1006

IOP Conf. Series: Journal of Physics: Conf. Series **1011** (2018) 012014 doi:10.1088/1742-6596/1011/1/012014

Effect of Temperature and Time of Sintering to Doping Ag On Microstructure of Perovskite Material $(La_{1-x}Ag_x)_{0.8}Ca_{0,2}MnO_3$

B Kurniawan[1*], S D Rosanti[1], R Kamila[1], N B Sahara[1], D S Razaq[1], E P Yandra[1] and T Komala[1]

Department of Physics, Faculty of Mathematics and Natural Sciences (FMIPA), Universitas Indonesia, Depok 16424, Indonesia

*Corresponding author: bkuru@fisika.ui.ac.id

Abstract. Particle synthesis has been performed on perovskite material $(La_{1-x}Ag_x)_{0.8}Ca_{0,2}MnO_3$ using the sol-gel method. Bulk samples were varied in temperature and time of sintering and doping variations of Ag x=0 and x=0.5. Calculation of stoichiometry was done to get the mass of each precursor. Precursors were included La_2O_3, $AgNO_3$, $Ca(NO_3)_2.4H_2O$ and $Mn(NO_3)_2.4H_2O$. Sintering temperature variations were done at 650°C for 5 hours and 900°C for 24 hours. Scanning Electron Microscopy (SEM) was done to see sample morphology with secondary electron mode (SE), homogeneity with backscattered electron mode (BSE) and specify elements contained in the sample with semi-quantitative EDX properties. The SEM-EDX results show the characterization of samples depend on density, porosity, and granular morphology. Microstructures of perovskite materials have a better characterization at doping x=0.5 and at the highest temperature and sintering time.

1. Introduction

Perovskite manganites are ones of the fascinating materials in condensed-matter research, because of their electrical and magnetic properties [1]. Substituting the Ln site by alkali earth or alkali element (*A*) in order to form $Ln_{1-y}A_yMnO_3$ (La = trivalent rare earth element and A = alkaline earth) compounds exhibits unusual physical effects around their ferromagnetic (FM)–paramagnetic (PM) phase-transition temperature (T_C). The unusual effects were such as microstructure, transport properties, colossal magnetoresistance and magnetocaloric (MCEs) [2-6]. The ionic radii of La^{3+} and Ag^+ can incorporate into the lattice and substitute La ions, and the Ln site disorder induce by Ag substitution gives effect on microstructure.

The application of perovskite LCMO is in the fields of magnetic memory devices, refrigeration, sensor, biology, medicine, and catalyst [7]. Physical properties of manganites may be modified by the chemical composition as well as by various treatments affecting their microstructure [8-9].

In this paper, perovskite $(La_{1-x}Ag_x)_{0.8}Ca_{0,2}MnO_3$ particle with different doped was synthesized by sol-gel processes. We study the effect of the sintering temperatures and time on the morphological of the samples.

2. Experimental

The material $(La_{1-x}Ag_x)_{0,8}Ca_{0,2}MnO_3$ has been prepared by sol-gel method where x=0 and x=0.5. Precursors were used including of La_2O_3, $AgNO_3$, $Ca(NO_3)_2.4H_2O$ and $Mn(NO_3)_2.4H_2O$. Each precursor was dissolved with nitric acid to obtain a transparent solution. Citric acid ($C_6H_8O_7.H_2O$) with a mole ratio of citric acid 1:1.2 total metal ions. The ammonia solution is added to adjust the pH of the solution to 7. The resulting mixture is homogenized using 400 rpm magnetic stirrer and temperature at 80-90 ° C to form a gel. The resulting gel was heated to 120°C in the oven for 3 hours to dry the sample to be gel highly viscous. After obtaining gel highly viscous, the sample was heated to a 500°C calcining temperature for 5 hours in the furnace for the decomposition of the organic precursor. Sintering temperature variations were done at 650°C for 5 hours and 900°C for 24 hours. Samples have characterization by using Scanning Electron Microscopy (SEM) and the composition of elements was examined by using X-ray Spectroscopy (EDX).

3. Result and Discussion

Figure 1. SEM image of $(La_{1-x}Ag_x)_{0,8}Ca_{0,2}MnO_3$, (x = 0; T = 900 °C, t = 24 h)

Figure 2. SEM image of $(La_{1-x}Ag_x)_{0,8}Ca_{0,2}MnO_3$, (x = 0.5; T = 900 °C, t = 24 h)

Figure 3. SEM image of $(La_{1-x}Ag_x)_{0,8}Ca_{0,2}MnO_3$, (x = 0; T = 650 °C, t = 6 h)

Figure 4. SEM image of $(La_{1-x}Ag_x)_{0,8}Ca_{0,2}MnO_3$, (x = 0.5; T = 650 °C, t = 6 h)

In Figure 1 until Figure 4, it is showed the SEM patterns of the $(La_{1-x}Ag_x)_{0,8}Ca_{0,2}MnO_3$ when x=0 and x=0.5 at 900 °C for duration of 24 h, samples were sintered at different temperatures and time duration. Figure 1 showed that the grain size of LCMO (x=0) were in the range 59 – 111 nm with an average grain size of 78.4 nm. Figure 2 showed that the grain size of LACMO (x = 0.5) were in range 71 – 119 nm with an average grain size of 100.6 nm. It is found that the grain sizes are increased with increasing the Ag-doped. Meanwhile, at Figure 3 and Figure 4,.for the sample of $(La_{1-x}Ag_x)_{0,8}Ca_{0,2}MnO_3$, when x=0 and x=0.5 at 650 °C for duration of 5h, it is found that the samples are agglomerated because of short sintering time. The grain size growth by increase the sintering time, the grain have more time to grow up and poros can be released gradually [10].

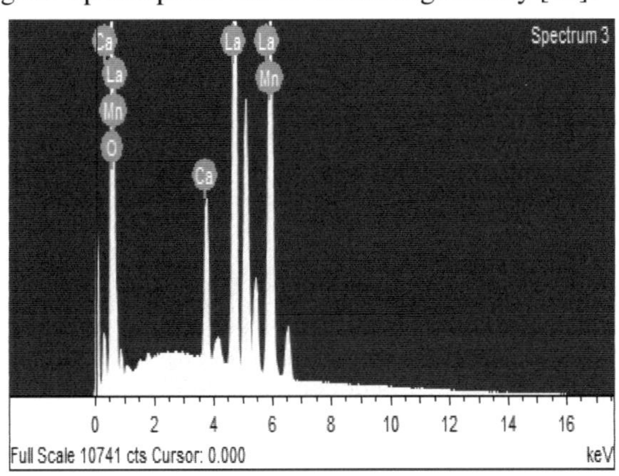

Figure 5. EDX image of $(La_{1-x}Ag_x)_{0,8}Ca_{0,2}MnO_3$, (x = 0; T = 900 °C, t = 24 h)

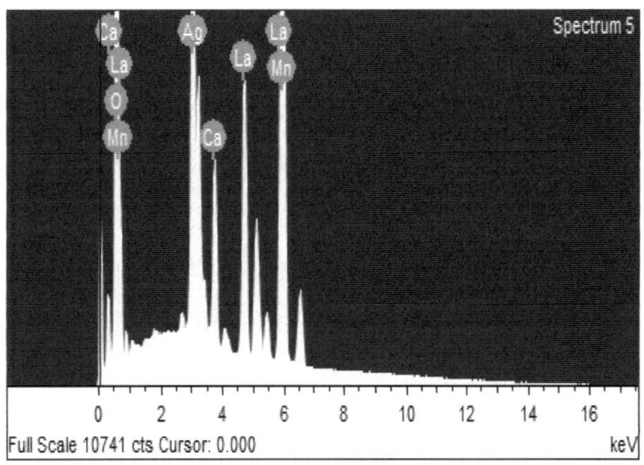

Figure 6. EDX image of $(La_{1-x}Ag_x)_{0,8}Ca_{0,2}MnO_3$, (x = 0.5; T = 900 °C, t = 24 h)

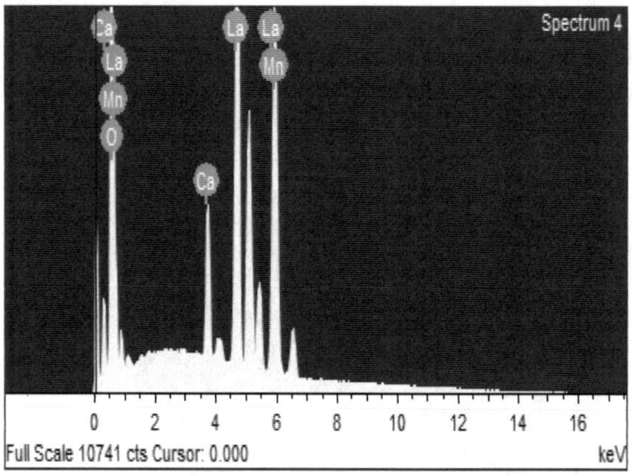

Figure 7. EDX image of $(La_{1-x}Ag_x)_{0,8}Ca_{0,2}MnO_3$, (x = 0; T = 650 °C, t = 6 h)

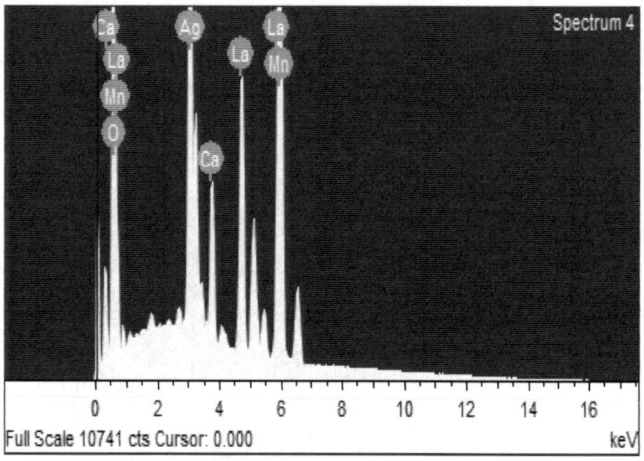

Figure 8. EDX image of $(La_{1-x}Ag_x)_{0,8}Ca_{0,2}MnO_3$, (x = 0.5; T = 650 °C, t = 5 h)

In Figure 5 until Figure 8 showed that there is no quantifiable loss of each elements. the composition of $(La_{1-x}Ag_x)_{0,8}Ca_{0,2}MnO_3)$ are presented in Table 1. The composition almost same as a quantitative value between designated composition and measured composition, it happened because the EDX method is semi-quantitative analysis.

Table 1. Compositional Analysis of $(La_{1-x}Ag_x)_{0,8}Ca_{0,2}MnO_3$, a. ($x$ = 0; T = 900 °C, t = 24 h); b. (x = 0.5; T = 900 °C, t = 24 h); c. (x = 0; T = 650 °C, t = 6 h); d. (x = 0.5; T = 650 °C, t = 6 h)

Doping concentration	Element	Weight (%)	Atomic (%)
x = 0	La	50.52	15.75
T = 900 °C	Ag	0	0
t = 24 h	Ca	3.52	3.81
	Mn	22.59	18.10
	O	23.01	63.32
x = 0.5	La	18.53	4.32
T = 900 °C	Ag	19.22	5.77
t = 24 h	Ca	2.96	2.39
	Mn	22.58	13.30
	O	36.71	74.23
x = 0	La	47.78	13.65
T = 650 °C	Ag	0	0
t = 6 h	Ca	3.39	3.36
	Mn	21.69	15.67
	O	47.78	13.65
x = 0.5	La	20.95	5.29
T = 650 °C	Ag	18.78	6.10
t = 6 h	Ca	3.01	2.63
	Mn	25.43	16.23
	O	31.83	69.75

Figure 5 until Figure 8 and Tabel 1 showed the chemical compositions and semi-quantitative EDX measurements. It is revealed that the cation atomic ratio of La: Ag: Ca: Mn is approaching to the nominal values in the composition of $La_{1-x}Ag_xCa_{0.2}MnO_3$ [7].

Figure 9. BSE image of $(La_{1-x}Ag_x)_{0,8}Ca_{0,2}MnO_3$, (x = 0; T = 900 °C, t = 24 h)

Figure 10. BSE image of $(La_{1-x}Ag_x)_{0,8}Ca_{0,2}MnO_3$, (x = 0.5; T = 900 °C, t = 24 h)

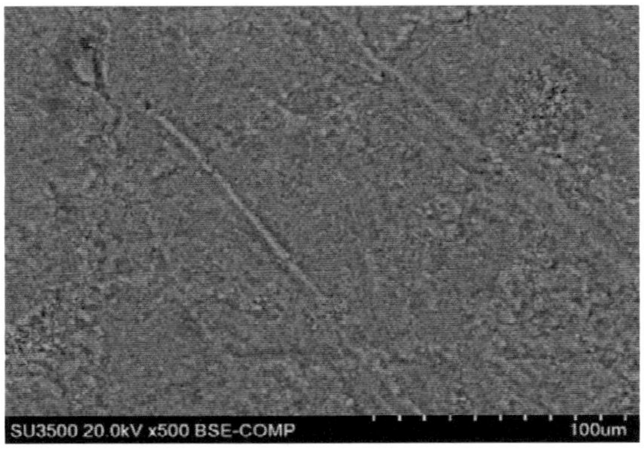

Figure 11. BSE image of $(La_{1-x}Ag_x)_{0,8}Ca_{0,2}MnO_3$, (x = 0; T = 650 °C, t = 6 h)

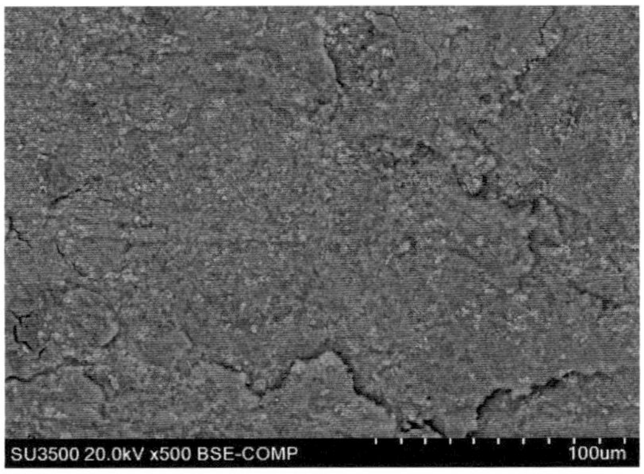

Figure 12. BSE image of $(La_{1-x}Ag_x)_{0,8}Ca_{0,2}MnO_3$, (x = 0.5; T = 650 °C, t = 6 h)

Figure 9 until Figure 12 showed that the backscattered electron mode (BSE) of LACMO (x = 0 and x = 0.5) at 900 °C for duration of 24h are homogenous. LACMO (x = 0 and x = 0.5) at 650 °C for the duration of 5 h are agglomerated due to that some of Ag react with oxygen. They can be seen in silver color. Besides, we also found that there are perovskite phase and Ag.O phase at LACMO (x = 0 and x = 0.5) at 650 °C for the duration of 5 h.

Summary

$(La_{1-x}Ag_x)_{0.8}Ca_{0.2}MnO_3$ were synthesis by a sol-gel process. The samples were varied in temperature and time of sintering and doping variations of Ag x = 0 and x = 0.5. Sintering temperature variations were done

IOP Conf. Series: Journal of Physics: Conf. Series **1011** (2018) 012014 doi:10.1088/1742-6596/1011/1/012014

at 650°C The SEM-EDX results show the characterization of samples depend on density, porosity, and granular morphology. Microstructures of perovskite materials have a better characterization at doping x = 0.5 and at the highest temperature and sintering time.for 5 hours and 900°C for 24 hours. The grain sizes are increased with increasing the Ag-doped and at the higher temperature and longer duration of sintering will make the nucleation processes in the sample better.

Acknowledgment
This work is financially supported by Universitas Indonesia under research grant "Hibah PITTA (Hibah Publikasi Internasional Terindeks untuk Tugas Akhir Mahasiswa Universitas Indonesia" with grant contract number is 634/UN2.R3.1/HKP.05.00/2017. The authors are also grateful to Department of Physics, Universitas Indonesia for research facilities.

References
[1] Smari M, Hamdi R, Dhahri E, Hlil E K, and Bessais L 2016 Correlation between magnetic and electric properties of $La_{0.5}Ca_{0.3}Ag_{0.2}MnO_3$ based on critical behavior of resistivity *Ceram. Int.*, vol. **42** no. 8 pp 10405–10409,
[2] Kalyana Lakshmi Y and Reddy P V 2010 Influence of silver doping on the electrical and magnetic behavior of $La_{0.7}Ca_{0.3}MnO_3$ manganites *Solid State Sci.* vol. **12** no. 10 pp 1731–1740.
[3] Thanh T D, Linh D C, Manh T V, Phan T L, and Yu S C 2016 Magnetic and Magnetocaloric Properties of $La_{0.8-x}Ag_xCa_{0.2}MnO_3$ Exhibiting the Crossover of First-and Second-Order Phase Transitions *IEEE Trans. Magn.* vol. **52** no.7 pp10–13
[4] Riahi K *et al* 2016 Effect of synthesis route on the structural, magnetic and magnetocaloric properties of $La_{0.78}Dy_{0.02}Ca_{0.2}MnO_3$ manganite: A comparison between sol-gel, high-energy ball-milling and solid state process *J. Alloys Compd.* vol. **688** pp1028–1038
[5] Ye S L *et al* 2002 Effect of Ag substitution on the transport property and magnetoresistance of $LaMnO_3$ *J. Magn. Magn. Mater.* vol. **248** no.1 pp 26–33
[6] Debnath J C and Strydom A M 2015 Large low field magneto-resistance and temperature coefficient of resistance in $La_{0.8}Ca_{0.2}MnO_3$ epitaxial thin film *J. Alloys Compd.* vol. **621** pp 7–11
[7] Xia W, Li L, Wu H, Xue P, and Zhu X 2017 Structural, morphological, and magnetic properties of sol-gel derived $La_{0.7}Ca_{0.3}MnO_3$ manganite nanoparticles *Ceram. Int.* vol. **43** no. 3 pp 3274–3283
[8] Pekała M and Drozd V 2008 Magnetocaloric effect in nano- and polycrystalline $La_{0.8}Sr_{0.2}MnO_3$ manganites *J. Non. Cryst. Solids* vol. **354** no 47–51 pp 5308–5314
[9] Pekała M., Drozd V, Fagnard J F, and Vanderbemden P 2010 Magnetocaloric effect in nano- and polycrystalline manganites $La_{0.5}$Ca0.5MnO_3 *J. Alloys Compd.* vol. **507** no. 2 pp 350–355
[10] Ma J *et al* 2014 Enhancement of temperature coefficient of resistivity in $La_{0.67}Ca_{0.33}MnO_3$ polycrystalline ceramics *Ceram. Int.* vol. **40** no. 3 pp 4963–4968

Characterization of SnO_2 Film with Al-Zn Doping Using Sol-Gel Dip Coating Techniques

A Doyan[1], Susilawati[1], N Ikraman[1], M Taufik[2]

[1]Magister Science, Post Graduate of Mataram University, Mataram, Indonesia
[2]Physics Education Mataram University, Indonesia

Email: arisdoyan@yahoo.co.id

Abstract. $Sn_{1-2x}Al_xZn_xO_2$ film has been developed using sol-gel dip coating technique. The materials $SnCl_2.2H_2O$, $AlCl_3$ and $ZnCl_2$ dissolved in water and ethanol with 5:95 volume ratio. Variations dopant concentration x = 0.000, 0.005, 0.0025, and 0.050. The film was grown with sol concentration 0.4 M, the withdrawal speed of 12 cm/min and sintering at 600 °C for 30 minutes. The characteristics $Sn_{1-2x}Al_xZn_xO_2$ films with various doping concentration phase were characterized by XRD. The morphological characteristics and the composition of the constituent elements of the film were characterized by SEM-EDX. The characteristics of the shape, structure, and size of the particles were characterized by TEM. The XRD results show that all films have a tetragonal SnO_2 rutile phase without any secondary phase with an average particle size in the range 5.14 – 2.09 nm. The SEM results show that the film grown has a smooth morphology with a striped texture (x = 0.00), and there is a crack (x = 0.050). The EDX results show that the composition and distribution of the constituent elements of the film are uniformly distributed. TEM results show that the particle films has tetragonal rutile structure, orthorhombic and amorphous with a spherical shape.

Keywords: Tin oxide, Zinc doping, Al doping, Sol-Gel, Dip Coating

1. Introduction

The tin oxide (SnO_2) is one n-type transparent semiconductor material with a wide energy band gap (~ 3.6 eV), combines high optical transparency and low resistivity that makes this material applied to solar cells, Liquid Crystal Display and other optoelectronic devices. In addition, SnO_2 is very sensitive to the presence of surrounding gas which makes it is applied as gas sensors [1]. Due to its nature and application, the study of SnO_2 engineering is very intensive. The preparation and growth of SnO_2 film can be done with various techniques such as Chemical Vapor Deposition (CVD) [2], DC and RF Sputtering [3], Pulsed Laser Deposition [4], Chemical Bath Deposition [5], and Sol-Gel Spray Pyrolysis [6], Sol-Gel Spin Coating [7], and Sol-Gel Dip Coating [8]. Of all these methods, Sol-Gel Dip Coating is a very good technique, because it is the simplest one that can grow the film on various forms of the substrate, ease in controlling particle growth and doping, and economical addition.

The characteristics of SnO_2 films are known to be enhanced by the addition of doping. Dopants generally have larger or smaller valence electrons with Sn^+, and have a similar ionic radius. Dopants may be of a single element such as Sb, F, Zn, Al, Mn or double-shaped elements such as Sb-F, Zn-Co, and others that depend on the application of SnO_2 itself. The SnO_2 films doped with Zn^{2+} and Al^{3+} ions are known to affect SnO_2 properties themselves, including stabilizing SnO_2 particles and changing energy band gaps, increasing optical transmittance and electrical conductivity, altering structural, increasing responsiveness, selectivity and stability in sensor gas. However, it is not yet known how the effect of these two dopants combined on the characteristics of the SnO_2 film. So it is necessary to further investigate the effect of these two dopants on the characteristics of the SnO_2 film. Therefore, in this study the growth of the SnO_2 thin film with combined doping of Zn^{2+} and Al^{3+} was done using Dip Coating Sol-Gel.

2. Materials and Methods

The basic ingredients used as the films in this study were as follows Tin (II) chloride dihydrate ($SnCl_2 \cdot 2H_2O$) with molar mass 225.63gr/mol, 98% purity, Merck). The solvent used was Ethanol (C_2H_5OH) with a molar mass of 46.07gr/mol (98% purity, Merck) and Aquades. The dopant material used was Zinc dichloride ($ZnCl_2$) with molar mass 136.30 g/mol (98% purity, Merck) and Aluminum Chloride ($AlCl_3$) with molar mass 133.34 g/mol (98% purity, Merck). The substrate film used is glass preparations with dimensions of 2.54 mm x 76.2 mm x 1 mm. While the material for washing the substrate is Hydrochloride (HCl 1.0 M) and ceramic detergent. The procedure of $Sn_{1-2x}Zn_xAl_xO_2$ sol preparation is a modification of the sol-gel preparation of $SnCl_2.2H_2O$. The $ZnCl_2$ and $AlCl_3$, dissolved in 5 ml of distilled water using a magnetic stirrer for 30 minutes at 80 ° C until the mass composition ratio $SnCl_2.2H_2O$, $ZnCl_2$, $AlCl_3$ with varying concentrations x (m/m) = 0.000, 0.005, 0.025, 0.050. Then 45 ml of ethanol (C_2H_5OH) was added in the mixture solution to obtain a constant solubility concentration of 0.4 mol / L. The solution was then stirred at 80 °C for two hours and kept for two days. The glass substrate before the deposition was washed using detergent followed by substrate soaking for one day in a mixed solution of 100 ml of HCl 1M with 300 ml of distilled water for one day. Then the substrate was washed with flowing distilled water. The substrate is then dried using an oven at 100 °C for one hour [7,9]. Thin layer deposition is done using an instrument with a dip coater pull Speed (Withdrawal Speed) at 12 cm/minute, long soaking 5 seconds, in sintering the electric furnace at a temperature of 600 °C for 30 minutes. This step is done as much as 5 times repetition. The Characterization of $Sn_{1-2x}Zn_xAl_xO_2$ film sample structure was done using XRD Rigaku Smartlab type with Cu Kα radiation. The XRD operating conditions are set at 40.0 kV 30 mA, and the XRD spacer angle range is set at the angle range (2θ) 5°-90°. The morphological characteristics and sample composition are characterized using SEM paired with EDX. In this study, the samples were characterized by samples with doping composition x = 0,000 and 0.050. The regional characterization, size, and regularity of samples were observed using JEOL JEM 1400 TEM. The samples characterized by TEM were samples with dopant composition x = 0.050.

3. Results and Discussion

3.1. Phase and Structure Characteristics

Figure 1 shows the XRD results for $Sn_{1-2x}Zn_xAl_xO_2$ samples at various dopant concentrations. The dopant ions Zn^{2+} and Al^{3+} replaced Sn^{4+} at SnO_2 lattice caused ionic radius Zn^{2+} (0.74 Å) and Al^{3+} (0,5 Å) significantly different with the ionic radius Sn^{4+} (0.69) based of not found the diffraction peaks of Sn, SnO, ZnO and Al_2O_3.These results are similar to other researchers [3, 10]. All samples were oriented to 110, 101, and 211. The addition and increase of Al-Zn dopants increased the intensity of diffraction, except at x = 0.050 which indicated the dopant solubility limit in the lattice SnO_2.
The lattice parameter of the diffraction graph is determined by the equation

$$\frac{1}{d^2} = \left(\frac{h^2 + k^2}{a^2} + \frac{l^2}{c^2} \right) \quad [1]$$

The estimated particle size of the film is determined using the Scherer formula

$$D = \frac{K\lambda}{\beta \cos\theta} \quad [2]$$

Where D is the average size of the crystal, K is a constant associated with the crystal form and has a value of 0.9, λ is the X-ray wavelength used (nm), and β represents the peak width at half the maximum diffraction peak (FWHM) (radians).

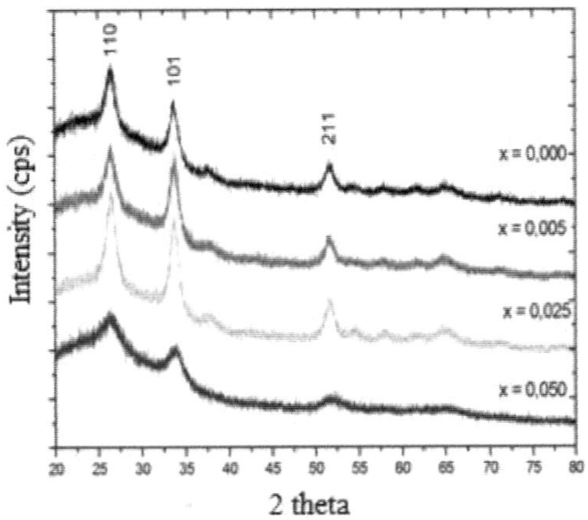

Fig. 1. Graph of the relationship between the diffraction angle with the intensity for variation of doping concentration.

The lattice parameter data and thin film particle size estimation on dopant concentration variations are shown in Table 1.

Table 1. The crystal lattice parameters and particle size of $Sn_{1-2x}Zn_xAl_xO_2$ film

No.	Sample	*a* values (Å)	*c* values (Å)	D (Å)
1.	$x = 0.000$	4.78444	3.18134	51.41
2.	$x = 0.005$	4.75640	3.18336	56.94
3.	$x = 0.025$	4.73352	3.16838	54.78
4.	$x = 0.050$	5.19802	2.93922	20.99

Table 1. shows that the lattice constant a decrease with increasing dopant, but increases drastically at dopant concentration x = 0.050. While the value of c grid constant increased at dopant concentration x = 0.005 which then decreased for dopant concentration x = 0.025 and x = 0.050. This indicates that the substitution of dopant ions Zn^{2+} and Al^{3+} causes SnO_2 lattice distortion caused by the ionic difference of Sn^{4+} radius. The lattice expansion is caused by a Zn^{2+} ionic radius greater than the Sn^{4+} ionic radius, whereas the lattice shrinkage caused by the ionic radius of Al^{3+} is smaller than that of the Sn^{4+} ionic radius which is in agreement with Khan, et al. [11]. This pattern of random lattice constant changes is caused by the substituted dopant replacing the SnO_2 lattice randomly. While at dopant concentration x = 0,050 seen increase of lattice constant a = 5.19802 Å and c = 2.93922 Å close to ZnO lattice constant value. These results indicate that on the axis axes Sn^{4+} ions are replaced with larger Zn^{2+} ions. The particle size of SnO_2 with x = 0.000 is 5.14 nm in the range of 2.67 - 7.76 nm. At dopant concentration x = 0.0050 there was an increase in particle size with an average size of 5.69 nm with a size range of 0.87 - 8.72 nm. At dopant concentration x = 0.025 the particle size decreased with an average size of 5.48 nm with a range of 2.60 - 7.51 nm. At dopant concentration x = 0.050 particle size decreased with mean size 2.09 nm with

IOP Conf. Series: Journal of Physics: Conf. Series **1011** (2018) 012015 doi:10.1088/1742-6596/1011/1/012015

range 1.01 - 3,69 nm. The increase in average particle size at low doping concentrations has grown SnO_2 film with Zn doping [12].

3.2. Characteristics of Morphological

The characteristic Film surface morphology is characterized using Scanning Electron Microscope (SEM). FIG. 2 (a) shows the $Sn_{1-2x}Zn_xAl_xO_2$ film image at dopant concentration x = 0,000 textured lines, whereas for dopant concentration x = 0.050 (Fig. 2 (b)) has a smooth surface with two crack types Size 3.27 µm and 0.80 µm and the presence of crater (cater). A crack with a size of 3.27 µm is applied to the previous coating which has been covered by other coatings, whereas a crack with a size of 0.80 µm was observed on the final coating. This crack is due to the high sol high concentration [13] and the high rate of heating at the time of sintering by the electric furnace [14]. Figure 2 (c) shows the transverse incision image of SEM. The films grown have varying thicknesses with a range of 4.37 - 5.00 µm. This thickness is categorized as a thin film [15].

(a) (b) (c)

Fig. 2. The surface morphological image of the film at magnification 10000 times for (a) x = 0.000, (b) x = 0.050, and (c) the transverse incision image of SEM

3.3. Characteristics of Composition

Distribution characteristics and composition of the embedded film material were characterized using the Energy Dispersive X-Ray Spectroscopy (EDX) coupled with the SEM test apparatus. The composition of the film sampling elements for x = 0.000 is indicated by the presence of the energy peaks shown in Fig. 3 (a). The peak at 3.44 and 3.68 eV energy emission from Sn and peak at 0.52 eV represents the energy emission of element O. The pattern of element distribution for a film with x = 0.050 by Fig. 3(b). The peak at energy 0.37, 3.44, 3.66, and 3.94 eV shows the spectra of Sn element. The peak at 0.52 eV shows the spectra of element O. The peak at 1.01, and 1.48 eV shows the dopant elements of Zn and Al. The peaks at 0.25 and 1.74 eV show the spectra of C and Si elements. The presence of C element indicates that the solvent does not evaporate completely, while the Si element in the substrate element (SiO_2).

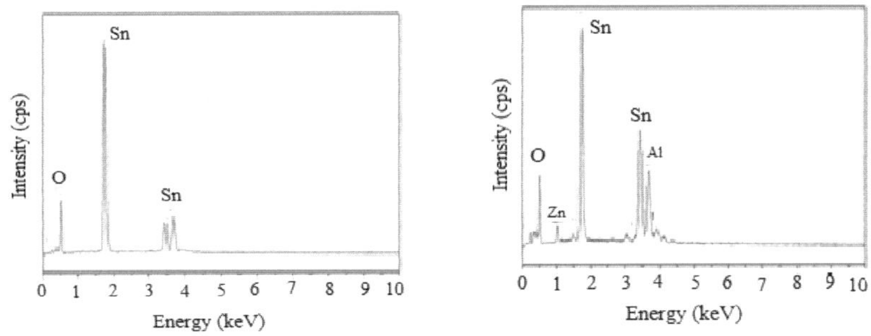

Fig.3. The EDX graph of the film at dopant concentration x = 0.000 dopant and concentration x = 0.050

3.4. Characteristics of Shape and Particle Size

The characteristics of particle shape and size were characterized using TEM. The samples characterized by TEM are samples with doping concentration x = 0.050. The particles are grown spherical with a size range of 3.72 - 6.16 nm. This result differs from the result of particle size estimation with Scherer formula on XRD with particle size range 1.01-3.69 nm (figure 4(a)). TEM can also be used to determine the particle structure through electrons diffraction future in the selected area (Selected Area Diffraction

Electron - SAED). Figure 4(b) shows an irregular pattern of electron diffraction that indicates that some particles have polycrystalline structures with a tetragonal rutile phase with various orientations. This diffraction pattern results in spot irregularities in electron diffraction indicating a lattice disturbance by the Al-Zn dopant corresponding to the XRD results [16]. Figure 4(c) shows that some particles also have polycrystalline structures with orthorhombic phases. The electron diffraction pattern is similar to that of Cd Doped SnO_2 diffraction film grown by Chemical Bath Deposition (CBD) method. The existence of an orthorhombic structure is caused by a lattice distortion caused by a substituted Zn doping ion replacing the SnO_2 lattice, while image 4(d) shows the particles are amorphous. This is due to the nucleation and Growth of imperfect film crystals due to the shortness of sintering time [17].

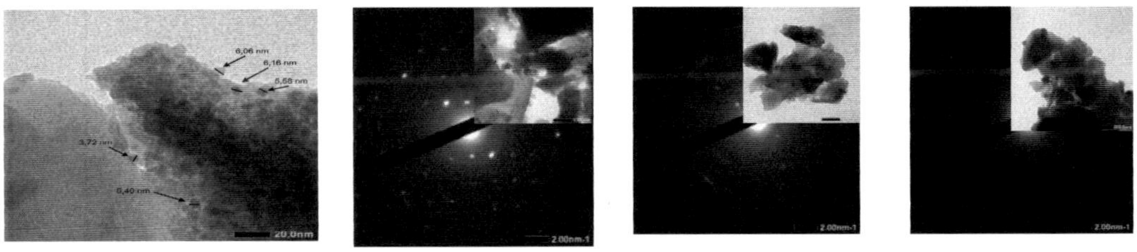

(a) (b) (c) (d)

Fig. 4. Electron Diffraction Pattern Image $Sn_{1-2x}Zn_xAl_xO_2$

4. Conclusion

Results and data analysis can be concluded that the addition of Al-Zn doping does not change the phase SnO_2. The SEM results show that the grown film has a smooth morphology with a striped texture (x = 0,000) and there is a crack (x = 0.050). EDX results show that the constituent elements of the film are spread evenly with the composition according to the calculation. The TEM results show that the film particles have three tetragonal, orthorhombic and amorphous, polycrystalline rutile structures

Acknowledgement

Thank you to all staff and laboratory analytical chemistry Mataram University and laboratory physics Department Institute Technology Bandung and A big thank you to the students the material physics research groups have tried to help prepare samples and data retrieval. We also thanks to the education department of research and technology for the basic research grant in 2017.

References

[1] Batzill M, and Diebol U 2005 The Surface and Materials Science of Tin Oxide *Progress in Surface Science* **79** pp 47–154

[2] Jeong Jin, Choi,Seung Pyung and Hong, Kwang Joon 2006 Structural and Optical Properties of SnO2 Thin Films Deposited by Using CVD Techniques *Journal of the Korean Physical Society*, Vol. **48**, No. 5, May 2006, pp. 960-963

[3] Xu Bo, Ren Xiao-Guang, Gu Guang-Rui, Lan Lei-Lei and Wu Bao-Jia 2016 Structural and optical properties of Zn-doped SnO_2 films prepared by DC and RF magnetron co-sputtering *Superlattices and Microstructures* Volume **89** pp 34-42

[4] Weng Min-Hang, P, Cheng-Tang Y, Ru-Yuan,H, and Chun-Chih, 2011 Structure Optical and Electrical Properties of ZnO Thin Films on The Flexible Substrate by Cathodic Vacuum Arc

Technology with Different Arc Currents *Ceramics International Journal 05,* **43**: 1-6. DOI: 10.1016/j.ceramint.2011.05.043.

[5] Maddu, Akhiruddin, Hasiholan, Rodo Tua, and Kurniati, Mersi 2009 The growth of Nanocrystal SnO_2 film with Chemical Bath Deposition Method (CBD) *Journal of Nanoscience & Nanotechnology* ISSN 1979-0880.

[6] Sriram S and Thayumanavan, A 2013. Effect of Al Concentration on the Optical and Electrical Properties of SnO_2 Thin Films Prepared by Low Cost Spray Pyrolysis Technique *International Journal of Chem. Tech Research* **5**(5): 2204 – 2209.

[7] Doyan, Aris., Susilawati, Imawanti, and Yanika Diah. 2017 Synthesis and characterization of SnO_2 thin layer with a doping aluminum is deposited on quartz substrates *AIP Conference Proceedings* **1801**, 020005 doi: 10.1063/1.4973083

[8] Carvalho, D H Q, Schiavona, M A, Raposoa M T, de Paivaa R, Alvesa, J L A, Paniagob, and Roberto, M 2012 Synthesis and characterization of SnO_2 thin films prepared by dip-coating method *Physics Procedia* **28** 22–27.

[9] Doyan Aris, Susilawati, Fitri S.A Ahzan, Sukainil 2017 Crystal Structure Characterization of Thin Layer Zinc Oxide *IOP Publishing* Ltd, doi: 10.1088/1757-899X/196/1/012016.

[10] De Arijit, 2015 Study of Sol-Gel Derived Spin Cated Cd-Sn Oxide Films on Glass *J. Thin. Film. Sci. Tec.* **4**, 83-87. DOI: 10.12785/ijtsf/040204.

[11] Khan Z, Raza K M, Shoeb Z, Mohammad K M, Shahid, 2010 Optical and Structural Properties of Thermally Evaporated Cadmium Sulphide Thin Films on Silicon (100) Wafers *Materials Science and Engineering B.*, **174**, 1-3: 145-149. DOI: 10.1016/j.mseb.2010.03.006.

[12] Rajwali, K and Ming-Hu F 2015 Dielectric and magnetic properties of (Zn, Co) co-doped SnO_2 nanoparticles *Chinese Physics B* **24**(12), 127803

[13]Uno U E, Emetere M E and Aplha M 2014 Crystalline Grain Size Effects On The Conductivity Of The Doped Tin Dioxide (SnO_2) With Zinc (Zn) *Journal of Ovonic Research* Vol. **10** No. 3 pp 83-88

[13] Miller T A, Bakrania S D, Perez C and Wooldrige M S 2006 Nanostructured Tin Dioxide Materials for Gas Sensor Application *J. F. Nanomaterials* pp 1-24

[15] Singh, Karnail B and Tirumkudulu Mahesh S 2007 Cracking in Drying Colloidal Films *Phys. Rev. Lett.* **98**, 218302

[16] Pawar B G, Pinjari P, Divak V K, Sanjay S P, Aniruddha B, and Sung H 2012 *Chemical Enginering* ISRN, 954869, 1-7. DOI: 10.542/2012/954869

[17] Zhao Jun-Hua, Tan Rui-Qin, Yang Ye, Xu Wei, Li Jia, Shen Wen-Feng, Wu Guo-Qiang, Yang Xu-Feng, Song Wei-Jie 2015 Synthesis mechanism of heterovalent Sn_2O_3 nanosheets in oxidation annealing process *Chinese Physics* B Volume **24**, Issue 7, article id. 070505

IOP Conf. Series: Journal of Physics: Conf. Series **1011** (2018) 012016 doi:10.1088/1742-6596/1011/1/012016

Effect of annealing temperature on optical properties of TiO_2 18 NR-T type thin film

Nandani[1)], A Supriyanto[1)], A H Ramelan[1)], F Nurosyid[1)]
[1)]Physics Department of Mathematics and Natural Science Universitas Sebelas Maret, Jl. Ir. Sutami No. 36 A Kentingan, Surakarta 57126

E-mail: agusf22@staff.uns.ac.id

Abstract
It has succeeded in growing TiO_2 18NR-T thin film using T-49 screen printing method with an annealing treatment temperature variation of 200°C, 450°C, 500°C and 550°C for 1 hour. The result of TiO2 18NR-T thin film will be characterized using UV-visible Spectrophotometry (UV-*Vis*). Characterization with UV-*Vis* obtained a result that treatment with 4 variations of temperature influence absorbance value and transmittance. The greatest absorbance value was obtained in the sample with temperature treatment 550°C while the largest transmittance value was obtained at temperature 200°C. The absorbance peak is at the wavelength of ultraviolet light. Optical band values in each temperature variation are obtained at 3.53 eV; 3.46 eV; 3.38 eV and 3.2 eV.

1. Introduction

Titanium dioxide (TiO_2) is a semiconductor material having a large optical band gap of 3.2 eV to 3.8 eV [1], making TiO_2 effective for photovoltaic [2] because of its stable properties. TiO_2 can be used in solar cell applications [3-5] because TiO_2 absorbs light at the wavelength of the ultraviolet region [6-8]. The TiO_2 material is a non-toxic, cheap, non-corrosive and easily obtainable material [9-11].

The increasing need encourages scientists to develop transparent materials to produce more advanced technology. There has been much fabrication of thin layers of semiconductor material to obtain new devices from modifying the thin films one of them for transparent solar cell applications [12]. Therefore research on TiO_2 thin films continues to be developed so that the deficiencies they have can be minimized.

Deposition methods have been widely used to produce desired thin films, such as spin coating [13]; sol-gel [14] and screen printing [9]. Each method of deposition has its advantages and disadvantages. In this research using screen printing method, in addition to the effective method is also easy to do.

In this study TiO_2 used in the form of pasta is TiO_2 18NR-T, one of the materials for making a solar window. The deposition was performed using a 49-type screen with a mesh size of 49 per cm. Annealing temperature treatment was provided to determine the effect of temperature rise on the optical band gap of the TiO_2 18NR-T layer. The TiO_2 18NR-T thin film was then characterized by absorption value and its transmittance value to determine the optical band value due to the variation of annealing temperature.

2. Methods

The research begins with preparing the ingredients, TiO_2 which is used is brand dyesol with the type of 18NR-T in the form of a yellow paste, the substrate (glass preparation), ethanol, quads, acetone, and alcohol. While the tool used is ultrasonic cleaner, digital scales, micropipette, stirrer, cellophane tape, screen printing, oven, crucible, furnace and characterized by UV-*Vis* spectroscopy.

Pasta TiO_2 18NR-T, one gram, and ethanol, one mL, were mixed and stirred homogeneously using a magnetic stirrer for 1 hour with a rotation speed of 300 rpm at room temperature. The next stage is deposition of the TiO_2 18NR-T solution in the substrate with screen printing method. The substrate to be used was first washed using ultrasonic cleaner using soap water, quads, acetone and alcohol for 10 min respectively.

The screen printing process is done after the substrate is dry, the screen used is a T-49 type affecting the amount of TiO_2 18NR-T solution to be deposited. In the substrate there is a limitation of the area with the size of 1.5 x 1 cm as the deposition of the TiO_2 18NR-T solution using scouth as:

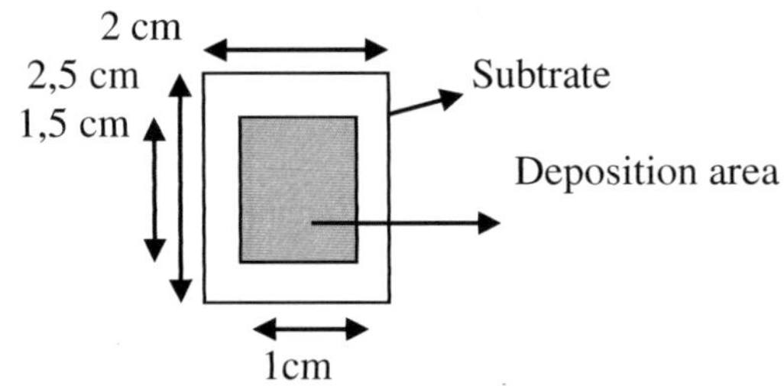

Figure 1: Schematic area of deposition of TiO_2 18NR-T solution

Deposition of TiO_2 18NR-T solution yielding 1 thin layer TiO_2 18NR-T using specification 3 drops of TiO_2 18NR-T solution and 5 times pressure. The result of deposition of theTiO_2 18NR-T solution by screen printing method was then carried out drying using the oven with temperature 130°C for 10 minutes. The best samples were then annealed using a furnace with variations of 200°C, 450°C, 500°C and 550°C aging for 1 hour. The provision of temperature variations aims to compare the results of crystal structure which influence absorbance value, transmittance and optical band gap energy.

The result of thin layer annealing TiO_2 18NR-T is then characterized using UV-*Vis* to measure absorbance value and its transmittance so that it can be processed to produce value of optical band gap energy using the Tauc method.

3. Results And Discussion

Annealing Temperature Dependence of Optical Properties of Thin Film TiO_2 18NR-T

TiO_2 18NR-T thin films have been grown on a glass substrate. The thin growth films prepared using a screen printing method with T-49 mesh size. Annealing treatments with temperature variations of 200°C, 450°C, 500°C, and 550°C were given to determine the effect of absorbance values, transmittance and optical band gap energy in the manufacture of TiO_2 18NR-T thin films. The result of TiO_2 18NR-T thin film above the transparent glass substrate.

The Results of absorbance and reversed transmittance Characterization can be shown with graphs in figure 2 and figure 3. The results of absorbance measurements showed that the treatment of variations in annealing temperature affects the absorbance rate of TiO_2 18NR-T thin film. The higher the given temperature value, the higher the absorbance value of the TiO_2 18NR-T thin film. This suggests that high temperatures accelerate the evaporation of oxygen during the annealing process.

A high absorbance value results in a low transmittance value, which means the TiO_2 18NR-T thin film with low transmittance is more absorbing than high transmittance. The results of the transmittance measurements can be shown in figure 3. The temperature of annealing affects the transmittance which results in the higher the given temperature decrease. In the wavelength range of 250 nm to 380 nm, there is a significant decrease in transmittance due to TiO_2 18NR-T over absorbing the ultraviolet wavelength region.

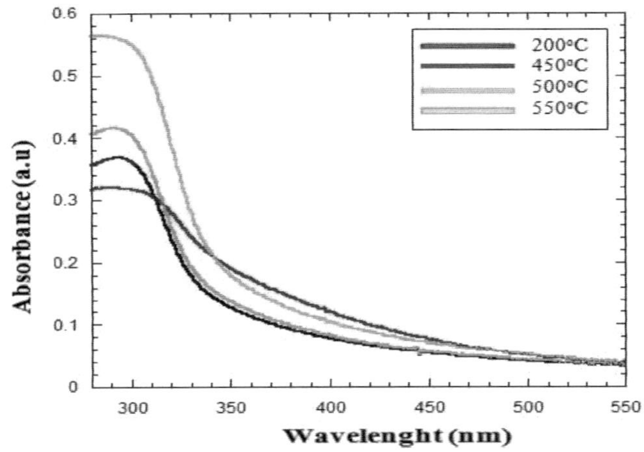

Figure 2. TiO_2 18NR-T thin film absorbance curve with annealing temperature variation of 200°C, 450°C, 500°C and 550°C

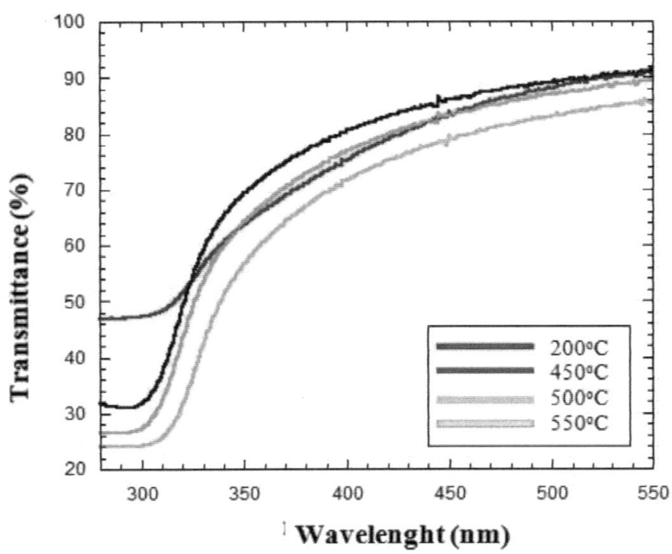

Figure 3. TiO_2 18NR-T thin film transmittance curve with annealing temperature variation of 200°C,

450°C, 500°C and 550°C

Determination of Optical Gap Values of TiO_2 18NR-T Thin Film of Variation of Annealing Temperature

The optical gap value of the TiO_2 18NR-T thin film is obtained using the Tauc Plot method by drawing a straight line on the relationship graph $\alpha h\upsilon^{\frac{1}{2}}$ and $h\upsilon$ to cut the abscissa axis of the energy axis. The calculation uses the equation 1.

$$\alpha h\upsilon = A(h\upsilon - E_g)^2 \quad (1)$$

A is the absorption coefficient, hυ is energy (eV), and E_g is the optical band gap energy.And the calculation of α is calculated by the equation:

$$\alpha = -\frac{1}{d}\ln T \quad (2)$$

The treatment of annealing temperature variations given in the TiO_2 18NR-T thin film produces an optical band gap energy which can be shown in the following figure 4.

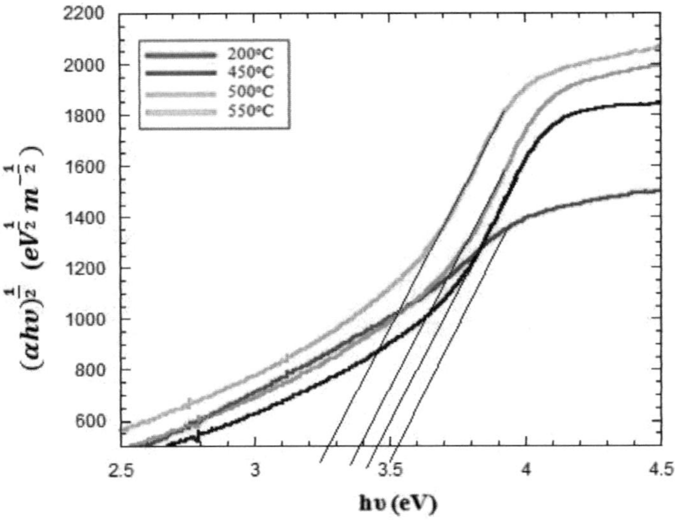

Figure 4. Determination of optical band gap TiO_2 18NR-T using Tauc Plot method.

The effect of annealing temperature on the TiO_2 18NR-T thin film results in different optical band gaps. At annealing temperature treatment of 200°C resulted in the optical band gap of 3.53 eV, while at 550°C maximum given temperature produces optical band gap of 3.27 eV. Increasing annealing temperatures result in a decreased optical band gap.

4. Conclusion

The results are the annealing temperature affects the optical band gap value of the TiO_2 18NR-T thin film; this is accompanied by changes in absorbance and transmittance values. Characterized by UV-*Vis* spectroscopy show that the higher temperature annealing causes the TiO_2 18NR-T thin film more absorbance. The maximum absorbance of the TiO_2 18NR-T thin film appeared at annealing temperature 550°C. This suggests that high temperatures accelerate the evaporation of oxygen during the annealing process. The effect of annealing temperature is increasing annealing temperatures cause a decreased optical band gap. Minimum band gap energy occurs in the TiO_2 thin film with annealing temperature 550°C is 3.27 eV.

5. Acknowledgments

This research is supported by the grant of Penelitian Unggulan - Universitas Sebelas Maret (PU-UNS), 2017.

6. References

[1] Gratzel, M 2003 *Journal of Photochemistry and Photobiologi Photochemistry.* **4** 145-153
[2] Macak J M, Zlamal M, Krysa J and Schmuki P 2007 *Small.***3** 300–304
[3] Pugliese D, Lamberti A, Bella F, Sacco A, Bianco S, Tresso E 2014 *Electron.***15** 3715–3722
[4] Roy P, Kim D, Lee K, Spiecker E, Schmuki P 2010 *Nanoscale.***2**45–59
[5] Gao X, Li J, Baker J, Hou Y, Guan D, Chen J, Yuan C 2014 *Commun.***50** 6368–6371
[6] Fuyuki T, Matsunami H 1986*J Appl Phys.* **25**128
[7] Gan J Y, Chang Y, Wu T 1998*Appl Phys Lett.*72.
[8] Miao L 2003 *Appl Surf Sci.***212** 255
[9] Hoffmann M R, Martin S, Choi W, Bahnemann D W, e*nvironmental applications of semiconductor photocatalysis, Chem. Rev.***95**69–96
[10] Fujishima A, Rao T N, Tryk D A 2000 *Journal Photochem. Photobiol.* **C1**1–21
[11] Tryk D A, Fujishima A, Honda K 2000*Electrochim Acta.***45**2363–2376
[12] Nurosyid F, FurqoniL, Supriyanto A and Suruyana R 2016. *Journal of Physics.* **776** 1-7
[13] Khan M I, Bhatti K A, RabiaQindeel, Hayat S A and Norah Alonizan 2017 *Result in Physics.***7** 1437-1439
[14] Zareen A, Ali S and Irfan M 2014 *European Scientific Journal.* **2** 1857-7431

IOP Conf. Series: Journal of Physics: Conf. Series **1011** (2018) 012017 doi:10.1088/1742-6596/1011/1/012017

The Effect of Pluronic 123 Surfactant concentration on The N_2 Adsorption Capacity of Mesoporous Silica SBA-15: Dubinin-Astakhov Adsorption Isotherm Analysis

Donanta Dhaneswara[1], , Siti Agustina A. A. A.[1], Dewantoro Adhy P.[1], Farhan Delayori[1,3] and Jaka Fajar Fatriansyah[1,2]
[1]Metallurgical and Materials Engineering, Faculty of Engineering, Universitas Indonesia, ok, Jawa Barat, 16424
Email: [2]fajar@metal.ui.ac.id, [3]fdelayori@gmail.com

Abstract. Mesoporous SBA-15 has been successfully synthesized at various concentration of Pluronic 123 surfactant (7mM, 50 mM, 54 mM, 60 mM and 66 mM) and the effect of these various concentrations on the N_2 adsorption capacity has been investigated. The adsorption analysis was conducted using Dubinin-Astakhov isotherm model for multilayer adsorption phenomenon. It was found that etryat low concentration of Pluronic 123, the system exhibits type I adsorption isotherm while at high concentration, the system exhibits type IV adsorption with H1 hysteresis curve which indicates the existence of pores with cylindrical geometry, relatively uniform pore size and possibility of pore network effects. It also was found that, by using D-A isotherm model fitting, at 60 mM concentration of Pluronic 123, SBA-15 has the highest adsorption capacity which stands at 421 cm^3/gram.

1. Introduction

In recent years, mesoporous materials draw the attention due to their applications in many fields such as: catalysis [1], adsorbent [2], drug delivery [3], biosensor [4] and separation [5]. Mesoporous material is a material with pore of diameters in the range of 2-50 nm as defined by IUPAC [6]. This material has highly ordered mesostructure which yields high surface area. This advantage allows diffusion and adsorption of large molecules which are applicable in wide applications. There are compounds of mesoporous materials such as: silica, alumina, carbon and transition metal oxides. Other Micropores such as has limited application to chemical catalysis and separation due to bulky nature of adsorbate [7]. One of the widely used compound is silica due to its abundance, inexpensive, thermally stable, chemically inert and harmless. Zhao et. al [8] successfully produced Santa Barbara Amorphous 15 (SBA-15) which has the larger pores and thicker walls and has exceptional thermal and mechanical stability and chemical resistance which make this SBA-15 a promising material for many applications.

Pure SBA-15 has ordered hexagonal structure with diameter of pores ranging from 4.6 nm to 30 nm and was synthesized using triblock copolymer in acid environment [9]. Surface of pure SBA-15 can be as high as 650 m^2/gram although theoretically this kind of structure can only has surface area of 204^2 m/gram. Ravikovitchet. al [10] reported that SBA-15 can also have regular cylindrical arranged in hexagonal structure. The synthesis of SBA-15 using triblockc opolymer as template has gained interest due to the ease of process. In this paper we used Pluronic 123 (P123), a triblock copolymer due to its capability to form long cylindrical micelles which are specifically suitable to synthesis SBA-15 with precursor TEOS. The suitable concentration of template/surfactant is crucial to synthesis SBA-15 since the pores size and structure are heavily dependent on template structure. The use of SBA-15 as adsorbent is related to its adsorption capacity. Higher adsorption capacity towards selective adsorbate is preferred. In order to study the possibility the use of SBA-15 to adsorb compounds such as inert gases, heavy metals and etc, the isotherm adsorption study is important. In this paper we investigate the effect of P123 on adsorption capacity of SBA-15 using Dubinin- Astakhov (DA) isotherm model. This model was developed to explain the multilayer adsorption and volume filling.

2. Experimental Methods

2.1 Synthesis and Characterization

In this synthesis, the precursor used was Tetraethyl Orthosilicate (TEOS) and the template used was P123. The process is called sol-gel solution and the solution is in acidic condition. First step, 5 ml ethanol was added to 31.25 gram TEOS and stirred for 30 minutes at 25°C. Second, 5 ml ethanol was mixed with 10 ml HCl. This solution then was mixed with ethanol (10 ml) + 50 ml distilled water solution and was stirred for 30 minutes at room temperature. The mixed solution was then refluxed for 2 hours at temperature of 60 C. Next the surfactant template solution was made. The concentration of surfactant was varied at 7 mM, 50 mM, 55 mM, 60 mM and 66 mM. This surfactant then was added into mixture of HCl (10 ml) and ethanol (25 ml). This surfactant solution was then added drop wise into TEOS solution and was continuously stirred until the solution turned into gel. The gel then was dried at 100°C for an hour and calcined for 5 hours. The product was then ready for adsorption measurement. The adsorption characterization was conducted using Quantachrome equipment at 77 K. N_2 gas was used as adsorbate. The specific surface area was determined by the Brunauer-Emmer-Teller (BET). Small angle X-ray scattering (SAXS) measurements (Shimadzu XRD-6000) was carried out to determine the structure of the synthesized samples.

2.2 Dubinin-Astakhov isotherm model of adsorption

Freundlich [11] has recognized empirical model of an adsorption isotherm which was derived from Langmuir equation [12] and was extended to heterogeneous surfaces. Later, BET equation was derived from a simpification of Langmuir and can explain multilayer adsorption phenomena. Before BET equation was widely used, Dubinin [13] developed adsorption equation based on Weibull distribution of adsorption potential. The BJH, HK and density functional theory (DFT) were also developed but the calculation was complicated and time consuming. On the other hand, DA model by Dubinin is simpler and can be used to analyze adsorption on mesoporous and can be shown as

$$C = C_0 \exp\left[-\left(A/\beta E_0\right)^n\right] \quad (1)$$

where C is the volume of adsorbate filling mesopores, C_0 is the maximum capacity of adsorbent to adsorb adsorbate, $A = RT\ln\left(P_0/P\right)$ is the differential of molar work of adsorption, β is the affinity coefficient, E_0 is the characteristic energy, R is the Boltzmann constant, T is the equilibrium temperature, P/P_0 is relative pressure. Fitting can be done by linearizing eq. (1) which is written as:

$$\ln C = \ln C_0 - \left[RT/\beta E_0 \ln\left(P_0/P\right)\right]^n \quad (2)$$

DA model is valid over a range of N_2 relative partial pressure up to 0.02.

IOP Conf. Series: Journal of Physics: Conf. Series **1011** (2018) 012017 doi:10.1088/1742-6596/1011/1/012017

3. Results and Discussion

First, SAXS results are shown in Figure 1. All measured samples show three main peaks (100), (110) and (200). These results are in agreement with the results obtained by Su et. al [14]. These three peaks show that synthesized SBA-15 has 2D hexagonal structure. As concentration of surfactant increases, there is tendency that the peaks shift to the lower angle. This shifting is expected as the increase of surfactant concentration can result in the expansion of lattice parameters thus increasing pore size [15]. In addition, at concentration 60 mM, the (100) peak is much broader than other concentration. This may indicate that more non-uniform mesopores were produced. Second, The N_2 adsorption-desorption isotherm (bold-dash lines) of SBA-15 with various concentration of P123 (7 mM, 50 mM, 54 mM, 60 mM and 66 mM) is shown in figure 2. It can be seen from figure 2 that there are two kinds of adsorption type. At low concentration of P123 (7mM), the curve exhibits type I isotherm without hysteresis loop. This means that the adsorption is limited to a single monolayer of adsorbate at the surface of SBA-15, which in this case is N_2. This curve indicates that at low concentration of surfactant, SBA-15 has small external surface. The limitation of uptake (maximum filling volume of adsorption) is governed by the accessible pores and not by internal surface area. This indication resonates with the fitting results which show that at concentration 7 mM, the maximum filling capacity is lowest. At high concentration of P123 (50 mM, 54 mM, 60 mM and 66 mM) curves exhibit type 4 isotherm. This translates to an adsorption which proceeds via multilayer adsorption. This is followed by a capillary condensation. Their curves reveal an H2 type hysteresis loops. This which indicates the existence of pores with relatively non uniform pore size.

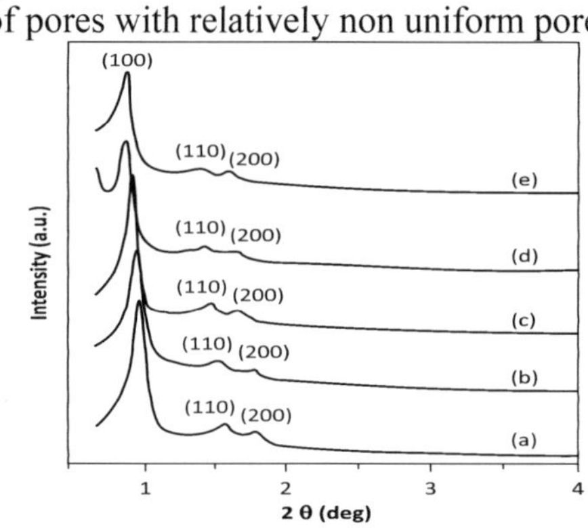

Figure 1. SAXS patterns for P123 concentration at (a)7 mM (b)50 mM (c)54 mM (d)60 mM and (e)66 mM..

DA isotherm model fitting is shown in figure 3. All curves were fitted up to relative pressure up to 0.3. The fitting results of maximum filling capacity/maximum adsorption capacity (C_0) are shown in figure 4. The lowest adsorption capacity is at 7 mM which stands at 175 cm3/gram and the highest adsorption capacity is at 60 mM which stands at 421 cm3/gram. The quadratic nature of relation between surfactant concentration and maximum adsorption capacity can be explained as follows. At optimum concentration, the hexagonal template which is formed in self-assembly manner by surfactant reach its maximum surface area. Further addition of surfactant leads to decrease of surface area. At maximum surface area of hexagonal template, the pore size will be optimum thus yields highest adsorption capacity. The behavior of optimum surfactant concentration value for producing specific geometry (such as hexagonal lamellar) may be attributed to the effort of the system to minimize free energy. The lamellar structure may form branches as concentration is increased. The characteristic energy adsorption for all samples are given in figure 5. At high concentration of P123 (50 mM, 54 mM, 60 mM and 66 mM), the value of their characteristic energy are around 200-250 J mol^{-1}, much lower in comparison with the value of characteristic energy for low concentration of P123 at 7 mM which has value 503.638 J mol^{-1}. The characteristic energy of adsorption is related to the isosteric heat of adsorption. These results show that at high concentration of surfactant, the system needs much lower energy to adsorp gas yielding high uptake. On the other hand, at low concentration of surfactant, the system needs much higher energy to adsorp gas thus yielding low uptake. We suggest that the characteristic energy can be used as a parameter to differentiate between two types of adsorption (type 1 and type 4) as well as two types of pore structure and geometry.

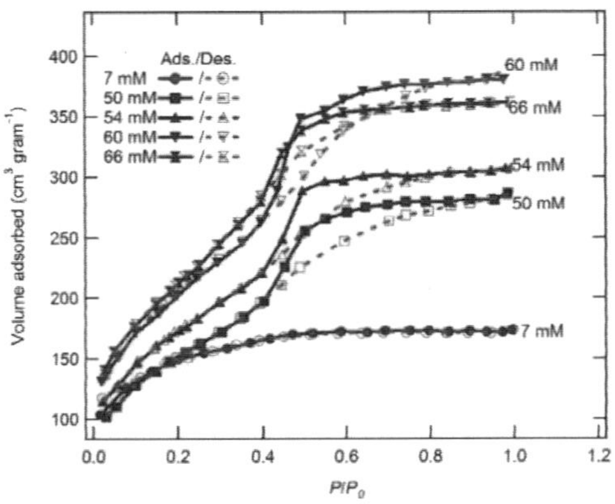

Figure 2. The N_2 adsorption-desorption isotherm (bold-dash lines) of SBA-15 with various concentration of P123.

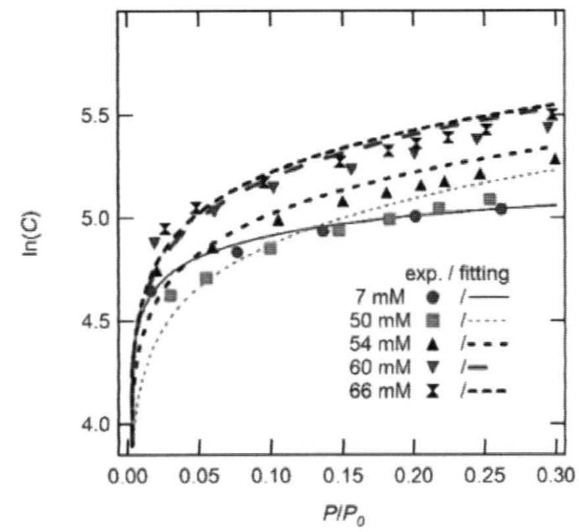

Figure 3. DA isotherm model fitting for various concentration of P123. All curves were fitted up to relative pressure up to 0.3.

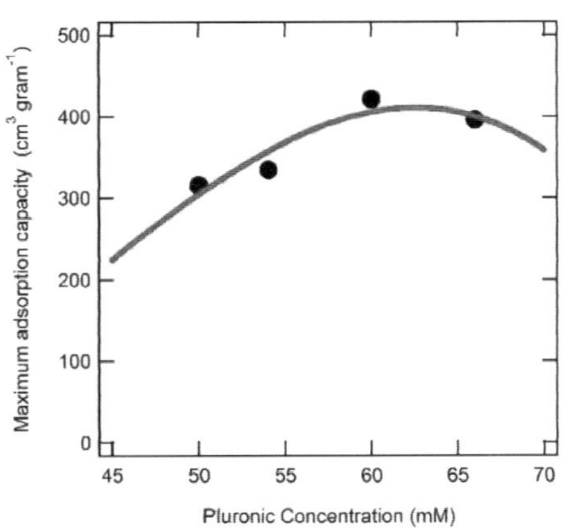

Figure 4. The fitting results of maximum filling capacity/maximum adsorption capacity for various concentration of P123.

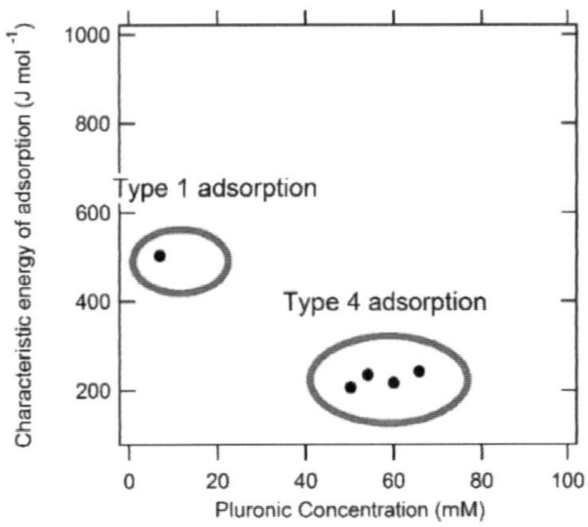

Figure 5. Characteristic energy of adsorption for various concentration of P123.

4. Conclusions

SBA-15 has been successfully synthesized with various concentration of P123 (7mM, 50 mM, 54 mM, 60 mM and 66 mM). DA model was found to be fitted well up to relative pressure of 0.3 and can be used to estimate adsorption capacity of SBA-15. It was found that at concentration 60 mM, SBA-15 has highest adsorption capacity. At this optimum concentration, the hexagonal template which is formed in self-assembly manner by surfactant reaches its maximum surface area, thus yields highest adsorption capacity. In addition it was found that the characteristic energy is strikingly different for low and high concentration. All of this results can be used as guidance for the use of mesopores materials to the applications such as study of adsorption for other gases or heavy metals.

5. Acknowledgement

This writing is supported by DRPM (Universitas Indonesia Research Directorate) under PITTA 2017 grant. There is no conflict of interest between the authors and the institution regarding the publication of this article.

References

[1] Abdalla, Z. E. A., And Baoshan L., Chemical Engineering Journal, 200, 113-121, 2012.
[2] Chang, C. Y., Chau, L. K., Hu, W. P., Wang, C. Y., & Liao, J. H., Microporous And Mesoporous Materials, 109(1), 505-512. 2008.
[3] Pasqua, L., Cundari, S., Ceresa, C., & Cavaletti, G., Recent Development, Current Medicinal Chemistry, 16(23), 3054-3063, 2009.
[4] Xu, X., Lu, P., Zhou, Y., Zhao, Z., & Guo, M., Materials Science And Engineering: C, 29(7), 2160-2164, 2009.
[5] Zornoza, B., Irusta, S., Téllez, C., & Coronas, J., Langmuir, 25(10), 5903-5909, 2009.
[6] Mccusker, L. B., Liebau, F., & Engelhardt, G., Nomenclature Of Structural And Compositional Characteristics Of Ordered Microporous And Mesoporous Materials With Inorganic Hosts (Iupac Recommendations 2001), Pure And Applied Chemistry, 73(2), 381-394, 2001.
[7] Bérubé, F., Kleitz, F., Kaliaguine, S., Journal Of Materials Science, 44.24: 6727, 2009.
[8] Zhao, D., Feng, J., Huo, Q., Melosh, N., Fredrickson, G. H., Chmelka, B. F., & Stucky, G. D., Science, 279.5350: 548-552, 1998.
[9] Zhao, D., Sun, J., Li, Q., & Stucky, G. D., Chemistry Of Materials, 12.2: 275-279, 2000.
[10] Ravikovitch, P. I., And Alexander V. N., The Journal Of Physical Chemistry B 105.29, 6817-6823, 2001.
[11] Clayton, W., Capillary And Colloid Chemistry. By Prof. H. Freundlich. Translated By H. Stafford Hatfield, B. Sc., Ph. D. Pp. Xv+ 883. London: Methuen And Co., Ltd., 1926.
[12] Langmuir, I., The Adsorption Of Gases On Plane Surfaces Of Glass, Mica And Platinum. Journal Of The American Chemical Society, 40.9: 1361-1403, 1918.
[13] Saeidi, N., Parvini, M., & Sarsabili, M. R., International Journal Of Engineering (Ije), Transactions A: Basics, 27(10), 1511-1518, 2014.
[14] Su, F., Zeng, J., Bao, X., Yu, Y., Lee, J. Y., & Zhao, X. S., Chemistry Of Materials, 17(15), 3960-3967, 2005.
[15] Dhaneswara, Donanta, and Nofrijon Sofyan., *International Journal Of Technology* 7.6 (2016): 1009-1015.

The Effect of UVC Irradiation on the Mechanical Properties of Chitosan Membrane in Sterilization Process

N N Rupiasih[1, 2, *], **M Sumadiyasa**[1] **and I K Putra**[1]

[1]Department of Physics, Faculty of Mathematics and Natural Sciences, Udayana University, Denpasar, Bali 80361, Indonesia

*Corresponding author e-mail: rupiasih@gmail.com

Abstract. The present study, we report about the effect of UVC irradiation on the mechanical properties of chitosan membrane in the sterilization process. The membrane used was chitosan membrane 2% which prepared by a casting method using chitosan as matrix and acetic acid 1% as a solvent. The UVC source used was germicidal ultraviolet (UVG) which widely used for sterilization purposes. Variation doses were done by the varying time of irradiation, e.g. 5 min, 15 min, 30 min, and 60 min. Those samples are named as S1, S2, S3, and S4, respectively. Chitosan membrane before irradiation namely S0 also used for comparative study. The effect of UVC irradiation on the mechanical properties of membranes has been examined by different techniques including FTIR, DMA, and the water uptake capability. The results showed that ultimate tensile strength (UTS) and moduli of elasticity (E) were increased by increasing the irradiation time. From FTIR analysis obtained that no new molecules were formed in irradiated membranes. The water uptakes capability of the membranes after irradiation was smaller compared with before irradiation, and among the irradiated membranes, the water uptake capabilities were increased by increasing the exposure time. These observations suggested that more care should be taken during the sterilization process and outdoor used of the membrane. The other side, the UVC irradiation can improve the mechanical properties of the membranes.

Keywords: chitosan membranes; short-wavelength ultraviolet (UVC); sterilization; mechanical properties; water uptake capability.

1. Introduction

Chitosan is a natural polymer derived by deacetylation of chitin. Chitin is a polysaccharide existing in the exoskeleton of crustaceans and insects, mollusk shells and fungous biomass [1, 2]. Chitosan is derived from the partial or full replacement of an N-acetyl group in chitin by an amine group; the process is called de-acetylation degree (DD) with value varying from zero to unity [3]. The amine and hydroxyl groups in chitosan polymer act as potential sites for intermolecular interactions take place. This characteristic has been extensively developed in various processes, for example, controlled drug release, sorption phenomena, and metal complexation [3-5].

Currently, UVG has been used widely for sterilization purposes. UVG irradiation is a disinfection method that uses short-wavelength ultraviolet (UVC) light to kill or inactivate microorganisms. It was used widely in medical sanitation, sterile work facilities, drinking, wastewater treatments, and air sanitization. Those facilities were enclosed and could be circulated, so the possibility of exposure to high radiation is large enough.

Many studies about photodegradation of structural modifications in chitosan-based systems induced by UV irradiation have been reported [3, 6]. The main photodegradation mechanisms involved rupture of glycosidic linkage and formation of a carboxylic group in the absence or presence of oxygen, and production of the hydroxyl group. The measurements of the static contact angle of chitosan films exposed to low-intensity UV lamps have revealed an enhancement in the surface energy with the exposure time [7]. This behavior has been associated with the increase of the surface polarity of films caused by the scission of glycosidic bonds and pyranose rings during the photo-oxidation process [3, 8]. A significant reduction of UV sensitivity has been detected in the study of UV effects on polymeric blends of chitosan matrix based. Chitosan-based blends show a lowering in the mechanical properties upon UV irradiation compared with pure chitosan films [3, 6]. The other sides, photo-irradiation-based techniques can be very useful to improve the performance of polymeric membranes in technical separations such as micro and ultra-filtration, gas separation, and as membrane adsorbers. Based on those backgrounds, thus it is importance to study about the effect of UVC irradiation on the mechanical properties of chitosan membrane which has been given during sterilization.

2. Materials and methods

2.1 Materials and membranes Preparation
Chitosan powder used was extracted from shrimp with a DD of 87.9% and solubility in acetic acid 1% is 99.4%. Acetic acid and NaOH were analytical grades (p.a.), and demineralized water was used in preparing solutions. Chitosan membrane 2% was prepared by a casting method using chitosan as matrix and acetic acid 1% as a solvent. The membrane solution was prepared by dissolving 5 g of chitosan powder in 250 ml of acetic acid 1%. The complete casting method described in Ref.1 and 2.

2.2 UVC irradiation and characterizations
Dry membrane was cut into different sizes for various characterization techniques. Some membranes were kept before irradiation as a comparative study, named as S0 and some membranes were irradiated by UVC radiation. The UVC source used was UVG. Variation doses were done by the different time of irradiation, e.g., 5 min, 15 min, 30 min, and 60 min. Those samples are named as S1, S2, S3, and S4, respectively. Various methods have been used to characterize the effect of UVC irradiation on the mechanical properties of membranes including FTIR, DMA (dynamic mechanical analysis), and the water uptake capability. FTIR spectra were recorded using an IR Prestige-21 FTIR spectrometer, Shimadzu. Spectra were taken with a resolution of 4 cm^{-1} in the range of 400 to 4000 cm^{-1}. The mechanical strength of membranes was measured using DMA following the standard method ASTM D3039. The sample is rectangular with a cross-section of 5 x 0.5 mm^2 and length 20 mm. All measurements were conducted at room temperature and in dry condition.

Water uptake experiment was done by following protocol described in Ref. 1. The water uptake of the membrane was calculated using the following formula, where m_{dry} is the mass of dry membranes and m_{wet} is the mass of wet membranes [1].

$$Water uptake\,(\%) = \frac{(m_{wet} - m_{dry})}{m_{dry}} \, x\, 100\% \quad (1)$$

3. Results and discussion
Figure 1 show IR spectra of chitosan membranes S0, S1, S2, S3, and S4. All spectra showed a similar pattern; in the figure, only S0 is labeled, and other spectra are written in Table 1. The S0 spectrum obtained was similar to the spectrum obtained in our previous study and also has been reported by others [1-3].

For membrane before irradiation (S0, Fig. 1), IR spectrum shows the absorption bands at around 3475, 2920 and 2880 cm^{-1} are characteristic of stretching vibrations of $-OH$, $-CH_2$, and $-CH_3$ groups, respectively. The absorption bands at 3370 and 1580 cm^{-1} represent of stretching and bending vibrations of $-NH_2$ amine group. The absorption band at around 1630 cm^{-1} corresponds to $-C{=}O$ stretching of carbonyl group which is related to the presence of N-acetyl units due to the

DD value around 87.9%. The peaks at 1409 and 1328 cm^{-1} related to the vibrations of –OH and –CH groups in the pyranose ring. The strong peak at 1047 cm^{-1} associates to the anti-symmetric stretching of C–O–C glycosidic linkage. This result is by earlier studies [1-3, 8, 9].

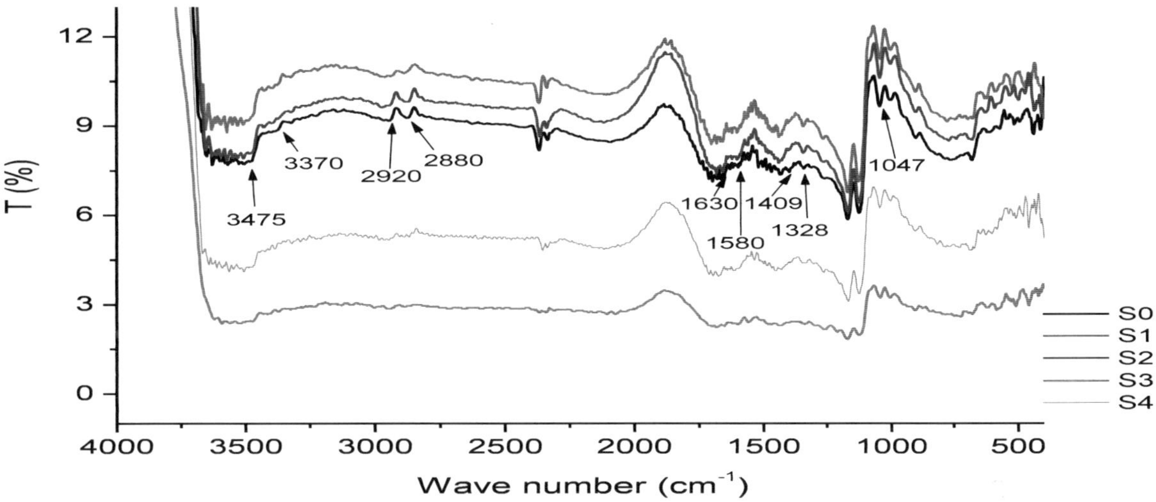

Figure 1. IR spectra of chitosan membranes: S0, S1, S2, S3, and S4

For membranes after irradiation (S1, S2, S3, and S4), it observed some changes in the absorption bands of the stretching –OH group and the bending $-NH_2$ group. Both bands were shifted to higher or lower frequencies. More can be seen in the data Table 1. These observations indicated that photodegradation and photo-cross-linking or photo-grafting process occurred in the membranes simultaneously [3, 8, 10] which depended on the irradiation time exposures. This also can be described from the ratio A_{3475}/A_{1580} and A_{1047}/A_{1580} IR spectra of chitosan membranes as shown in Table 2, where A_{3475}, A_{1580}, and A_{1047} are the absorbances of hydroxyl-stretching, amine-bending bands and glycosidic-stretching respectively [3, 8]. From this analysis, we observed an increase around 1% to 2.3% in the ratio A_{3475}/A_{1580} and increase or decrease around 0.7% to 2.3% in photodegradation process occurs along the entire membranes as a consequence, a change in membrane properties found.

Table 1. The main bands observed in chitosan membranes: S0, S1, S2, S3, and S4

Functional Groups	Wavenumbers Observed (cm^{-1})				
	S0	S1	S2	S3	S4
–OH stretch	3475	3482	3479	3471*	3475
$-CH_2$ stretch	2920	2934	2934	2934*	2943
$-CH_3$ stretch	2880	2887	2878	2878*	2869
$-NH_2$ amine group stretch and bend	3370 and 1580	3395 and 1581	3388 and 1581	3374* and 1555*	3388 and 1537
–C=O stretch of carbonyl group	1630	1630	1630	1630*	1630
–OH and –CH groups in the pyranose ring	1409 and 1328	1393 and 1337	1393 and 1337	1393* and 1328*	1383 and 1337
The anti-symmetric stretch of C–O–C glycosidic linkage	1047	1047	1047	1045*	1047

Ref. 1-3. *: Peak is very weak.

The UTS, modulus of elasticity and elongation-at-break are presented here to characterize the mechanical properties of the chitosan membranes, as shown in Fig. 2. UTS is the maximum tension that can be supported by the membranes until the moment it breaks. Elongation-at-break is a measure of flexibility of the membranes before breaking. These measurements are important once the mechanical properties of membranes or films depend on the filmogenic nature of the material used, which is directly related to its structural cohesion [11].

Table 2. The ratio A_{3475}/A_{1580} and A_{1047}/A_{1580} IR spectra of chitosan membranes: S0, S1, S2, S3, and S4

Sample	A_{3475}/A_{1580}	A_{1047}/A_{1580}
S0	1,000	0,916
S1	1,000	0,917
S2	1,007	0,896
S3	1,010	0,923
S4	1,023	0,893

Figure 2a and b showed that UTS and E moduli increased by increasing the irradiation time, where ε is increased up to irradiation of 15 min then decreased. These mean that UVC irradiation affected the mechanical properties of chitosan membrane in doses dependent manner. There has been membrane modification (functionalization) during the sterilization process, which includes photo-cross-linking of membranes and photo-grafting of membrane surfaces, either via "grafting-from" or "grafting-to" ways [10]. The crosslinking reactions occurred in competition with chain scission. Thus the modifications of the chemical structure will cause the changes of the mechanical property of the membrane; they will also contribute to the development of entirely novel membrane-based materials [10, 12].

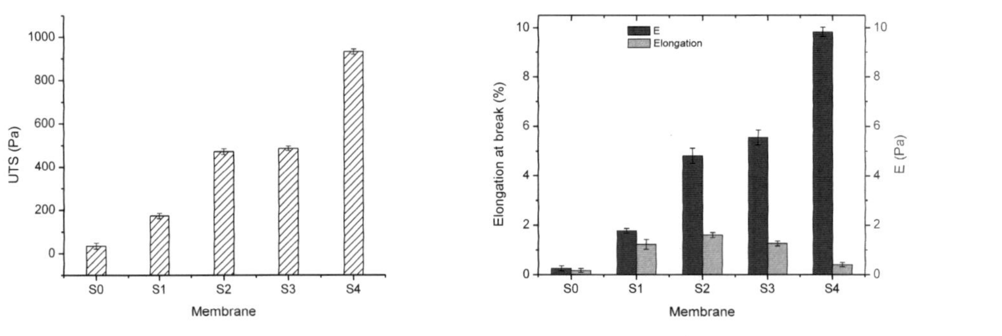

a) b)

Figure 2. a) Ultimate tensile strength (UTS) and b) elongation-at-break (ε) and elastic moduli (E)

The water uptakes of all membranes, at various times, were investigated and the results obtained as shown in Fig. 3a and 3b. It showed that the dynamic of water uptake is highly affected by the UVC irradiation which depending on the irradiation time given. The dynamical of water uptake is due to varies changes in the chemical structure of the entire membrane [3, 13] which showed by IR spectra. The water uptake of S0 is reached a maximum after 24 h while the water uptake of S1, S2, and S3 reached a maximum after 1 h and S4 after 6 h. They were reached maximum faster compared with S0. Figure 3b showed the water uptake capabilities of irradiated membranes were smaller compared with S0, but among the irradiated membranes, the water uptake capability was increased by increasing the exposure time. The UVC irradiation affected the water uptake capability of chitosan membranes in doses dependent manner. This result is in agreement with the tensile modulus obtained by DMA test. It can be described that might occur the competition between chain scission, which initiated by UVC irradiation and crosslinking reactions that changed the chemical structure of irradiated membranes. The crosslinking have increased the strength of the membrane, so it reduced the water uptake capability.

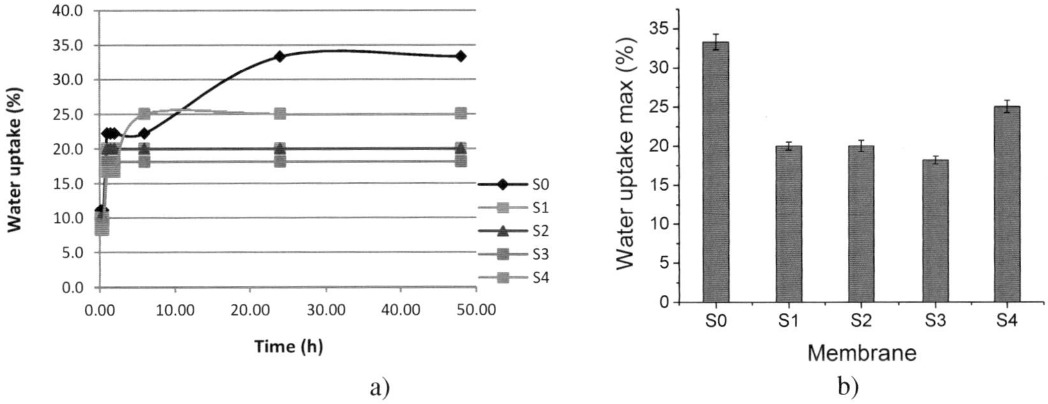

Figure 3. Water uptake of all membranes: S0, S1, S2, S3, and S4. a) Water uptake as a function of times and b) The maximum percentage of water uptake (%) of each membrane

4. Conclusion

This study demonstrated that the mechanical properties of chitosan membranes were greatly affected by UVC irradiation within the short time of irradiation (i.e., 5 min). It affects the mechanical properties of chitosan membranes which strongly depends on the time of irradiation. These observations suggested that more care should be taken during the sterilization process and outdoor use of the membrane. The other side, the UVC irradiation can improve the mechanical properties of the membranes.

Acknowledgements

Author thankful to the Ministry of Research, Technology and Higher Education, Republic of Indonesia, Research Grant No: 415.73/UN14.4.A/PL/2017 for financial supported.

References

[1] Rupiasih N N, Sumadiyasa M, and Putra I K 2017 The Effect of Variations in the Ratio of Matrix/Solvent on the Physical and Mechanical Properties of Chitosan Biopolymer Membranes *IOP Conf. Series: Materials Science and Engineering* **196** (012039) 1-9

[2] Rupiasih N N *et al* 2017 Utilization of Shrimp Skin Waste (Sea Lobster) As Raw Material for the Membrane Filtration, *IOP Conf. Series: J. Phys.: Conf. Ser.* **846** (012003) 1-5

[3] Praxedes A P P *et al* 2012 Effects of UV irradiation on the wettability of chitosan films containing dansyl derivatives *Journal of Colloid and Interface Science* **376**, 255-261

[4] Rupiasih N N 2016 Effects of electrolytes on ion transport in Chitosan membranes *Journal of Physics: Conference Series* **776** (012045) 1-6

[5] Rupiasih N N, *et al* 2015 Transport phenomena in chitosan synthetic membranes with emphasis on the effect of variations in the ratio of matrix/solvent, *J. Phys.: Conf. Ser.* **622** (012004), 1-7

[6] Rupiasih N N, *et al* 2017 The Study of the Effect of UV-C Radiation on the Current-Voltage Characteristics of Chitosan Membranes, in *Radiation in Medicine and Biology*, edited by Pandit B. Vidyasagar, et al., Pan Stanford, 175-185

[7] M. Rinaudo 2006 Chitin and chitosan: Properties and applications, *Progress in Polymer Sciences* **31** (7), 603-632

[8] Rupiasih N N and P.B. Vidyasagar 2008 Comparative study of effect of low and medium dose rate of γ irradiation on microporous polysulfone membrane using spectroscopic and imaging techniques *Polymer Degradation and Stability* **93** 1300-1307

IOP Conf. Series: Journal of Physics: Conf. Series **1011** (2018) 012018 doi:10.1088/1742-6596/1011/1/012018

[9] Jaroslaw M W *et al* 2005 Degradation of chitosan and sodium alginate by gamma radiation, sonochemical and ultraviolet methods, *Radiation Physics and Chemistry* **73** (5), 287-295
[10] Dongming He, *et al* 2009 Photo-irradiation for preparation, modification and stimulation of polymeric membranes *Progress in Polymer Science* **34** (1) 62-98
[11] Souza B W S, *et al* 2010 Influence of electric fields on the structure of chitosan edible coatings *Food Hydrocolloids* **24** 330-335
[12] Andrady A L *et al* 1996 Spectral sensitivity of chitosan photodegradation *J. Appl. Polym. Sci.* **62** (9), 1465-1471
[13] Larché J F *et al* 2012 Photooxidation of polymers: Relating material properties to chemical changes *Polymer Degradation and Stability*, **97** (1) 25-34

IOP Conf. Series: Journal of Physics: Conf. Series **1011** (2018) 012019 doi:10.1088/1742-6596/1011/1/012019

Study and Fabrication of Super Low-Cost Solar Cell (SLC-SC) Based on Counter Electrode from Animal's Bone

D R Fadlilah[1], M N Fajar[1], A N Aini[1], R I Haqqiqi[1], P R Wirawan[1], and Endarko[1*]

[1]Physics Department, Institut Teknologi Sepuluh Nopember Kampus ITS Sukolilo-Surabaya, 60111, Indonesia

[*]Corresponding email address: endarko@physics.its.ac.id

Abstract, The synthesized carbon from bones of chicken, cow, and fish with the calcination temperature at 450 and 600°C have been successfully fabricated for counter electrode in the Super Low-Cost Solar Cell (SLC-LC) based the structure of Dye-Sensitized Solar Cells (DSSC). The main proposed study was to fabricate SLC-SC and investigate the influence of the synthesized carbon from animal's bone for counter electrode towards to photovoltaic performance of SLC-SC. X-Ray Diffraction and UV-Vis was used to characterize the phase and the optical properties of TiO_2 as photoanode in SLC-SC. Meanwhile, the morphology and particle size distribution of the synthesized carbon in counter electrodes were investigated by Scanning Electron Microscopy (SEM) and Particle Size Analyzer (PSA). The results showed that the TiO_2 has anatase phase with the absorption wavelength of 300 to 550 nm. The calcination temperature for synthesizing of carbon could affect morphology and particle size distribution. The increasing temperature gave the effect more dense in morphology and increased the particle size of carbon in the counter electrode. Changes in morphology and particle size of carbon give effect to the performance of the SLC-SC where the increased morphology's compact and particle size make decreased in the performance of the SLC-SC.

1. Introduction

The oil and gas supplies in Indonesia will not be available all time. Solar Cell is one of the solutions to this problem because of its ability to convert sunlight into electric current. Along with the development of a solar cell, the most vigorous research conducted now is DSSC (Dye-Sensitized Solar Cell). DSSC is the 3rd generation in Solar Cell which has inexpensive fabrication compared to conventional solar cells (silicon-based) [1]. DSSC is composed of nanoporous TiO_2 photoanode and Pt or carbon counter electrode separated by iodide/triiodide electrolyte [2]. Based on statistical data in 2015 from Indonesia's farm and animal's health department, meat production levels increased compared to 2014, so it makes new problem due to increased waste animal's bone [3]. In 2012 [4], there was research about comparative analysis and characterization of animal's bone as an adsorbent substance, and it can be used to synthesize carbon. In 2017 [2], carbon could be applied to be a counter electrode which had electronics conductivity, corrosion resistant to tri-iodide, high reactivity for reducing tri-iodide, and low cost. From these statements, it can be said that animal's bone can replace Pt as the counter electrode. From this issue, the utilization of animal's bone to synthesis carbon as a counter electrode in the component of DSSC as Super Low-Cost Solar Cell (SLC-SC) will be

IOP Conf. Series: Journal of Physics: Conf. Series **1011** (2018) 012019 doi:10.1088/1742-6596/1011/1/012019

investigated. The morphology and the particle size distribution of the synthesized carbon from animal's bone will be observed and characterized too.

2. Research Method

2.1. Preparation of TiO_2 paste

TiO_2 powders were synthesized by previous research method [2,5]. The mixture of 20 mL of Titanium (III) chloride ($TiCl_3$) and 100 mL of distilled water was stirred for an hour. NH_4OH solution was dropped slowly to the mixed solution until the rate of the acidity reached to 9 and then the result of the solution was stirred until a white precipitate appeared. The precipitate was filtered and washed several times with distilled water; this process would be stopped at the rate of the acidity equal to 7. Removal process of residual organics and the stabilization of the materials were carried out by calcination for 3 hours at 450 °C. The TiO_2 pastes were prepared by the mixture of distilled water, ethylcellulose, terpineol, and ethanol was then added and kept stirring for 10 min. The mixed oxide pastes were obtained from 0.7 mL of acetic acid, 0.3 g of PEG, and 0.7 mL of Triton X-100.

2.2. Preparation of Carbon

Bones dried at 170 °C for 4 h. After dried, the bones were calcination at 450 and 600 °C for 1 h. The solids carbon that obtains from calcination then crushed in a mortar to make carbon powder. The same way prepares the bones of chicken, fish, and cow.

2.3. Preparation of Photoanode

TiO_2 paste was deposited by *screen printing* method with a dimension of 1×1 cm^2 onto ITO glass with size 2.5×2.5 cm^2. The deposited TiO_2 paste onto the ITO glass was then heated for 5 min at 100 °C. The photoanode was immersed overnight in the dye that extracted from grape skin at room temperature

2.4. Preparation of Counter Electrode

Carbon paste was prepared by mixing carbon powder with ethanol and was then stirred for 1 h. The carbon pastes were deposited onto a conductive glass substrates-ITO glass using doctor blade method with the thickness around ~10 μm. The electrode was heated for 30 min at 250 °C to remove the residual organic and stabilize the material.

2.5. SLC-SC fabrication

SLC-SC was fabricated using previous research [2] with the dye-sensitized TiO_2 photoanode, mostly as an electrolyte, and the synthesized carbon as the counter electrode.

3. Result and Discussion

3.1. TiO_2 phase analysis

The match!software was used for qualitative analysis of all data from X-Ray Diffraction (XRD). Figure 1 shows the TiO_2 anatase has been successfully formed. The crystal plane 101 situated at 2θ around 25.3° is the crystal plane with the highest intensity in the TiO_2 powder [5]. Another crystal field also identified namely (004), (200), (105), (211), (204), (116), (220) and (215) that are located at 2θ for 38.3 °, 48°, 54°, 55°, 62°, 69°, 71° and 75°, respectively. The results are consistent with those obtained by S. Mahshid et al.[6].

3.2. Optical characteristics of Dye and Photoanode

UV-Vis performed the optical properties of dye and photoanode. Figure 2 shows the absorbance graph of TiO_2 thin film, a dye that extracted from the grape skin, and TiO_2 that immersed in the dye for 24 hours. It can be analyzed from Figure 3 that the TiO_2 thin film can absorb 300 – 550 nm wavelength with maximum wavelength range at 300 – 380 nm whereas the dye has two peaks wavelength range that can be absorbed in the range of 200 – 300 nm and 550 – 800 nm. After immersed the TiO_2 thin

IOP Conf. Series: Journal of Physics: Conf. Series **1011** (2018) 012019 doi:10.1088/1742-6596/1011/1/012019

film in the dye, it was found that the wavelength that can be absorbed by the thin film layer is in the range of 200 – 800 nm with maximum absorbable wavelength at 300 – 800 nm.

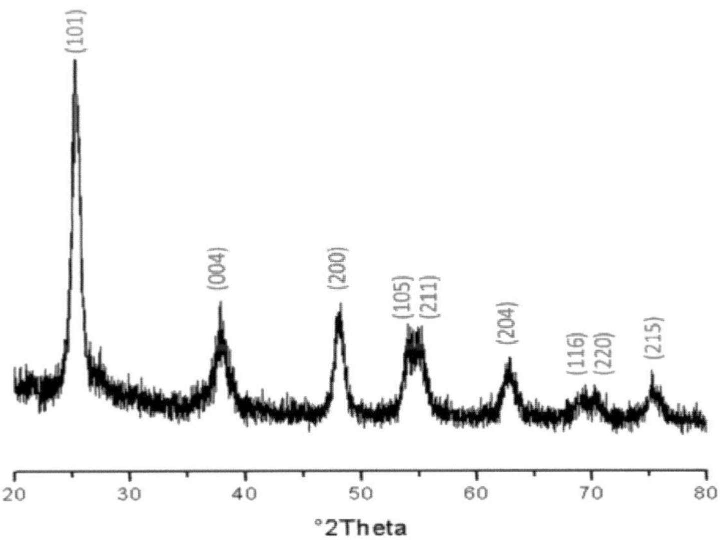

Figure 1. Graphic XRD of TiO_2 powder shows anatase phase

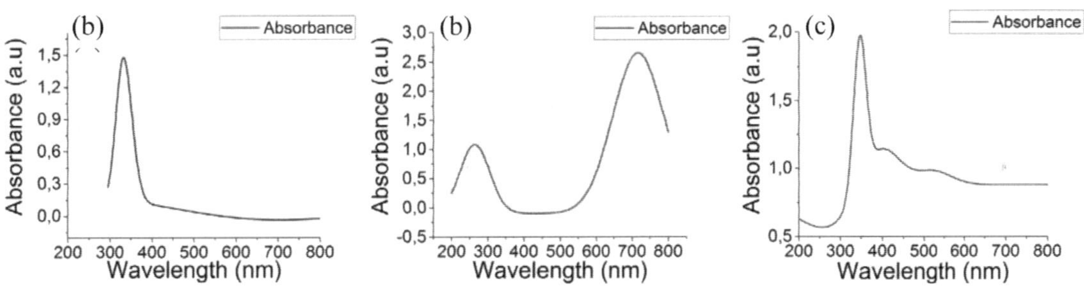

Figure 2. UV-Vis shows the absorbance graphs of (a) TiO_2 thin film, (b) dye that extracted from the grape skin, (c) TiO_2 thin film that immersed in dye for 24 hours

3.3. Study of Morphological and Particle Size Analyzer in Counter Electrode

The morphology of each type of counter electrode shows the same form of flakey or in pieces and there are pores of different particle sizes as shown in Figure 3. Measurement of the particle size of carbon from SEM micrograph was performed using the ImageMif software. Figure 4 shows the particle size of the synthesized carbon powder with calcination temperature at 450 °C from a fish, chicken, and cow bones. It can be analyzed that the synthesized carbon powder from fish bone has the biggest size compared to the others with dominant particle size found at 1, 0.8 and 0.4 μm, respectively. Meanwhile, the synthesized carbon with calcination temperature at 600 °C showed that each carbon powder is denser, and increased their particle size.

To obtain the result more accurate, we used *Particle Size Analyzer* device so that the measurement results are in the form of particle distribution and overall condition of the sample. Figure 4 shows the particle size distribution of the synthesized carbon with calcination temperature at 450 and 600 °C,

IOP Conf. Series: Journal of Physics: Conf. Series **1011** (2018) 012019 doi:10.1088/1742-6596/1011/1/012019

respectively. The results revealed that the calcination temperature could be increasing the particle size. This fact also had been reported by [7].

(b)

Figure 3. The SEM micrograph shows the surface morphology of animal's bone with calcination temperatures: a) 450 °C (fish (a), chicken (b), and cow (c)), and b) 600 °C (fish (a), chicken (b), and cow (c))

The Polydispersity Index (PdI) and average particle size of the synthesized carbon from each calcination temperatures are summarized in Table 1. It can be concluded from Table 1 that the addition of temperature during the manufacturing process could enhance in particle size whereas the average Polydispersity Index (PdI) in each sample also varies. PdI shows the uniformity of size, the smaller the PdI value indicates an increasingly uniform size. The synthesized carbon can achieve the lowest and highest average PdI value from the bones of cow and chicken with calcination temperature at 600 °C, respectively.

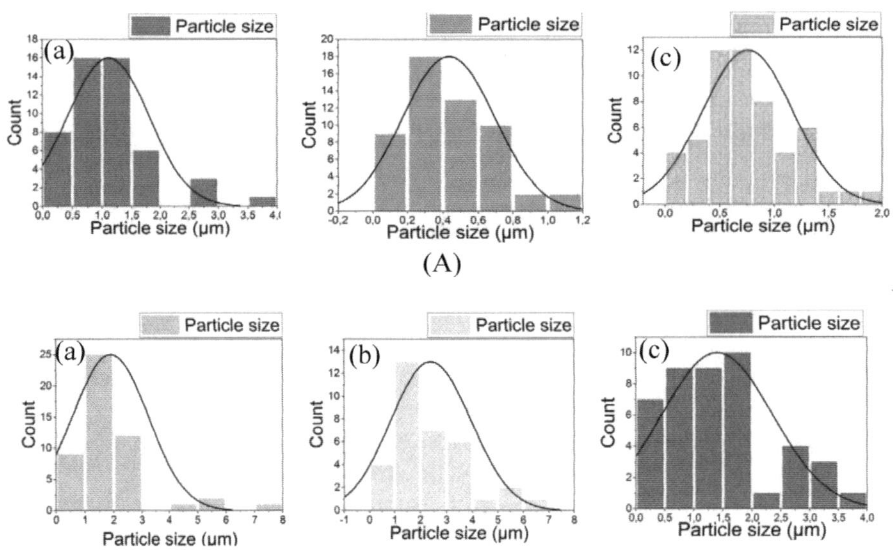

(B)

Figure 4. Particle size distribution of the synthesized carbon using imageMif with calcination temperatures: (A) 450 °C (fish (a), chicken (b), and cow (c)), and (B) 600 °C (fish (a), chicken (b), and cow (c))

Table 1. Particle size from *Particle Size Analyser* measurement

Sample	Average Particle Size	Average PdI
Fish 450°C	1455.00	0.712
Fish 600°C	4702.67	0.832
Chicken 450°C	705.90	0.562
Chicken 600°C	2781.33	0.911
Cow 450°C	1528.00	0.750
Cow 600°C	5902.00	0.374

3.4. I-V characterization

Characterization of current, voltage and efficiency of Super Low-Cost Solar Cell was measured with the solar simulator under illumination at 100 mW/cm^2 were summarized in Table 2. Table 2 shows the highest efficiency (0.063%) can be achieved by the synthesized carbon from the fishbone with calcination temperature of 450 °C. Decreased efficiency can be seen for all sample due to increasing calcination temperature from 450 to 600 °C except for the chicken bone. The fact due to increased in particle size. The particle size affects the contact area of the electrode counter, the wider the contact of the electrode the more electrons can be passed on the electrode counter in the electron transport process from the outer circuit of the DSSC to electrolyte [5]. Meanwhile, the photovoltaic performance of SLC-LC with the synthesized carbon from animal's bone as a counter electrode has the better performance compared to the SLC-SC with commercial black carbon as the counter electrode. The results show that the synthesized carbon from animal's bone has a potential application in dye-sensitized solar cells (DSSC) and also can be one of the solutions for animal's bone waste problems. Based on the results from [8], the counter electrode based on composite carbon black and Pt, it had given the efficiency value of 0.43%. This fact shows that the synthesized carbon from animal's bone has a potential to be applied to the counter electrode for DSSC with very low cost in the future.

Table 2. Performance characteristics of the SLC-SC

	J_{SC} (mAcm^{-2})	V_{oc}(V)	FF	η(%)
Fish 450°	0.05739	1.58475	0.69186	0.06292
Fish 600°	0.05718	1.56553	0.41677	0.03730
Chicken 450°	0.08092	1.55064	0.34468	0.04325
Chicken 600°	0.06587	1.61891	0.475714	0.05072
Cow 450°	0.06768	1.63598	0.512190	0.05671
Cow 600°	0.06656	1.67014	0.461573	0.05131
Commercial black carbon	0.01029	0.42144	0.424568	0.00184

4. Conclusion

The calcination temperature on synthesized carbon from animal's bone gave effect to morphology and particle size distribution. The increased temperature gave the effect more dense in morphology and increased the particle size of carbon. Changes in the morphology and particle size of carbon give effect to the performance of the SLC-SC where the increased morphology's dense and particle size make decreased in the performance of the SLC-SC. The synthesized carbon from animal's bone has a potential application in dye-sensitized solar cells (DSSC) and also can be one of the solutions to animal's bone waste problems.

References

[1] Wongcharee K, Meeyoo V, and Chavadej S 2007 Dye-sensitized solar cell using natural dyes extracted from rosella and blue pea flowers *Solar Energy Materials and Solar Cells* vol. **91**, no.7 pp 566–571

[2] Muhammad Noer Fajar and Endarko 2017 Study on conventional carbon characteristics as counter electrode for dye sensitized solar cells *Journal of Physics: Conference Series*, vol. **853** no.1 012044

[3] K. Pertanian 2015 Statistik Peternakan dan Kesehatan Hewan 2015 *Direktorat Jenderal Peternakan dan Kesehatan Hewan. Kementerian Pertanian RI. Jakarta*

[4] Mohammed A, Aboje A A, Auta M, and Jibril M 2012 A comparative analysis and characterization of animal bones as adsorbent *Adv. Appl. Sci. Res.* vol. **3** no. 5 pp 3089–3096

[5] Adawiyah S R and Endarko 2017 Structural and morphological characterization of TiO_2-SnO_2 thin film prepared by combining doctor-blade and sol-gel techniques in *IOP Conference Series: Materials Science and Engineering* vol. **188** 012062

[6] Mahshid S, Askari M, Ghamsari M S, Afshar N, and Lahuti S 2009 Mixed-phase TiO_2 nanoparticles preparation using sol–gel method *Journal of Alloys and Compounds*, vol. **478** no. 1 pp 586–589

[7] Rosanti S D and Puryanti D 2015 Pengaruh Temperatur Terhadap Ukuran Partikel Fe_3O_4 Dengan Template PEG-2000 Menggunakan Metode Kopresipitasi *Jurnal Ilmu Fisika* vol. **7** no. 1 pp 39–44

[8] Kang G, Choi J, and Park T 2016 Pt-Free counter electrodes with carbon black and 3D network epoxy polymer composites *Sci. Rep.* vol. **6**, 2016

IOP Conf. Series: Journal of Physics: Conf. Series **1011** (2018) 012020 doi:10.1088/1742-6596/1011/1/012020

Efficient and Stable Photovoltaic Characteristics of Quasi-Solid State DSSC using Polymer Gel Electrolyte Based on Ionic Liquid in Organosiloxane Polymer Gels

H Pujiarti[1], W S Arsyad[2], Shobih[3], L Muliani[3] and R Hidayat[1*]

[1] Physics of Magnetism and Photonics Research Division, Faculty of Mathematics and Natural Sciences, Institut Teknologi Bandung, Jl.Ganesha 10, 40132, West Java, Indonesia
[2] Physics Department, Faculty of Mathematics and Natural Sciences, Halu Oleo University, Kendari, Indonesia
[3] Electronics and Telecommunication Research Center LIPI, Jl. Sangkuriang Komp. LIPI Gd. 20, Bandung, Indonesia

E-mail: rahmat@fi.itb.ac.id

Abstract. Dye-Sensitized Solar Cell (DSSC) is still one of the promising solar cell types among the third generation of solar cells because of easiness of fabrication and variety of available materials. In this type of solar cell, the electrolyte is one of the important components for regenerating excited dyes and transporting electric charge carriers to the counter electrode. Indeed, the power conversion efficiency of DSSC can be then significantly affected by the chemical and physical properties of the electrolyte. The simplest electrolyte system of an I^-/I_3^- redox couple in an organic solvent, however, has some drawbacks due to corrosive properties, volatile and leakage problem. Use of solid phase or gel phase electrolyte may overcome those problems, but it is often considered to suppress the efficiency due to low ion diffusion. Here, we report the photovoltaic characteristics of DSSC using polymer gel electrolyte (PGE), which is composed of ionic liquid and an organosiloxane polymer gel. The better cell performance with power conversion efficiency of about 6% has been obtained by optimizing the mesoporous size of the TiO_2 layer and the PGE viscosity.

1. Introduction

Dye-Sensitized Solar Cells (DSSCs) is organics solar cells with dye molecules as the light absorbers. Since the first report in 1991 by O'Regan and Gratzel [1], DSSCs have received much attention for several decades because of the easiness of fabrication and low-cost production [2-3]. In a conventional DSSC, regeneration of excited dyes was performed by using I_3^-/I- redox couple dissolved in an organic solvent. Such solution may cause several problems including the leakage and volatilization of the liquid solvent, possible desorption of the adsorbed photosensitizers, and corrosion of the Pt counter electrode. Those critical effects may hamper the long-term performance of DSSC [4]. In order to solve those problems, other types of electrolytes have been introduced, such as solid state and quasi-solid state (QSS) electrolyte. In the last decade, quasi-solid state electrolytes have been much investigated because of their stability and large ionic conductivity. However, the overall energy conversion efficiencies of quasi-solid DSSCs are usually lower than liquid DSSCs, which commonly to be assigned due to poor ion diffusion and poor interfacial contact with its TiO_2 mesoporous layer [5,6]. Tao *et al.* fabricated QSS-DSSC with PCE of 9.61 % by using electrolyte based on a low molecular mass organo-gelator (LMOG) material [7]. The best record efficiency for QSS-DSSC was about 10.1% as reported by Suzuka's group by using TEMPO polymer gel electrolyte [8]. In this report, we present the improvement on the performance of DSSC using polymer gel electrolyte (PGE), which is composed of ionic liquid and an organosiloxane hybrid polymer gel (HPG), by optimizing the mesoporous structure of the TiO_2 layer and controlling the viscosity by adjusting the crosslinking degree of HPG. In the previous work, we have shown that the performance and efficiency of this kind of quasi-solid-state DSSC may crucially be affected by scarce penetration of the PGE into the mesoporous TiO_2 layer [9].

2. Experiments

2.1. Fabrication of photoelectrode

For preparing DSSC working electrode, FTO glass (15Ω/sqr, Solaronix) was firstly washed with detergent. Then it was followed by cleaning in an ultrasonic bath with acetone and isopropanol (1:1) for 15 minutes and drying. Afterwards, a series of TiO_2 layers were deposited onto the top of FTO by the same method as in the previous work [9,10]. In this work, the layer was then treated in 40 mM Ti-isopropoxide solution in isopropanol and sintered at 500°C for 30 min. After natural cooling to 80°C, the TiO_2 electrodes were immersed in 0.7 mM N719 dye solution in ethanol for 24 h in dark condition.

2.2. Electrolyte

PGE used here is consisted of HPG and Imidazolium Ionic Liquid (IIL) (Mosalyte TDE 250, Solaronix). The HPG was synthesized from 3(trimethoxysilyl)propyl methacrylate (TMSPMA, Sigma-Aldrich) via the sol-gel method in a mix solution of ethanol, distilled water, and acetic acid with 1:10:0.1:0.25 in a volume ratio. The solution was then stirred and heated at 45 °C for 17h and 24h. The mixing process of HPG and IIL was carried out at room temperature with 1:1 in a volume ratio.

2.3. DSSC assembly and Characterization

The photoanode and the Pt drilled counter electrode (Dyesol) were sealed together with a 25 μm Surlyn (Solaronix) spacer. The PGE was then injected into assembled cells. In this research three differents electrolytes were used, namely pure ionic liquid (DSSC L), DSSC PGE 17h and DSSC PGE 24h. The morphology of TiO_2 nanoparticles and the thickness of the layer were observed by SEM, and its porosity analyzed by PoreDizM70 Matlab program. The current-voltage characteristics and EIS measurements were carried out by the same instruments as in our previous works. [9-11] The viscosity measurements of electrolytes by using a Brookfield viscometer.

3. Results and Discussion

3.1. Porosity of TiO_2

The performance of the fabricated DSSC was strongly influenced by the mesoporous structure of TiO_2 layer. Figure 1 shows a typical SEM image of the mesoporous TiO_2 layer prepared in this work and its porosity analyzed by PoreDizM70 Matlab program. From this SEM image, the mean pore diameters are estimated to be about 60 nm while the particle sizes are about 30-40 nm.

3.2. Photovoltaic Performances of DSSC

Figure 2 shows the current density-voltage (*J-V*) curves of DSSCs fabricated in this work. The parameter performances of DSSC including PCE are summarized in Table 1. It can clearly be seen that the performance of DSSC using ionic liquid is slightly better than the DSSCs using PGE with PCE of about 6%. This efficiency reduction is commonly observed in DSSCs with PGE, which often attributed due to the high viscosity and low ion diffusion as well as poorer electrolyte interfacial contact TiO_2. Recently, we have also empahsized again the crucial effect of scarce electrolyte penetration into the mesoporous layer due to that electrolyte rheology. In order to investigate the stability, the cells were also measured after one month (34 days) after the cell fabrication.

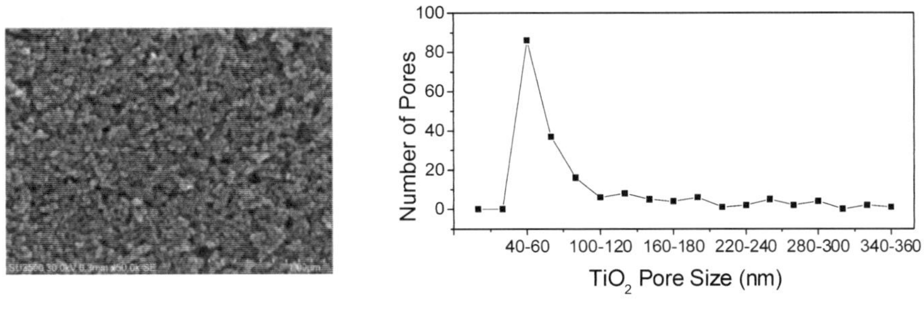

Figure 1. Typical SEM image of TiO_2 mesoporous film used in this work.

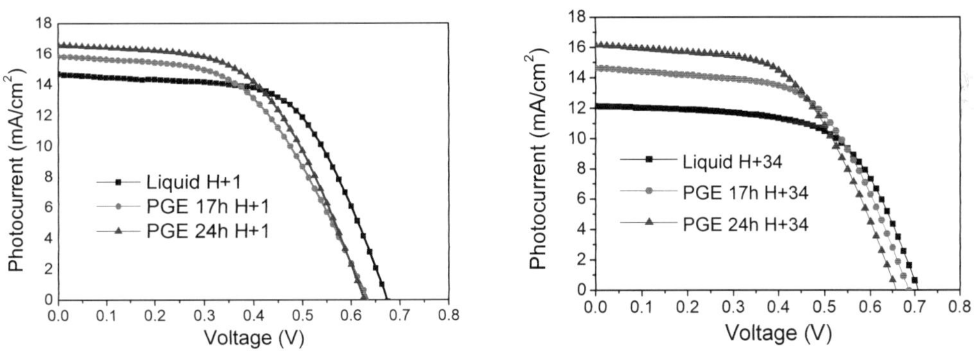

(a) (b)

Figure 2. The photocurrent density-voltage (*J-V*) characteristics of the fabricated DSSCs measured at different times of (a) one day (fresh condition), and (b) one month (34 days) after the fabrications.

Table 1. The parameter photovoltaic performances of DSSC using PGE and IL (as comparison)

Sample	TiO_2 thickness	J_{sc} (mA/cm^2)		Voc (V)		Efficiency (%)	
		H+1	H+34	H+1	H+34	H+1	H+34
DSSC L	17	14.7	11.8	0.675	0.74	6.04	5.21
DSSC 17h	17	15.8	14.6	0.638	0.68	5.25	5.65
DSSC 24h	17	16.6	16.2	0.625	0.66	5.70	5.73

3.3. Impedance Characteristics

Analysis of electrochemical impedance spectroscopy (EIS) spectra is important to be done as it can provide some information on the processes of transport and transfer, diffusion, and accumulation of charge carriers as well as loss recombination of charge carriers. [2, 6,11] In order to explain and analyze the EIS spectra of DSSC, we must estimate the parameter from EIS spectra by fitting with its equivalent circuit model. From this analysis, we can deduce the physical processes and the related parameters of the device. Fig. 3 represents the Nyquist plot of DSSC measured under 0.6V DC bias voltage and light condition at one day (fresh) and one month after fabrication for (a) DSSC using liquid electrolytes, (b) DSSC with PGE using HPG synthesized for 17h, (c) DSSC with PGE using HPG synthesized for 24h. The Nyquist plot can be well fitted with an equivalent circuit as shown in Fig. 3 (d) and the corresponding parameters from fitting results are tabulated in Table 2. The impedance spectrum of each solar cell exhibit three semicircles with almost similar impedance magnitude. Each semicircle can be described, from high frequency to low frequency, as followings. The first semicircle corresponds to the charge transfer processes at the counter electrode/electrolyte interfaces, the second semicircle corresponds to transport electron in the TiO_2 mesoporous, while the third semicircle corresponds to diffusion process from electrolyte.

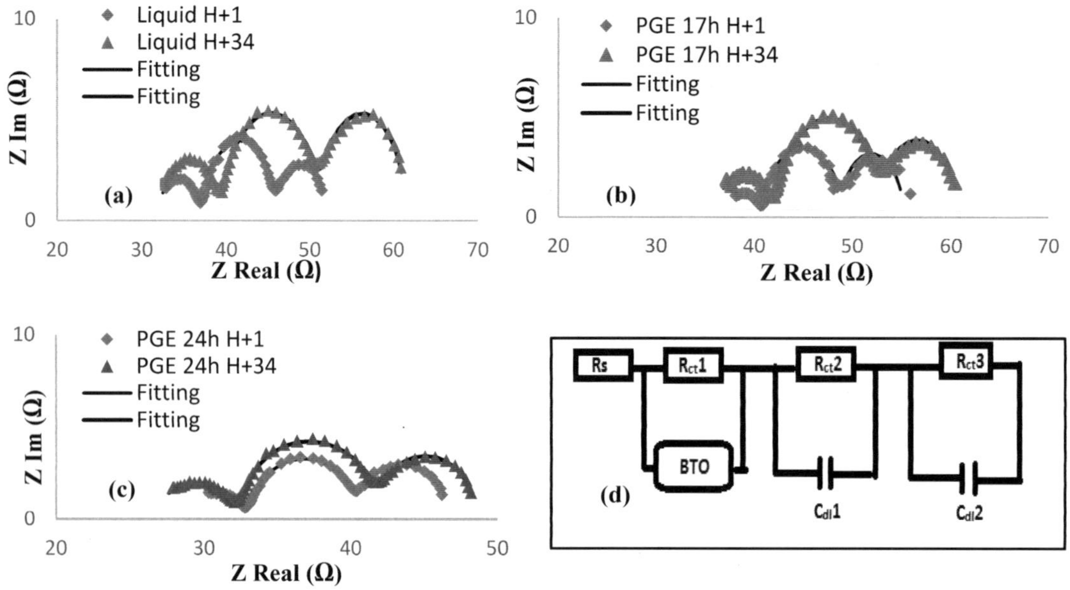

Figure 3. Nyquist plots of DSSC with different electrolytes measured at the 1st day after the fabrication and measured at the 34th days after the fabrication. (a) DSSC with liquid electrolyte, (b) DSSC with PGE synthesized for 17h, (c) DSSC with PGE synthesized for 24h, (d) equivalent circuit

Table 2. The equivalent circuit parameters of DSSCs using PGE and IL, which were measured at two different times after preparation, namely after one day (H+1) and after ariund one month (H+34).

	DSSC Liquid		**DSSC 17h**		**DSSC 24h**	
Condition	H+1	H+34	H+1	H+34	H+1	H+34
Rs [Ω]	29.4	29.4	35.5	33.3	26.8	24.0
$R_{ct}1$ [Ω]	10.17	30.3	5.8	10.4	6.7	9.2
$R_{ct}2$ [Ω]	8.5	10.6	7.5	10.2	6.1	6.8
$R_{ct}3$ [Ω]	5.8	11.1	6.4	7.6	6.8	8.6
$C_{dl}1$ [μF]	403.2	390.3	473.4	456.8	519.5	467.8
$C_{dl}2$ [mF]	65.8	33.6	51.5	42.9	52.3	44.7
L	0.3	0.8	1.6	0.3	0.4	1.1
r k [Ω]	4.9	3.8	180.5	16.2	243.3	255
rm [Ω]	227.6	61.9	447.8	347.8	350.6	164.5
Ym[S*s^a]	26.6	44.9	27	34.6	12.4	6.9
a	0.9	0.8	1	0.9	1	1

3.4. Viscosity

Figure 4 shows the curve of viscosity *vs.* shear stress of the ionic liquid, PGE, and HPG. The viscosity of the ionic liquid is relatively larger than the others. The PGE shows similar viscosity characteristics with the ionic liquid. However, HPG used here shows a Newtonian behavior where its viscosity does not significantly influence by the changing of shear rate. The PGE using HPG synthesized for 17h have larger viscosity than PGE using HPG synthesized for 24h. This result shows that the viscosity characteristics in these PGEs are dominated by the ionic liquid content, which has strong ionic interaction, and are not by the HPG cross-linking degree. Based on their photovoltaic characteristics in the present work, the DSSC with PGE 24h produces better working performance than the DSSC with PGE 17h, which has less cross-linking and higher viscosity in contrary to our previous report before [9]. It should be noted, however, the TiO_2 layer has in the present work has a more porous structure by avoiding preparing the layer in a multi-step process.

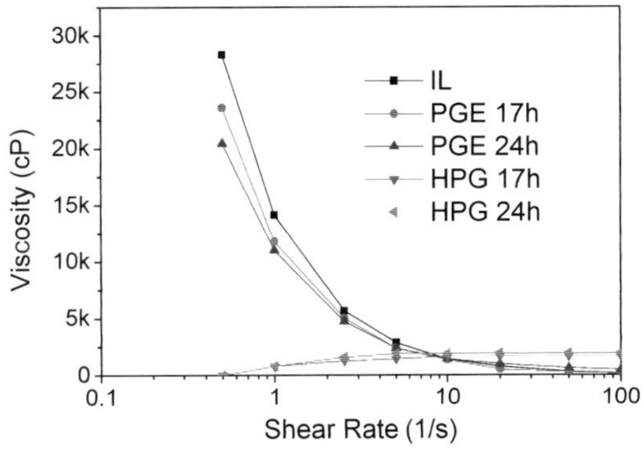

Figure 4. The viscosity characteristics of PGEs and HPG that were synthesized in 17h and 24h.

4. Conclusion

The performance enhancement of quasi-solid state DSSC using PGE studied here was achieved by optimizing the mesoporous structure of TiO_2 layer and controlling the crosslinking degree of HPG as indicated by its viscosity. The PCEs of DSSCs with PGEs are almost close to the PCE of the reference cell with pure ionic liquid, which is about 6%. The performance of DSSC with PGE 24h, which has smaller viscosity and has a presumably higher crosslink degree, shows better performance than that of PGE 17h because the TiO_2 layer in the present work has more porous structure. It should be also noted that DSSC with PGE has better stability than the DSSC with pure IL because of less electrolyte leakage.

5. References

[1] O'Regan and Gratzel M 1991 *Nature* **353** 737-740
[2] S Sarker, A J S Ahammad, H W Seo and D M Kim, 2014 *Int. J. of Photoenergy* http://dx.doi.org/10.1155/2014/851705
[3] K. J. Hwang, D. W. Park, S. Jin, S. O. Kang and D. W. Cho 2015 *Mat. Chem. & Phys.* **150** 594
[4] S Zahra, M Rasoul, W Hui-Pung, S Jia-Wei and D E W Guang 2015 *J. of App. Mat. & Interface*, DOI: 10.1021/acsami.5b01519
[5] J. H. Yang, C. W. Bark, K. H. Kim and H. W. Choi 2014 *Materials* **7** 3522-3532
[6] Bisquert J and Fabregat-Santiago F, 2009 *J. Phys.Chem. C* **113**.17278-17290
[7] L. Tao, Z. Huo, Y. Ding, Y. Li, S. Dai, L. Wang, J. Zhu, Xu. Pan, B. Zhang, J. Yao, M. K. Nazeruddin and M. Gratzel 2015 J. *of Mat. Chem. A* **3** 2344-2354
[8] Suzuka M, Hayashi N, Sekiguchi, T, Sumioka K, Takata N H, Ikeda H, Oyazu K and Nishida H 2016 *Scientific Reports* DOI: 10.1038/srep28022
[9] Arsyad, W S, Bahar H, Prijamboedi B and Hidayat R 2017 *Ionics* DOI:10.1007/s11581-017-2230
[10] Pujiarti H, Arsyad W S, Hidayat R 2014 *AIP Conference Proceesings* **1617** 39-42
[11] Hidayat R, Handayani Y S and Wulandari P 2015 *Material Science Forum* **827** 162-167

Acknowledgement
The authors would like to appreciate the financial support from Riset KK B ITB 2016.

Analysis of Impact of Tropical Cyclone Blance on Rainfall at Kupang Region Based on Atmospheric Condition and Satellite Imagery

S ROGUNA[1,2,3], I J A SARAGIH[1,3], P S SIREGAR[1,3], A M JULIUS[3]

[1] Meteorology Major, Sekolah Tinggi Meteorologi Klimatologi dan Geofisika, Jl. Perhubungan 1 no. 5 Pondok Betung, Pondok Aren, Tangerang Selatan
[2] Stasiun Meteorologi Majene, Jl. Lettu Muh Yamin no.13, Majene, Sulawesi Barat
[3]Badan Meteorologi Klimatologi dan Geofisika, Jl. Angkasa 1 No.2, Kemayoran, Jakarta

Email: satrianaroguna1996@gmail.com

Abstract. The Tropical Depression previously identified on March 3, 2017, at Arafuru Sea has grown to Tropical Cyclone Blance on March 5, 2017. The existence of Tropical Cyclone Blance gave impacts like increasing rainfall for some regions in Indonesia until March 7, 2017, such as Kupang. The increase of rainfall cannot be separated from the atmospheric dynamics related to convection processes and the formation of clouds. Analysis of weather parameters is made such as vorticity to observe vertical motion over the study area, vertical velocity to see the speed of lift force in the atmosphere, wind to see patterns of air mass distribution and rainfall to see the increase of rainfall compared to several days before the cyclone. Analysis of satellite imagery data is used as supporting analysis to see clouds imagery and movement direction of the cyclone. The results of weather parameters analysis show strong vorticity and lift force of air mass support the growth of Cumulonimbus clouds, cyclonic patterns on wind streamline and significant increase of rainfall compared to previous days. The results of satellite imagery analysis show the convective clouds over Kupang and surrounding areas when this phenomena and cyclone pattern moved down from Arafuru Sea towards the western part of Australia.

Keywords*: Rainfall, Vorticity, Vertical Velocity, Streamline*

1. Introduction

A tropical cyclone is a form of extreme weather disorder that begins the emergence of a low-pressure center over the ocean that triggers the process of convection and the formation of clouds intensively. Generally, the formation of tropical cyclones is effective in latitudes above 10° north latitude and southern latitudes (Haryani and Zubaidah, 2012). Approximately 65% of tropical cyclones are formed in the region between 10o and 20o of the equator. Tjasyono in Suryantoro (2008) reveals that there are some necessary conditions (geographical and climatological conditions) in the formation of tropical cyclones, i.e., sea surface temperature is quite hot (> 26°C). Coriolis parameters must be greater than the minimum value found in the surrounding latitude 5° northern and southern hemispheres, and the presence of vertical wind shear is weak in the thick troposphere.

Tjasyono, (2004) mentions the growth of tropical storms in general divided into three stages. The birth stage is characterized by the arrangement of clouds and storm lines relating to the motion of the wind disturbance. In the adult stage, characterized by a strong circulation with symmetrical conditions and regular cloud patterns with cyclone eyes that have low pressure. The dead stage is characterized by the widening of the circulation so that its size and shape become less symmetrical. In the northern hemisphere, the average age of tropical cyclones is 6 days and 5 days in the southern hemisphere (Asrianti et al., 2013).

This research aims to determine the impact of tropical cyclone Blance formed on March 5, 2017, which was previously identified as a low-pressure center in the Arafuru Sea to change the pattern and intensity of rainfall in Kupang region on 5 - 7 March 2017.

2. Methods

The data used in this research is observation data in the form of rainfall from Kupang Meteorological Station to see the fluctuation of rainfall in Kupang area before until the time of the cyclone. In this study we also uses weather parameter data taken from ECMWF (European Center for Medium-Range Weather Forecast) with a resolution of 0.25x0.25 for selected layers, in the form of streamline data to see the wind patterns and air mass distribution generated by the occurrence of cyclones so as to affect weather conditions in the Kupang region, parameter of vorticity to see the existence of vertical air movement that raises the mass of air and vertical velocity to see the speed of lifted air masses, as well as using IR satellite channel data taken from BMKG. The method used in this research is a descriptive analytic method of the results of the processed data parameters before until the cyclone in Kupang region especially on March 5-7, 2017. For observation data, time series graph is used to see the fluctuation of rainfall. While the weather parameter data is processed by using GrADS software (The Grid Analysis and Display System) and satellite data is processed by using software GMSPLD / SATAID (Satellite Animation and Interactive Diagnosis).

3. RESULTS AND DISCUSSION

3.1. Analysis of Vorticity

Based on Figure. 1. parts A1-B1-C1, A2-B2-C2, A3-B3-C3 sequently shows layers of vortices of 850 hpa, 700 hpa and 500 hpa on March 5-7, 2017. Seen from A1 until C3, the vortices are relatively negative between $-1e^{-05}$ to $-3e^{05}$ although in A1 and C1, the value still shows positive which indicates on March 5, 2017, has not been yet too affected of the cyclone. These results show strong vortices with negative values in the southern hemisphere, thus indicating the presence of a strong enough air mass increase on the 3 dates up to 500 hpa layer.

3.2. Analysis of Vertical Velocity

It is an upward movement due to a mechanical process (not thermodynamic), where a positive valu(+) indicates no vertical upward movement. While the negative value (-) indicates a vertical upward movement (Zakir, 2010). Based on figure 2. the value of vortices either on March 5, March 6 or March 7, 2017 shows a negative value of speed. On March 5, 2017, vertical velocity in the region of Kupang is indicated by a value between -0.1 to -0.15 and on March 6, 2017 vertical velocity in the region of Kupang is indicated by a value between -0.3 to -0.35 which each of both is at an altitude of 850, so indicates that at this altitude enables the formation of Cumulonimbus clouds. Likewise on March 7, 2017, the value of vertical velocity is valued with a value of between -0.15 to -0.2 at an altitude of 850mb where at that height, this energy has been able to allow the air mass lifted so as to form Cumulonimbus clouds.

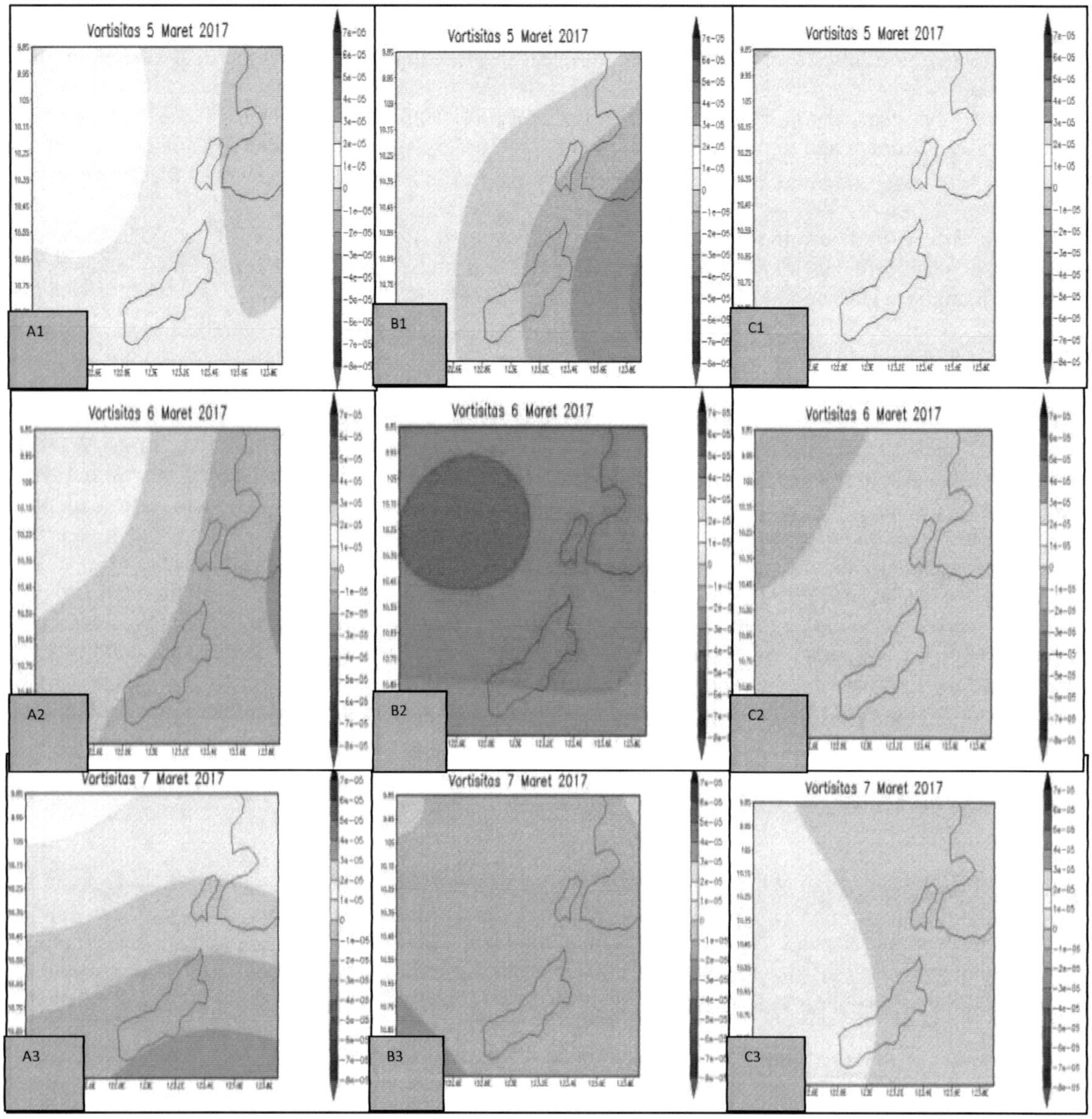

Figure 1. Vorticity sequently (from left-right) from 850 hpa, 700 hpa, and 500 hpa on March 6-7, 2017

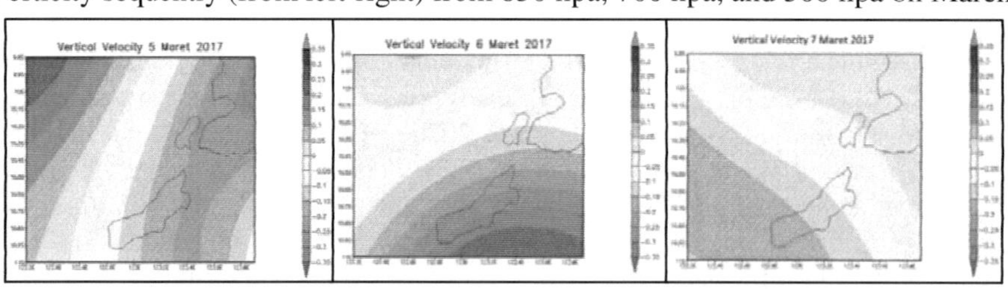

Figure 2. Vertical Velocity in 850 hpa layer sequently on March 5-7, 2017

3.3. Analysis of Surface Wind

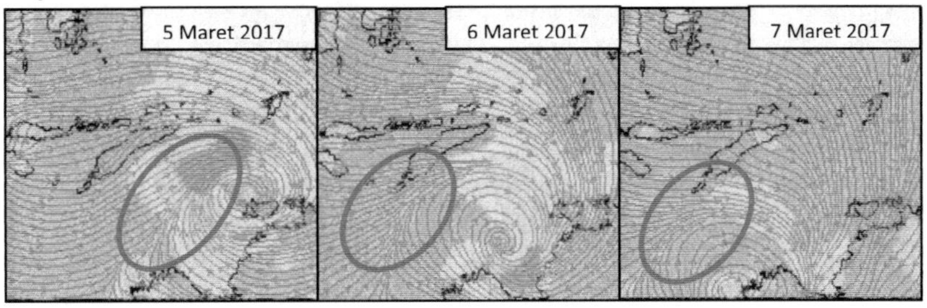

Figure 3. Surface Wind Pattern on March 6 -7, 2017 around Kupang and surrounding areas

Based on figure 3 on the surface layer, the cyclone pattern that has formed was previously originated from the Arafuru Sea which has moved to the southeastern region of Kupang on March 5, 2017, March 6, 2017, until March 7, 2017, has reached to the western part of Australia. The air mass coming from the west forms a convergence pattern around the southern part of Kupang. The air mass that moves in this pattern typically slows down the speed so that it triggers a lot of air masses gathered in the area. This has the potential to affect large parts of Kupang in the presence of an increase in convection activity and thus the potential for the growth of convective clouds such as Cumulonimbus clouds.

3.4. Analysis of Rainfall Data

Table 1. Rainfall in Kupang region on March 3, 2017, until March 7, 2017

Dates	Number of rainfall
March 3, 2017	No Rain
March 4, 2017	No Rain
March 5, 2017	12.0 mm
March 6, 2017	15.5 mm
March 7, 2017	20.8 mm

An increase rainfall indicates an increase of convective clouds, as shown in table 1. Based on the table, there is an increase in rainfall between March 3, 2017, until March 7, 2017, in the Kupang region. On March 3-4, 2017, there was no occurrence of rain events that occurred on that day, followed by the incident on March 5, 2017, with a measured amount of 12.0 mm. On the next day, rainfall increased to 15.5 mm and increased on March 7, 2017, to 20.8 mm. This significant increase in the value of rainfall indicates that the impact brought by the presence of this tropical cyclone "Blance" is sufficient to affect convection activity over the Kupang region resulting in the growth of convective clouds which produces rain such Cumulus and Cumulonimbus.

3.5. Analysis of Satellite Imagery

Based on the Figure. 4, originated from a Tropical Depression located on Arafuru Sea on March 3, 2017. The Tropical Depression was identified as a 1-day continuous growth and growth on March 4, 2017. Until March 5, 2017, the Tropical Depression has grown Into a Tropical Bluff Cyclone that continues to experience movement and growth until it reaches its mature stage at 18:00 UTC on the same day. On March 6, 2017, Tropical Cyclones began to experience a weakening stage until on March 7, 2017 the Tropical Cyclone was slowly extinct. From the time of the Tropical Depression to the Tropical Cyclone Blance, the direction of the movement is from the Arafuru sea descending towards the southwest past Kupang and ending in western Australia.

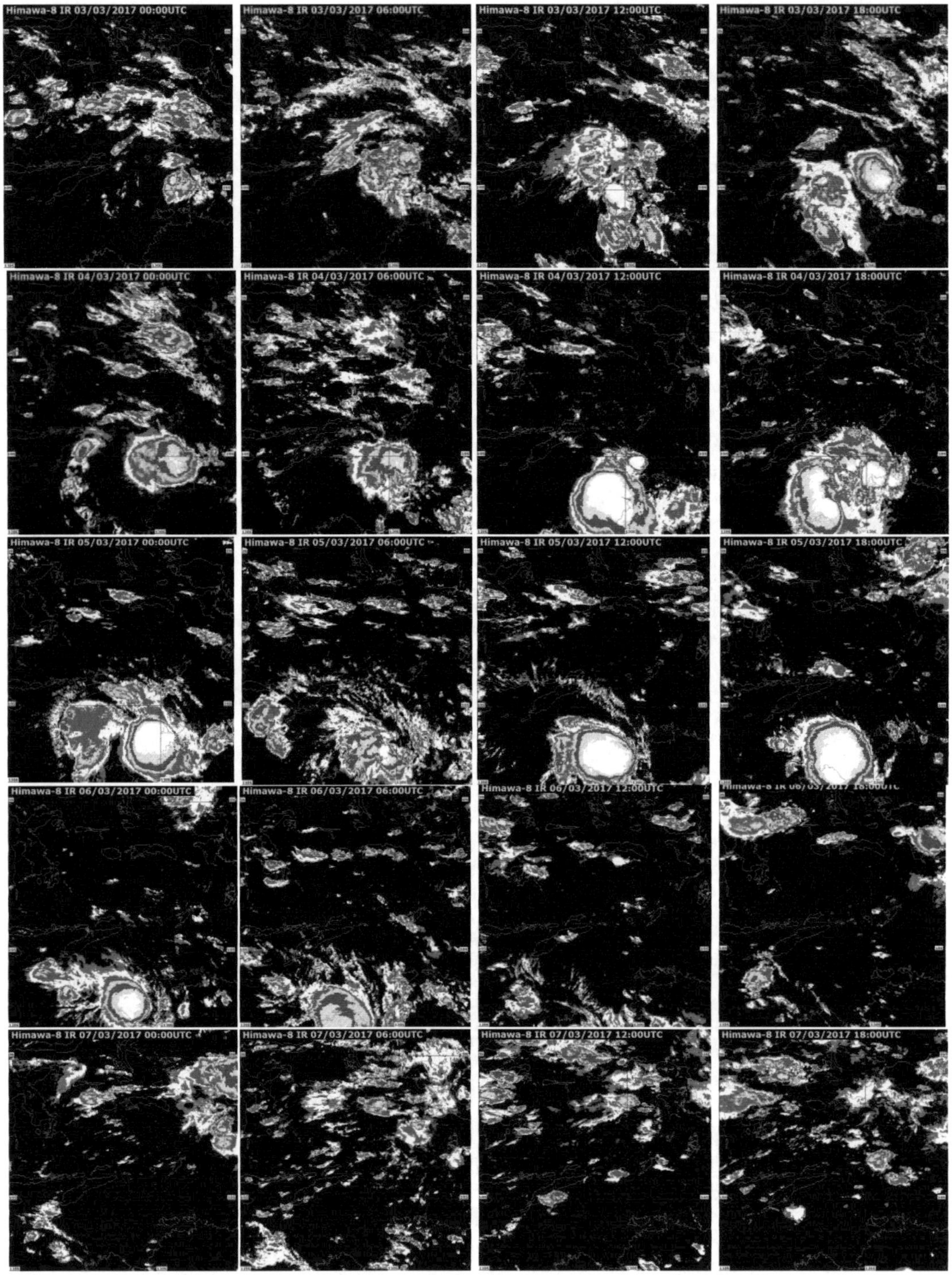

Figure 4. HIMAWARI-8 Satellite Imagery of Tropical Cyclone "Blance" motion every 6 hours on March 5-7, 2017

IOP Conf. Series: Journal of Physics: Conf. Series **1011** (2018) 012021 doi:10.1088/1742-6596/1011/1/012021

4. Conclusion

The presence of tropical cyclone "Blance" gave impacts such as significant increase of rainfall in Kupang region from March 5, 2017, to March 7, 2017, based on the observation data, where on March 3-4, 2017 there has been no rain event since it was still a Tropical Depression and not yet approaching Kupang. The parameters that run into change as impacts of the occurrence of cyclones which resulted an increase rainfall are vorticity that shows a negative value (strong vortices in the southern hemisphere) which indicates a strong lifting of air masses over Kupang, vertical velocity with a negative value on 850 hpa layer indicates the occurrence of upward vertical movement in the southern hemisphere with a strong enough energy which enabled to occur the growth of convective clouds, convergence pattern that causes the buildup of air masses resulting an increase of convection activity, and the movement of tropical cyclone "Blance" which previously located in the Arafuru Sea declined southwestward across the Kupang region and finally stopped in west of Australia.

5. References

[1] Haryani NS, Zubaidah A 2012 *Dinamika Siklon Tropis di Asia Tenggara Menggunakan Data Penginderaan Jauh.* (Widya, 29 Vol **324)** p 54-58

[2] Suryantoro A 2008 *Siklon Tropis di Selatan dan Barat Daya Indonesia dari Pemantauan Satelit TRMM dan Kemungkinan Kaitannya dengan Gelombang Tinggi dan Putting Beliung* (Majalah Sains dan Teknologi Dirgantara Vol **3** chapter **1)** p 21-32

[3] Tjasyono BHK. 2004. Klimatologi. (Bandung: Institut Teknologi Bandung)

[4] Zakir, A., Khotimah, M.K., dan Sulistia, W., 2010 *Prespektif Operasional Cuaca Tropis* (Jakarta: BMKG)

IOP Conf. Series: Journal of Physics: Conf. Series **1011** (2018) 012022 doi:10.1088/1742-6596/1011/1/012022

Potential Fungus surface resistance of the silica/acrylic coated leaves waste composite

Masturi[1,*], **WN Jannah**[1], **RM Maulana**[1], **T Darsono**[1], **Sunarno**[1], **S Rustad**[2]
[1] Physics Department of Universitas Negeri Semarang
D7 Building 2nd Floor, Sekaran, Gunungpati, Semarang, Indonesia
[2] Dian Nuswantoro University, Semarang, Indonesia

*Email: masturi@mail.unnes.ac.id

Abstract. The composite coated by some materials coaters have been made. This coating was done to isolate the fungus possibly growing on the composite. The composite was made from a mixture of teak leaves waste and polyurethane polymer using a simple mixing method; then the mixture was pressed at a pressure of 3 metric-tons for 15 minutes. The composite produced then was coated with acrylic only and acrylic-silica using spray method. The coated samples then were characterized using scanning electron microscopy (SEM) to determine the surface pores. Further, it was obtained the average surface pore size of acrylic coater is 1.24 μm, while the acrylic-silica pore forms an oval shape with a length and a width of 0.75 μm and 0.38 μm, respectively. In comparison with the fungus size of 2-7 μm, it can be concluded that the composite is proper as home appliance application.

Introduction

The wood demands are increasing for several decades as people growth. It challenges the researcher and industrialist to develop a material replacing the rare woods, and one of the ways is through the composite engineering producing a lignocellulosic material having properties like the woods [1-6]. These composites have been made from natural fibers and agricultural wastes mixed with a polymer to strengthen bonding with pressing technology [7-8].

Some research relating to the lignocellulosic composite have been done, such as Mango leaves waste composite [9], composite using cellulose fibers [10], tea waste composite [11], and peanuts-shell composite [12]. However, most of them have not performed the safety aspect of its usage yet, especially in handling the fungus that enables grows on the composite as the hygroscopic environment. Further, as a hygroscopic environmental condition where the water easily penetrates into the composite network, the composite will be targeted for the growth of fungi and other biological agents [13-14]. Therefore, it is very important to ensure the fungus that may grow on the composite must be isolated, and one of the ways is by a coating process.

Several works of composite coatings have also been done, such as Iswanto *et al.* [8] that have treated liquid paraffin, liquid silicone and waterproof coating on composites. Further, Lesar and Humar [15] used waxes to coating on wood surfaces and to know their resistance to fungi. However, these studies have not examined the possibility of fungus resulted from the composite passing to the surface, regarding the contents of the leaf waste usually produce fungus.

In this works, we have made a polyurethane/teak leaf waste composite coated with some material coaters. By using the coating process, the composite produced has properness and safety from fungus to be used as several home appliance. We have also measured the surface pores to ensure that the fungus possibly growing cannot pass to the surface.

Material and Methods

The leaf waste used is Javanese teak leaf waste, while polyurethane, acrylic, and silica were purchased from Indrasari, Semarang, Indonesia.

For first, polyurethane was mixed with crushed-teak leaves at the certain fraction. The mixture was stirred until homogeneously and then pressed at 3 metric tons pressure for 15 minutes to produce polyurethane/teak leaf waste composite. The composite then was coated using spray method with several material coaters, including acrylic only and acrylic-silica, with the mass of silica and acrylic is 5 g and 30 g, respectively. The mixing process of silica and acrylic was done using magnetic stirrer for 5 minutes to ensure the mixture is homogeneous. The spray method was operated with a pressure of 10 bars.

Scanning electron microscopy (SEM) test was performed to determine the average pore magnitude of the composite surface [16]. This pore then was compared to the several sizes of fungi to judge closure performance of the coater to the fungi that may grow from the teak leaf waste inside.

Result and Discussion

Cellulose and lignin contents in the teak leaf waste can trigger fungi growth when there is water infiltrating into the composite as a hygroscopic environment. The coating treatment is to prevent water to insert into the composite and also ensure the composite has resistance to fungi grown from teak leaf waste inside.

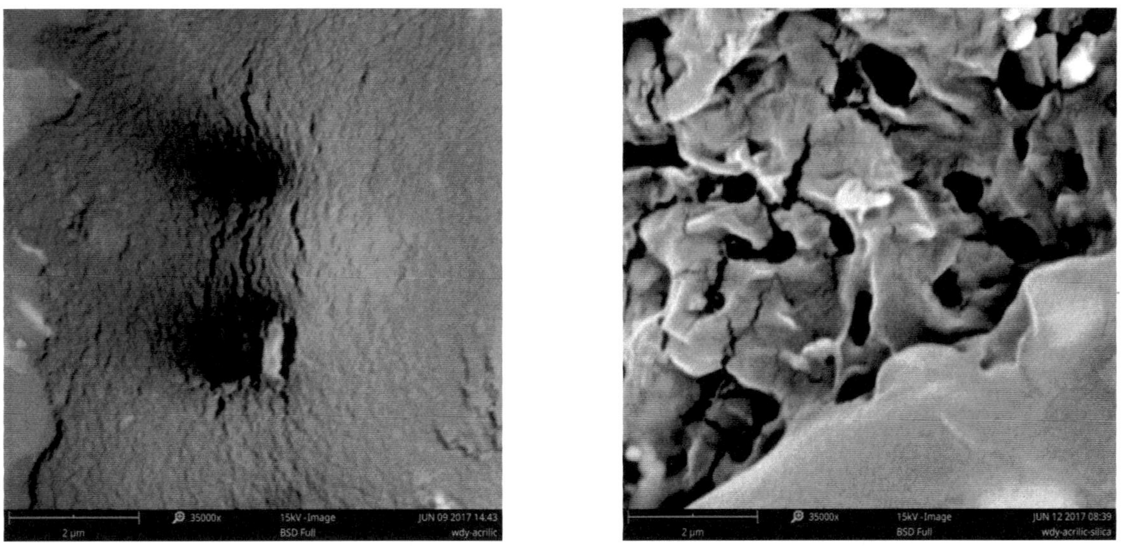

a) b)

Figure 1. SEM image of composite for a) acrylic, b) acrylic-silica coatings

Using SEM measurement (Figure 1), it was obtained the composite surface pore size as shown in Table 1. For the composite coated by acrylic only, the average pore size is 1.24 μm with very rare distribution and has a ball shape (Figure 1(a)). This pore may occur due to an interaction between acrylic and the teak leaf waste particle underneath. As this interaction, some of the acrylic chains are attracted inside and forms some pores. This suggestion is based on the acrylic's active group, i.e., the vinyl group attracted by some particle, especially teak leaf waste particle on the composite surface [9,16]; Masturi et al., 2017). Meanwhile, the silica/acrylic coater tends to form more oval shapes having average length and width of 0.76 μm and 0.38 μm, respectively (Figure 1(b)). The smaller size of the pore is suggested due to stronger interaction occurs between acrylic and silica than that of acrylic and teak leaf waste, especially between slightly electropositive silicon and electronegative oxygen of acrylic's vinyl group to form dipole-dipole interaction, while the large density of the pores is due to the strong interaction. However, we have not examined this interaction yet in this work.

Table 1 shows both pore sizes of acrylic and acrylic-silica are fairly small if compared to some wooden fungi having average pore size 3-7 μm [17]. It means that the fungus possibly growing on the polyurethane/teak leaf waste composite cannot get out passing the pores and further may stay for a long time under the coater to die.

Table 1. The pore size on the composite surface for several materials

Coating materials	Pore size (μm)
Acrylic	1.24
Acrylic/silica	Length = 0.76 , width = 0.38

As the pore size above, it can be concluded that the composite produced is safe as home appliance raw material in competing for the woods to reduce the woods usage and deforestation.

CONCLUSION

We have synthesized polyurethane/teak leaves waste composite coated with acrylic and acrylic-silica. The surface morphology of the acrylic coated and acrylic-silica coated composites have a pore size of 1.24 μm and 0.38-0.76 μm, respectively. These pore sizes are relatively smaller compared to wooden fungi having average size 3-7 μm. It means so that the composite is reasonably safe from the fungi may grow from the teak leaf waste inside, and further, it can be used as home appliance material competing for the woods.

ACKNOWLEDGEMENT

We would like to thanks to Research, Technology and Higher Education Ministry in financing this work via the Mainstay Research Grant for the fiscal year 2017.

Reference

[1] Zivkovic I, Pavlovic A, Fragassa C 2016 Improvements in wood thermoplastic matrix composite materials properties by physical and chemical treatments *Int. J. Quality Res.* **10**(1) 205–218
[2] Kowaluk G 2014 Properties of lignocellulosic composites containing regenerated cellulose fibers *Bioresour.* **9**(3) 5339-5348
[3] Ashori A 2008 Municipal solid waste as source of lignocellulosic fiber and plastic for composite industries *Polym-Plastics Technol Eng.* **47**(8), 741-744
[4] Paukszta D, Borysiak S 2013 The influence of processing and the polymorphism of lignocellulosic fillers on the structure and properties of composite materials - a review *Mater.* **6** pp 2747-2767
[5] Rosa S M L, Santos E F, Ferreira CA, S.M.B 2009 Nachtigall. Studies on the properties of rice-husk-filled-PP composites – effect of maleated PP *Mater. Res.* **12**(3) pp 333-338
[6] Sumaila M, Ugheoke B I, Timon L, Oloyode T 2006 A Preliminary mechanical characterization of polyurethane filled with lignocellulosic material *Leonardo J. Sci* **9** pp 159-166
[7] Kose C, Terzi E, Buyuksari U, Ever E, Ayrilmis N, and Kartal S N 2011 Particleboard and MDF panels made from a mixture of wood and pinecones : resistance to decay fungi and termites under laboratory conditions *J. Bioresour.* **6** (2) pp 2045–2054
[8] Iswanto A H, Sucipto T, Nadeak S S D, and Fatriasari 2015 Post-treatment effect of particleboard on dimensional stability and durability properties of particleboard made from sorghum bagasse *IOP Conf. Ser. Mater. Sci. Eng.* **180** pp 1 – 8

[9] Masturi, Sunarno, Jannah W N, Alighiri D, Purwinarko A, Susilawati, Amri U, Rustad S 2017 Effect of fiberglass reinforcement on compressive strength of teak (Tectona grandis L.f.) leaves waste composite *J. Phys. Conf. Ser*, **824**, 012006-9
[10] Maafi E M, Malek F, Tighzert L, Dony P 2010 Synthesis of polyurethane and characterization of its composites based on alfa cellulose fibers *J.Polym. Environ.* **18** pp 638-646
[11] Mattos B D, Misso A L, De Cademartori P H G, De Lima E A, Magalhães W L E, and Gatto D A 2014 Properties of polypropylene composites filled with a mixture of household waste of mate-tea and wood particles *Constr. Build. Mater.* **61** pp 60–68
[12] Pirayesh H and Khazaeian A 2012 Using almond (Prunus amygdalus L.) shell as a bio-waste resource in wood based composite *Compos. Part B Eng.* **43** (3) pp 1475–1479
[13] Sumarni G and Muslich M 2004 Keawetan 52 jenis kayu Indonesia *J. Peneltian Hasil Hutan* **22** (1) pp 1–9
[14] Benthien J T, Thoemen H, Maikowski S, and Lenz M T 2012 Resistance of flat-pressed wood – plastic composites to fungal decay: effects of wood flour content, density, and manufacturing technology *J. Wood Fiber Sci.* **44**(4) pp 422–429
[15] Lesar B and Humar M 2011 Use of wax emulsions for improvement of wood durability and sorption properties *J. Wood Prod.* **69** pp 231–238
[16] Wyart Y, Georges G, Deumie C, Amra C, Moulin P 2008 Membrane characterization by microscopic methods: multiscale structure *J Membrane Sci.* **315** pp 82-92
[17] Akbari S, Eslahi N, Kish M H 2017 Evaluation of hydrphiic properties of acrylonitrile/acrylic acid copolymer films dendrigrafted with citric acid. *Polyolefins J.***4** pp 215-242

Numerical Modeling of 3D Seismic Wave Propagation around Yogyakarta, the Southern Part of Central Java, Indonesia, Using Spectral-Element Method on MPI-GPU Cluster

Sudarmaji[1], Indra Rudianto[2] and Budi Eka Nurcahya[3]

Department of Physics, Faculty of Mathematics and Natural Sciences, Universitas Gadjah Mada, Yogyakarta, Indonesia

E-mail: [1]ajisaroji@ugm.ac.id, [2]indrarudianto.official@gmail.com, [3]budiekanurcahya@yahoo.com

Abstract. A strong tectonic earthquake with a magnitude of 5.9 Richter scale has been occurred in Yogyakarta and Central Java on May 26, 2006. The earthquake has caused severe damage in Yogyakarta and the southern part of Central Java, Indonesia. The understanding of seismic response of earthquake among ground shaking and the level of building damage is important. We present numerical modeling of 3D seismic wave propagation around Yogyakarta and the southern part of Central Java using spectral-element method on MPI-GPU (Graphics Processing Unit) computer cluster to observe its seismic response due to the earthquake. The homogeneous 3D realistic model is generated with detailed topography surface. The influences of free surface topography and layer discontinuity of the 3D model among the seismic response are observed. The seismic wave field is discretized using spectral-element method. The spectral-element method is solved on a mesh of hexahedral elements that is adapted to the free surface topography and the internal discontinuity of the model. To increase the data processing capabilities, the simulation is performed on a GPU cluster with implementation of MPI (Message Passing Interface).

1. Introduction

On 26 May 2006 at 22:54 UTC, the Yogyakarta earthquake occurred with moment magnitude M_w 6.4 [1]. The earthquake has caused severe damage to the densely populated area. About 6,000 people were killed, 50,000 were injured, and between 0.5 and 1 million homeless. The total loss was estimated at over 3 billion U.S. dollars [2]. Due to its huge damages on human casualties, buildings and materials, the active fault system must be well investigated and characterized to prevent future seismic hazards.

In recent years, numerical modeling using GPU computing have been successfully used to simulate the seismic wave propagation, such as study of fluid saturated porous media [3, 4], high-order finite-difference simulation [5], and finite-difference numerical modeling in realistic topography [6]. However, numerical modeling of 3D seismic wave propagation in Yogyakarta basin has not yet been available. For Yogyakarta basin, it poses several challenges. The notable topography around the city makes the issue more difficult. It also covers a large area, so that it needs more data processing power capability.

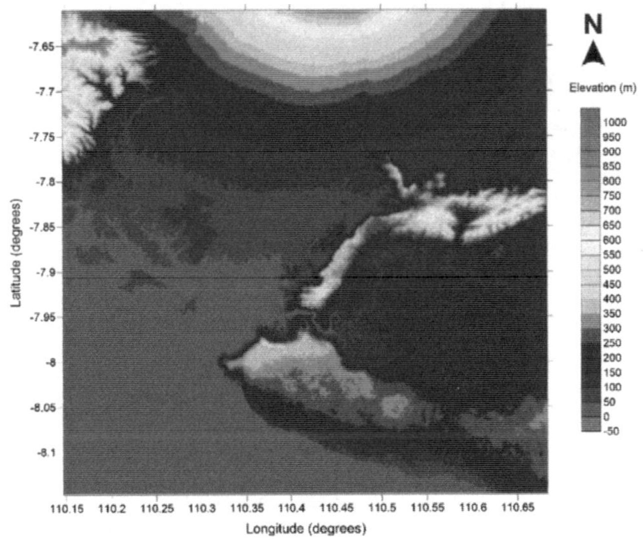

Figure 1. Topography map of Yogyakarta basin. It spans from -8.15° to -7.61° latitude and from 110.15° to 110.68° longitude. Surface topography is imported using NASA Shuttle Radar Topographic Mission (SRTM) with an original resolution of 3 arcsec (~90 m).

To accommodate a considerable surface topography, we used the spectral-element method (SEM) to simulate the seismic wave propagation in Yogyakarta basin. It used a weak formulation for solving elastic wave equation and naturally incorporates free surface conditions. The method has been successfully applied in many areas of seismology [7, 8].

In this study, we present numerical modeling of 3D seismic wave propagation in Yogyakarta basin to observe its seismic response due to varying surface topography and its implication on earthquake hazards.

2. 3D Model and Mesh Implementation

The model we used for simulation covers most of the southern part of Central Java with dimension of 59.3 km × 59.4 km horizontally and from +0.96 to -30 km vertically. It spans from -8.15° to -7.61° latitude and from 110.15° to 110.68° longitude (Figure 1). Surface topography was imported using NASA Shuttle Radar Topographic Mission (SRTM) with an original resolution of 3 arcsec (~90 m). For meshing purposes, we converted the geographic coordinate system into the UTM (Universal Transverse Mercator) coordinate system. Our model lies in the 49S UTM projection zone.

To incorporate high-resolution topography, we used mesh doubling as a function of depth [9]. Figure 2 shows mesh implementation of our model. Three consecutive mesh doublings facilitate in an increase in element size with depth. We also added three additional flat interfaces, which correspond to the three mesh doubling layers. We used this technique in order to accommodate a finer mesh near topography and decrease the number of elements in the model which lead to a reduction of the computational burden. The first interface was added at the depth of 750 m to facilitate a finer mesh near the surface. Note that resolution of the mesh in the surface is not controlled by shear-wave, but by surface-wave because surface-wave speed is lower than shear-wave speed. The second interface was added at the depth of 1750 m to facilitate a transition layer between two mesh doublings. The third interface was then added at the depth 3750 m to facilitate the third double layer. It has constant element size to the bottom of the model which could significantly decrease the number of elements in the model.

We used 3D homogeneous wave velocity model based on the one-dimensional (1-D) spherically symmetric Earth models [10] as a preliminary study. The compressional- and shear-wave speeds is 5800 m/s and 3200 m/s respectively. The density in the model is 2600 kg/m^3.

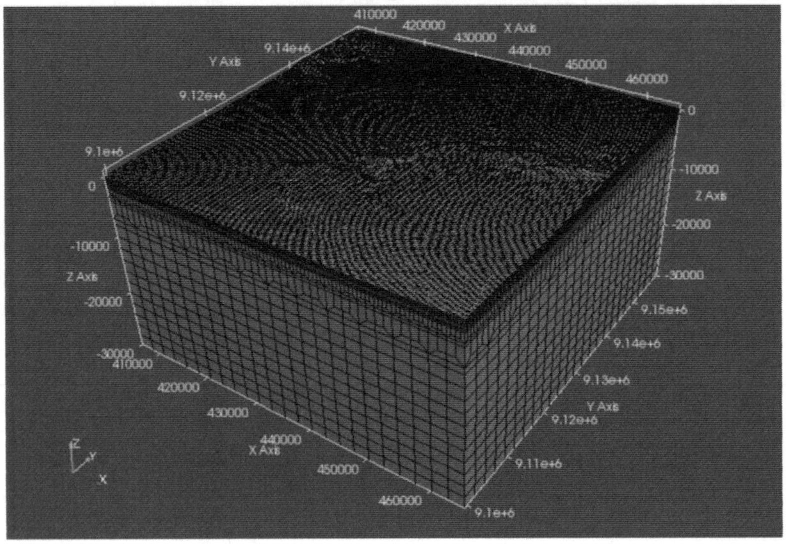

Figure 2. Spectral-element mesh for Yogyakarta basin. The size of the model is 59.3 km × 59.4 km horizontally and from +0.96 to -30 km vertically. There are three consecutive mesh doubling facilitate in an increase in element size with depth.

3. Simulation of the 26 May 2006 Yogyakarta Earthquake (M_w 6.4)

The focal mechanism and properties of the source were obtained from The Harvard Centroid Moment Tensor (CMT) solution with M_w 6.4. The epicenter of the earthquake is located at -8.03° latitude and 110.54° longitude and the hypocenter is 21.7 km. The 3D moment-tensor inversion indicated the fault striking 323.11°, dipping 76.62°, and raking -176.02°.

We used the SPECFEM3D software package [11] and decomposed the model into 213696 hexahedral elements. In each spectral element we used the polynomial degree of $N = 4$, and thus each element contains $(N + 1)^3 = 125$ Gauss-Lobatto-Legendre (GLL) integration points. Total number of GLL points in the model is 14,124,456. The minimum and maximum distances between Gauss-Lobatto-Legendre integration points in the model are 43.17 m and 991.42 m respectively. We used a Gaussian source time function with a half-duration of 3.7 s, and therefore a maximum frequency is 0.27 Hz. The time step used was $\Delta t = 2$ ms, and we propagated the signal for 30000 time steps, thus total simulation time is 60 s.

The simulation was carried out on a GPU cluster with 4 NVIDIA Quadro K4000 graphics cards and 24 core processors Intel Xeon CPU E5-620 @2.4 GHz using MPI (Message Passing Interface) and CUDA (Compute Unified Device Architecture) at Computational and Seismic Laboratory, Geophysics Sub-Department, Universitas Gadjah Mada.

4. Results and Discussion

Snapshots of the simulation for the norm of all three components of the velocity wave field are displayed in Figures 3. We observed two consecutive wave fields with a strong amplitude travel toward the Yogyakarta basin in the northwest direction at t = 20 s (Figure 3d). The mountainous areas in the northwestern part of the model caused the waves to reflect back to the basin. After t = 30 s (Figure 3f), some wave fields still trapped and reflected in the basin and travel to the north, northeast and southwest direction follows the basin geometry.

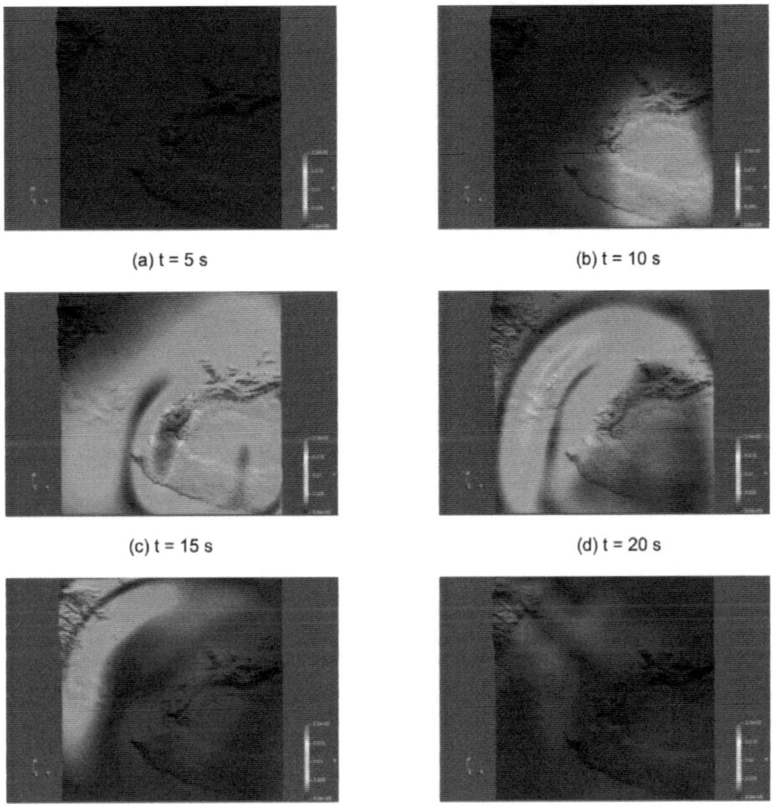

(a) t = 5 s (b) t = 10 s

(c) t = 15 s (d) t = 20 s

(e) t = 25 s (f) t = 30 s

Figure 3. Snapshots of the simulation of the 26 May 2006 M_w 6.4 Yogyakarta earthquake for the norm of all three components of the velocity wave field. There are two consecutive wave fields with a strong amplitude travel toward the Yogyakarta basin in the northwest direction at t = 20 s (d). After t = 30 s (f), some wave field still trapped and reflected in the basin and travel to the north, northeast and southwest direction follows the basin geometry.

Figure 4 shows resulting synthetic seismograms recorded at the surface located at -7.74° latitude and 110.28° longitude for East (a), North (b), and Vertical (c) component of velocity. All the synthetic waveforms are low-pass filtered with a corner frequency of 1 Hz. The station has a horizontal distance of 43 km from the source.

As a preliminary study, we aim to analyze how varying topography in our model governs the propagation of seismic wave. Although we have accomplished much, there are some limitations in our implementation. For example, mesh implementation on the surface still insufficient to access the full effect of topography. In our future works, we intended to develop our mesh implementation. To increase mesh resolution and access the full effect of topography, we use an unstructured hexahedral meshing technique. We also enlarge the dimension of the model we wish to take into account of possible earthquake and/or seismic damage, and include all currently available relevant seismological information available in our model.

5. Conclusions

We presented the numerical modeling of 3D seismic wave propagation around Yogyakarta using spectral-element method on MPI-GPU cluster. We used a mesh doubling technique to incorporate high-resolution topography and accommodated a finer mesh near the surface. We observed two

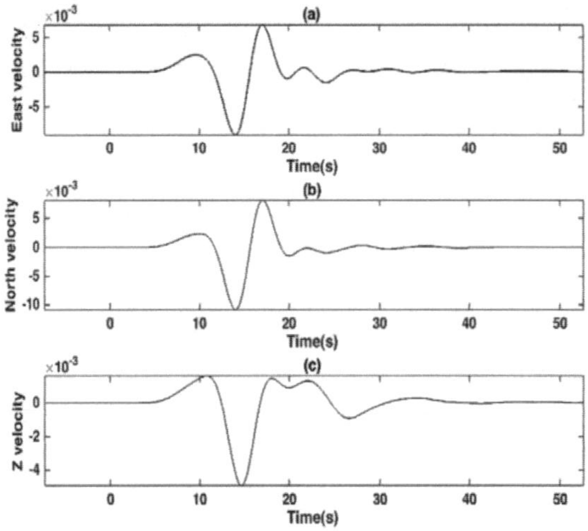

Figure 4. Resulting synthetic seismograms recorded at -7.74° latitude and 110.28° longitude for East (a), North (b), and Vertical (c) component of velocity. All the synthetic waveforms are low-pass filtered with a corner frequency of 1 Hz.

consecutive wave fields with a strong amplitude travel toward the basin. Furthermore, the basin and surrounding mountains significantly increase the duration of shaking in the Yogyakarta area.

In our future works, we intend to develop our mesh implementation, enlarge the dimension of the model, and include all currently available relevant seismological information in our model.

Acknowledgments
We would like to express our gratitude to PT. Chevron Indonesia for hardware donations given to Geophysics Sub-Department, Universitas Gadjah Mada.

References

[1] Harvard-CMT 2006 Focal mechanism of the 2006 Yogyakarta earthquake, The website of the Harvard Global CMT, id#20062253A, (available at http://www.globalcmt.org)
[2] Consultative Group on Indonesia 2006 Preliminary damage and loss assessment, Yogyakarta and central Java natural disaster: A joint report of BAPPENASS, the provincial and local governments of D.I Yogyakarta, the provincial and local governments of central java, and international partners, in *The 15 Meeting of the Consultative Group on Indonesia (CGI) Jakarta, June 14, 2006 140 pp.*, Jakarta
[3] Sudarmaji, Sismanto and Waluyo 2014 *Int. Conf. on Phys.*
[4] Sudarmaji, Sismanto, Waluyo and Soedijono B 2016 *AIP Conf. Proc.* **1755** 100001-1–7
[5] Hasym I B, Sudarmaji, Sismanto and Waluyo 2014 *Int. Conf. on Phys.*
[6] Permana T and Sudarmaji 2014 *Int. Conf. on Phys.*
[7] Komatitsch D, Liu Q, Tromp J, Süss P, Stidham C and Shaw J H 2004 *Bull. Seismol. Soc. Am.* **94** 187206
[8] Komatitsch D, Ritsema J and Tromp J 2002 *Science* **298** 1737–42
[9] Komatitsch D and Tromp J 1999 *Geophys. J. Int.* **139** 806–22
[10] Dziewonski A M and Anderson D L 1981 *Phys. of the Earth and Planetary Interiors* **25** 297–356
[11] Komatitsch D, Erlebacher G, Göddeke D and Michéa D 2010 *J. Comp. Phys.* **229** 7692–7714

IOP Conf. Series: Journal of Physics: Conf. Series **1011** (2018) 012024 doi:10.1088/1742-6596/1011/1/012024

Delineation of The Sumatra Fault in The Central Part of West Sumatra based on Gravity Method

R D Saragih[1] and K S Brotopuspito[1]

[1]Geophysics Laboratory, Department of Physics, Universitas Gadjah Mada, Sekip Utara, Yogyakarta, Indonesia 55281

Corresponding author email : [1]rikadesrinasaragih@ymail.com, [1]kirbani@yahoo.com

Abstract. The Sumatra Fault System is elongated across the Sumatra Island, Indonesia, Southeast Asia including the central part of West Sumatra, Indonesia, Southeast Asia. The Sumatra Fault and subsurface structure on the Central Part of West Sumatra had been analyzed using gravity method. Bouguer anomaly data were obtained from GRDC (Geological Research and Development Centre) maps, Bandung, Indonesia (i.e. without terrain correction). In this study, terrain correction had been applied to these Bouguer data. Bouguer anomaly in a horizontal plane at 3000 meters high and equivalent depth of mass point 7000 meters were obtained using Dampney Method. Residual and regional anomalies were separated using upward continuation method at 8000 meters high. The result of the SVD on residual anomaly shows two negative anomalies on northwest – southeast. The zero miligal per meter square quantity coincides remarkably well with trace faults which is a part of the Sumatra Fault System. Two negative anomalies are located around the Sianok Segment and Sumani Segment.

1. Introduction

The study area is concentrated in the central part of West Sumatra which is part of West Sumatra, Indonesia, Southeast Asia. This study area lies between lattitudes 0°N – 1°S and longitudes 99,76°E – 101,50°E. The Sumatra fault system is elongated across the Sumatra Island, Indonesia, including the central part of West Sumatra, Indonesia. The Sumatra fault is one of the active faulting in Indonesia. The fault zone of the Sumatra Fault runs along the entire 1600-km long Bukit Barisan mountain range on the west side of the Sumatra island [1]. From topographic maps and stereographic aerial photographs shows that the Sumatran Fault is highly segmented [2]. Topographic depressions, volcanoes, and east-west aligned magnetic anomalies are the signs found at the junction of the segments [1]. In addition to this, we can prove segmentation of the Sumatra Fault with subsurface identification. We can use geophysical method for subsurface identification.

Gravity method is one of the several methods in geophysics for subsurface identification of geological structure including faults. Local variations in the densities of rocks cause minute changes in the gravity field [3]. The gravity method is a passive method so does not need source to use and this is one of the advantage of the gravity method. The another advantage of gravity method is the gravity field relatively easy to interpretate because variations of the earth's gravity is relatively uniform. Gravity field in the central part of West Sumatra has never been spesifically discussed. In this study, we had been discussed the gravity field in the central part of West Sumatra and its correlation with the

Sumatra Fault. To enhance gravity anomalies, we have applied terrain corrections, upward continuation method and Second Vertical Derivative (SVD) to gravity data.

The geological information of the study area were obtained from the geological maps of Padang [4] and Solok Quadrangle [5]. The main fault of the study area is a part of the Great Sumatra Fault which trends to the northwest – southeast and represents as dextral fault [4]. The Great Sumatra Fault is associated with the formation of volcanoes and the other fault directions are northeast – southwest and north – south [4]. The simplification of the geological map of the study area is shown in Figure 1.

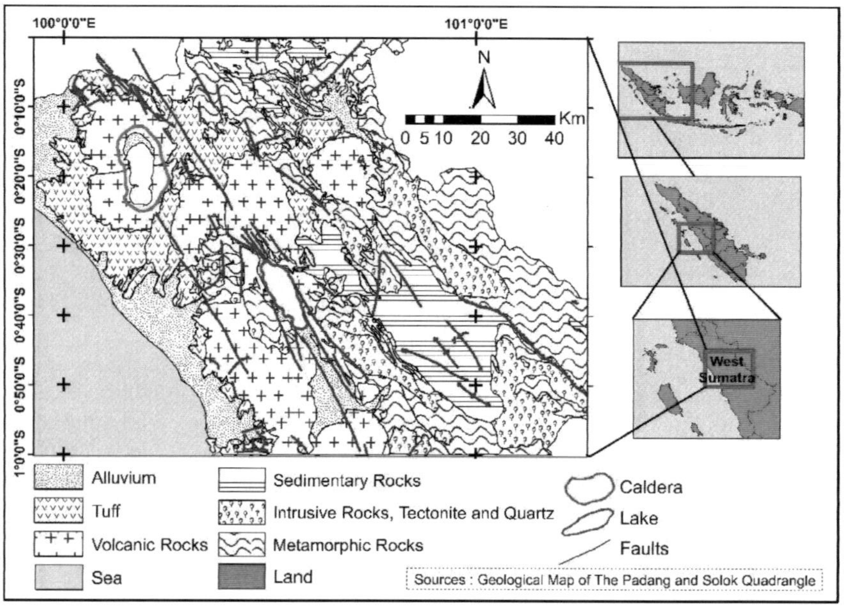

Figure 1. Simplification of the geological map of the study area.

2. Method

Gravity data were obtained from Geological Research and Development Center (GRDC) Bandung, Indonesia, Southeast Asia in the form of contour maps for Padang [6] and Solok Quadrangle [7]. Gravity data in these contour maps were acquired with LaCoste and Romberg model G land gravity meters [6] and [7]. Lattitude corrections were made using Geodetic Reference Systems, 1967. Bouguer corrections were computed using a density of 2,67 tm^{-3}. Terrain corrections have not been applied because sufficiently reliable topographic maps were unavailable [6] and [7].

Gravity data has obtained from Bouguer contour maps with 1,5 km of spacing grid. After the Bouguer contour map is extracted into gravity data, we have applied terrain corrections [8]. For terrain corrections, we have obtained topographic maps which include land via http://earthexplore.usgs.gov/ and marine via http://www.marine-geo.org/portals/gmrt/about.php. For inner and outer zone, we had been used 5 km [9] and 167 km [10], [11], [12], [13] and [14], respectively.

2.1. Reduction Bouguer anomalies onto a horizontal plane

Bouguer anomalies are measurements on an irregular grid with variety of elevations so, we have been projected onto a horizontal reference plane with Dampney Method [15]. Bouguer anomaly measurements on an irregular grid and at a variety of elevations can be synthesized by an equivalent source of discrete point masses on a plane of arbitrary depth below the surface [15]. By keeping the depth of the plane within certain limits relative (shown in Equation 1) to the stasion spacing, the synthesized field closely approximates the true gravity field above the topography.

$$2.5\Delta x < (h - z_i) < 6\Delta x \quad (1)$$

IOP Conf. Series: Journal of Physics: Conf. Series **1011** (2018) 012024 doi:10.1088/1742-6596/1011/1/012024

2.2. Upward continuation

Bouguer anomalies in a horizontal plane, resulted from reduction onto a horizontal plane, is mixing of regional and residual anomaly. Residual and regional anomalies were separated with upward continuation method [16]. Upward continuation transforms the potential field measured on one surface to the field that would be measured on another surface farther from all sources [17]. The potential field is completely known if over another surface and no masses are located between the two surfaces. The fact that gravity fields obey Laplace's equation permits us to determine the field over an arbitrary surface. By Fourier transforming the measured data, multiplying by the exponential term of equation, and inverse Fourier transforming the product, a level to level continuation can be achieved. Fourrier transform of upward continuation is given by Equation 2 [17].

$$F[\psi_U] = e^{-\Delta z|k|}, \quad \Delta z > 0 \tag{2}$$

2.3. Second Vertical Derivative (SVD)

Deliniation of the Sumatra Fault were made using Second Vertical Derivative method. The second vertical derivative can be obtained from the horizontal derivatives because the gravity field satifies Laplace's equation [3]. If ϕ is measured on a horizontal surface, then Laplace's equation can be transformed to the Fourier domain [17] as shown in Equation 3.

$$F\left[\frac{\partial^2 \phi}{\partial z^2}\right] = |k|^2 F[\phi] \tag{3}$$

Hence, the second vertical derivative of a potential field measured on a horizontal surface is framed as a three-step filtering operation: Fourier transform the potential field, multiply by $|k|^2$, and inverse Fourier transform the product [17].

3. Result and Discussion

Terrain correction had been applied to Bouguer data and results values –50.71 to 44.04 mgal. Gravity field in the central part of West Sumatra which is applied terrain correction is shown in Figure 2.

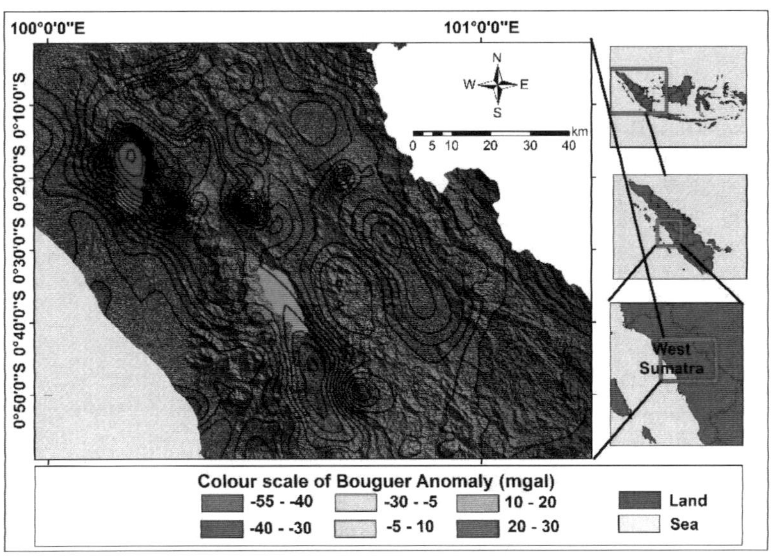

Figure 2. Bouguer anomaly have been applied terrain correction in the central part of West Sumatra based on data from GRDC maps. Low to high anomaly is marked by colour scale (mgal).

We do not make detailed interpretation in these anomalies to avoid ambiguity of interpretation. As we have explained in method that Bouguer anomalies are measurements on an irregular grid with variety of elevations. Because of that, we have continued our data processing. Bouguer anomaly has been projected onto a horizontal reference plane with Dampney method. The result of these method is Bouguer anomaly in a horizontal plane at 3000 meters high and equivalent depth of mass point 7000 meters. Bouguer anomalies in a horizontal plane is mixing of regional and residual anomaly. Because of they are mixing, we need to separated them. Regional and residual anomaly have been separated with upward continuation method with 8000 meters high. The result of Upward Continuation is regional and residual anomaly. To delineation of the Sumatra Fault in the study area, we have been used Second Vertical Derivative (SVD) Method to residual anomaly. The SVD Map on residual anomaly in the study area is shown in Figure 3.

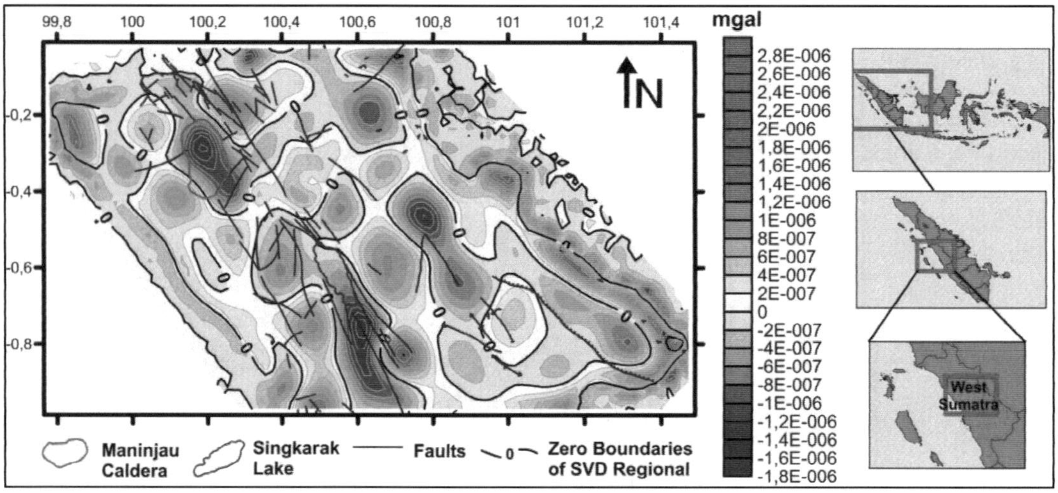

Figure 3. The SVD Map of residual anomaly in the central part of West Sumatra. Low anomalies are concentrated in the surrounding area of Maninjau and Singkarak. SVD Map forms elongated zero contours which correspond to the faults.

Low anomaly in the surrounding area of Maninjau Caldera in Figure 3 is assumed as a deficit massa of the large eruption. Maninjau is a large collapsed caldera that was resulted from a large eruption of silicic pyroclastic material (220 – 250 km^3) [18]. Low anomaly in the sorrounding area of Singkarak Lake is assumed by subsidence of the lake. At present, all of the faults in the Singkarak Lake are active, therefore subsidence of the lake still continues [19]. The lowest anomaly in the southwestrn part of the lake is correlated with a structural graben. Sumani Segment runs northwestward from the volcanic terrane of Lake Diatas to the southwestern flank of Lake Singkarak which occupies a structural graben [2].

The resultant Second Vertical Derivative (SVD) gravity map forms elongated zero contours which correspond to the edges of local geologically anomalous density distributions features [20]. The zero miligal per meter square quantity coincides remarkably well with most lithological boundaries when compared with the major geologic contacts [20]. In this study area, the zero miligal per meter square quantity has no correlated with litology boundaries [21]. The SVD map forms elongated zero countours which correspond to the faults which is a part of the Sumatra Fault. The zero miligal per meter square quantity coincides remarkably well with fault segmentation. When compared with Map of geometrically defined segments of the Sumatra Fault System [2], two negative anomalies are located around the Sianok and Sumani Segment.

4. Conclusion

We had been spesifically discussed the gravity field in the central part of West Sumatra and its correlation with the Sumatra Fault. To enhance gravity anomaly, we have been used terrain correction, upward continuation and SVD method. From gravity method, we can prove segmentation of the Sumatra Fault in the central part of West Sumatra. Deliniation of the Sumatra Fault in the central part of the West Sumatra with gravity method shows that two negative anomalies are located around the Sianok and Sumani Segment.

References

[1] Tjia D H Active Faults in Indonesia 1978 *Bulletin Geol. Soc. Malaysia* **10** pp 73-92
[2] Sieh K and Natawidjaja D Neotectonics of the Sumatran Fault, Indonesia 2000 *J. Geophys. Research* **105** pp 28 295–28 326
[3] Telford W M, Geldart L P and Sheriff R E 1990 *Applied Geophysics Second Edition* (Australia: Cambridge University Press) Chapter 2 pp 6–60
[4] Kastowo and Leo W G 1996 *Geological Map of The Padang Quadrangle, Sumatera* (Bandung: Geological Research and Development Centre)
[5] Silitonga H P and Kastowo 2007 *Geological Map of The Solok Quadrangle, Sumatera* (Bandung: Geological Research and Development Centre)
[6] Buyung N., Subagio and Walker D S A 1992 *Bouguer Anomaly Map of The Padang Quadrangle, Sumatera* (Bandung: Geological Research and Development Centre)
[7] Buyung N, Mirnanda E and Walker D S A 1992 *Bouguer Anomaly Map of The Solok Quadrangle, Sumatera* (Bandung: Geological Research and Development Centre)
[8] Hinze J W, Frese V B R R and Saad H A 2008 *Gravity and Magnetic Exploration, Principles, Practices and Applications* (Australia: Cambride University Press) p 136
[9] Hammer S Terrain Corrections for Gravimeter Stations 1939 *Geophysics* **4** p 184–194
[10] Karl J H The Bouguer Correction for The Spherical Earth 1971 *Geophysics* **36** p 761–762
[11] LaFehr R T Standardization of Gravity Reduction 1991a *Geophysics* **56** p 1170–1178
[12] LaFehr R T An Exact Solution for The Gravity Curvature (Bullard B) 1991b *Geophysics* **56** p 1179–1184
[13] Kane M F A Comprehensive System of Terrain Corrections Using a Digital Computer 1962 *Geophysics* **27** p 455–462
[14] Nagy D The Gravitational Attraction of a Right Rectangular Prism 1966 *Geophysics* **31** p 362–371
[15] Dampney C N G The Equivalent Source Technique 1969 *J. Geophysics* **34** pp 39–53
[16] Grant F S and West G F 1965 *Interpretation Theory in Applied Geophysics* (New York: McGraw-Hill Book Company) p 217
[17] Blakely R J 1996 *Potential Theory in Gravity and Magnetic Applications* (Australia: Cambridge University Press) chapter 12 pp 312–325
[18] Pribadi A, Mulyadi E, Pratomo I Mekanisme Erupsi Ignimbrit Kaldera Maninjau, Sumatera Barat 2007 *J. Geologi Indonesia* **2** pp 31–41
[19] Sidarto Pembentukan dan Perkembangan Danau Singkarak Sumatra Barat 2011 *JSDG* **21** 293–305
[20] Aku M O Application of Second Vertical Derivative Analytical Method to Bouguer Data For the Purpose of Delineation of Lithological Boundaries 2014 *J. Science World* **9** 27–32
[21] Saragih R D Identifikasi Sesar Sumatra Di Sumatra Barat Bagian Tengah Berdasarkan Analisis Data Gravitasi 2016 *Thesis* (Yogyakarta : Universitas Gadjah Mada)

Acknowledgments

Authors would like to thank Dr.rer.nat Sintia Windhi Niasari, Dr. Eddy Hartantyo, Dr.Ing. Ari Setiawan and Mauludin Kurniawan, M.Sc., for their kind helps and supports during data processing and writing this paper.

The Study of Fault Lineament Pattern of the Lamongan Volcanic Field Using Gravity Data

K N Aziz[1], E Hartantyo and S W Niasari

[1]Geophysics Laboratory, Physics Departement, Universitas Gadjah Mada, Sekip Utara, Yogyakarta, Indonesia 55281

[1]Corresponding author email: khafidh.KNA@gmail.com

Abstract. Lamongan Volcano located in Tiris,East Java,possesses geothermal potential energy. The geothermal potential was indicated by the presence of geothermal manifestations such as hot springs. We usedsecondary gravity data from GGMplus. The result of gravity anomaly map shows that there is the lowest gravity anomaly in the center of the study area coinciding with the hot spring location.Gravity data were analyzed using SVD method to identify fault structures. It controls the geothermal fluid pathways. The result of this research shows thatthe type of fault in hot springsisanormal fault with direction NW-SE. The fault lineament pattern along maaris NW-SE.Maar indicates anormal fault.As the result we know that gravity data from GGMplus which analyzed with SVD can be used to determine the type and trend of fault.

1. Introduction

Indonesia is located between three tectonic plates, i.e. the Eurasian, the Indo-Australian, and the Pacific Plates [1]. These plates formed convergent plate margin namely subduction zone. Consequently, Indonesia has many active volcanoes [2]. Active volcanism in convergent margin becomes a potential target for geothermal explorations. These geothermal potentials are indicated by manifestations such as hot springs [3].

The Lamongan Volcano is one of the 76 active volcanoes in Indonesia [4]. The Lamongan Volcano, which is stratovolcano type, is located in 7.983°S and 113.342°E [5]. It is also situated in the Sunda arc and in between three volcanic complexes, that is Bromo, Semeru, and Argopuro volcano, as shown in Figure 1 [6]. Moreover, the Lamongan Volcano has 61 basaltic cinder or spatter cones and 29 maars[4, 7].Tiris, a village located on the eastern of Lamongan Volcano, has a geothermal potential characterized by hot springs, with temperatures between 35°C and 45°C. There are also zeolite veins on rocks, indicating a boiling zone along the hydrothermal outflow zone [3].

Gravity methodwas conducted to identify subsurface structures in Lamongan Volcano. The basic principle of the gravity method was to measure the difference value of the gravitational field due to variations in rocks density [8]. In this research, we usedgravity method to identify faultstructures that controlling geothermal fluid pathways using Second Vertical Derivative (SVD) [9].We usedsecondarygravity data from Global Gravity Model Plus (GGMPlus).

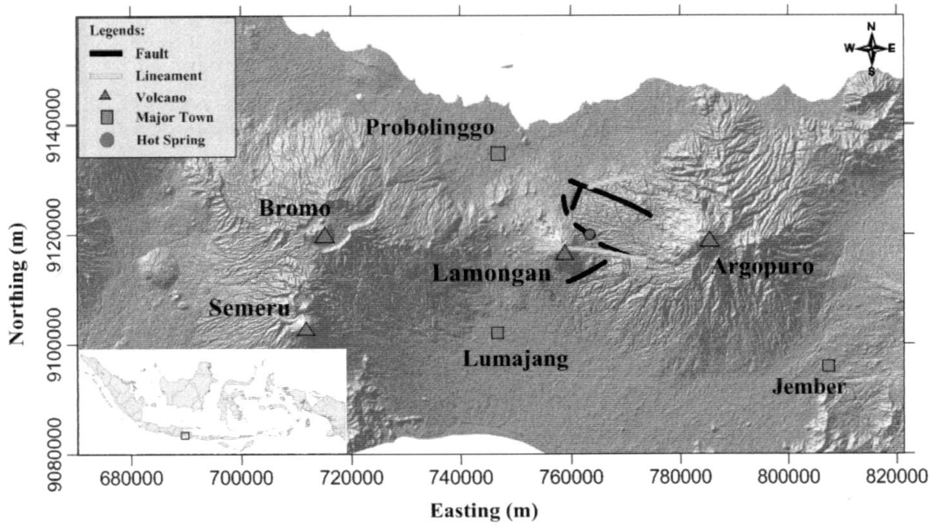

Figure 1. East Java map shows thelocation of theresearch area. There are 3 major towns around, i.e. Probolinggo, Lumajang, and Jember. Four volcanoes i.e. Bromo, Semeru, Argopuro, and Lamongan showed by ared triangle. The fault on hot springs shown by the black line [10, 11]

2. Materials and Methods

In this research, gravity data was obtained from GGMPlus in radial derivative of disturbing potential form.Gravity data satellite was downloaded fromhttp://ddfe.curtin.edu.au/gravitymodels/GGMplus. GGMPlus provides gravity data covering all continents and coastal zones with a total of 3 billion points data. Gravity data was available in grid form with space between points ~200 m [12]. The coordinates of this research in the 49S UTM zone elongated from 681731 m to 806625 m and 9085476 m to 9137909 m.

The data obtained from GGMPlus was available in the form of 5°x5° areas, so we have to select data according to the research area [12]. The utilized GGMPlus data was the gravity anomaly.Gravity anomaly was subsequentlyseparated into regional anomaly using upward continuation and subtracted to produce theresidual anomaly. The residual anomaly was analyzed with SVD to generate SVD anomaly.

Structural analysis was conducted by slicing SVD anomaly. SVD was used to delineate the contact of lithology with contrast density and enhancing local anomaly and to identify the types of fault and the estimated dipping fault [13, 14]. The slice was made by cross-section of the fault boundary. The fault boundary was classified by a zero value of SVD anomaly [15]. The slice on the contour anomaly of SVD resultedin a curve which then analyzed to determine the type of fault in the area. There were criteria used to determine the type of faults, as follows [14, 16]

$$\left|\frac{\partial^2 \Delta U}{\partial z^2}\right|_{\max} > \left|\frac{\partial^2 \Delta U}{\partial z^2}\right|_{\min} \quad (1)$$

For normal faults, and

$$\left|\frac{\partial^2 \Delta U}{\partial z^2}\right|_{\max} < \left|\frac{\partial^2 \Delta U}{\partial z^2}\right|_{\min} \quad (2)$$

For thrust faults

3. Results and Discussion

Gravity anomaly response was obtained from GGMplus gravity data. The gravity anomaly ranges from -20 mGal to 380 mGal(shown in Figure 2).

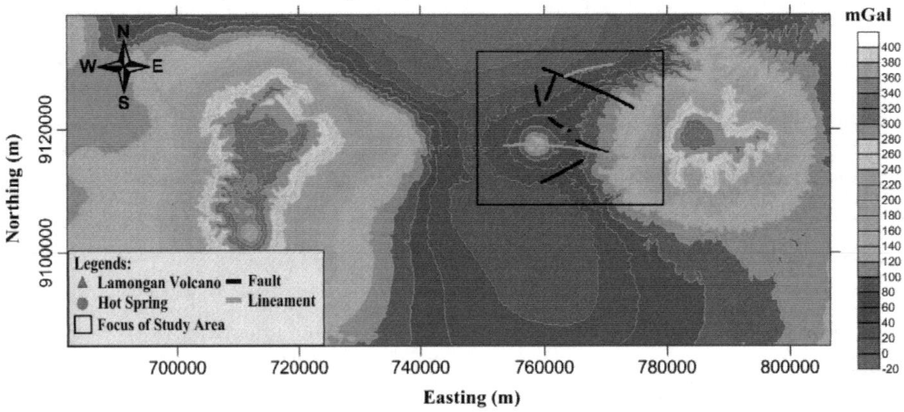

Figure 2. Contour map of gravity anomaly in the study area was obtained from GGMplus and overlaid with fault, lineament, and hot springs. The focus of our study shown by the black rectangular and will be explained further in Figure 3.

Figure 2shows the response of gravity anomaly in the research area affected by variations of rock density. The contour of gravity anomaly shows thathighest anomalyassociatedbythe Bromo, the Semeru, and Argopuro volcano, whilelowest anomalyconformed to the length of thenorthern-southern in center of the study area. The results show that there is the lowest gravity anomaly in the center of the study area coinciding with the hot spring location.

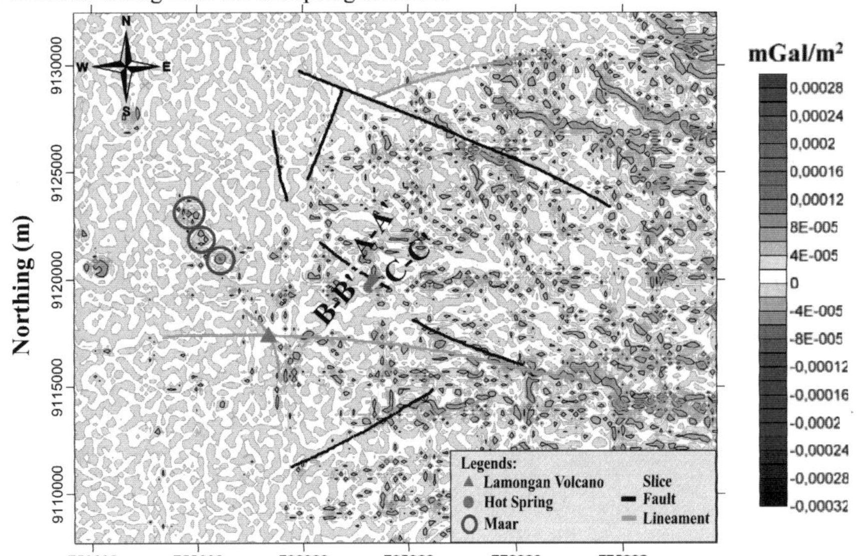

Figure 3. Contour map of SVD anomaly surroundsLamongan Volcanic Field. The line A-A', B-B', C-C', and D-D' show position ofslice. The east of Lamongan Volcano is dominated by highest and lowest anomaly.The black line and green line show the position of fault and lineament based on geological data.

Contour map of SVD anomaly (Figure 3) shows anomaly value between -0.00032 mGal/m^2 to 0.00028 mGal/m^2. East side was dominated by highest and lowest anomaly which corresponds with Argopuro Volcano.In the north-west of Lamongan, there is alow anomaly that located in the center area and thehigh anomaly was locatedaround the center. This result corresponds with maar (blue circle) [4, 7]. The trend of fault lineament pattern directed in NW-SE. This direction indicatesa weak zone which was assumed as thefluid pathways from the body of volcano to the surface. This weak zone can be defined as a maar.The fault lineament was known from the pattern of maar distribution. In the north-east of Lamongan Volcano, zero value of SVD anomaly indicates the contact boundary of thefault plane. It is shown in Figure 3 that the trend of fault lineament pattern was directedNW-SE. Thisdirectioncorresponds tothe geological data which also displayed NW-SE.In addition, the fault in anarea that located in north-east of Lamongan Volcano coinciding with hot springs location.

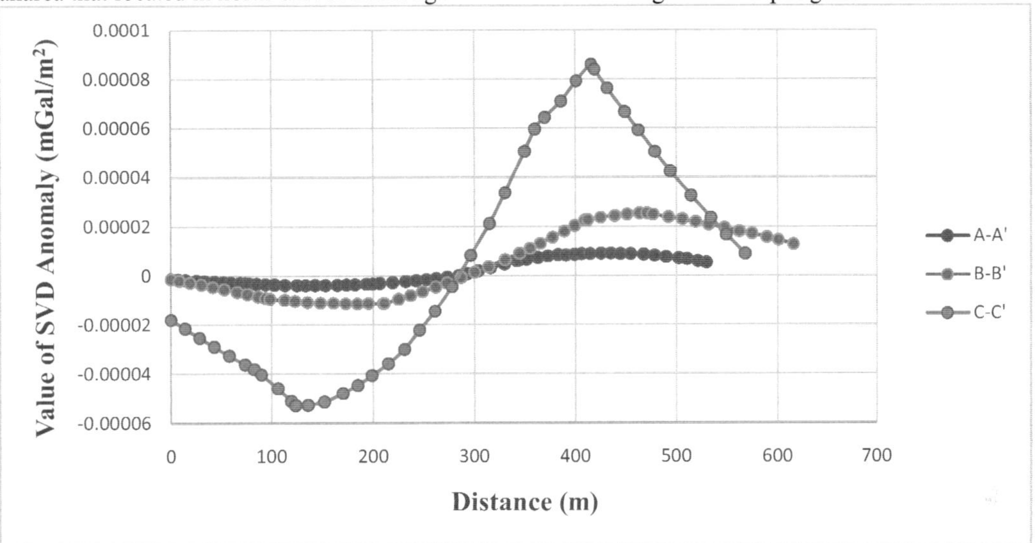

Figure 4. Curveprofile of slice resultof SVD anomaly contour map. a) Slice A-A', b) Slice B-B' and c) Slice C-C'

Overall from this curve, the value of SVD anomaly is ranged between -0.000053mGal/m^2 to 0.000084mGal/m^2.The value of SVD anomaly of slice A-A', B-B' andC-C' has formed the curve with the maximum amplitude which is higher than minimum amplitude. Based on the value of SVD anomaly and equation 1 and 2, the result can be classified as normal fault. This normal fault has a coinciding with hot springs location.

4. Conclusion

From SVD anomaly contour map and curve profile of slicing result, we can conclude that there are two faults. The type offaultwhich corresponds with hot springsis anormal fault with trend direction NW-SE. Moreover, the trend of fault lineament patternalongmaar can beindicated by normal fault. This lineament trends in NW-SE.

5. Acknowledgements

We would like to thank to "Kementerian Riset, Teknologi, dan Pendidikan Tinggi" for the PUPT funding no. 2452/UN1.P.III/DIT-LIT/LT/2017. We are grateful to Western Australian Geodesy Group, Curtin University for data providing the gravity data used in this study. Wewould like to thank the anonymous reviewers for important suggestions and comments.

References

[1] Putra R R, Kiyono J, Ono Y and Parajuli H R 2012 Seismic hazard analysis for Indonesia *Journal of Natural Disaster Science* **33** 59-70

[2] Deon F, Förster H-J, Brehme M, Wiegand B, Moeck I, Scheytt T, Jaya M S, Putriatni D J and Supoyo 2015 Greenfield exploration of hidden magmatically driven geothermal systems in active subduction zones: case study Lamongan (Eastern Java, Indonesia) *Proc. on World Geothermal Congress*19-25 April 2015 Melbourne

[3] Deon F, Förster H-J, Brehme M, Wiegand B, Scheytt T, Moeck I, Jaya M S and Putriatni D J 2015 Geochemical/hydrochemical evaluation of the geothermal potential of the Lamongan volcanic field (Eastern Java, Indonesia) *Geothermal Energy***3**

[4] Carn S A 2000 The Lamongan volcanic field, East Java, Indonesia: physical volcanology historic activity and hazard *Journal of Volcanology and Geothermal Research***95** 81-108

[5] Global Volcanism Program 2013 Lamongan (263320) in volcanoes of the world v.4.6.0Venzke E (ed) Smithsonian Institution Available at http://volcano.si.edu/volcano.cfm?vn=263320 (accessed 17.05.17)

[6] Deon F, Moeck I, Sheytt T and Jaya M S 2012 Preliminary assessment of the geothermal system of the Tiris Area, East Java, Indonesia *74th EAGE Conf. and Exhibition incorporating EUROPEC 2012*

[7] Carn S A and Pyle D M 2001 Petrology and geochemistry of the Lamongan Volcanic Field, East Java, Indonesia: primitivesunda arc magmas in an extensional tectonic setting? *Journal of Petrology***42** 1643-83

[8] Dentith M and Mudge S T 2014 *Geophysics for The Mineral Exploration Geoscientist* (Cambridge: Cambridge University Press) chapter 3 p 85

[9] Muin F, Retnowati E, Kadir W G A and Suryanto W 2013 Gravity and passive seismic application in exploration and development Bunyu field Indonesia *Tyumen 2013-New Geotechnology for the Old Oil Provinces*

[10] Suharsono and Suwarti T 1992 *Geological Map of the Probolinggo Quadrangle, Jawa*(Bandung: Geological Research and Development Centre)

[11] U.S. Department of the Interior U.S. Geological Survey 2017 EarthExploler at https://earthexplorer.usgs.gov (accessed 17.05.17)

[12] Hirt C, Claessens S, Fecher T, Kuhn M, Pail R and Rexer M 2013 New ultra-high resolution picture of earth's gravity field *Geophysical Research Letters* **40**4279-83

[13] Gupta V K and Ramani N 1982 Optimum second vertical derivatives in geological mapping and mineral exploration *Geophysics***47** 1706-15

[14] Sastranegara T, Nainggolan S S and Raharjo I B 2015 The application of a triangular mesh for gravity inversion to reconstruct subsurface geological structures in the Hululais geothermal prospect, Bengkulu *Proc. onWorld Geothermal Congress* 19-25 April 2015 Melbourne

[15] Saragih R D 2016 *Identification of Sumatra Fault in the Central Part of West Sumatra Based on Gravity Data Analysis* thesis (Yogyakarta: Universitas Gadjah Mada)

[16] Sari E P and Subakti H 2015 Identification of Baribis faults – West Java using second vertical derivative method of gravity*4th Int. Symp. on Earthquake and Disaster Mitigation 2014 (ISEDM 2014)*ed A Saepuloh, I Meilano and Z Zulhan (Bandung: AIP *Conf. Proc.*) 1658 pp 030016-1 – 030016-9

Verification Modal Summation Technique for Synthetic and Observation Seismogram for Pidie Jaya Earthquake M6.5

Irwandi Irwandi[1*], Fashbir[1], Daryono[2]

[1]Physics Department, Syiah Kuala University, Jl. Syech Abdurrauf, Banda Aceh, 23111, Indonesia
[2]Indonesian Agency for Meteorology, Climatology and Geophysics, Jakarta, Indonesia
*email: irwandi@unsyiah.ac.id

Abstract. Neo-Deterministic Seismic Hazard Assessment (NDSHA) method is a seismic hazard assessment method that has an advantage on realistic physical simulation of the source, propagation, and geological-geophysical structure. This simulation is capable on generating the synthetics seismograms at the sites that being observed. At the regional NDSHA scale, calculation of the strong ground motion is based on 1D modal summation technique because it is more efficient in computation. In this article, we verify the result of synthetic seismogram calculations with the result of field observations when Pidie Jaya earthquake on 7 December 2016 occurred with the moment magnitude of M6.5. Those data were recorded by broadband seismometers installed by BMKG (Indonesian Agency for Meteorology, Climatology and Geophysics). The result of the synthetic seismogram calculations verifies that some stations well show the suitability with observation while some other stations show the discrepancies with observation results. Based on the results of the observation of some stations, evidently 1D modal summation technique method has been well verified for thin sediment region (near the pre-tertiary basement), but less suitable for thick sediment region. The reason is that the 1D modal summation technique excludes the amplification effect of seismic wave occurring within thick sediment region. So, another approach is needed, e.g., 2D finite difference hybrid method, which is a part of local scale NDSHA method.
Keyword: Synthetic Seismogram, Pidie Jaya Earthquake

1. Introduction

Regarding the problem of ground shaking estimation lies in generic attenuation relations for PSHA procedure, the alternative NDSHA (Neo-Deterministic Seismic Hazard Assessment) procedure has advantages in calculating the strong ground motion from realistic model of synthetic seismograms respect to the source of specific properties. NDSHA may immediately and effectively accommodate any reliable information from those approaches in order to adequately compute the ground shaking maps[1][2][3]. Simplify NDSHA procedure can be divided into four steps as shown in Figure 1. The first step was identification of the fault or geographical areas that produce earthquakes and earthquake characteristics. The second step was determination of earthquake source definition for each cellular

IOP Conf. Series: Journal of Physics: Conf. Series **1011** (2018) 012026 doi:10.1088/1742-6596/1011/1/012026

area in seismogenic zones. The third step was computation of synthetic seismograms' base of sources definition and structural characteristic sites. Finally, the fourth step was picking the maximum value from the synthetic seismograms as important parameters to be found which represent the strong ground motion, such as PGA (Peak Ground Acceleration) and PGV (Peak Ground Velocity).

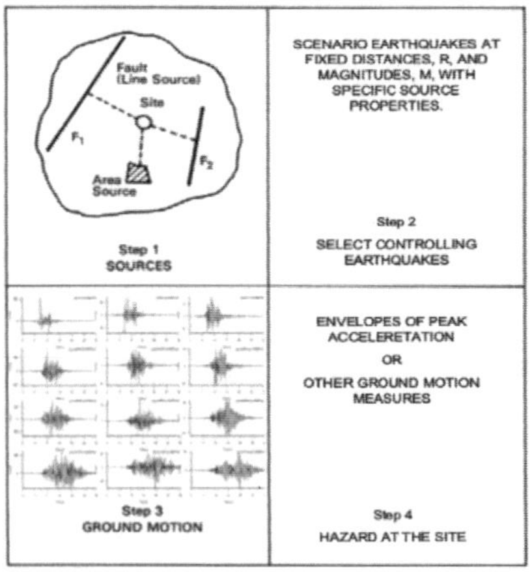

Figure 1. The procedures of seismic hazard assessment in general can be divided into four steps. NDSHA procedure is starting from
(1) source identification,
(2) source definition,
(3) compute the synthetic seismograph,
(4) and pick up the importance strong ground motion parameters for mapping.
The parameters PGA and PGV are sensitively depend on the computation method of synthetic seismograph because it depend on the algorithm and sites of the structural model data.

2. Pidie Jaya (Pijay) Earthquake M6.5

In the early morning at 5.03 local time (UTC+7) there was a strong earthquake happened in Pidie Jaya area that shook Pidie Jaya, Pidie, Bireun. The earthquake happen in an area that was not expected to be an active fault previously and earthquake rarely happen there, so that the buildings in that area have a high vulnerability. The earthquake was felt to Banda Aceh, Medan and several other cities in the north of Sumatra Island. As reported by the BNPB, the dead victims were 104 people and 19,130 houses were reportedly damaged where 17,673 of them are in Pidie Jaya. A total of 387 people have been staying at the hospitals in Pidie, Pidie Jaya and Bireun for medical treatment due to injuries[4].

The great fault of Sumatra has been well known and it has produced many destructive earthquakes[5]. But the Pijay earthquake was a surprise one because it was far from the Sumatra fault zone[6] and there was an effort to make an additional fault line (black line in Figure 2) connected to the Sumatran fault [7].

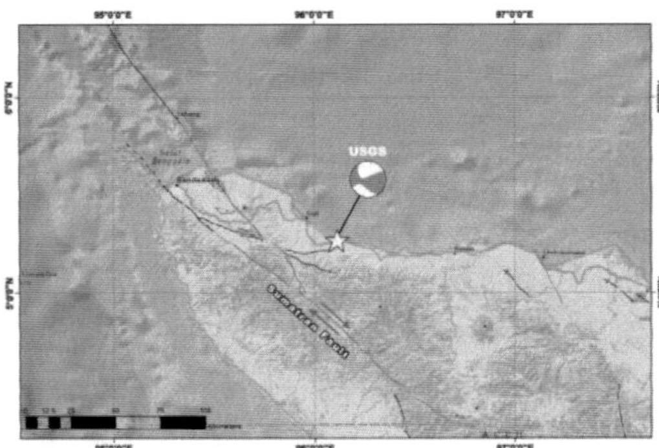

Figure 2. Active fault in Aceh [7] represented by the red lines

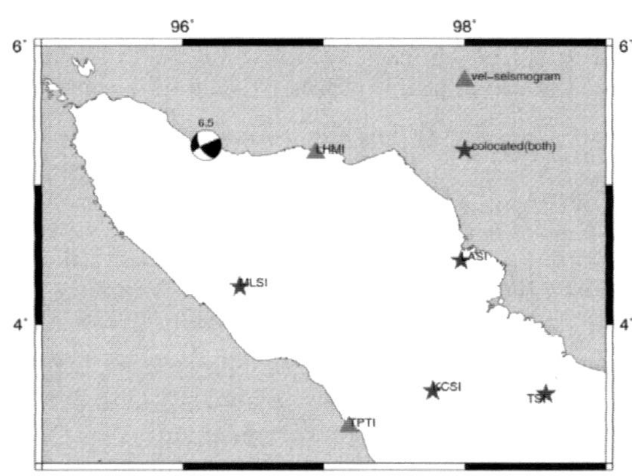

Figure 3. Epicenter location, focal mechanisms, and available data for seismograph and accelerograph when the Pidie Jaya earthquake occurs.

3. Method

The computational algorithm is based on the modal summation technique for the generation of Rayleigh wave synthetic seismograms (radial and vertical component of motion). Equation 1 describes the asymptotic expression of the Fourier transform of the displacement for the radial (u_x) and the vertical (u_z) components of motion at a distance r from the source[8]. The radial component is described by

$$u_r^R(r,z,\omega) = \sum_{m=1}^{\infty} \frac{e^{-i\frac{3}{4}\pi}}{\sqrt{8\pi\omega}} \frac{e^{-ik_m^R r}}{\sqrt{r}} \frac{\chi_m^R(h_s,\omega)}{\sqrt{c_m^R v_m^R I_m^R}} \frac{F_r^R(z,\omega)}{\sqrt{v_m^R I_m^R}} \quad (1)$$

The beauty of the modal summation technique computer codes for layered, inelastic structural models lies in the separation of the calculation of the spectral quantities related with the medium properties (phase velocity *c*, group velocity *v*, energy integral *I* and complex wave number *k*) from those that describe the source and its position w.r.t. the sites of interest. Detail information and detail governing mathematical formula for modal summation technique were explained by Panza[9].

Parameters involved for double-couple calculations are origin time, hypocenter magnitude and focal mechanics information. We used Pidie Jaya earthquake data obtained from USGS[10]. The earthquake epicenter was located at 5.283°N, 96.168°E with a focal depth of 13.5 km, which could be categorized as a shallow earthquake. The seismic mechanism was the right lateral strike slip with magnitude = 6.5 Mww, strike = 243°, dip = 81°, rake = 33°.

The synthetic seismograph from modal summation technique will be verified with the observation seismographs for Pidie Jaya earthquake M6.5 on 7 December 2016 from BMKG seismometer stations. BMKG has used the seismometers with type Trillium-120 (Nanometrics), BBVS-120 (Geodevices), and STS2. It is important to know the true instrumental response for applying the instrument correction and filtering the data at the adopted threshold of frequency used in the seismograms computation, before comparing the computed seismograms with the real recorded seismic data. The data of Pidie Jaya earthquake have been recorded and also available, as shown in Figure 3.

4. Comparison Synthetic and Observation Seismogram

Synthetic and field observations data are gotten from different sources. Synthetic data are gotten from simulation while observation data are gotten from field observation from BMKG station recording ground shaking when the earthquake occurs. Figure 3 shows the available data from six stations, whereas four stations have seismometer and accelerometer while another two stations have seismometer only. Due to the limitations in this writing, we only discuss several seismograms from two stations KCSI and LASI, as shown in Figure 4.

Figure 4a, 4c, and 4e show the seismograms recorded by the seismometer that measure velocity for the KCSI station. The waveform package between syntheses of simulation results 1 Hz (red) and 10 Hz (green) have similarities. Verification with observational data is compatible with its wave envelope especially for the horizontal components EW (east-west) and NS (north-south). The conformity of this maximum amplitude is very useful for validating the PGV (peak ground velocity).
Waveform data for the LASI station (Figure 4b, 4d, and 4f) show that the compatibility between synthetic data with observations waveform is significantly different, relative to KCSI station. The waveform envelope of observation is larger than the waveform from synthetic seismogram and followed by the significant coda wave effect. This significant difference can happen because the LASI station is located in thicker sedimentary layer compared with the KCSI station which is located in a thinner sedimentary layer or bedrock area. The seismograms waveforms are more dispersed than those of synthetic seismograms because the synthetic seismograms are computed on a 1D rock structure without considering the scatter, site effect, and another factors.

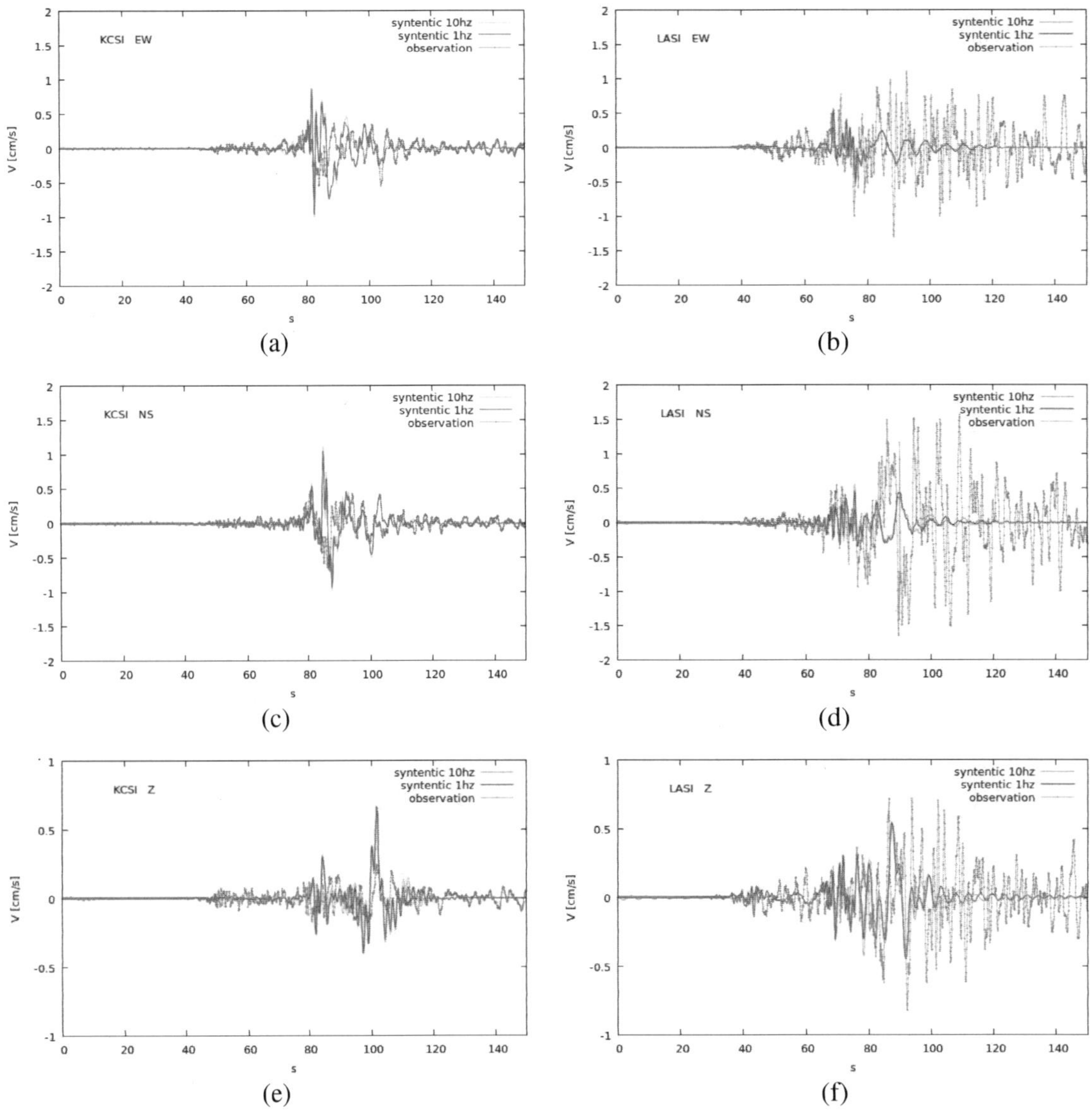

Figure 4. The recorded seismograms are plotted with a gray line while the synthetic seismograms computed at 10Hz and 1Hz cutoff frequency are plotted with green and red lines, respectively. The Left side are taken from the station KCSI (Kutacane) 3.522°N 97.772°E, with distance of 263 km, while the right side are taken from the station Langsa Aceh (LASI) 4.457°N 97.970°E, with epicenter distance of 219.78 km.

5. Conclusions and Suggestion

Modal summation technique has successfully generated synthetic seismogram and has adequate verification for waveform envelope with observational data from BMKG seismometer which is located on bedrock area. Considering phase wave, there is a slight difference between synthetic and observation. However, the verification of waveform envelope is sufficiently precise for NDSHA purpose which we concern on PGV. For regions which have significant thick sedimentary layer, the 1D model of summation technique must be consolidated with 2D finite different model so that the amplification effects can be taken into account. Research on the synthetic seismogram needs to be compared with other methods and developed in order to increase the precision of NDSHA map.

6. References

[1] Panza, G. F., Irikura, K.,Kouteva, M., Peresan, A., Wang, Z., &Saragoni, R. (2011). Introduction to the special issue on advanced seismic hazard assessments. Pure and Applied Geophysics, Vol. 168, 1–2. doi:10.1007/s00024-010-0179-9.

[2] Hassan, Hany M., Giuliano F. Panza, Fabio Romanelli, Mohamed N. ElGabry, Andrea Magrin (2017). Insight on seismic hazard studies for Egypt. Eng Geology 220 (2017) 99–109.

[3] Irwandi (2017). Advantages of Realistic Model Based on Computational Method: NDSHA versus Standard PSHA. IOP Conf. Series: Earth and Environmental Science 56 (2017) doi:10.1088/1755-1315/56/1/012007.

[4] AHA Center, Bachtiar Andy Musaffa. Form 2-Situation Update No.9 Aceh Pidie Jaya Earthquake. Monday, 19 December 2016.

[5] Tabei, T., Kimata, F., Ito, T., Gunawan, E., Tsutsumi, H., Ohta, Y., Yamashina, T., Soeda, Y., Ismail, N., I. Nurdin and Sugiyanto, D. (2015). Geodetic and Geomorphic Evaluations of Earthquake Generation Potential of the Northern Sumatran Fault, Indonesia.

[6] Irwandi, Marwan, Muksin, and Fashbir. Applications of the VLF-EM method for rapid Sumatran fault identification in Leuser national park, Aceh. AIP Conference Proceedings 1861, 030050 (2017); doi: 10.1063/1.4990937

[7] Bradley, Kyle Edward, Wang Yu, Shireen Federico (2016). Mw 6.5 Earthquake Strikes Northeast Aceh, Sumatra, Indonesia. Earth Observatory Singapure. http://www.earthobservatory.sg/news/mw-65-earthquake-strikes-northeast-aceh-sumatra-indonesia.

[8] Vaccari, F. (2016). A web application prototype for the multiscalemodelling of seismic input. In Earthquakes and Their Impact on Society (pp. 563-584). Springer International Publishing.

[9] Panza, G. F., F. Romanelli, F. Vaccari, (2001), Seismic wave propagation in laterally heterogeneusanelastic media: theory and applications to seismic zonation, AdvGeophys 43, 1–95.

[10] USGS (United Stated Geological Survey) (2016). Earthquake Hazard program. M 6.5-14km WNW of Reuleuet, Indonesia. https://earthquake.usgs.gov/earthquakes/eventpage/us10007ghm#moment-tensor.

Acknowledgments

I wish to acknowledge assistance from Department of Geoscience, Trieste University, and International Centre for Theoretical Physics (ICTP) and BMKG. I wish to encourage members of Physics Department, Geophysics Engineering Department, and TDMRC.

IOP Conf. Series: Journal of Physics: Conf. Series **1011** (2018) 012027 doi:10.1088/1742-6596/1011/1/012027

Subsurface Structure Mapping Using Geophysical Data in Candi Umbul-Telomoyo, Magelang, Central Java, Indonesia

A P Affanti*, E Prastyani, P D Maghfira, S W Niasari

Geophysics sub-department, Universitas Gadjah Mada

*adellaaffanti@gmail.com

Abstract. Candi Umbul warm spring is one of the manifestations in the Telomoyo geothermal prospect area. A geophysical survey had been conducted using VLF (Very Low Frequency) EM, VLF R and magnetic methods in the Candi Umbul-Telomoyo. VLF EM, VLF R and magnetic data were aimed to image the conductivity and magnetic anomalies distribution of the subsurface beneath the Candi Umbul-Telomoyo. VLF EM data had been mapped with Karous-Hjelt filter and analysed by tipper analysis, VLF R data had been modelled using 2layinv and analysed using impedance analysis. On the other hand, magnetic data processing was done with upward continuation. The Karous-Hjelt filter and 2layinv models show the highest conductivity distribution that located at 4800-5000 m were correlated with tipper and impedance analyses. In addition, the high-low magnetic contrast from the quantitative magnetic data interpretation indicates a fault (which could be a fluid pathway) which is closed to the Candi Umbul warm spring manifestation.

Introduction

Indonesia is a country with high geothermal prospect potential [1]. Telomoyo volcano, as one of geothermal prospect area locates in Magelang, Central Java, Indonesia. One of geothermal manifestation is the Candi Umbul warm spring, temperature of $(35 - 36)^0$ C [2]. Warm spring indicates that there are fluids beneath the earth. A conductivity distribution of VLF data and susceptibility distribution of magnetic data are able to map the fault.

Previous study found that upflow zone located around the Telomoyo caldera [3]. In the contrary, the outflow zone was interpreted around the Candi Dukuh, Candi Umbul and Pakis Dadu manifestations [4]. A previous magnetic model showed a normal fault with strike N 60^0 E/45^0 [5]. Stratigraphy of research area consists of Kaligetas Formation, Gunungapi Gilipetung rock, Gunungapi Andong and Kendil rock, Gunungapi Telomoyo rock, Gunungapi Merbabu rock, Kerucut Gunungapi sedimentation and alluvium [6]. Four main structures in Candi Umbul-Telomoyo has direction N 100-110^0 E/N 280-290^0 E, N 60-70^0 E/N 240-250^0 E, N 20-30^0 E/N 200-210^0 E, and N 150-160^0 E/N 330-340^0 E, where the major direction of the structures are northwest-southeast and north-south [2]. Whereas, VLF and magnetic data were aimed to identify the major structure which controls Candi Umbul warm spring.

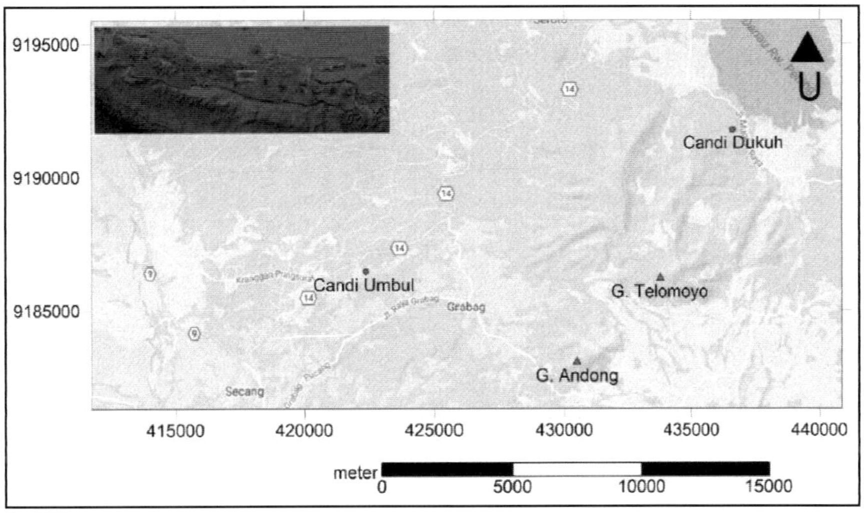

Figure 1. The research area was located in Candi Umbul-Telomoyo, Magelang Regency, Central Java, Indonesia. The Candi Umbul warm spring manifestation is one of the three warm springs manifestation of Telomoyo geothermal area. The Candi Umbul is located on the west side of Telomoyo volcano.

Method

Acquisition of VLF EM data was performed on March 18-19, 2017 while VLF R data acquisition conducted was on March 25-26, 2017. Magnetic data, acquisition was done on April 6-9, 2017. Based on geological map Magelang-Central Java [7], the direction of the main structure is north-south. For that reason, VLF's line had west-east direction using the transmitter from NWC (19800 Hz) and VTX3 (18200 Hz). VLF data consist of 54 points in a line of 10.3-kilometres length. Magnetic data acquisition consist of in 3 lines (105 points).

VLF EM data were processed using spline and moving average. Tilt data also filtered using Fraser equation [8]. In order to get the current density distribution of VLF EM, the data were also filtered using Karous-Hjelt [9]. The presence of conductor is indicated by low resistivity [10]. The relation between magnetic field represented by tipper (A, B) which used as VLF EM data analysis. Tipper A is parallel to x coordinate (east), while tipper B parallel to y coordinate (north) and transmitter direction [11]. The ratio of the horizontal magnetic field and the orthogonal magnetic field (called impedance) were used in VLF R data analysis [12]

On the other hand, magnetic data processing started with diurnal and IGRF correction. The result from the total anomaly magnetic field was reduced to pole and upward continuation 1500 m. Upward continuation is a process that transforms anomalies measured on a surface to those that have been measured on some higher surface [13].

Result and Discussion

Tipper analysis was done by calculating the ratio between the value of the vertical magnetic field and the horizontal magnetic field. The horizontal magnetic field is the primary magnetic field derived from the transmitter. While the vertical magnetic field is a secondary magnetic field resulting from induction of the primary magnetic field. The variation of the tipper value shows variations in lateral conductivity beneath the surface [14]. Based on tipper analysis result, the existence of conductive anomaly was indicated by the increasing tipper values at distances of 1200 m, 3200 m, 4200 m, 5600 m and 8250 m in local coordinate (from west direction of the profile line survey).

The first processing on VLF EM data was the interpolation of tilt data using a spline to make the same measurement spacing (100 metres). Karous-Hjelt filter gave the result of high current density distribution in red colour which associated with the conductive zone. The conductive zone found at distance of 4000 m to 5600 m of local coordinates. While the resistive zone (blue colour) appears at 5700 m-7000 m of local coordinates. In VLF EM data, a consistent conductive anomaly appears at 5600 metres distance (shown in Figure 2). The conductive anomaly is estimated to be a fault. While the fault at distance of 5600 metres is a relatively north-south fault that indicated as a river.

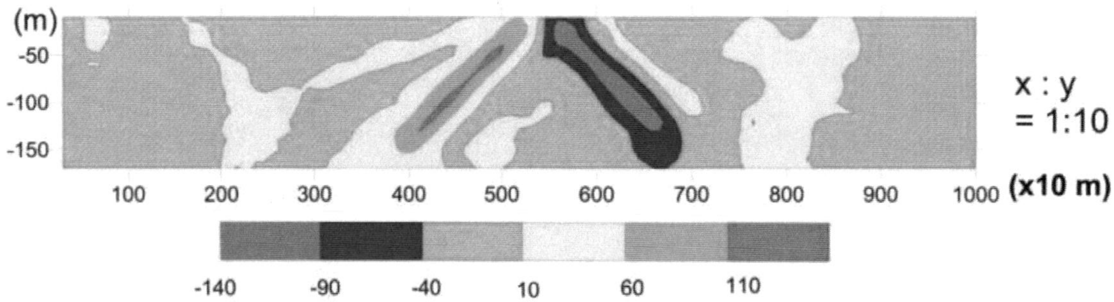

Figure 2. Current density distribution map. The conductive zone was characterized by a high current density (red colour). The conductive zone was at distance of 4000 m to 5600 m local coordinates. While the resistive zone showed at a distance of 5700 m-7000 m local coordinates.

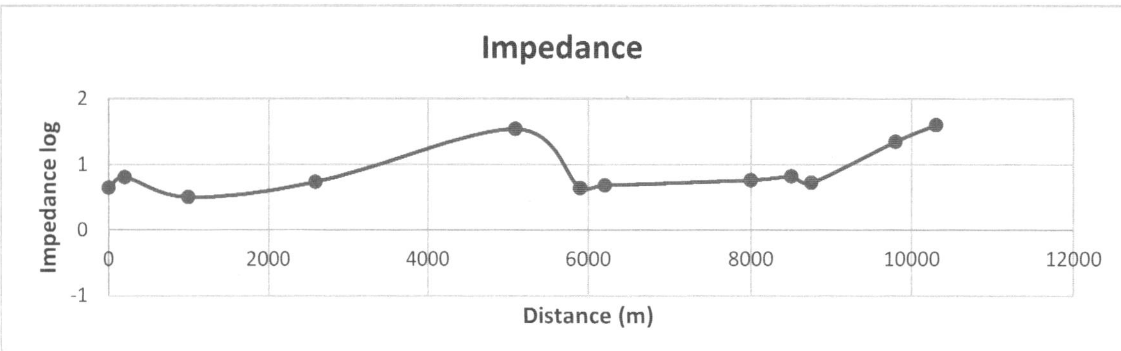

Figure 3. The result of the sorted impedance log data versus distance at frequency 18200 Hz. The approximate location of the conductive anomalies was at a distance of 1000 m, 5200 m and 8750 m.

For VLF R, the impedance log data were sorted based on the quality bar of electromagnetic signal that received and the transmitter signal direction. Since the apparent resistivity value is proportional to the impedance value, the existence of the conductor was also characterized by the small impedance value measured. Based on Figure 3, the approximate location of the conductive anomaly shown by the decrease of impedance which located at distances of 1000 m, 5200 m and 8750 m local coordinate.

The input for VLF R modelling were apparent resistivity, phase and distance. Weighting data was also used to get the fit model. Weighting parameter was quality bar and noise (electric noise, magnetic noise and magnetic electric noise). The higher quality bar and the absence of noise would give the higher data weight. Based on the result of 2layinv modelling, the conductive zone of the model was found at distances of 1000 m-2500 m that indicated by red colour The green colour in the east of survey area showed a resistive zone which correlated to andesite-basaltic rock in Telomoyo volcano.

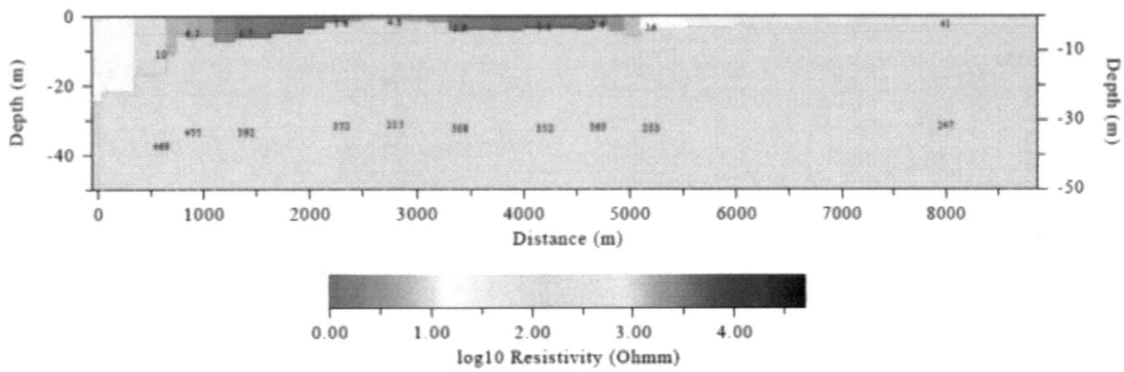

Figure 4. Resistivity model of Candi Umbul-Telomoyo volcano using the 2layinv software. The conductive zone is shown at distance of 1500 m-3000 m (red colour) while the resistive zone located in the east of survey area represented by green colour.

In magnetic processing, total magnetic field anomaly which got from diurnal and IGRF correction had the distribution value of -681 nT to 271 nT. The wide range of this anomaly caused by variation of lithology in the survey area. In order to change dipole effect become monopole, total magnetic field anomaly was reduced to pole. The reduce to pole map still consisted of local anomaly and regional anomaly which should be separated with upward continuation. Upward continuation made the anomalous feature clearer. Figure 5 has a high anomaly with red to pink colour while the low anomaly has the blue colour. The presence of high-low anomaly indicated the structure beneath the surface.

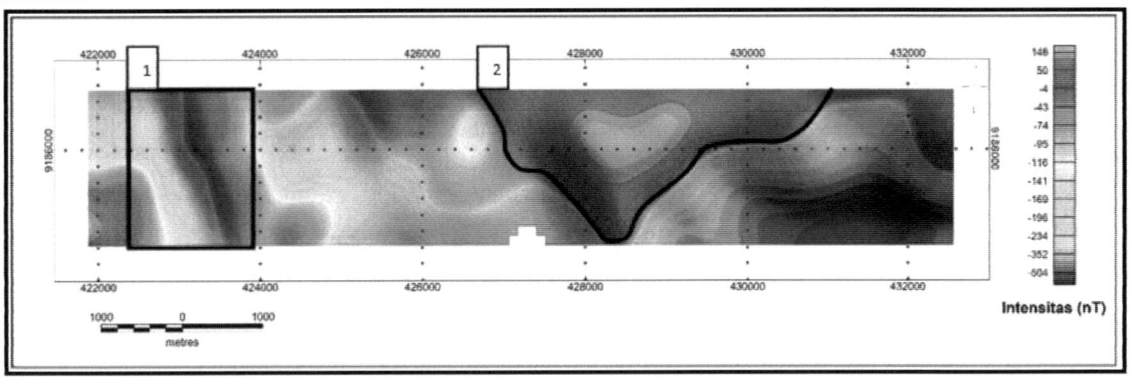

Figure 5. Map of magnetic field anomaly from 1500 m upward continuation of geothermal prospect area of Candi Umbul-Telomoyo volcano. The number 1 anomaly pattern was identified as the normal fault structure and the number 2 anomaly pattern was identified as the lithologic contact between the Gilipetung andesite rocks and the Merbabu volcano pyroclastic deposits.

Conclusion

A geophysical survey can be a brief method to reveal a structure beneath Candi Umbul-Telomoyo geothermal prospect area. A geophysical survey using VLF and magnetic method show the fault structures which control Candi Umbul warm spring located at distance of 1500 m and 5600 m. At distance of 1500 m, the fault has northwest-southeast direction while at distance of 5600 m, the fault has north-south direction.

Acknowledgement

Thank you for Chevron UPP for the VLF instrument. For geophysics Universitas Gadjah Mada laboratory instrument pool, thank you for instrument that had been used for the survey. We would like to thank to Faculty of Mathematic and Natural Science for the funding.

References

[1] Hochstein M P 2008 Indonesia: Geothermal Prospect and Developments-Preface *Geothermics* **37** 217-219

[2] Hermawan D and Rezky Y 2011 *Deliniasi Daerah Prospek Panas Bumi Berdasarkan Analisis Kelurusan Citra Landsat di Candi Umbul-Telomoyo, Provinsi Jawa Tengah* (Bandung: Pusat Sumber Daya Geologi)

[3] Hermawan D, Widodo S and Mulyadi E 2012 *Sistem Panasbumi Daerah Candi Umbul-Telomoyo Berdasarkan Kajian Geologi dan Geokimia* (Bandung: Pusat Sumber Daya Geologi)

[4] Praromadani Z S 2013 *Pemodelan Sistem Geothermal Daerah Telomoyo dengan Menggunakan Data Magnetotellurik* (Depok: Universitas Indonesia)

[5] Kharisa N A, Narendratama R, Wulandari I, Faisal M I, Kirana K, Zipora R, Arfiansyah I and Suyanto I 2015 *Analysis of Magnetic Anomaly Data for Identification Structure in Subsurface of Geothermal Manifestation at Candi Umbul Area Magelang Central Java Province Indonesia* (Jakarta: International Geothermal Convention and Exhibition)

[6] Prameswari M *2014 Distribusi Anomali Gas Udara Tanah CO2, Mercury (Hg) Tanah, dan Suhu Udara Tanah untuk Mengetahui Distribusi Zona Panas di Kompleks Gunung Telomoyo, Jawa Tengah* (Yogyakarta: Universitas Gadjah Mada)

[7] Thaden R E, Sumadirja H and Richards P W 1975 *Geologic Map of the Magelang and Semarang Quadrangles, Java* (Bandung: Geological Survey of Indonesia)

[8] Fraser DC 1969 Contouring of Vlf-Em *Geophysics* **6** 34 958-967

[9] Karous M, Hjelt S E 1983 Linear Filtering of VLF Dip-Angle Measurements *Geophysical Prospecting* **31** 782-794

[10] Parasnis D S 1986 Principles of Applied Geophysics (New York: Chapman and Hall Ltd)

[11] Pedersen L B 1998 Tensor VLF Measurements: Our First Experience *Exploration Geophysics* **29** 52-27

[12] Cagniard L 1953 Basic Theory of the Magneto-telluric Method of Geophysical Prospecting *Geophysics* **18** 605-635

[13] Blakely R J 1995 Potential Theory in Gravity and Magnetic Applications (Cambridgeshire: Cambridge University)

[14] Oskooi B and Pedersen L B 2005 Comparison between VLF and RMT methods. A combined tool for mapping conductivity changes in the sedimentary cover *Journal of Applied Geophysics* **57** 227-241

IOP Conf. Series: Journal of Physics: Conf. Series **1011** (2018) 012028 doi:10.1088/1742-6596/1011/1/012028

Spatial Analysis of Gravity Data in the California, Nevada, and Utah (US)

NA Ferani[1], E Hartantyo[1], and SW Niasari[1]

[1]Geophysics Laboratory, Department of Physics, Universitas Gadjah Mada, Sekip Utara, Yogyakarta, Indonesia 55281

Corresponding author email: tniaannisaferani@gmail.com

Abstract. The geological condition of western North America is very complex because of the encounter of three major plates namely North America, Juan de Fuca, and Pacific Plate. The process of Juan de Fuca subduction and Pacific transform against North America plate created many mountains and produced Great Basin that we can see extending across California, Nevada, and Utah. The varied natural condition causes the varied value of gravity anomaly distribution. Using Topex free-air anomaly analyzed with second vertical derivative (SVD), we can analyze the fracture structures that occur in the Great Basin. The results show that the maximal SVD anomaly value is higher than the minimal SVD anomaly value at the western and eastern border of Great Basin. This explains that the two of Great Basin border are normal faults with trend direction NW-SE in the western boundary and NE-SW trending in the eastern boundary. This research result corresponds with the high seismicity data along the fault. Through this research, we can know that topex free-air anomaly data can be used to determine the type and trend of fault on a regional scale.

1. Introduction

Subduction of Farallon plate begun in the Mesozoic era had dominated the geological history of western North America [1]. At the late of Mesozoic, the Farallon plate subducting the North America plate moved upward to form a horizontal slab beneath North American Plate [2]. About 30 million years ago, end of the Farallon plate produced triple junction in western California and divided the Farallon plate into two parts, namely Juan de Fuca in the north, and Cocos in the south [3]. Relic of the Farallon plate is moving down towards the North America plate until now, but a drastic decrease in slope occurred to a depth of 400 km. It was caused by the presence of the plume about 17 million years ago [4, 5]. The occurrence of this plume produced upwelling of the asthenosphere, thinning of the lithosphere, extending zone and normal fault in the western North America to form Great Basin and Range that extended throughout Nevada to Utah [6].

The Gorda plate is the southernmost part of the Juan de Fuca plate system, which subducts the North America plate towards N 50^0E – N 55^0 E [7]. The encounter of 3 plates between the Gorda, Pacific, and North America plate greatly affected the geological condition of the western North America region. This encounter was referred to as Mendocino Triple Junction (MTJ) where the Pacific plate that moved to the northwest pushes the Gorda Plate. The area with high seismic is located in the San Andreas Fault which marks the boundary of the Pacific-North America plate [8]. The change in the movement of the Pacific plate from the azimuth N60^0W to the N37^0W caused an extension between the Sierra Nevada and the Colorado Plateau. This was related to the stretching of the Basin and Range areas [9]. Basin and Range's eco region spans an area of 343,169 km^2, mostly in Nevada (65.4%), Utah (25.1%) and the rest are scattered in several other states [10].

The varied natural appearance is highly correlated with its gravity value. In this research, we used secondary data obtained from http://topex.ucsd.edu/cgi-bin_data.cgi to analyze spatially the region of western North America. The accuracy of topex data always increases significantly through GPS to get a more precise orbit point [11]. Our research area map can be seen in Figure 1.

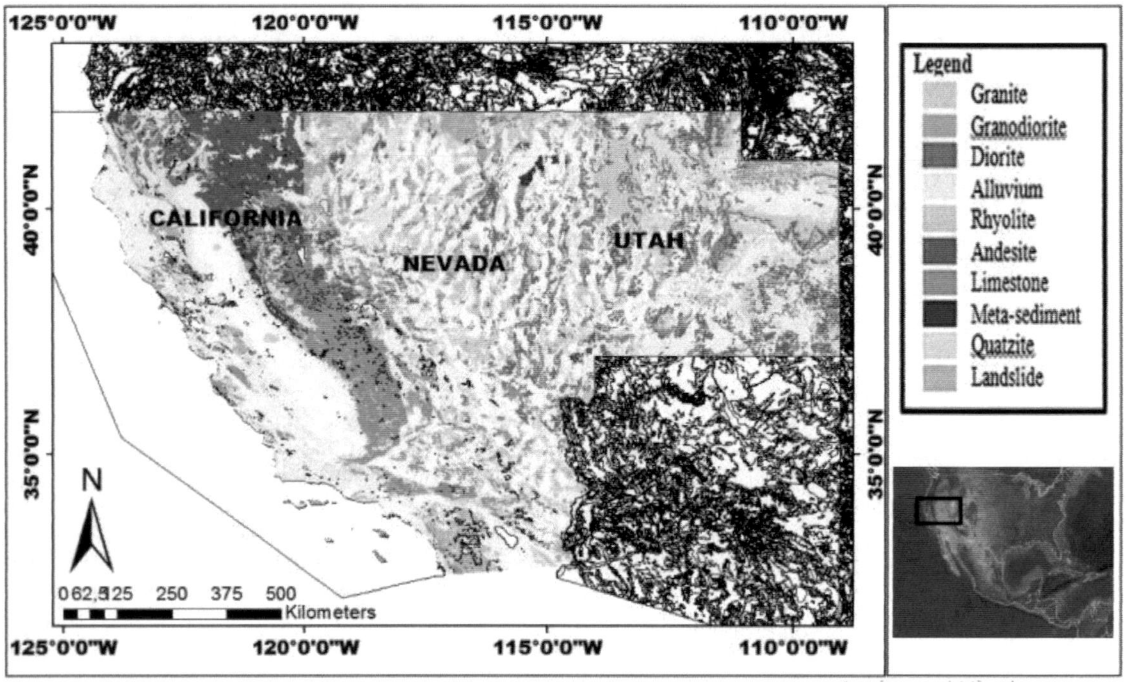

Figure 1. Geological map of California, Nevada, and Utah. Generally, surface appearance in the California is dominated by igneous rocks in the Sierra Nevada and sedimentary rocks in the Great Valley. Nevada and Utah are dominated by sedimentary and metamorphic rocks throughout the Basin and Range region [12, 13, 14].

2. Method

In this research, Topex free-air anomaly data that were used to map faults in the Great Basin area were obtained via http://topex.ucsd.edu/cgi-bin/get_data.cgi. The free-air data were better used to delineate deeper basin than shallow dip [15]. The coordinates of this research area were 112^{0}W-127^{0}W and 37^{0}N-42^{0}N. To separate residual and regional anomalies, there were several frequently used methods such as Continuation, Moving Average, Polynomial, Griffin, and visual methods. We used upward continuation to separate residual and regional anomalies up to 50km. This method applied the Blakely equation [16].

$$U\,(x, y, z_0 - \Delta z) = \frac{\Delta z}{2\pi}\int_{-\infty}^{\infty}\int_{-\infty}^{\infty}\frac{U(x',y',z_0)}{[(x-x')^2+(y-y')^2+\Delta z^2]^{3/2}}\,dx'dy' \qquad (1)$$

$U\,(x, y, z_0 - \Delta z)$ is the potential value in the continuation result field, Δz is the value of upward, and $U(x', y', z_0)$ is the potential value in the observation field. After doing upward continuation, regional anomaly data were analyzed using Second Vertical Derivative (SVD) to see the fracture structure. SVD is a method that can be used to see the effect of gravitational gradient on mapping [17]. By analyzing the maximum and minimum values of SVD, we can determine the type of fault occurring at that location [18]. With the equation :

For normal faults :

$$\left(\frac{\partial^2 \Delta g}{\partial z^2}\right) \max > \left|\left(\frac{\partial^2 \Delta g}{\partial z^2}\right)\right| min \tag{2}$$

For reverse faults :

$$\left(\frac{\partial^2 \Delta g}{\partial z^2}\right) \max < \left|\left(\frac{\partial^2 \Delta g}{\partial z^2}\right)\right| min \tag{3}$$

With $\left(\frac{\partial^2 \Delta g}{\partial z^2}\right)$ is the second vertical derivative of Δg, and Δg is the value of gravity anomaly.

3. Results and Discussion

Based on free-air anomaly data from Topex, the gravity response anomaly results in the range of -160mGal/θ^2 to 280mGal/θ^2, as shown in Figure 2.

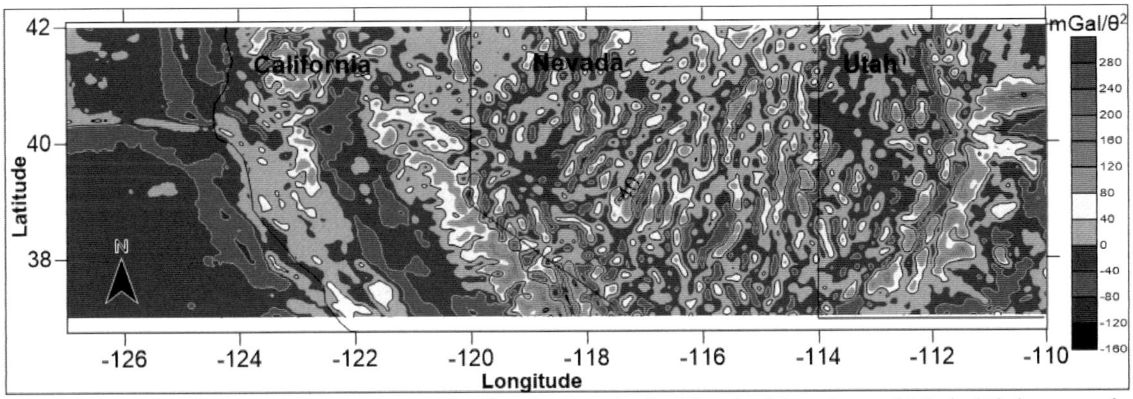

Figure 2. Map of free-air gravity anomaly distribution in California, Nevada, and Utah. High anomaly values are marked with red color and low anomaly values are marked with blue to purple color. The black line is the boundary of the three countries.

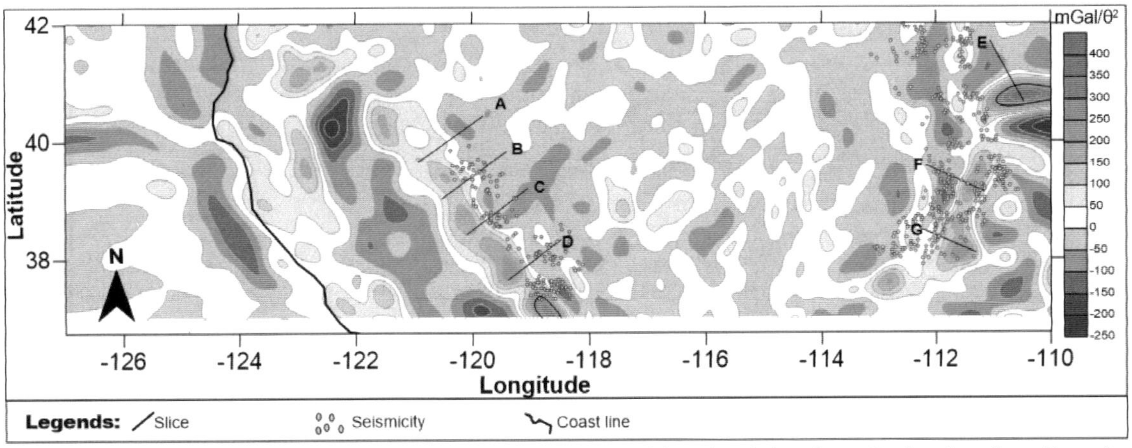

Figure 3. Mapping of SVD analysis results in overlay with seismicity data. Red color describes the high anomaly response and blue color illustrates the low anomaly response. A, B, C, D, E, F, and G slices are made along the area that is assumed a fault because having a high anomaly contrast.

Figure 2 shows the gravity anomaly response caused by the variation of rock density in the research area. The high anomalies associated with Cascadia volcanic arc and the Sierra Nevada mountains located in California and the western rocky mountains in Utah. In the Nevada state, we can see that there are repetitive high-low anomalies. These are related to the Great Basin area along Nevada country. After doing upward continuation to eliminate local anomalies, we used SVD analysis with results in Figure 3.

The SVD anomaly contour that is shown in Figure 3 shows the anomaly values between -250 to 400mGal/θ^2. The white color on the map illustrates the zero anomaly value that correlates with the lineament of fault. This corresponds with another study that in the eastern Great Basin there is a Wasatch fault that extends along 370km and has 6 large segments such as Bringham City Segment, Weber Segment, Salt Lake City Segment, Provo Segment, Nephi Segment, and Levan Segment [19]. In the western Great Basin bordering on the Sierra Nevada mountains, there is Sierra Nevada Fault [20]. The cross-section pattern on SVD anomaly map in figure 3 is made of curves can be seen in Figure 4.

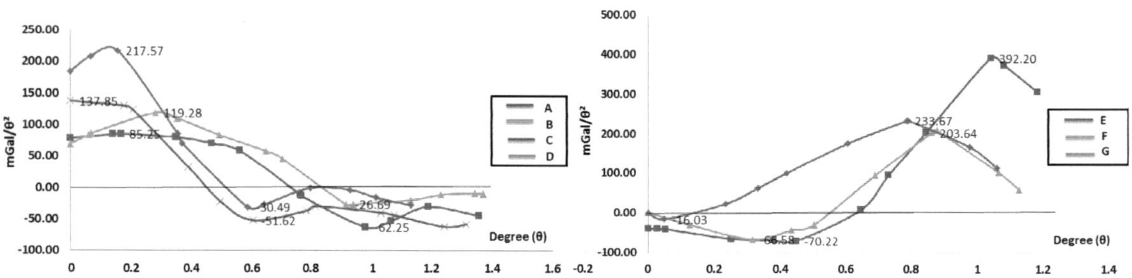

Figure 4. A, B, C, and D slices illustrate the SVD values at the western Great Basin boundary while E, F, and G Slices represent the eastern Great Basin boundary. A black line extending on the zero Y-axis signifies the high and low anomaly boundary of the SVD anomaly contour pattern.

From figure 4, we can see that from all slices, both of western and eastern frontier of Great Basin indicate that the maximum value of SVD is higher than the minimum value of SVD. Based on equation (2) and (3), we can know that the fault on the western and eastern Great Basin is a normal fault.

4. Conclusion

Through this research, we can conclude that topex free-air data can be analyzed using SVD analysis for regional coverage area. With SVD analysis we can also get the direction of the fault lineament. It can be seen from the lineament of normal fault in the western Great Basin with NW-SE trending and in the eastern Great Basin with NE-SW trending.

5. Acknowledgement

We are grateful to NASA and UCSD for providing topex free-air anomaly data so that this research can be done. We would like to thank the reviewers who have provided suggestions and comments for this paper to be better.

References

[1] Sigloch K, Mcquarrie V and Nolet G 2008 Two-stage subduction history under North America inferred from multiple-frequency tomography *Nature Geoscience* **1** 458-462

[2] Bunge H P and Grand S P 2000 Mesozoic plate-motion history below the northeast Pacific Ocean from seismic images of the subducted Farallon slab *Nature* **405** 337-340

[3] Stoffer P W 2006 *Wheres the San Andreas Fault?: a guidebook to tracing the fault on public lands in the San Fransisco Bay Region* (Reston, VA: U.S. Geological Survey) p 3

[4] Pavlis G L, Sigloch K, Burdick S, Fouch M J and Vernon F L 2012 Unravelling the geometry of the Farallon plate: Synthesis of three-dimensional imaging results from USArray

Tectonophysics **532-535** 82-102

[5] Xue M and Allen R M 2007 The fate of the Juan de Fuca plate: Implications for a Yellowstone plume head *Earth and Planetary Science Letters* **264** 266-276

[6] Zoback M L 1983 Structure and Cenozoic tectonism along the Wasatch fault zone, Utah *Geological Society of America Memoirs Tectonic and Stratigraphic Studies in the Great Basin* **157** 3-28

[7] Chaytor J D, Goldfinger C, Dziak R P and Fox C G 2004 Active deformation of the Gorda plate: Constraining deformation models with new geophysical data *Geology* **32** 353-356

[8] Oppenheimer D 2013 A policy for Rapid Mobilization of USGS OBS (RMOBS) *Mendocino Triple Junction Offshore Northern California* available at https://woodshole.er.usgs.gov/operations/obs/rmobs_pub/html/mendocino.html (accessed 20.04.17)

[9] Atwater T and Stock J 1998 Pacific-North America Plate Tectonics of the Neogene Southwestern United States: An Update *International Geology Review* **40** 375-402

[10] Benjamin M. Sleeter, Tamara S. Wilson, and William Acevedo, editors Status and Trends of Land Change in the Western United States-1973 to 2000 *USGS Publications Warehouse* available at https://pubs.usgs.gov/pp/1794/a/ (accessed 19.07.17)

[11] Schwintzer P, Kang Z, Reigber C and Zhu S 1995 GPS satellite-to-satellite tracking for TOPEX/Poseidon precise orbit determination and gravity field model improvement *Journal of Geodynamics* **20** 155-166

[12] California geologic map data *Interactive maps and downloadable data for regional and global geology, geochemistry, geophysics, and mineral resources; products of the USGS Mineral Resources Program* available at https://mrdata.usgs.gov/geology/state/state.php?state=CA (accessed 18.05.17)

[13] Nevada geologic map data *Interactive maps and downloadable data for regional and global geology, geochemistry, geophysics, and mineral resources; products of the USGS Mineral Resources Program* available at https://mrdata.usgs.gov/geology/state/state.php?state=NV (accessed 18.05.17)

[14] Utah geologic map data *Interactive maps and downloadable data for regional and global geology, geochemistry, geophysics, and mineral resources; products of the USGS Mineral Resources Program* available at https://mrdata.usgs.gov/geology/state/state.php?state=UT (accessed 18.05.17)

[15] Simpson R W and Jachens R C 1989 Chapter 3: Gravity methods in regional studies *Geological Society of America Memoirs Geophysical Framework of the Continental United States* pp 35-44

[16] Blakely R J 2009 *Potential theory in gravity and magnetic applications* (Cambridge: Cambridge Univ. Press) p 316

[17] Grant F S and West G F 1965 *Interpretation theory in applied geophysics* (New York: McGraw-Hill) p 248

[18] Achil B S 2016 *Subsurface Structure Identification from Gravity Anomalies Supported by Focal Mechanism Data in Nusa Tenggara Timur Province Sheet of Kupang-Atambua* thesis (Yogyakarta: Universitas Gadjah Mada) p 103

[19] Wu L C and Robert B S 2002 Integrated seismic-hazard analysis of the Wasatch font, Utah *Bulletin of the Seismological of America* **92** 1904-1922

[20] Oldow J S, Geissman J W and Stockli D F 2008 Evolution and strain reorganization within late Neogene structural stepovers linking the central walker lane and northern eastern California shear zone, western great basin *International Geology Review* **50** 270-290

IOP Conf. Series: Journal of Physics: Conf. Series **1011** (2018) 012029 doi:10.1088/1742-6596/1011/1/012029

Identification the geothermal system using 1-D audio-magnetotelluric inversion in Lamongan volcano field, East Java, Indonesia

N Ilham[1] and S W Niasari[2]

[1]Undergraduate Program of Geophysics, Universitas Gadjah Mada, Yogyakarta, Indonesia
[2]Geophysics sub-Department, Universitas Gadjah Mada, Yogyakarta, Indonesia

Corresponding author: nandailham23@gmail.com

Abstract. Tiris village, Probolinggo, East Java, is one of geothermal potential areas in Indonesia. This area is located in a valley flank of Mount Lamongan and Argopuro volcanic complex. This research aimed to identify a geothermal system at Tiris area, particularly the fluid pathways. The geothermal potential can be seen from the presence of warm springs with temperature ranging 35-45^{0}C. The warm spring locations are aligned in the same orientation with major fault structure in the area. The fault structure shows dominant northwest-southeast orientation. We used audio-magnetotelluric data in the frequency range of 10 Hz until 92 kHz. The total magnetotelluric sites are 6. From the data analysis, most of the data orientation were 2-D with geo-electrical direction north-south. We used 1-D inversion using Newton algorithm. The 1-D inversion resulted in low resistive anomaly that corresponds to Lamongan lavas. Additionally, the depth of the resistor are different between the area to the west (i.e. 75 m) and to the east (i.e. 25 m). This indicates that there is a fault around the aligned maar (e.g. Ranu Air).

1. Introduction

Indonesia is the largest archipelago in the world located between three major plates i.e. Eurasian, Pacific, and Indo-Australian plate. The collision between those plates makes Indonesia hosts 200 volcanoes that located along Sumatera, Java, and the island of eastern part of Indonesia [1]. These volcanoes provide a huge geothermal energy potential in Indonesia, which about 40% of the world's geothermal energy, equivalent to ~29.215 MWe. However, it has been little developed. It is just about 4% of ~29.215 MWe is already installed as the source energy [2]. One of the geothermal potential in Indonesia located on Mount Lamongan which can be seen from geothermal manifestations that appear in Tiris village, Probolinggo. Tiris, which was located on the flank of Mount Lamongan, has geothermal potentials equivalent to 92 MWe from the total 1206.5 MWe in East Java [3].

Mount Lamongan (alt. 1651 m) is a basaltic stratovolcano located in the eastern spur of Java, Indonesia [4]. It lies between two large mountains, Mount Tengger to the west and Mount Argopuro to the east. Mount Lamongan is surrounded by numerous prehistoric eruptive centers comprising up to 61 cinder cones and 27 maars [5]. The lithology of Mount Lamongan consists of Lamongan lavas and

Lamongan volcanic rocks overlying Argopuro volcanic rocks. These lithologies were formed in the Quarternary period [6].

Mount Lamongan was an active volcano in East Java. It has three vents i.e. Tarub the older, Tjupu, and Lamongan itself. Lamongan was detached from other vents by a fault aligned northwest-southeast which extends to the northwest into a series of aligned maars [7] shown in figure 1.

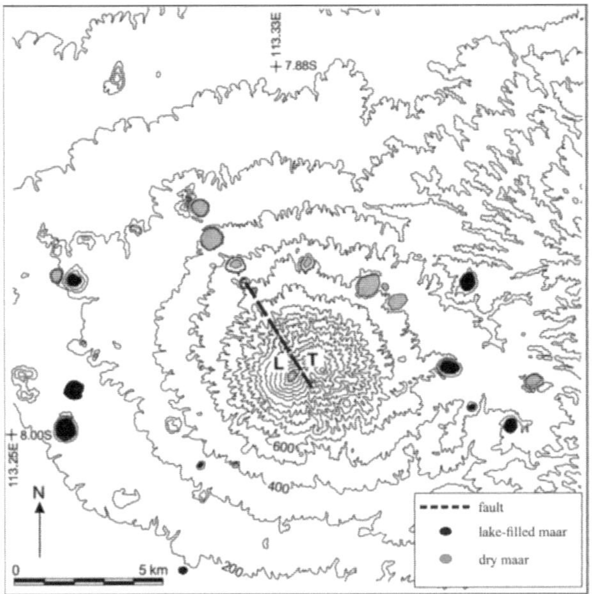

Figure 1. The dash line shows the fault which detached Lamongan (L) with Tarub vent (T). The fault has same orientation with aligned maars at the northwest. The black dot is lake-filled maars, meanwhile the grey one is dry maars (modified from [7]).

The geothermal manifestations in Tiris area consist of warm springs and deposits of iron oxide [3]. They appear through fractures in Tancak river with northwest-southeast orientation. Hydro-geochemical studies classified the thermal water as bicarbonate water with high HCO_3 concentrations. Bicarbonate water type indicates that the appearance of hot water is an outflow process from the reservoir. Moreover, in Tiris area, there are zeolite-bearing veins, indicative of satellite boiling zones along the hydrothermal outflow zone. These veins are also indicative of deep-seated geothermal reservoirs [8]. Thus, Tiris geothermal area assumed as an outflow area with the Argopuro being the upflow area of this geothermal system.

This research conducted the primary data of audio magnetotelluric. We used 1-D inversion to imaging subsurface resistivity. Thus, the fault as part of geothermal system can be identified.

2. Materials and methods

Magnetotelluric (MT) is a passive geophysical method that measures the time-varying orthogonal horizontal electric (**E**) and magnetic (**H**) field components by utilizing electromagnetic waves for imaging the subsurface conductivity distribution [9]. Correlation between the magnetic field (**H**) and the electric field (**E**) can be described as an impedance tensor (**Z**) [10].

$$\mathbf{E} = \mathbf{ZH} \tag{1}$$

Electromagnetic waves are sourced from the solar wind (with frequencies <1 Hz) and lightning activity (with frequencies >1 Hz) [11]. This wave will then reach the surface of the earth and propagates down to the subsurface. The amplitude of the wave propagating to the subsurface will be attenuated by 1/e. The depth at which the amplitude is attenuated by 1/e called skin depth [12].

$$\delta \approx 500\,(T\rho_a)^{1/2} \tag{2}$$

where ρ_a is the apparent resistivity (Ωm) and T is the period (s). The depth of penetration will be directly proportional to the resistivity and period [13]. The resistance layer and high period will make the wave penetrate deeper into the earth.

Resistivity is related to the characteristic impedance corresponds to homogeneous and isotropic earth. The apparent resistivity is the magnitude of the impedance as a function of the period [14]. The value of the apparent resistivity and the magnetotelluric phase in the homogeneous medium is shown in the equations below.

$$\rho_a = 1\ (\omega\mu)^{-1}\ |\mathbf{Z}|^2 = 0.2T\ |\mathbf{Z}|^2 \tag{3}$$

$$\Phi = X^{-1}Y \tag{4}$$

where X and Y are the real and imaginary parts of impedance [15].

The 1-D inversion is using Newton algorithm. The inversion problem can be solved by looking for a zero value of a function $\frac{\partial E}{\partial m}$ iteratively using Newton method [16]. The inversion model in the (n+1) iteration can be expressed in the formula below. Where the function

$$m_{n+1} = m_n - \left(\frac{\partial^2 E}{\partial m^2}\right)^{-1}_{m=m_n} \left(\frac{\partial E}{\partial m}\right)_{m=m_n} \tag{5}$$

where E is the data prediction error and m is the model. The estimate of the inversion solution requires a second derivative of the data prediction error function.

This research used the audio magnetotelluric method to identify the fault as the fluid pathways. There are 6 sites on the north Mount Lamongan eastward with azimuth line N90^{0}E. The distance between each sites is 2000 m, with the distance between AMT2 and AMT11 sites is 5000 m as exceptional. There is a site i.e. AMT8 between AMT2 and AMT11. However, the data component obtained from AMT8 differs from other sites so that the site is not included in the inversion. The data component of the AMT8 is the xy component, while at the other sites is yx which is then used for inversion. The line passed 2 faults i.e. the fault around aligned maars and the fault around warm springs. The sites and the geological features of the research area shown in figure 2.

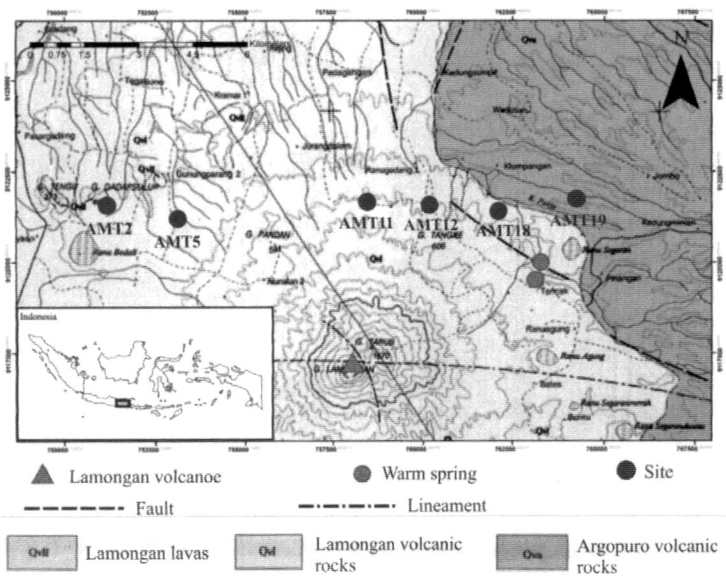

Figure 2. The map of the research area. The points of measurement are shown as blue dots with azimuth line N90^{0}E. The red dots and triangle is the warm spring points and the peak of Mount Lamongan respectively. (Inzet: The black square is the research area which located on the eastern of Java).

IOP Conf. Series: Journal of Physics: Conf. Series **1011** (2018) 012029 doi:10.1088/1742-6596/1011/1/012029

3. Results and discussion

In this research, we used the primary data which acquired at the line northern Mount Lamongan using Stratagem, 36716-01 REV.D version. 1-D inversion was carried out to imaging the subsurface conductivity of the research area using IPI2Win software. The inverse problem is solved using variant of the Newton algorithm of the least number of layers. The period time, resistivity, and phase used for regularizing the process of the fitting error minimizing by each sites. Thus, the inverse problem is solved for each sounding curve separately shown in figure 3.

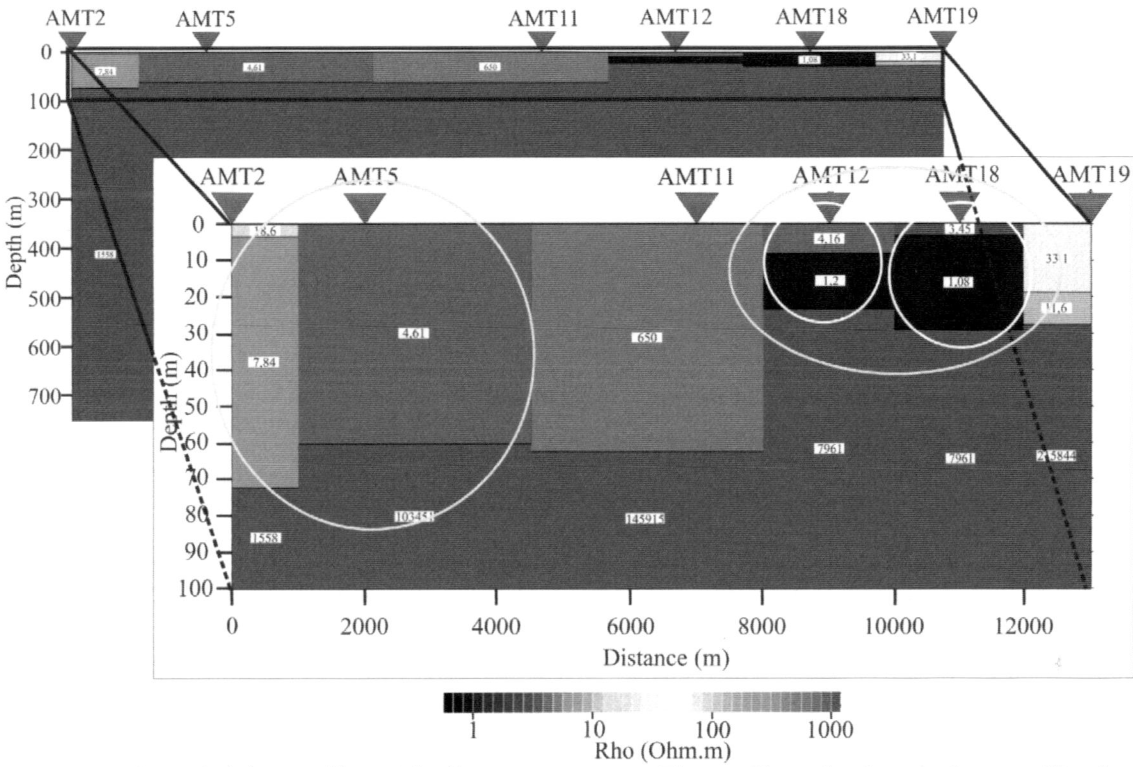

Figure 3. The resistivity profiles of the line measurement. The profile at the front is the magnification of the profile at the back with 100 m depth. The yellow circle shows the depth difference of low resistive anomaly. It indicates that there is a fault located in the high resistive anomaly. Meanwhile, the white circle indicates the fault too which there is an uncomformity of the resistivity block.

The resistivity profile shows that there is low resistivity anomaly in the depth up to 70 m. AMT2 and AMT5 sites have a resistivity value ranging 4.65-18.6 Ωm, which means a conductive anomaly at the depth 60-75 m. At AMT11, there is a resistive structure with a resistivity value of 643 Ωm. Three sites to the east, show the existence of low resistive anomaly in the depth ranging 25-30 m. The low resistive anomaly interpreted as Lamongan lavas.

Carn [4] identified a fault in the research area which is align with the maars located in the northwest. It is the continuation fault from the peak of Mount Lamongan. The fault and the lineament of the maars have northwest-southeast orientation. This structure can be proved by the existence of a resistive structure in the resistivity profile. The resistive structure located in the zone between AMT5 and AMT11 sites.

In warm spring area, there is a fault that is oriented to the northwest-southwest as well. The fault located at the eastern research area, between AMT12 and AMT18 sites. The resistivity profile shows the depth difference between the same resistivity block. The difference indicates there is a fault there as fluid pathways. The fluid emerges to the surface as warm springs.

From the resistivity profile, it can be seen that there is an anomaly depth difference between the western and the east. The western conductive zone has a greater depth than the east. This indicates a fault, most likely related to normal fault. Despite that, we do not find the dip of the fault from the geological mapping.

4. Conclusion

In this study, it can conclude that there is two faults in the research area. It can be seen from the depth difference of the resistivity value in the resistivity profile. Both faults are part of the geothermal system of Mount Lamongan as the fluid pathways, which the fault on the east is the pathways of the warm springs. Moreover, the low resistive anomaly at the profile interpreted as Lamongan lavas.

5. Acknowledgments

We would like to thank to "Kementrian Riset, Teknologi, dan Pendidikan Tinggi" for the PUPT funding no. 2452/UN1.P.III/DIT-LIT/LT/2017. Our thanks go to Mr. Joko, Arfaq, Rita, Dheny, Hardi, and Alvian as acquisition team, Rama for important suggestions and comments, and the instrument that used in this research from the geophysics lab.

References

[1] Darma S, Harsoprayitno S, Setiawan B, Hadyanto, Sukhyar R, Soedibjo A W, Ganefianto N, and Stimac J 2010 *Proc. World Geothermal Congress (April 25-29 Bali)*
[2] Rozaq K, Rahayu D and Bramantio B 2015 *Proc. World Geothermal Congress (April 19-25 Melbourne)*
[3] Utama W and Putra T M K 2012 *Proc. 1st ITB Geothermal Workshop (March 6-8 Bandung)*
[4] Carn S A 2000 *J. of Volcanology and Geothermal Research* **95** 81-108
[5] Carn S A and Pyle D M 2001 *J. of Petrology* **42** 1643-1683
[6] Suharsono and Suwarti T 1992 *Geological map of the Probolinggo quadrangle, Jawa* (Bandung: Geologicar Research and Development Centre)
[7] Carn S A 1999 *Bull Volcanol* **61** 92-105
[8] Deon F, Förster H J, Brehme M, Wiegand B, Schyett T, Moeck I, Jaya M S and Putriatni D J 2015 *J. Geothermal Energy* **3** 20
[9] Bakker J, Kuvshinov A, Samrock F, Geraskin A, and Pankratov O 2015 *Earth, Planets and Space* **67** 160
[10] Booker J R 2014 *Surv. Geophys.* **35** 7-40
[11] Simpson F and Bahr K 2005 *Practical magnetotellurics* (Cambridge: Cambridge University Press) pp 3-4
[12] Unsworth M 2007 *Encyclopedia of Geomagnetism and Paleomagnetism* ed D Gubbins and E Herrero-Bervera (Dordrecht: Springer) p 670
[13] Kyzhnyak M 2014 *Geoelectric strike and its application in magnetotellurics* (Reykjavik: Faculty of Earth Science, University of Iceland) p 7
[14] Naidu G D 2012 *Deep crustal structure of the Son-Narmada-Tapti* (Berlin: Springer) p 21
[15] Caldwell T G, Bibby H M and Brown C 2004 *Geophys. J. Int.* **158** 457-469
[16] Grandis H 2009 *Pengantar pemodelan inversi geofisika* (Bandung: Himpunan Ahli Geofisika Indonesia (HAGI)) pp 77-79

IOP Conf. Series: Journal of Physics: Conf. Series **1011** (2018) 012030 doi:10.1088/1742-6596/1011/1/012030

How strong was the 2015/2016 El Niño event?

Iskhaq Iskandar[1,*], DeniOkta Lestari[2], PutriAdia Utari[2], QurniaWulan Sari[2], Dedi Setiabudidaya[1], Wijaya Mardiansyah[1], Supardi[1] and Rozirwan[3]

[1]Department of Physics, Faculty of Mathematics and Natural Sciences, University of Sriwijaya, Inderalaya Campus, South Sumatra, Indonesia
[2]Graduate School of Environmental Sciences, University of Sriwijaya, Palembang Campus, South Sumatra, Indonesia
[3]Department of Marine Sciences, Faculty of Mathematics and Natural Sciences, University of Sriwijaya, Inderalaya Campus, South Sumatra, Indonesia

Corresponding author iskhaq@mipa.unsri.ac.id

Abstract. On the interannual timescale, the Indonesian climate is strongly influenced by a coupled ocean-atmosphere modes in the tropical Pacific Ocean. During a warm phase (El Niño event), negative sea surface temperature anomalies (SSTA) in the western tropical Pacific lead to suppress convection activities causing reduce precipitation over the maritime continent. The situation is reverse during the cold season(La Niña event). In this study, the evolution of 2015/2016 El Niño event is evaluated based on the collected data by the Tropical Atmosphere Ocean/ Triangle Trans-Ocean Buoy Network (TAO/TRITON) buoys. The results show that the evolution of the event has started in boreal spring (April – May 2015). It came to thepeak in boreal fall/winter (October – December 2015). The event lasted until boreal spring before it terminated in April/May 2016. In addition, the intensity of the event is classified as a strong event, and it is one of the strongest El Niño events during the last three decades.

1. Introduction

On interannual time-scale, the Indonesian climate is strongly influenced by a coupled ocean-atmosphere mode in the tropical Indo-Pacific region. From the equatorial Pacific, a swing of warm and cold phase, refers to the El Niño and La Niña phase, significantly modulates the rainfall intensity and distribution over the Indonesian region [12,3,4]. The significant reduce of rainfall intensity was observed during the occurrence of the El Niño event, while excess rainfall was observed over most of the Indonesian region during the La Niña event [5].

Theintensity of theEl Niño and La Niña eventscan bemeasured using oceanicindex as well as atmospheric index. The oceanic index describes the sea surface temperature anomaly (SSTA) in the equatorial Pacific Ocean [5]. Positive SSTA exceeding one standard deviation and lasting for at least five consecutive months that indicates an El Niño event. The situation is reversed for the La Niña event [4,6]. Meanwhile, the atmospheric index is expressing the sea level pressure gradient between the eastern and western tropical Pacific[5]. This index is well known as the Southern Oscillation Index (SOI) that is calculated as a gradient in the sea level between Tahiti and Darwin, Australia. During El Niño (La Niña) event, the sea level pressure in the western (eastern) tropical Pacific is higher than that observed in the eastern (western) counter part.

During 2015/2016, the Niño3.4 index indicated a positive value and most part of the Indonesian regions have experienced deficit rainfall. Note that the Niño3.4 index is a sea surface temperature(SST) anomaly index representing the El Niño conditions. It is calculated as an average SST anomaly in the equatorial Pacific Ocean bounded by a box of 170°W - 120°W, 5°S - 5°N. This study is designed to describehow strong was the 2015/2016 El Niño eventcompared to previous El Niño events.

2. Data and Methods

2.1. Data

This study relies on the observed oceanic and atmospheric parameters based on the Tropical Atmosphere Ocean/Triangle Trans-Ocean Buoy Network (TAO/TRITON) buoys project [7,8].The TAO/TRITON buoys provide various oceanic and atmospheric dataset. However, in this study only oceanic temperature atmospheric winds are used. The TAO/TRITON buoys recorded the oceanic temperature with various vertical resolutions from 1 m in the upper 10 m to 250 m in the deeper layers [8]. The TAO buoys recorded the oceanic temperature from 1 m depthdown to depth of 750 m, while the TRITON buoys recorded the temperature from 1.5 m down to depth of 750 m, The TAO buoys measured the winds at height of 4 m above sea level, while the TRITON buoys recorded the winds at height of 3.5 m above sea level. In addition, this study also used the Niño3.4 index obtained from the Observing System Monitoring Center, National Oceanic and Atmospheric Administration (NOAA).

2.2. Method

The monthly value of the Niño3.4 index was used to show the time evolution of 2015/2016 El Niño event. The Niño3.4 is defined as an averaged SSTA in the region bounded by 5°N to 5°S, from 170°W to 120°W [8]. Note that the anomaly fields are calculated as a deviation from the monthly climatology of all fields for a period of January 2000 – December 2015.

3. Results

The time evolution of the 2015/2016 El Niño is represented by the time series of the Niño3.4 index shown in Figure 1. It is shown that the evolution of 2015/2016 El Niño was started in boreal spring 2015 (April-May) when the index exceeding one standard deviation. The index gradually increased from boreal summer to boreal fall, and it reached the peak in October/November. The termination of the 2015/2016 El Niño was started as the index gradually decreasing in December 2015. The event was completely terminated in boreal spring 2016 (April/May) when the index fall below one standard deviation. Note that the 2015/2016 El Niño was lasting for about one year from boreal spring 2015 to boreal spring 2016.

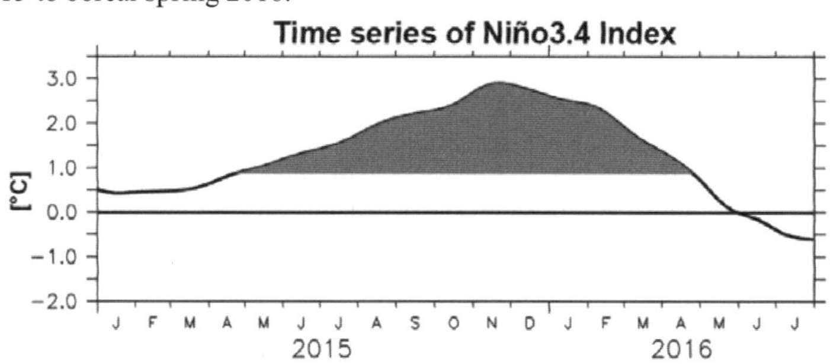

Figure 1. Time series of Niño3.4 index from January 2015 to July 2016 and wasdefined as anaveraged sea surface temperature anomaly between 5°N– 5°S, 170°–120°W.

Spatial evolution ofthe 2015/2016 El Niño event can be seen in Figure 2. At the initial evolution of the event, the easterly surface winds in the eastern tropical Pacific were weakened, while the westerly winds in the western part were strengthened (Figure 2a). As a result, the SST in the eastern tropical Pacific anomalously warm, while in the western the SST anomalously cold. By October 2015, the easterly winds in the eastern tropical Pacific were terminated and anomalous westerly winds were observed along the equatorial Pacific (Figure 2d). The maximum warm SSTA in the eastern Pacific was observed in November 2015 (Figures 2e) as the El Niño came to its peak phase (see Figure 1). The anomalous westerly winds gradually weakened in the following months, and the easterly winds replaced the westerly anomalies in the eastern Pacific since January 2016 (Figure 2g). As the result, the SSTA in the eastern (western) Pacific got cooling (warming). By February 2016, strong easterly winds were observed in the eastern Pacific, while in the western Pacific the anomalous westerly winds were terminated and maximum warm SSTA remained in the central Pacific (Figure 2h).

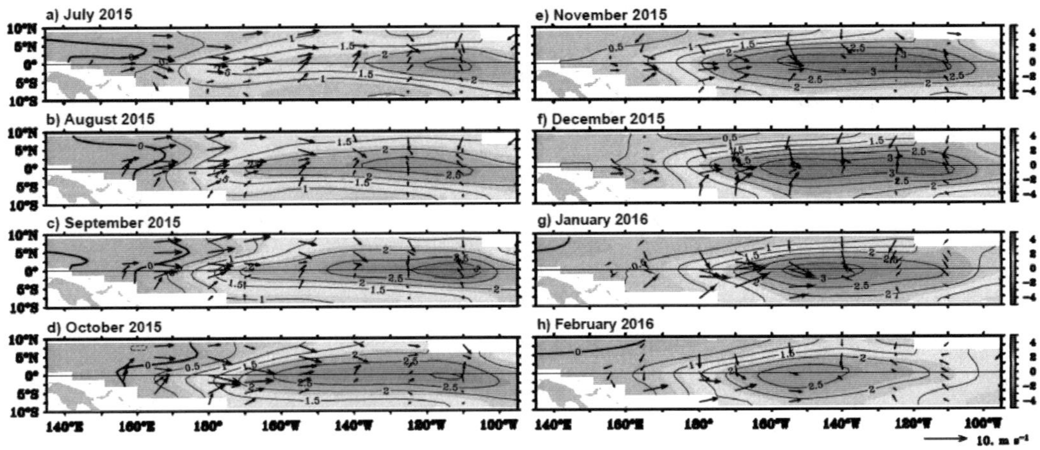

Figure 2. Ocean-atmosphere evolution associated with the 2015/2016El Niño is represented by SSTA (*shaded* in °C) and surface wind anomalies (*vectors* in Nm/s) from July 2015 through February 2016.

In order to examine the dynamics underlying of the evolution on 2015/2016El Niño event, we then examined time-longitude evolution of theanomalous zonal wind, dynamic height, SST, and the upper layer heat content (Figure 3).From early 2015, a series of strong intraseasonal westerly wind anomalies was observed in western equatorial Pacific (Figure 3a). The westerly wind anomalies were strengthened in boreal spring and expanded eastward. These intraseasonal westerly wind anomalies excited intraseasonaldownwelling Kelvin propagating eastward across the equatorial Pacific as shown by intraseasonal variations in the dynamic height (Figure 3b). The series of these wind-focedintraseasonal Kelvin waves warmed the eastern Pacific (Figure 3c) and accumulated the heat in the eastern Pacific (Figure 3d). In the western Pacific, the SST anomalously negative due to the upwelling of cold subsurface water, so that the heat content anomalously is negative. During the peak phase of the 2015/2016El Niño, the SSTA in the eastern Pacific could reach >3°C (Figure 3c).

At the end of 2015, strong westerly wind anomalies were observed in the western equatorial Pacific, while easterly wind anomalies already occupied the eastern Pacific (Figure 3a). These strong westerly wind anomalies forced downwelling Kelvin waves with amplitude smaller than that of the previous downwelling waves (Figure 3b). Starting from early 2016, the westerly wind anomalies in the western equatorial Pacific were weaken even reversed their direction, while the easterly wind anomalies in the eastern Pacific were strengthened (Figure 3a). The easterly wind anomalies in the eastern Pacific induced upwelling causing cool SSTA and reducing heat content in the eastern Pacific(Figures 3c-d). In June/July 2016, the western Pacific warm pool was returned as positive

SSTA occupied the western Pacific, while the eastern Pacific cold tongue replaced the positive SSTA in the eastern Pacific (Figure 3c).

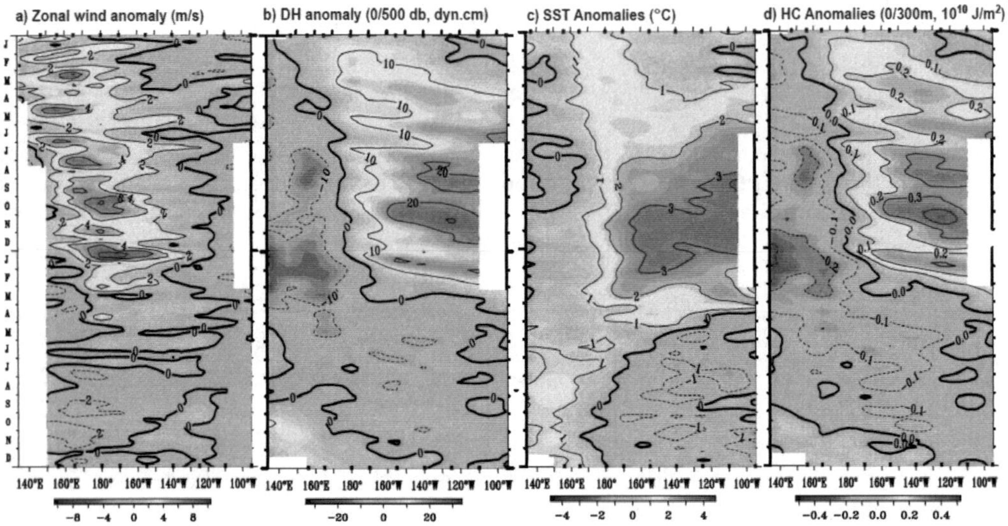

Figure 3. Time-longitude diagram of a) zonal wind anomaly (m/s), b) dynamic height anomaly (0/500 db, dyn.cm), c) sea surface temperature anomaly (°C), and d) heat content in the upper 300 m depth ($\times 10^{10}$ J/m^2) from January 2015 – December 2016.

The intensity of the 205/2016 El Niñowas compared with that of the previous El Niñoevents during the last three decades from 1980 to 2016. Note that the Niño3.4index was used to represent each El Niño event. There were 6 strong El Niño events observed during 1980 – 2016, namely the 1981/1982, 1991/1992, 1997/1998, 2002/2003, 2009/2010 and 2015/2016 El Niño event (Figure 4). It is clearly shown that 3 out of 6 events were categorized as a strong El Niño event in which the Niño3.4 index exceeding 2.5°C. Those 3 events are the 1981/1982, 1997/1998 and 2015/2016 El Niño events. In particular, the 2015/2016 El Niño has a comparable magnitude with the El Niño on 1997/1998. However, the initial development of the 2015/2016 El Niñowas slightly slower than that the 1997/1998 El Niño.

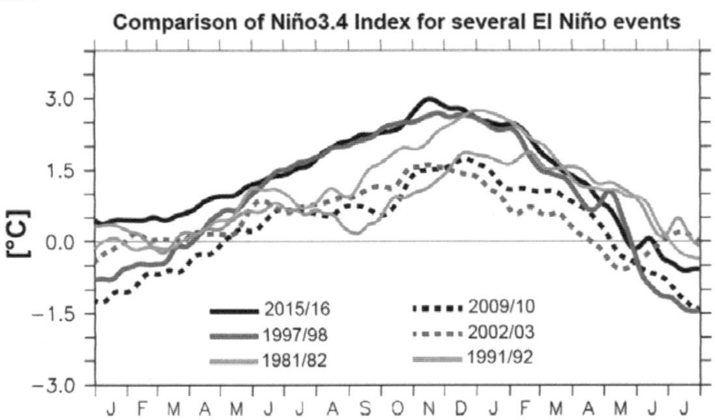

Figure 4. Time series of the Niño3.4index for several El Niño events during the last three decades.

4. Conclusion

During boreal spring 2015 until boreal spring 2016, El Niñoevent took place in the tropical Pacific. This extreme climate event has caused catastrophic climate event around the globe. In particular, most part of the Indonesian regions has experienced deficit precipitation leading to a severe drought event. This study was designed to evaluate the dynamics of the evolution of the 2015/2016 El Niño event and compared with previous El Niño events.

The results show that the development of 2015/2016 El Niño was induced by a series of strong intraseasonal westerly winds observed in the western Pacific from the early spring to late autumn 2015. These intraseasonal westerly winds forced intraseasonaldownwelling Kelvin waves that accumulating oceanic heat content in the eastern equatorial Pacific. These intraseasonaldownwelling Kelvin waves together with a weakening of easterly winds in the eastern equatorial Pacific warmed SST there, leading to a further weakening of easterly winds. As the El Niño came to its peak phase, the positive (negative) SSTA occupied the eastern (western) equatorial Pacific. The termination of 2015/2016 El Niño was associated with the weakening of westerly winds in the western equatorial Pacific and the strengthening of easterly winds in the eastern equatorial Pacific in early 2016. The easterly winds in the eastern Pacific induced upwelling of cool subsurface water leading to a negative SSTA and reducing heat content in the eastern equatorial Pacific. By spring 2016, positive SSTA occupied the western Pacific, while the eastern Pacific cold tongue replaced the positive SSTA there.

The 2015/2016 El Niño was one of the strongest El Niño events in the last three decades. It has comparable amplitude with the 1997/1998 El Niño, and it was lasting a relatively longer, from boreal spring 2015 until boreal spring 2016.

Acknowledgment

The authors thank the Global Tropical Moored Buoy Array Project Office, NOAA/Pacific Marine Environmental Laboratory for providing the TAO/TRITON datasets. This study is supported by the Ministry of Research, Technology and Higher Education through a *PenelitianBerbasisKompetensi2017* (No. 454/UN9.3.1/LT/2017) and by the University of Sriwijaya through *HibahProfesi 2017* (No. 1011/UN9.3.1/PP/2017).

References

[1] McPhaden MJ, Zebiak SE, and Glantz MH. 2006. ENSO as an integrating concept in Earth science,*Science*, **314**, 1740–1745, doi:10.1126/science.1132588.
[2] Trenberth KE, Caron JM, Stepaniak DP, Worley S. 2002. The evolution of ENSO and global atmospheric surface temperatures,*J. Geophys. Res.*, **107**, 4065, doi:10.1029/2000JD000298.
[3] Trenberth KE and Stepaniak DP. 2001. Indices of El Niño evolution, *J. Clim.*, **14**, 1697– 1701.
[4] Trenberth, K. E. (1997), The definition of El Niño. Bull. Amer. Meteor. Soc., 78, 2771−2777.
[5] Philander SGH.1990. El Niño, La Niña, and the Southern Oscillation. *Academic Press*, 289 pp.
[6] Iskandar I. et.al. 2017. Evolution of 2015/2016 El Niño and its impact on Indonesia, *AIP Conference Proceedings,***1857**, 080001 (2017); doi: http://dx.doi.org/10.1063/1.4987095.
[7] McPhaden MJ. 1995, The Tropical Atmosphere Ocean (TAO) Array is Completed. *Bulletin of the American Meteorological Society*, **76**(5):739-741.
[8] McPhaden MJ. et al.1998. The Tropical Ocean-Global Atmosphere (TOGA) observing system: A decade of progress. *J. Geophys. Res.*, **103**, 14,169-14,240.

IOP Conf. Series: Journal of Physics: Conf. Series **1011** (2018) 012031 doi:10.1088/1742-6596/1011/1/012031

Investigation of Aceh Segment and Seulimeum Fault by using seismological data; A preliminary result

U. Muksin[1,2], Irwandi[2,3], I. Rusydy[2], Muzli[4], K. Erbas[5], Marwan[3], Asrillah[3], Muzakir[3], N. Ismail[2,3]

1Department of Physics, Syiah Kuala University, Banda Aceh, Indonesia
2Tsunami and Disaster Mitigation Research Center, Syiah Kuala University, Banda Aceh
3Geophysics Engineering, Syiah Kuala University, Banda Aceh
4Badan Meteorologi, KlimatologidanGeofisika Indonesia, Jakarta
5 Research Center for Geosciences, GFZ-Postdam, Potsdam-Germany

Abstract. The Seulimeum Fault has not generated large earthquake after last large earthquake with magnitude of M 7.3 occured in 1936. The Seulimeum Fault is accompanied by the Seulawah volcano that reported to be active in 1839, 1975 and 2010. The activity of the Seulimeum Fault could be related with the existence of the Seulawah volcano and the Seulawah volcano activity could also triggered by the Seulumeum Fault activity. The objective of the longterm research is to investigate the relation between the Seulimeum Fault and the Seulawah Volcano. The aim of this paper is to present the first result of the investigation of the Seulimeum Fault based on the seismicity and geomorphology. A seismic network consisting of 17 seismometers (Trilium Compact) and data logger (DSS Cube) were deployed in Aceh Besar. The seismic network was installed for 3 months to record earthquakes along the Seulimeum and the Aceh Faults. The Seulimeum Fault is considered to be active as several local earthquakes were recorded. The Seulimeum Fault is much more active in the region of the bifurcation of the The Aceh Segment and the Seulimeum Fault. The mechanisms of earthquakes along the Seulimeum Fault were mostly strike slip following similar to the Sumatran Fault characteristics.

1. Introduction

Aceh is one of the very seismically active region in the world that has generated more than 2200 earthquakes with magnitude larger than M 6.0 since 1990 to November 2004. The Aceh region has become much more active after the Mw 9.3 2004 Aceh earthquakes indicated by more earthqukes with magnitudes larger than M 6.0 occurred[1]. The sources of large earthquakes could be from the subduction zone, the Wharton Ridge, the Sumatran Fault and even from the secondary faults. Although the magnitudes of inland earthquakes in Aceh were smaller than those along the subduction zone, inland large earthquakes with magnitude between M 6.0 and M 7.0 could be much more destructive than those along the subduction zone since the inland earthquakes are located close to the population. Among secondary faults in Aceh that has generated destructive earthquakes included the Lampahan Fault in Central Aceh on July 2nd, 2013 and in Pidie Jaya on December 6th, 2016. One of the secondary faults that has not released large earthquake is the Seulimeum Fault which is located in the northernmost Aceh (Figure 1).

The Seulimeum Fault is one of the secondary faults in Sumatra that could generate destructive earthquake with magnitude larger than M 7.0 as a large earthquake occurred along the fault in 1936 [2]. Although there is no report of large earthquake along the Seulimeum Fault, the Seulimeum Fault has been generating small scale earthquakes [3]. Small scale earthquakes often generated along secondary faults that close to volcanoes e.g. in Tarutung[4] and in Central Aceh[5]. The Seulimeum Fault is accompanied by the Seulawah Volcano which was reported to be active in 1975, 2010, and 2013 indicated by tremor earthquakes.

The seismic activities along the Seulimeum Fault could be related to the Seulawah Volcano and also the Seulawah Volcano activity could be caused by the seismic activities along the Seulimeum Fault. Therefore the objective of the research is to investigate the characteristics of the Seulimeum Fault from the seismic activities and its relation to the Seulawah Volcano. The seismicity distribution along the Seulimeum Fault could delineate the Seulimeum Fault complemented by geomorphology.

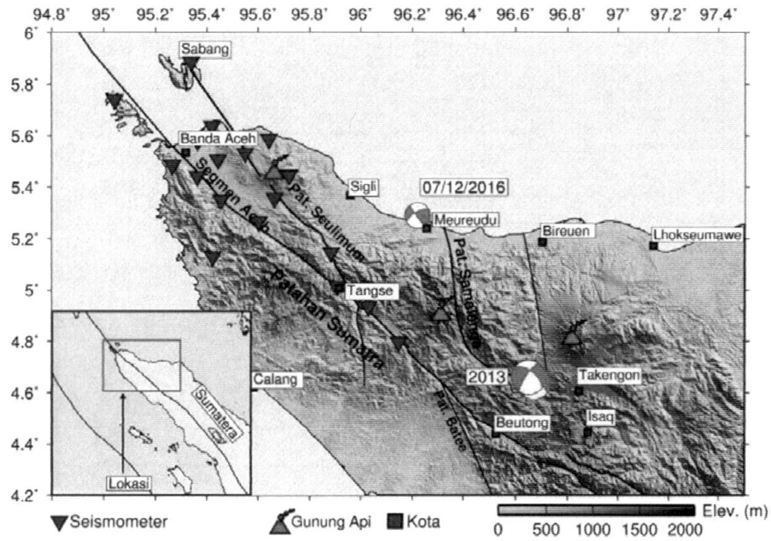

Figure 1: The study area covering the Seulimeum Fault, the Aceh Segment, and the Seulawah Volcano. The white triangles represent the seismic stations.

2. Data and Methods

A seismic network consisting 17 seismometers and DSS Cube data logger provided by GFZ-Potsdam was installed in Aceh Besar, the northernmost Sumatra. We used broadband seismometers and recorded the waveforms in the DSS Cube datalogger with the sampling rate of 100 sps. The seismic network covered the main geological features including Seulawah Volcano, The Seulimeum Fault, and the Aceh Segment in the region of 4.8 N- 5.6 N and 95.4 E – 96.0 E as shown in Figure 1. The network was deployed for three months because of the limitation of the instrument availability since January 2017. This kind of seismometer distribution in general could detect earthquakes with magnitude down to M 2.0 depending on the noise level around the seismometers.

The seismic waveforms from all the stations were downloaded manually and the P and S arrivals were observed visually (picked manually). The first preliminary locations of the earthquakes were determined by using HYPO71 [6]. The earthquakes were then relocated by using a simultaneous earthquake relocationand 1D velocity determination [7]. Several 1D velocity modelswere employed then the inversion is conducted for each velocity model. The RMS values of each result are compared and the velocity model with the smallest velocity model is chosen as the best 1D model. The 1D velocity model could then be used for seismic imaging to study the structure of the beneath the area of interest including the Seulawah Volcano.

3. Preliminary Results and Discussion

The northernmost region of Aceh is relativelyseismically active indicated by several earthquakes recorded along the Seulimeum Fault and the Aceh Segment. Figure 2 shows the waveforms of an earthquake along the Seuliemum Fault in the Krueng Raya region. The waveforms indicate that even within 40 seconds (Figure 2) two earthquakes occurredalong the Fault.

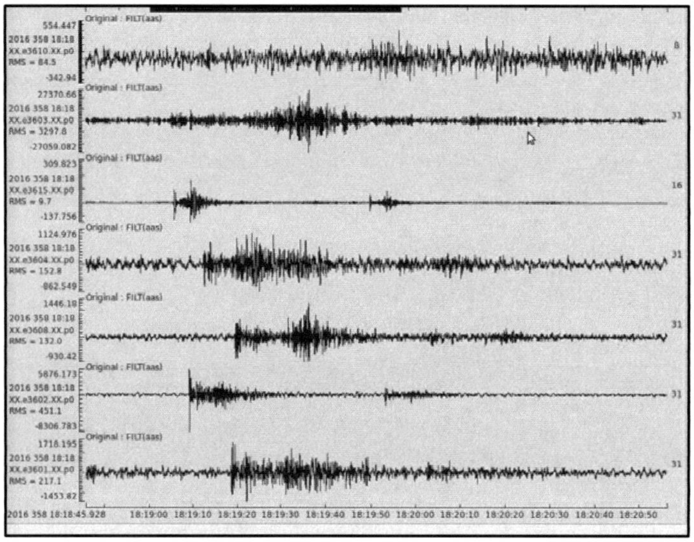

Figure 2:An example of recorded seismic waveforms of an earthquake that occured in Kreung Raya, Aceh Besar

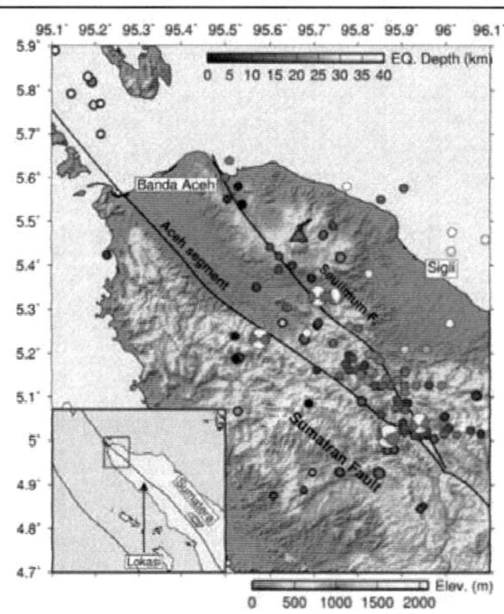

Figure 3: The distribution of earthquakes in Aceh along the Seulimeum Fault and the Aceh segment.

The distribution of earthquakes is shown in Figure 3 indicating that both, the Seulimeum and the Aceh Faults are currently active. Both faults are much more active in the region close to the bifurcation the two faults. The Seulimeum Fault is also active in the Krueng Raya region. The historical earthquake also indicated that the Kreung Raya region wasactive. A large earthquake occurred around Krueng Raya that affected the region significantly. The local earthquake data has not been fully analyzed but we could suggest the pattern of the fault from the focal mechanisms provided by International Seismological Center[1]. The focal mechanisms of the earthquakes indicate that the Seulimeum Fault is rigth lateral strike slip which is almost similar to the Sumatran Fault. The Seulimeum Fault strikes NNW-SSE while the Sumatran Fault similar to the fault linewe derived from the earth elevation model (ASTER-GDEM).

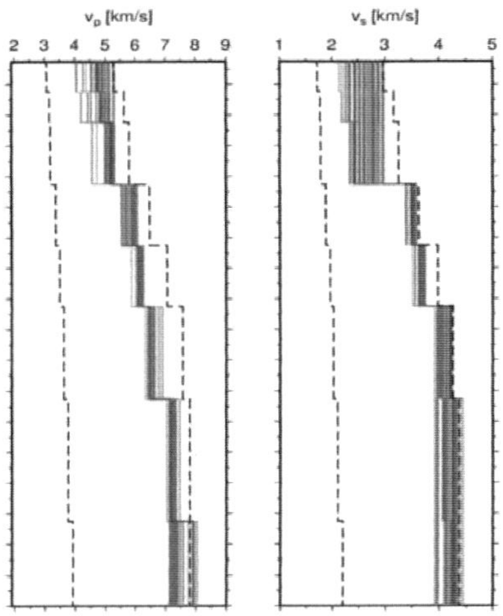

Figure 4: The 1D velocity model after 30 iterations (a) Vp model and (b) Vs model. The red lines indicate the models with the smallest RMS values. The dash lines represent the minimum and the maximum boundaries of the imposed velocity models.

The locations of the earthquakes need to be relocated by using a simultaneous earthquake relocation and velocity determination. We have experimented to obtain the 1D velocity model as shown in Figure 4. We imposed 64 initial velocity models and then the 1D inversion was applied for each initial model with 30 iterations. The initial models were chosen arbitrarilyto be within the boundary corridor indicated by dashed line in Figure 4. All imposed models converged to one best velocity model indicated by red lines in Figure 4. The 1D velocity model will be used for imaging the structure beneath the study area.

4. Conclusion

The Seulimeum Fault and the Aceh segment seem to be active as several local earthquakes recorded along the faults. Both segments are more active in the region close to the bifurcation of the two faults. The Seulimeum Fault is also active in the Krueng Raya region. We have determined the 1D velocity model of the northernmost Sumatra (Aceh Besar) that could be used for further analysis to image the seismic structure of the Seulawah Volcano and the structure beneath the faults. The Seulimeum Fault is right lateral strike-slip similar to that of the Sumatran Fault.

5. References

IOP Conf. Series: Journal of Physics: Conf. Series **1011** (2018) 012031 doi:10.1088/1742-6596/1011/1/012031

[1] International Seismological Centre. 2017 In: Centre IS. Thatcham, United Kingdom.
[2] Sieh K, &Natawidjaja D. 2000. Journal of Geophysical Research: Solid Earth (1978–2012). 2000. *105:* 28295-28326.
[3] Ryberg T, Muksin U, &Bauer K. 2016. Journal of Volcanology and Geothermal Research. 2016. *321:* 73-84.
[4] Muksin U, Haberland C, Nukman M, Bauer K, &Weber M. 2014. Journal of Asian Earth Sciences. 2014. *96:* 123-131.
[5] Muksin, Bauer K, Muzli, Masturyono, Trond Ryberg, &Haberland C. 2015: 75th Annual Meeting of the German Geophysical Society. Hannover, Germany.
[6] Lee WHK, &Valdes C. 1985. HYPO71PC: A personal computer version of the HYPO71 earthquake location program: US Geological Survey.
[7] Kissling E, Ellsworth W, Eberhart-Phillips D, &Kradolfer U. 1994. Journal of Geophysical Research: Solid Earth (1978–2012). 1994. *99:* 19635-19646.

Acknowledgments
The research is funded by the Indonesian Government under the scheme of PenelitianUnggulanUniversitas 2017-2019. The seismic instruments were provided by Research Center for Geosciences, GFZ-Potsdam. All maps were created by using generic mapping tools (GMT-PLOT).

Spatial Gravity Analysis of the Cascadia Subduction Zone using Satellite Data

A Hanatan[1], E Hartantyo[2], S W Niasari[3]

Geophysics Laboratory, Department of Physics, Faculty of Mathematics and Natural Sciences, Universitas Gadjah Mada, Sekip Utara, Yogyakarta, Indonesia 55281

[1] Corresponding author email: ardiana.hanatan@ugm.ac.id
[2] hartantyo@ugm.ac.id , [3] sintia_windhi@ugm.ac.id

Abstract. Cascadia Subduction Zone is a subduction zone elongated about 1000 km length. The remnants of Farallon plate subduct the North American plate and form this subduction area. One of Farallon plate remnants, i.e. Juan de Fuca plate, subducts dominantly the North American plate. We focused on the observation of three states, i.e. Oregon, Idaho, and Wyoming. This research aims to determine the direction, the shape, and the initial coordinates of subduction in our study area. We obtained free air corrected gravity data from TOPEX. Then we visualized data to get contour map and found that Cascadia Subduction Zone has direction from west to east that can be proofed by increasing of gravity anomaly. The gravity anomaly ranges from -140 mGals until 320 mGals. We applied upward continuation and got the result that the subduction is elongated from north to south. Initial coordinate detail of subduction shown by SVD result. The subduction starts from coordinate 46.811° Northern Hemisphere and Longitude of 123.436° into 41.260° Northern Hemisphere and longitude of -123.204°. This coordinate appropriate with the result of magnetotelluric research that shows a high resistivity. We can conclude that from gravity satellite data, we can visualize the contour map then take several steps to get details information of subduction.

1. Introduction

A geological condition in Northern American plate, particularly in the Northwestern United States shows complex geological features. This complexity was started on 150 Ma resulted in the Farallon plate subduction. It shaped the Pacific Northwest and affected its upper mantle and crust [1]. The convergence between the subduction zone of the North American west coast and the East Pacific Rise caused a fragmentation of Farallon plate [2,3]. Since Oligocene era, Farallon plate was divided into Juan de Fuca plate, Cocos plate, Nazca plate, and several microplates [4].

Juan de Fuca plate is an oceanic plate that subducts the Northern American plate along the Cascadia Subduction Zone. The subduction of the Juan de Fuca Plate towards Northern America plate had 50° of dipping, ~ 400 km of depth, and ~75 km of thickness [5]. This subduction moves 30-45 mm/year [6]. Cascadia Subduction Zone that elongated about 1000 km can be included as a vulnerable area and has a high possibility of earthquakes [7]. One of the influencing factors of seismic activities is the geometry of subduction slab [8]. Oceanic plates usually brought chemical and physical bound water into subduction zones [9]. The result of this subduction emerges some geological features in the Northwestern United States, i.e. Oregon, Idaho, and Wyoming.

Previous research studied about the model of lithospheric structure because of Juan de Fuca Plate's influence towards the Cascadia Subduction [10]. This gravity inverse study used a density-depth relation in the subducted slab and oceanic asthenosphere beneath Vancouver Island and Oregon. The result showed that the down going slab is steeper beneath Oregon than beneath Vancouver Island.

In this research, we used secondary data. We analyzed the Cascadia Subduction Zone in the Northwestern United States spatially. The spatial analysis involves several factors, i.e. direction, shape, and the initial coordinate of Cascadia Subduction Zone in our study area.

2. The Data Sets and Methods

In geophysics, the spatial analysis can be used for many applications. Spatial analysis is the previous step before modeling. Modeling of gravity anomaly data aimed to estimate the geometry, depth, and densities of an object [11]. In this research, we used gravity secondary data from http://topex.ucsd.edu/cgi-bin/get_data.cgi in free-air corrected format data with ±1.25 km in space. These data were a satellite data based which is already resulted in free air anomaly map [12]. The longitude-latitude coordinate taken was 40° to 48° Northern Hemisphere and -126° until -103°. Our research focus is in the Cascadia Subduction Zone, particularly subduction from Juan de Fuca plate towards North America plate through Oregon, Idaho, and Wyoming.

The next step was creating a contour map. We separated the regional and residual anomaly used upward continuation. We used 230.000 m of upward continuation. Upward continuation will eliminate the local anomaly gravity data in our contour map. The principle of this separation used low pass filter to get the low frequency. In this upward continuation step, we used equation (1) [13]:

$$U(x, y, z_0 - \Delta z) = \frac{\Delta z}{2\pi} \int_{-\infty}^{\infty} \int_{-\infty}^{\infty} \frac{U(x, y, z_0)}{\left[(x - x`)^2 + (y - y`)^2 + (\Delta z)^2\right]^{\frac{3}{2}}} dx`dy` \quad (1)$$

The general problem in the interpretation of anomaly map is that it is hard to handle the effect of shallow structure from the deeper-seated ones. The result of removal regional field effect is the indeterminate and non-unique set of residuals. Second Vertical Derivative (SVD) can be used to determine the shallow and deeper structures [14]. SVD is often used to enhance the structures of localized near-surface [15].

3. Results and Discussion

Oregon, Idaho, and Wyoming are three states located in the Northwestern United States. For prior information, the rock arrangement in Oregon is dominated by young rocks. Idaho has the most varied rock types and unit ages. Wyoming has the oldest rock age than Oregon and Idaho. So we have a prior hypothesis that Wyoming is the state that has the longest distance from the Juan de Fuca subduction zone boundary.

We got topography and gravity data from TOPEX. We used topography data to know the space line of gravity data. From gravity data downloaded, we got the anomaly gravity response in the observational area. The difference color of contour map shows the difference value of gravity anomaly. The highest value of gravity anomaly showed by dark red color whereas the lowest showed by black color. It ranged from -140 mGals until 320 mGals as shown in Figure 1. As we know, one mGals is equal to 1 cm/s^2.

The initial area of subduction zone showed by a specific increase of gravity anomaly. Further, we can confirm it by upward continuation step. There are weak zones around the subduction that dominated by purple color. These weak zones have a high possibility of earthquakes and hazards around Cascadia Subduction Zone.

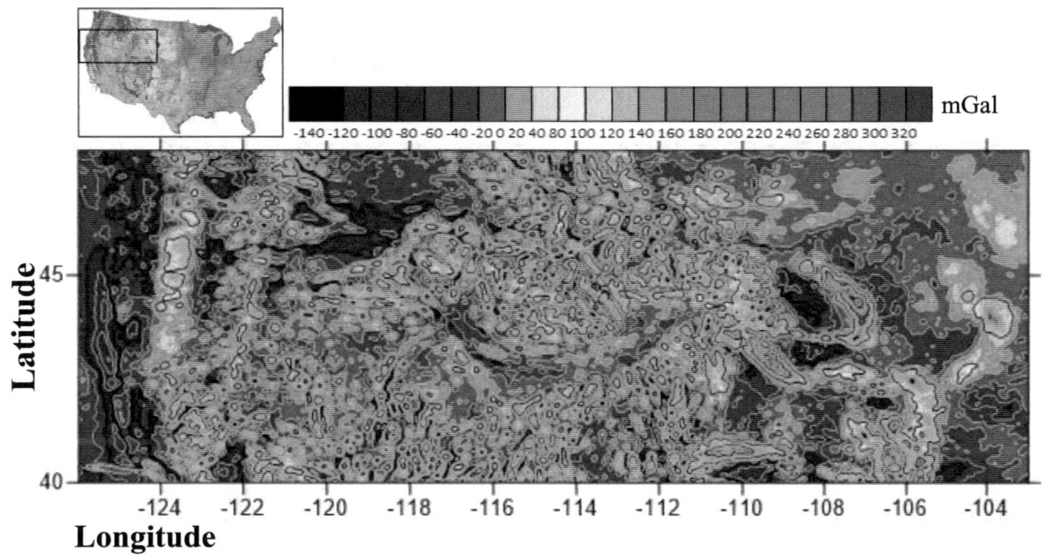

Figure 1. The picture shows gravity anomaly of the study area. The map shows gravity anomaly in the coordinate 40° until 48° and -126° until -103°. In the west area, the dominant color of the map is dark purple. This color indicates low gravity anomaly. In the east area the red color that indicates high anomaly gravity. These value can be associated with the geological features, i.e. Yellowstone plume. Inset: American continent (Source: [16]).

Figure 1 shows the differences of gravity anomaly that illustrated by different color scales. High gravity anomaly lies dominantly in the east. Our hypothesis is the gravity anomaly increased from west to east. This fact can be attributed to the direction of Juan de Fuca plate subduction towards North American plate. Juan de Fuca subduction stretched from west to east, particularly in our observation is from Oregon to Wyoming. The highest gravity anomaly lies around the coordinate -110^{0}BB because there are geothermal areas of Yellowstone.

After we got the gravity anomaly map, we used upward continuation to separate the regional and residual gravity anomaly. In this case, we took the regional gravity anomaly because the study area is large. Upward continuation process was done by software with several limits (i.e. 13.000 m, 20.000 m, 30.000 m, 40.000 m, 50.000 m, 60.000 m, 70.000 m, 80.000 m, 90.000 m, 100.000 m, 200.000 m, and 230.000 m). From the visualization of several limits of upward continuation, we finally decided to used 230.000 m. The reason for this decision is because the regional anomaly looks clearest in height 230.000 m than another limitation (Figure 2).

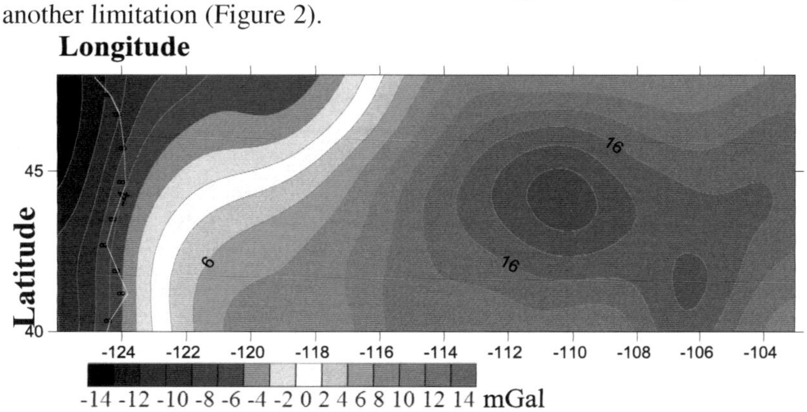

Figure 2. Gravity anomaly map of 230.000 m upward continuation. The gravity anomaly increased from black into red. The gravity anomaly value increased from west to east. It means that our early hypothesis is true. The white line shows the coastline.

IOP Conf. Series: Journal of Physics: Conf. Series **1011** (2018) 012032 doi:10.1088/1742-6596/1011/1/012032

Figure 2 shows the difference of gravity anomaly regionally. The result of upward continuation step confirmed our hypothesis before. There is an increase of gravity anomaly from west to east that can be correlated with the Juan de Fuca oceanic plate subduction towards North American plate. In the study area, i.e. Oregon, Idaho, and Wyoming, there is an increase of gravity anomaly too. Further, we can used SVD (Second Vertical Derivative) to know the initial coordinates of the subduction zone in the gravity anomaly map from 230.000 m of upward continuation. Gravity anomaly map after SVD process is shown in Figure 3.

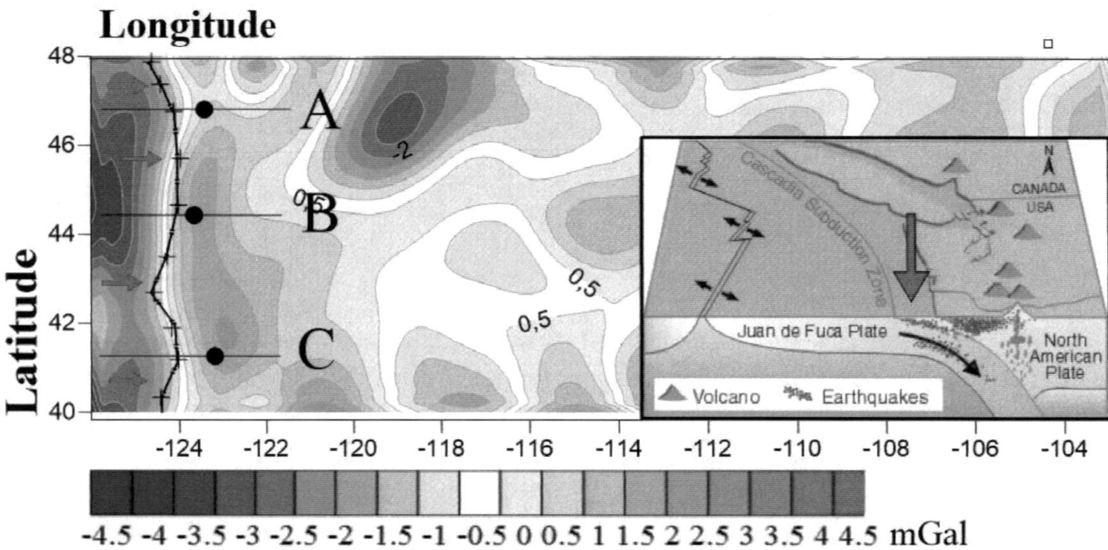

Figure 3. Gravity anomaly map after SVD process. There is contrast color from one area and others. The Juan de Fuca oceanic plate dominated by blue color. It shows a low gravity anomaly. The boundary of Juan de Fuca plate and the North American plate is dominated by white color that shows higher gravity anomaly. The increase of gravity anomaly can be correlated with the beginning of Cascadia Subduction Zone. Line A, B, and C show the position of slice. Black line shows the coastline whereas black dots show the beginning of subduction zone in our slices. Inset: Cascadia Subduction Zone (Source: [17]).

Figure 3 displays increasing gravity anomaly from the west coast to the east, as indicted by blue color (low anomaly) and red color (high anomaly). From the figure, we can conclude that the subduction has an elongated shape from North to South. The details coordinate of beginning subduction zone can be known by slicing the SVD map. We slice in three lines (i.e. line A, line B, and line C). From three slices, we got the graphic as shown in Figure 4.

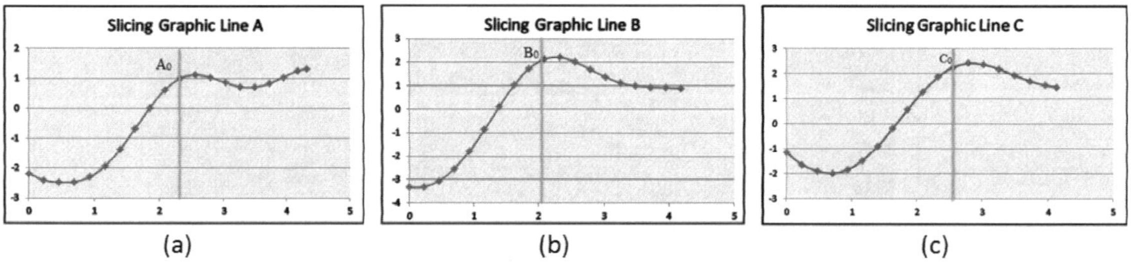

Figure 4 (a, b, c). The picture shows the slicing graphic of SVD anomaly from three lines. There is a similarity from these graphic. Similarity can be shown by the graphics that start with the low value and then increased. The orange line shows the beginning of subduction.

Similarity of three SVD curves can be correlated with the gravity anomaly, which shows the increasing density from Juan de Fuca oceanic plate, the beginning of subductionto the North American Plate. The orange line of each curve in Figure 4 indicates the beginning of subduction from North to South in 46.811° to 41.260° Northern Hemisphere and longitude of -123.436° until -123.204°.

4. Conclusion

Cascadia Subduction Zone formed by the Juan de Fuca oceanic plate subduction towards the North American continental plate. The subduction is elongated from north to south that starts from coordinate 46.811° to 41.260° Northern Hemisphere and longitude of -123.436° into 123.204° in the study area. Cascadia Subduction Zone has the direction from west to east that can be proofed by the increase of gravity anomaly from -140 mGals into 320 mGals.

Acknowledgments

We thank TOPEX data for free air corrected gravity data and to the reviewers of International Conference on Theoritical and Applied Physics (ICTAP) 2017 for the suggestions. The authors would like to thank to the Faculty of Mathematics and Natural Sciences for the funding to attend the ICTAP 2017 conference.

References

[1] Obrebski M, Allen R M, Pllitz F, and Hung S H 2011 *Geophysical Journal International* 185 (London: Oxford) p 1003-1021
[2] Atwater T 1989 *The Geology of North American* (America: Geological Society of America) p 21-72
[3] Lonsdale P 1991 *AAPG* Mem 47 (California: The Gulf and Peninsular Province of the Californias) p 87-124
[4] Govers R and Meijer P Th 2001 *Earth and Planetary Science Letters* (Amsterdam: Elsevier) p 115-131
[5] Xue M and Allen R M 2007 *Earth and Planetary Science Letters* (Amsterdam: Elsevier) p 264
[6] Chen C, Zhao D and Wu S 2015 *Tectonophysics* (Amsterdam: Elsevier) p 73-88
[7] Clague J J 1997 *Geophysics* (British: Geological Survey of Canada) p 439-460
[8] Campione M and Capitani G C 2013 *Nature Geoscience* vol 6 p 847-851
[9] Nediamovic M R, Bohnenstiehl D R, Carbotte S M, J P Canales, R P Dziak 2009 *Earth and Planetary Science Letters* (Amsterdam: Elsevier)
[10] Romanyuk T V, Blakely R and Mooney W D 1998 *Phys. Chem. Earth.* vol 23 no 3 (California: U.S. Geological Survey) p 297-301
[11] Grandis H and Dahrin D 2014 *J. Math. Fund.* Sci. vol 46 no 1 (Bandung: ITB Journal Publisher) p 1-13
[12] Harabaglia P and Doglioni C 1998 *Geophysical Research Letters* vol 25 no 5 (Oklahoma) p 703-706
[13] Blakely R J 1995 *Potential Theory in Gravity and Magnetic Applications* (United States: Cambridge University Press)
[14] Nurpratama M I and Darusman C A 2015 *Proceedings in World Geothermal Congress 2015* (Melbourne, Australia)
[15] Gupta V K and Ramani N 1982 *Geophysics* vol 47 no 12 p 1706-1715
[16] American continent map data 2006 *National Land Cover Database (NLCD) Delivers Update* available at https://eros.usgs.gov/lir/usgs (accessed 08.08.17)
[17] Cascadia Subduction Zone 2010 *U.S. Geological Survey- Reducing the Risk from Volcano Hazard* available at https://pubs.usgs.gov/fs/2000/fs060-00/ (accessed 08.08.17)

IOP Conf. Series: Journal of Physics: Conf. Series **1011** (2018) 012033 doi:10.1088/1742-6596/1011/1/012033

Seismic Imaging in Complex Geological Area Using Pre-Stack Depth Migration (PSDM) and Tomography Method

A B Prabowo[1] and Sudarmaji[1]

[1] *Geophysics Universitas Gadjah Mada*

Sekip Utara, Bulaksumur, Yogyakarta, Indonesia 55281

Agung.budi.p@mail.ugm.ac.id

Abstract. Pre-Stack Depth Migration (PSDM) method is required for seismic imaging in the area with complex geological structures such as fault, anticline, or syncline. Result of the PSDM depends on the interval velocity model as its input. A valid interval velocity model can yield appropriate seismic section and vice versa. One way to get an accurate interval velocity model is tomography method. This method can refine interval velocity model until it has minimum error. In this research, it's used 2 kinds of tomography method, i.e. horizon based tomography and grid based tomography. Horizon based tomography considers the depth model in interval velocity refinement process, while grid based tomography doesn't. Tomography is done till 6th iteration where the result of 6th iteration show the best quality of seismic section. The results show that the best quality section is PSDM using horizon based tomography. It has more geological sense than the PSDM section using grid based tomography and PSDM without tomography. Furthermore, PSDM using horizon based tomography is capable to correct pull-up effect with better continuity.

Keywords: PSDM, horizon based tomography, grid based tomography.

1. Introduction

Reflection seismic method is one of the geophysical methods to subsurface imaging. This method can yield section that have higher resolution and wide area coverage than other geophysical methods. One of the stages that determines the quality of section on seismic method is seismic data processing. In such case of simple geological structures and lateral velocity variations, the data processing is sufficient until PSTM. But, in complex geological area such as fault and presence of salt dome or reef carbonate with high lateral heterogeneity, it is necessary to apply pre-stack depth migration (PSDM) using interval velocity model as its input.

Previous PSDM research was conducted Atmaja and Sudarmaji by correcting the pull-up effect in seismic data using PSDM. The PSDM result indicates that the pull-up effect which is still visible on the PSTM section can be identified and corrected [1]. Then, Nugroho and Sudarmaji compared the horizon based tomography and grid based tomography methods in the PSDM on seismic data. The results show that horizon based tomography has better quality than grid based tomography [2].

2. Method

This research used 2D reflection seismic data located in Alberta foothills, Canada which was acquired by Husky Oil and Talisman Energy. Alberta Foothills has complex geological structure. On the southern part of the region is dominated by thrust faults, while in the north is dominated by folds [3]. The complexity of geological structure in Alberta Foothills can be seen in geological section that parallel to the seismic line as shown in figure 1. The seismic data has high velocity variation because it is located in complex geological area.

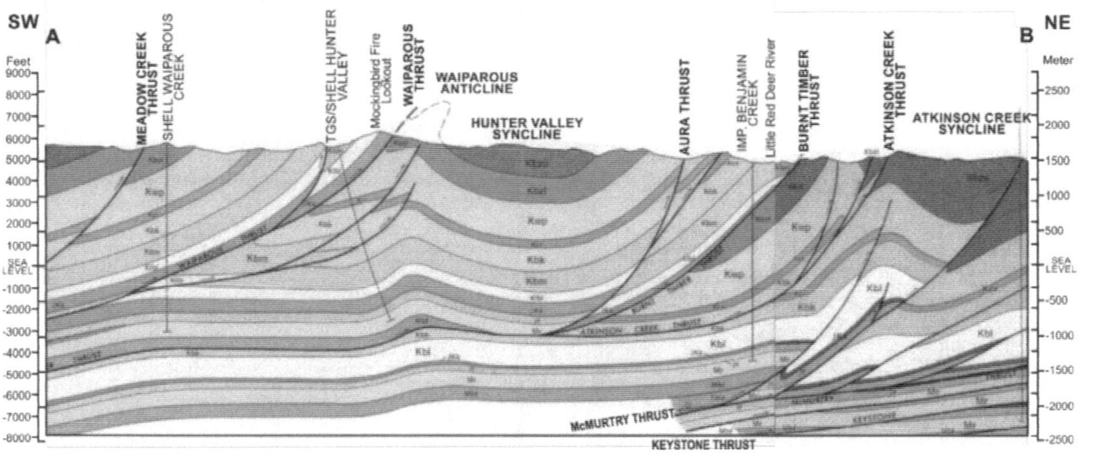

Figure 1. Geological section in Alberta Foothills that parallel to the seismic line. It shows geological complexity such as folds and faults in Alberta Foothills (Modified from [4] and [5]).

The seismic data consists of 143 shot gather, 4 ms sampling rate, trace number 39764, and 3000 ms recording length [6]. This seismic data processing used software Paradigm Echos and Geodpeth. Steps in seismic data processing is divided into 2 parts, first is pre-stack time migration (PSTM). The second step is pre-stack depth migration (PSDM). On PSDM steps include of interval velocity model refinement process using horizon based tomography and grid based tomography methods. The tomography process is done iteratively to get the best section. Determination of section quality is seen from flat or not the gather, reflector continuity, and validated by geological section (figure 1).

3. Results and Discussion

3.1. Velocity model analysis

In this research, 6th iteration tomography shows the best result, so the tomography process stopped at 6th iteration. The result of the 6th iteration of horizon based tomography and grid based tomography are shown by figure 2 and 3. Based on these images, it is known that the result of horizon based tomography has a lower interval velocity at shallow depth. In addition, distribution of interval velocity on the horizon based tomography looks smoother than the grid based tomography results. It is caused horizon based tomography consider depth model in interval velocity model refinement process, whereas grid based tomography only refines velocity using statistical calculation considering the depth model [7].

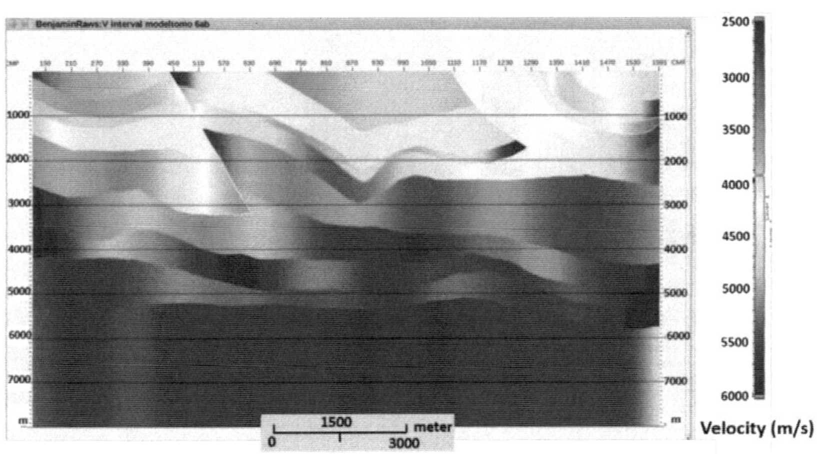

Figure 2. Interval velocity model resulted from horizon based tomography 6th iteration. Blue colour indicates high velocity zone, while yellow colour indicates low velocity zone.

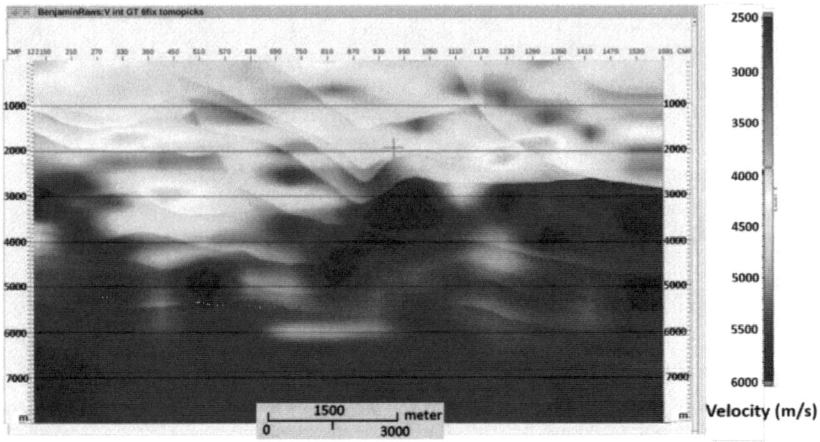

Figure 3. Interval velocity model resulted from grid based tomography 6th iteration. Blue colour indicates high velocity zone, while yellow colour indicates low velocity zone.

3.2. *PSDM section analysis*

Initial PSDM section (without tomography) is shown in figure 4, while PSDM using horizon based tomography section and PSDM using grid based tomography section shown by figures 5 and 6. Based on those sections, PSDM using tomography section shows a significant difference compared to the initial PSDM section. The first difference is shown by the anticline structure in figures 4 (a), 5 (a), and 6 (a). Geometry structure of anticline on PSDM horizon based tomography section looks more suitable with the geological section than the PSDM using grid based tomography section. It shows that horizon based tomography has more geological sense. The next difference lies in the reflector continuities shown by figure 4 (b), 5 (b) and 6 (b). It shows that the reflector on PSDM using horizon based tomography section have better continuities than reflector on the others section. Horizon based tomography is also capable in correcting the pull-up effect shown in figure 4 (c) on the initial PSDM section. The pull-up effect is characterized by the presence of an upward-looking reflector resembling an anticline. The pull-up effect is caused by a significant velocity difference between the Rundle Group and the Blairmore formation. The carbonate-dominated in Rundle group causes the seismic wave velocity in the area to be relatively higher than the seismic wave velocity in the sandstone-dominated in Blairmore Formation. On the PSDM using horizon based tomography section (figure 5 (c)), the pull-up effect is reduced. In contrast, on the PSDM using grid based tomography section, the pull up effect still appears as shown in figure 6 (c).

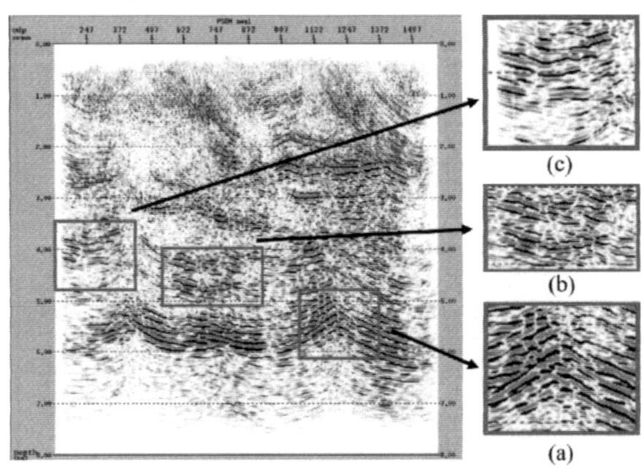

Figure 4. The result of PSDM initial. Input for PSDM initial uses velocity model without tomography (initial interval velocity model).

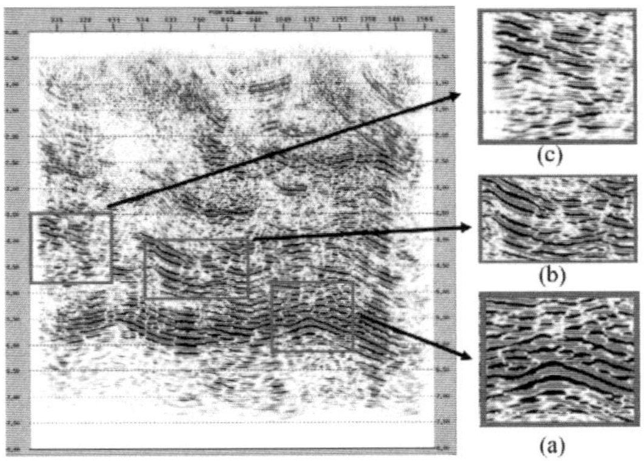

Figure 5. The result of PSDM horizon based tomography. This seismic section uses interval velocity model resulted from 6th iteration of horizon based tomography.

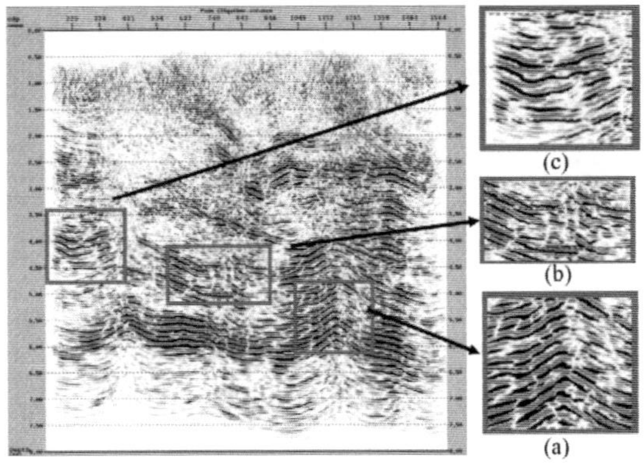

Figure 6. The result of PSDM grid based tomography. This seismic section uses interval velocity model resulted from 6th iteration section of grid based tomography.

3.3. Comparison of PSDM section and PSTM section

PSDM horizon based tomography section has several advantages compared with the PSTM section. PSDM horizon based tomography is capable to reduce the pull-up effect seen in the PSTM section as shown in figure 7 and 8 (a). It is characterized by the appearance of flat reflector. The next difference is shown in figure 7 (b) and 8 (b). The reflector on the horizon based tomography PSDM section has better continuity than the reflector on the PSTM section. It is caused by the accurate velocity model on the PSDM. The Velocity model that used in the PSDM section is generated from horizon based tomography method. This method uses ray tracing algorithm so that it can accurately calculate the velocity. Figure 7 (c) and 8 (c) show the structural imaging differences on the PSDM and PSTM sections. In the PSTM section, the reflector's appearance looks straight, while the horizon based tomography PSDM section shows the anticline structure. The appearance of the anticline structure is corresponding to a geological section shown in figure 1.

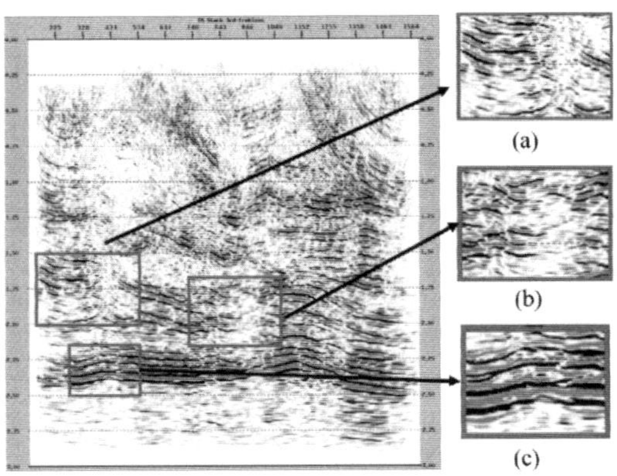

Figure 7. PSTM section. This seismic section uses RMS velocity model as input.

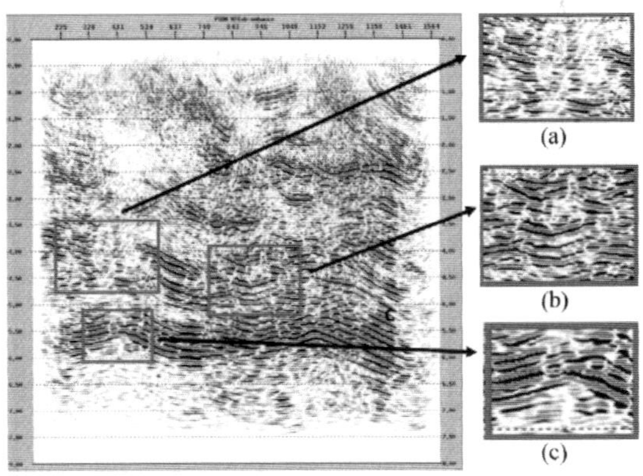

Figure 8. PSDM horizon based tomography. It was compared to PSTM section.

4. Conclusion

Based on the result, we know that PSDM section is better than PSTM section. It shows that the PSDM can handle lateral heterogeneity and high velocity variation in complex geological area. However, tomography method need to be applied on PSDM because this method needs accurate velocity to yield valid seismic section. On the resulted section shows PSDM horizon based tomography yields better seismic imaging than either PSTM or PSDM grid based tomography. In addition, PSDM horizon based tomography yields seismic section close to geological appearance. Finally, the best appropriate method applied in complex geological area such as Alberta foothills is horizon based tomography.

5. Acknowledgement

We would like to say thanks to Paradigm Geophysical inc. for software donation and Chevron for hardware donation given to laboratory of geophysical Universitas Gadjah Mada, so we can finish this research.

References

[1] Atmaja I S and Sudarmaji 2013 *Proc. Pertemuan Ilmiah XXVIII HFI Jateng dan DIY* (Maret 23 *Surakarta*)
[2] Nugroho Y D and Sudarmaji 2014 *J. of Applied Physics* **4** 63
[3] Wright G N, McMechanm M E, and Potter D E G 1994 *Structure and Architecture of Western Canada Sedimentary Basin* (Alberta: Canadian Society of Petroleum Geologists and the Alberta Research Council) pp 25-40
[4] Anonim 1972 *Geology Lake Minnewanka (East Half) West of Fifth Meridian Alberta* (Alberta: Geological Survey of Canada Department of Energy, Mines and Resources)
[5] Anonim 1972 *Geology Wildcat Hills (West Half) West of Fifth Meridian Alberta* (Alberta: Geological Survey of Canada Department of Energy, Mines and Resources)
[6] Anggara R *2014 Pengolahan Data Seismik 2D Lintasan "Eda Calosa" Kaki Pegunungan Alberta, Kanada Berbasis Metode Prestack Time Migration (PSTM) dan Prestack Depth Migration (PSDM) Isotropy dengan Menggunakan Paradigm Epos 4.0* (Yogyakarta: Universitas Gadjah Mada)
[7] Anonim 2009 *EPOS 4.0 Help* (Houston: Paradigm Geophysical Co)

IOP Conf. Series: Journal of Physics: Conf. Series **1011** (2018) 012034 doi:10.1088/1742-6596/1011/1/012034

Spectral-element simulation of two-dimensional elastic wave propagation in fully heterogeneous media on a GPU cluster

Indra Rudianto and Sudarmaji

Department of Physics, Faculty of Mathematics and Natural Sciences, Universitas Gadjah Mada, Yogyakarta, Indonesia

E-mail: indrarudianto.official@gmail.com

Abstract. We present an implementation of the spectral-element method for simulation of two-dimensional elastic wave propagation in fully heterogeneous media. We have incorporated most of realistic geological features in the model, including surface topography, curved layer interfaces, and 2-D wave-speed heterogeneity. To accommodate such complexity, we use an unstructured quadrilateral meshing technique. Simulation was performed on a GPU cluster, which consists of 24 core processors Intel Xeon CPU and 4 NVIDIA Quadro graphics cards using CUDA and MPI implementation. We speed up the computation by a factor of about 5 compared to MPI only, and by a factor of about 40 compared to Serial implementation.

1. Introduction

Over the past several years, numerical modeling using graphics processing units (GPUs) have been successfully used to simulate seismic wave propagation. One of the most widely used approach is the finite-difference method [1, 2] with varying degree of sophistication. It has been successfully applied to simulate seismic wave propagation in fluid saturated porous media [3, 4], high-order simulation [5], and heterogeneous model [6]. Unfortunately, significant difficulties arise in the present of topography [7]. Pseudospectral methods have become popular for acoustic and elastic simulations [8, 9], but are restricted to models with regular variations.

The spectral-element methods have been used for more than 30 years in computational fluid dynamics [10]. The method easily accommodates topography and naturally incorporates free-surface condition. It uses a weak formulation for solving elastic wave equation and lends itself well to be applied in parallel computation with distributed memory [11] and GPU cluster [12].

In this article, we present an implementation of the spectral-element method for simulation of two-dimensional elastic wave propagation in fully heterogeneous media. Based upon this implementation, a realistic topography, a complex subsurface structure, and wave-speed heterogeneity can be efficiently incorporated within the spectral-element mesh. The effects of heterogeneity, influence of realistic surface topography and analysis of simulation time by the implementation of Message Passing Interface (MPI) and Compute Unified Device Architecture (CUDA) are investigated.

IOP Conf. Series: Journal of Physics: Conf. Series **1011** (2018) 012034 doi:10.1088/1742-6596/1011/1/012034

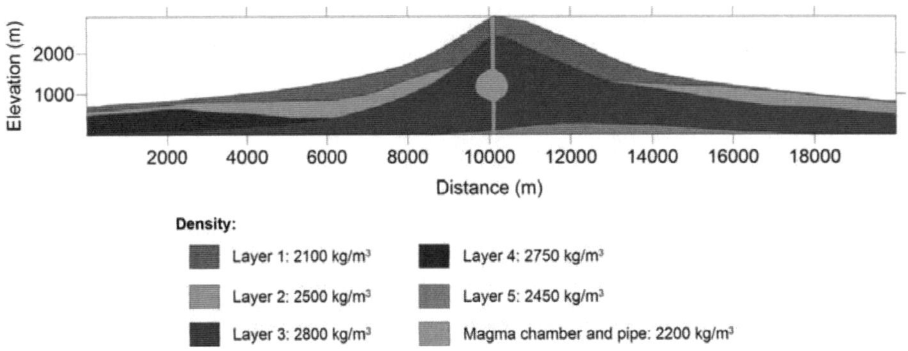

Figure 1. Subsurface structures of the model used for simulation based on Merapi gravity data inversion in 2011 [13].

2. Mesh Implementation

2.1. Model Setting

We define a model for simulation based on Merapi gravity data inversion in 2011 [13] as shown in Figure 1. It had a dimension from 0 to 20 km horizontally, and from 0 to 2912 m vertically. Throughout the entire model, P-wave speed was approximated based on the value of density [14], which may be written in the form

$$V_P = 108.28\,\rho^4, \tag{1}$$

with V_P denotes P-wave speed in m/s and ρ denotes density of medium in g/cm^3. The S-wave speed was approximated using the ratio of P- and S-wave speed given by [15], which has the value of 1.86. The summary of material properties of the model is given in Table 1.

Table 1. Material properties of the model.

Property	Density (kg/m^3)	P-wave speed (m/s)	S-wave speed (m/s)
Layer 1	2100	2106	1132
Layer 2	2500	4230	2274
Layer 3	2800	6656	3578
Layer 4	2750	6193	3329
Layer 5	2450	3901	2098
Magma chamber and pipe	2200	2537	1364

2.2. Mesh Generation

To incorporate a considerable surface topography, complex subsurface interfaces, and 2-D wave speed heterogeneity, we used an unstructured quadrilateral meshing technique with an initial element size of 30 m. We used a paving mesh scheme algorithm provided by TRELIS meshing software to produce a fairly uniform mesh shape and accommodated curved interfaces in the model (see Figure 2). The total number of elements in the model is 30999.

2.3. GPU Cluster

Simulation was carried out on a GPU cluster, which consists of 24 core processors Intel Xeon CPU E5-620 @2.4 GHz and 4 NVIDIA Quadro K4000 graphics cards using CUDA and MPI implementation

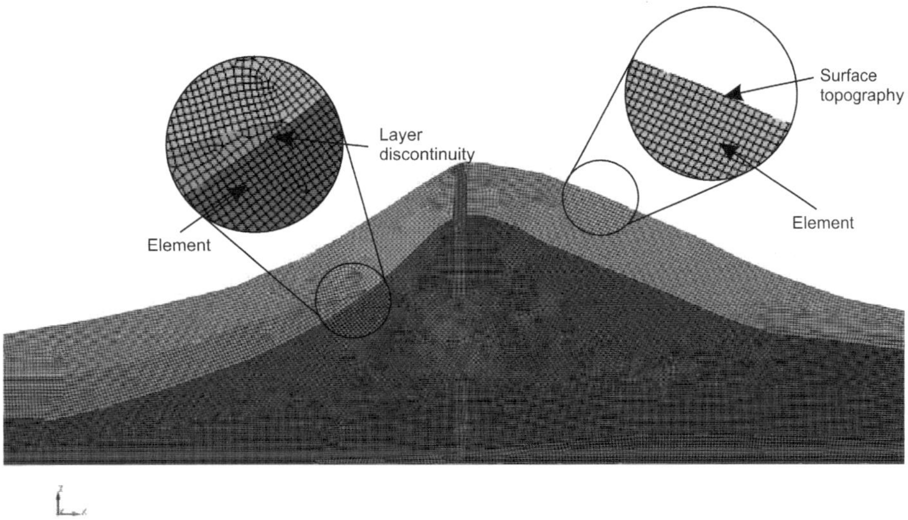

Figure 2. Unstructured quadrilateral mesh generation of our model. Total number of elements in our model is 30999. Each small rectangle in the model denotes one element. By our mesh implementation, we can observe that surface topography and curved layer discontinuities can be easily accommodated.

provided by SPECFEM2D package [12] at the Computational and Seismic Laboratory, Geophysics Sub-Department, Universitas Gadjah Mada.

3. Results and Discussion

3.1. Snapshots of Wave Propagation

We decomposed the model into 30999 quadrilateral elements. The minimum and maximum distances between grid points in the model are 14.07 m and 71.71 m respectively. In each spectral element, we interpolated a function with the polynomial degree of $N = 4$, and thus each element contains $(N+1)^3 =$ 125 Gauss-Lobatto-Legendre (GLL) integration points. The source of the earthquake is an elastic point source placed at the coordinates $\mathbf{x}_s = (10009, 1912)$. We used a Ricker source time function with a corner frequency of 8 Hz. The time step used is $\Delta t = 0.1$ ms, and we propagated the signal for 100,000 time steps, thus total simulation time is 10 s.

Snapshots of the simulation for the vertical component of velocity wave field are displayed in Figure 3. We have normalized the color scale to the maximum of all time steps. We observed at $t = 1$ s, the wave field begins to propagate on each layer. We also observed a fairly dense wave field in the uppermost layer, magma chamber and pipe. At $t = 3$, 5, and 8 s, a relatively dense wave field is still quite visible at the layer 1, magma chamber and pipe compared to the layer 3 which is concentrated around the top, as well as the right and left of the slope. At $t = 10$ s, only at the layer 1, magma chamber and pipe, the wave field is still quite visible despite its small amplitude.

3.2. Performance of Simulation

We performed the simulation using three different schemes. The first scheme is a Serial simulation (1 CPU, 1 Process). This scheme requires 1 hour, 25 minutes and 31 seconds of the wall-clock time. The second scheme is an MPI only simulation (48 CPU, 48 Processes). This scheme requires 10 minutes and 50 seconds. The last scheme is MPI and CUDA simulation (4 GPU, 4 Processes). This scheme requires only 2 minutes and 8 seconds. By using MPI only implementation, we can speed up the computations by a factor of about 7 compared to Serial implementation. By using MPI and CUDA, we can speed up

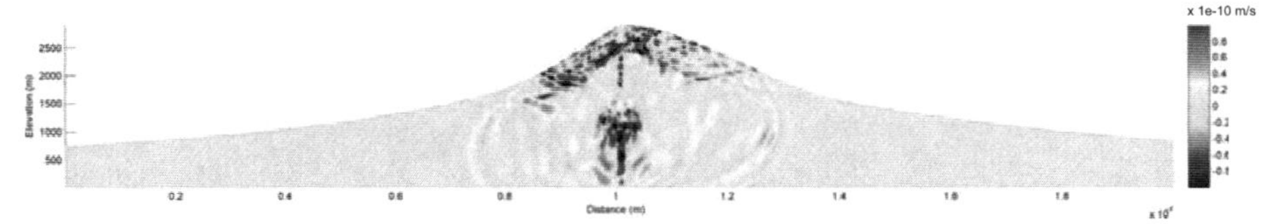

(a) t = 1 s

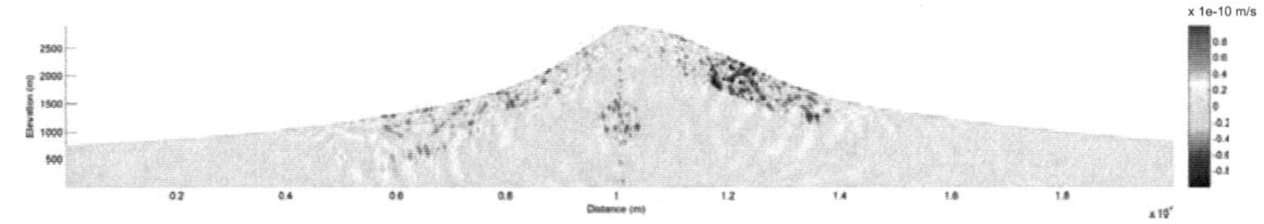

(b) t = 3 s

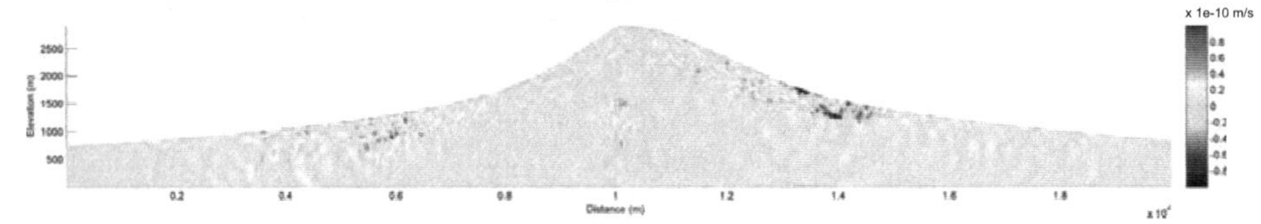

(c) t = 5 s

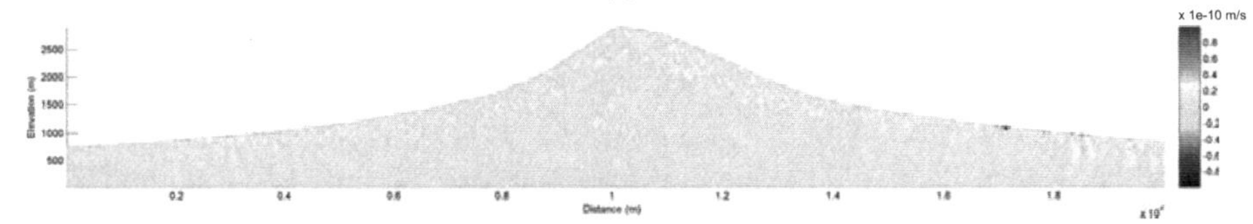

(d) t = 8 s

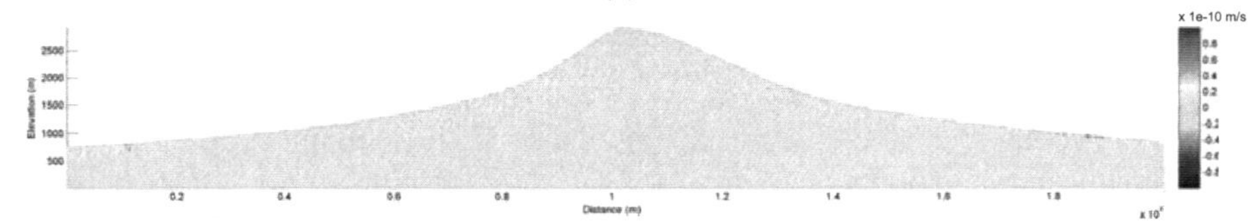

(e) t = 10 s

Figure 3. Snapshots of the simulation for the vertical component of velocity wave field subsequently for $t = 1$ s (a), $t = 3$ s (b), $t = 5$ s (c), $t = 8$ s (d), $t = 10$ s (e).

the computations by a factor of about 5 compared to MPI only, and by a factor of about 40 compared to Serial implementation.

4. Conclusions

We presented an implementation of the spectral-element method for simulation of two-dimensional elastic wave propagation in fully heterogeneous media. Based upon this implementation, a realistic geological feature including surface topography, a complex subsurface interface and 2-D wave-speed heterogeneity can be easily accommodated.

By using MPI only implementation, we can speed up the computation by a factor of about 7 compared to Serial implementation. By using MPI and CUDA implementation, we can speed up the computation by a factor of about 5 compared to MPI only, and by a factor of about 40 compared to Serial implementation.

Acknowledgments

We would like to express our gratitude to PT. Chevron Indonesia for hardware donations given to Geophysics Sub-Department, Universitas Gadjah Mada. The authors would also like to say thanks to

IOP Conf. Series: Journal of Physics: Conf. Series **1011** (2018) 012034 doi:10.1088/1742-6596/1011/1/012034

Yoyok Adisetya for his help building and configuring a GPU cluster at the Computational and Seismic Laboratory.

References

[1] Virieux J 1986 *Geophysics* **51** 889–901
[2] Graves R W 1996 *Bull. of the Seismol. Soc. of Am.* **86** 1091–1106
[3] Sudarmaji, Sismanto, Waluyo and Soedijono B 2016 *AIP Conf. Proc.* **1755**
[4] Sudarmaji, Sismanto and Waluyo 2014 *Int. Conf. on Phys.*
[5] Hasym I B, Sudarmaji and Waluyo 2014 *Int. Conf. on Phys.*
[6] Permana T and Sudarmaji 2014 *Int. Conf. on Phys.*
[7] Robertsson J O A 1996 *Geophysics* **61** 1921–1934
[8] Reshef M, Koslofft D, Edwards M and Hsiung C 1988 *Geophysics* **53** 1175–1183
[9] Reshef M, Koslofft D, Edwards M and Hsiung C 1988 *Geophysics* **53** 1184–1193
[10] Patera A T 1984 *J. Comput. Phys.* **54** 468–488
[11] Komatitsch D and Vilotte J P 1998 *Bull. of the Seismol. Soc. of Am.* **88** 368–392
[12] Komatitsch D, Erlebacher G, Göddeke D and Michéa D 2010 *J. Comp. Phys.* **229** 7692–7714
[13] Aziza L 2014 *Subsurface modelling of Merapi volcano using gravity method based on 2011 data* Bachelor's thesis Universitas Gadjah Mada
[14] Gardner G H F, Gardner L W and Gregory A R 1974 *Geophysics* **39** 770–780
[15] Ratdomopurbo A and Poupinet G 2000 *J. of Volc. and Geotherm. Research* **100** 193–214

IOP Conf. Series: Journal of Physics: Conf. Series **1011** (2018) 012035 doi:10.1088/1742-6596/1011/1/012035

MT2D Inversion to Image the Gorda Plate Subduction Zone

Y K Lubis[1], S W Niasari[1], and E Hartantyo[1]

[1] Department of Physics, Universitas Gadjah Mada, Yogyakarta, Sekip Utara, Yogyakarta, Indonesia 55281

[1] Corresponding author e-mail: yennikhairanilubis@gmail.com

Abstract. The magnetotelluric method is applicable for studying complicated geological structures because the subsurface electrical properties are strongly influenced by the electric and magnetic fields. This research located in the Gorda subduction zone beneath the North American continental plate. Magnetotelluric 2D inversion was used to image the variation of subsurface resistivity although the phase tensor analysis shows that the majority of dimensionality data is 3D. 19 MT sites were acquired from EarthScope/USArray Project. Wepresent the image of MT 2D inversion to exhibit conductivity distribution from the middle crust to uppermost asthenosphere at a depth of 120 kilometers. Based on the inversion, the overall data misfit value is 3.89. The Gorda plate subduction appears as a high resistive zone beneath the California. Local conductive features are found in the middle crust downward Klamath Mountain, Bonneville Lake, and below the eastern of Utah. Furthermore, mid-crustal is characterized by moderately resistive. Below the extensional Basin and Range province was related to highly resistive. The middle crust to the uppermost asthenosphere becomes moderately resistive. We conclude that the electrical parameters and the dimensionality of datain the shallow depth(about 22.319 km) beneath the North American platein accordance with surface geological features.

1. Introduction

The Gorda plate is a southernmost microplate of the Juan de Fuca plate system. The Gorda plate covers about 45,000 km^2 as estimated by Chaytor [1]. The western frontier of the Gorda plate is the Gorda Ridge, and the south frontier is the Mendocino Fracture Zone. Both of them restrict the Gorda plate from the Pacific plate. The half-spreading rates of the Gorda Ridge are 27.5 mm/yr near the Blanco Fracture Zone on the north and 14.0 mm/yr to the south near Mendocino Fracture Zone since 2 Ma [2]. Moreover, the relative movement rate of the North American plate is about 58 mm/yr, and it contrasts to the Pacific plate [3].The Gorda plate moves relatively to the east and subducts the North American plate since late Mesozoic and early Tertiary[4].

The subduction of the Gorda plate, as a part of the largest of Juan de Fuca plate system, beneath the North America plate formed several geological features such as the Klamath Mountain, the Cascadia Range, the Central Basin and Range, and the Wasatch fault. Furthermore, the unit of individual rock of the North American plate generally is younger to the eastward [4].The geological structures denote the variation of the subsurface physical properties namely resistivity, density, seismicity, susceptibility, etc. These parameters are commonly employed to identify the subsurface structure includingto imagethesubductionzone.The Central Basin and Range (CBR) showed a high resistivity value as a result of thermal releasing and basaltic hydrating [9]. The contrast resistivity around CBR is a representation of the upper crust resistivity value [8].The study area is located in the western North American plate (Figure 1). This research aims to delineate the Gorda plate subduction zone beneath the North American plate using 2D inversion magnetotelluric data. Magnetotelluric is a geophysical method that usenatural electromagnetic wavesto describe the subsurface resistivity distribution [10][11].This method informs the subsurface resistivity structure by simultaneous measuring of an electrical field and magnetic field components [12]. In this study, we also analyze the dimensionality

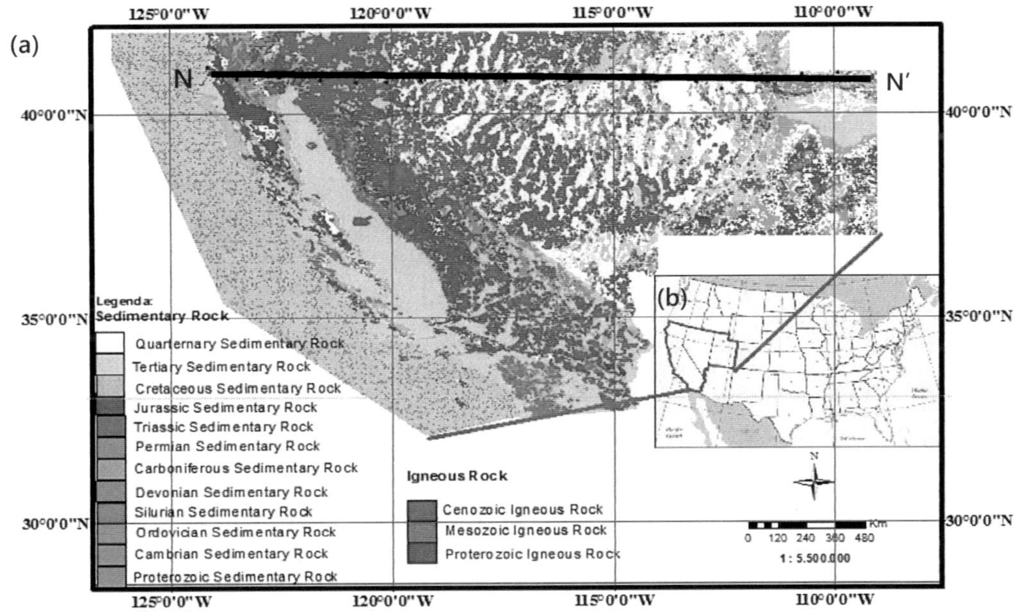

of magnetotelluricdata by using phase tensor analysis.

Figure 1: (a)TheSurface Geology Map of California, Nevada, and Utah shows the distribution of lithologies. A bold line green N-N' with black dots denote MT sites that extend ~1300 kilometers [5, 6, 7]. (b) Inset: The North American Map (Modified from USGS Geology Map, 2017)[8]

2. Materials and Methods

This research employs the magnetotelluric method. We have taken 19 MT sites as data. These data were acquired from the openaccess datafrom National Science Foundation in Earth Scope USArray Project [13]. The site spacing is 70 kilometers and the range frequency used is 10^{-1} – 10^{-4} Hz. These MT sites elongated about 1300 kilometers from Redwood Park of California to Brown Park of Utah. These data consist of the coordinate, the impedance, and the frequency of the MT sites.

The two-dimensionality of MT data inversion was implemented using WinGlink software. The MT data were inverted applying the nonlinear gradient conjugate.The first step that we didwasmasking process. Subsequently, the homogeneous half-space model was used to createthe initial model in resolution setting of the model. The next process wastodetermineinversion parameters.We selectXY and YX components, error floor 5%,and the weight function for horizontal and vertical orientation. Moreover, we conduct the tau test by plotting rms value until L-curve is formed. Then, the tau value optimum of the tau test (τ) chosen here is $\tau = 10$. The tau value optimum is the most balanced value of the L-curve. Thistau test controls the relative weight and the roughness [14]. Furthermore, the last stage wasthe inversion process to produce a convergent model whichwas indicated by the absence of changes to the model although the iteration is still conducted.

Furthermore, the dimensionality of magnetotelluric data wasconducted using the skew angle value (β), and the ellips shape of tensor phase was done employing tensor phase analysis [15]. This analysis results the characteristic of data. The circle shape with skewness value (β)distributed within $-3^0 > \beta > 3^0$indicated1D character. The ellips shape signified 2D character when skewness value (β) range is -3^0 up to 3^0 or 3D character when skewness value (β) is less than -3^0 or more than 3^0. Mathematically, the formula of tensor phase can be written as follows:

$$\Phi = R^T(\alpha - \beta)\begin{bmatrix} \Phi_{max} & 0 \\ 0 & \Phi_{min} \end{bmatrix} R(\alpha + \beta) \quad (1)$$

$$\alpha = \frac{1}{2}\arctan\left(\frac{\Phi_{12} + \Phi_{21}}{\Phi_{11} + \Phi_{22}}\right) \quad (2)$$

$$\beta = \frac{1}{2}\arctan\left(\frac{\Phi_{12} + \Phi_{21}}{\Phi_{11} + \Phi_{22}}\right) \quad (3)$$

R^T is transposed matrix, β is skewness angle, and α is an angle that express the dependency of tensor toward coordinate system. The geoelectrical strike orientation is shown by the remaining of $\alpha - \beta$[14].

3. Results and discussion

The result of 2D inversion model of magnetotelluric is shownin Figure 2. The root mean square (rms) value of this model generally is 3.889. This rms value indicates the suitability of the data and the electrical properties model. Then, the phase tensor analysis is given by Figure 3 in period 19.6928 s. The phase tensor model describes the majority of magnetotelluric data character is 3D. The characters of the data dimensionality candescribesthe relevancy of the subsurface resistivity properties with geological structures on surface.

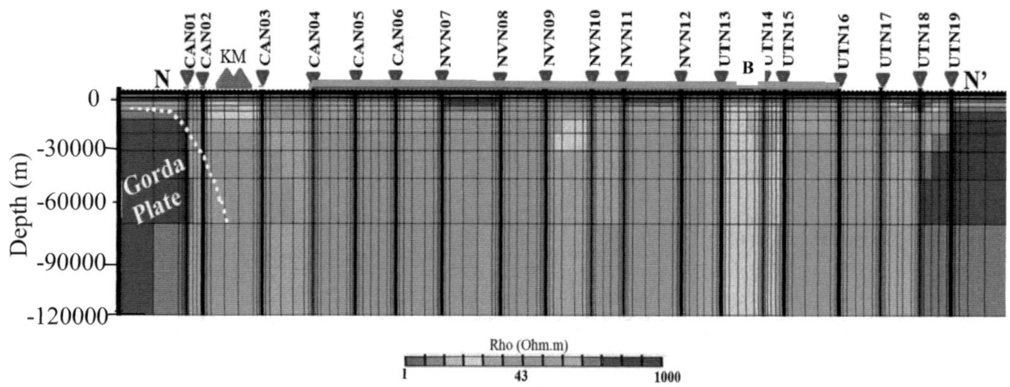

Figure 2:2D resistivity Model of profile N-N'. The Gorda plate subduction is assigned by white dashes and associated to the high resistivity zone. The shallow resistivity values exist below the Klamath Mountain (red triangles) and the Bonneville Lake (BL). The shallow conductivity values are found in the Central Basin and Range province (orange bolt line).

The highest rms value found is 13.237 at UTN18 site in the eastern Utah. The data analysis described at this site was3D character. Moreover, the local conductive anomaly are found in the upper crust (~22 km) underneath this site, but the geological feature on the surface can't be connected. This might be caused by the incompatibility of the 3D data of UTN18 with the 2D inversion magnetotelluric used at shallow depth. Otherwise, the lowest rms value is 1.047 located in Nevada at NVN12 site with 2D character. This site has a low resistivity at shallow/deep depththat relates to the addition of the widespread extension of Central Basin and Range (CBR) province. CBR is part of a great basin filled with streams flowing into the lakes(e.g Lahontan Lake and Bonneville Lake) or drowning in the valleys [13]. Subsequently,low resistivities(10-100Ωm) in the mid-crustal extend from western Nevada to Western Utah at depths > 20 km. These zones are related to the addition of the widespread extension of the Central Basin Range. Further below the CBR,the low resistivity zones extend to a depth 120 km. These can be explained by magma mobilization from the upper asthenosphere.

Thesubduction zone of the Gorda plate as shown in Figure 2 is characterized by high resistivity (more than 500 Ωm). Thisresistoris continuous from shallow depth to greater depth, and itis interpreted as the body of theGorda plate.Atthe east, the shallow conductorwasfound belowKlamath Mountain (KM)estimatedas a resultof arc magmatism and molten products due to plate subduction[16]. The presence of the melt product is decided as the explanation for the high conductivitiesbeneath fore arc and volcanic arc in the state of California [10]. High conductive crustalalso appears in northeastern Utah that passes the Bonneville Lake. This conductor is associated to fluid content emplaced in shallow depth and magmatism activity from the greater depth (more than

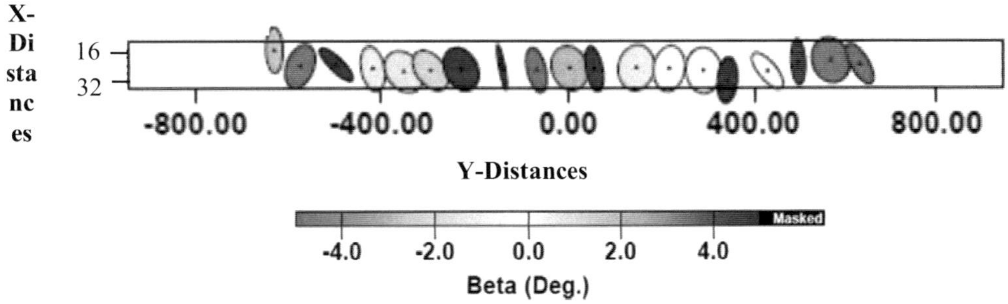

120 km).

Figure 3. The result of Phase Tensor analysis replects the dimensionality of magnetotelluric data in periode 19,6928 s. The 3D characters of data are assigned by the ellips shape with skewness value (β) is less than -3^0 or more than 3^0. The 2D characters of data is showed by the ellips shape with skewness value range is -3^0 up to 3^0. The 1D characters of data has the circle shape and the skewness value is same with 2D character. The dimensionality characters of UTN18 and NVN12 are showed by the red rectangular and the blue rectangular.

3. Conclusion
In summary, the magnetotelluric 2D inversion model has been applied to delineate the subsurface resitivity distribution underneath the western North American plate in the Gorda plate zone. This 2D inversion magnetotelluric model and the characteristic of the short period magnetotelluric data (19s)are correspond to the geological structure on the surface.

4. Acknowledgments
This research has been possible due to the availability of magnetotelluric data from EarthScope USArray Project. The authors are very thankful to the reviewers for their constructive suggestionto improve this paper.

5. References
[1] Chaytor J D, Goldfinger C, Dziak R P and Fox C G 2004 Active deformation of the Gorda plate: Constraining deformation models with new geophysical data *Geology*__32__353
[2] Wilson D S 1988 Tectonic history of the Juan de Fuca Ridge over the last 40 million years *Journal of Geophysical Research*__93__11863

[3] Atwater T 1970 Implications of Plate Tectonics for the Cenozoic Tectonic Evolution of Western North America *Geological Society of America Bulletin* **81** 3513

[4] Jachens R C and Griscom A 1983 Three-dimensional geometry of the Gorda Plate beneath northern California *Journal of Geophysical Research: Solid Earth* **88** 9375–92

[5] California geologic map data *Interactive maps and downloadable data for regional and global geology, geochemistry, geophysics, and mineral resources; products of the USGS Mineral Resources Program* available at https://mrdata.usgs.gov/geology/state/state.php?state=CA (accessed 18.05.17

[6] Nevada geologic map data *Interactive maps and downloadable data for regional and global geology, geochemistry, geophysics, and mineral resources; products of the USGS Mineral Resources Program* available at https://mrdata.usgs.gov/geology/state/state.php?state=NV (accessed 18.05.17)

[7] Utah geologic map data *Interactive maps and downloadable data for regional and global geology, geochemistry, geophysics, and mineral resources; products of the USGS Mineral Resources Program* available at https://mrdata.usgs.gov/geology/state/state.php?state=UT (accessed 18.05.17)

[8] Anon 2017 Geologic maps of US states *Interactive maps and downloadable data for regional and global geology, geochemistry, geophysics, and mineral resources; products of the USGS Mineral Resources Program*

[9] Wannamaker P E, Booker J R, Jones A G, Chave A D, Filloux J H, Waff H S and Law L K 1989 Resistivity cross section through the Juan de Fuca Subduction System and its tectonic implications *Journal of Geophysical Research: Solid Earth* **94** 14127–44

[10] Bedrosian P A and Feucht D W 2014 Structure and tectonics of the northwestern United States from EarthScopeUSArraymagnetotelluric data *Earth and Planetary Science Letters* **402** 275–89

[11] Naidu G D 2012 Deep Crustal Structure of the Son-Narmada-Tapti Lineament, Central India *Springer Theses*

[12] Chen C, Zhao D and Wu S 2015 Tomographic imaging of the Cascadia subduction zone: Constraints on the Juan de Fuca slab *Tectonophysics* **647-648** 73–88

[13] Kapinos G, Montahaei M, Meqbel N and Brasse H 2016 Three-dimensional electrical resistivity image of the South-Central Chilean subduction zone *Tectonophysics* **666** 76–89

[14] Rodi W and Mackie R L 2001 Nonlinear conjugate gradients algorithm for 2-D magnetotelluric inversion *Geophysics* **66** 174–87

[15] Caldwel T G, Bibby H M and Brown C 2004 *Geophys. J. Int.* **158** 457-69

[16] Meqbel N M, Egbert G D, Wannamaker P E, Kelbert A and Schultz A 2014 Deep electrical resistivity structure of the northwestern U.S. derived from 3-D inversion of USArraymagnetotelluric data *Earth and Planetary Science Letters* **402** 290–304

IOP Conf. Series: Journal of Physics: Conf. Series **1011** (2018) 012036 doi:10.1088/1742-6596/1011/1/012036

Geo-electrical and geological strikes of the Mount Lamongan geothermal area, East Java, Indonesia – preliminary results

L R Nugraheni[1], S W Niasari[2] and M Nukman[2]

[1]Undergraduate Program of Geophysics, Universitas Gadjah Mada, Yogyakarta, Indonesia 55281

[2]Geophysics sub-Department, Universitas Gadjah Mada, Yogyakarta, Indonesia 55281

Email: nugrahenirita@gmail.com

Abstract. Geothermal manifestations located in the Tiris, Mount Lamongan, Probolinggo, consist of warm springs. These warm springs have temperature from 35° until 45°C. Tiris fault has NW-SE dominant orientation, similar to some lineaments of maars and cinder cones around Mount Lamongan. The Mount Lamongan geothermal area is situated between Bromo and Argapura volcanoes. This study aims to map the geo-electrical and geological strikes in the study area. Phase tensor analysis has been performed in this study to determine geo-electrical strike of study area. Geological field campaign has been conducted to measure geological strikes. Then, orientation of geo-electrical strike was compared to geological strike. The result presents that the regional geological strike of study area is NW-SE while the orientation of geo-electrical strike is N-S.

1. Introduction

Mount Lamongan is a unique basaltic stratovolcano at East Java, Indonesia. Lamongan Volcanic Field is surrounded by 61 basaltic cinders, 29 prehistoric maars, 2 prehistoric vents, and 1 active vent [1] (see figure 1). This mountain is situated between Bromo volcano to the west and Argapura volcano to the east. Mount Lamongan belongs to volcanic belt zone as a result of subduction between the Indo-Australian Oceanic Plate and the Eurasian Continental Plate.

Warm springs as an indication of a geothermal system and geological structures as one component of a geothermal system are found in the Tiris. Warm springs temperatures range between 35 and 45°C [2]. The orientation of Tiris fault is NW-SE similar to some maars and cinder cones lineaments [3].

Position of Mount Lamongan geothermal system is still debated whether Mount Lamongan has independent geothermal system or a part of the Argapura geothermal system. Several geothermal researches have been conducted at Mount Lamongan to solve this problem. Geochemical evaluation of the geothermal potential of the Lamongan Volcanic Field was investigated [4]. Distribution of ground temperature at Mount Lamongan was analyzed with remote sensing [5]. The result shows that north side Mount Lamongan have ground temperature distribution which generated by geothermal potential Mount Lamongan [5].

IOP Conf. Series: Journal of Physics: Conf. Series **1011** (2018) 012036 doi:10.1088/1742-6596/1011/1/012036

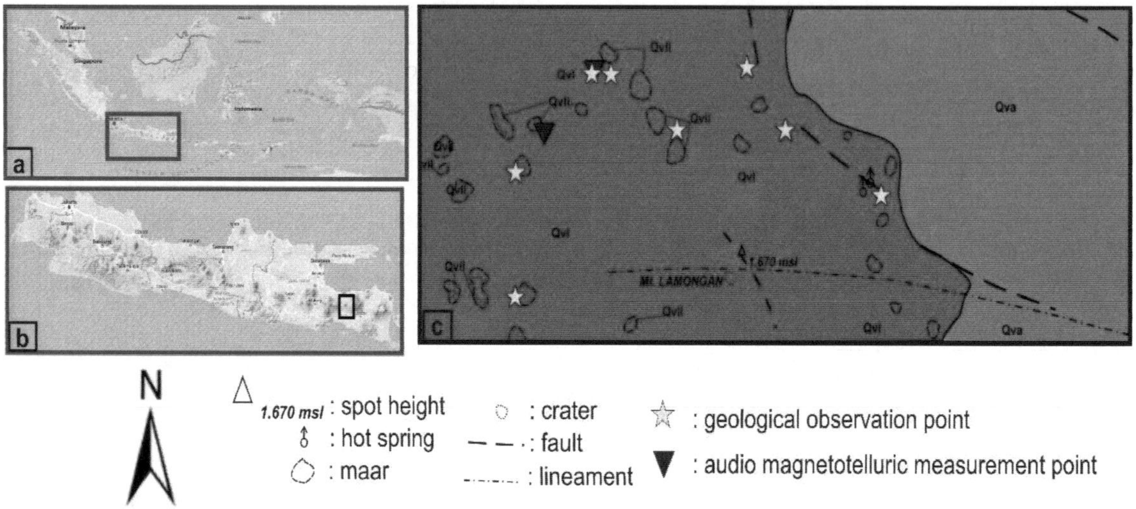

Figure 1. (a) The Indonesian map. Area in purple rectangle is Java Island. (b) The Java Island map. Area in black rectangle is study area. Study area located on volcanic belt zone. (c) Geological map of study area with 2 audio magnetotelluric measurement points (blue rectangles) and 8 geological observation points (yellow stars). Dominant lithology of study area is volcanic rocks from Mount Lamongan consist of Qvl and Qvll (red color) and small part from Mount Argapura consists of Qva (orange color) (modified from [6]).

2. Materials and Methods

Magnetotelluric (MT) is a passive electromagnetic method utilize fluctuation of natural electric (**E**) and horizontal magnetic fields (**H**) in orthogonal directions for determining resistivity structure of the Earth [7]. This method involves the comparison of the horizontal components of the magnetic and electric fields associated with the flow of telluric currents [8]. Element correlates **E** and **H** in frequency domain is impedance tensor (**Z**) [9]. It also reflects the conductivity distribution in some volume below the measurement point [10]. Impedance tensor consists of diagonal (Z_{xx} and Z_{yy}) and off-diagonal components (Z_{xy} and Z_{yx}). Correlation **E** and **H** can be written in matrix form.

$$\mathbf{E} = \mathbf{Z}\,\mathbf{H} \tag{1}$$

$$\begin{bmatrix} E_x \\ E_y \end{bmatrix} = \begin{bmatrix} Z_{xx} & Z_{xy} \\ Z_{yx} & Z_{yy} \end{bmatrix} \begin{bmatrix} H_x \\ H_y \end{bmatrix} \tag{2}$$

Magnetotelluric signals are generated from two sources. Sources from low frequency (less than 1 Hz) are generated from interaction of the solar wind with earth's magnetic field. Meanwhile sources from high frequency (greater than 1 Hz) are generated by thunderstorm activity [11].

Penetration estimate of electromagnetic waves can be determined crudely by skin depth (δ). In electromagnetic, skin depth refers to specific depth where the amplitude of the EM wave reduces to 1/e of its initial amplitude at the surface [12].

$$\delta \approx 503 \sqrt{\frac{\rho}{f}} \tag{3}$$

Where ρ is resistivity (ohm.m) and f is frequency (Hz).

Galvanic distortion is one of the key problems in magnetotelluric [13]. It can distort the regional MT responses. Information of conductivity variations at depth is contained in the phase and shape of the amplitude response as a function of the period. Since phase response is free of the galvanic distortions, it can be used to visualize variations of magnetotelluric data at depth without distortion from local heterogeneities. Phase response or phase tensor (ϕ) is ratio of impedance tensor real (X) and imaginary (Y) parts [14].

$$\phi = X^{-1}Y \quad (4)$$

Phase tensor can be represented graphically in ellipse. Major and minor axes of ellipse depict the principal axes and values of the orientation of the major axes specified by the angle ($\alpha - \beta$). α expresses tensor's dependence on the coordinate system. β or skew angle can be thought of as a rotation and is a measure of tensor's asymmetry [14]. The ellipse axis direction is the electrical strike in 2D data [15]. Information of magnetotelluric data can be defined completely in phase tensor by variations of three coordinate invariants consist of ϕ_{max}, ϕ_{min}, and β [14-15].

The area of study lies at the Mount Lamongan geothermal area. The Mount Lamongan geothermal area is interesting geothermal volcanic system area because its position lies between complex volcanoes. Orientation of geo-electrical strike from phase tensor analysis was compared to regional geological strike. Information of the geo-electrical strike, geological strike, and dimensionality of magnetotelluric data are important in subsurface modeling process.

3. Results

Geological field campaign results show that dominant orientation of geological strike is NW-SE with majority value is about N165°E. This result consistent with Carn's research. From Carn's research in Mount Lamongan, some of the maars have NW-SE orientation comparable with the strike of the Tiris fault [3]. The NW-SE strike orientation of geological features in study area may be caused by regional compression N-S from subduction of the Indo-Australia Plate beneath the Eurasia Plate, which also controls the volcanic belt zone in Sunda Arc.

Phase tensor analysis presents that the dimensionality of data from AMT 2 and AMT 4 are two dimensional for frequency higher than 0.1 Hz. 2-D magnetotelluric data refer to resistivity variation occurring to the depth and one of the horizontal axes direction [7]. These data are graphically represented as ellipses with major axis orientation is N-S (N0°E and N180°E) (see figure 2). The orientation of major axis shows the geo-electrical strike direction of study area. The geo-electrical strike refers to the direction representing the orientation of electric current flow in the subsurface due to lateral inhomogeneity of electric conductivity on Earth [16].

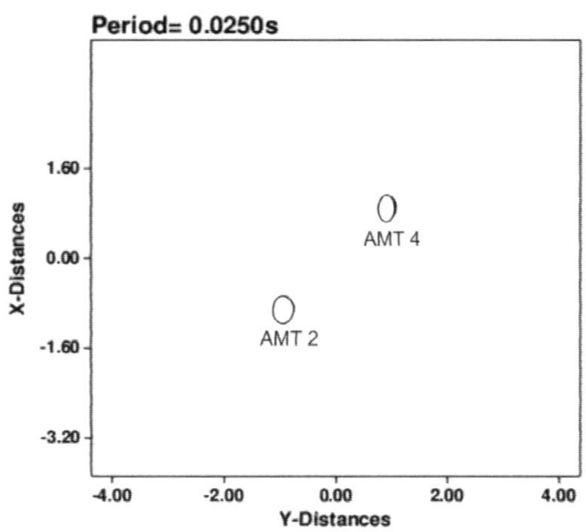

Figure 2. Phase tensor plot from AMT 2 and AMT 4 data in frequency 40 Hz. Dimensionality of data is 2-D. Major axis of ellipse as the geo-electrical strike of area study has orientation N-S.

The results show that orientation of geological and geo-electrical strikes in study area are different (see figure 3). Orientation difference is about 15°. However, the orientation geo-electrical strike still shows the direction of regional compression from N-S, similar with compression which controls the geological strike. This different orientation can be caused by complexity of lithology resistivity and geological structures of study area.

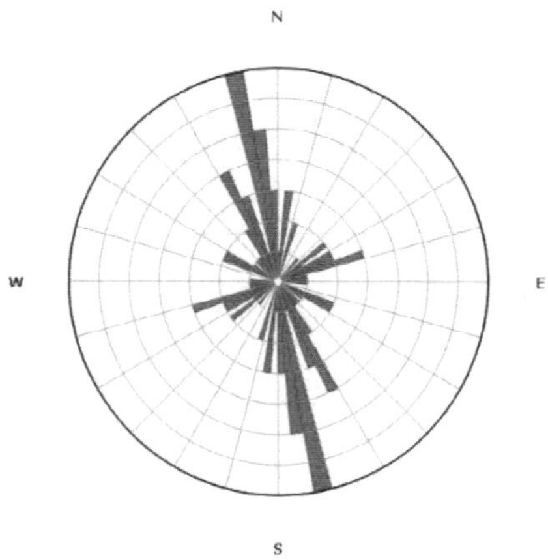

(a)

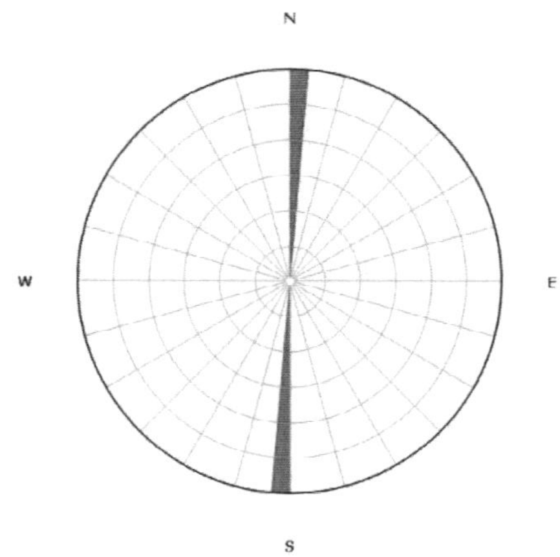

(b)

Figure 3. (a) Rose diagram of geological strikes in study area with NW-SE as dominant orientation. (b) Rose diagram of geo-electrical strikes in study area with N-S as dominant orientation.

4. Conclusion

The orientations geo-electrical and geological strikes of the Mount Lamongan are different. However, their orientations show regional compression from N-S as controller. Information about orientation of geological and geo-electrical strikes in study area are important to get a suitable and accurate model with a minimum error value in subsurface modelling process of magnetotelluric data.

Acknowledgments

We would like to thank to "Kementrian Riset, Teknologi, dan Pendidikan Tinggi" for the PUPT funding no. 2452/UN1.P.III/DIT-LIT/LT/2017. We want to thank Geophysical Laboratory Universitas Gadjah Mada for the permission of use STRATAGEM 26716-01 REV.F and geological tools. We thank Yuni Indrawati for her helps during geological field campaign. Our gratitude to Nanda Ilham and Arriqo' Fauqi Romadlon for their help during geophysical acquisition.

References

[1] Carn S A 2000 *J. Volcanology and Geothermal Research,* **95** 81-108
[2] Deon F, Moeck I, Scheytt T, Jaya M S 2012 *Geophysical Research Abstracts* **14**
[3] Carn S A 1999 *J. Bull Volcano* **61** 92-105
[4] Deon F, Förster H, Brehme M, Wiegand B, Scheytt T, Moeck I, Jaya M S, Putriatni D J 2015 *J. Geothermal Energy* **3** 1-20
[5] Faridah and Krisbiantoro 2014 *J. Berkala Fisika* **17** 67-72
[6] Suharsono and Suwarti T 1992 *Geological map of the Probolinggo quadrangle, Jawa* (Bandung: Geological Research and Development Centre)
[7] Shimpson F and Karsten B 2005 *Practical magnetotellurics* (Cambridge: Cambridge University Press)

IOP Conf. Series: Journal of Physics: Conf. Series **1011** (2018) 012036 doi:10.1088/1742-6596/1011/1/012036

[8] Cagniard L 1953 *J. Geophysics* **18** 605-635
[9] Vozoff K 1972 *J. Geophys.* **37** 1 98-141
[10] Becken M and Burkhardt H 2004 *J. International. Geophys* **159** 69-82
[11] Naidu G D 2012 *Deep crustal structure of the Son-Narmada-Tapti* (Berlin: Springer) p 14
[12] Khyzhnyak M 2014 *Geoelectric strike and its application in magnetotellurics* (Reykjavik: Háskólaprent) p 6-7
[13] Chave A D and Jones A G 2012 *The magnetotelluric method (theory and practice)* (Cambridge: Cambridge University Press) p 134
[14] Caldwell T G, Bibby H M and Brown C 2004 *J. International. Geophys.* **158** 457-469
[15] Booker J R 2014 *J. Surveys in Geophysics* **35** 1 7-40
[16] Niasari S W 2016 *AIP Conference Pro.* **1755** (Maryland: American Institute of Physics) p 100002

IOP Conf. Series: Journal of Physics: Conf. Series **1011** (2018) 012037 doi:10.1088/1742-6596/1011/1/012037

Resistivity Image from 2D Inversion of Magnetotelluric Data in the Northern Cascadia Subduction Zone (United States)

F B Gultom[1*], S W Niasari [2], E Hartantyo [2]

[1] Master of Physics Study program, Faculty of Mathematics and Natural Science, Universitas Gadjah Mada, Sekip Utara, Yogyakarta, 55281 Indonesia

[2] Geophysics Laboratory, Physics Department, Faculty of Mathematics and Natural Science, Universitas Gadjah Mada, Sekip Utara, Yogyakarta, 55281 Indonesia

[*] Corresponding author email: fades.br.gultom@mail.ugm.ac.id

Abstract. Cascadia Subduction Zone (CSZ) lies between Pacific margin and North America plate. The purpose of this research is to identify the CSZ along Oregon, Idaho, Wyoming from conductivity (σ) contrast in the subsurface by using the magnetotelluric (MT) method. MT is an electromagnetic method that use frequency between 10^{-4} Hz and 10^{4} Hz. We obtained the MT data from the EarthScope USArray in the form of EDI-File (five components of the electromagnetic field). We analyzed the MT data using phase tensor and modeled the data using 2D inversion. From the phase tensor analysis, the 3D data dominated the eastern regions. Global data misfit is 6,88, where WYI18 (close to Yellowstone) contributes misfit of 29,3. This means that the model response does not fit the data, which implies the data is not fully 2D. The 2D inversion results are found high resistivity anomalies (more than 500 ohm.m) at shallow depth beneath Oregon and Wyoming, which coresspond to high density anomalies. This high resistivity anomalies might correspond to the north American plate. Thus, it can be concluded that 2D inversion model can be used for most 3D MT data to illustrate the resistivity distribution in the Cascadia Subduction Zone.

1. Introduction

The north western America has more active tectonic than the eastern regions that tend to be more stable [1]. One of the most complex geological structures in North America is the Cascadia Zone [2]. In Cascadia, there is a subduction between the Juan de Fuca (JdF) and the Northern plate [3]. The slab thickness is 75 km that subducts the North American plate with a dip 50° and a rate of 32-44 mm/year [3]. This subduction beneath the North America plate occured in the early Cenozoic [4]. This subduction process is certainly related to the fact of the emergence of the Cascadia Volcanic Arc in the Oregon. Meanwhile, Idaho is surrounded by more varied rock. The emergence of a geological feature found in Idaho such as Columbia river basalt and Snack River Plain [5]. In the North eastern Wyoming revealed the dominance of oldest rock.

In the previous research, it had been shown that the 1D and 3D model of resistivity distribution due to JdF plate subduction using magnetotelluric method [1,6]. In addition, the density anomaly over the research area had been studied using gravity methods [2]. Both of the previous studies results show similarity responses, i.e high resistivity and high-density anomaly beneath the JdF plate subduction area. In the measured MT of the eastern Cascadia detected a very low resistivity to the south [6].

Reviews of The Magnetotelluric (MT) method are provided in some papers and electromagnetic books [7,8]. Magnetotelluric method used natural electromagnetic signals to reveal resistivity distribution in the subsurface [8]. The penetration depth of MT soundings depend on the period and the resistivity [7]. The magnetic field and electric field that are transmitted in the XYZ direction resulted in Ex, Ey, Hx, Hy, Hz components [9].

In this study, we focus on identifying the Cascadia Subduction Zone along Oregon, Idaho, Wyoming from resistivity (ρ) contrast in the subsurface. We used phase tensor method to analyze the MT data. And then, we performed 2D inversion to found out. If the 2D inversion could not resolve fully 3D data, but resolve 1D or 2D data. We also hypotheses that there is an increase in subsurface resistivity throughout the study area beneath the North American plate.

2. The Data Sets and Methods

We downloaded the EDI-file from EarthScope USArray (http://ds.iris.edu/spud). The total stations were 20 points shown in figure 1. At each point of measurements, we needed to insert the coordinates in the website to obtain the data in the form of impedance and frequency (EDI-file). The coordinates in this research are Latitude coordinates from 43,68^0 to 44,34^0 and Longitude coordinates from -124,16^0 to -107,11^0. Measurement points distance was 70 km for each site from Oregon to Wyoming (figure 1).

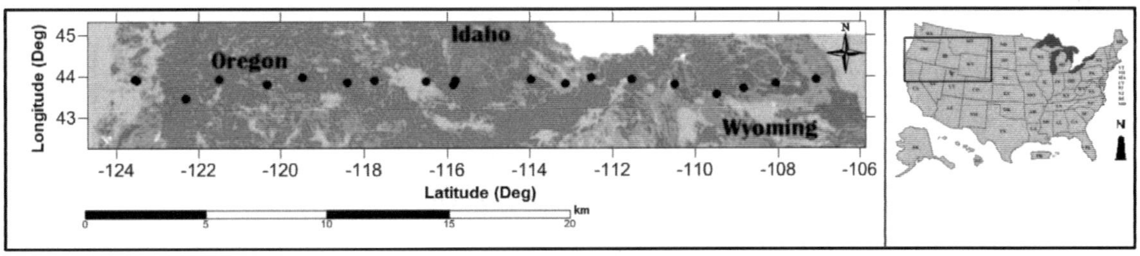

Figure 1. Left: Geological map of Oregon, Idaho, Wyoming (United States). The rocks consist of Cenozoic Igneous rock and Cenozoic Sedimentary rock. The MT data are shown by black dots. [10,11,12]. Right: US territory map. The research study is shown by black rectangle [13].

First, we performed the dimensionality data analysis using phase tensor (Φ) [14]. This used to determine the characteristic of data in the research area. This result will reveal the inversion modeling type that corresponds to characteristic data. Data with a 3D character represented by a red or blue ellipse (with a value of $\beta > 3^o$ or $\beta < -3^o$). The 2D character data represented by a white ellipse with β values between 3° and -3°. The 1D character data represented by a white circle with the value of β between > -3° and <3° [14]. The phase tensor equation could be written as:

$$\phi = X^{-1}Y \quad (1)$$

where X^{-1} is the inverse of the X component (the real component), while Y is the imaginary component of the phase tensor.

Then, we performed 2D inversion using NLGC (Non-Linear Gradient Conjugate) [15]. This serves to minimize the residual effects and spatial data of resistivity. The inverse problem could be written as:

$$d = F(m) + e \quad (2)$$

where d is an observation data, e is an error value, m is calculated data and F is a forward modeling function.

3. Results and Discussion

Figure 2 shows 2D inversion model result after 208 iteration. The red and the blue blocks represent the respectively condition in the subsurface. The RMS value at the last iteration is 6,88. There is a significant contribution of RMS value from the WYI18, which is 29,3. The contribution of a large error value is due to the 3D character data of the WYI18.

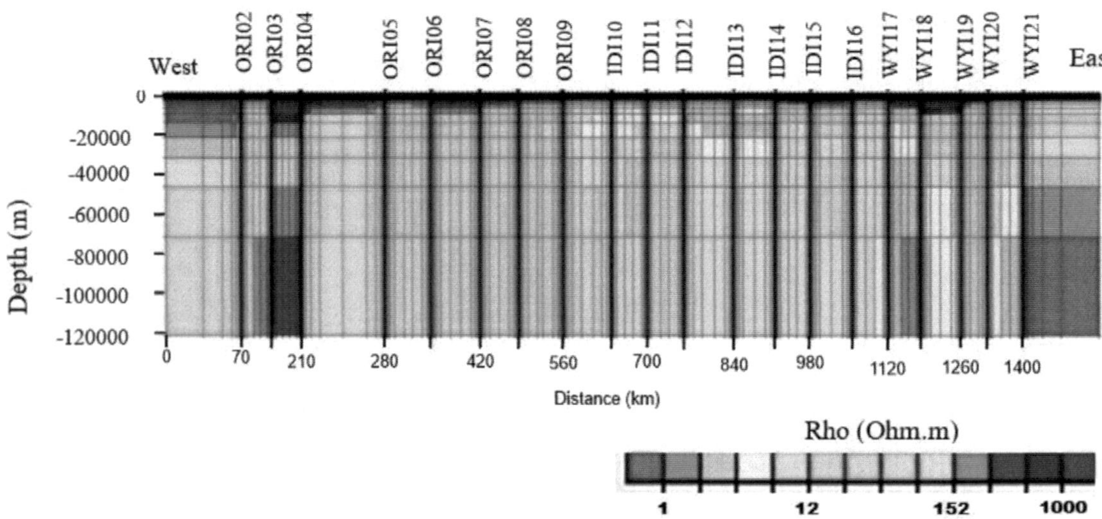

Figure 2. Resistivity cross section model along Oregon, Idaho, Wyoming. This model image 2D MT inversion after 208 Iteration. Blue color represent the resistive area. ORI03 and WYI18 reveal the high resistivity. The resistivity anomalies are correlates with emerge geology features such as Cascadia Volcanic Arc in Oregon and Yellowstone in Wyoming.

The result of 2D inversion show that the Cascadia Subduction Zone has a high resistivity distribution in the subsurface of Oregon. We found high resistivity (more than 500 ohm.m) at shallow depths of < 20 km and also at 70-120 km depth (beneath ORI03). These results are similar with the previous studies where at a depth of 20-50 km there is a low-velocity anomaly along the western Cascadia boundary [16]. Gravity studies also show an increasing of density anomalies at the depth of JdF plate subduction which is close to high resistivity anomaly beneath ORI03 [2]. Beneath Wyoming, high resistivity anomaly is found near to the surface (about 10 km depth). At WYI17 and WYI 18, below the resistor, there is a conductive anomaly (i.e. < 7 ohm m), which might corresponds to the Yellowstone geothermal area [1].

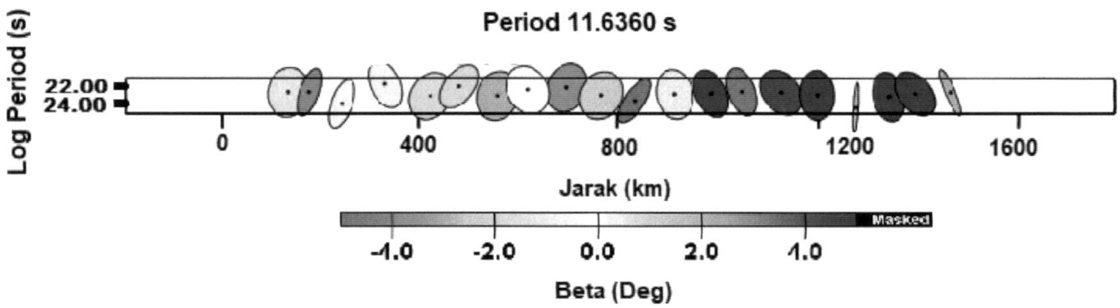

Figure 3. The Phase Tensor of 20 sites on period of 11,6 s. The ellipse colour represent beta (β) value. This value describes the character of the data. The 3D data is represented as red (β <-3) or blue (β>3).

The dimensionality analysis results are shown in the period of 11.6360 s and beta variation (β) (see figure 3). The short period corresponds to a shallow depth, which correlates to the geological features in the surface. Additionally, the skew variation shows the data characteristics. The phase tensor over a given period illustrates the lateral changes of the conductivity structure at a depth corresponding to that period. From this analysis, the red phase tensor and blue ellipses are found in the eastern region, which correspond to the 3D character, including MT sites of WYI 18. Although at the short period the data is relatively 2D, but at the long period the data are 3D. As in WYI18 there is a failure of 2D inversion particulary at depths > 50 km, which is indicated by a high RMS error value (i.e 29.3) and orange ellips phase tensor. That is indicates the data at this MT site is 3D.

4. Conclusion

From MT studies in three countries in Northwestern America (Oregon, Idaho, and Wyoming), we found that the high resistivity anomalies in two sites (Oregon and Wyoming). This high resistivity value is more than 500 Ohm.m is seen at shallow depths (< 20 km). In the western regions tend to be more resistive than the east. At the east of the study area, there is a high conductive anomaly that corresponds to Yellowstone geothermal area.

Acknowledgments
Our special thank to IRIS data that provides MT data along Oregon to Wyoming district. Many thank to anonymous reviewers for their comments and suggestions to improve this paper. We also want to thank to Faculty of Mathematics and Natural Science for the funding to attend the ICTAP 2017.

References

[1] Meqbel N M, Egbert G D, Wannamaker P E, Kelbert A and Schultz A 2014 *Earth and Planetary Science Letters* p 290-304
[2] Romanyuk T V, Blakely R and Mooney W D 1998 *Phys. Chem. Earth.* **23** no 3 (California: U.S.Geological Survey) p 297-301
[3] Xue M and Allen R M 2007 *Earth and Planetary Science Letters* (Amsterdam: Elsevier) p 264
[4] Atwater 1989 *The Geology Of North America* **72** (America: Geological Society of America) p 21
[5] Darold A and Humphreys E 2013 *Earth and Planetary Science Letters* (United States: Elsevier) p 232-242
[6] Wannamaker P E, Booker J R, Jones A G, Chave A D, Fillox J H, Waff H S and Law L K 1989 *Journal of Geophysical Research* **94** p 14,127-14,144
[7] Simpson F and Bahr K 2005 *Practical Magnetotelluric* Cambridge University Press Cambridge
[8] Vozoff K 1972 *Geophysics International Journal* **37** p 98-141
[9] Niasari S W, Munoz G, Kholid M, Suharto E and Ritter O 2015 *Australia Proceedings World Geothermal Congress* p 1-5
[10] Walker G W and MacLeod N S 1991 *Geologic map of Oregon U.S. Geological Survey* available at https://mrdata.usgs.gov/geology/state/state.php?state=OR (accessed 25.05.17)
[11] Bond J G, Kauffman J D, Miller D A, and Venkatakrishnan R 1978 *Geologic map of Idaho: Moscow, Idaho, Idaho Bureau of Mines and Geology with contributions from U.S. Geological Survey* available at https://mrdata.usgs.gov/geology/state/state.php?state=ID (accessed 25.05.17)
[12] Green, Gregory N, Drouillard, and Patricia H 1994 *The Digital Geologic Map of Wyoming in ARC/INFO Format: U.S. Geological Survey* available at https://mrdata.usgs.gov/geology/state/state.php?state=WY (accessed 25.05.17)
[13] United State map *State Minerals Statistics and Information; product of the USGS mineral resources program* available at https://minerals.usgs.gov/minerals/pubs/state/ (accessed 04.07.17)

[14] Caldwell T G, Bibby H G and Brown C 2004 *Geophysics Journal* **158** p 457-469
[15] Rodi W and Mackie R L 2001 *Geophysics* **66** p 174-187
[16] Chen C, Zhao D, and Wu S 2015 *Tectonophysics* ***647-648*** (Amsterdam: Elsevier) p 73-88

IOP Conf. Series: Journal of Physics: Conf. Series **1011** (2018) 012038 doi:10.1088/1742-6596/1011/1/012038

Subsurface structure identification uses derivative analyses of the magnetic data in Candi Umbul-Telomoyo geothermal prospect area

U Septyasari[1], S W Niasari[1], and P D Maghfira[1]

[1]Geophysics Sub-Department, Universitas Gadjah Mada, Yogyakarta, Indonesia 55281

E-mail: ufairaseptyasari@gmail.com

Abstract. Telomoyo geothermal prospect area is located in Central Java, Indonesia. One of the manifestations around Telomoyo is a warm spring, called Candi Umbul. The hydrothermal fluids from the manifestation could be from the subsurface flowing up through geological structures. The previous research about 2D magnetic modeling in Candi Umbul showed that there was a normal fault with strike/dip N60^0E/45^0 respectively. This research aims to know the distance boundary and the kind of the geological structure in the study area. We also compared the geological structure direction based on the geologic map and the derivative maps. We used derivative analyses of the magnetic data, i.e. First Horizontal Derivative (FHD) which is the rate of change of the horizontal gradient in the horizontal direction. FHD indicates the boundaries of the geological structure. We also used Second Vertical Derivative (SVD) which is the rate of change of the vertical gradient in the vertical direction. SVD can reveal normal fault or thrust fault. The FHD and SVD maps show that the geological structure boundary has the same direction with the north west-south east geological structure. The geological structure boundary is in 486 m of the local distance. Our result confirms that there is a normal fault in the study area.

1. Introduction

Indonesia is a country that located in the ring of fire zone. This condition makes Indonesia has about 127 active volcanoes around it [1]. These volcanoes are the sources of some geothermal prospect areas in Indonesia, especially in Java Island that dominated by volcanism activities. One of the geothermal areas in Java is Telomoyo.

One of the manifestations around Telomoyo is a warm spring, called Candi Umbul. This warm spring has a chloride water type which the temperature is about 36^0C and pH 7,6 [2]. Based on the research of structural lineaments satellite imagery data, the geothermal prospect area in Candi Umbul is correlated with Telomoyo activities [3]. Besides that, there were VLF-EM and 2D magnetic modeling research. The VLF-EM result showed that there was a high conductivity zone. It associated with the subsurface structure. The high conductivity zone was located in 5600 m of the local coordinate [4]. The 2D magnetic modeling in Candi Umbul showed that there was a normal fault with strike/dip N60^{0}E/45^0 respectively [5].

This research aims to know the distance boundary and the kind of the geological structure in Candi Umbul-Telomoyo geothermal prospect area. We used derivative analyses of the magnetic data, i.e. first horizontal derivative and second vertical derivative. We also compared the geological structure direction based on the geologic map and the derivative maps.

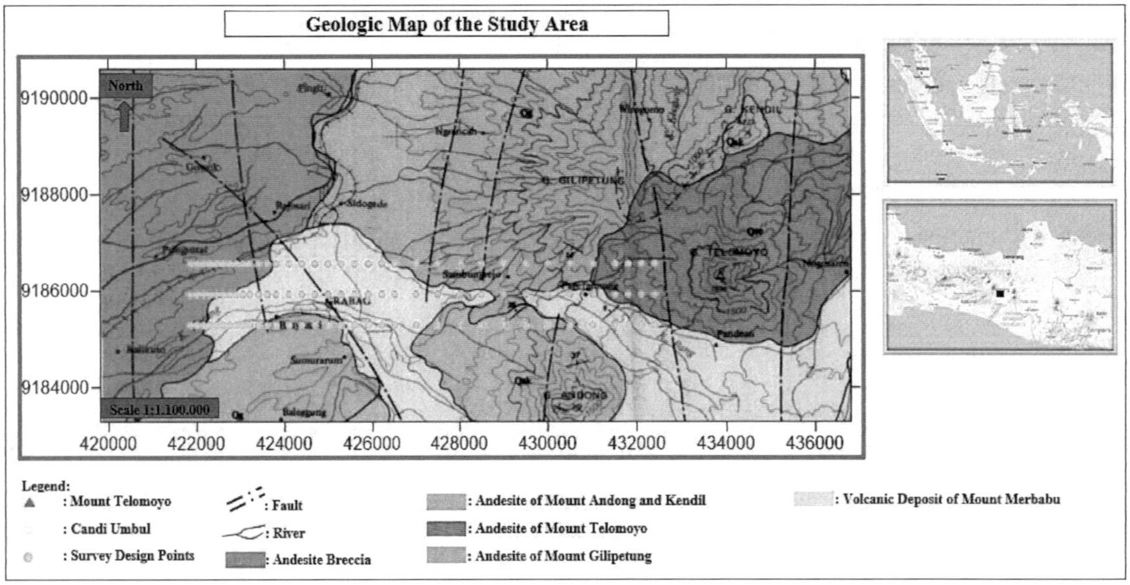

Figure 1. The red rectangle map is a part of the geologic map of the Magelang and Semarang quadrangles [6]. The survey design points overlay the geologic map [7]. The yellow rectangle map is the map of Indonesia, while the green rectangle map is the map of Java Island. The black rectangle in Java Island's map shows the research location.

2. Materials and Methods

This research is an advance part of the previous research. The method and the data are same, but we used different analyses. In this research, we used the magnetic method. The magnetic method is a potential method which measures the earth's magnetic field intensities. The magnetic field intensities are obtained from the magnetic properties of the underlying rocks and the environment where the rocks are in it [8].

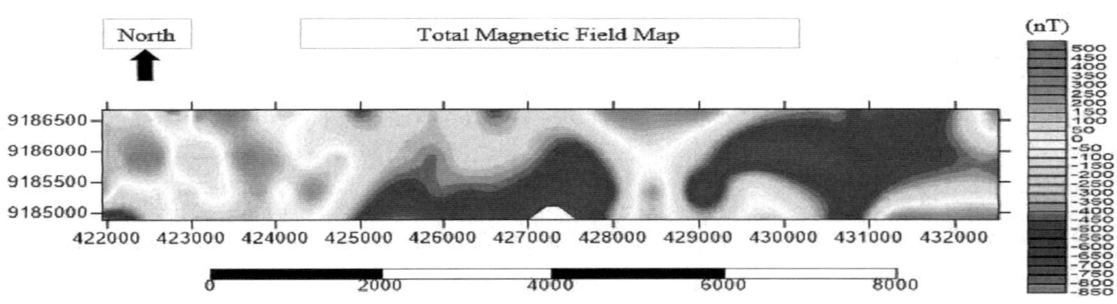

Figure 2. The map shows the total magnetic field of the study area. It still has the dipole data. The largest anomalies are in the west and in the middle of the map [7].

To delineate and analyze the structures beneath the surface in Candi Umbul-Telomoyo geothermal prospect area, we used First Horizontal Derivative (FHD) and Second Vertical Derivative (SVD) to the magnetic data. Both analyses aim to know the margins of the magnetic sources [9].

Before doing FHD and SVD, the data were transformed into a Reduction to Pole (RTP) because the study area still produced dipole data (Figure 2). RTP is a filter to make the data easier to be interpreted. It provides a simple approach to improve realistic estimations of the source of anomalies [10]. The dipole data have an asymmetric pattern [11]. This asymmetric pattern depends on the shape of perturbing body, the direction of the magnetic field and the inclination angle of the study area [12].

The First Horizontal Derivative (FHD) is the rate of change of the horizontal gradient in the horizontal direction. FHD indicates the boundaries of the geological structure. This derivative analysis is used to delineate high-frequency features clearly [13]. The gradient amplitude of first horizontal derivative can be defined by [14]:

$$\mathrm{FHD} = \sqrt{\left(\frac{\partial H}{\partial x}\right)^2 + \left(\frac{\partial H}{\partial y}\right)^2} \tag{1}$$

while the Second Vertical Derivative (SVD) is the rate of change of the vertical gradient in the vertical diretion. Theoretically, from the Laplace's equation, if H is a potential so $\nabla^2 H$ is equal to 0 [15]:

$$\frac{\partial^2(\Delta H)}{\partial z^2} = -\frac{\partial^2(\Delta H)}{\partial x^2} - \frac{\partial^2(\Delta H)}{\partial y^2} \tag{2}$$

the SVD can reveal normal fault or thrust fault. If the maximum absolute value is greater than the minimum absolute value, we conclude it as a normal fault. Vice versa, we conclude it as a thrust fault [16].

3. Results and Discussions

Figures 3 and 4 show the geological structure boundary that causes the anomalies. The geological structure boundary in figure 3 can be correlated with 0 value in the number scale range of figure 4. Besides that, we also used the geologic map from figure 1 to determine the geological structure in the study area. The FHD and SVD maps show that the geological structure boundary has the same direction with the north west-south east geological structure.

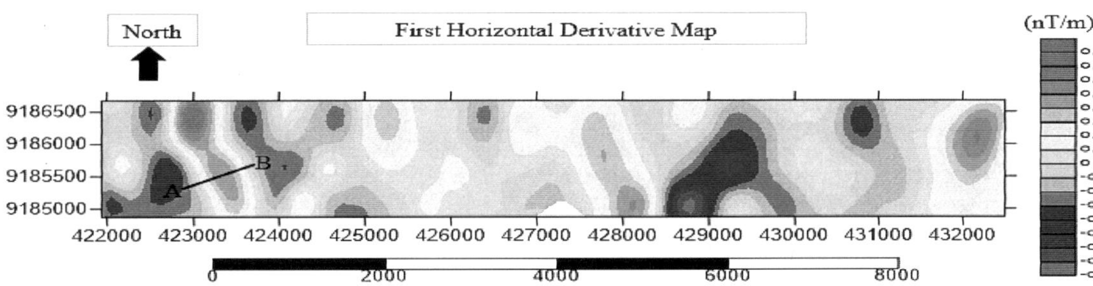

Figure 3. The map shows the first horizontal derivative in the X direction with A-B slicing line. Line A-B slices the boundary of the geological structure perpendicularly.

The geological structure that appears in the FHD and SVD maps is only the west side. The other geological structures do not appear on the map because the lithology is dominated by andesite in the middle and east side. Andesite is an igneous rock that contains some magnetic minerals. Those minerals will have a strong response to the magnetic anomalies. So, the lithology response will be stronger than the geological structure response.

We made two graphs of FHD and SVD from the A-B slicing line of figures 3 and 4. The first horizontal derivative tends to have characteristics with maximum or minimum value in the area with the geological structure. In figure 5a, the peak of the graph shows the maximum value. This maximum value indicates the geological structure boundary. From the graph, we know that the geological structure boundary is located in the local distance 486 m with the FHD value is 0,489 nT/m.

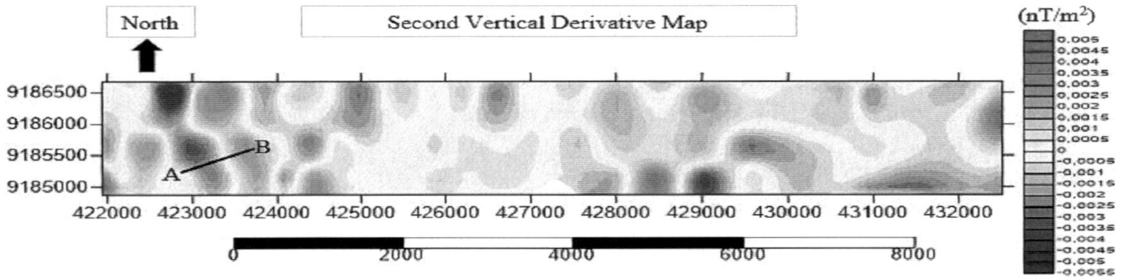

Figure 4. The map shows the second derivative in the vertical direction with A-B slicing line. Line A-B slices the boundary of the geological structure perpendicularly.

To determine the kind of the geological structure in the study area, we used the SVD graph in the figure 5b. In this graph, the anomaly that is caused by the geological structure will have maximum absolute value and minimum absolute value. The maximum absolute value is 0,00261 nT/m^2 and the minimum absolute value is 0,00239 nT/m^2. Since the maximum absolute value is greater than the minimum absolute value, we conclude it as a normal fault.

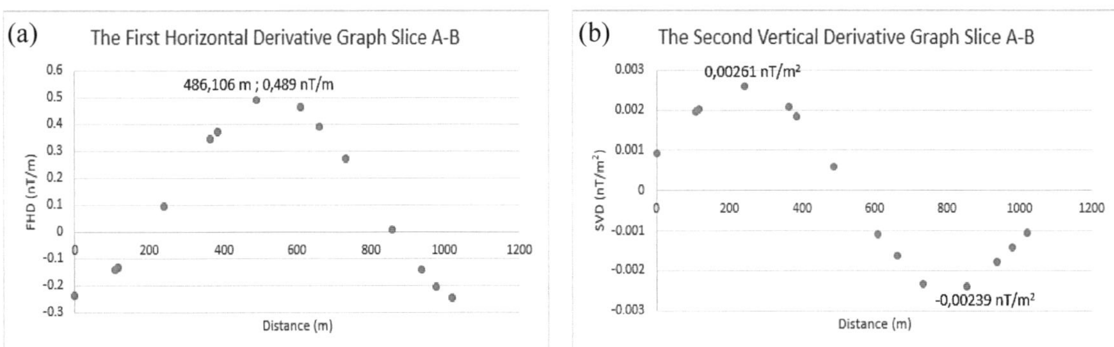

Figure 5. (a) The graph of the first horizontal derivative in the A-B slicing line. (b) The graph of the second vertical derivative in the A-B slicing line.

This research has confirmed the previous 2D magnetic modeling research. The kind of the geological structure in the study area is a normal fault. Besides that, this research has confirmed the geological map about the direction of the west side geological structure is north west-south east.

4. Conclusions

Based on the results, only the west side geological structure boundary appears in the FHD and SVD maps. The FHD and SVD maps show that the geological structure boundary has the same direction with the north west-south east geological structure. This geological structure boundary correlates with the maximum value of the FHD map and it has 0 value in the number scale range of the SVD map.

The FHD graph shows that the geological structure boundary is in 486 m of the local distance. It has maximum FHD value 0,489 nT/m. While the SVD graph has maximum absolute value at about 0,00261 nT/m^2 and minimum absolute value at about 0,00239 nT/m^2. Since the maximum absolute value is greater than the minimum absolute value, we conclude that the kind of the geological structure in the study area is a normal fault.

5. Acknowledgements

We would like to thank to "Fakultas Matematika dan Ilmu Pengetahuan Alam, Universitas Gadjah Mada" for the Hibah Penelitian Dosen funding no. 0088/J01.1.28/PL.06.02/2017 and thanks to "Geophysics Laboratory Instrument Pool, Universitas Gadjah Mada" for the instruments that had been used in the survey.

References

[1] Badan Geologi 2017 *Laporan Evaluasi Tingkat Aktivitas Gunungapi Bulan Maret 2017* [online] Available at: http://www.vsi.esdm.go.id/index.php/gunungapi/aktivitas-gunungapi/1509-laporan-evaluasi-tingkat-aktivitas-gunungapi-bulan-maret-2017 [Accessed 4 Jul. 2017]

[2] Geology and Geochemistry Integrated Survey Team 2010 *Laporan Survei Terpadu Geologi dan Geokimia Daerah Panas Bumi Candi Umbul-Telomoyo, Provinsi Jawa Tengah* (Bandung: Pusat Sumber Daya Geologi)

[3] Hermawan D and Rezky Y 2011 *Deliniasi Daerah Prospek Panas Bumi Berdasarkan Analisis Kelurusan Citra Landsat di Candi Umbul-Telomoyo, Provinsi Jawa Tengah* (Bandung: Pusat Sumber Daya Geologi)

[4] Affanti A P 2017 *Subsurface Conductivity Mapping Using VLF-EM and VLF-R in Candi Umbul-Telomoyo, Magelang* (Yogyakarta: Universitas Gadjah Mada)

[5] Kharisa N A, Narendratama R, Wulandari I, Faisal M I, Kirana K, Zipora R, Arfiansyah I and Suyanto I 2015 *Analysis of Magnetic Anomaly Data for Identification Structure in Subsurface of Geothermal Manifestation at Candi Umbul Area Magelang Central Java Province Indonesia* (Jakarta: International Geothermal Convention and Exhibition)

[6] Thaden R E, Sumadirja H and Richards P W 1975 *Geologic Map of the Magelang and Semarang Quadrangles, Java* (Bandung: Geological Survey of Indonesia)

[7] Maghfira P D 2017 *Identification of Subsurface Imaging Using Magnetic Method in Candi Umbul Area, Grabag, Magelang, Central Java* (Yogyakarta: Universitas Gadjah Mada)

[8] Mariita N O 2007 *The Magnetic Method* (Kenya: Kenya Electricity Generating Co Ltd)

[9] Hinze W J, Von Frese R R and Saad A H 2013 *Gravity and Magnetic Exploration: Principles, Practices, and Applications* (Cambridge: Cambridge University Press) p 323

[10] Blakely R J 1995 *Potential Theory in Gravity and Magnetic Applications* (Cambridge: Cambridge University Press) p 330

[11] Chandra P C 2016 *Groundwater Geophysics in Hard Rock* (Boca Raton: CRC Press/Taylor & Francis Group) p 72

[12] Pakiser L C and Mooney W D 1989 *Geophysical Framework of the Continental United States* (Boulder (Colo.): The Geological Society of America) p 56

[13] Feumoe A N S, Ndougsa-Mbarga T N, Manguelle-Dicoum E, Fairhead J D 2012 Deliniation of Tectonic Lineaments Using Aeromagnetic Data for the South-East Cameroon Area *Geofiz.* **29** 175-192

[14] Grauch V J S and Cordell L 1987 Limitations of Determining Density or Magnetic Boundaries from the Horizontal Gradient of Gravity or Pseudogravity Data *Short Note Geophys.* **52** 118-121

[15] Elkins T A 1951 The Second Derivative Method of Gravity Interpretation *Geophys.* **23** 29-50

[16] Reynolds J M 1984 *An Introduction to Application and Environmental Geophysics* (West Sussex: John Wiley and Sons Ltd)

IOP Conf. Series: Journal of Physics: Conf. Series **1011** (2018) 012039 doi:10.1088/1742-6596/1011/1/012039

Zonation of Landslide-Prone Using Microseismic Method and Slope Analysis in Margoyoso, Magelang

Muchamad Reza Aditya*[1], Arriqo' Fauqi Romadlon[1], Reymon Agra Medika[1], Yosua Alfontius[1], Zukhruf Delva Jannet[1], Eddy Hartantyo[1]

[1]Geophysics Study Program, Department of Physics, Faculty of Mathematics and Natural Sciences, Universitas Gadjah Mada, Yogyakarta, Indonesia

*Corresponding author: mochamadrezaaditya@gmail.com

Abstract. Margoyoso Village, Salaman Sub-district, Magelang Regency, Central Java is one of the villages that were included in landslide prone areas. The steep slopes and land use in this village were quite apprehensive. There were fractures with 5 cm in width and a length of 50 m. Moreover, these fractures appeared in the home residents. Although the local government has established a disaster response organization, this village is still not getting adequate information about the landslide prone areas. Based on the description before, we conducted research with geophysical methods and geotechnical analysis to minimize the danger of landslides. The geophysical method used in this research was microseismic method and geotechnical analysis. The microseismic measurement and slope stability analysis at Margoyoso village was a step in analysing the landslide-prone zone boundary. The results of this research indicated that landslide potential areas had a low peak ground acceleration values with a range from 36 gal to 46 gal. Measurement of slope stability indicated that a slope angle values between 55°-78° are a potential landslide slope because the soil in this village has very loose properties so it is very easy to move.

1. Introduction

Landslide is one of the most often natural disasters that happen in Indonesia. Almost all regions in Indonesia are risky from the landslide. According to *Data dan Informasi Bencana Indonesia 2016-Badan Nasional Penangulangan Bencana* (Disaster Data and Information in Indonesia - National Board for Disaster Management) Even until mid-2016, there have been 394 landslides that caused hundreds of victims, dozens of missing people, and up to thousands of people displaced[1]. Therefore, the landslide disaster must be taken seriously and society must be educated about prevention act so that nothing bad happens.

Margoyoso Village, Salaman District, Magelang Regency, Central Java is one of the villages in Indonesia which are included in landslide prone areas. Villagers have complained about lots of fractures that appear in this village, even that fracture also appeared around the villager's house with an average fracture of 5 cm with a length up to 50 m and also appeared cracks in the wall of villager's house. This fracture is suspected as a result of land-use change due to the new buildings that not supported by proper construction. Although the local government has established a disaster response organization such as the Tanggap Ing Sasmito forum, this village is still not getting adequate information about landslide prone areas.

Landslide disaster can occur because of many factors, such as geomorphology (slope, height, drainage, etc), geological (soil type, fracture) and seismic activity (earthquakes)[2]. Landslides that caused by earthquakes can increase the shear stress due to horizontal acceleration, or decrease in soil strength. [3]. A microseismic method for zonation landslide prone areas, from this method can be analyzed a

correlation between the thickness and type of the soft sediment cover [4]. In this case, the area with high potential of landslide and low potential of landslide have different seismic vulnerabilities [5]. Slope stability analysis is a geotechnical method that can measure soil safety factors caused by geological aspects such as fractures, pores, and lithology.
By conducting this research, it is expected to provide information about landslide-prone zones so that it has an impact on the reduction of potential victims and the loss of the community. This research is also a contribution from the geosciences especially geophysics in its application to the environment.

2. Methods and Materials

This research area located in Margoyoso, Salaman sub-district, Magelang regency. The time of research is about 2 weeks (April 30th - May 9th, 2017). The method used in this research is microcosmic method and slope stability analysis. In general, the research flow diagrams were conducted as follows:

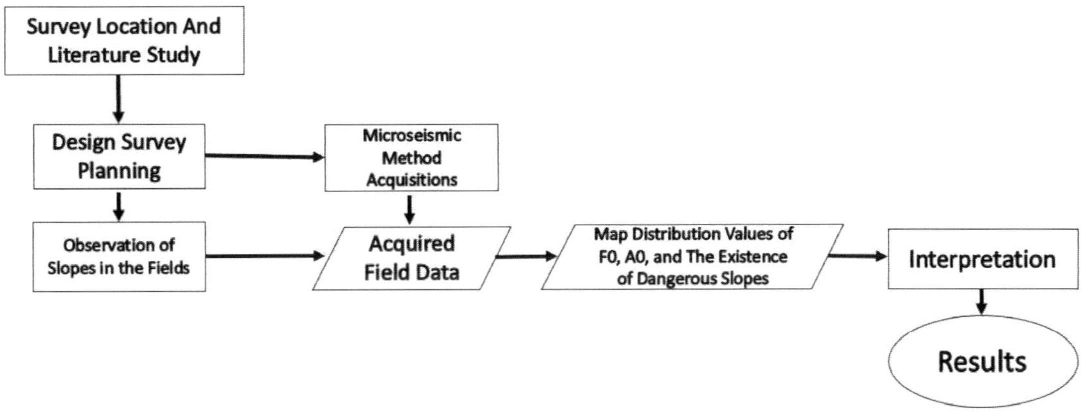

Figure 1. Research process flow diagram. This research was started with surveying of research location and it was finished with data interpretation.

2.1. Microseismic Method

The method used in the study of landslide-prone zonation was microseismic. The microseismic device can identify the vibration decay contained in the medium, along with the broadband seismometer measuring in real time. The main sources of vibrations in microseismic method are ocean wave, traffic, industrial and human activities[6]. Then, the ground movement can be recorded as a function of time in local site[7]. Site response analysis is important in seismic hazard assessment in earthquake prone zones. This research used H/V method, also called the Nakamura technique[8].

2.1.1. Instrumentation

In this study used seismometer Lennartz Electronic with brand LE-3Dlite. This instrument consists of one seismometer unit, one Data Logger unit, also other accessories such as LAN cable and connecting cable. This seismometer unit has an Eigen frequency 1 Hz and the acquired frequency limit is 100 Hz. While data logger is an instrument that can record data within a certain time. This instrument converts the analogue signal into a digital signal that can be processed by laptops. Other equipment required are a 12 volt accumulator, gps handheld, and compass.

2.1.2. Acquisition

In this research, the microseismic method focused on horizontal to vertical spectral ratio (HVSR) method. Ambient seismic noise is a natural vibration arising from nature or commonly referred to as small vibrations of nature (microtremor). Spectral analysis was done to obtain domain frequency of microtremor activity. In this analysis, we used fast fourier transform (FFT) with bandwidth frequency 1 Hz - 10 Hz to obtain natural frequency (f_0) and natural amplification (A_0) values from the research area[9]. Nakamura[10] explains that the HVSR Method can compare the ratio of spectral amplitudes between vertical and horizontal components. Mathematically, it can be expressed in the equation:

$$R(f) = \frac{\sqrt{H_{EW}{}^2(f) + H_{NS}{}^2\ (f)}}{V_{UD}\ (f)} \tag{1}$$

Which:

$R(f)$: HVSR spectrum ratio
H_{EW} : Horizontal component spectrum (East-West)
H_{NS} : Horizontal component spectrum (North-South
V_{UD} : Vertical component spectrum

The HVSR method is used on a three-component passive seismic (microtremor). Microzonation can predict the response and behaviour of the soil or sediment layer to external energy that affects the surrounding soils. Microtremor used very small amplitude ranging from 0.1 to 1.0 μm and the period between 0.1 seconds and 1.0 seconds[11].
Peak ground acceleration (PGA) is used to describe strong ground motion. PGA is a parameter that represents the acceleration of earthquake vibration on the ground. The PGA value is not included in the measurement of the energy of an earthquake[12].

2.2. Slope Analysis
Slope stability analysis is an important part to prevent the disturbances that potentially lead to disaster that can endanger the surrounding community. The increased of pore pressure causes decreasing of effective stress on the soil. Afterward, reducing soil shear strength and resulting slope failure[13].
According to Karnawati[14], there are several factors to control the occurrence of slope failure, slope failure with movement of soil mass type has a slope for the range of 20^0 to 40^0.
The measurement method applied to soil slope analysis (homogeneous) generally is an analysis method with slope angle. A slope will slide if the driving force is greater than its holding force[15].

3. Result and Discussion

3.1. Geological Study
The geological study shows that this area had steep slopes. Those slopes were worsened by poor land use. Moreover, in area with steep slopes there were not planted by plants with strong roots so that potentially occur landslides.

3.2. Microseismic Method Measurement

3.2.1. Natural Frequency (F_0)
Natural frequency characterizes the ground surface character when subjected to external vibrations (such as earthquakes). If the vibrations that hit it have a frequency greater than or equal to the natural frequency then there will be a resonance which will result in the amplification of seismic waves in the area affected by the vibration. The distribution of natural frequency values is shown in Figure 2A. The frequency of the region with blue color will resonate with earthquake frequency and this area is dangerous.

3.2.2. Amplification (A_0)
Amplification is defined as the magnitude of wave reinforcement that occurs in an area. Reinforcement of this wave is proportional to the ratio of the horizontal spectrum and the vertical spectrum of a medium. Soft sedimentary rocks are known to strengthen the movement of the soil during the earthquake and therefore the average damage caused more severe than the hard layer[6]. The distribution of amplification can be seen at Figure 2B. The area with high amplification value is dangerous, because in that area can be earthquake wave strengthening.

3.2.3. Peak Ground Acceleration
The calculation of Peak Ground Acceleration with Kanai method used Yogyakarta earthquake 2006 (6.2 Richter Scale). When the PGA value is smaller, the potential for landslide disaster will be greater.

The observed earthquake should be noted the time range of the earthquake because the frequency of the occurrence of an old earthquake can cause a large earthquake as well. In Figure 2C, the PGA values with a range of 36-52 gal is low enough so that the area is prone to land movement.

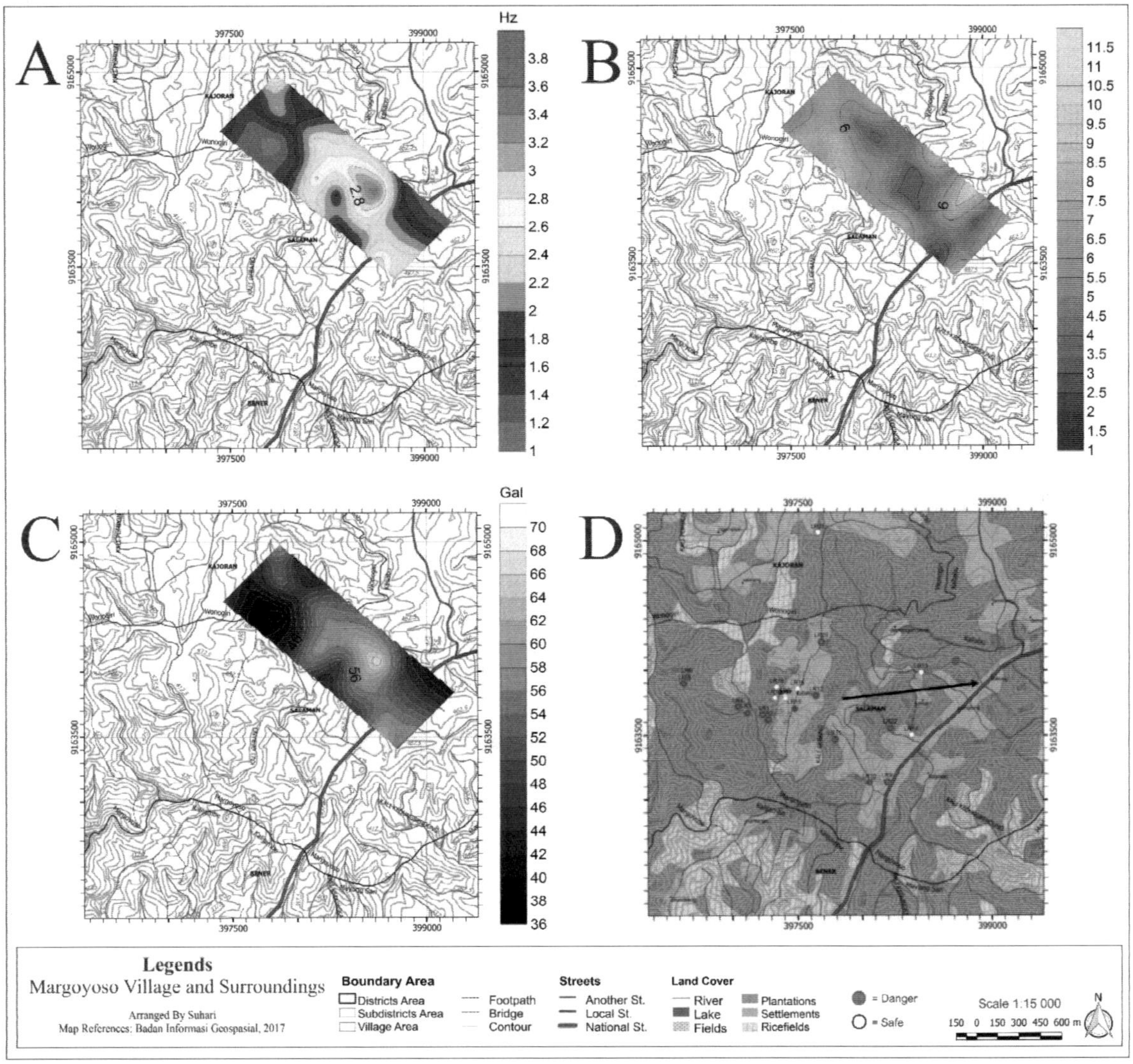

Figure 2. The result of landslide-prone zonation using microseismic method and slope analysis in Margoyoso, Magelang. **A**. Distribution map of natural frequency value (F_0). The blue colour indicated low natural frequency value. The read one indicated high natural frequency value. **B.** Distribution map of amplification value. The violet colour indicated low amplification value. The light green colour indicated high amplification value. **C.** Distribution map of PGA values. The black colour indicated low PGA value. The white colour indicated high PGA value. **D.** Distribution map of slope measurement point. The red circle indicated dangerous slopes. The white circle indicated safe slopes.

3.3. Slope Analysis

There were 22 observed slopes. 14 of 22 (63.63%) observed slopes have potential of landslides. The distribution of slope stability analysis point was shown in Figure 2D. There are several areas that vulnerable because the slope is unstable. The red circle indicated dangerous slopes. The white circle indicated safe slopes.

4. Conclusion

Zonation of landslide-prone with microseismic method and analysis of slope stability in Margoyoso Village, Salaman sub-district, Magelang had quite dangerous landslide potential. This research was a step in analysing the limits of landslide prone zone based on predetermined parameters.

Based on the parameters that have been described in this study (natural frequency, amplification, peak ground acceleration, and stability analysis of slope), it was found that the potential landslide area is in the eastern and western of the research area that had a low peak ground acceleration value with a range of 36-52 gal and slope failures.

5. References

1. Data dan Informasi Bencana Indonesia [Internet]. 2016 [cited 2017 Feb 5]. Available from: http://dibi.bnpb.go.id/data-bencana
2. Arnous MO 2011 Integrated remote sensing and GIS techniques for landslide hazard zonation : a case study Wadi Watier area *J Coast Conserv.* **15**(4) 477–97.
3. Lei CI 2012 Earthquake-Triggered Landslides *1st Civil and Environmental Engineering Student Conference* (London: Imperial College London) p. 1–6.
4. Torgoev A, Lamair L, Torgoev I, Havenith H 2013 A Review of Recent Case Studies of Landslides Investigated in the Tien Shan Using Microseismic and Other Geophysical Methods *Earthquake-Induced Landslides* (Berlin: Heidelberg) p. 285–94.
5. Gallipoli MR, Mucciarelli M, Vona M 2009 Empirical estimate of fundamental frequencies and damping for Italian buildings *Earthq. Eng. Struct. Dyn.* **38**(8) 973–88.
6. Tuladhar R, Yamazaki F, Warnitchai P, Saita J 2004 Seismic microzonation of the greater Bangkok area using microtremor observations *Earthq. Eng. Struct. Dyn.* **225**(October 2002) 211–25.
7. Sulistiawan H, Supriyadi, Yulianti I 2017 Seismic Hazard Analysis based on Earthquake Vulnerability and Peak Ground Acceleration using Microseismic Method at Universitas Negeri Semarang *Journal of Physics* p. 1–6.
8. Kind F, Kohler A, Ohmberger M, Richter G, Savvaidis A, Scherbaum F, et al 2002 *Site Effects Assessment Using Ambient Excitations SESAME.*
9. Sabrianto L, Awali A 2017 Seismic Microzonation Using Microtremor Measurement for Natural Disasters Mitigation of Earthquake at Regions Singaraja City the Province of Bali Indonesia *J Earth Sci.* **6**(1):1–6.
10. Nakamura Y, Nakamura T 1998 Stiffness Design for Specified Nonexceedance Probability of Seismic Response *J. Earthq. Spectra* **14**(1) 165–88.
11. Ishiyama Y, Hasanuddin 2006 Mikrotremor *Antologi Paper Pusat Vulkanologi dan Mitigasi Bencana Geologi.*
12. Douglas J 2003 Earthquake ground motion estimation using strong-motion records : a review of equations for the estimation of peak ground acceleration and response spectral ordinates *J. Earth-Science Rev* **61** 43–104.
13. Regmi RK, Lee G, Jung K 2013 Analysis on Failure of Slope and Landslide Dam *J. Civ. Eng.* **17** 1166–7.
14. Karnawati D 2007 Mekanisme Gerakan Massa Batuan Akibat Gempabumi; Tinjauan dan Analisa Geologi Teknik *Din. Tek. Sipil* **7**(2) 179–90.
15. Firmansyah, Feranie S, Tohari A, Latief F D E 2016 Prediction of landslide run-out distance based on slope stability analysis and center of mass approach *IOP Conf. Ser. Earth Environ. Sci* **29** 1–6.

6. Acknowledgments

The authors would like to thank Ministry of Research, Technology and Higher Education (Kemenristek Dikti) for research funding, Geological Disaster Technology Development and Research Center (BPPTKG) for permission of using seismometer, and also to Geophysics study program Universitas Gadjah Mada for providing HSE equipments.

IOP Conf. Series: Journal of Physics: Conf. Series **1011** (2018) 012040 doi:10.1088/1742-6596/1011/1/012040

Severe Drought Event in Indonesia Following 2015/16 El Niño/positive Indian Dipole Events

D O Lestari[1], E Sutriyono[2], Sabaruddin[3], and I Iskandar[4*]

[1]Graduate School of Environmental Sciences, University of Sriwijaya, South Sumatra, Indonesia
[2]Departement of Geology, Faculty of Engineering, University of Sriwijaya, South Sumatra, Indonesia
[3]Departement of Soil Sciences, Faculty of Agriculture, University of Sriwijaya, South Sumatra, Indonesia
[4]Departement of Physics, Faculty of Mathematics and Natural Sciences, University of Sriwijaya, South Sumatra, Indonesia

*Email: iskhaq@mipa.unsri.ac.id

Abstract. During boreal fall and winter 2015/16, Indonesia experienced catastrophic drought event causing many environmental problems. This study explored dynamical evolution of drought event in Indonesia associated with those two climate modes. Based on the Niño3.4 index, the evolution of the El Niño has started in April 2015, reached its peak in January 2016 and terminated in April 2016. Meanwhile, the Dipole Mode Index (DMI) revealed that the evolution of positive Indian Ocean Dipole has started in August, reached its peak in September and terminated in November 2015. It is shown that during those two events, Indonesia experienced severe drought in which the precipitation was extremely decreased. During the peak drought condition co-occurring with the dry season, the anomalous of precipitation reached –450 mm/month in September 2015. The peak of the drought was associated with the El Niño and positive Indian Ocean Dipole sea surface temperature anomaly (SSTA) patterns, in which negative SSTA covered the eastern tropical Indian Ocean and the western Pacific Ocean including Indonesia seas. Meanwhile, positive SSTA observed in the western tropical Indian Ocean and Eastern Pacific Ocean.

1. Introduction

In 2015/16, Indonesia experienced catastrophic drought. It was expected that was influenced by coupled ocean-atmosphere phenomenon in the tropical Pacific Ocean and the tropical Indian Ocean. Previous studies have found that during dry season precipitation in Indonesia region has high correlation with the El Niño Southern Oscillation (ENSO) and the Indian Ocean Dipole (IOD) events [1–3]. The ENSO is one major mode of climate variability that takes place in the tropical Pacific Ocean [4]. Meanwhile, the IOD is a coupled ocean-atmosphere phenomenon in the tropical Indian Ocean that has strong impact on the Indian Ocean rim countries [5,6]. A positive IOD is characterized by low sea surface temperature (SST) anomaly in the southeastern tropical Indian Ocean (off west Sumatra and south Java) and high SST anomaly in the western/central tropical Indian Ocean, associated with southeasterly (easterly) wind anomalies along the coast of Sumatra and Java (equator). As a consequence of change in the ocean-atmosphere circulation is an excess precipitation over the

IOP Conf. Series: Journal of Physics: Conf. Series **1011** (2018) 012040 doi:10.1088/1742-6596/1011/1/012040

east Africa, India, and southeast Asia, and deficit precipitation over the Indonesian region and Australia [7–9]. The purpose of this study is to analyze the dynamical evolution of drought event in the Indonesian region during 2015 associated with those two climate modes.

2. Data and Method

This study used three datasets, namely precipitation, SST, and low-level wind (850 mb). The precipitation data were obtained from the Tropical Rainfall Measuring Mission (TRMM) with horizontal resolution of 0.25° × 0.25° (https://mirador.gsfc.nasa.gov/). The data are available from January 1998 until present. The SST data were from the National Oceanic and Atmospheric Administration (NOAA)'s Optimum Interpolation Sea Surface Temperature (OISST) Version 2 (https://www.esrl.noaa.gov/psd/). The data are available from September 1981 until present, having horizontal resolution of 0.25° × 0.25°. Meanwhile, the low-level wind data were from the ERA-Interim of the European Centre for Medium-Range Weather Forecasts (ECMWF) with a constant horizontal resolution of 0.5° for period of January 1979 until present (http://apps.ecmwf.int/datasets/data/).

Monthly value of the Nino3.4 index and the Dipole Mode Index (DMI) were used to represent the evolution of two types of climate mode in the Indo-Pacific region. The Nino3.4 index is defined as an average SST anomaly in the equatorial Pacific Ocean bounded by 5°N to 5°S and 170°W to 120°W [4]. Meanwhile, the DMI which is defined as a difference in SST anomaly between western the tropical Indian Ocean region (10°S to 10°N and 50°E to 70°E) and its eastern region (0°S to 10°S and 90°E to 110°E) [5]. Note that, the anomalies of all variables are defined as the deviation from the monthly climatology over a period of 2001-2015.

3. Result and Discussion

3.1. Evolution of ENSO and IOD events 2015/16

We first calculated the SST anomaly based on Niño3.4 index and DMI region to identify the evolution of ENSO and IOD events. It is shown that during 2015 to 2016 the positive IOD event was co-occurred with the El Niño event. The evolution of El Niño event started from April 2015. It became stronger and reached the peak in January 2016 with a maximum Niño3.4 index reaching >2.5 °C. The event decayed in April 2016 (Figure 1a). The DMI revealed that the evolution of positive IOD has started in June/July, reached its peak in September and terminated in November 2015 (Figure 1b).

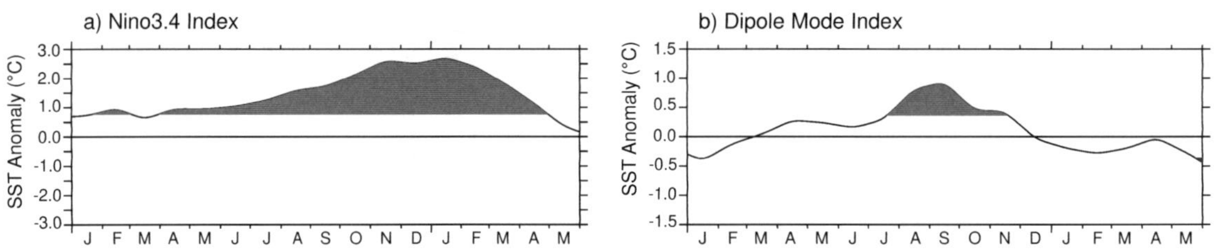

Figure 1. Time series of (a) Niño3.4 Index and (b) DMI during January 2015 to May 2016. Positive anomaly (shading) shows El Niño and positive IOD events.

3.2. Spatial Patterns of Precipitation during El Niño/Positive IOD 2015/16

Precipitation over the Indonesian region is generally decreased during 2015/16 (Figure 2). This condition associated with the El Niño and positive IOD events are concurrently occurred [10,11]. The significant negative precipitation anomalies first appeared over Sumatra, Kalimantan, and Papua Islands in July 2015. This drought condition continued and reached its peak in September-October

2015. During the peak of positive IOD event, almost all of area in the Indonesia experienced severe drought. The precipitation in the Indonesian region is decreased up to –450 mm/month (Figure 2e).

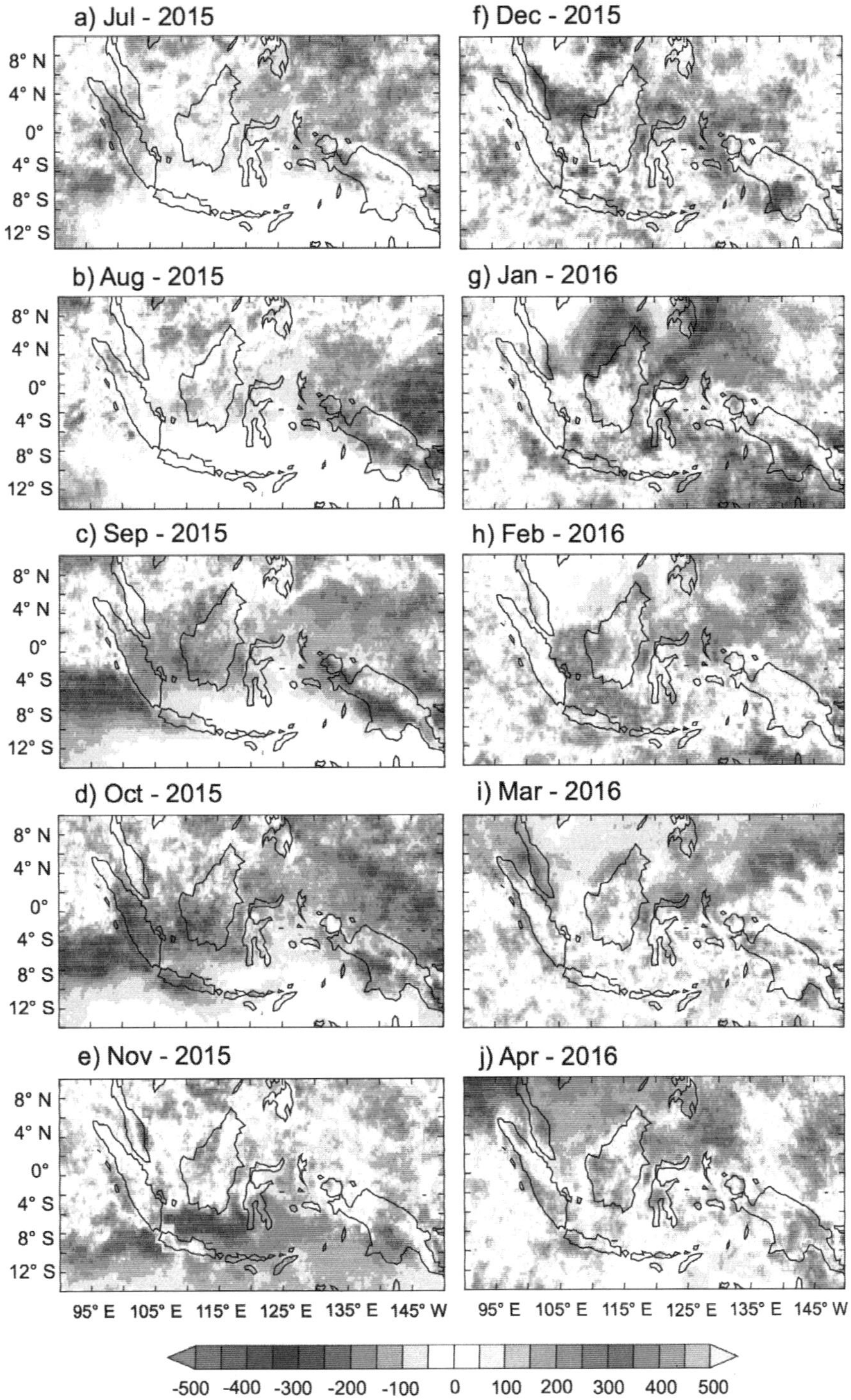

Figure 2. Spatial patterns of anomalous precipitation (mm/month) over the Indonesian region. Figures 2a to 2j indicate anomalies during July 2015 to April 2016.

In November 2015, while the positive IOD has decayed, the positive precipitation anomalies were observed in northern part of Sumatra, northeastern part of Kalimantan, west Java, and northern Papua Islands. However, the drought condition in southeastern Papua Island was persisting until the peak of El Niño event in January 2016. It is interesting to note that the intensity of negative precipitation anomalies in this area was higher during the peak of positive IOD and approach the peak of El Niño events (September and December 2015).

3.3. Ocean-Atmosphere Interaction in the Indo-Pacific Region

In order to explain the severe drought condition in Indonesian region during 2015/16, we evaluated the ocean-atmosphere interaction in the Indo-Pacific region. Figure 3 shows the spatial patterns of SST and horizontal wind anomalies in the Indo-Pacific region. The negative precipitation anomalies over the Indonesian region associated with negative SST anomalies in the Indonesian seas, eastern tropical Indian Ocean and western tropical Pacific Ocean.

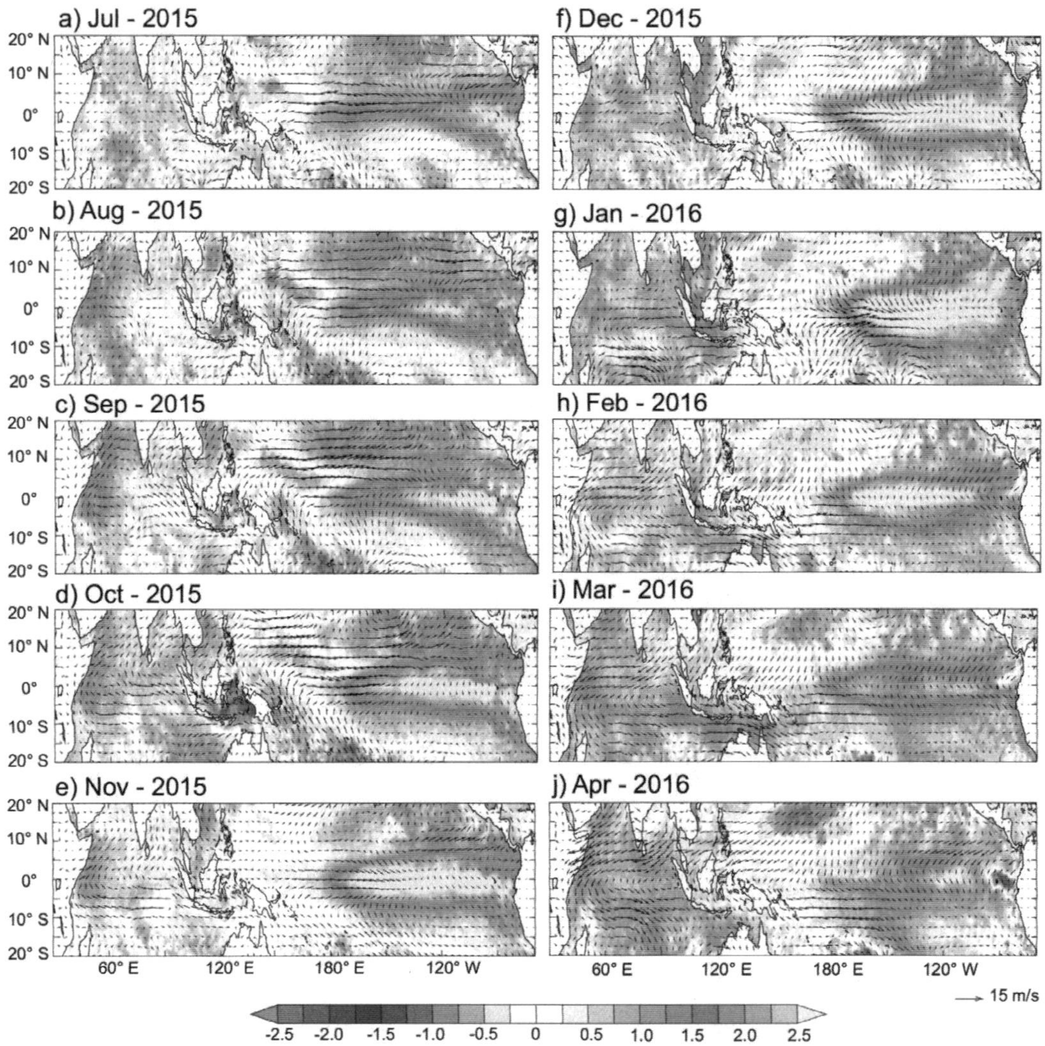

Figure 3. Spatial patterns of the low-level wind anomalies (m/s; *vector*) superimposed on the SST anomalies (°C; *shading*) over the Indo-Pacific region. Figures 3a to 3j indicate anomalies during July 2015 to April 2016.

During the El Niño and positive IOD events, the Indonesian seas experienced cooling phase. The most striking feature was observed during the maturing phase of the positive IOD. Meanwhile, the warmer SST anomalies was observed in the western tropical Indian Ocean and in the eastern to central tropical Pacific Ocean. The cooling effect of SST anomalies in Indonesian Seas and the warming effect of SST anomalies in the western tropical Indian Ocean and the eastern to central tropical Pacific Ocean have caused divergence wind originated from the Indonesia area. Therefore, the strong westerly wind anomalies were observed in the western tropical Pacific Ocean and strong southeasterly wind anomalies were observed in the eastern tropical Indian Ocean. This divergence circulation causes a descending atmospheric motion over the Indonesian region leading to a suppressed convection. As the positive IOD event terminated, the western Indonesian seas started to warm (Figure 3d).

4. Conclusion

The severe drought condition in the Indonesian region during 2015/16 is caused by the El Niño and positive IOD events that simultaneously occurred. The precipitation anomalies in the Indonesian region decreased drastically in dry season and continued until wet season. This condition is associated with SST and horizontal wind anomalies in the tropical Pacific Ocean and the tropical Indian Ocean. During the dry season, the SST in the Indonesian seas and its surrounding water was anomalously negative. Meanwhile, positive SST anomalies have covered the western (eastern) tropical Indian (Pacific) Ocean. This caused a divergence winds over the Indonesian region, leading to a descending atmospheric motion and a suppressed convection.

Acknowledgement

This research is supported by the PMDSU Scholarship from the Ministry of Research, Technology and Higher Education, Indonesia for the first author. The last author is supported by the Ministry of Research, Technology and Higher Education, Indonesia through *Penelitian Berbasis Kompetensi 2017* and by the University of Sriwijaya through *Hibah Profesi 2017.*

References

[1] Hendon H H 2003 Indonesian Rainfall Variability: Impacts of ENSO and Local Air-Sea Interaction *J. Clim.* **16** 1775–1790
[2] Aldrian E and Susanto R D 2003 Identification of Three Dominant Rainfall Regions within Indonesia and Their Relationship to Sea Surface Temperature *Int. J. Climatol.* **23** 1435–1452
[3] Saji N H and Yamagata T 2003 Possible Impacts of Indian Ocean Dipole Mode Events on Global Climate *Clim. Res.* **25** 151–169
[4] Trenberth K E 1997 The Definition of El Niño *Bull. Am. Meteorol. Soc.* **78** 1–7
[5] Saji N H, Goswami B N, Vinayachandran P N and Yamagata T 1999 A Dipole Mode in the Tropical Indian Ocean *Nature* **401** 360–363
[6] Webster P J, Moore A M, Loschnigg J P and Leben R R 1999 Coupled Ocean-Atmosphere Dynamics in the Indian Ocean during 1997-98. *Nature* **401** 356–360
[7] Yamagata T, Behera S K, Luo J, Masson S, Jury M R and Rao S A 2004 Coupled Ocean-Atmosphere Variability in the Tropical Indian Ocean *Earth's Clim. Ocean. Interact. Geophys. Monogr. Ser.* **147** 189–210
[8] Luo J-J, Sasaki W and Masumoto Y 2012 Indian Ocean Warming Modulates Pacific Climate Change *Proc. Natl. Acad. Sci.* **109** 18701–18706
[9] Nur'utami M N and Hidayat R 2016 Influences of IOD and ENSO to Indonesian Rainfall Variability : Role of Atmosphere-Ocean Interaction in the Indo-Pacific Sector *Procedia Environ. Sci.* **33** 196–203
[10] Luo J J, Zhang R, Behera S K, Masumoto Y, Jin F F, Lukas R and Yamagata T 2010 Interaction between El Niño and Extreme Indian Ocean Dipole *J. Clim.* **23** 726–742
[11] Annamalai H, Xie S P, McCreary J P and Murtugudde R 2005 Impact of Indian Ocean Sea Surface Temperature on Developing El Niño* *J. Clim.* **18** 302–319

The Use of Sparse Direct Solver in Vector Finite Element Modeling for Calculating Two Dimensional (2-D) Magnetotelluric Responses in Transverse Electric (TE) Mode

Lisa' Yihaa Roodhiyah[1], Tiffany Tjong[1], Nurhasan[1], D Sutarno[1]
1 Electromagnetic Induction Lab., Physics of Complex System Div., Faculty of Mathematics and Natrural Sciences, Institut Teknologi Bandung, Indonesia

Email: lisa_yihaa@s.itb.ac.id

Abstract. The late research,linear matrices of vector finite element in two dimensional(2-D) magnetotelluric (MT) responses modeling was solved by non-sparse direct solver in TE mode. Nevertheless, there is some weakness which have to be improved especially accuracy in the low frequency (10^{-3} Hz-10^{-5} Hz) which is not achieved yet and high cost computation in dense mesh. In this work, the solver which is used is sparse direct solver instead of non-sparse direct solverto overcome the weaknesses of solving linear matrices of vector finite element metod using non-sparse direct solver.Sparse direct solver will be advantageous in solving linear matrices of vector finite element method because of the matrix properties which is symmetrical and sparse. The validation of sparse direct solver in solving linear matrices of vector finite element has been done for a homogen half-space model and vertical contact model by analytical solution. Thevalidation result of sparse direct solver in solving linear matrices of vector finite element shows that sparse direct solver is more stable than non-sparse direct solver in computing linear problem of vector finite element method especially in low frequency. In the end, the accuracy of 2D MT responses modelling in low frequency (10^{-3} Hz-10^{-5} Hz) has been reached out under the efficient allocation memory of array and less computational time consuming.

1. Introduction

MT is one of the passive electromagnetic methods which measure natural variations of electric and magnetic vector field at the earth surface to map resistivity structure in the subsurface of the earth [1]. One of the methods which is used to model the complex subsurface electrical resistivity structure in MT is vector finite element method. The method can model the electric and magnetic vector fields conceptualy correct (divergence-free) based on the Maxwell's equation so the result is not a spurious solutionwhich is generally attributed to lack of enforcement of the divergence condition [2,3,8]. In vector finite element mehod, there is a linear problem of matrices which have to solve. It can be written as follow[2],

$$[K']\{A'\} = \{b\} \tag{1}$$

where $[K']$ is stiffness matrices, $\{A'\}$ is vector field which is modeled and $\{b\}$ is boundary condition.

In late research in 2015, the linear matrices of 2D magnetotelluric modeling using vector finite element method (eq. 1) was solved by non-sparse solver [4]. However, the result showed an unstable responses in low frequency (10^{-3} Hz-10^{-5} Hz) especially in phase respon which is not achieved yet. It is caused by the stiffness matrices ($[K']$) properties of vector finite element in magnetotelluric modeling which is ill-conditioned and sparse. Besides, the computational parameters (allocation memory of array and computational time consuming) of non-sparse solver was not efficient in dense mesh due to the huge allocation memory of matrices. Therefore, high cost computation will become the problem of modeling in dense mesh [9].

To overcome these issues, the robust efficient sparse direct solver PARDISO [5] is usedhere. Basically,It will use the properties of the matrices which is symmetrical and sparse to solve ill-conditioned matrices and also decrease the allocation memory of the matrices [7]. Therefore, It is assumed more efficient than the non-sparse direct solver.

2. Numerical Method

A computer program modeling in FORTRAN using vector finite element method has been developed which is capable of solving second order Maxwell differential equation in 2D MT problem inTE mode which can be written as follows,[6]

$$\nabla \times \left(\frac{1}{\sigma} \nabla \times \overrightarrow{H} \right) + i\omega\mu_0 \overrightarrow{H} = 0 \qquad (2)$$

where $\overrightarrow{H}$ is magnetic field, i is imaginary number, ω is angular frequency, μ_0 is magnetic permeability of free space, σ is conductivity. After magnetic field value is calculated, then electric field value can be calculated using equation as follow,

$$\frac{\partial H_z}{\partial y} - \frac{\partial H_y}{\partial z} = \sigma E_x \qquad (3)$$

where E_x is x-component of electrical field, H_z is z-component of magnetic field, H_y is y-component of magnetic field. The boundary condition which is used is dirichlet boundary condition. The magnetic field value is set by 1 in the surface andset by 0 in the other sides.Basically, sparse direct solverwhich is used to solve vector·finite element in equation (2)is LU decomposition which is solved in sparse calculation [5].In the end, responses which are modeled are apparent resistivity and phase response.

3. Result and Discussion

The program validation of 2D magnetotelluric response modeling has been done forhomogen half-space model and vertical contact model. Figure 1shows thatfor apparent resistivity of homogen half-space model both non-sparse solver and sparse solverhave a good agreement with true resistivity (10ohm.m).Nevertheless, for phase response of homogen half-space model it shows that only sparse direct solver has a good agreement with theoritical phase response of homogen half-space (45°) infrequency below 10^{-4} Hz. Several phase response datawhich is calculated by non-sparse solver has been scattered in those frequency.

On the other hand, response of vertical contact modelshows in figure 2. It shows that apparent resistivity of non-sparse solver has a good agreement withsparse solver.Both of them are not scattered. Based on the roughly calculated error of apparent resistivity using sparse direct solver is 0.78 in frequency 10^{-5} Hz. Then on the other side, error of apparent resistivity using non-sparse direct solver is 0.83in the same frequency 10^{-5} Hz.However, It shows that the phase response of sparse solverand non-sparse solver does not have a good agreement in frequency below 10^{-4} Hz. The sparse solver is not scattered while non-sparse solver is scattered in those frequency around the boundary of different resistivity.Next, based on the roughly calculatederror of phase using sparse direct solver is 0.16 in frequency (10^{-5} Hz). Then on the other side, error of phase using non-sparse direct solver is 3.27 in the same frequency (10^{-5} Hz).

It is caused by ill-conditioned properties of stiffness matrices in vector finite element which can not be overcome by non-sparse solver. The phase response is more sensitive than apparent resistivity in the boundary of different resistivity due to imaginary value.

Then, The advantagesof using sparse direct solver shows in table 1. It shows that sparse solver is more efficient in allocation memory of array and computation time than non-sparse solver.

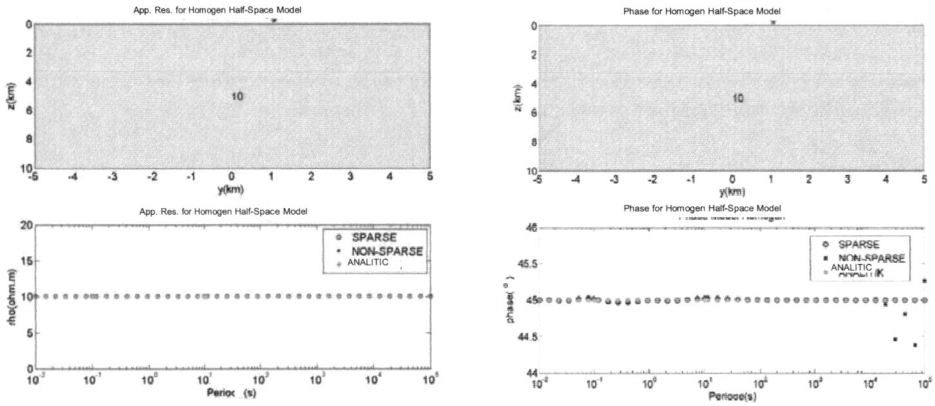

(a) **(b)**

Figure 1. (a) Apparent resistivity and (b) phase responses of homogen half-space model using sparse direct solver and non-sparse direct solver

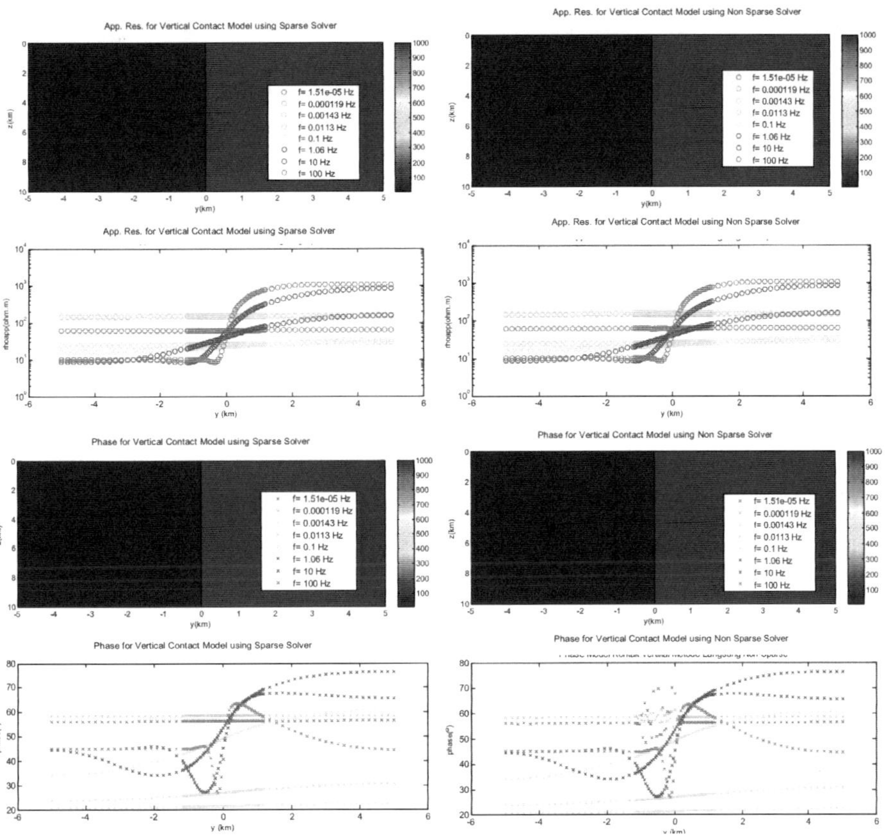

(a) (b)

Figure 2. (a)Apparent resistivity (upper-left) and phase (bottom-left) responses of vertical contact model using sparse direct solver, (b) Apparent resistivity (upper-right) and phase (bottom-right) responses of vertical contact model using non-sparse direct solver

Table 1. Computational parameters of sparse direct solver and non-sparse direct solver

Methods	**Sparse**	**Non-sparse**
Matrix dimension	1x74668	13544x13544
Allocation memory of array (*cells*)	74668	1,83E+08
Computation time (*s*)	175.6	1406.2

4. Conclusion

A computer program in FORTRAN to compute numerical solution of 2D MT responses based on vector finite element method utilizing sparse direct solver has been developed. It shows that using sparse direct solver in solving linear matrices of vector finite element has improved the accuracy of apparent resistivity and phase responses of 2-D MT modellingin low frequency (10^{-3} Hz-10^{-5} Hz).Additionaly, using sparse direct solver have decreased allocation memory of array and computational time consuming. Thus, 2D MT response modeling for low frequency can be developedunder the efficient allocation memory of array and less computational time consuming.

Acknowledgement
This work is supported by Directorate General of Human Resource for Science, Technology and Higher Education (DIKTI) for the financial support through PMDSU program.

References

[1] Cagniard, L. (1953):Basic theory of the magnetotelluric method of geophysical prospecting,*Geophysics*, **18**, 605-635.
[2] Jian, J. (2002): The finite element method in electromagnetics (second edition), John Wiley & Sons. Inc., New York.
[3] Sukarman, E. P. (2015): *Pemodelan struktur resistivitas 2D magnetotelurik menggunakan metode elemen hingga berbasis vektor (studi kasus : data MT sesar Sumatera)*, Thesisof Magister Program, Institut Teknologi Bandung, 19-33.
[4] Farquharson, C.G. dan Miensopust, M.P. (2011): Three-dimensional finite element modelling of magnetotelluric data with a divergence correction, *Journal of Applied Geophysics*,**75**, 699-710.
[5] Schenk, O and Gärtner, K 2004 Solving unsymmetric sparse systems of linear equations with PARDISO *Future Gener. Comput. Syst.* **20** 475-487
[6] Schaa, R., Gross, L., & Du Plessis, J. (2016): PDE-based geophysical modelling using finite elements: examples from 3D resistivity and 2D magnetotellurics,*Journal of Geophysics and Engineering*, ***13*(2)**, S59.
[7] Prihantoro, R., & Sutarno, D. (2016, August). The Use of Direct Solver in Vector Finite Element Modeling for Calculating 3-D Magnetotelluric Responses. In *Journal of Physics: Conference Series* (Vol. 739, No. 1, p. 012110). IOP Publishing.
[8] Prihantoro, R., Sutarno, D., & Nurhasan. (2015, September). Numerical solution of 3-D magnetotelluric using vector finite element method. In *AIP Conference Proceedings* (Vol. 1677, No. 1, p. 060018). AIP Publishing.
[9] Ledo, J., Queralt, P., Martí, A., & Jones, A. G. (2002). Two-dimensional interpretation of three-dimensional magnetotelluric data: an example of limitations and resolution. *Geophysical Journal International*, *150*(1), 127-139.

Two Dimensional Finite Element Based Magnetotelluric Inversion using Singular Value Decomposition Method on Transverse Electric Mode

Tiffany Tjong[1], Lisa Yihaa' Roodhiyah[1], Nurhasan[1], Doddy Sutarno[1]

[1] Electromagnetic Induction Lab., Physics of Complex System Div., Faculty off Mathematics and NaturalSciences, Institut Teknologi Bandung, Indonesia

E-mail : tiffany11002@students.itb.ac.id

Abstract. In this work, an inversion scheme was performed using a vector finite element (VFE) based 2-D magnetotelluric (MT) forward modelling. We use an inversion scheme with Singular value decomposition (SVD) method toimprove the accuracy of MT inversion.The inversion scheme was applied to transverse electric (TE) mode of MT. SVD method was used in this inversion to decompose the Jacobian matrices. Singular values which obtained from the decomposition process were analyzed. This enabled us to determine the importance of data and therefore to define a threshold for truncation process. The truncation of singular value in inversion processcould improve the resulted model.

1. Introduction

Magnetotelluric (MT) is an electromagnetic method whichis used to map electrical resistivity distributionin the subsurface. MT method involves measuring electric field and magnetic field on the surface of the earth. The data obtained usually presented as apparent resistivity and phase data [1].To derive electrical resistivity value from MT data, inversion method can be used. Inversion inMT usually are ill-posed where small changes in data could drastically affecting the inversion results [2]. Finite element discretization method is a good method to be used in electromagnetic field problems such as MT.The solution obtained from finite element method satisfying the physical concept of electromagnetic field [3].The usage of singular value decomposition (SVD) has been used to the ill-posed problems to stabilize the iteration process within the inversion [2].SVD is able to produce more precise and robust inversion result [4].There are some works which have applied SVD to the MT inverse problems such as [5],[6],[7].The previous inversion works mostly usefinite difference discretization. Within this work, we applied SVD inversion method to a vector based finite element (VFE) MT modelling for transverse electric (TE) mode.

2. Inversion with Singular Value Decomposition

The inversion scheme described in this work isheavily based on the work of [1].In general, the relationship of model parameter and data can be described as :

$$\mathbf{d} = \mathbf{g}(\mathbf{m}) \tag{1}$$

where **d** is a vector containing several data which are apparent resistivity and phase as a function of frequency and the location of measurement.The vector**m** is the model parameters vector. The model parameter in this case is true electrical resistivity for each cell in VFE discretization. **g(m)**is the forward modelling which produce a set of calculated data for a certain set of **m**.Basically, the inversion problem is to obtain the correct set of **m** which generates a set of **g(m)** that fit with the measurement data **d**.In simple inversion scheme, **m** is determined by using iterative method which involves estimating initial value and a perturbation of the value for each iteration. The perturbation of **m**is defined as follow :

$$\Delta \mathbf{m} = \left(\mathbf{J}^T\mathbf{J}\right)^{-1} \mathbf{J}^T \boldsymbol{\varepsilon} \qquad (2)$$

where **ε** is the error between measured data and calculated data and **J** is Jacobian matrix. Jacobian matrix **J** is also termed as the sensitivity matrix. Equation (2) is used when **J** is not singular. For inversion with SVD, Δ**m** is obtained by directly inverting **J** :

$$\Delta \mathbf{m} = \mathbf{J}^{+} \boldsymbol{\varepsilon} \qquad (3)$$

$\mathbf{J}^{+}$ is the generalized inverse of Jacobian matrix and is defined by :

$$\mathbf{J}^{+} = \mathbf{V}\, \mathbf{S}^{+}\, \mathbf{U}^{T} \qquad (4)$$

V, **S**, and **U** are the matrices which is a result of matrix **J** decomposition using SVD method [8]. Matrix **S** is a diagonal matrix with the components are the singular value of **J**. $\mathbf{S}^{+}$ is the invert matrix of **S**.

3. Results and Discussion

The inversion was performed on a homogenous model with several different singular value threshold. We use a synthetic model to test the inversion program. The resistivity value of the synthetic model is 100 Ohm-m.Initial resistivity value was set to 5 Ohm-m. Figure 1 shows the inversion result for singular value threshold ranging from 0.001 to 1. The lower singular value threshold (0.001 and 0.01) producing model with mean percentage errorvalue of 3.7657e+10 % and 685.8591 %. The higher threshold value also producing less accurate model with error value 59.0965%. The optimal model is attained with the threshold ranging from 0.02 to 0.1. The threshold for truncating the singular value influences the result of the inversion.

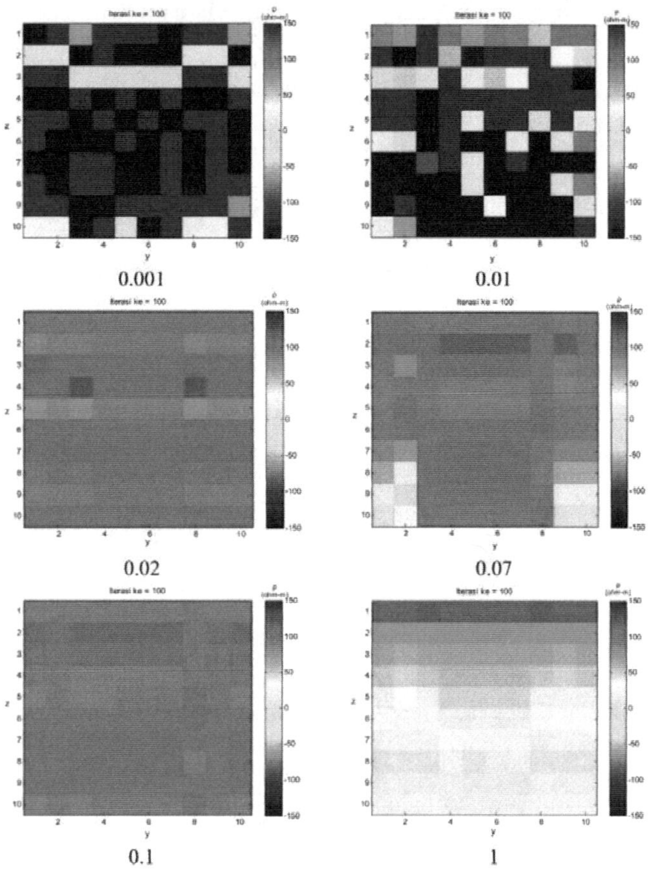

Figure 1. Inversion result with different singular value threshold for homogenous model

4. Conclusion

We have developed an inversion program using singular value decomposition method to solve the ill-posedness of magnetotelluricproblem.The test using homogenous model shown a similarity with the synthetic model. Truncation of singular values also affecting the inversion result. For homogenous model shown in this work, the optimum value to obtain the appropriate result is between 0.02 to 0.1.

Acknowledgments

This work is supported by Directotrate General of Human Resource for Science, Technology and Higher Education (DIKTI) for the financial support through Program Pendidikan Magister Menuju Doktor untuk Sarjana Unggul (PMDSU) program.

References

[1] Cagniard, L., 1953. Basic theory of the magneto-telluric method of geophysical prospecting. *Geophysics*, *18*(3), pp.605-635.
[2] Jupp, D.L.B. and Vozoff, K., 1975. Stable iterative methods for the inversion of geophysical data. *Geophysical Journal International*, *42*(3), pp.957-976.
[3] Kordy, M., Wannamaker, P., Maris, V., Cherkaev, E. and Hill, G., 2016. 3-D magnetotelluric inversion including topography using deformed hexahedral edge finite elements and direct

solvers parallelized on SMP computers–Part I: forward problem and parameter Jacobians. *Geophysical Journal International*, *204*(1), pp.74-93.

[4] Lines, L.R. and Treitel, S., 1984. Tutorial: A review of least-squares inversion and its application to geophysical problems. *Geophysical prospecting*, *32*(2), pp.159-186.

[5] Jupp, D.L. and Vozoff, K., 1977. Two-dimensional magnetotelluric inversion. *Geophysical Journal International*, *50*(2), pp.333-352.

[6] Kalscheuer, T. and Pedersen, L.B., 2007. A non-linear truncated SVD variance and resolution analysis of two-dimensional magnetotelluric models. *Geophysical Journal International*, *169*(2), pp.435-447.

[7] Candansayar, M.E., 2008. Two-dimensional inversion of magnetotelluric data with consecutive use of conjugate gradient and least-squares solution with singular value decomposition algorithms. *Geophysical Prospecting*, *56*(1), pp.141-157.

[8] Jackson, D.D., 1972. Interpretation of inaccurate, insufficient and inconsistent data. *Geophysical Journal International*, *28*(2), pp.97-109

Development of low cost and accurate homemade sensor system based on Surface Plasmon Resonance (SPR)

F D Laksono, Supardianningsih, M Arifin and K Abraha

Department of Physics, Universitas Gadjah Mada, Sekip Utara BLS 21 Yogyakarta, Indonesia

Email: muhammad.arifin@ugm.ac.id

Abstract. In this paper, we developed homemade and computerized sensor system based on Surface Plasmon Resonance (SPR). The developed systems consist of mechanical system instrument, laser power sensor, and user interface. The mechanical system development that uses anti-backlash gear design was successfully able to enhance the angular resolution angle of incidence laser up to 0.01°. In this system, the laser detector acquisition system and stepper motor controller utilizing Arduino Uno which is easy to program, flexible, and low cost, was used. Furthermore, we employed LabView's user interface as the virtual instrument for facilitating the sample measurement and for transforming the data recording directly into the digital form. The test results using gold-deposited half-cylinder prism showed the Total Internal Reflection (TIR) angle of 41,34°± 0,01° and SPR angle of 44,20°± 0,01°, respectively. The result demonstrated that the developed system managed to reduce the measurement duration and data recording errors caused by human error. Also, the test results also concluded that the system's measurement is repeatable and accurate.

1. Introduction

Surface Plasmon Resonance (SPR) can be described as a collective resonance on the oscillations of electrons in the valence band in a material by light [1]. The light coming through the medium boundary with the angle comes greater than the critical angle when it concerns the boundary surface resulting in the total internal reflection (TIR) optical phenomenon. When the TIR occurs, the refractive electromagnetic waves have a field waveform with amplitude decaying exponentially away from the boundary plane of two mediums called evanescent waves. The resonance condition is achieved if the frequency of the light photon corresponds to the natural frequency of the electrons of the surface oscillating against the atomic restoring force. When the resonance takes place on a planar surface, there is a surface plasmon (SP) excitation which is a surface electromagnetic wave with the direction of propagation parallel to the metal / dielectric boundary plane[2].

When the SP wave vector and the evanescent wave vector corresponding to the magnitude, then the resonance is eliminated due to the phase difference. As a result, at a certain angle of light occurs attenuated total reflection (ATR) is marked by a drastic reduction in the intensity of reflected light [3]. Wave oscillations in the boundary plane of the metal and the outer medium are very sensitive to any changes in this boundary, for example, the change due to absorption by molecules on the metal surface. If the change occurs then the ATR angle shifts. The magnitude of the ATR angle and its shift is a parameter of the magnitude measured by the SPR instrument [4].

The commercial instrument has several advantages such as easy to operate and precise measurement results. On the other hand, however, commercial instruments are usually treated as a black box and

IOP Conf. Series: Journal of Physics: Conf. Series **1011** (2018) 012043 doi:10.1088/1742-6596/1011/1/012043

only produce results without knowing the actual process [5]. As for the SPR homemade instrument has many advantages, including researchers, have full access to the operation of the system, and it is possible to modify the configuration according to the research scheme[6,7]. SPR homemade is also cheaper and can be modified both hardware and software.

The SPR instrument mechanical system enhanced by anti-backlash gear based on the Torsional Spring-Loaded Anti-backlash model [8–11].The laser detector acquisition system and stepper motor controller utilizing Arduino Uno which is easy to program, flexible, and low cost, was used [12]. LabView's user interfaces as the virtual instrument for facilitating the sample measurement and for transforming the data recording directly into the digital form. Also, LabVIEW is simplified application development and have been customized for data acquisition systems, software simulations, embedded electronics, and real-time controls [13].

2. Instrument Design

The basic design of homemade and computerized sensor system based on SPR is shown in Figure 2. The main component is a Mechanical system, optic system, and computerized system. The optic system includes a low power laser (HeNe 632.8 nm, 3 mW), two detectors (one and two), a polarizer, a beam splitter. The computerized system consists Main Unit and Computer. The half-cylinder prism was deposited by gold and mounted on prism holder.

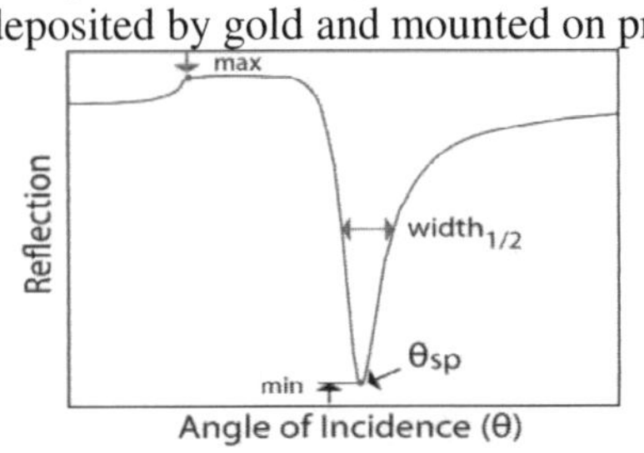

Figure 1. Ideal SPR spectra.

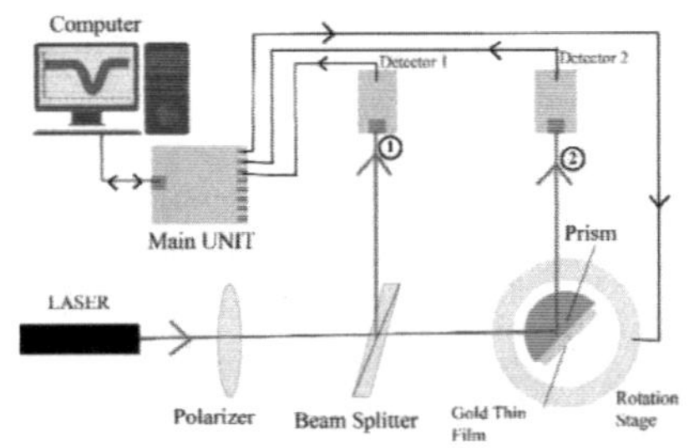

Figure 2. block diagram of the SPR instrument.

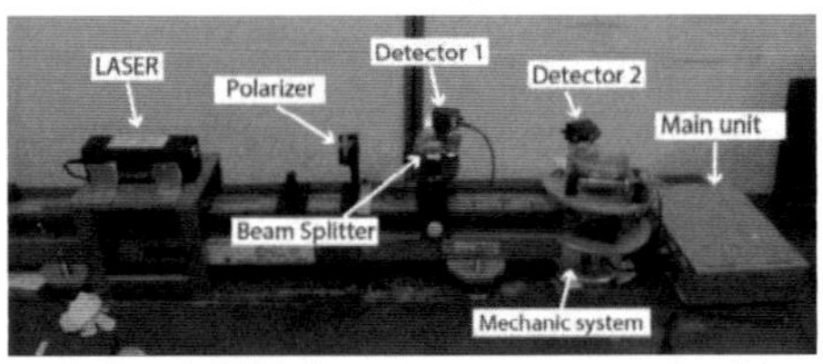

Figure 3. Real SPR instrument set-up in this paper.

Angle scanning to find SPR angle is performed by measuring reflected light intensity as the prism and detector rotated in the θ-2θ mechanism. Prism and detector are rotated by angle θ and 2θ respectively. The data plotted on the SPR graph is shown in Figure 1 are angles and reflections. The experimental SPR spectra measurement of the gold thin film shown in Figure 2. The light from LASER pass through a polarizer to enhance the overall SPR signal by removing the light from a laser that's not polarized in TM mode or parallel to the plane of incidence. The P polarized laser beam then divided into two parts with the same wavelength and intensity by the beam splitter. One part of the laser beam to the first laser detector as the reference laser intensity. The other part of the light is reflected by the prism toward the second laser detector as the laser intensity. Reflection data obtained by dividing intensity from detector two (2) with detector one (1). Real SPR instrument set-up is shown in Figure 3.

2.1 Mechanical system

The mechanical system shown in Figure 5, includes two motor stepper, two torsional spring-loaded anti-backlash gear, detector holder, and prism holder. The hybrid stepper motor is used to obtain good torque and angular, which the angular resolution is 1.8°/step. Wheel gear that connected to the worm gear is designed with 180 teeth to obtain 1:180 ratio. Both motor steppers utilize a worm drive that

rotates the table 1° for 100 full steps of the stepping motor. This defines the accuracy of the instrument which is ±0.01° for both the resonance and critical angle.
The repeatability measurement of SPR spectra enhances by a torsional spring-loaded anti-backlash gear as wheel gear. The anti-backlash gear used to reduce backlash of the mechanical system. Design of torsional spring-loaded anti-backlash gear shown in Figure 6. Anti-backlash gear has two wheel gear which connected by spring. Mechanism of anti-backlash gear to reduce backlash shown in Figure 4. In Figure 4 (a) wheel gear g1 and wheel gear g2 interact, but having a space between the interconnected teeth causes a backlash. In Fi 4 (b) the anti-backlash gear g1 has two-wheeled gears whose direction of emphasis differs due to the spring effect. Therefore, when wheel gear g1 interacting with the wheel gear g2 there is no space then minimize the backlash. The torsional spring-loaded anti-backlash model has a very good accuracy level on small loads.

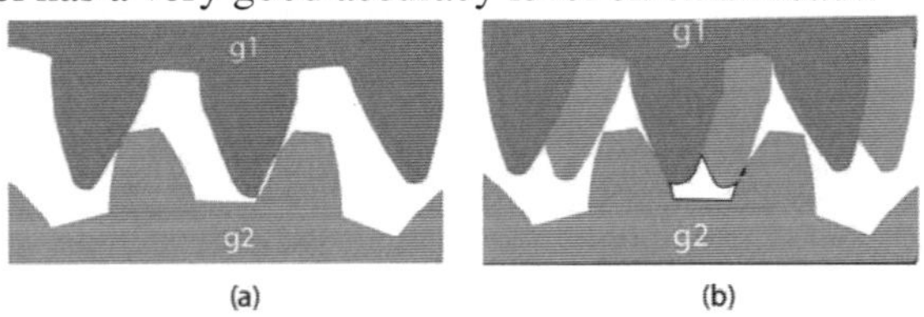

Figure 4. (a) Wheel gear with backlash and (b) anti-backlash gear

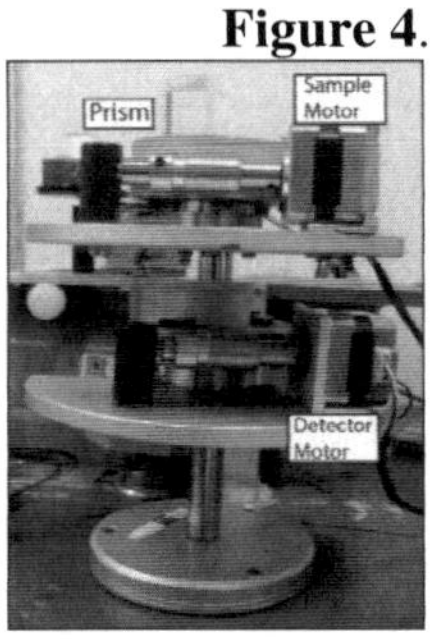

Figure 5. Mechanical system in this paper

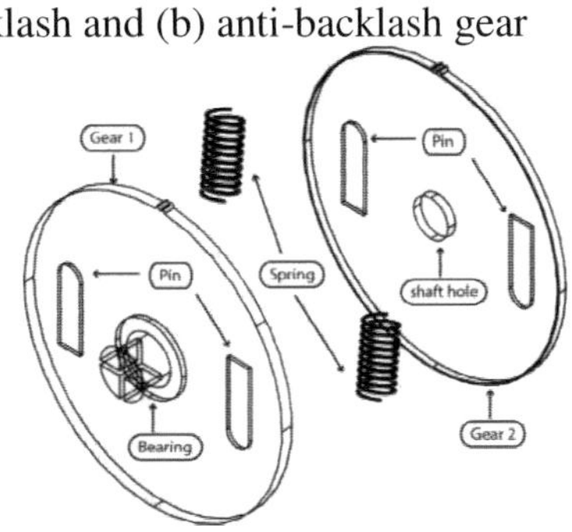

Figure 6. Design of torsional spring-loaded anti-backlash gear

3. Electronic System
The electronic system has three tasks. First, to drive two motor stepper, each one with a bipolar stepper motor. Second, to convert the analogical response of the photodiode to a digital value. Third, to respond any command and send data to the computer. These three tasks executed through USB commands as a serial signal.
The electronics of the main unit of the SPR instruments consists of a microcontroller based system (Arduino UNO), which was selected because it has USB connectivity, cheap, and facilitates the control of the system by synchronizing with LabView Software. This electronic system is composed of a microcontroller, a power supply, motor driver, and a computer. There is two analog input; these are the main photodetector and the reference photodetector. Also, the microcontroller gives us threaten digital output to control the stepping motors via motor drivers.
The photodetector used is Si photodiode (OPT101, Texas Instrument). This photodiode selected because has fast responses, built in with trans-impedance amplifier, high linearity light power, and voltage. The photodetector directly connected to ADC device (ADS1115) to reduce possible noise causes of long analog cable. Besides, this ADC device has 16bit resolution and I2C communication protocol which support with Arduino UNO. When ADS1115 gain amplifier set to ±4.096V, then the resolution is 125μV. The light reflected or transmitted by SPR instrument detect by photodetector converted to a voltage. This voltage converted to digital form and sent to Arduino UNO by ADS1115.
In the movement routing, the photodetector motor rotated two steps (0.02°), and the sample motor rotated one step (0.01°), then the ADC is started to sample the sensors. The main value measured by SPR instrument corresponds to the angle when the minimal intensity of the light occurs. This angle determined from the measured angle dependence on the reflected light of the voltage form, where

voltage proportional to incident light. Software counters are used to know the actual position and determine when the system has to stop. This software uploaded inside the microcontroller Arduino UNO. The Arduino UNO labeling each data to aid LabVIEW software to organize all data collected.

4. LabVIEW Software Interface

The LabVIEW software to be the interface between SPR instrument and user. The state machine diagram is shown in Figure 7 and the block diagram shown in Figure 8. Each program made in LabVIEW is called a virtual instrument and construct of three component. A front panel which is the user interacts a block diagram which is the program that controls the front panel and icon/connector which is connecting between programs (represent with the box in Figure 8).

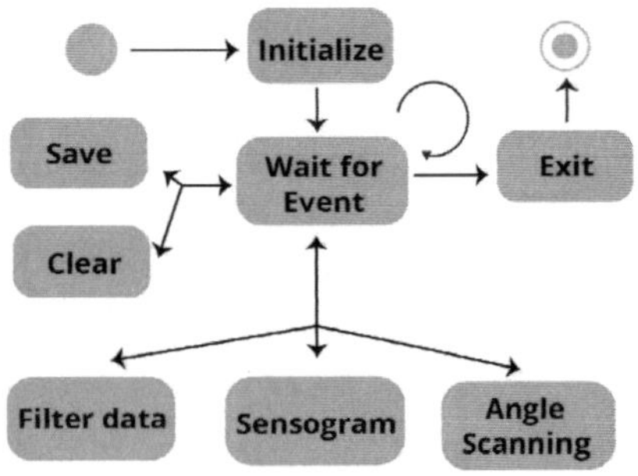

Figure 7. State machine diagram of user interface

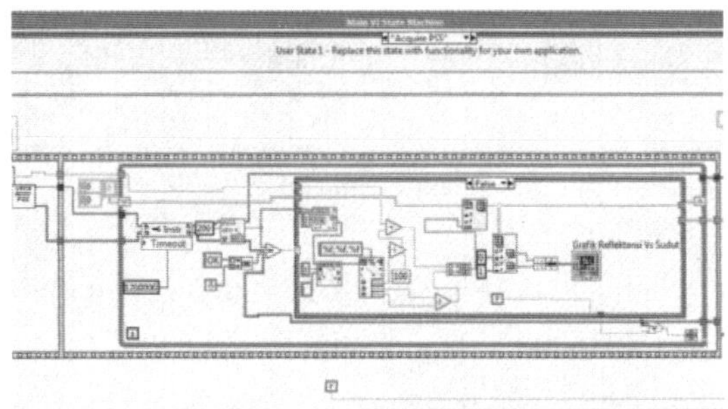

Figure 8. Partial source code of LabVIEW program

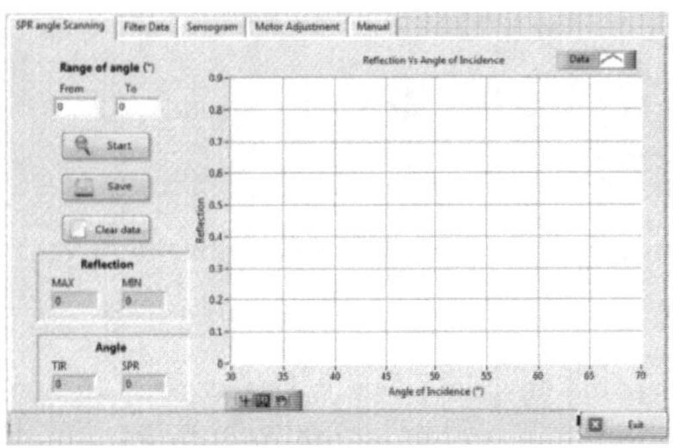

Figure 9. The front panel of the LabVIEW program is showing the user interface of the "SPR angle scanning" tab.

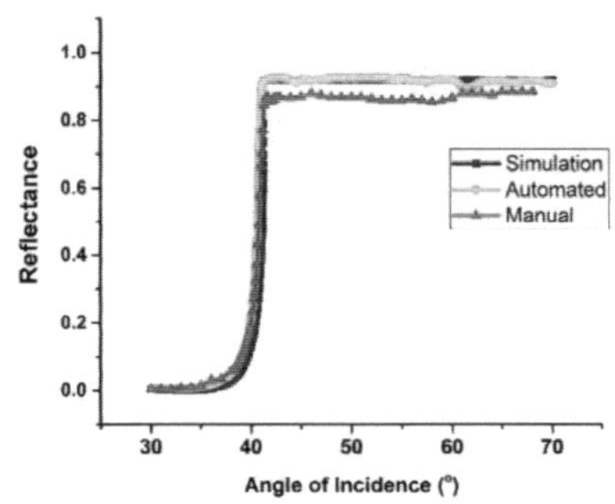

Figure 10. Comparison of resonance angle measurement of half-cylinder prism for the manual system developed a system (Automated), and theory-based simulation.

The program begins with the circle in the top left corner, then initialize all electronic and program component. "loop" cycle (above wait for event diagram) serves to make selection state (action) in the user menu. The Software interface consists of one pop-up window and five tabs. The following describes the function of each tab.

- *Setting port*. This Pop-up window shows a box to determine the serial port used for communication between LabVIEW and SPR instrument. The communication port uses a serial signal passed over USB so that the port name is prefixed by COM. The port number used is the serial port connected to the Arduino UNO microcontroller.
- *SPR angle scanning*. The signal of the light received by detector one and detector two is graphed in real time. In this tab, the initial and final degrees should be specified. When the scanning is started, sample motor is moved one step which leads to change the angle of prism 0.01° (θ). The detector motor is moved two step which leads to change the angle of detector 0.02° (2θ). When the scanning process is finished, each motor rotated back to lead the prism and detector back to the initial position. In the reflection box corresponding to the maximum

and minimum intensity of light. In the angle box corresponding to the angle of TIR (total internal reflection) and SPR (surface plasmon resonance). This screen of the "SPR angle scanning" tab is shown in Figure 9.

- *Filter data*. In this tab, a filter data obtained from the last SPR angle scanning. Filter method to smoothing is carried on through the "Savitzky and Golay function" from LabView software. This function is a digital filter that can be applied to a set of digital data points to increase the signal-to-noise ratio without greatly distorting the original signal by convolution. First, the polynomial order and the number of data points is specified to each side of the current data point to use for the least squares minimization. After filter process finished, a graph which smoothed is displayed. In the reflection box corresponding to the maximum and minimum intensity of light. In the angle box corresponding to the angle of TIR (total internal reflection) and SPR (surface plasmon resonance).
- *Sensorgram*. In this tab, the signal of the light received by detector graphed in real time with the x-axis as time and y-axis as reflection. First, SPR angle, duration, and time interval should be specified. When the process is started, the motor rotates to SPR angle then sampling the signal of the light every time interval until duration time run out. When the scanning process is finished, each motor rotated back to original position.
- *Motor Adjustment*. In this tab, setting the motor to adjust the position of the sensor and prism to get the position 'zero'.
- *Manual*. In this tab, contain a manual guide for operating the SPR instrument.

5. Test and result

5.1 Test using half-cylinder prism as sample

This test is intended to determine whether the angle of measurement results has been matching with the reference. References are obtained via simulation using the Winspall program and related variables of the datasheet. There are two parameters to be considered in this test, i.e., the angle of the prism when there is a phenomenon of total internal reflectance and the graph form from scanning.

The first test compared the measurement results using a manual system, an automated system and simulated from an angle of 30° to 70°. The simulation was performed based on the glass datasheet (BK7) which used to have a prism refractive index of 1.5151 when subjected to a laser beam with a wavelength of 632.8 nm. Figure 10 shows a comparison graph between manual system measurement data, automated systems, and simulations. The theoretical simulation using winspall software with prism refractive index constant parameters at 632.8 nm wavelength based study is 1.5151. The manual system measurement has same construction with automated instrument, but it is using manual rotating table despite motorized rotary stage. Manual system measurement uses manual light intensity measurement using portable photodetector and record data manually. The test results using half-cylinder prism showed the Total Internal Reflection (TIR) angle of 41.31°± 0.01°. The time taken to perform a single measurement in this test is for four minutes.

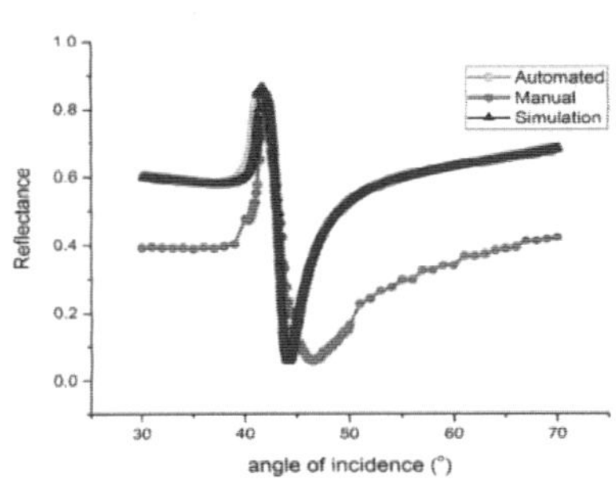

Figure 11. Comparison of resonance angle measurement of gold-deposited half-cylinder prism for the manual system developed a system (Automated), and theory-based simulation.

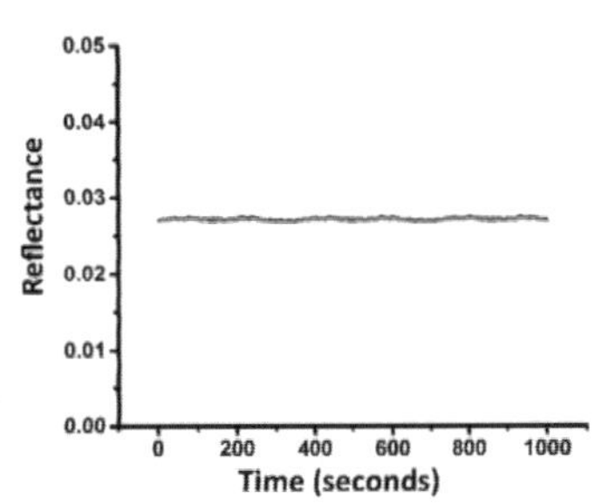

Figure 12. Sensorgram graph at SPR angle of the gold-deposited half-cylinder prism.

5.2 Test using gold-deposited half-cylinder prism

The first test was performed on a half-cylinder prism which has a thickness of 30 nm gold layer by comparing the measurement results using manual and automated SPR instruments plus theoretical simulations. Tests using manual and automated SPR instruments are done from an angle of 30° to 70°. The theoretical simulation using winspall software with gold dielectric constant parameters at 632.8 nm wavelength based study is ε_{Au} = -12.033 + 1.163i [14]. The test is done to know the difference of graph form from each measurement method. Figure 11 shows graphs of manual, automated, and simulated measurements. Test results using gold-deposited half-cylinder prism showed the Total Internal Reflection (TIR) angle of 41.34°± 0.01° and SPR angle of 44.20°± 0.01°, respectively.

5.3 Real-time Test measurement using sensorgram

Real-time measurements on SPR instruments can be used to obtain sensorgrams. There are three ways to get the sensorgram, one of which is measuring the reflectance of time shift [15]. Reflectance is measured at a fixed angle. Usually, the selected angle is the SPR angle because it is a unique parameter of measurement and can also describe its refractive index. Therefore, this test aims to test the ability of SPR instruments in performing measurements in real time. Measurements were made on gold-deposited half-cylinder prism at the SPR angle every second for 1000 seconds. The result of the test shown in Figure 12 demonstrated a reflectance value that tends to be constant even if actual data is seen up and down reflectance value, but because the value of up and down is small then not visible on the graph. The reflectance value is small because the measurement angle is done at an angle of 44.20 ° which is the SPR angle of sample A. Due to the non-treatment of the gold layer the reflectance value remains constant.

6. Conclusion

Development of low cost and accurate homemade sensor system based on Surface Plasmon Resonance (SPR) have been successful. The result demonstrated that the developed system managed to reduce the measurement duration, data recording errors caused by human error, and easy to use. Also, the test results also concluded that the system's measurement is repeatable and accurate.

7. Acknowledgement

This work was supported by *Hibah Peningkatan Kapasitas Peneliti Dosen Muda 2017* coordinated by Directorate of Research, Universitas Gadjah Mada.

8. References

[1] Jensen T R, Malinsky M D, Haynes C L, and Van Duyne R P 2000 Nanosphere Lithography: Tunable Localized Surface Plasmon Resonance Spectra of Silver Nanoparticles *J. Phys. Chem. B* **104** 10549–56

[2] Simon H J, Mitchell D E and Watson J G 1975 Surface plasmons in silver films—a novel undergraduate experiment *Am. J. Phys.* **43** 630–6

[3] Orfanides P, Buckner T F, Buncick M C, Meriaudeau F and Ferrell T L 2000 Demonstration of surface plasmons in metal island films and the effect of the surrounding medium—An undergraduate experiment *Am. J. Phys.* **68** 936–42

[4] Piliarik M and Homola J 2006 SPR Sensor Instrumentation *Engineering* vol 4 (Berlin Heidelberg: Springer-Verlag)pp 95–116

[5] Miller L S, Nakhleh M B, Nash J J and Meyer J A 2004 Students' Attitudes toward and Conceptual Understanding of Chemical Instrumentation *J. Chem. Educ.* **81** 1801–8

[6] Lavine B K, Westover D J, Oxenford L, Mirjankar N and Kaval N 2007 Construction of an inexpensive surface plasmon resonance instrument for use in teaching and research *Microchem. J.* **86** 147–55

[7] Luna-Moreno D, Espinosa Sánchez Y M, Ponce de León Y R, Noé Arias E and Garnica Campos G 2015 Virtual instrumentation in LabVIEW for multiple optical characterizations on the same optomechanical system *Opt. - Int. J. Light Electron Opt.* **126** 1923–9

[8] Lindsay D 2012 Anti-backlash resistant gearing for a seat mechanism **2** 6

[9] Yang Z, Shang J, Yu N, Luo Z and Wang X 2013 Effects of shrink range and preload on dynamics characteristics of anti-backlash geared rotor-bearing system with composite mesh stiffness *J. Vibroengineering* **15** 1642

[10] Yang Z, Shang J, Luo Z, Wang X and Yu N 2015 Nonlinear Dynamics Modeling and Analysis of Torsional Spring-Loaded Antibacklash Gear with Time-Varying Meshing Stiffness and Friction *Adv. Mech. Eng.* **5** 203438–203438

[11] Deng X, Wang J, Wang Q, Liu Q Q and Zhang J 2010 The study and analysis of new Anti-backlash worm gear *2010 Int. Conf. Mech. Autom. Control Eng. MACE2010* 2984–8

[12] D'Ausilio A 2012 Arduino: A low-cost, multipurpose lab equipment *Behav. Res. Methods* **44** 305–13

[13] Sumathi S and Surekha P 2007 *LabVIEW based Advanced Instrumentation Systems* (Berlin, Heidelberg: Springer Berlin Heidelberg)

[14] Olmon R L, Slovick B, Johnson T W, Shelton D, Oh S-H, Boreman G D and Raschke M B 2012 Optical dielectric function of gold *Phys. Rev. B* **86** 235147

[15] Schasfoort R B M and McWhirter A 2008 SPR Instrumentation *Handbook of Surface Plasmon Resonance :* (Cambridge: Royal Society of Chemistry) pp 35–80

The Correlation of Blood Glucose Concentration and the Movement of Laser Secondary Speckle Pattern of the Artery

M A Saputra , P Prajitno

Department of Physics, Faculty of Science and Mathematics,
Universitas Indonesia, Kampus UI Depok, Depok 16424, Indonesia

Corresponding author's e-mail: prawito@sci.ui.ac.id

Abstract. Blood glucose is the molecule needed for human life, it usually measured invasively (by taking blood). but that measurement is still very vulnerable. The alternative method namely the non-invasive method is very interesting. In addition, the article [1] explains the relationship between the movement of the arterial pulse with glucose concentration, therefore the research study to investigate the correlation between the blood glucose and the movement of laser speckle pattern resulted from the arterial movement will be promising as the non-invasive method for measuring the blood glucose concentration. In this study, the laser speckle pattern imaging method, where the microscopically movement of the object is illuminated by a laser beam and recorded by the high-speed camera in a certain interval time, are used to identify the movement patterns of the artery. From the image processing, the graphs such as electrocardiograph (ECG) can be extracted. The average of the maximum peaks of the graph can be correlated with the blood glucose concentration in the blood, as the same as shown in the article [2]. From the data that has been obtained in this research, the movement of the speckle tends to increase in accordance with the rise of blood glucose concentration.

1. Introduction

Blood glucose or glucose (C6H12O6) in blood is a molecule that's very important which provides energy to all cells in the body. If the number of glucose not stable, it can cause health problems. Therefore, blood glucose checking is really needed. Blood glucose checking can be done by invasive and non-invasive way. Invasive way means that the blood glucose will be checked by taking blood sample of the patient to be analyzed in the laboratory or by using glucometer.

The other way is the one that has been developed by researchers [1] [2] which is non-invasive method where the blood sample is not taken to check the blood glucose.

In this research, the experiment such as the previous research which is explained in the interconnected articles [3], [4], [5] has been conducted. Those articles study about the microscopically movement of reflection from laser beam which illuminate the skin near artery. From the movement, it can be correlated with the arterial movement which later can be correlated to many things but more specifically to blood glucose concentration and dehydration. On article [3], researchers try to develop a new method to see the movement of arterial pulse that can estimate blood pressure from movement of speckle which can be read from around the pulse. Later on, speckle pattern reading method is developed more and more on the article [4] for other things, which are to be correlated with changes of blood glucose. This article doesn't show how the data can be processed to be certain kind of data such as ECG.

In this research, more detail methods in the data processing have been done so data like ECG can be obtained and later on will be correlated with blood glucose. In the other side, the camera that will be used is pocket camera with lower frame rate than the frame rate from the previous experiment.

2. Theoretical explanation

Blood flows through all part of the body transmits substance and oxygen that are needed by body tissues. One of the transmitted substances is blood glucose. The amount of blood glucose will influence the blood parameter like how L. S. Jefferson explained on reference [4].

$$C_v(t) = \frac{(1-\varepsilon)\cdot q_0 \cdot h(t)}{F} \tag{1}$$

Where C_v (t) is the concentration of glucose on venous at certain t time, and F is blood flow which is represented as glucose content. The glucose pulse is symbolized by q_0, where that symbol represents the amount of glucose in milligram (mg) in kilograms of blood per one heartbeat. ε is part of the glucose pulse that is extracted by blood system and metabolism that will not pass the vein. h(t) is a factor from organ that causes delay and distortion on the glucose pulse at the vein.

Based on article [6], if h(t) = 1, and total of all q_0 is C_{artery}, F and total of all $C_v(t)$ is C_v, where C_{artery} and C_{vena} is concentration of glucose on artery and vena. From that, ε can be written like the equation below:

$$\varepsilon = \frac{(C_{Arteri} - C_{Vena})}{C_{Arteri}} \tag{2}$$

Concentration of glucose on artery and vena is proportional with glucose pulse (q_0) and blood flow (F).

Based on article [3], the relation between the radial movement of blood vessels (dr) with pressure change on vessel is shown by the equation below:

$$dr \approx \frac{r}{E} dP \tag{3}$$

Where r is the value of radius of vessel and E is elastic modulus of vessel wall. So the radial relative changes are linear to the pressure of vessel. This research uses laser as the medium to see the small changes.

When laser is illuminated to rough area, granularity pattern will be formed. These patterns will be formed when laser which is coherent touches the area of rough surface so the reflection will disperse to all directions and will result laser speckle patterns.

The speckle in this research is obtained subjectively because the obtained speckle structure is based on parameter of reading system like if the size of lens aperture is bigger, the size of speckle will change, and if the imaging position is changed, the obtained pattern will also change to be unsuitable for the initial pattern.

The size of area can be determined by diffraction-limited resolution from Airy disk lens which diameter can be obtained by formula (4) like the equation below [5]:

$$d = \frac{2.4\lambda u}{D} \tag{4}$$

Where λ is wavelength from laser beam, u is the distance from object to lens and D is diameter of aperture.

A. Ullal [7] have investigated about how to process the movement of laser speckle. When the object does tilting process with α as the angle like in Figure 1, sensor will read the laser speckle pattern which also moves or shifts.

IOP Conf. Series: Journal of Physics: Conf. Series **1011** (2018) 012044 doi:10.1088/1742-6596/1011/1/012044

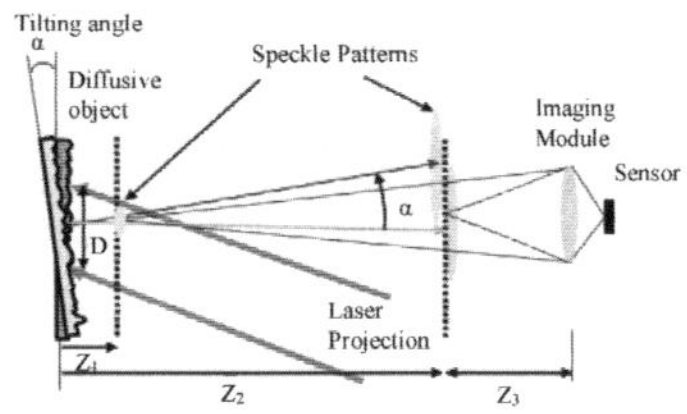

Figure 1. Schematic laser speckle pattern system [5]

Figure 2. Process of data taking and the obtained laser speckle

Figure 3. Block diagram of laser speckle movement data processing

Based on article from Z Zalevsky [4], it is explained that the tilting movement has correlation with shifting movement from camera reading so the equation below can be adaptable.

$$\beta = \frac{4\pi \tan\alpha}{\lambda} \approx \frac{4\pi\alpha}{\lambda} \tag{5}$$

From the equation (5), β is relative shifting from laser speckle pattern to object changes, α is a tilting angle from an object and λ is wavelength from the laser.

$$\tan\alpha \approx \alpha \approx d \tag{6}$$

where d is the changes that happen on the vibrating object. Based on [4], the change of glucose flow is proportional to shift changes on laser speckle pattern

$$C_v(t) \propto \beta(t) \tag{7}$$

3. Experimental results

The system that's used in this research can be seen on Figure 1. From this figure, it can be seen that there's a green laser (532 nm) which is directed to the part of artery. Gypsum is applied around the wrist to prevent any movement, but a hole with the size of laser beam is made on part of the pulsating artery. After that, camera will be positioned to the side with an angle of 45 degree from the direction of the illuminated beam. Other than that, there's 10 centimeters distance from the lens to the reflection and this is based on Airy disk pattern which can be obtained from Airy disk formula (4) above by determining the aperture of the used lens, laser and the distance from the beam source. Camera that's used has size of 512x384 pixel with frame rate of 240 fps and have focus lens 4.5-45.0 mm with aperture lens opening 1:3.5 - 6.5. Recording process and data taking can be seen on Figure 2.

There are some steps which are needed to be followed like on Figure 3, to change the laser speckle pattern so pulse like ECG can be obtained. The result of recording is in form of video recording with duration of 10 seconds which has 512x384 pixel and frame rate of 240 fps which can be seen on Figure 2.

After the image is taken, then the preprocessing and image segmentation will be conducted. The purpose of this is to make processing becomes easier to obtain pulse like ECG. For the process itself, there are several steps, which are, for the first, image will be rotated with the same direction of laser speckle movement, after that from the rotation, vertical movement of the speckle will be obtained. Then, it will be separated from its background. Object separation on this phase is conducted to get an image that will be able to be cut and it also helps the processing becomes easier, then the value of the green one will be taken because the green one is more dominant. These process will be done for every frame.

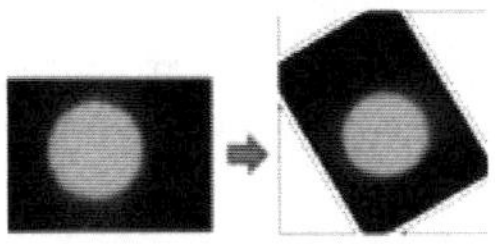

Figure 4. before (left) and after (right) rotation

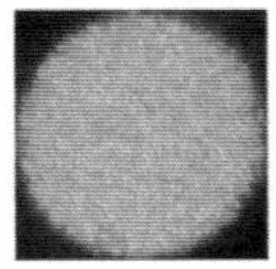

Figure 5. The result of image cropping

Figure 6. The result of RGB separation

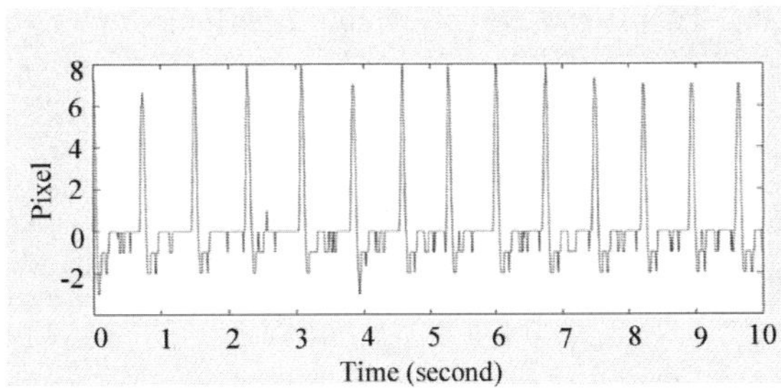

Figure 7. The result of processing like ECG

After being processed in the previous step, the next process is image correlation processing [6] that will be conducted to determine the movement of speckle. The first thing that's needed to do in this image processing is taking a small part of speckle pattern on initial frame, 81x81 pixel of every small pattern in the middle of the picture that has the initial coordinate.

The next step is searching for the highest correlation value between patterns in previous 81x81 pixel square and the picture in the next frame. Every pixel on the frame which is tested with 81x81 pixel square will be analyzed based on its highest correlation. The coordinate of the area that has the highest correlation will be used as the new coordinate origin and this is called movement of laser speckle. Based on article [8] below, the best method to determine the correlation on laser speckle pattern is C3 method.

$$C3 = \frac{\left\{\sum_{i=1}^{n}\sum_{j=1}^{m}\left[f(x_i, y_1) - \bar{f}\right]\cdot\left[g(x_i, y_1) - \bar{g}\right]\right\}^2}{\sum_{i=1}^{n}\sum_{j=1}^{m}[f(x_i, y_1) - \bar{f}]^2 \cdot \sum_{i=1}^{n}\sum_{j=1}^{m}[g(x_i, y_1) - \bar{g}]^2} \cdot \tag{8}$$

$\bar{f}$ is the average intensity of laser speckle pattern on the first frame, $f(x_i, y_i)$ is intensity of every pixel with coordinate (x_i, y_i), $\bar{g}$ is the average intensity of laser speckle pattern on the second frame (after movement), and $g(x_i, y_i)$ is intensity of every pixel with coordinate (x_i, y_i).

After obtaining the highest correlation, the changes are called movement of laser speckle. Then, every frame from the pictures is processed and the obtained result from the experiment is shown on Figure 7.

A series of tests need to be carried-out to test whether the obtained data has correlation with blood glucose. Data from every test is compared with a measurement instrument (glucometer) which has high accuracy based on article [7]. There are 2 types of tests, which are stability test and dynamic test.

Stability test is a test to see whether the used instrument has measurement value be relatively the same as when there's no changes in parameter. This test is done to one healthy non diabetic person with duration of 30 minutes and time interval of 5 minutes. The result can be seen on Figure 8(a).

Dynamic test is a test to see changes that happen when there are parameter changes in subject which in this matter is the blood glucose concentration. This test is done to 5 healthy non diabetic people with duration of one hour and time interval of 10 minutes. But before doing this, these 5 peoples will have to drink a high glucose drink.

In the stability test data, the data that have been compared are the blood glucose concentrations which have the mostly constant values with average values of the maximum peaks from the movement of laser speckles. This shows that the developed instrument is stable when there are no changes in parameter.

In the dynamic test data as on Figure 9-12. have a mostly good level of correlation between the average amplitudes of the like ECG graph resulted from the camera and blood glucose concentration with correlation value of about 80%. But the Figure 13. subject has low level of correlation with value of 63%. This is caused by the significant movement of the subjects' arms.

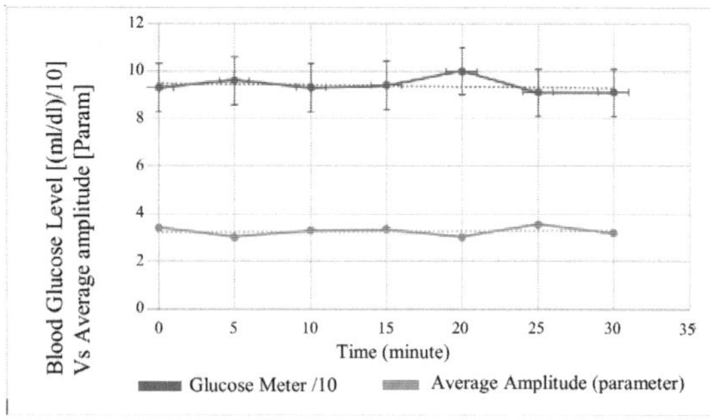

Figure 8. Stability test data

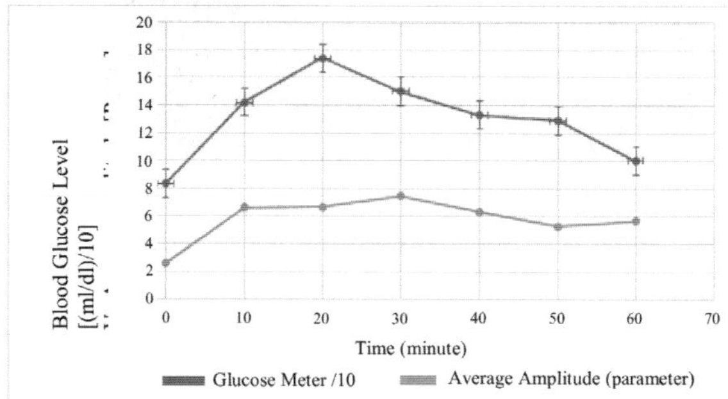

Figure 9. Subject 1, Corr.81%

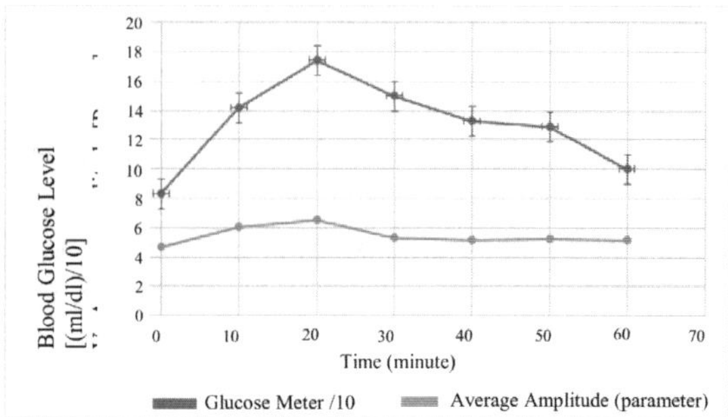

Figure 10. Subject 3, Corr.84%

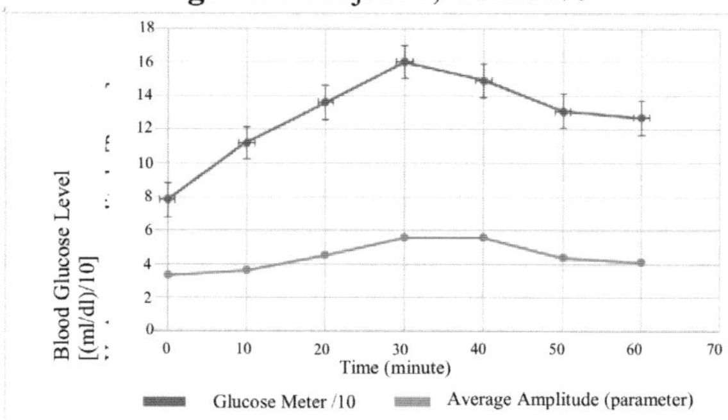

Figure 11. Subject 2, Corr.92%

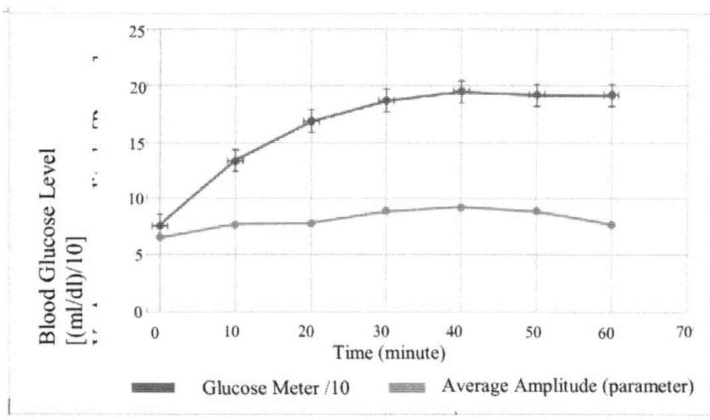

Figure 12. Subject 4, Corr.84%

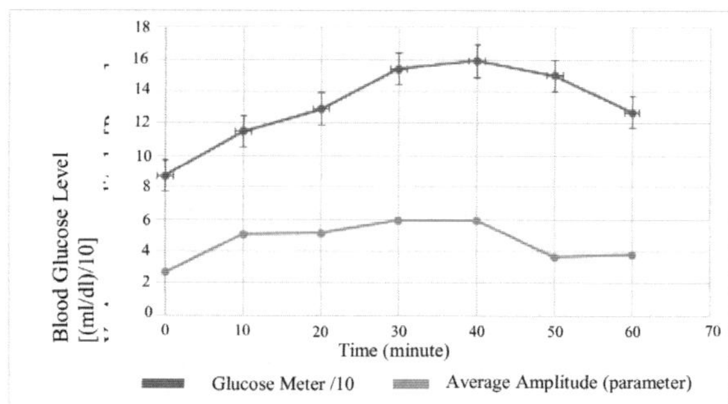

Figure 13. Subject 5, Corr.63%

4. Conclusions

From the experiment that had been done, it can be seen that the movement of laser speckle pattern has high correlation to changes of blood glucose concentration and the used instrument is stable which is showed by the barely changed result of the stability test. In the future measurement blood glucose using low budget camera is possible. By using different way to take data and the use of a pocket camera, it can also be generated changes in the peak of laser speckle pattern.

5. Acknowledgements

This work is supported by research grants of Indexed International Publication of Student Final Project (Hibah Publikasi Internasional Terindeks untuk Tugas Akhir (PITTA) Mahasiswa), Universitas Indonesia, Grant No. 672/UN2.R3.1/HKP.05.00/2017.

References

[1] Esenaliev R, Larin K, Larina I, Motamedi M, and Opt Lm 2001 Noninvasive monitoring of glucose concentration with optical coherence tomography

[2] MacKenzie H, Ashton H, Spiers S, Shen Y, Freeborn S, Hannigan J, Lindberg J, Rae P, and Clin C 1999 Advances in photoacoustic noninvasive glucose testing pp 1587–1595

[3] Beiderman Y, Horovitz I, Teicher M, Vicente J G, Mico and Zalevsky Z 2010 Remote estimation of blood pulse pressure via temporal *Biomedical Optics* pp 1-7

[4] Beiderman Y, Blumenberg R., Rabani N, Teicher M, Garcia J, Mico V and Zalevsky Z 2011 Demonstration of remote optical measurement configuration that correlates to glucose concentration in blood *Medical optics and biotechnology* pp 1-13

[5] Ozana N, Arbel N, Beiderman Y, Mico V, Garcia J, Anand A, Javidi B, Epstein Y and Zalevsky Z 2014 Improved noncontact optical sensor for detection of glucose concentration and indication of dehydration level *Medical optics and biotechnology* pp 1-15

[6] Jefferson L S, Chernington A D and Goodman H.M 2001 *The endocrine system, Volume 2, the endocrine In Handbook of Physiology* (Oxford University Press)

[7] Limitations on Resolution and Contrast: The Airy Disk edmund optics, [Online]. Available: https://www.edmundoptics.com/resources/application-notes/imaging/limitations-on-resolution-and-contrast-the-airy-disk/. [Accessed 14 August 2017].

[8] Zalevsky Z, Beiderman Y, Margalit I, Gingold S, Teicher M, Mico V and Garcia J 2009 Simultaneous remote extraction of multiple speech sources and heart beats from secondary speckles pattern Optical Society of America **21566-21580** 17

[9] Gubarev F, Li L, Klenovskii M, and Glotov A 2016 Speckle pattern processing by digital image correlation *EDP Sciences* **04003** p 48

[10] Ullal A, Parmar G M, and Chauhan P H 2013 Comparison of glucometers used in hospitals and in outpatient settings with the laboratory reference method in a tertiary care hospital in Mumbai *Indian J. Endocrinol. Metab*. **17** pp 1-4

IOP Conf. Series: Journal of Physics: Conf. Series **1011** (2018) 012045 doi:10.1088/1742-6596/1011/1/012045

Online Vibration Monitoring of a Water Pump Machine to Detect Its Malfunction Components Based on Artificial Neural Network

P Rahmawati, P Prajitno

Departemen Fisika, Fakultas Matematika & Ilmu Pengetahuan Alam, Universitas Indonesia, Kampus UI Depok, Depok 16424, Indonesia

Corresponding author's e-mail : prawito@sci.ui.ac.id

Abstract. Vibration monitoring is a measurement instrument used to identify, predict, and prevent failures in machine instruments[6]. This is very needed in the industrial applications, cause any problem with the equipment or plant translates into economical loss and they are mostly monitored component off-line[2]. In this research, a system has been developed to detect the malfunction of the components of Shimizu PS-128BT water pump machine, such as capacitor, bearing and impeller by online measurements. The malfunction components are detected by taking vibration data using a Micro-Electro-Mechanical System(MEMS)-based accelerometer that are acquired by using Raspberry Pi microcomputer and then the data are converted into the form of Relative Power Ratio(RPR). In this form the signal acquired from different components conditions have different patterns. The collected RPR used as the base of classification process for recognizing the damage components of the water pump that are conducted by Artificial Neural Network(ANN). Finally, the damage test result will be sent via text message using GSM module that are connected to Raspberry Pi microcomputer. The results, with several measurement readings, with each reading in 10 minutes duration for each different component conditions, all cases yield 100% of accuracies while in the case of defective capacitor yields 90% of accuracy.

1. Introduction

Indonesia is a developing country that has many great natural resources. One of developments that Indonesia has been doing in industrial field is about how to use and manage natural resources effectively. Based on data from Ministry of Industry in 2015, non-petroleum refinement industrial sector have grown about 5.04 percent, which is higher than economic growth that have grown about 4.79 percent [9].

Indonesian industries have been using high technological machines to shorten the natural resources processing to become more effective. These machines may be broken because of their age or failure in use. Therefore, machine's condition monitoring is needed, because it's an important part of industrial processing maintenance. Sometimes a machine may break down unexpectedly from operating, it may cause relatively big financial loss and also may bring hazards that can harm work safety.

There are 2 ways that can be used to monitor the condition of machine, which are with vibration or lubricant. Vibration works by analyzing the damage through vibration from the machine, because every vibration from the machine indicates that there's something broken in the machine. Lubricant is a type of analysis that works by investigating the fluid from the inside of the machine such as lubricant.

Nowadays, there have been many researches that try to develop instruments to do vibration monitoring method, started from using wire to wireless, and each research also has used different sensor. Most of the researches use Micro-Electro-Mechanical System (MEMS) accelerometer and do processing data by applying Fast Fourier Transform (FFT). From the previous research of this year, a new instrument had been developed by using Field-Programmable Gate Array (FPGA) [1]-[6].

In this research, a system had been developed to detect any damage on water pump which happen to components that are bearing, capacitor and impeller by using MEMS accelerometer with data processing that uses Relative Power Ratio (RPR). And Artificial Neural Network (ANN) method is used to classify the malfunction. As for the result of this analysis will be transmitted by GSM network.

2. Methodology

The main component of this result consists of MEMS accelerometer, Raspberry Pi and GSM module. The type of MEMS accelerometer that is used in this research is MPU-6050 which has 3 axis. Then, the result data of MEMS accelerometer acquisition are sent by using I2C communication line which is located on Raspberry Pi and later on will be processed by using Power Spectral Density (PSD), Relative Power Ratio (RPR) and Artificial Neural Network (ANN) algorithms that are embedded on Raspberry Pi. The programming language that's embedded on Raspberry Pi is Python programming language. The result of Raspberry Pi processing will be transmitted by GSM module as text message to cellular phone. USART is used for Raspberry Pi and GSM module interfacing. The overall system and work flow are shown as Figure 2.1, while the general flow diagram is shown by Figure 2.2.

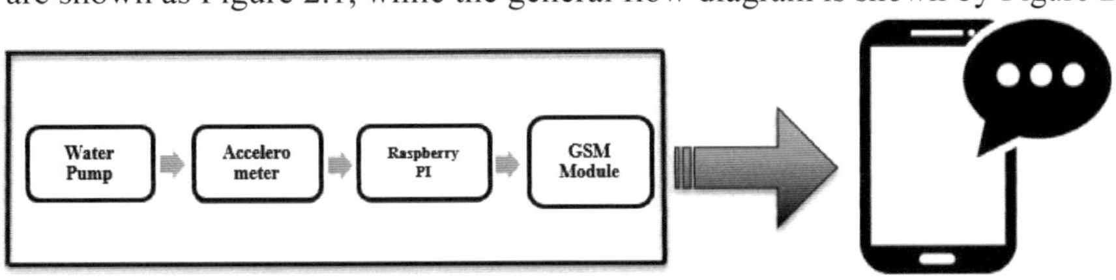

Figure 2.1 General System Diagram

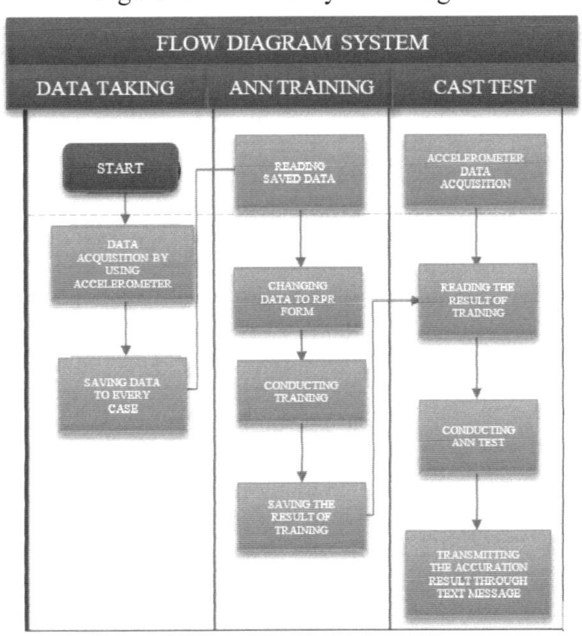

Figure 2.2 The General Flow Diagram

2.1. Vibration Data Acquisition

Sensor that's used to take vibration data is MEMS accelerometer type MPU-6050 which is triaxial accelerometer with 16 bit output resolution in every axis and sampling rate frequency is 1000 Hz. In addition, measurement range of this accelerometer is ± 2g, so theoretically the resolution is $4g/2^{16}$ = 61.04 µg/bit. MPU-6050 supports 400 kHz Fast Mode I2C serial interface for communication with other devices. Sensor is installed on the top surface of the water pump body by using rubber tape 3m.

The first vibration data taking is done to make it as training data on ANN within 5 minutes duration. Training data is data that consists of any malfunction on components which are bearing, impeller and capacitor. After that, training data will be processed to be training data on ANN which is embeddded on Raspberry Pi.

Figure 2.3 Components which malfunction will be detected (Capasitor,Impeller, and Bearing)

2.2. Data Processing by Using Relative Power Ratio (RPR) Method

Vibration data that is taken by using accelerometer has a relatively big number of data, so RPR method is needed to minimize the number of data to make training on ANN become easier. In this RPR method, vibration data are divided into 3000 data per one set of data. Every set of data is calculated by Power Spectral Density (PSD) algorithm, so the power value can be known in every frequency on every set of data. Then, with 10 Hz frequency band, the RPR value can be calculated by following equation:

$$RPR(df) = \frac{Power(df)}{Sum(PSD)} \quad (1)$$

with RPR(df) is the value of RPR on frequency band, Power(df) is the total power of frequency band and Sum(psd) is the overall total power on every frequency.

2.3. Malfunction Classification by Using Artificial Neural Network (ANN)

Artificial Neural Network (ANN) is an artificial intelligence that can analyze data by comparing data to other data which had been chosen as reference. On ANN, first thing that's needed to be done is doing training by giving data which represent every component's malfunction. ANN that's used in this research is a type of feed-forward network with backpropagation training. On ANN, there are certain value of input nodes, hidden nodes and output nodes that have to be set. The chosen value of input nodes is 50 which is the data number of one data set, and the chosen value of hidden layer is 25 nodes and the chosen value of output nodes is 6 based on the number of malfunction that wants to be classified into several classes which are normal, broken capacitor, broken impeller, broken bearing, broken capacitor & impeller, and broken capacitor & bearing.

Besides training data on ANN, testing process is also conducted by taking real time and online data that later will be compared with the result of the training.

2.4. Result of Analysis Transmission using GSM Module

The result of analysis from ANN is transmitted through text message by using GSM Module. The result that will be transmitted is just the result of malfunction analysis. Text message transmission is done by every one minute with duration of 10 minutes.

3. Results

3.1. ANN Training

Data taking to get ANN training data is conducted for about 5 minutes to every malfunction case on water pump. So the number of data which is 300.000 data is obtained from sampling rate of MEMS accelerometer MPU 6050 is 1 kHz. Then, data from every malfunction is converted into the form of set of data with value of 3000 data per set of data. From this data set, RPR calculation is conducted in frequency band with value of 10 Hz which later will be projected into a graphic like in Figure 3.1.

On Figure 3.1., there's RPR graphic from every case where each case has different pattern. The graphic from broken impeller and broken capacitor case show that RPR value is almost the same in every frequency band but RPR from broken impeller is bigger than RPR from broken capacitor. However, graphic from broken bearing case shows that the RPR value is smaller, so there's a visible bold line in the lower part. The case that gives the biggest value of RPR is the broken capacitor and impeller case.

From these different patterns of malfunction, by using human's visual skill, the malfunction of water pump component can be known, but it takes a long time. Therefore, ANN training is needed to be done by giving the result of RPR data to ANN. The chosen ANN is the feed-forward network type and backpropagation trainer with hidden layer which value is 25 and activation function that uses Tanh layer and Softmax Layer.

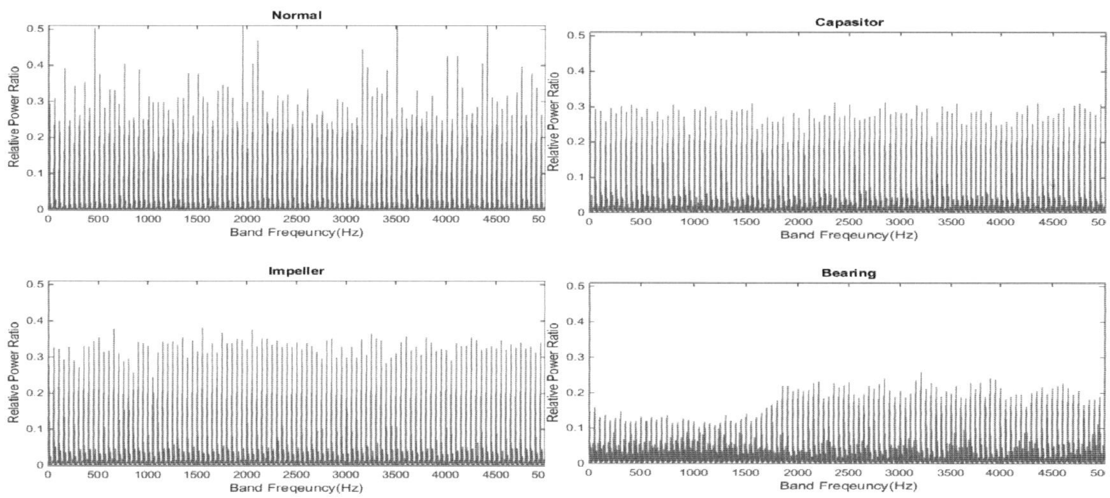

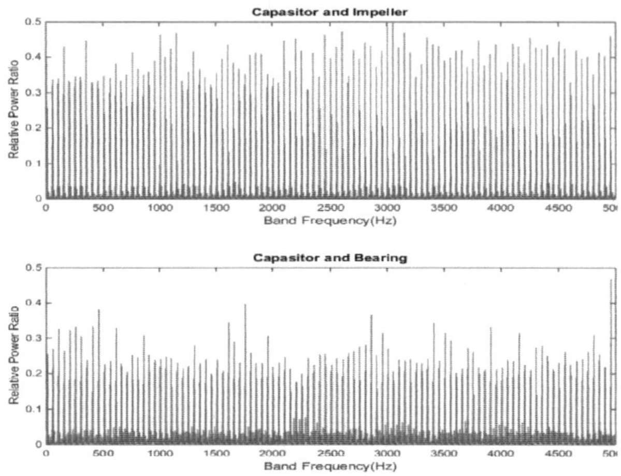

Figure 3.1. Relative Power Ratio (RPR) Graphic from Every Case

3.2. Testing

Testing for each case is conducted in every minute for about 10 minutes where test data is taken by online. Then the result of analysis will be transmitted by using GSM module through text message. The result from testing system shows that there's a match between the analyzed case and there is an error on the 4th minute on capacitor testing (based on Figure 3.2)

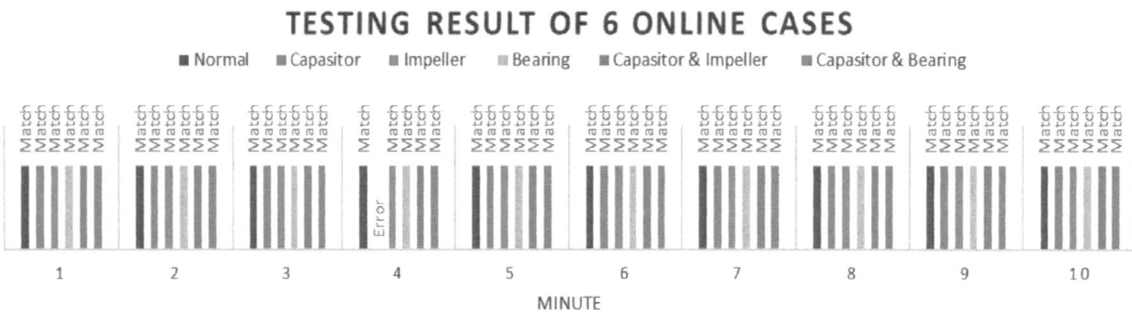

Figure 3.2 The Testing Result System

4. Conclusion

From the result of component's malfunction testing on water pump that's conducted by the artificial system, it can be concluded that vibration monitoring system can be built by acquisitioning vibration data through MEMS MPU-6050 accelerometer by using Raspberry Pi microcomputer. This microcomputer had been embedded with Artificial Neural Network (ANN) that uses feed-forward dan backpropagation. The training uses hidden layer that uses Tanh Layer and also uses Softmax Layer as the output with the chosen value of hidden layer is 25. The result of testing data for every case shows that there's a match between the result of test and types of cases that has been testing, except for capacitor case that has 1 error from 10 times of testing. This result may give simplicity for industrial workers to know the machine malfunctions fastly in certain period remotely.

Suggestions for further research are the data taking for training data and testing data must taken in same electrical condition because the different voltage source used will change the vibration pattern of each case and it is also necessary to vary the experimental conditions, for example the duration of the use of the water pump machine should be more than 30 minutes and then taking training data for each case.

IOP Conf. Series: Journal of Physics: Conf. Series **1011** (2018) 012045 doi:10.1088/1742-6596/1011/1/012045

Acknowledgments
This research is a final assignment research under the support of the Instrumentation Physics Division at the Department of Physics of the Universitas Indonesia.

References

[1] Bengherbia B, Zmirli M O, Toubal A, & Guessoum A 2017 FPGA- based wireless sensor nodes for vibration monitoring system and fault diagnosis *Measurement* **101** 81-92.
[2] Chikuruwo M N, Maregedze L, & Gankayi T 2016 Design of an Automated Vibration Monitoring System for Condition Based Maintenance of a Lathe Machine(Case Study) *IEEE.*
[3] Gicrlak P, Burghardt A, Szybicki D, Szuster M, & Muszynka M. 2016 On-line manipulator instrument condition monitoring based on vibration analysis *Mechanical System and Signal* **89** 14-26
[4] Helal I A, Vuong T-H, David J, Bellaaj N M, & David M P 2015 Vibration Monitoring Based on MEMS Accelerometers *16th International Conference on Science and Techniques of Automatic Control & Computer Engineering* Monastir
[5] Hong K, Huang H, Fu Y, & Zhou J 2016 A vibration measurement system for health monitoring of power transformers *Measurement* **93** 135-147
[6] Ikram W, Chen S -L, Harvei T, Olsen T, Mikalsen E, Svoen G, Myhre B 2014 Vibration- Based Wireless Machine Condition Monitoring System *IEEE*
[7] Pratama R A 2015 Perancangan Pengendali Neural Network pada Gerak YAW dan Sudut Slip Samping Kendaraan Roda Empat (Depok: University Indonesia)
[8] Stoica P, & Moses R 2005 *Spectral Analysis of Signals* (New Jersey: Prentice Hall)
[9] Indonesia K P 2016 Laporan Kinerja Kementrian Perindustrian Tahun 2015 (Jakarta)
[10] Fausett L 1994 *Fundamentals of Neural Networks (*New Jersey: Prentice-Hal)
[11] Stoica P & Moses R 2005 *Spectral Analysis of Signals* (New Jersey: Prentice Hall)
[12] Cheung V & Cannons K (n.d.) *An Introduction to Neural Network* (Canada: University of Monitoba)
[13] Randall R B 2011 *Vibration-based Condition Monitoring* (Vol. III) (Australia: Wiley)

IOP Conf. Series: Journal of Physics: Conf. Series **1011** (2018) 012046 doi:10.1088/1742-6596/1011/1/012046

Automatic Quadcopter Control Avoiding Obstacle Using Camera with Integrated Ultrasonic Sensor

Hanafi Anis, Ahmad Haris Indra Fadhillah, Surya Darma* and Santoso Soekirno*

Department of Physics, Faculty of Mathematics and Natural Sciences, University of Indonesia, Kampus UI Depok,
Depok 16424, Indonesia
E-mail : suryadarma@sci.ui.ac.id, santoso.s@sci.ui.ac.id

Abstract. Automatic navigation on the drone is being developed these days, a wide variety of types of drones and its automatic functions. Drones used in this study was an aircraft with four propellers or quadcopter. In this experiment, image processing used to recognize the position of an object and ultrasonic sensor used to detect obstacle distance. The method used to trace an obsctacle in image processing was the Lucas-Kanade-Tomasi Tracker, which had been widely used due to its high accuracy.Ultrasonic sensor used to complement the image processing success rate to be fully detected object. The obstacle avoidance system was to observe at the program decisions from some obstacle conditions read by the camera and ultrasonic sensors. Visual feedback control based PID controllers are used as a control of drones movement. The conclusion of the obstacle avoidance system was to observe at the program decisions from some obstacle conditions read by the camera and ultrasonic sensors.

1. Introduction

Using a computer vision is another alternative, and some of these kind of works have been done. Barron et al. [8] provides a survey of optical flow computation algorithms. Gosiewski et al.[9] using an optical flow based on a lucas-kanade method for reading collisions. Sagar and Visser [10] using an optical flow method that combined with a background substractor to knowing the distance between the UAV and the collision. Computing optical flow and performing three dimensional reconstruction are very similar mathematically. Using this method, data from the camera will be reconstructed to form the coordinates of obstacle presence. The image coordinates are connected to the Robot Operating System (ROS), so the drone can navigate automatically avoid obstacles. But there are some drawbacks using such a method. One of which is the difficulty of getting the appropriate programming to achieve goals. A high degree of programming difficulty becomes a major drawback. Therefore here the author tries to use this method with the program as easy as possible.

One other challenge is the condition when the camera cannot distinguish between obstacles or not. Under current conditions the drone cannot be stopped because it is automatic so it can cause accidents and damage to the drone. Therefore here the author wants to provide additional features with the use of ultrasonic sensors to help the drones to know the existence of obstacles with a higher level of accuracy. The authors more preferred to uses a computer vision method as the main thing. Like Ohya et al. [12] said that a primary focus of machine intelligence and advanced robotics is to capture human faculties in a computer and since humans use vision for obstacle avoidance while navigating. The ultrasonic sensor is used as a backup, if an error has occurs. Here the author aims to create an automatic navigation system on drones to pass through obstacles using a camera that integrated with ultrasonic sensor. The limitations of the coupled problem consist of how to implement the Image Processing method using a single camera on the drone, transferring the object readings data by ultrasonic sensors, as well as the use of PID-based control ROS as a drone flying control.

2. The Proposed Method

The method studied in this paper uses image processing and ultrasonic sensors. Image processing method contains feature detection and optical flow. The theory behind those techniques is worked out in the following subsections.

2.1. Image Processing

The Image Processing method will produce the output of a path planning for quadcopter movement. The process consists of several stages, such as Pre-processing, Grouping, and Decision making. This stage consists of filtering, feature detection, and optical flow.

The feature detection method will define and track the important points of an image, such as the corner and edge. The Shi and Tomasi feature detection method outperforms the other three methods based on both, the detection range and the ratio of frames containing a detected target. Based on that experiment the feature detection method used in this paper is Shi-Tomasi method.

$$R = \min(\lambda 1 \lambda 2)$$

The Shi-Tomasi method uses a minimum value of R that acts as a threshold value for the quality level of a corner and edge. The eigenvalues $\lambda 1$ and $\lambda 2$ represent the bidirectional axis value of the ellipse to determine the corner [12].

Next, the optical flow method is used in object tracking through feature comparison of two frames. A feature found in the first frame compared to the same features found in the second frame. The vector is obtained by the location of an object on a particular specific pixel (feature) from one point in the frame, to another point in a different frame is what is known as the optical flow of the object. Here the author uses the Lukas-Kanade method as a guide for optical flow.

$$h \approx \frac{\sum_x F'(x)[G(x)-F(x)]}{\sum_x F'(x)^2}$$

The h value for the gradient of the two frame images is the vector of the optical flow of the area. The Lucas-Kanade method makes it possible to produce unusual movements of certain environments from our destination points [13]. The downside of this method is if there is a large undetected movement in the search area. To solve the problem, this method is compensated by developing LK algorithm into pyramid shape [18]. This process involves a pictures with different resolutions, starting at the highest position on the pyramid (low resolution) and moving gradually to the lowest position of the pyramid with higher resolution (high definition image).

From Pre-processing, we are grouped into the Grouping stage, where the existing features will be grouped by finding the average value of each feature and the frame will be split into four different quadrants. Decision Making is the controller of the drone movement to go to the quadrant without obstacles.

2.2. PID Controller

Ultrasonic sensors are used as an additional feature on the drone to help the drone determine the presence of obstacles or not. The principle is to detect the time that the ultrasonic wave is required from the transmitter to the receiver. To detect an obstacle from ultrasonic sensor, it depends on many factors, such as the orientation, reflectively, curvature, etc., of the surface of the obstacle toward the sensor and on the threshold used for the detection of the received echoes [15].

IOP Conf. Series: Journal of Physics: Conf. Series **1011** (2018) 012046 doi:10.1088/1742-6596/1011/1/012046

2.3. *Decision Making*

Specifications of the camera with image processing and ultrasonic sensors, integrated into a decision-making system based on information obtained from both sensors. The decision making algorithm uses two sensors as a reference movement. The essence of avoiding obstacle is how quickly the sensor recognizes that there is an obstacle in front of the drone. In the flowchart figure 1 below, there is a program decision-making flow that states the linkage and relationship between image processing and ultrasonic sensors in some conditions.

The decision-making algorithm method begins with obstacle detection by the camera and the ultrasonic sensor calculates the distance in front of the drone, and we set the minimum distance of the drone near the obstacle to be 50 cm. The program is divided into several conditions in avoiding obstacles, can be seen in Figure 1. If the image processing detects an obstacle, the program identifies whether the distance is 50 cm or less, if more then the image processing information will be used as a reference for the obstacle-avoiding drone, but if the distance is less than 50 cm, then the reference distance from ultrasonic is used as a reference drone to avoid. In other situations, if image processing can not detect an obstacle, it will be divided into two conditions, if the distance read by the ultrasonic sensor is still above 50 cm, then the obstacle detection process is still in progress. But if the obstacle is still not detected, but the distance on the ultrasonic sensor is below 50 cm, then the drone uses this information to dodge.

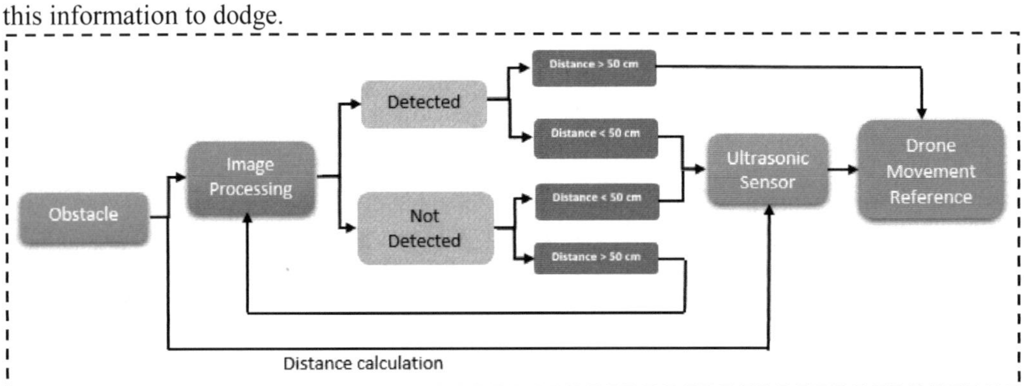

Figure 1. Flowchart Decision Making Algorithm

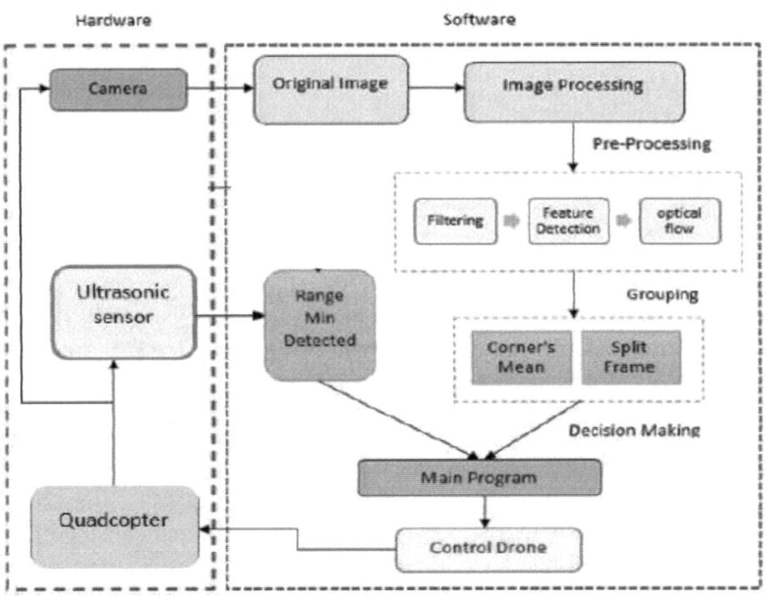

Figure 2. Flowchart of drone work process.

The whole avoiding obstacle work system, can be seen in figure 2, with data processing and ultrasonic sensors running simultaneously. After the decision making algorithm is processed, the information reference is obtained and used as the control drone reference. After the drone moves, then the process will come back from beginning.

3. Evaluation and Experimental Result

Before test the quadcopter, the authors have to know the characterization of each sensor that used in this paper. The process of ultrasonic sensor characterization done by the author to get the precision value of the sensor. The result show that

$$S\,(cm) = 0.0176\ \times t(s) - 0.1625$$

The percentage of obstacle detection success using the Lucas-Kanade-Tomasi method has a success rate above 85% with the highest yield at 100 centimeters.

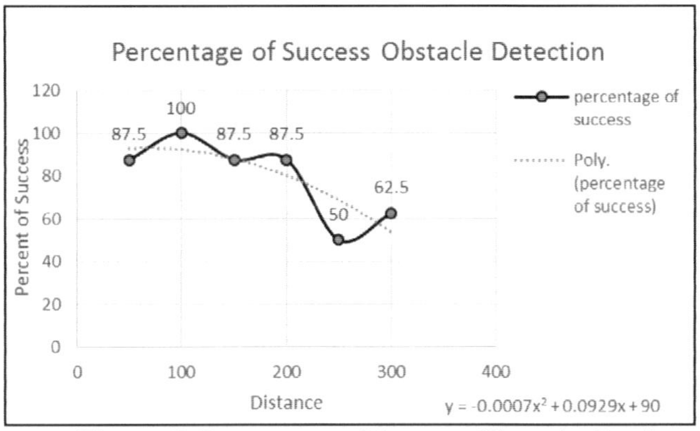

Figure 3. Percentage of Success Obstacle Graph

The result of the plot of the line with the quadratic equation had a decreasing trendline proportional to the greater distance. This was because the ability of Lucas-Kanade-Tomasi was algorithm decrease to trace the features of an obstacle, as the resolution of the obstacle decreases proportionally to the obstacle away from the camera. we get the equation from trendline this chart is

$$y = \ -0.0007x^2 + 0.0929x + 90$$

Figure 4. Feature detection with 500 point (left) and 50 point (right).

In figure 3, there is a comparison of the number of features detected by the Lucas-Kanade-Tomasi algorithm. In the left figure, the maximum number of features tracked is 500 points, which the algorima can choose as many as 500 points in the frame. While in the picture to the right, the maximum number of features that can be tracked is 50 points. The limit points are set by changing the threshold value of the feature. The reduction in the number of features indicates the location of the obstacle is more specific than if the number of features more, because if the number of features reproduced, then the algorithm will look for more features than in the obstacle, so the background can identify there is a feature.

Figure 5. Average value point in different quadrant and different distance.

Further data obtained results will be used for testing to find the optimal distance of obstacle readings using the optical flow method. This test is performed to determine the distance from the obstacle object to the optimal camera for the program to get information on the obstacle presence of the image. In the program as you can see in figure 4, the author divides the image into 4 quadrants with each having the same size. The existence of an average value determines where the obstacle is in a particular quadrant.

4. Conclusion and Future Work

The conclusions, we can see the avoiding obstacle system integration between image processing with complementary ultrasonic sensors. Image processing in detecting obstacle with 85% success rate at 50-200 cm distance, where there is a relation between distance and image processing capability in recognizing obstacle. From these results it can also be seen that the resolution of the camera affects the results of obstacle tracking, because the far more obstacle, the obstacle resolution will be reduced so that the lucas-kanade-tomasi algorithm reduced effectiveness.

This drone framework is designed to be as simple as possible, in order to meet the purpose of the author as well as many people who can use this author's proposal easily. This aspect has great flexibility properties, because it can be used not only on quadcopter, but other flying robots or ground robots. For that the authors hope this paper can be implemented to the upcoming human life, where all the difficult work can be done by robots that are automatic.

From this paper the authors expect a lot of development that can be done, such as the need for an algebraic program to read two different obstacles but are at the same distance. The main developmental hope of the author is, the drone can avoid the obstacles that are dynamic. By calculating the approximate movements of the object, the drone can avoid such moving obstacles, but higher complexity programming is required. This is still a shortcoming of this paper, so it is expected that many future can develop this paper to solve the problem.

5. Acknowledgment

This research was funded by Hibah PITTAUI, under contract number 679/UN.2.R3.1/HKP.05.00/2017

References

[1] Frew E and Raja S 2004 Obstacle avoidance with sensor uncertainty for small unmanned aircraft *43rd IEEE Conf. Decision Control* **1** pp 614-619

[2] Jian L and Li X 2011 Vision-based navigation and obstacle detection for UAV *Int. Conf. Electron. Commun. Control* pp 1771–1774

[3] Delin L, Fei W, Biao W, and Ben M C 2012 Implementation of Obstacle Avoidance Technique for Indoor Coaxial Rotorcraft with Scanning Laser Range Finder *Proc. of the 31st Chinese Control Conference (Hefei)* pp 5135-5140

[4] Richard E Z and Larry D A 1997 Design and Flight Test of 35-Gigahertz Radar for Terrain an Obstalce Avoidance *Journal of Aircraft* **34** pp 261-263

[5] Richard E Z, Raymond F C, Andre Z, Ron S, and Robert G B 1996 Development and Flight Test of Terrain-Referenced Guidance with Ladar Forward Sensor *Journal of Guidance Control and Dynamics* **19** pp 823-828

[6] Jeffrey B S, Brandon C, Andrew C, Randal W B, and Timothy W M 2005 Static and Dynamic Obstacle Avoidance in Miniature Air Vehicles *AIAA Infotech at Aerospace (Virginia)* pp 823-828

[7] Meng G and Pah H 2016 The Application of Ultrasonic Sensor in the Obstacle Avoidance of Quadrotor UAV *Proc. of 2016 IEEE Chinese Guidance, Navigation, and Control Conference (Nanjing)* pp 976-981

[8] Barron J L, Fllet D J, and Beauchemin S S 1994 Performance of Optical Flow techniques *International Journal of Computer Vision* **12** pp 43-77

[9] Zdzislaw G, Jakub C, and Leszek A 2011 Vision-based obstacle avoidance for Unmanned Aerial Vehicles *4th Internationa Congress on Image and Signal Processing (Shanghai)* pp 2020-2025

[10] Jaysinh Sagar and Arnoud Visser 2014 Obstacle avoidance by combining background substraction, optical flow, and proximity estimation *International Micro Air Vehicle Conference and Competition 2014* pp 142-149

[11] Han S B, Lee S G, and Lim J K 2006.The study on map building algorithm and traveling algorithm of the mobile robot with the ultrasonic sensor *Journal of the Korean Society of Marine Engineering* 2006

[12] Akihisa O, Akio K, and Avinash K 1998 Vision-Based Navigation by a Mobile Robot with Obstacle Avoidance Using Single-Camera Vision and Ultrasonic Sensing *IEEE Transaction on Robotic and Automation* **14** pp 969-978

[13] Shi J and Carlo T 1994 Good features to track *Proceedings Computer Vision and Pattern Recognition* pp 593-600.

[14] Bruce D L and Takeo K 1981 An iterative image registration technique with an application to stereo vision *Proceedings of the DARPA Imaging Understanding Workshop* pp 21-130.

[15] Dan T, Nabil B, Fazel F, and Kristopher 2014 Experimental Evaluation of Four Feature Detection Methods for Close Range and Distant Airborne Targets for Unmanned Aircraft Systems Applications *International Conference on Unmanned Aircraft Systems (Orlando)* pp 1267-1273

[16] Beauchemin S S and Barron J L 1995 The computation of optical flow *ACM Computing Surveys (New York)* **27** pp 433-466

IOP Conf. Series: Journal of Physics: Conf. Series **1011** (2018) 012047 doi:10.1088/1742-6596/1011/1/012047

Transmittance and Tunneling Current through a Trapezoidal Barrier under Spin Polarization Consideration

F A Noor*, E Nabila, H Mardianti, T I Ariani and Khairurrijal

Physics of Electronic Materials Research Division, Faculty of Mathematics and Natural Sciences, Institut Teknologi Bandung, Bandung 40132, Indonesia

*fatimah@fi.itb.ac.id

Abstract. The transmittance and tunneling current in heterostructures under spin polarization consideration were studied by employing a zinc-blended structure for the heterostructures. An electron tunnels through a potential barrier by applying a bias voltage to the barrier, which is called the trapezoidal potential barrier. In order to study the transmittance, an Airy wave function approach was employed to find the transmittance. The obtained transmittance was then utilized to compute the tunneling current by using a Gauss quadrature method. It was shown that the transmittances were asymmetric with the incident angle of the electron. It was also shown that the tunneling currents increased as the bias voltage increased.

Introduction

It was known that spin-polarized electron transport in semiconductors attracts great attention due to its application in spintronic devices [1-3]. The development is focused on the possibility and methods of spin injection into semiconductors [4]. In order to obtain the desired spin orientation, spin-polarized carriers from magnetic materials are injected [4]. However, the conductivity mismatch of the metal-semiconductor structure is a general difficulty for electrical injection of ferromagnetic semiconductors, as reported by Schmidt et al. [4-5]. On the other hand, a spin transistor can also be achieved by using only nonmagnetic materials that exploit the unique characteristics of bulk inversion asymmetry in (110)-oriented semiconductor heterostructures, as reported by Hall et al. [6-7]. Furthermore, Voskoboynikov et al. have proposed that a non-magnetic semiconductor material can be produced by Rashba spin-orbit coupling due to the Dresselhaus effect occurring in zinc-blende [8-10].

Several methods have been proposed to study the tunneling current in heterostructures under spin polarization [11-13]. Suryamas et al. have studied the tunneling process in a heterostructure with a zinc-blende semiconductor under spin polarization consideration [14-15]. However, they did not consider the effect of bias voltage on the heterostructure. In this research, we studied the electron transmittance and tunneling current through a nanometer-thick zinc-blende semiconductor with spin consideration by considering the bias voltage applied to the barrier (called the trapezoidal barrier). The transmittance was derived by considering the Dresselhaus effect and solving the Schrödinger equation. The effects of the electron's incident angle and bias voltage were investigated.

Theoretical Model

IOP Conf. Series: Journal of Physics: Conf. Series **1011** (2018) 012047 doi:10.1088/1742-6596/1011/1/012047

The potential profile of the heterostructures when applying a bias voltage to the barrier is depicted in Fig. 1. Here, V_0 is the barrier height, d is the barrier width, e is the electronic charge, V_b is the bias voltage, and E is the energy. We consider an electron tunneling through a zinc-blende semiconductor; in this study, GaSb was used. The potential profile is therefore expressed as:

$$V(z)=\begin{cases} 0 & z<0 \\ V_0-e\dfrac{V_b}{d}z & 0\le z<d \\ -eV_b & z\ge d. \end{cases} \quad (1)$$

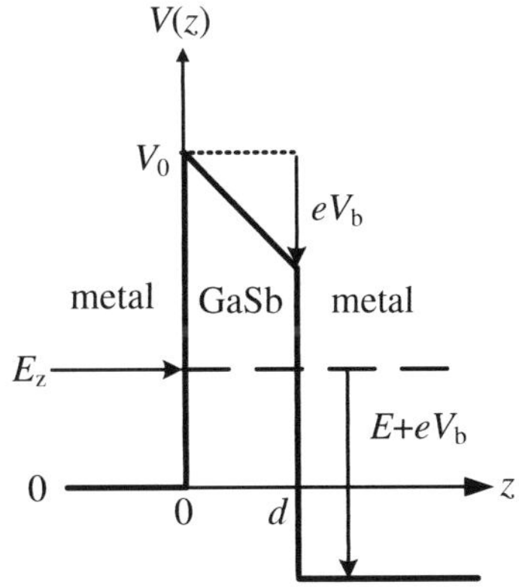

Figure1. Potential profile of heterostructures under a bias voltage applied to the barrier.

The model begins from the Hamiltonian equation, which describes an electron in a material as given by:

$$H\Psi = E\Psi \quad (2)$$

where E is the total electron energy, Ψ is the electron wave function, and H is the Hamiltonian. The Hamiltonian, H, is composed of H_O for the heterostructure without spin polarization and H_D for considering the spin polarization (Dresselhauss term) and is written as $H = H_0 + H_D$. Here H_D is expressed by [16]:

$$H_D = \gamma\left(\alpha_x k_x - \alpha_y k_y\right)\left(\partial^2/\partial z^2\right) \quad (3)$$

where α_i are Pauli matrices ($i\in\{x,y\}$) and γ is a Dresselhauss constant.

By solving the Schrödinger equation and applying the boundary condition at $z = 0$ and $z = d$ [17], it is easy to find the formulation of the transmission coefficient, f. The formulation is [18]

$$f_\pm = -2i\frac{k_{1\pm}\delta_{1\pm}}{m_1}\exp\left(-ik_{3\pm}d\right)\times\left\{\left(\frac{2eV_b}{m_2{}^2\eta^2 d}\right)^{\frac{1}{3}}\delta_{2\pm} + i\left(\frac{k_{1\pm}\delta_{3\pm}k_{3\pm}\delta_{4\pm}}{m_1 m_3}\right) - \left(\frac{k_{1\pm}k_{3\pm}}{m_1 m_3}\right)\left(\frac{2eV_b}{m_2{}^2\eta^2 d}\right)^{-\frac{1}{3}}\delta_{5\pm}\right\}. \quad (4)$$

Here,

$$k_{1\pm} = \left(\frac{2m_1 E_{z\pm}}{\eta^2}\right)^{\frac{1}{2}} \Bigg/ \left(1 \pm \frac{2\gamma m_1 k_\rho}{\eta^2}\right)^{\frac{1}{2}}, \quad (5)$$

$$k_{3\pm} = \left(\frac{2m_1}{\hbar^2}(E_z + eV_b)\right)^{\frac{1}{2}}, \tag{6}$$

$$\eta_\pm(z) = \left(\frac{2m_2 eV_b}{\hbar^2 d}(E_z + eV_b)\right)^{\frac{1}{3}}\left\{(\Phi_0 - E_z)\frac{d}{eV_b} - z\right\}, \tag{7}$$

$$\delta_{1\pm} = Ai'(\eta(d))Bi(\eta(d)) - Ai(\eta(d))Bi'(\eta(d)), \tag{8}$$

$$\delta_{2\pm} = Ai'(\eta(d))Bi'(\eta(0)) - Ai'(\eta(0))Bi'(\eta(d)), \tag{9}$$

$$\delta_{3\pm} = Ai(\eta(0))Bi'(\eta(d)) - Ai'(\eta(d))Bi(\eta(0)), \tag{10}$$

$$\delta_{4\pm} = Ai(\eta(d))Bi'(\eta(0)) - Ai'(\eta(0))Bi(\eta(d)), \tag{11}$$

$$\delta_{5\pm} = Ai(\eta(0))Bi(\eta(d)) - Ai(\eta(d))Bi'(\eta(0)), \tag{12}$$

From Eq. (4), we can obtain the transmittance as follows:

$$T_\pm = f_\pm^* f_\pm, \tag{13}$$

where the signs '+' and '–' denote spin up and spin down, respectively, $f_\pm^*$ is the conjugate of $f_\pm$.

Finally, the tunneling current is computed by using the following equation [19]:

$$J_z = \frac{em_1 kT}{2\pi^2\hbar^3} \times \int_0^\infty T(E_z)\ln\left\{\frac{(1+\exp[(E_F - E_z)/kT])}{1+\exp[(E_F - E_z - eV_b)/kT]}\right\}dE_z, \tag{14}$$

where k is the Boltzmann constant, T is the temperature, E_F is the Fermi energy of metal, and $T(E_z) = T_+(E_z) - T_-(E_z)$ is the total transmittance. The tunneling current in Equation (14) is easily evaluated by using the Gauss-Laguerre Quadrature method [20].

Calculated Results and Discussion

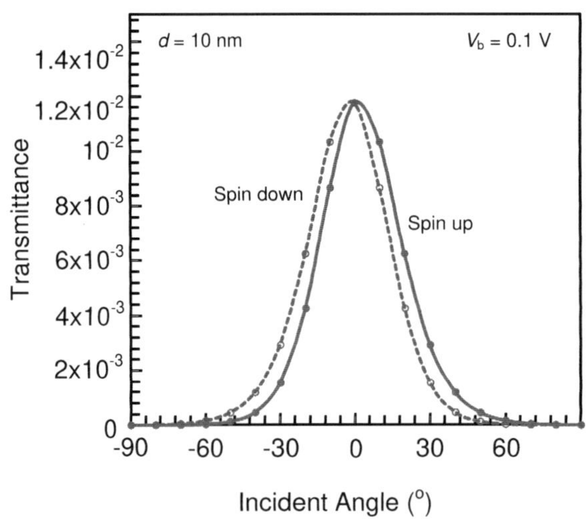

Figure 2. The transmittance depends on the incident angle of the electron.

The transmittances and tunneling currents for electrons in metal/GaSb/metal structures can be calculated on the basis of Equations (13) and (14). The parameters in this study were: $\gamma_1 = 0$, $\gamma_2 = 187$ eV/Å3, $V_0 = 0.2$ eV, $E = 0.1$ eV, $d = 10$ nm. Figure 2 presents electron transmittance versus incident angle of the electron for a barrier width and bias voltage of 10 nm and 0,1 V, respectively. The transmittances show the highest value at an incident angle of 0° (z-direction) for a spin up and spin down polarization respectively. This gives the same results as the calculation without considering the bias voltage [14-15], in which the transmittance for each state is quasi-symmetric with the incident angle of the electron because of the bulk inversion asymmetry properties of the zinc-blende.

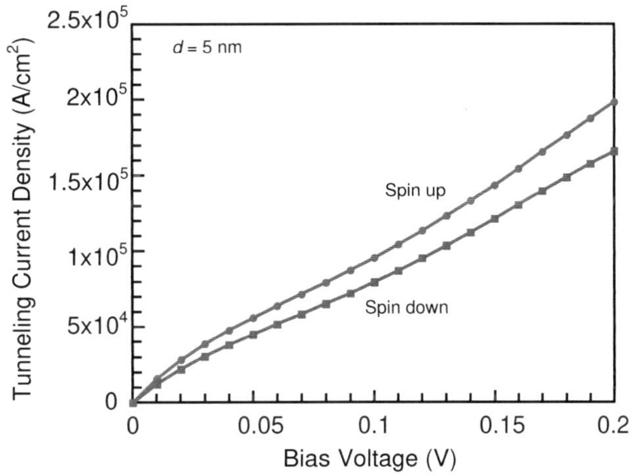

Figure 3. Electron tunneling current dependence on bias voltage.

The bias voltage dependence of the transmittance in metal/GaSb/metal heterostructures for a spin up and spin down states is depicted in Figure 3. It can be seen that as the bias voltage increased, the tunneling current increased for both states. However, the tunneling current for electron spin up was larger than that for electron spin down. This means that the electron spin upstate is more dominant than the spin down one.

Conclusions

The electron transmittance and tunneling current through metal/GaSb/metal heterostructures have been calculated by taking into account the spin polarization. The transmittance and tunneling current were evaluated for the up and down states. It was shown that the transmittances are quasi-symmetric with the incident angle of the electron. It was also shown that the tunneling current increased as the bias voltage increased. Moreover, the tunneling current for the spin-up state was larger than that for the spin down one.

Acknowledgments

This research was supported by *Riset Desentralisasi* research grants for the fiscal year of 2017.

References

[1] Zutic I, Fabian J, Das Sarma S 2004 *Rev. Mod. Phys.* **76** 323.
[2] Fabian J, Matos-Abiague A, Ertler C, Stano P, Zutic I 2007 *Acta Phys. Slov.* **57** 566.
[3] Lu J W, Chen E, Kabir M, Stan, M R, Wolf S A 2016 *Int. Mater. Rev.* **61** 456.
[4] Perel V I, Tarasenko S A, Yassievich I N 2003 *Phys. Rev. B* **67** 201304.
[5] Schmidt G, Ferrand D, Molenkamp L W, Filip A T, and Wees J 2000 *Phys. Rev. B* **62** R4790.
[6] Hall K C, Lau W H, Gundogdu K, Flatte M E *2003 Appl. Phys. Lett.* **83** 2937.
[7] Flatte M E, Byers J M, Lau W H 2002 Spin Dynamics in Semiconductors *Semiconductor*

Spintronics and Quantum Computation (Berlin: Springer-Verlag).
[8] Voskoboynikov A, Liu S S, Lee C P 1998 *Phys. Rev. B* **58** 15397.
[9] Boda A, Boyacioglu B, Erkaslan U, Chatterjee A 2016 *Physica B Condens Matter* **498** 43.
[10] Tutunculer H, Koc R, Olgar E 2004 *J. Phys. A* **37** 11431.
[11] Dakhlaoui H, Jaziri S 2005 *Physica B* **355** 401.
[12] Lebedeva N, Kuivalainen P 2003 *J. Appl. Phys.* **93** 9845.
[13] Michaeli K, Varade V, Naaman R, Waldeck D H 2017 *J. Phys : Condens Matter* **29** 103002.
[14] Suryamas A B, Abdullah M, Khairurrijal 2006 *Indonesian J. Phys.* **17** 43.
[15] Suryamas A B, Abdullah M, Khairurrijal 2006 Transmittance Coefficient of Electron Tunneling through a Nanometer Thick Square Barrier with Spin Polarization Consideration *International Conference on Mathematics and Natural Sciences (ICMNS)* (Bandung, 29-30 November 2006) pp. 941.
[16] Dresselhauss G 1955 *Phys. Rev.* **100** 580.
[17] BenDaniel D J, Duke C B 1966 *Phys. Rev.* **152** 683.
[18] Noor F A, Nabila E, Sustini E, Khairurrijal 2016 Influence of Incident Angle of Electron on Transmittance and Tunneling Current in Heterostructures with Bias Voltage by Considering Spin Polarization Effect *2nd Material Research Society of Indonesia (MRS-id) Meeting* (Bandung, 24-26 October 2016) published in IOP Conf. Series: Materials Science and Engineering 214 (2017) 012037
[19] Noor F A, Abdullah M, Sukirno, Khairurrijal, Ohta A, Miyazaki S 2010 *J. Appl. Phys.* **108** 093711.
[20] De Vries P L 1993 *A first course in computational physics* (New York: John Wiley & Sons).

Repeatability and Reliability Characterization of Phonocardiograph Systems Using Wavelet and Backpropagation Neural Network

Sumarna, J Astono, A Purwanto, D K Agustika*)

Department of Physics Education, Universitas Negeri Yogyakarta,
Jl. Colombo No. 1, Yogyakarta, Indonesia

E-mail: dyah_kurniawati@uny.ac.id

Abstract. Phonocardiograph (PCG) system consisting of an electronic stethoscope, mic condenser, mic preamp, and the battery has been developed. PCG system is used to detect heart abnormalities. Although PCG is not popular because of many things that affect its performance, in this research we try to reduce the factors that affecting its consistency To find out whether the system is repeatable and reliable the system have to be characterized first. This research aims to see whether the PCG system can provide the same results for measurements of the same patient. Characterization of the system is done by analyzing whether the PCG system can recognize the S1 and S2 part of the same person. From the recording result, S1 and S2 then transformed by using Discrete Wavelet Transform of Haar mother wavelet of level 1 and extracted the feature by using data range of approximation coefficients. The result was analyzed by using pattern recognition system of backpropagation neural network. Partially obtained data used as training data and partly used as test data. From the results of the pattern recognition system, it can be concluded that the system accuracy in recognizing S1 reach 87.5% and S2 only hit 67%.

Keywords: Phonocardiograph, wavelet, backpropagation, neural network

1. Introduction

Heart disease is one of the main causes of death [1]. One method to monitor heart abnormalities is auscultation technique by using a stethoscope. This is usually done by medician and requires special skills and very difficult to do. Furthermore, a stethoscope cannot hear the low-frequency component of heartbeat while it sometimes carries important information of heart condition [1][2]. One method to overcome the problems of the stethoscope is by using Electrocardiograph. Electrocardiograph (ECG) shows heart condition from heart muscle relaxation and contraction. ECG output analysis is used to determine the existence of heart abnormalities [3]. However, ECG is quite expensive and cannot be afforded by many health facilities in remote areas. Therefore, in this research PCG system that is cheaper than ECG but easier to use than stethoscopes was built.

Phonocardiography (PCG) is a method for recording heart sound from mechanical pulses of heartbeat using stethoscope and recording of heart sounds is in the form of graphics [4]. PCG system in this research consists of a stethoscope, mic condenser, mic preamp, and PC as can be seen in Figure 1.

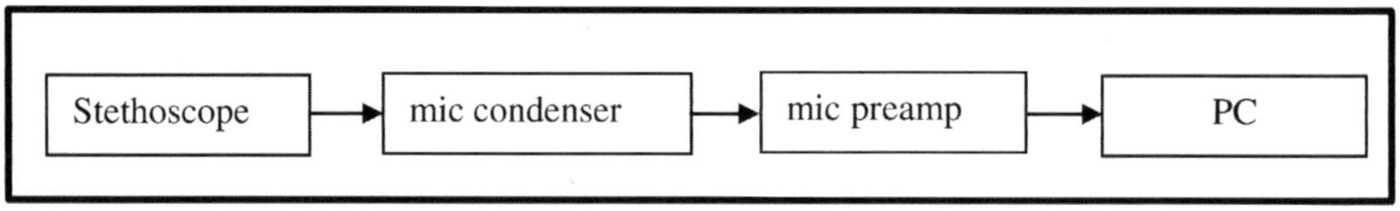

Figure 1. Block diagram of PCG system

Although PCG has advantages compared to other cardiovascular recording devices, PCG is rarely used because of many things that affect its consistency such as the position of chest pieces, the type of chest pieces, the length of the filter pipe, the working point of the transistor on the amplifier and so forth. The novelty in this study is the use of PCG as a cardiac device and we try to overcome its weakness by aligning the components that cause instrument inconsistencies.

To determine whether the PCG system designed and constructed in this research is repeatable and reliable, the PCG system output results should be analyzed. PCG output result is a graph depicting a person's heartbeat. Heartbeat comes from opening and closing of heart valves. The closing of the mitral and tricuspid valves at the beginning of the ventricular systole phase causes the first heart sound (S1) to occur. The closing of the pulmonic and aortic valves occurring at the beginning of the ventricular diastole phase results in a second heart sound (S2) [4]. If the output signal from PCG is consistent then for the heartbeat of the same person the results should remain the same. Therefore, to demonstrate the repeatability and reliability of the PCG system that has been made, recording of the heartbeat was cut into sections S1 and S2 and then processed by the pattern recognition system. If the repeatability and reliability of the system are decent then the value of S1 and S2 in various data retrieval with the same conditions will be recognizable by the pattern recognition system.

The signal processing of PCG system is done in three stages which are preprocessing, feature extraction and pattern recognition. Preprocessing is performed on the PCG's output response with the aim of eliminating noise obtained from the surrounding environment and also preparing signals for the next stage by using Discrete Wavelet Transform. The next step is feature extraction to select the signal character that represents the original signal, hence it can improve pattern recognition system efficiency [3]. The last step is the signal pattern recognition into S1 or S2.

The first stage of ECG signal processing is preprocessing, where it is necessary to remove noise from the input signal using Wavelet Transform. Wavelet transforms decompose signals into a number of scales related to the frequency components and analyzes each scale by a certain resolution. Wavelet Transform uses short time intervals to evaluate higher frequencies and longer time intervals for lower frequencies. One of the advantages of Wavelet Transform is that it is capable of deciphering signals at various resolutions [3]. Signal decomposition of wavelet transform can be seen as a process of signal filtering. In decomposition, the signal is passing a low-pass filter and high-pass filter. The output of low pass filter is called approximation and those come from high pass filter is called detail. The approximation is considered as original signal while detail usually comes from noise [5].

Pattern recognition system used in this research is Backpropagation Neural Network (BPNN). The neural network consists of a group of artificially linked neurons. This research uses the artificial neural network in the classification of S1 and S2, where the input is the approximation coefficient range of wavelet transformation results from the feature extraction and the output unit representing the class of S1 and S2. Each hidden unit counts the weighted amount of its input. BPNN allows the acquisition of practical input/output mapping information within multilayer networks. BPNN executes gradient descent searches to minimize mean square error (MSE) between the desired output and actual network output by adjusting the weights [3].

2. Research Methodology

This research investigated the ability of the PCG system in recognizing S1 and S2 from the patient's heart condition. Heartbeat recording was performed for 15 seconds. The signals then cut on the part of S1 and S2 for data analyzing. For data analysis 9 data of S1 and 9 data of S2 was used. Data S1 and S2 are then pre-processed by using DWT (Discrete Wavelet Transform). DWT divided signals into approximations and details. The next step is to extract the feature by selecting the range of highest and lowest of approximation coefficient. This range of values is then used as training data and test data on the pattern recognition system by using BPNN. The block diagram of S1 and S2 detection method in PCG system is depicted in Figure 2.

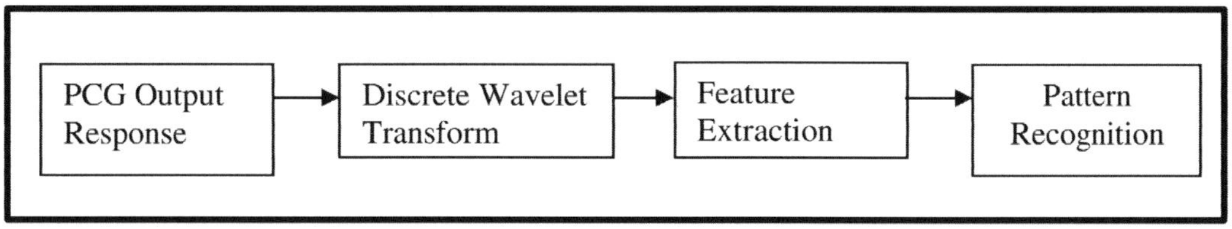

Figure 2. Block diagram of PCG system S1 and S2 detection method

3. Results and Discussion

3.1. Pre-processing by using Discrete Wavelet Transform and Feature Extraction

The recorded heartbeat from PCG system can be seen from Figure 3. The signal than cut at S1 and S2 part. S1 and S2 from different measurement almost show the same result. The cutting signal started from the beginning of the first peak and ended at the lowest point of the latest peak.

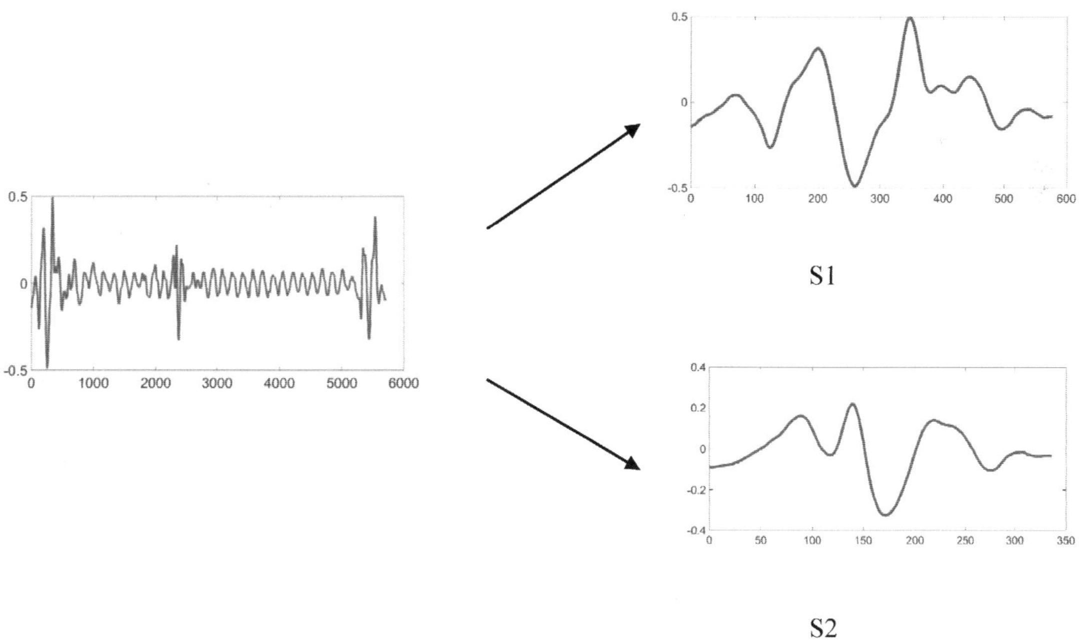

Figure 3. Cutting Signal at S1 and S2

After cutting the signal, S1 and S2 than transform into Discrete Wavelet Transform of Haar level 1 where they filtered into low pass filter and high pass filter as can be seen in Figure 4. The low pass filter part called approximations and considered as the original signal while the high pass one considered as noise and called details. The coefficient approximation then used as input in BPNN.

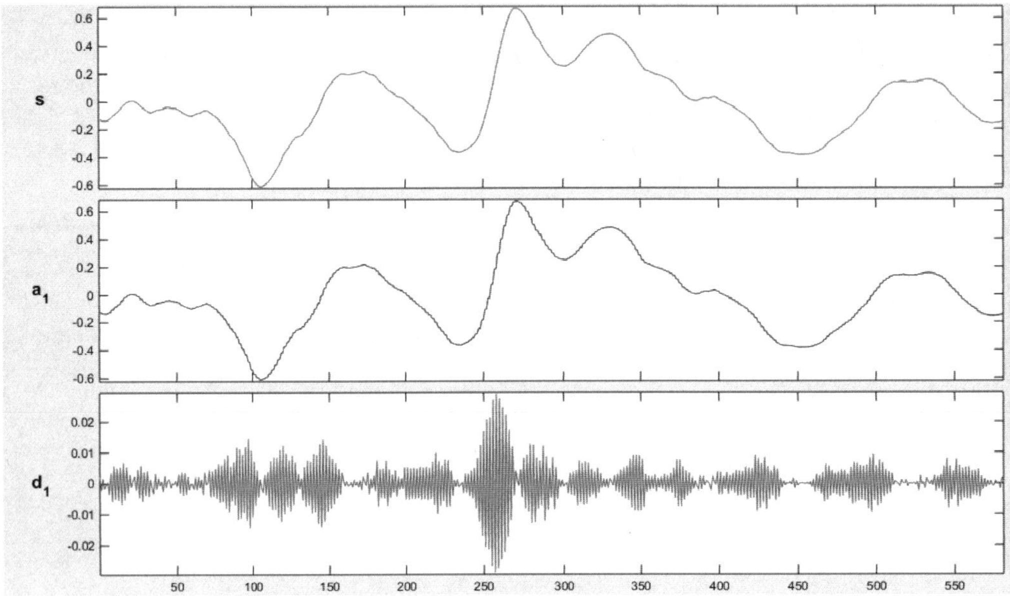

Figure 4. DWT processing from signal (s) into approximation (a_1) and detail (d_1)

However to improve the efficiency of pattern recognition, feature extraction stage has to be done and it was done by selecting the range of approximation coefficient. Table 1 shows the feature extraction result of nine data retrieval.

Table 1. Feature Extraction Result of Nine Data Retrieval.

No.	Coefficient of approximation data range of S1	Coefficient of approximation data range of S2
1	1.810	1.730
2	1.995	1.230
3	2.078	1.573
4	2.095	1.635
5	2.248	1.804
6	2.140	1.376
7	1.589	1.366
8	1.860	1.259
9	1.934	0.992

3.2. Pattern Recognition by using BPNN

The signal is divided into two separate classes that are S1 and S2. Each file of nine data retrieval in 15 seconds recording was selected. The approximation coefficients are given as an input to the BPNN. To train the network, 4 data (two from S1 and two from S2) are utilized. while 14 (seven from S1 and seven from S2) data are used for testing the network. One input from feature extraction result, two hidden layers and 25 numbers of neurons in each hidden layer is used for training and testing the PCG signal. At the output layer two neurons were used, and the target matrix is (1,0) for S1 and (0,1) for S2. After several training steps the weight of matrix that gave the highest accuracy was saved. The picture of BPNN architecture is depicted in Figure 5 .

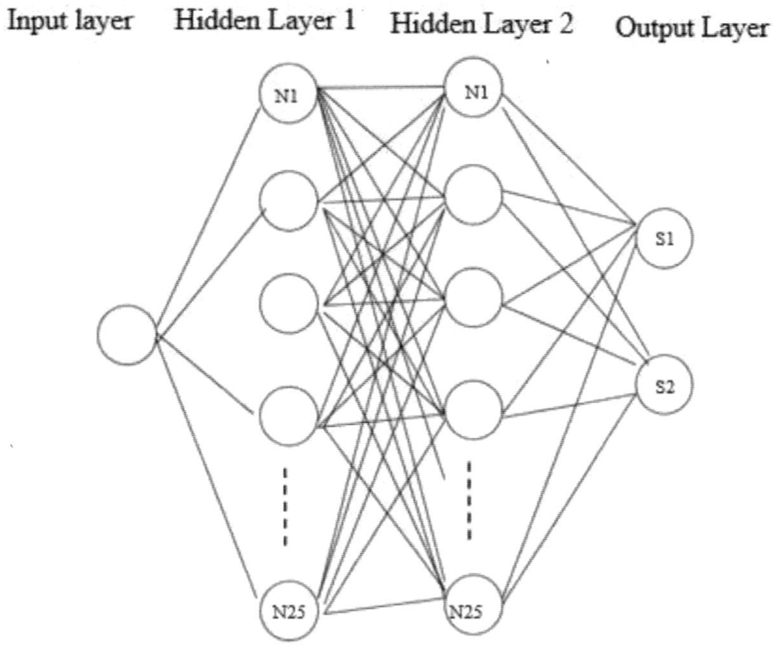

Figure 5. BPNN architecture

From pattern recognition by using BPNN, the results show that the PCG system recognition of S1 reaches 87.5% while S2 only hit 67% as can be seen in Table 2.

Table 2. BPNN result

Output/Target	Accurcy
S1	87.5%
S2	67%

As can be seen in Table 2, the accuracy of the system is not high. It is because the wavelet that was used in analyzing process not suitable for the signal waveform. Therefore, the research to determine kind of wavelet is still needed/

4. Conclusion

The S1 and S2 detection of the PCG signal based on DWT, feature extraction and BPNN shows that the recognition of S1 reaches 87.5% while S2 only hit 67%. The accuracy can be improved by using different wavelet in DWT, different method in feature extraction and different architecture in BPNN.

References

[1] Kao W C and Wei C C 2011 Automatic phonocardiograph signal analysis for detecting heart valve disorders *Expert Syst. Appl.* **38** (6) pp 6458–6468
[2] Abbas A K and Bassam R 2009 *Phonocardiography Signal Processing* **4** (1)
[3] H. M. Rai, "ECG Signal Classification using Wavelet Transform and Back Propagation Neural Network," vol. 3, pp. 212–215.
[4] Chen T, Yang S, Ho L, Tsai K, Chen Y, Chang Y, Lai Y, Wang S, Tsao Y, and Wu C 2017 S1 and S2 Heart Sound Recognition Using Deep Neural Networks **64** (2) pp 372–380

[5] Agustika D K and Triyana K Application of Principal Component Analysis and Discrete Wavelet Transform in Electronic Nose for Herbal Drinks Classification pp. 6–11.

Quadcopter Control Using Speech Recognition

H Malik, S Darma* and S Soekirno*

Department of Physics, Faculty of Mathematics and Natural Sciences, University of Indonesia, Kampus UI Depok,
Depok 16424, Indonesia

E-mail : suryadarma@sci.ui.ac.id, santoso.s@sci.ui.ac.id

Abstract. This research reported a comparison from a success rate of speech recognition systems that used two types of databases they were existing databases and new databases, that were implemented into quadcopter as motion control. Speech recognition system was using Mel frequency cepstral coefficient method (MFCC) as feature extraction that was trained using recursive neural network method (RNN). MFCC method was one of the feature extraction methods that most used for speech recognition. This method has a success rate of 80% - 95%. Existing database was used to measure the success rate of RNN method. The new database was created using Indonesian language and then the success rate was compared with results from an existing database. Sound input from the microphone was processed on a DSP module with MFCC method to get the characteristic values. Then, the characteristic values were trained using the RNN which result was a command. The command became a control input to the single board computer (SBC) which result was the movement of the quadcopter. On SBC, we used robot operating system (ROS) as the kernel (Operating System).

1. Introduction

Speech recognition is an algorithm that makes the machine can recognize the voice spoken by humans. Research on speech recognition had been started since the 1930s. But the development began to look significant around the 1960s.

In speech recognition there are two major stages, they are feature extraction and classification. Each stage has its own methods. Some researchers have tried several methods such as cross-correlation, Mel frequency cepstral coefficient, hidden Markov model, artificial neural network and so on. However, for a relatively crowded environment, the success rate of the speech recognition system was relatively decreased [1]. Therefore, some researchers such as Yanmin Qian et al. in his paper proposed ways to improve speech recognition for environments with relatively more noise with the Deep Neural Network (DNN) method [1]. In addition, various methods of feature extraction stage in speech recognition have various degrees of success rate [2]. The Mel Frequency Cepstral Coefficient (MFCC) method had an error rate of 7.21-17.8%, while with Mel Filter-Bank (FBANK) had an error rate of 1.35-6.82% [2].

The methods used to create speech recognition systems had various types. Such as used a combination of Mel frequency cepstral coefficient (MFCC) as feature extraction with Dynamic Time Wrapping (DTW) as shifting features and K-Nearest Neighbor (KNN) as a classification [3], combined method between Mel frequency cepstral coefficient (MFCC) as feature extraction with Vector quantization (VQ) and Gaussian Mixture Models (GMM) [4], combined method between Mel

IOP Conf. Series: Journal of Physics: Conf. Series **1011** (2018) 012049 doi:10.1088/1742-6596/1011/1/012049

frequency cepstral coefficient (MFCC) as feature extraction with Linear Discriminant Analysis (LDA) as a classification [5], and the method of combination of Discrete wavelet Transform (DWT) as feature extraction with Linear Discriminant Analysis (LDA) as a classification [5]. Each method had its own success rate, as in [3] with a success rate of 98.4%, and at [4] with a success rate of 94.31%.

In this research, the authors designed a speech recognition system as a quadcopter control tool using Mel Frequency Cepstral Coefficient (MFCC) method as feature extraction and Recursive Neural Network (RNN) as classification that is expected can produce better results, and then made a comparison from two types of databases, they are existing databases with new databases.

2. The Proposed Method

MFCC, as described by Davis and Mermelstein, was as a profitable approach to speech recognition. MFCC is one of the most common and familiar algorithms, used in signal applications for feature extraction. The following steps were used in this method which was summarized in **Figure 1**. An explanation of the MFCC already existed in the paper [3] [4] [5].

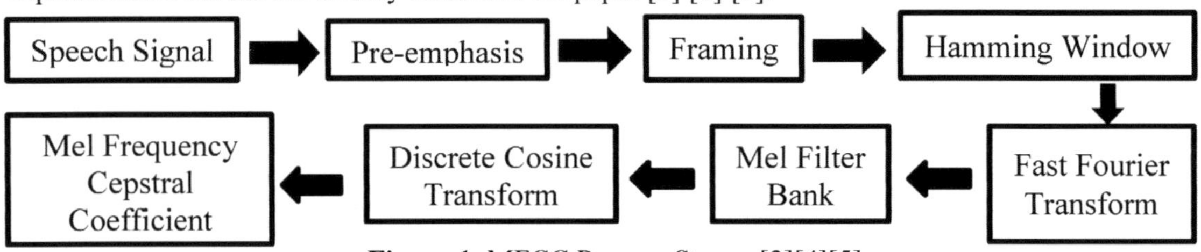

Figure 1. MFCC Process Stages [3][4][5]

In this research, the quadcopter was controlled using voice commands. The sound signal was captured by the microphone which was processed using the DSP module that had been available on the laptop by using the software.

Then the characteristic value that was obtained from MFCC was trained using Recursive Neural Network [1] [6] [7]. In this research, Recursive Neural Network type that used was Recurrent Neural Network [6].

After the training process, then we did the testing process to see the results when inputted voice command which output was worth close to the target value of each word. The output value of the test result was the input condition for moving the quadcopter, 'naik' command represented 'take off', 'bawah' command represented 'landing', 'maju' command represented 'positive pitch', and 'balik' command represented 'positive yaw'. Here it was a flowchart of the created system.

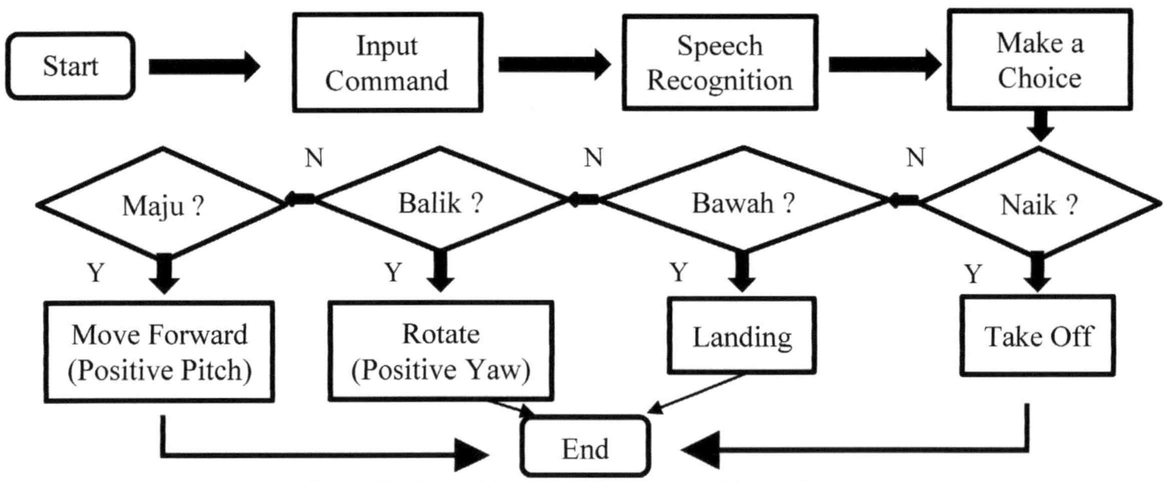

Figure 2. Speech Recognition System Flow Chart

The quadcopter movement was controlled by using the PID controller [8] created by removed the Integral component into a PD controller. In this research, the PD controller served to adjust the amount of distance that was taken by the quadcopter. The PD controller was a controller that the output signal was directly proportional to the error signal and error rate. The equation for the PD controller was,

$$PD = k_p e_p + k_d e_d \, , \tag{1}$$

where k_p was a proportional constant, while k_d was a derivative constant, e_p was an error of a proportional component, and e_d was an error of a derivative component.

3. Evaluation and Experimental Result

In this research, the sound database was needed from every word of the command used. We used an existing database and new database then compared both of them. The new database that created then was trained using the Recurrent Neural Network with the target value '1' for 'maju', '2' for 'mundur', and '3' for 'kanan'. After the database was trained, the calibration data testing process did ten times for each word. From the testing process we obtained the following results.

Table 1. Success rate of characterization test

Command	Maju	Mundur	Kanan	Succes Rate (%)
Maju	4	6	0	40
Mundur	5	5	0	50
Kanan	8	2	0	0

After calibration data processing, then the data was tested twenty times with four types of words that were 'maju', 'mundur', 'naik', and 'balik'. After that, we compared existing databases with new databases that were created in this research, and the result was shown in **Figure 3**.

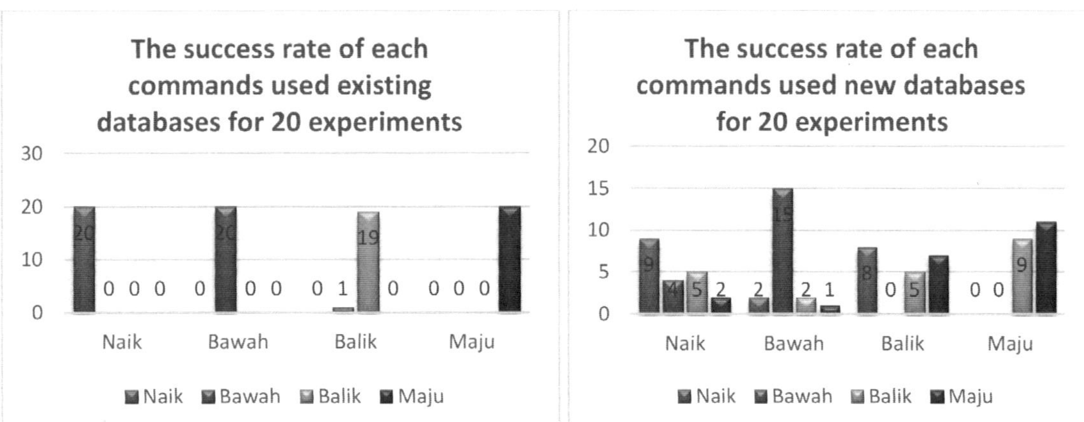

(a) (b)

Figure 3. (a). The success rate of each command used existing databases for 20 experiments. **(b).** The success rate of each command used new databases for 20 experiments

Table 2. Fine tuning data results

No	Kp	kd	Target Distance (cm)	Reached Distance (cm)	Time (second)
1	0	1	200	245	2.844
2	0.02	0.9	200	235	3.32

3	0.04	0.8	200	160	3.231
4	0.06	0.7	200	190	3.045
5	0.08	0.6	200	190	3.172
6	0.1	0.5	200	195	3.21
7	0.12	0.4	200	220	3.377
8	0.14	0.3	200	200	2.766
9	0.16	0.2	200	200	3.752
10	0.18	0.1	200	205	3.24
11	0.2	0	200	185	6.213

The quadcopter movement was controlled by using the PID controller. So, we had to calibrate the controller to find k_p and k_d. Here, it was the results of calibration data to find k_p and k_d values that had precision distance and the least travel time.

The k_p and k_d values that yield the best value were $k_p = 0.14$ and $k_d = 0.3$. From the value of k_p and k_d, then tested PID controller with three tests. Here it was the results of PID testing.

Table 3. PID test results

No	Target Distance (cm)	Reached Distance (cm)	Time (second)
1	200	200	2.766
2	200	194	3.263
3	200	190	3.345

4. Conclusion and Future Work

Based on the test and discussion data, we could conclude that, Speech Recognition System created could already recognize the spoken word, the success rate for each word used existing databases was 100% except for the word 'balik' by 95%. While, the success rate for each word used new databases was 45% for 'naik', 75% for 'bawah', 25% for 'balik', and 55% for 'maju'. From that, we could conclude that a new databases still need more improvement. Based on table 2, the value of k_p and k_d that produced the best value was when $k_p = 0.14$ and $k_d = 0.3$ with distance reached = 200 cm and time = 2.766 seconds.

In making speech recognition system there was still much to be improved. Therefore, there were some suggestions that could be did for the development of the next system such as, making databases had to be from the more varied responders in order to increase the quality of the databases created, the sound of the quad-spin propeller motion could increase the disturbance for the system made, so it should be added filter to resolve this, as well as PID controller that made could still be refined again for the value of its response to be better.

5. Acknowledgment

This research was funded by Hibah PITTA UI, under contract number 679/UN.2.R3.1/HKP.05.00/2017

References

[1] Y Qian, T Tan, and D Yu 2016 *Neural Network Based Multi-Factor Aware Joint Training for Robust Speech Recognition* vol 24, no 12, pp 2231–2240 (IEEE/ACM Trans. Audio Speech Lang. Process)

[2] H B Sailor and H A Patil 2016 *Novel Unsupervised Auditory Filterbank Learning Using Convolutional RBM for Speech Recognition* no 99, pp 2341–2353 (IEEE/ACM Trans. Audio Speech Lang. Process)

[3] M A Imtiaz 2016 *Isolated Word Automatic Speech Recognition (ASR) System using MFCC , DTW & KNN* pp 106–110 (Asia Pacific Conference on Multimedia and Broadcasting (APMediaCast))
[4] U G Patil 2016 *Automatic Speech Recognition of Isolated Words in Hindi Language using MFCC* pp 433–438 (International Conference on Computing, Analytics and Security Trends (CAST))
[5] H Ali, X Zhou, and S Tie *Comparison of MFCC and DWT Features for Automatic Speech Recognition of Urdu* pp. 154–158
[6] X Chen, X Liu, Y Wang, M J F Gales, and P C Woodland 2016 *Efficient Training and Evaluation of Recurrent Neural Network Language Models for Automatic Speech Recognition* vol 24, no 11, pp 2146–2157 (IEEE/ACM Trans. Audio Speech Lang. Process)
[7] R Setiono, B Baesens, and C Mues 2008 *Recursive neural network rule extraction for data with mixed attributes* vol 19, no 2, pp 299–307 (IEEE Trans. Neural Networks)
[8] B. Yasmina and A. Zoubir 2017 *Dynamic Modeling , Simulation and PID Controller of Unmanned Aerial Vehicle UAV* no. Intech, pp. 1–6

IOP Conf. Series: Journal of Physics: Conf. Series **1011** (2018) 012050 doi:10.1088/1742-6596/1011/1/012050

10 GHz Standing-Wave Coplanar Stripline on $LiNbO_3$ Crystal for Radio to Optical-Wave Conversion

F Darwis[1], Y N Wijayanto[1], A Setiawan[1], D Mahmudin[1], A N Rahman[1] and P Daud[1]

[1]Research Center for Electronic and Telecommunication, LIPI, Kampus LIPI, Jl. Sangkuriang, Bandung 40135, Indonesia

E-mail: fajri.darwis@lipi.go.id

Abstract. Recently, X-band radar systems are used widely for surveillance and navigation applications. Especially in archipelago or maritime country, the surveillance/navigation radar systems are required to monitoring critical areas and managing marine traffic. Accurate detection and fast analysis should be improved furthermore to provide security and safety condition. Therefore, several radar systems should be installed in many places to coverage the critical areas within radar networks. The radar network can be connected using optical fibers since it has extremely low propagation loss with optical-wave to carry-out the radar-wave. One important component in the scenario is a radio to optical-wave conversion component. In this paper, we report a 10 GHz radio to optical-wave conversion component using standing-wave coplanar stripline (CPS) on $LiNbO_3$ optical crystal as the substrate. The standing-wave CPS electrodes with narrow slot are arranged in an array structure. An optical waveguide is located close to the narrow slot. The CPS electrodes were analysed using electromagnetic analysis software for 10 GHz operational frequency. Responses for slot width and electrode length variation are reported. As results, return loss of -14.580 dB and -19.517 dB are obtained for single and array CPS electrodes respectively. Optimization of the designed radio to optical-wave conversion devices was also done.

1. Introduction

Nowadays, radar systems are important to provide security and safety condition using electromagnetic-wave in radio-bands to identifying objects. The radar systems can be used in many applications such as for military and defend, traffic management, agricultural, medical, and so on. Especially in archipelago country with many islands, radar systems are required to monitoring and surveying remote or borderline areas on land and sea. We have developed radar systems for surveillance and navigation in X-band operational frequency [1]. The developed radar systems were installed in coastal area to surveillance of ship traffics and other objects in a strait. The developed radar systems have detection range of up-to 20 Nautical Mile with radiation power less than 1W. Based on these results, detection range should be improved to cover large areas.

Improvement of the detection area can be solved by adding and installing radar units in many places. It can be realized using a network of radar systems. Optical fiber links are promising to support the radar networks through radio-over-fiber (RoF) technology [2]. Optical fibers have low propagation loss characteristics of 0.2dB/km using lightwave to carry out the radar signals. In RoF technology, an interface to convert wireless radar signal to a lightwave signal is important [3]. The interface can be implemented by combining radar antennas and optical modulators [4]. An interface using patch antennas embedded with a gap was reported [5-6].

In this paper, a design of an interface between microwave and lightwave using Coplanar Stripline (CPS) on $LiNbO_3$ substrate was proposed. The design consists of a substrate with two pieces of conductors arranged to operate in parallel and separated by a narrow gap. The advantage of the CPS is that it can be installed in series or in shunt and CPS is the balanced transmission line [7].

In the following sections, we will discuss the proposed interface structure and analyse its characteristics. Performance of the proposed interface to detect wireless microwave polarization are also reported and discussed.

2. Device Structure

The coplanar strip line (CPS) structure consists of a pair of symmetric electrodes with certain length which is used in a ground–signal configuration [8-9]. The structure of CPS is depicted in Figure 1.

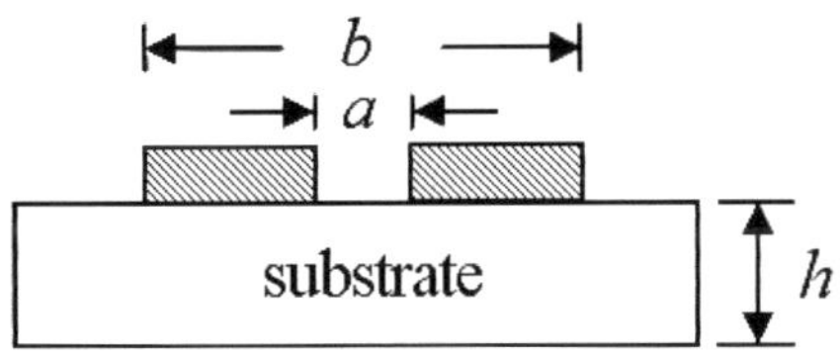

Figure 1. Structure of coplanar strip line.

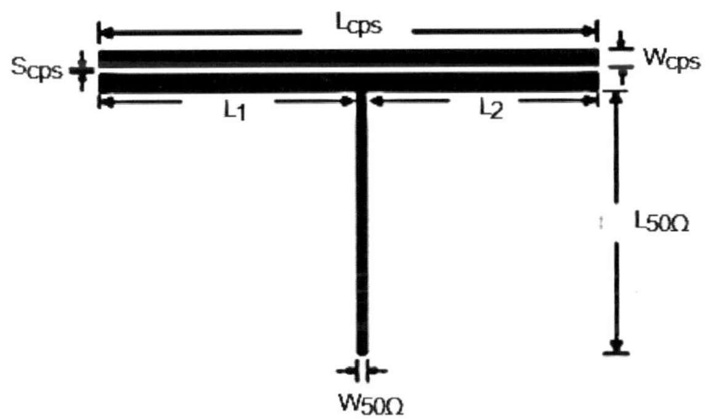

Figure 2. The proposed design of CPS.

In Figure 1, the gap between two conductors is define by variable a and the sum of total width of two conductors and a is defined by b. Variable h is height of substrate. Figure 2 shows the proposed design of CPS with feeding. The CPS uses 50 Ω of feeding. First position of the feeding is placed in center of CPS. The feeding will be connected to connector as input of microwave signals. Lithium Niobate ($LiNbO_3$) is used as a substrate. The substrate has a good characteristic for light wave [10].

To design the CPS, first we must determine the values of the width of electrode (w = 0.142 mm), gap between two electrodes (a = s = 0.031 mm), height of substrate (h = 0.3 mm), relative permittivity of substrate (ε_r = 28), and working frequency (f = 10 GHz). The characteristic impedance of CPS is calculated using equations (1) till (5).

$$k = \frac{a}{b} \qquad k' = \sqrt{1-k^2} \tag{1}$$

$$k_1 = \frac{\sinh(\pi a/4h)}{\sinh(\pi b/4h)} \tag{2}$$

$$\varepsilon_{eff} = 1 + \frac{\varepsilon_r - 1}{2}\,\frac{K'(k)}{K(k)}\,\frac{K(k_1)}{K'(k_1)} \tag{3}$$

$$\frac{K'(k)}{K(k)} = \begin{cases} \left[\frac{1}{\pi} ln\left(2\frac{1+\sqrt{k'}}{1-\sqrt{k'}}\right)\right] & for\ 0 \le k \le 0.707 \\ \left[\frac{1}{\pi} ln\left(2\frac{1+\sqrt{k}}{1-\sqrt{k}}\right)\right]^{-1} & for\ 0.707 \le k \le 1 \end{cases} \tag{4}$$

$$Z_0 = \frac{\eta}{\sqrt{\varepsilon_{eff}}}\,\frac{K(k)}{K'(k)} \tag{5}$$

From the equations above, the value of characteristic impedance (Z_0) is 42.65 Ω.

After the design of CPS is done, we need to design the feeding of CPS. The feeding of CPS use microstrip line with characteristic impedance 50 Ω. LineCalc software is used to calculate the length and width of microstrip line. We obtain the length of microstrip line is 1.951 mm and the width of microstrip line is 0.058 mm.

The proposed design will be simulated and optimized using ADS 2011.10 software to find the numerical characteristic. From the software we obtain the S(1,1) characteristic. The graph of S(1,1) characteristic is depicted in Figure 3.

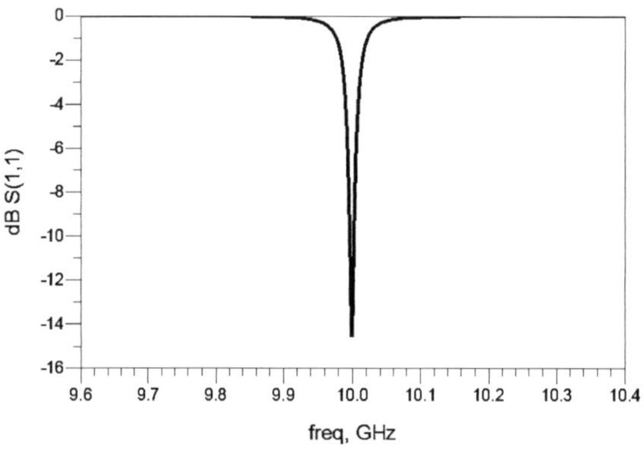

Figure 3. S(1,1) characteristic of single CPS.

In Figure 3, the value of return loss is -14.580 dB. This result is acquired by optimization using ADS 2011.10. The positions of feeding affect the return loss value. The best position of feeding after optimization are L_1 = 1.7405 mm and L_2 = 1.5905 mm and the length of CPS is 3.392 mm. For the width of feeding is 0.061 mm.

3. Operational Principle

In previous section, we know that the CPS consists of two electrodes. The electrodes are separated with certain gap. One electrode is used as a signal and the other as a ground. In this design, we consider gap value between electrodes. It is to get the good electric field. To make connection between microwave signals to CPS, the feeding is used. The size of feeding is 50 Ω due to the connector impedance we use for this design is 50 Ω. So, matching of connector and feeding line is needed to have the maximum power transfer.

In order to see the influence of gap between frequencies, we make this design with 3 difference values of gap. The response frequency is showed in Figure 4. From Figure 4 we see that the frequency is shifted in gap value 41 μm and 51 μm. It occurs because effective the electric constant (ε_{eff}) is reduced as shown in equation (1) till (4). The return loss of proposed design better than the others, but all of them still have good return loss. Plot of the electric field (normalized to the voltage) with gap is depicted in Figure 5. The high electric field is obtained from the smallest gap between electrodes.

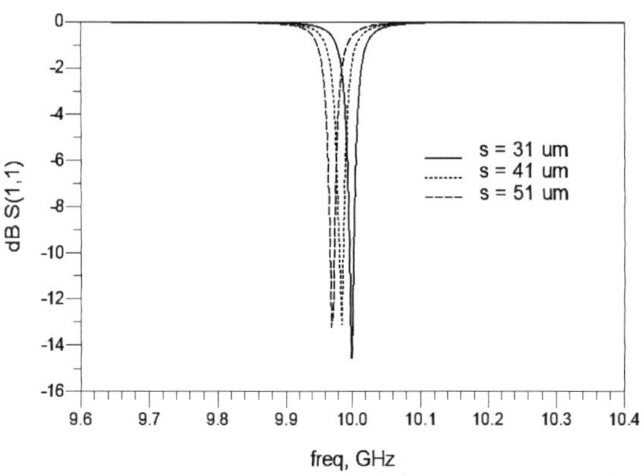

Figure 4. Response frequency with several values of gap.

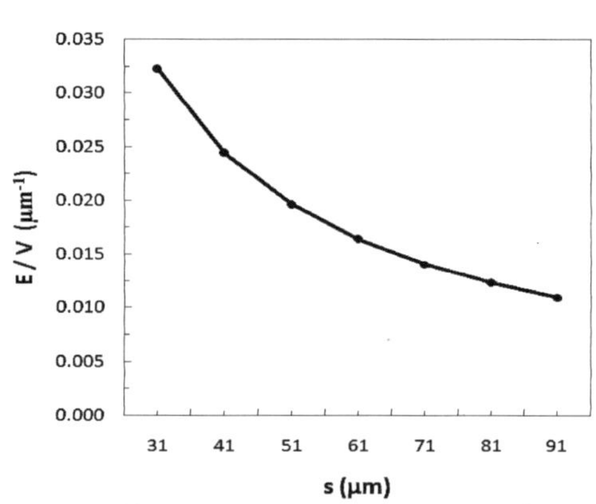

Figure 5. Electric field with gap.

4. Array Structure

To see the characteristic of CPS with array structure, we use four CPS with feeding that are arranged in horizontal position. The proposed design of array structure is depicted in Figure 6.

In Figure 6, we use wilkinson power divider technique to divide the power from input port [11]. The characteristic impedance in input port is 50 Ω. This design uses three of 2-way power divider to connect four CPS. Two of 2-way power divider is used to connect the two CPS in each power divider. The last power divider is used to connect the two of 2-way power divider. The power divider is designed using microstrip line. The characteristic impedance after 50 Ω of transmission line is $Z_0\sqrt{2}$ equal with 70.71 Ω.

Since the CPS is designed for microwave – light wave conversion, range between CPS must consider the light wave transmission. By using the equation below, the range from starting point first CPS to starting point second CPS (*d*) can be calculated.

$$F(\theta) = \sum_{h=1}^{N} A \exp\left[-j(h-1)k_m\left(n_0 d \sin\theta - n_g d\right)\right] \quad (6)$$

Where *A*: amplitude, *N*: number of array, k_m: wave number, *d*: range CPS, *n*: refractive index and $\sin\theta = 0$. The value of *d* is 13.94 mm calculated with 1 wavelength.

The characteristic of array CPS is depicted in Figure 7. In the figure, the return loss before optimization is about -5.71 dB. To find the best result of return loss, we do the optimization. The optimization was done by change the impedance Z_0 (width of feeding) in feeding. The best widths of feeding that we have are 0.1 mm in input, 0.02 mm in CPS feeding and 0.03 mm in between input and CPS. The return loss after optimization is -19.517 dB. It shows that the performance of array CPS is good.

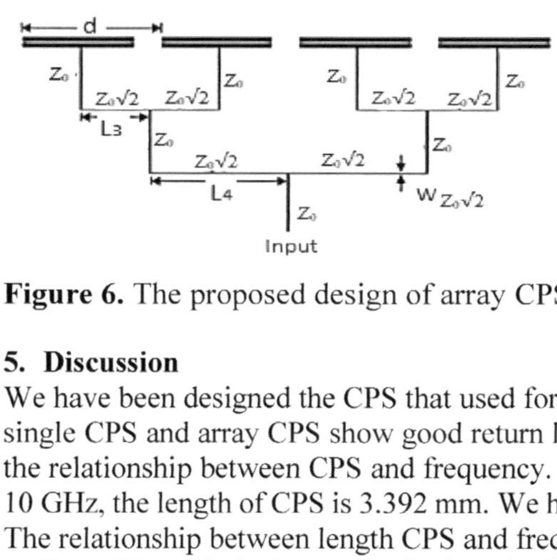

Figure 6. The proposed design of array CPS.

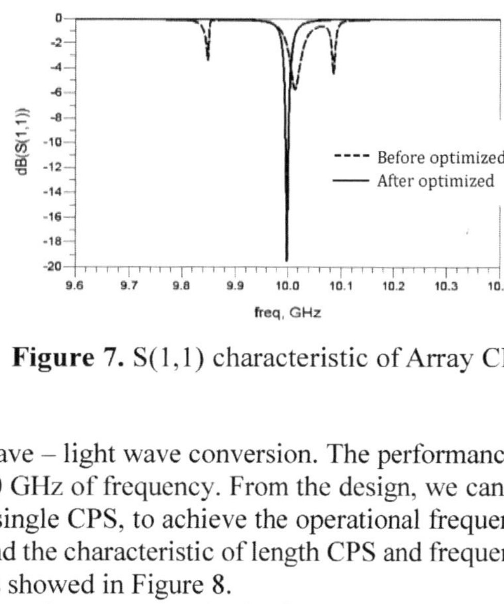

Figure 7. S(1,1) characteristic of Array CPS.

5. Discussion

We have been designed the CPS that used for microwave – light wave conversion. The performance of single CPS and array CPS show good return loss at 10 GHz of frequency. From the design, we can see the relationship between CPS and frequency. For the single CPS, to achieve the operational frequency, 10 GHz, the length of CPS is 3.392 mm. We have found the characteristic of length CPS and frequency. The relationship between length CPS and frequency is showed in Figure 8.

From Figure 8, it shows that the length of CPS inversely proportional with frequency. To know the relationship between feeding position with electric field magnitude, we change the position of feeding, L_1 and L_2, to get the values of electric field. Figure 9 shows the effect of 3 condition of feeding position to electric field.

On position of $L_1 = 1.7405$ mm and $L_2 = 1.5905$ mm shows the maximum of electric field in 10 GHz of frequency. It's mean that the standing wave in that position of feeding better than the others. From the return loss value, we find the VSWR is about 1.46.

The strong microwave electric field can be generated using the proposed CPS structure. Since ligthwave propagates into an optical waveguide where it is located close to the slot of CPS structure. The electric fields between microwave and lightwave are interacted each other's. The interaction of them can be calculated by considering transit-time effect. As results, optical modulation can be obtained by considering transit-time and interaction between microwave and lightwave electric fields.

IOP Conf. Series: Journal of Physics: Conf. Series **1011** (2018) 012050 doi:10.1088/1742-6596/1011/1/012050

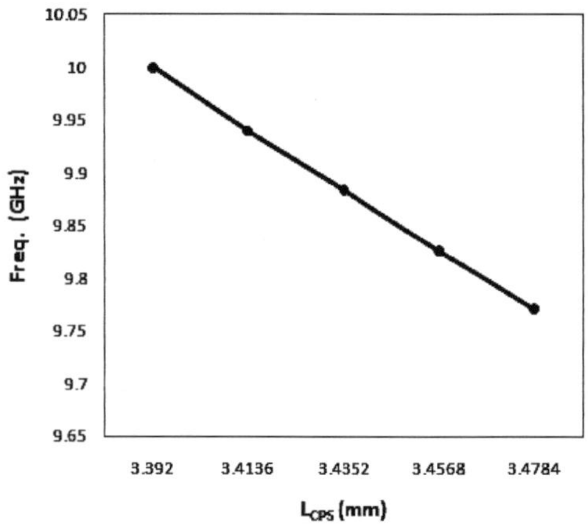

Figure 8. Length CPS vs Frequency.

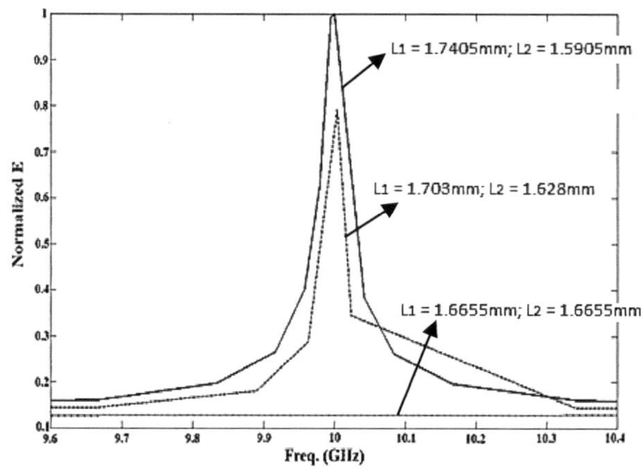

Figure 9. Normalized Electric field with feeding position.

6. Conclusion

The radio to optical-wave conversion using standing-wave CPS electrodes on a $LiNbO_3$ crystal as the substrate was proposed and reported for 10 GHz operational frequency. The standing-wave CPS electrodes can be generated strong radio-wave electric field when radio-wave signal is coupled to the feeding of the CPS electrodes. The strong radio-wave electric field can be used for optical modulation by considering interaction of the radio and optical-wave electric fields. Analysis of the CPS electrodes was reported in 10 GHz operational frequency. Calculation results for return loss, VSWR, and radio-wave electric field in single and array CPS electrodes was reported and discussed. The proposed device is promising for radio to optical-wave conversion in the ROF technology. The ROF technology can be applied for high-speed communication and high-resolution imaging systems.

Acknowledgments

This research activity was conducted through The Excellent Research Grant from Indonesian Institute of Sciences (LIPI), Indonesia and through Thematic Program, funded by Indonesian Institute of Sciences (LIPI), Indonesia.

References

[1] Sulistyaningsih, Wahab M, Mahmudin D, Permana D, Maulana Y Y, Estu T T 2016 *TELKOMNIKA* **14** 507

[2] Wijayanto Y N, Mahmudin D, Daud P 2012 *Proc. Of The 6th National Radar Seminar and The 1st Int. Conf. On Radar, Antenna, Microwave, Electronics and Telecommunication (Bali, Indonesia)* p 73

[3] Wijayanto Y N, Murata H, Hermida I D P 2016 *Jurnal Elektronika dan Telekomunikasi* **16** 7

[4] Shinada S, Kawanishi T, Sakamoto T, Andachi M, Nishikawa K, Kurokawa S, Izutsu M 2007 *IEEE Photonics Technology Letters* **19(10)** 735

[5] Wijayanto Y N, Murata H, Okamura Y 2011 *IEICE Electronics Express* **8(2011)** 491

[6] Wijayanto Y N, Murata H, Shiomi H, Okamura Y 2011 *CLEO: 2011 (Optical Society of America, Washington, DC, 2011)* paper JWA122

[7] Simons R N 2001 *Coplanar Waveguide Circuits, Components, and Systems* New York John Wiley & Sons, Inc.

[8] Wadell B C 1991 *Transmission Line Design Handbook* Norwood MA: Artech House

[9] Lee T H 2004 *Planar Microwave Engineering: A Practical Guide to Theory Measurement, and Circuits* Cambridge Cambridge University Press

[10] Bemal M P, Roussey M, Amet J, Baida F I, Burr G W 2007 *2007 Europ. Conf. On Lasers and Electro-Optics and Int. Quantum Electronics Conf. Munich*

[11] Pozar D M 2011 *Microwave Engineering (Fourth Edition)* USA John Wiley & Sons, Inc.

IOP Conf. Series: Journal of Physics: Conf. Series **1011** (2018) 012051 doi:10.1088/1742-6596/1011/1/012051

Characteristic Exponent of Normal and Oblique Rolls in Homeotropically Aligned Nematic Liquid Crystal

V Saraswati and F Nugroho

Department of Physics, Universitas Gadjah Mada, Sekip Utara BLS 21 Yogyakarta, Indonesia

Corresponding author: fakhrud@ugm.ac.id

Abstract. Soft-mode turbulence (SMT) is one of an experimental example of spatiotemporal chaos, observed in electroconvection system of homeotropically aligned nematic liquid crystal (NLC), due to a non-linear interaction between Nambu-Goldstone mode denoted by the **C(r)**-director and the convective mode **q(r)**. There are two types of stripe patterns in the SMT, namely normal rolls (NR) and oblique rolls (OR) which separated by a point of applied frequency, called the Lifshitz frequency (f_L). We report a study of the phase transition from normal to oblique rolls by observing the patterns with an applied frequency below and beyond of f_L. The temporal fluctuations of the pattern images had been analyzed using autocorrelation function. It fits with Kohlrausch Williams Watts (KWW) function, showing there is a dynamical glass-forming liquid in the transition of NR-OR regime. Also, we found a new type of defect in the NR regime which never been reported before, a dynamic defect which takes the shape of a ring first to a spot in the end.

1. INTRODUCTION

The changing of natural phenomenon that occurs around us is an extremely unstable system. A system in far from equilibrium which infinitely rich in a change of the dynamics sometimes exhibits a variety of pattern structure [1]. The most well-known phenomenon that triggers a pattern formation is turbulence in the fluid system. Recently, significant development related to the detailed classification of turbulence has been obtained; one of them is spatiotemporal chaos (STC) which are spatially and temporally disordered structure [2]. Very recently, one of the experimental models of STC that attracts a good deal of researcher's attention is Soft-Mode Turbulence (SMT) which shows a special scenario for the transition to turbulence, because it directly appears from a quiescent state via single supercritical bifurcation [3].

The SMT has been found in electroconvection system of homeotropically aligned nematic liquid crystal (NLC). In the homeotropic alignment, the director $\bar{n}$ of an NLC is perpendicular to the electrodes which same as the *x-y* plane. When the ac voltage V applied to the homeotropic NLC system, the first transition called the Fréedericksz transition occurs at $V \geq V_F$, below the critical threshold for convection V_C. The director $\bar{n}$ tilts with respect to the *z*-axis by the transition, and its projection on the *x-y* plane is called the $\bar{C}\,(r)$-director, where $\bar{r} = (x\,,y)$. The direction of $\bar{C}\,(r)$ which behaves as Nambu-Goldstone mode can be freely rotated, because of the same energy [4]. If the ac voltage V applied to the system beyond the point of second transition V_C, the electroconvection occurs and exhibits a stripe pattern because of the existences of a homogenous wave vector $\bar{q}(r)$. The nonlinear interaction between the resulting convective modes $\bar{q}$ and the Nambu-Goldstone modes $\bar{C}$ leads to the SMT [5]. To inform how far the system of SMT moves from the critical point (V_C), some researchers [4, 6, 7] proposed the normalized voltage ε as a control parameter, defined by $\varepsilon \equiv (V^2 - V_C^2)/V_C^2$.

There are two types of stripe patterns in the electroconvection of the NLC in the homeotropic system, respectively to the ac frequency f of the applied voltage, namely oblique rolls (OR) and normal rolls (NR). Both patterns are separated by a point of frequency called the Lifshitz frequency (f_L). Below the f_L lies OR regime, while beyond f_L is for NR regime. For distinguish both regimes, a line defect called blacklines are usually used. Blacklines found only in the OR regime.

We are interested in investigating the phase transition between the regime of NR and OR by extracting the statistical properties of both patterns relaxation using autocorrelation function. In the SMT, the autocorrelation function shows a decay behavior, and usually, it is approached by a simple exponential decay as a type of the chaotic system [6]. But, recently some studies [7] found that simple exponential equation cannot perfectly fit the autocorrelation function of the SMT relaxation. According to that reason, we also tried to use an alternative fitting form for the whole ε using the Kohlrausch-William-Watts (KWW) function, a generalization equation of the simple exponential decay by introducing the Kohlrausch exponent (β).

2. EXPERIMENTAL SETUP

The experimental set up is similar to that in ref. [6, 7]. We carried out our experiments by using the nematic liquid crystal *p*-methoxybenzilidene-*p'*-*n*-butylaniline (MBBA) with doping of tetra-*n*-butyle-ammonium bromide (TBAB). MBBA is filled between two parallel glass plates whose surfaces are coated with transparent electrodes, indium tin oxide (ITO). We used rectangular electrodes with different size; there are 3 x 1.5 cm^2 and 2 x 2.5 cm^2 which we set up into a sandwiched cell. The space between the electrodes is maintained with polymer spacers of 50 μm. To realize homeotropic alignment surfaces of the glass plates are treated with a surfactant (*n*-*n'*-dimethyl-*n*-octadecyl-3-aminopropyl-trimethoxy silyl chloride: DMOAP). Statistical data were measured at the stabilized temperature 30.00 ± 0.05^{0}C. The CCD camera (Panasonic BD400) mounted on the microscope, and the software (Pixel View) were used to capture pattern images on the *x*-*y* plane. Then, the image analysis was performed by custom software.

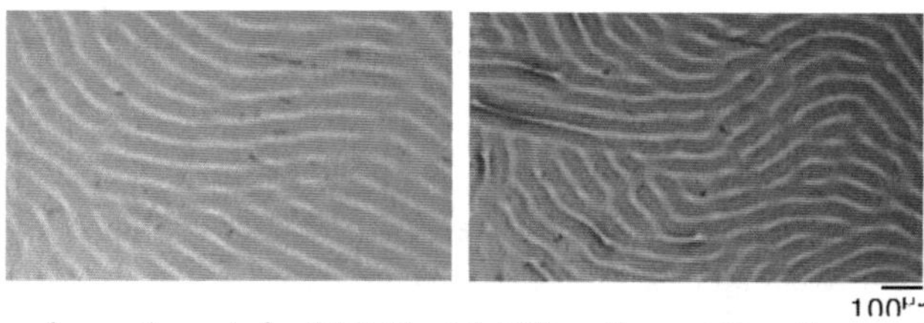

Figure 1. The pattern formation at f = 950 Hz (a) NR pattern when the control parameter is ε = 0.050 (b) the transition regime between NR and OR is at ε = 0.075, indicated by the appearance of blacklines.

By observing the pattern with the change of frequency and the voltage simultaneously, we obtained f_L for our cell is 750 Hz which separates the OR ($f < f_L$) and the NR regimes ($f > f_L$). We cautiously jumped the voltage until we obtained the transition regime between OR and NR as shown in Figure 1. Moreover, we obtained the phase diagram of the applied ac voltage – frequency (V–f), respectively. For extracting the statistical properties of the transition NR-OR regime, we set the observation frequency in the Lifshitzt frequency f = 750 Hz. The experimental procedure is as mention in Ref [7]. We set the desired positive control parameter ε (0.025-0.075) by jumping to the voltage above a convective threshold V_C. We captured the electroconvection images in 20 min. Then, we calculated the autocorrelation function $\tilde{Q}(\boldsymbol{r}, \tau)$ of the pattern fluctuation in the Eulerian picture using formula as mentioned in Ref [7].

IOP Conf. Series: Journal of Physics: Conf. Series **1011** (2018) 012051 doi:10.1088/1742-6596/1011/1/012051

3. RESULT AND DISCUSSION

The first experimental result of our procedure is the phase diagram for the normal and oblique regime. Figure 2 shows the phase diagram of the applied ac voltage – frequency ($V-f$), respectively for our cell. The resulting phase diagram from our observation is similar to the schematic phase diagram proposed in the Ref [2]. From Figure 2 we can conclude that the state of NR regime is narrow, proofed by when we set the applied frequency at $f = 950$ Hz the NR regime appeared below the control parameter $\varepsilon = 0.075$ and experienced a transition to OR regime at $\varepsilon = 0.075$ as shown in

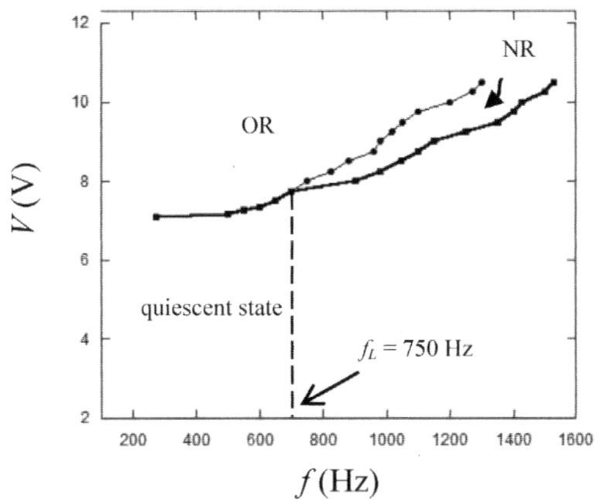

Figure 2. Phase diagram of $V-f$ obtained from observation. The Lifshitz frequency f_L is about 750 Hz.

While for the experimental results of the temporal autocorrelation analysis for $0.025 \leq \varepsilon \leq 0.075$ are shown in Figure 3, where $\tilde{Q}(\tau)$ denotes the spatially average $\tilde{Q}(\boldsymbol{r},\tau)$. All the obtained autocorrelation functions fall of rapidly to zero. Also the decay goes faster by increasing of the control parameter ε, meaning that the dynamical of pattern fluctuation in NR regime is chaotic. Although the autocorrelation function at $\varepsilon = 0.050$ and 0.075 are coincide. Similar with Ref [7], we also found that the simple exponential is not suitable for describing $\tilde{Q}(\tau)$ at low ε. Because of that, we used the KWW equation for fitting function [7] which defined:

$$\tilde{Q}(\boldsymbol{r},\tau) = \alpha \exp\left[-(|\tau|/\tau_{KWW})^{\beta}\right] \qquad (1)$$

where each α, β, and τ_{KWW} denotes a normalized constant, the Kohlrausch exponent and the relaxation time of the KWW function. The KWW relation is empirical for relaxation in disorder system and usually used to explain the dynamic in colloidal system. As illustrated in Figure 3, the KWW function well describes the autocorrelation function of SMT for low ε.

There are two kinds of parameters that we used to describe the effect of the fitting by the KWW relation quantitatively; β and τ_{KWW}. Figure 4 (a) shows an exponent β dependence with increasing of ε. If $\beta > 1$ the KWW equation is called the compressed exponential, while for $\beta < 1$ is called the stretched exponential. As illustrated in Figure 4 (a), the β value for the transition regime from NR to OR when $\varepsilon \to 0$ is greater than 1. Implying that at low ε the dynamical of the SMT's temporal fluctuation is analog with relaxation which appears in glass forming liquid (GFL) dynamics [7]. This dynamical heterogeneity in the GFL has been discussed in Ref. [8] before.

Then, we discuss τ_{KWW} parameter or the correlation time proposed by Nugroho ***et al.*** [7] as a correction of the correlation time τ_s by Hidaka ***et al.*** [6] regarding fitting the autocorrelation function to the simple exponential equation. We obtained that the KWW fitting satisfied the relationship $\tau_{KWW} \propto \varepsilon^{-1}$ as illustrated in Figure 4; namely $c_0 = -0.00$ s^{-1} defined in $\tau_{KWW}{}^{-1} = c_0 + c_1\varepsilon$, showing the bifurcation of the

Also, this was employed as evidence that softening of irregular modes occurs at critical point $\varepsilon = 0$. SMT that occurs supercritically and the bifurcation point coincides with electroconvection threshold $\varepsilon = 0$.

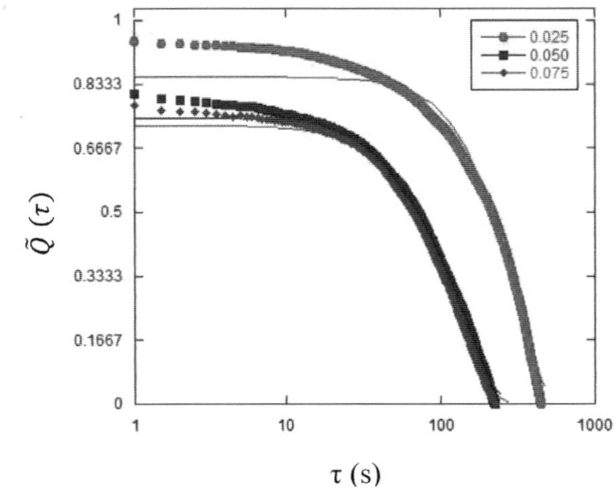

Figure 3. The plot of the scaled autocorrelation function with the KWW fitting (shown by a line that intersects with the plotted data) for several values of the control parameters; $\varepsilon = 0.025$ (red circle), 0.050 (blue square), and 0.075 (green diamond). Each normalization constant was obtained by fitting. The R value of the fitting is around 0.99, representing perfect fit between the data and the KWW function.

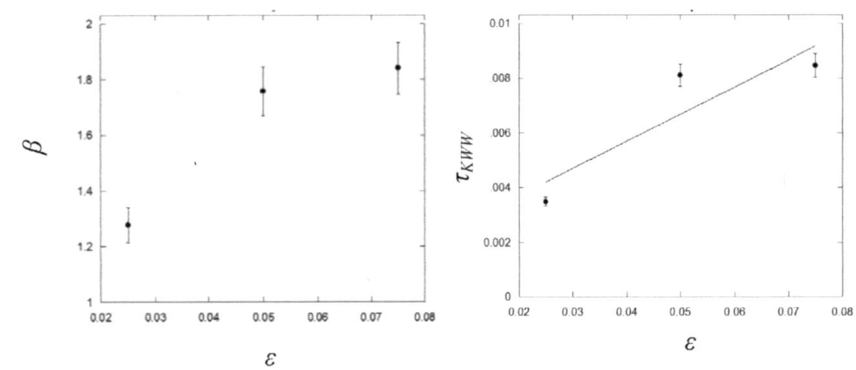

Figure 4. An ε dependence of β (a) and τ_{KWW} (b) in Eq (2). Standard deviation is used as an error bar.

To get more information about the phase transition from NR to OR we tried to do observation in the NR at higher fixed frequency $f = 1500$ Hz with a convective threshold V_C is 15.9 volt. We expected that we would see another transition from NR to OR. But something different has happened. We didn't find a sign of transition from NR to OR even we applied a high control parameter $\varepsilon = 1.0$ in the system. Furthermore, it is really surprising that we observed a new type of defect appears in the NR regime starts from $\varepsilon = 0.8$. There are no studies before that reported a discovery of defects in the NR regime. Defects in the SMT frequently find in the OR regime which called by blacklines. The existence of blacklines is related to the local interaction between **C(r)** and **q(r)** that have been discussed before in Ref [5]. Blacklines can be observed for the whole ε, and more blacklines will appear by increasing of the control parameter ε. It is different with the defects we found in the NR regime. Those defects are very dynamic. In the first appearance the defects take a ring shape, but after several seconds their shape is changed into a point before they begin to disappear. We called those new type of defects by blackspot as shown in Figure 5.

Blackspot only appears when the SMT system is applied by a high control parameter, in random positions and has a short lifetime.

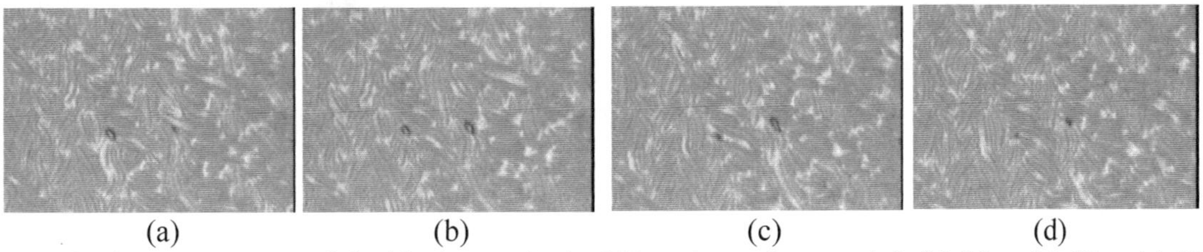

Figure 5. The appearances of the blackspots in the NR regime when $\varepsilon = 1.0$, (a) 85 s, (b) 87 s, (c) 90 s, (d) 92 s from initial observation time. The blackspots have a short lifetime.

4. CONCLUSION

We investigated the phase transition in the SMT from NR to OR regime by examined the temporal autocorrelation function of its pattern fluctuation. Our research revealed that in the relaxation of the SMT, there is a GFL-dynamics at low ε indicated by the compressed exponential of KWW which is suitable for fitting the autocorrelation function. Furthermore, at a higher frequency beyond of f_L, we found that there is no transition from NR to OR. On the other hand, in the NR regime, we observed new kind of defects called blackspot when the systems applied by a high control parameter. The future researchers are needed to conduct a broader investigation of blackspots properties.

ACKNOWLEDGMENT

The author would like to thank for the support from our partners in the Computational and Nonlinear Laboratories, Department of Physics, Faculty of Mathematics and Natural Sciences, Gadjah Mada University, Indonesia. This work is funded by Indonesia Directorate General of Higher Education (DIKTI) with contract number 119/LPPM/2015 and 0204/J01.1.28/PL.06.02/2017.

References

[1] Mori H and Kuramoto Y 1994 *Dissipative Structures and Chaos*, translated by Paquette (G.C., Springer: German)

[2] Anugraha R 2008, New Type of Order-Disorder Transition in Soft-Mode Turbulence of Electroconvection in Nematics (*Dissertation*: Graduate School of Engineering, Kyushu University, Kyushu, Japan)

[3] Anugraha R, Hidaka Y, Oikawa N and Kai S 2008 A Transition to Spatiotemporal Chaos under a Symmetry Breaking in a Homeotropic Nematic System *Journal of the Physical Society of Japan* **77** 073001.

[4] Hidaka Y, Tamura K, and Kai S 2005 Soft-mode Turbulence in Electroconvection of Nematics *Progress of Theoretical Physics Supplement* No.161.

[5] Anugraha R, Hidaka Y, Ueki T, and Kai S 2008 Symmetry-dependent defect structures in soft-mode turbulence *Phys. Rev. E* **80** 041701

[6] Hidaka Y, Huh J H, Hayashi K, and Kai S 1997 Soft-mode turbulence in electrohydrodynamic convection of homeotropically aligned nematic layer *Phys. Rev. E* **56** (6)

[7] Nugroho F, Narumi T, Hidaka Y, Yoshitani J, Suzuki M, and Kai S 2012 Glassy dynamics in relaxation of soft-mode turbulence *Phys. Rev. E* **85** 030701

[8] Weeks E R, Crocker J C, Andrewm Levitt C, Schofield A, and Weitz D A 2000) Three-dimensional direct imaging of structural relaxation near the colloidal glass transition *Science* **287**

Physical Evaluation of PVA/Chitosan Film Blends with Glycerine and Calcium Chloride

A D Nugraheni[1, a)], D Purnawati[1], and A Kusumaatmaja[1,2]

[1]Department of Physics, Universitas Gadjah Mada
Jl. Sekip Utara, Yogyakarta, 55281 Indonesia
[2]Nanomaterials Research Group, Universitas Gadjah Mada
Jl. Kaliurang Km 4, Sekip Utara, Yogyakarta, 55281 Indonesia

[a)]ari.dwi.n@ugm.ac.id

Abstract. PVA/chitosan film has been fabricated by using drop casting method. PVA/chitosan film is produced by dissolving 2% (w/v) PVA solution and 2% (w/v) chitosan solution. PVA/chitosan film is produced with weight ratio variation (w/w) 100/0, 75/25, 50/50 and 0/100. The film is fabricated using drop casting method in Petry dish with diameter 11 cm at room temperature and RH 50%–60% during seven days. The mechanical properties were characterized by using Universal Technical Machine (UTM) and UV-Vis to understand the physical properties of weight ratio (w/w) of PVA/Chitosan film by addition of plasticizer and calcium chloride. The film thickness tends to decrease with PVA content. The addition of chitosan will increase film thickness, and it will decrease swelling index, elongation (%), and transmittance of UV rays. The additions of plasticizer to PVA/Chitosan film will increase film thickness and elongation (%), and it will decrease swelling index, tensile strength and transmittance of UV rays. The crosslink of PVA/Chitosan film with calcium chloride will decrease film thickness, swelling index, elongation (%) and transmittance of UV rays, and increase tensile strength.

Keywords: chitosan, PVA, Drop casting, plasticizer, calcium chloride

1. Introduction

In the few years, development of food packaging with the antibacterial system has been increased. It could happen because of the consumer response and modern trend by produced the healthy and fresh food. The research was focused on the development of biopolymer film which is safe for health and environment. The biopolymer film was used as synthetic food packaging which is antibacterial and environment-friendly. This film was investigated to know the capability of keeping the food longer and increased the quality of food by avoided vapor movement, oxygen, and fat [1]. Biopolymer film was investigated by using a polymer with protein and polysaccharide, for example, chitosan, cellulose and cassava flour [2]. Chitosan is a synthetic polymer with antibacterial properties [3]. Chitosan was blended with other material for increasing the mechanical properties. Addition of PVA into Chitosan polymer was investigated. It showed the decreasing of UV transmittance which is given antibacterial properties [4]. Chitosan was added by silica and PEG to increased mechanical properties [5].

PVA is a polymer with low toxicity, biocompatible, hydrophilic, chemically stable and easy to fabricate [5]. PVA and Chitosan blends were developed and characterized in biology application. The film was accessible to biodegradable, and antioxidant by mixing PVA and Chitosan with different weight ratio [2]. Addition of chitosan into PVA solution will be decreasing mechanical strain of film. It will be increasing the mechanical stiffness of film [4]. The chemical synthesis between chitosan and PVA could be improved in mechanical strength thermal stability and swelling capability in chitosan film [6]. Addition of glycerin into PVA will be decreased transition temperature and thermal decay [7].

Modified PVA by addition of glycerin made tensile film decrease, but the elongation at break is higher. PVA/chitosan could be modified by added another material or fabricated by other methods. The purpose of modification in mechanical and optic properties to the blend film will be applicable. Modification by addition of calcium chloride to chitosan will be an influence to decreased the depreciation mass of tomatoes and extend the age of tomatoes 5-8 days longer than control sample [7]. The film decreases by the addition of calcium chloride to sodium alginate solution [8].

In these research, PVA/Chitosan film was made by drop casting method. The weight ratio of PVA/chitosan will be made by addition with glycerin and calcium chloride. The film will be characterized by UTM and UV vis spectroscopy. The prospect of the film for food packaging will be investigated by test the film with fresh vegetable.

2. Experimental Procedure

2.1 Materials

In conducting this research, chitosan medium molecular weight with deacetylation percentage (DD) of 75-85% deacetylated, and poly(vinyl alcohol) (PVA) molecular weight 89.000-98.000 (+99% degree hydrolysis) were purchased from Sigma Aldrich (St.Louis, MO, USA). The solvent of chitosan was 98% acetic acid which purchased from a local company of PT.Brataco (Indonesia). Glycerol and calcium chloride provided by Sigma Aldrich, respectively.

2.2 Preparation of film

Preparation of PVA/Chitosan with glycerine: The chitosan was soluble acid thus a blend of chitosan solution 2% was prepared by dissolving chitosan in acetic acid 2wt%. After stirring and heating at 80^0C during 5 hours, chitosan 2wt% was mixed with PVA 2wt% solution and stirred for 4 hours at 60^0C until producing a homogeneous solution. The weight ratio for PVA/chitosan was 100/0, 75/25, 50/50 and 0/100. Glycerin 10% (w/w) from chitosan mass was added. The solution was stirred for one hour at 70^0C.

Preparation of PVA/Chitosan with calcium chloride: PVA/chitosan at 75/25 before addition of glycerin was dropped in Petri dish by drop casting method. This film was put in 1 % (w/v) calcium chloride in ethanol solution. The solution with the film was stirred for 30-40 minute at room temperature.

Characterization: Mechanical characterization was used *Universal Testing Machine* (UTM) ZWICK/ZOS. The output of this characterization included thickness, strain, and stress. Optical characterization was used *UV-visible spectrometer* SHIMADZU UV-1700.

3. Result and Discussion

Mechanical Properties

Table 3.1 shows the thickness of film with a different weight ratio of PVA/Chitosan and by addition of glycerine and calcium chloride. PVA without the addition of chitosan shows that decreasing of thickness. PVA/chitosan blend shows the increasing in thickness rather than PVA 100%. The thickness will be increasing by addition chitosan. Chitosan 100% shows the thickest film. The increase in thickness will decrease by addition PVA because the thickness of film depends on PVA composition. Chitosan was a positive charge, and it has wide hydration film. Chitosan could hold water molecule that could avoid polymer chain and produce the thick film. When PVA concentration increased, the thickness will be decreasing.

Table 3.1 Thickness (a) and wide (A)

Weight ratio	a (mm)	A (mm^2)	U_A (mm^2)
50/50	0.029	0.145	0.029428
75/25	0.038	0.19	0.038328
PVA 100%	0.016	0.08	0.016763
75/25 + glycerin	0.056	0.28	0.056223
75/25 (0,1 gr/10 ml)	0.025	0.125	0.025495
75/25 (0,2 gr/20 ml)	0.021	0.105	0.021587
75/25 (0,3 gr/30 ml)	0.022	0.11	0.022561

Addition glycerin will influence the polymer film. Also, glycerin will be increased the thickness of the film. It is caused by hygroscopic properties of glycerin. Hygroscopicity is the capability for attracting and the bond water molecule in the air by absorption process. This property allows glycerin to attract water molecule from the air to the film. Water molecule will be attached and soak in the polymer film. Addition glycerin will increase the thickness of the film. Decreasing in thickness by adding calcium chloride is caused by breaking in function chain of PVA/Chitosan film because of the reaction of calcium chloride [8]. The decreasing of thickness is caused translation of positive charge. Calcium chloride is solved in ethanol. It forms calcium ion and chloride ion. The result is electrolyte concentration in solution will be increased. The film thickness became decreased.

Table 3.2 Stress

Weight ratio	Σ (MPa)	Uσ (MPa)
50/50	12.26	2.488
75/25	22.00	4.438
PVA 100%	17.44	3.655
75/25 + glycerin	4.84	0.973
75/25 (0,1 gr/10 ml)	26.10	5.125
75/25 (0,2 gr/20 ml)	12.41	2.552
75/25 (0,3 gr/30 ml)	33.53	6.877

PVA/chitosan blend film has high stress than pure chitosan film [2]. There is an interaction between PVA and chitosan molecules. This interaction will increase mechanical strength from PVA/chitosan blend film. The increasing of mechanical strength from PVA/chitosan film because of the interaction between –OH and NH_2 of both polymer [9].

Pure PVA has the higher strain (ε). Also PVA into the blend, the flexibility of film will be increased [2]. Addition of glycerin will be decreased intermolecular force between PVA and chitosan. It causes the distance between PVA and chitosan wider and sprawl. The film blend will be comfortable to be broken when the force was given. By the addition of plasticizer (glycerin), the strain became increasing [10]. Glycerin cause is decreasing intermolecular force between PVA and chitosan. Decreasing of intermolecular force will increase the flexibility of blend film. Addition of calcium chloride will influence the strain and stress of blend film. The strain indicates the formation of between carboxyl group in the interface of PVA with Ca2+ from calcium chloride [8; 11]. The increasing of calcium ion ((Ca2+) will decrease the strain of film.

Table 3.3 Strain

Weight ratio	ε (%)	Uε (%)
50/50	13.32	2.018
75/25	49.155	2.229
PVA 100%	61.67	2.350
75/25 with glycerin	53.60	2.270
75/25 *crosslink* 0,1 gr/10 ml	15.55	2.024
75/25 *crosslink* 0,2 gr/20 ml	2.704	2.001
75/25 *crosslink* 0,3 gr/30 ml	22.22	2.049

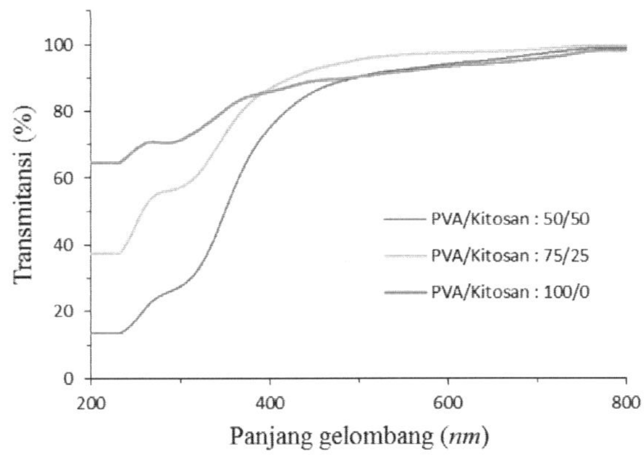

Figure 1. Transmittance spectrum of PVA/chitosan film

Picture 1 shows that pure PVA has the higher transmittance spectrum than another blend. It is at 60% - 90%. By the addition of chitosan, the transmittance spectrum will decrease. Transmittance spectrum is the capability to protect the film. The transmittance spectrum is related to the sum of radiation energy that could penetrate the film. The high transmittance spectrum shows the high radiation energy to penetrate the film.

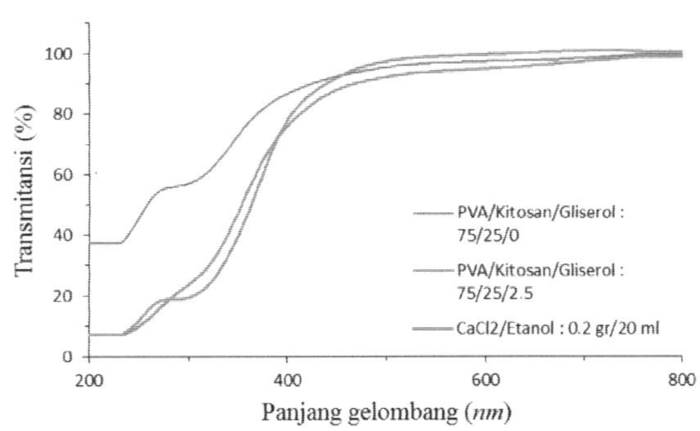

Figure 2. Transmittance spectrum of modified PVA/chitosan film

Picture 2 shows the PVA/chitosan 75/25 has the highest transmittance spectrum at 30% - 90%. After addition of calcium chloride, the transmittance spectrum continues to decrease and constant at wavelength 250 nm. By the addition of glycerin, the transmittance decrease but not stable. The low spectrum transmittance shows capability to protect the radiation. Picture 2 shows that modified PVA/chitosan with calcium chloride could resist the radiation.

4. CONCLUSION

The weight ratio of PVA/chitosan will influence the physical properties of blend film. Domination of PVA concentration will increase the swelling, spectrum transmittance, strain and stress of the blend film. Addition of plasticizer and crosslink with calcium chloride will decrease the swelling, transmittance spectrum, and stress of blend film.

Acknowledgements
Part of this work was supported by a research grant of "PUPT (Penelitian Unggulan Perguruan Tinggi) by the Directorate General of Higher Education (DIKTI), Ministry of Education and Culture, the Republic of Indonesia.

References
[1] Dutta P K, Tripathi S, Mehrota G K, Dutta J 2009 Perspectives for chitosan based antimicrobial film in food applications *J. Food Chem.* **114** 1173-1182
[2] Hajji S, Chaker A, Jridi M, Maalej H, Jellouli K, Boufi S, Nasri M 2016 Structural analysis, and antioxidant and antibacterial properties of chitosan-poly(vinyl alcohol) biodegradable film *J. Environ. Sci. Pollut. Res.* **23** 15310-15320
[3] Shakeel A, Ahmad M, and Ikram S, 2014 Chitosan: A Natural Antimicrobial Agent-A Review *J. Appl. Chem.* **3**(2) 493-503
[4] Bonilla J, Fortunati E, Atares L, Chiralt A, Kenny J M 2013 Physical, Structural and antimocrobial properties of poly vynil alcohol-chitosan biodegradable films *Food Hydro.* **35** 463-470
[5] Widhi Mahatmanti F, Nuryono, and Narsito 2014 Physical Characteristic of chitosan based film modified with silica and polyethylene glycol *Indo. J. Chem.* **14**(2) 131-137
[6] Yu Q, Song Y, Shi X, Xu C, Bin Y 2011 Preparation and properties of chitosan derivative /poly (vynil alcohol) blend film crosslinked with glutaraldehyde *J. Carbohydr. Polim.* **84** 465-470
[7] Rahmawati I S, Hastuti E D, Darmanti S 2011 Pengaruh Perlakuan Konsentrasi Kalsium Klorida (CaCl2) dan Lama Penyimpanan terhadap Kadar Asam Askorbat Buah Tomat (Lycopersicum esculentum Mill.) *Buletin Anatomi dan Fisiologi* **19**(1)
[8] Rhim J W and Lee J H 2004 Effect of $CaCl_2$ Treatment on Mechanical and Moisture Barrier Properties of Sodium Alginate and Soy Protein-based Films *J. Food Sci. Biotechnol.* **13**(6) : 728 – 732
[9] Chen R H, Hwa H D 1996 Effect of molecular weight of chitosan with the same degree of deacetylation on the thermal, mechanical, and permeability properties of ther prepared membrae *J. Carbohydr. Polym.* **29** 353 – 358
[10] Wirawan S K, Prasetya A dan Ernie 2012 Pengaruh plasticizer pada karakteristik edible film dari pektin *J. Reaktor* **14**(1) 61 – 67
[11] Pavlath A E, Gossett C, Camirand W, Robertson G H 1999 Ionomeric films of alginic acid *J. Food Sci.* **64** 61-63

The influence of curvature configuration on the characteristic of alcohol gel insertion jacket of polymer optical fiber liquid level sensor

N Arumnika, *H Kuswanto
Physics Study Program, Faculty of Mathematics and Natural Sciences, Universitas Negeri Yogyakarta, Jl. Colombo 1 Yogyakarta, Indonesia

Email: herukus61@uny.ac.id; arumnika@gmail.com

Abstract: This study aimed to determine the effect of curvature configuration to sensitivities and linearities of Polymer Optical Fiber (POF) water level sensor. POF type SH-4001-1.3 has been used in this study. The jacket of POF of 20 cm was removed. Transparent piped inserted by alcohol gel has been used to replace the jacket. This is head of a sensor. The head of a sensor is curved with variations of the specified path length, peel length, the width of curvature, the height of curvature and waveform. Configuration A (20 cm, 34 cm, 6 cm, 2 cm, 1 wave), configuration B (20 cm, 34 cm, 8 cm, 2 cm, 1 wave), configuration C (20 cm, 34 cm, 9 cm, 2 cm, ½ wave), configuration D (20 cm, 34 cm, 10 cm, 2 cm, ½ wave). The head of a sensor inserted into the water tank. The light source inserted to one end POF is a He-Ne laser light with a power of 5 mW and a wavelength of 632.8 nm. Power output at the other end received by the Optical Power Meter (OPM). The curvature configuration the head sensor of POF affects the output. Configuration A has good sensitivity, however good linearity given by configuration.

Keywords: POF, sensor, curvature

1. Introduction

Optical fiber technology has a very important role in the telecommunications industry. Optical fibers now have usability in remote sensing. The optical fiber is used as a sensor to measure level sensors petrol [1], liquid levels [2-4], the level of the liquid in a volumetric flasks as applications in industry [5], strain measurement [6], the humidity sensor in storage chemical [7,8], and temperature sensors [9,10].

On the other hand, optical fibers have been used for measuring water level in many forms, some are non-intrusive sensors based on light attenuation while passing tank walls as in [11], but being only useful in transparent tanks [7]. The industrial world needs a tool to measure the water level in a large tank. The optical fiber is the only part of the measuring device that introduced into the tank – that is, only light and plastic or glass interacts with the water – the method becomes safe and corrosion free, without electrical sparks that could cause a fire or explosion of a tank [12]. Plastic optical fibers (POF) have received increasing attention because of their clear technical advantages over glass fibers, such as flexibility and large core diameter, which enables efficient connection and coupling resulting in a low-cost system for a local area network [13]. POF has a higher numerical aperture than glass optical

IOP Conf. Series: Journal of Physics: Conf. Series **1011** (2018) 012053 doi:10.1088/1742-6596/1011/1/012053

fibers, and the system also has the advantage from a wider acceptance angle which provides a better light collecting conditions [14].

Fiber optic sensor used a modified behavior of light waves. Methods for modifying can be done against the amplitude (intensity), frequency, and phase [15]. Amplitude modification has the advantage of being easy to be processed [9,16-20]. This method is a simple method. Intensity changes occurred due to the leakage caused by the optical fiber subjected to pressure, temperature[10] changes, bending [20] or other physical treatment

This research examined the sensitivity of changes in the intensity of the POF sensor which has multiple bends and using water level sensor with fiber optic in a sinusoidal configuration which has been inserted gel. Here, the variations of fiber optic configuration are the length of the track, the length of peeling, the width of curvature, curvature height, and waveform, respectively.

2. Experimental Set-Up

The research was to find the effect of the curvature of the POF cladding and coating of optical power received by OPM. As is well known, not only used in communications systems, but also plastic optical fiber has been widely used as a sensor. In this study, POF was used as a water level sensor. POF type SH-4001-1.3 has a core refractive index of 1.49 and the cladding refractive index of 1.41. Design of fiber optic experiment contained in Figure 1.

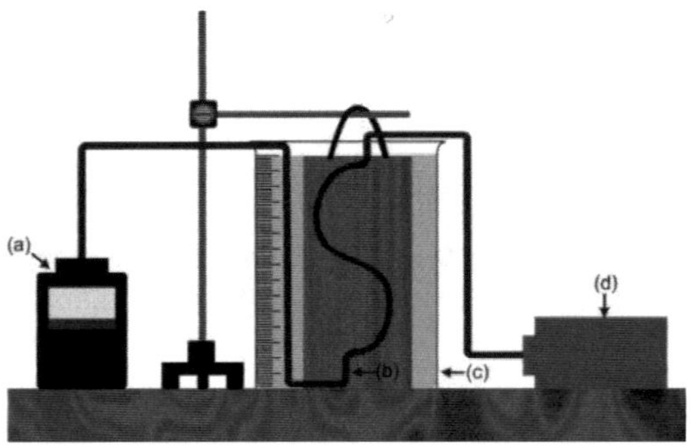

Figure 1. Design a research tool

Description: (a) Optical Power Meter (OPM); (b) Fiber Optics; (c) Water storage containers; (d) Helium-Neon Laser

Table 1 represents the configuration of the fiber optic specifications.

Table 1. Specifications configuration in points (b)

Specifications	Optical fiber configuration			
	A	B	C	D
Length Tracks (cm)	20	20	20	20
Long Exfoliation (cm)	34	34	34	34
curvature width (cm)	6	8	9	10
High-curvature (cm)	2	2	2	2
Form (wave)	1	1	½	½
Design Picture	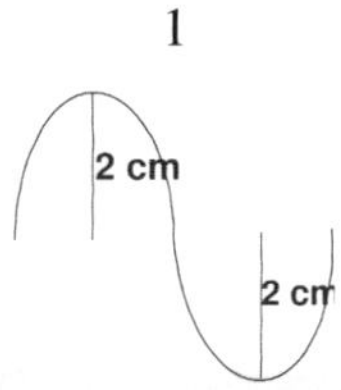	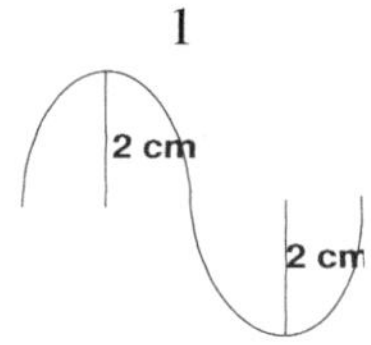	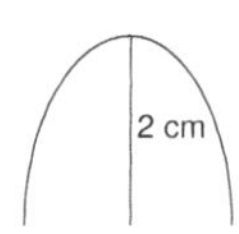	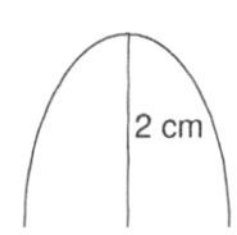

3. Result And Discussion

In the particular study, the optical mechanism was done on POF is to perform intensity modulation. Intensity modulation was done by measuring the optical power output fiber by modifying the state of the optical cladding. Research-based water level sensor sinusoidal-shaped POF included in the classification of intrinsic fiber optic sensors where the next POF as a guiding light once served as the sensing the water level in certain parts of the POF. The research result is presented in Figure 2 which is a graph in each configuration. The increase in altitude (independent variables) were conducted nine times, i.e 4 cm, 6 cm, 8 cm, 10 cm, 12 cm, 14 cm, 16 cm, 18 cm and 20 cm. The light source Helium-Neon laser used have $\lambda = 632.8$ nm and power of 5 mW.

Figure 2 shows that the optical power (dBm) that has been detected by the OPM to the water level (cm) in each normalized width curvature. Normalization was done by dividing the score of optical power at every level rises with the highest absolute score of optical power. This study measured the amount of optical power loss resulting from the different configuration curvature. The loss of power due to the treatment granted by the characteristics of optical fiber, which will be weakened when bending or coating on the cladding.

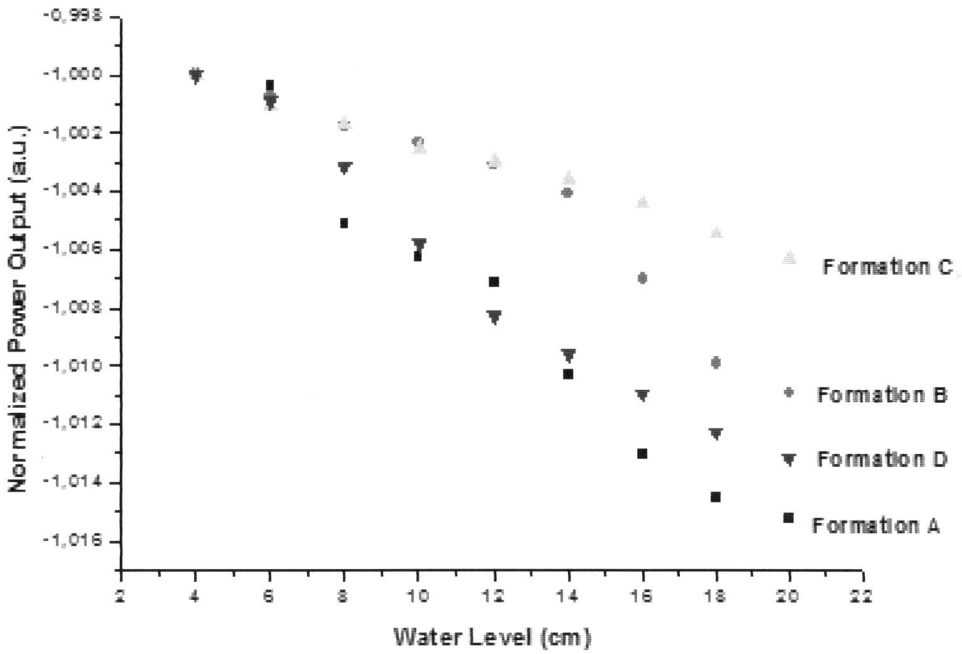

Figure 2. Graph normalized combined power output (a.u.) to the water level (cm) with a wide variety of curvature

Figure 2 shows when the water level changes, there will be a change in the optical power in spite of these changes are very small. Optimal data obtained at the time of configuration C. Figure 2 is a graph of the optical power normalized by the water level. The x-axis shows water level, with the unit of cm while Y-axis shows normalized power output with the unit of a.u.

The sensitivity and linearity of the liquid level sensor is a change in the output sensitivity to any changes in the smallest input. For linear elements, the determination of the sensitivity is the gradient (Δy / Δx) or the slope of the graph. Sensitivity analysis is a technique to assess the impact of various

IOP Conf. Series: Journal of Physics: Conf. Series **1011** (2018) 012053 doi:10.1088/1742-6596/1011/1/012053

changes of each key variables toward the result. Determination elements sensitivity is obtained by appeared gradient when applied to fitting using origin. The linearity test aimed to determine whether the two variables have a linear correlation. By using origin, it has obtained regression score of the linear fitting result. Simple regression analysis is a correlation between two variables; they are independent variable and the dependent variable. Linearity can be stated in a straight line equation. Table 2 shows a score of sensitivity and linearity.

Table 2. The value of the gradient at each configuration

Configuration	Sensitivity (dBm/cm)	Linearity
A	-0,056	0,98
B	-0,036	0,96
C	-0,018	0,99
D	-0,040	0,98

The negative sign indicates that the line on the graph tends to the left. The best sensitivity is presented in Figure 3 which shows a combination of graphic on each configuration.

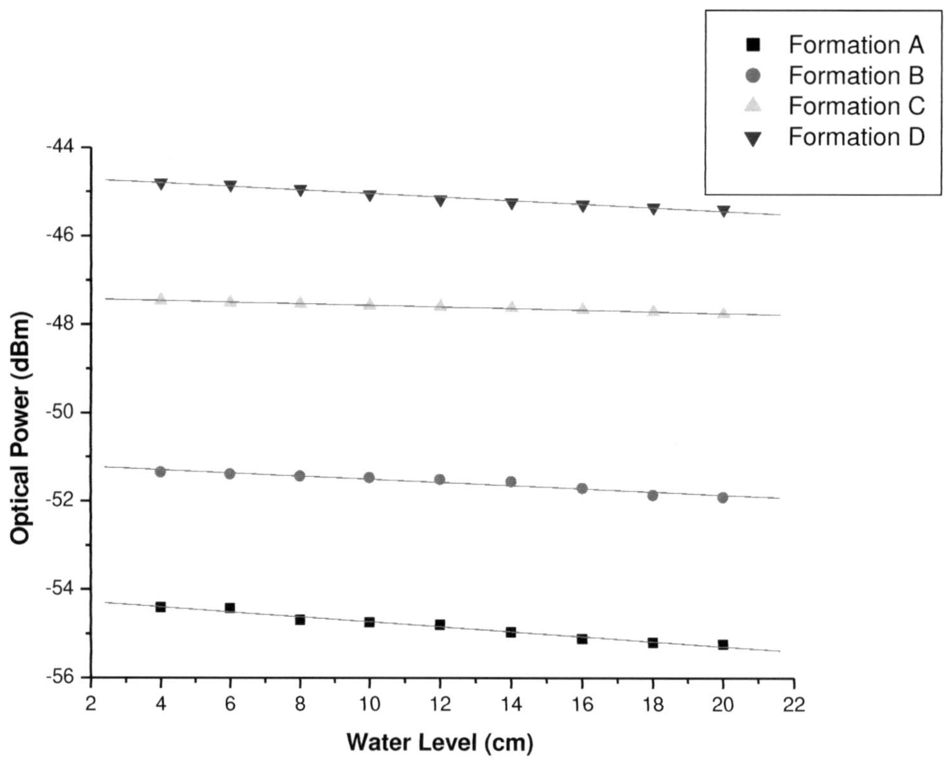

Figure 3. Combined graph of wide variations for each configuration

The best sensitivity obtained in Configuration A with the curvature of 6 cm width with a score of

0.056 dBm/cm. It means when there is a change in the water level, there will be a change in the optical power. Table 2, shows the linearity of the optical fiber in the form of sinusoidal, with the variation of Configuration A, Configuration B, Configuration C and Configuration D with the score close to 1. Best linearity obtained in Configuration C with a linearity coefficient of 0.99 which means that there is a linear correlation between the light intensity and the increasing of the water level.

4. Conclusions

The coating on the cladding on each optical fiber curvature configuration could provide the output of POF because of the differences of reflection on the coating material. Reflection on curved POF occurs when the light hits the boundary of core-cladding and cladding-water. The light intensity score as the output of sinusoid optical fiber in each curvature will be smaller as the water level increases. Best sensitivity obtained in Configuration A while best linearity is Configuration C.

REFERENCES

[1] Montero D S, Vazquez C 2012 *ISRN Sensor Net.*2012:7.
[2] Sameer MC, Nicolas A, Jaeger F. *J.* 2007 *Opt. Eng.* **46** 114401–7.
[3] Antonio-Lopez J E, Sanchez-Mondragon J J, and May-Arrioja DA. 2011. *Opt. Lett.* **36** 3425–7
[4] Lomer M, Quintela A, Lopez-Amo M, Zubia J, Lopez- Higuera JM. 2007. Meas. *Sci. Tech..* **18** 2261–7.
[5] Mamun A, Ahmed N, Ahamed NA, Rahman SAM, Ahmad B, Sundaraj K 2014 *Ind. J. of Sci. and Tech..* **7** 1321–6.
[6] Harith Z, Irawati N, Batumalay M, Rafaie HA, Yun G, Harun SW, Nor RM, Ahmad H. 2015 *Ind. J. of Sci. and Tech..* **8** 1–5.
[7] Montero DS, Vazquez C, Mollers I, Arrue J, Jager D. A. 2009 *Sens.* **9** 6446–55.
[8] Chang X, Li M, Han X 2009 *Front. of Optoelectr. Chi.* **2** 362.
[9] Arrue J, Zubia J, Fuster G, Kalymnios D 1998 IEE *Proc. Optoelectr.* **145** 313–8.
[10] Moraleda AT, Vazquez C, Zubia J, Arrue J 2013 *Sens.* **13** 13076–89.
[11] Chen YC, Chen LW, Chen PC. *Opt Letters.* 2005; 30(3):230–2.
[12] Chen YC, Kuang JH, Chen LW, Chuang HC. 2006 *Opt Letters.* **31** 879–81.
[13] Chen YC, Chen PC, Kuang JH. 2006 *Appl Opt.* **45** 6668–74.
[14] Chen YC 2007*Appl Opt.* **46** 4570–78.
[15] Lu WH, Chen LW, Xie WF, Chen YC 2012 *Sens.* **12** 7485–95.
[16] Kuang JH, Chen PC, Chen YC 2010 *Sens.* **10** 10198–10.
[17] Zhao J, Bao T, Kundu T 2016 *J. of Sens.* **2016** 5.
[18] Annepu VNV, Rao AB 2015 *Ind. J. of Sci. and Tech..* **8** 1–5.
[19] Montalvo J, Vázquez C, Montero S, 2006 *Opt. Exp.* **14** 4601-4610.
[20] Arifin A , Yusran, Miftahuddin, B. Abdullah, T 2017 *AIP Conf. Proc.* **1801**. 1-6

The honey insertion cladding to improve the sensitivity of temperature polymer optical fiber sensor

M Arwani and H Kuswanto

Physics Study Program, Faculty of Mathematics and Natural Sciences, Universitas Negeri Yogyakarta, Jl. Colombo 1 Yogyakarta, Indonesia

Email: herukus61@uny.ac.id

Abstract. The sensitivity of temperature polymer optical fiber (POF) sensor has been studied. Part of cladding (9 cm) was substituted with honey. Polymer cladding was stripped mechanically and the honey inserted into the tube. Plastic gel closed the two end sides of the tubes. The optical power output was detected by Optical Power Meter (OPM). Honey cladding and temperature changing effect to the internal reflection and optical fiber output intensity. Highest output intensity changing at 20°C was shown by optical fiber coated by longan honey as cladding. The range of 10-50° C, as the rise of surroundings temperature, the attenuation was getting smaller. Best sensitivity was fiber with sensing part coated by *Longan* honey. Best linearity was sensing fiber with sensing part coated by *Pracimantoro* honey.

Keywords: honey cladding, sensor, POF

1. Introduction

The optical fiber is a kind of wire that made from thin and smooth silica glass or polymer. It is different from copper wires which are conducting electrical current; the optical fiber transmitted an optical signal from one to another end of the wire. A kind of sensor that using optical fiber as an active component or only as an optical waveguide in sensing mechanism is known as optical fiber sensor [1].

The total internal reflection occurs when the incident light is reflected from an interface at a higher angle than the critical angle. However, its intensity does not abruptly decay to zero at the interface, and a small portion of light penetrates into the reflecting medium. This penetrated electromagnetic field is called the evanescent wave [2]. No energy is lost by the evanescent wave unless there is absorption in the medium in which it is traveling [3] This penetration depth of evanescent wave could be determined by Equation 1.

$$d_p = \frac{\lambda}{2\pi\left[\frac{sin^2\theta}{n^2} - 1\right]^{\frac{1}{2}}} \tag{1}$$

Honey is composed primarily of the sugars glucose and fructose, and third highest of honey's component is water. The viscosity of honey is affected by temperature, moisture content and floral source. Table 1 shows how the viscosity changes the temperature, moisture content, and floral source

IOP Conf. Series: Journal of Physics: Conf. Series **1011** (2018) 012054 doi:10.1088/1742-6596/1011/1/012054

change. The viscosity of honey decreases rapidly when the temperature rises. 1% moisture is equivalent to 3.5°C and affects viscosity [2].

Table 1. (a) Viscosity changes as temperature change. (b) Viscosity changes as water content change [3].

(a)

Temperature (°C)	Viscosity (P)
13.7	600.0
29.0	68.4
39.4	21.4
48.1	10.7
71.1	2.6

(b)

Water Content (%)	Viscosity at 25°C (P)
15.5	138.0
17.1	69.0
18.2	48.1
19.1	34.9
20.2	20.4

This particular research examined the influences of honey cladding and surrounding temperature changing toward the output intensity of optical fiber. Best kind of honey that has highest intensity changes, best sensitivity and linearity also has determined by the data result.

As light source entering the fiber core, it will keep traveling until reach the surface field between core and cladding medium (Figure. 1).

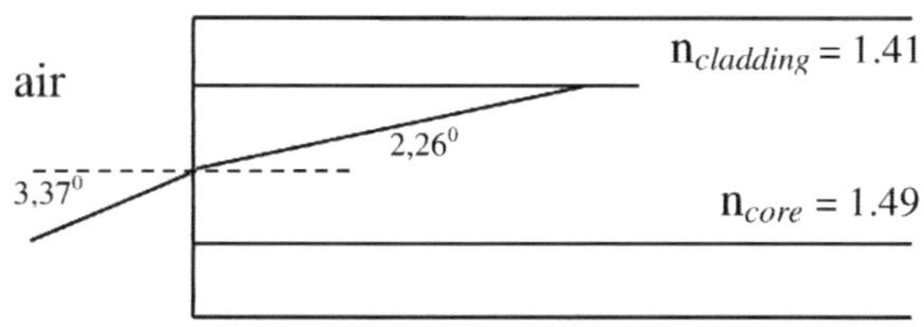

Figure 1. Laser light after entering the fiber core

The light will be reflected back to core only if the angle comes to surface of two medium (θ_i) were greater than critical angle (θ_{crit}). Critical angle (θ_{crit}) of fiber optic was obtained by subtituting the core and cladding refractive index to Equation 2.

$$\theta_{crit} = \arcsin\left(\frac{n_2}{n_1}\right) \quad (2)$$

Plastic optical fibers (POF) have flexibility and large core diameter, which enables efficient connection and coupling resulting in a low-cost system for a local area network. These fibers have attention because of their clear technical advantages over glass fibers [4-6]. POF has a higher numerical aperture than glass optical fibers, and the system also has the advantage from a wider acceptance angle which provides a better light collecting conditions [7].

Fiber optic sensor used a modified light waves. The intensity modification can be used as a parameter. Intensity modification is easy to be processed. Intensity changes occurred due to the leakage caused by the optical fiber subjected to physical treatment as pressure, temperature [8-9] changes, bending [10-12].

In this experiment, there was part of a fiber optic which was had removed the cladding and was replaced with honey which affected by the total internal reflection in fiber optic. Fiber optic that was using the fluorinated polymer as cladding has a critical angle at 71.14° will change its magnitude due to of different cladding refractive index.

This particular research examined the influences of honey cladding and surrounding temperature changing toward the output intensity of optical fiber. Best kind of honey that has highest intensity changes, best sensitivity and linearity also has determined by the data result.

2. Methods

The research was started by determining the refractive index of honey using refractometer at a temperature of 20°C. Refractometer was calibrated first using distilled water, and then some drops of honey were put into the prism or sample part of the refractometer. And then, cover the prism was gently closed until honey spreading all over the prism. The next step was reading the scale on refractometer that has been directed toward a light source by peep on it ocular lens. Brix scale that has read before was converted into a refractive index using Brix to refractive index conversion table.

In this research, fiber optic SH-4001-1.3 was cut off 4 m in length, and some cladding part of those wire was removed by 9 cm. Cladding part of fiber was peeled using acetone, and thus only main part of the fiber that remained. The peeled fiber was surrounded by a small tube which then was injected with honey. The part of fiber which had coated by honey was functioned as a probe and placed in a reservoir filled with water. Water in the reservoir had variated it temperature and functioned as surroundings. A tip of fiber was connected with a laser, and the other tip was connected with OPM. The data results were collected by observing the of the intensity of fiber optic using OPM for increasing temperature of surroundings from 10°C-50°C and decreasing temperature from 50°C-10°C.

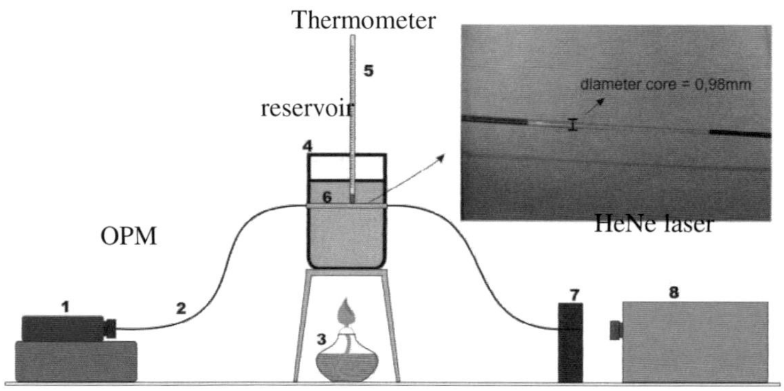

Figure 2. Scheme of experiment set up

3. Result and Discussion

Figure 3 presents the changes in the output intensity of the optical fiber as a function of changes in ambient temperature. Figures 3a, b, c, and d are for sensors with cladding of *madurasa, longan, madujago and pracimantoro* honey respectively. The data are taken at the time of increase and decrease in temperature.

The results show the consistency of the output intensity value when the temperature is raised and lowered. Fig.2 shows that highest attenuation at room temperature of 20°C was shown by fiber optic with *Madurasa* honey as cladding.

IOP Conf. Series: Journal of Physics: Conf. Series **1011** (2018) 012054 doi:10.1088/1742-6596/1011/1/012054

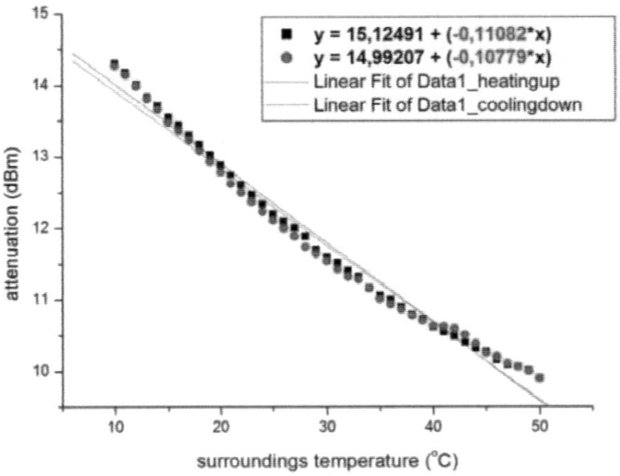

(a)

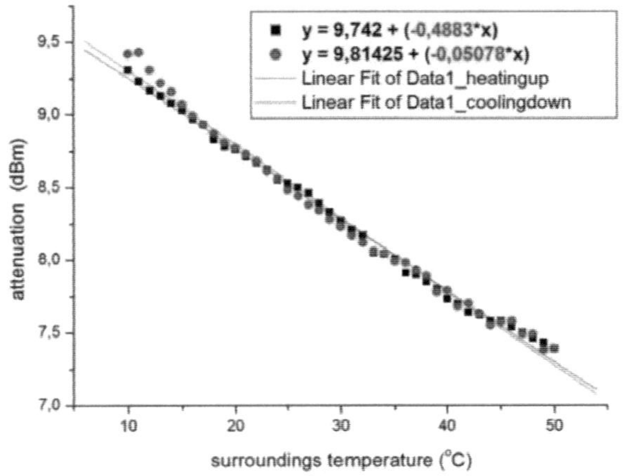

(b)

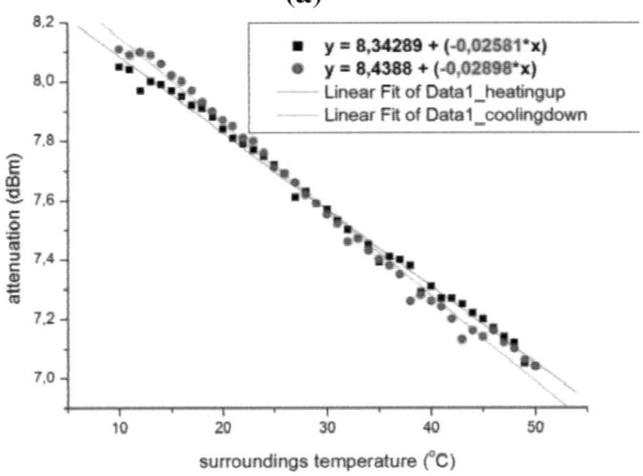

(c)

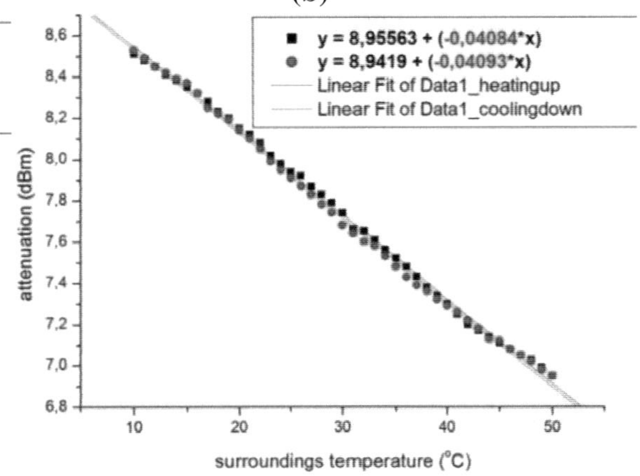

(d)

Figure 3. Output power attenuation of fiber optic with various honey cladding at various surroundings temperature. (a) madurasa honey, (b) longan honey, (c) madujago honey, (d) pracimantoro honey

Sensitivity is the changes in the output for every smallest input changes. For any linear element, sensitivity could be determined by it gradient ($\Delta y/\Delta x$). Table 2 showed the gradient value of graph which was obtained by linear fitting of *Origin* software. If the determination coefficient (R) approaching to 1, then both variables have a linear relationship. At Table 2, minus sign (-) show that the graph is decreasing linearly.

Table 2 shows that from all kind of honey in this experiment, the highest sensitivity was shown by madurasa honey with the gradient of (0,1108 ± 0,002) in the heating process, and -(0.1078 ± 0.003) of cooling down. However, best linearity was shown by pracimantoro honey with the determination coefficient of (-0.99909) in heating process and (0.99283) of cooling down.

To confirm the effect of this type of honey, we also measured its refractive index. Table 3 showed the refractive index of some honey that was used as cladding. Furthermore, the critical angle of fiber optic was compared to the angle of the incoming ray to know whether the ray through the cladding or reflected back to the core.

Table 2 Sensitivity and linearity of honey as fiber optic cladding

Honey	Sensitivity (dBm/°C)		Linearity	
	Heating	Cooling	Heating	Cooling
Madurasa	- (0.1108 ± 0,002)	- (0.1078 ± 0,0030)	- 0.99283	- 0.98956
Longan	- (0.0488 ± 0,0005)	- (0.0507 ± 0,0009)	- 0.99773	- 0.99455
Madujago	- 0.0258	- 0.0289	- 0.99726	- 0.99625
Pracimantoro	- (0.0408 ± 0,0003)	- (0.0409 ± 0,0003)	- 0.99909	- 0.99877

Table 3. Refractive index of some medium and critical angle when it was used as cladding

Medium	n_{core}	$n_{cladding}$	θ_i (°)	NA	θ_{crit} (°)
Fluorinated Polymer	1.49	1.4100	87.4	0.48	71.14
Madurasa honey	1.49	1.4901	87.4	-	-
Longan honey	1.49	1.4849	87.4	0.12	85.26
Madujago honey	1.49	1.4765	87.4	0.20	82.28
Pracimantoro honey	1.49	1.4821	87.4	0.15	84.10

In fiber optic with honey cladding, even the laser rays were reflected back to the core; however, still, there is attenuation of the output. This is because the rays which are traveled through fiber optic interact with honey cladding by the evanescent wave. Evanescent wave was penetrated the cladding medium and traveled along fiber optic. Moreover, with the changes of cladding refractive index, it will affect the penetration depth of evanescent wave through the cladding medium (d_p). Table 4 shows the penetration depth of evanescent wave through varied honey cladding

Table 4. Penetration depth of evanescent wave

Cladding	λ (nm)	$sin^2\theta$	n	Penetration depth (nm)
Fluorinated Polymer	632.8	0.998	0.946	297.77
Longan honey	632.8	0.998	0.997	1452.31
Madujago honey	632.8	0.998	0.991	789.46
Pracimantoro honey	632.8	0.998	0.995	1085.44

4. Conclusion

Honey cladding and temperature changes affect the internal reflection and optical fiber output intensity. Greatest output intensity changes at 20°C was shown by optical fiber coated by *Longan* honey as cladding. The range of 10-50° C, as the rise of surroundings temperature, the attenuation is getting smaller. Best sensitivity was fiber with sensing part coated by *Longan* honey. Best linearity was sensing fiber with sensing part coated by *Pracimantoro* honey.

5. References

[1] Yin S Y 2008 *Fiber Optic Sensors (Second Edition)*. CRC Press.
[2] National Honey Board 2005. Honey - A Reference Guide to Nature's Sweetener. www.honey.com ,Retrieved 14 March 2016.
[3] Supriyatin, 2010 Berk. *Fis.* **13** 119 – 124
[4] Harith Z, Irawati N, Batumalay M, Rafaie HA, Yun G, Harun SW, Nor RM, Ahmad H. 2015 *Ind. J. of Sci. and Tech.*. **8** 1–5.
[5] Montero DS, Vazquez C, Mollers I, Arrue J, Jager D. A. 2009 *Sens.* **9** 6446–55.
[6] Chang X, Li M, Han X 2009 *Front. of* Optoelectr. *Chi.* **2** 362.
[7] Arrue J, Zubia J, Fuster G, Kalymnios D 1998 IEE *Proc. Optoelectr.* **145** 313–8.
[8] Moraleda AT, Vazquez C, Zubia J, Arrue J 2013 *Sens.* **13** 13076–89.
[9] Chen YC, Chen LW, Chen PC. *Opt Letters*. 2005; 30(3):230–2.
[10] Kuang JH, Chen PC, Chen YC 2010 *Sens.* **10** 10198–
[11] Lu WH, Chen LW, Xie WF, Chen YC 2012 *Sens.* **12** 7485–95.
[12] Chen YC 2007 *Appl Opt.* **46** 4570–78. 10.

IOP Conf. Series: Journal of Physics: Conf. Series **1011** (2018) 012055 doi:10.1088/1742-6596/1011/1/012055

Detection of salts in soil using transversely excited atmospheric (TEA) carbon dioxide (CO_2) laser-induced breakdown spectroscopy (LIBS) by the aid of a metal mesh

N Idris[1,*], M Ramli[2], A Khumaeni[3], and K Kurihara[4]

[1]Department of Physics, Faculty of Mathematics and Natural Sciences, Syiah Kuala University, Jl. Syech Abdurrauf No. 3 Darussalam, 23111 Banda Aceh, Aceh, Indonesia
[2]Department of Chemistry, Faculty of Mathematics and Natural Sciences, Syiah Kuala University, Jl. Syech Abdurrauf No. 3 Darussalam, 23111 Banda Aceh, Aceh, Indonesia
[3]Department of Physics, Department of Physics, Faculty of Science and Mathematics, Diponegoro University, Tembalan, Semarang 50275 Indonesia
[4]Department of Physics, Faculty of Education and Regional Studies, University of Fukui 3-9-1, Bunkyo, Fukui 910-8507, Japan
* Email: nasrullah.idris@unsyiah.ac.id

Abstract. In this work, a nickel metal mesh was used to allow a direct detection of salt in soil sample by LIBS utilizing unique characteristics of a TEA CO_2. The metal mesh is placed in the front of the soil sample to prevent the soil sample from blowing off upon focusing the high pulsed laser beam irradiation. LIBS apparatus used in this work is a TEA CO_2 laser operated at wavelength of 10.6 μm with pulse energy and duration of 3J and 200 ns, respectively. The laser beam was focused using a ZnSe lens (f = 200 mm) onto soil sample after passing through the metal mesh. The emission spectrum from the induced plasma was detected using an optical multichannel analyzer (OMA) system consisting of a 0.32-m-focal length spectrograph with a grating of 1200 graves/mm and a 1024-channel photodiode detector array with a micro-channel plate intensifier. The soil sample used is a standard soil and ordinary soil containing several salts such as Ca, Mg at high concentration. The LIBS experiment was carried out at high pressure surrounding gas of 1 atmosphere. It was observed that by the aid of the metal mesh, strong breakdown gas plasma can be produced just after TEA CO_2 laser irradiation on soil sample without significant sample blowing off. It was found that emission lines from salts, Ca (Ca II 393. 3 nm, Ca II 396.3 nm, Ca I 422.5 nm), and also other salts including Mg and Na can clearly be detected with strong emission intensity and narrow spectral width. This result implies that a TEA CO_2 LIBS assisted by the metal mesh (metal mesh method) can be used for direct analysis several salts such as Ca, Mg, and Na in soil sample.

1. Introduction

Due to the physical and chemical properties namely soft matrix and containing many element of organics and inorganics at various levels including major, minor and trace elements, soil is considered as complex sample. Thus, it is basically difficult to make direct analysis even using laser induced breakdown spectroscopy (LIBS). One serious problem in analysing soil sample using LIBS is sample blowing off upon direct focusing a high power laser beam on the soil sample. This problem is actually general problem in LIBS when dealing with powder sample. Therefore, several schemes have been adopted in LIBS to suppress the sample blowing effect to allow direct analysis of powder sample including soil. General approaches are making the powder sample in the form of pellet and employing

binder materials such as potassium bromide, poly(vinyl alcohol), starch, silver and aluminium [1]. A neodymium yttrium aluminum garnet (Nd-YAG) laser with typical wavelength and pulse duration of 1064 nm and 10 ns, respectively, is normally adopted in LIBS as an excitation source.
On the other hand, in previous study it was found that LIBS using a TEA CO_2 shows very promising analytical performance for soil pollution analysis [2]. It is considered due to unique characteristics of the CO_2 laser pulse. The TEA CO_2 laser has a far longer wavelength and pulse duration of 10.6 m and 200 ns, respectively. Moreover, it was also found a unique phenomenon in laser induced breakdown spectroscopy (LIBS) using the TEA CO_2 laser on metal sample at atmospheric pressure, namely strong gas breakdown plasma can be induced when the TEA laser beam focused on the metal sample while leaving the metal sub target without any damage [3-4]. The unique phenomenon has been utilized for various unique applications of the TEA CO_2 LIBS [5]. Recently, the unique metal sub target effect in the TEA CO_2 LIBS was studied for detecting salt in soil sample [6]. In the study, for escaping from blowing off effect, the soil sample was mixed with silicon grease as binder material and then the soil mixed with silicon grease sample is attached on a nickel metal sub target. It was found that the soil sample can be fixed with high stability on the metal sub target without any blowing off effect. It was also found that strong plasma can be induced from the soil sample mixed with silicon grease (SMS) by attaching the SMS to the nickel sub target effect. The plasma produced is very strong, resulting in strong salt emission lines allowing a quantitative analysis, such as Ca, Na, Mg. In this study, the unique phenomenon in the TEA CO_2 LIBS on a metal sample at atmospheric pressure is utilized with different strategy, namely a metal mesh is used for covering the soil sample for avoiding the soil sample from blowing off while at the same time the metal mesh works to enhance plasma generation.

2. Experimental Procedure

The basic experimental setup used in this work is similar to the previous study [6]. However, instead of the metal mesh sub target, the nickel metal mesh was used in this study. The soil sample and the metal mesh configuration are shown in Figure 1. The LIBS apparatus consists of a TEA CO_2 laser (3 J, λ=10.6 μm, Δt= 200 ns) as a plasma excitation source and an optical multichannel analyzer (OMA) system as an optical detector for the plasma emission. The TEA CO_2 laser is a Shibuya SQ-4000 laser actually constructed for laser marking application. The OMA system (ATAGO Macs-320) consists of a spectrograph (0.32-m-focal length) with a grating of 1200 graves/mm, and a photodiode detector array (1024-channel) equipped with a micro-channel plate image intensifier. The spectral resolution of the OMA system is 0.2 nm at 500 nm.

There used in this study is the standard soil sample, the same as used in the previous study [6]. The standard soil sample is produced by Japan Industrial Test Powder (Japan Industrial Standard Test Powders 1 Class 7) with a specified chemical composition as displayed in Table 1. There are two kinds of the nickel metal mesh used in this study, namely metal mesh no. 40 (equivalent to 0.420 mm particle size) and no. 50 (equivalent to 0.297 mm particle size). The soil sample in amount of 3 mg is inserted in a hole with a diameter of 15 mm and a depth of 50 mm made on a metal plate as shown in Figure 1. The soil sample was pressed manually to put in the hole. The soil sample in the metal hole is then covered by the metal mesh. The sample on the metal holder is situated in a circular metal chamber with a diameter of 120 mm. The circular metal chamber is equipped with several windows and valves for filling different kinds of surrounding gases in the chamber and evacuating the gas. This experiment was carried out at high pressure of 1 atmosphere under helium surrounding gas flowing at a rate of 5 L/min.

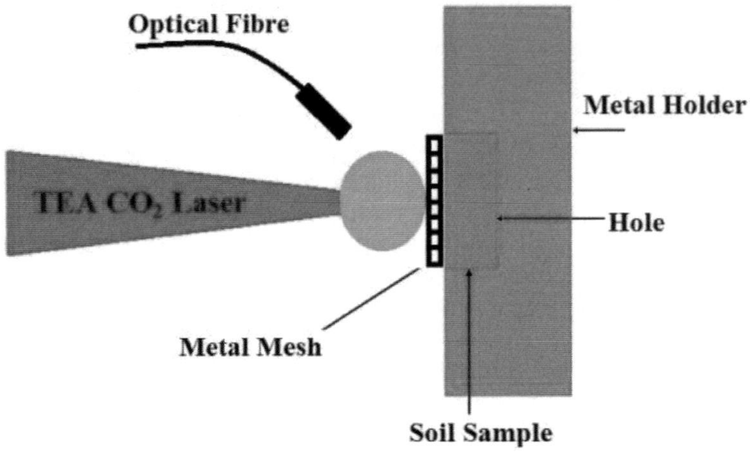

Figure 1 Experimental setup of LIBS apparatus using a TEA CO_2 laser using a metal mesh

In order to produce plasma from the soil sample, the CO_2 laser beam was focused by a zinc selenium (ZnSe) lens (f = 200 mm) through a ZnSe window onto the metal mesh covered-soil sample in the chamber. Small part of the laser beam was intercepted by a photon drag detector (Hamamatsu Photonics, B749) and sent to the OMA system for synchronizing the data acquisition. It is considered that most part of the focused laser beam passes through the mesh reaching the soil sample inducing the soil plasma. The plasma emission was collected and delivered to the spectrograph of the OMA system using an optical fibre. One end of the optical fibre was set at an inclined position of about 45° to the laser beam path, and around 7.5 cm apart from the centre of the plasma, as displayed in Fig. 1, to allow collection the entire plasma emission into the fibre optic core (θ= 27^0 in solid angle). The output optical signal from the other end of the optical fibre was then fed into the entrance slit of the spectrograph. The photodiode array was installed at the exit slit of the spectrograph for detecting the plasma emission spectrum. The OMA system was operated by a personal computer using SpectraView software. The emission spectrum was displayed on the computer monitor and was recorded in the computer memory. The detector was run in gated mode. The gate delay and gate width of the OMA system was set mostly at 1 μs and 50 μs, respectively during the experiment. The emission spectrum acquisition was carried out by 10 accumulations and was repeated 3 times. It was estimated the fluctuation of intensity of the emission lines is approximately 5%.

3. Results and Discussion

It was observed during this experiment that strong, luminous plasma can be produced from the soil sample covered by the nickel metal mesh after focusing the TEA CO_2 laser beam. The plasma generation was made under helium surrounding gas at atmospheric pressure, since preliminary experimental results show the emission intensity is boldly increased in case using helium as surrounding gas. Figure 2 shows emission spectrum taken from the plasma produced on the metal mesh covered-soil sample. The metal mesh used in this experiment was no. 40. This mesh is equivalent to a nominal sieve opening of 0.42 mm. It can clearly be seen the atomic and ionic emission lines due to salt, Ca, namely Ca I 422.67 nm, Ca II 393.36 nm, and Ca II 396.84 nm. The emission lines due to other soil constituents also appears together with the salt emission lines, namely Al atomic emission lines (Al I 394.40 nm and Al I 396.15 nm) and Si ionic emission lines (Si II 412.80 nm). The emission lines feature strong intensity especially Ca lines due to its lower excitation energy. It can also be seen the emission lines exhibit narrow spectral width with relatively low background. It is interesting since the emission spectra were acquired under almost time integrated mode, namely the delay and the width of the gate time of the OMA system were set at 1 μs and 50 μs, respectively as normally adopted in LIBS without a gating function. It is assumed that most part of the laser beam passes through the mesh and arrive on the soil sample, ablating the soil sample. Almost simultaneously, the long pulse duration of the TEA CO_2 laser interacts with the metal mesh inducing strong gas breakdown through initial electron liberation by multiphoton ionization from the metal mesh in the focusing region of the laser beam. This initial electrons promote avalanche or cascade ionization by collision and further absorption of the later part of the pulse energy of the laser light through free-free transition, inverse Bremsstrahlung, resulting strong gas breakdown plasma [3]. In the case of the sample covered by the metal mesh, strong helium gas breakdown plasma produced on the metal mesh, just in front of the soil sample. After the strong helium gas breakdown plasma produced, the laser energy is then reserved in helium metastable excited states. It is considered that the helium metastable excited states play important role in plasma excitation process [7-8]. Thus, the ablated atoms from the sample populated in the helium gas plasma will be excited effectively by the helium metastable excited states. In case of the soil sample covered by the metal mesh, the ablated soil atom moves into the region of helium gas breakdown plasma located just in front of the soil sample, and then are excited by the helium metastable excited states. This results in strong emission intensity with a narrower spectral width and low background since the emission takes place through a delayed excitation mechanism by the helium metastable excited states.

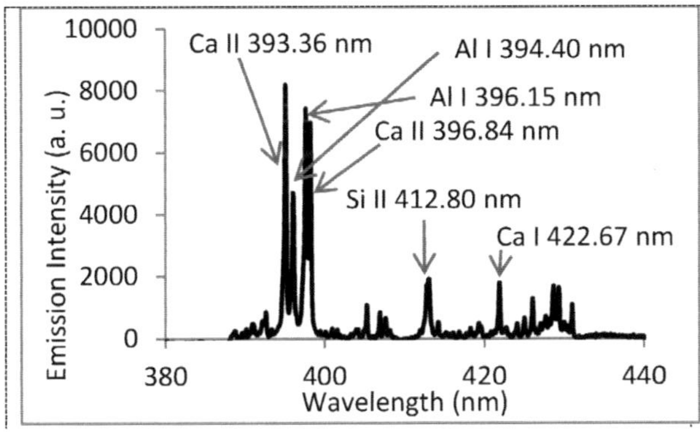

Figure 2. Emission spectrum taken from the plasma induced on the soil sample when the metal mesh no. 40 used.

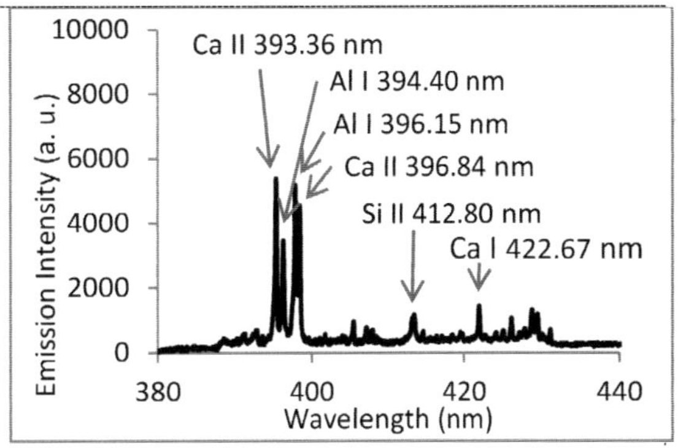

Figure 3. Emission spectrum taken from the plasma produced on the soil sample when the metal mesh no. 50 employed.

Figure 4. Photograph from the top view of the interior situation of the circular metal chamber including the sample on the metal holder and the optical fibre probe taken just after focusing the TEA CO_2 laser on the metal mesh covered-soil sample. The metal mesh used in this case is no. 40.

When using the metal mesh no. 40, the soil sample blowing off still occurs significantly as shown in Figure 4. As it can be seen in Fig. 4, the soil sample was scattered on the interior bottom of the sample chamber just after focusing the laser beam onto the soil sample covered by the metal mesh. We then tried to use the metal mesh no. 50 which its nominal sieve opening is 0.297, about 25% smaller than that of the metal mesh no. 40. It is expected the soil sample blowing off effect will be suppressed further while keeping an optimal laser pulse energy passing through the metal mesh for producing strong plasma. Figure 3 shows emission spectrum taken from the plasma produced on the soil sample when the metal mesh no. 50 employed. It can be observed in Figure 3 that the emission lines from Ca and other constituents of the soil sample appears with similar features to the spectrum detected when using the metal mesh no. 40. The difference is only in the emission intensity, which is around 25% lower than that of using the metal mesh no. 50. This is consistent with the difference in the nominal sieve opening between the two metal meshes, affecting to the portion of the laser pulse beam passed through the metal mesh. It is considered that in case of using the metal mesh no. 50, the laser pulse energy passed through the mesh is about 25% lower than that of using the metal mesh no. 40. Nevertheless, as shown in Figure 3 although the laser energy passes through the metal mesh no. 50 is lower, the produced plasma exhibits a useful emission spectrum with favorable features for spectrochemical analysis. The emission lines from other salts, namely Mg and Na can also clearly be detected at different wavelength measurement ranges with similar emission spectral line features. Thus, the TEA CO_2 LIBS employing the metal mesh method can be used for carrying out quantitative analysis of salt in soil sample.

4. Conclusions

The nickel metal mesh was employed to allow direct analysis of soil sample using a TEA CO_2 laser. The blowing off effect can be suppressed substantially by using the metal mesh for covering the soil sample. It is found that by applying the metal mesh on the soil sample, strong, luminous plasma can be produced, resulting in strong emission intensity of salt, Ca (ionic lines: Ca II 393.36 nm, and Ca II 396.84 nm, and atomic line: Ca I 422.67 nm) and other constituents. The emission lines feature a narrow spectral width and low background, implying the essential role of the delayed excitation through helium metastable excited states produced in the strong helium gas breakdown plasma. The dimension of the metal mesh must be carefully selected to obtain optimal condition namely it can passes through sufficient laser pulse energy to the soil sample surface while it is able to prevent the soil sample blowing off. This TEA CO_2 laser induced breakdown spectroscopy utilizing the unique metal mesh can be used for conducting a direct analysis on soil sample. A quantitative analysis is being conducted and will be reported elsewhere.

Acknowledgments
The authors (N. I., and M. R.) thank to Research and Society Service Directorate (DP2M), Directorate General of the Research Strengthening and Developing, Ministry of Research, Technology and Higher Education, Government of Republic Indonesia, for research funding through international collaboration and publication research grant at Institute for Research and Community Service, Syiah Kuala University, Aceh, Indonesia 2016 with contract grant no. 6/UN11.2/PP/SP3/2016 and no. 6/UN11.2/PP/SP3/2017.

References

[1] Gondal M A, Hussain T, Yamani Z H, Baig M A 2007 *Talanta* **72** (2) 642-9
[2] Idris N, Kagawa K, Sakan F, Tsuyuki K and Miura S 2007 *Appl. Spectrosc.* **61** (12) 1344-51
[3] Ramli M, Idris N, Fukumoto K, Niki H, Sakan F, Maruyama T, Kurniawan K H, Lie T J, and Kagawa K 2007 *Spectrochimica Acta Part B: Atom. Spect.* **62** (12) 1379-89
[4] Lee Y I, Idris N, Kurniawan K H, Lie T J, Tsuyuki K, Miura S, and Kagawa K 2007 *Curr. Appl. Phys.* 7(5) 540-46
[5] Khumaeni A, Lie Z S, Lee Y I, Kurihara K, Kagawa K, and Niki H 2011 *Appl. Spectro.* **65** 236-41
[6] Idris N, Ramli M, and Kurihara K 2016 Detection of salt in soil by employing the unique sub-target effect in a pulsed carbon dioxide (CO_2) laser-induced breakdown spectroscopy *Proc. International Seminar on Sensors, Instrumentation, Measurement and Metrology* (Malang, Indonesia, 10-11 August 2016) (Institute of Electrical and Electronics Engineers, IEEE) pp. 79-83
[7] Suliyanti M M, Hidayah A N, Pardede M, Jobiliong E, Abdulmadjid S N, Idris N, Ramli M, Lie T J, Hedwig R, Tjia M O, Kurniawan K H, Lie Z S, Niki H, and Kagawa K *Spectrochimica Acta Part B: Atom. Spec.* **62** (12) 1379-89
[8] Bišćan M and Milošević S 2013 *Spectrochimica Acta Part B: Atom. Spec.* **81** 20-25

The Performance of CO_2 Laser Photoacoustic Spectrometer In Concentration Acetone Detection As Biomarker For Diabetes Mellitus Type 2

F H Tyas, J G Nikita, D K Apriyanto, Mitrayana* and M N Amin
Department of Physics, Universitas Gadjah Mada, Yogyakarta, INDONESIA

*Corresponding author: mitrayana@ugm.ac.id

Abstract. Breath analysis is useful for the diagnosis of human diseases and monitoring of metabolic status. However, because of the low concentrations and the large numbers of compounds in the breath, the breath analysis requires highly sensitive and highly selective instruments to identify and determine the concentrations of certain biomarkers [1]. Various methods developed over the past 20 years to detect biomarker gases [2]. CO_2 laser photoacoustic spectroscopy offers a sensitive technique for the detection and monitoring of gas footprints at low concentrations [3]. The performance of photoacoustic spectrometer (PAS) examined with intracavity configuration. In this research, the highest observed intracavity power was (49,96 ± 0,02) W for active medium gas composition He: N_2: CO_2 at 30:50:50. The highest laser absorption line for standard acetone gas set at 10P20, and the lowest detection limit set at (30 ± 4) ppb. For application purposes, the photoacoustic spectrometer was used to measure the concentration of acetone gas in exhaled gases from a group of patients with type 2 diabetes mellitus and a group of healthy volunteers. Exhaled gas sampling method took manually, and the measurement result was examined using multicomponent analysis. The measurement showed that the highest acetone gas concentration for type 2 diabetes mellitus patients was (162 ± 3) × 10 ppb and the lowest one was (101 ± 3) × 10 ppb. Furthermore, for healthy volunteers, the highest acetone gas concentration was (85 ± 3) × 10 ppb and the lowest one was (15 ± 3) × 10 ppb.
Keyword: CO_2 laser photoacoustic spectrometer, acetone, diabetes mellitus, multicomponent analysis

1. Introduction

Breath analysis is useful for the diagnosis of human diseases and monitoring of metabolic status. However, the breath analysis requires highly sensitive and highly selective instruments to identify and determine the concentrations of certain biomarkers [1]. CO_2 laser photoacoustic spectroscopy offers a sensitive technique for the detection and monitoring of gas footprints at low concentrations. Because it ensures high output power in the wavelength region of 9 – 11μm where more than 250 gas molecules exhibit strong absorption of bands in atmospheric, industrial, medical, military and science environments [3]. CO_2 lasers have extensive tunability and high laser power that allows measurement of mixtures of several gases at concentrations (sub) parts per billion volume, 1 ppb = 1: 10^9 [4].

Results from Basic Health Research Reports stated that the prevalence of DM patients in Indonesia in 2013 (2.1%) is increased compared to 2007 (1.1%). The highest prevalence of DM found in Yogyakarta province with 2.6% prevalence followed by Jakarta with 2.5% and Sulawesi 2.4% [5]. This study examines the measurement of acetone gas concentration as a biological marker (biomarker) of type 2 diabetes mellitus through human breath. Type 2 diabetes mellitus is selected because, in adult age range, type 2 diabetes accounts for about 90% to 95% of all diagnosed diabetic cases [6]. The photoacoustic spectroscopy method of the CO_2 laser was applied in this study to measure the concentration of acetone gas (C3H6O) in the breath sample. This study conducted on healthy people volunteers and patients with type 2 diabetes mellitus who has a range of ages 25-80 years, amounting to 31 individuals each.

2. Experiment details

The research materials used in this research were exhalation gas samples of DSH patients and healthy volunteers with age range 25 - 80 years which consist of 31 people in each group. CO_2 gas, N_2, and He as the active medium of the CO_2 laser. Distilled water as a coolant for the CO_2 laser tube. N_2 gas to clean the FA cells and determine the background signal. The standard gas (ethylene, ammonia, and acetone) as a calibrator. KOH, and $CaCl_2$ to fill the scrubber tube and 75% alcohol for cleaning the hose covering device on the sample bag.

Equipment calibration was also carried out on three standard gases of ethylene, acetone and ammonia gas to obtain the relationship matrix of gas concentration to the selected laser line, 10P14 for ethylene, 10P20 for acetone and 10R14 for ammonia. In this research, CO_2 laser power optimization is done to get optimal laser output power. Some of the factors that affect the optimization of are: the ratio of the active ingredient laser (CO_2 gas, N_2, and He), the discharge arrangements in the form of variations of the power supply and voltage current and grating position settings.

In this study, every patient and healthy person volunteers are guided to take a deep breath; then breath is retained and then streamed to a hose connected to scrubbers and sample bag. The scrubber contains KOH and $CaCl_2$ which serves to filter CO_2 and H_2O in the volunteers' exhaled gas. Replacement of KOH and $CaCl_2$ on the scrubber is done periodically as the powders begin to appear solid and the volunteers feel heavy in blowing them. Hoses used for exhaling the gas washed with alcohol before being used by volunteers and hoses replaced for each volunteer. Sample bag that has been used then cleaned by using a vacuum pump and nitrogen gas. After the sample bag cleaned, it can reuse for sampling the next volunteer breath.

The process of measuring breath gas samples of patients with type 2 diabetes in this study was carried out by flowing gas into the sample bag with the flow pump into the photoacoustic cell. A photoacoustic cell is a place of interaction between breath gas with CO_2 laser beam radiation in which there is a microphone as a detector of the photoacoustic signal generated from the interaction. In this study, gas samples in photoacoustic cells were measured under conditions of gas flowing systems and recorded on the three laser absorption lines: ethylene (10P14), acetone (10P20) and ammonia (10R14) using LabVIEW software.

Results of signal and power recorded on processed using software LabView origin to obtain the normalized signal. Then put on a normalized signal matrix equation as follows :

$$\begin{bmatrix} \left(S/I\right)_1 \\ \left(S/I\right)_2 \\ \left(S/I\right)_3 \end{bmatrix} = \begin{bmatrix} k_{11} & k_{12} & k_{13} \\ k_{21} & k_{22} & k_{23} \\ k_{31} & k_{32} & k_{33} \end{bmatrix} \begin{bmatrix} C_1 \\ C_2 \\ C_3 \end{bmatrix}, \tag{1}$$

with k_{ij} searched through each gradient of gas standards used are ethylene, acetone, and ammonia analyzed on several variations of laser absorption lines are 10P14, 10P14, 10R14 in various concentration.

After each sample of breath was recorded and resulted in a normalized signal recording, data analysis was performed to calculate the amount of biomarker gas concentration in each sample by the inverse multicomponent matrix. In each new bag sample measurement, cell cleansing done by using nitrogen gas that passed on photoacoustic cells.

3. Result and discussion

The composition ratio of the laser's active medium gas CO_2 (He: N_2: CO_2) in this study was 30: 50: 50 with laser's operating voltage ranging from 8.18 to 9.13 kV and the current ranged from 11.15 to 13.53 mA. The adjustment of the composition of active medium gas generates a maximum intracavity power of (49.96 ± 0,02) W and producing two groups of CO_2 laser 10R and 10P.

The result of CO_2 laser scanning generates the laser line output spectrum with the strongest FA signal of acetone gas is found at line 10P20. The spectrum patterns of acetone absorption have a distinct characteristic than other gas spectrum patterns such as ethylene gases which have the strongest signals only on a single 10P14 laser line and ammonia gas on a single 10R14 laser line. Acetone gas absorption lies in almost all CO_2 laser lines, although it has a small signal.

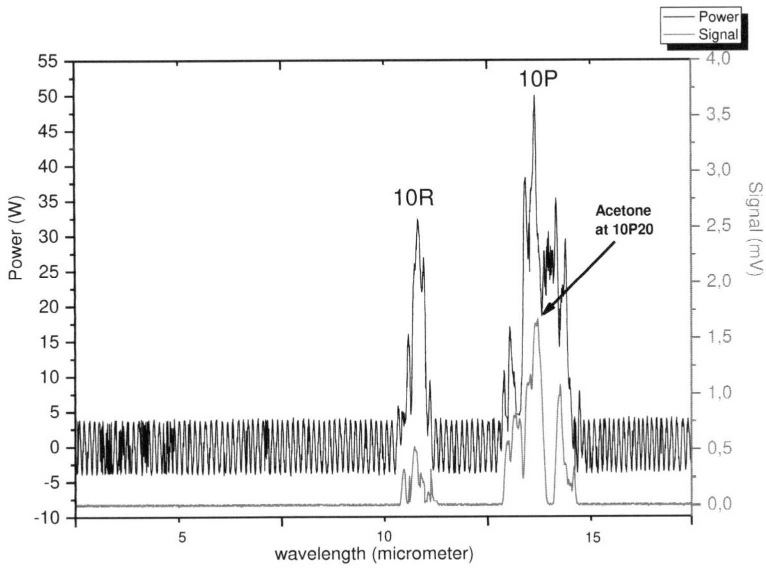

Figure 1. The strongest FA signal of acetone gas

In an axial flowing-type CO_2 laser, the turbulence in the active medium gas stream (He, N_2 and CO_2) simply affects the power output during the sample recording process. However, this problem solved by normalizing the photoacoustic signal to its laser power. Linearity curves for several variations of ammonia and ethylene gas concentrations analyzed by recording the photoacoustic signals on the three strongest absorption lines for ethylene, acetone and ammonia gas, laser lines 10P14, 10P20 and 10R14. The gradient of each linearity curve that obtained is then used as a multi-component matrix of analysis as shown in (2) :

$$\begin{bmatrix} \left(S/I\right)_1 \\ \left(S/I\right)_2 \\ \left(S/I\right)_3 \end{bmatrix} = \left[\begin{bmatrix} 0{,}023 & 0{,}002 & 0{,}0015 \\ 0{,}0114 & 0{,}0179 & 0{,}0059 \\ 0{,}001 & 0{,}0002 & 0{,}0256 \end{bmatrix}\right] \begin{bmatrix} C_1 \\ C_2 \\ C_3 \end{bmatrix}. \tag{2}$$

Then to find the concentration of the sample with a known amplitude of the photoacoustic signal, matrix inversion is applied to the equation (2) and resulted respectively being the concentration of ethylene gas, acetone gas and ammonia gas as shown in (3) :

$$\begin{bmatrix} C_1 \\ C_2 \\ C_3 \end{bmatrix} = \begin{bmatrix} 46.088 & -5.132 & -1.517 \\ -28.833 & 59.221 & -11.959 \\ -1.575 & -0.262 & 39.215 \end{bmatrix} \begin{bmatrix} \left(S/I\right)_1 \\ \left(S/I\right)_2 \\ \left(S/I\right)_3 \end{bmatrix}. \tag{3}$$

With C_1, C_2, and C_3 respectively being the concentration of ethylene gas, the concentration of acetone gas and the concentration of ammonia gas. While $(S/I)_1$, $(S/I)_2$ and $(S/I)_3$ are respectively normalized photoacoustic signals at line 10P14, 10P20 and 10R14.

Figure 2 shows that the concentration of acetone possessed by each DM type 2 patient is higher than that of healthy volunteers. In this research, the highest acetone gas concentration for diabetes mellitus type 2 patients was $(162 \pm 3) \times 10$ ppb and the lowest one was $(101 \pm 3) \times 10$ ppb. Furthermore, for healthy volunteers, the highest acetone gas concentration was $(85 \pm 3) \times 10$ ppb and the lowest one was $(15 \pm 3) \times 10$ ppb. The average value of the concentration of acetone gas of type 2 DM patients$(118 \pm 3) \times 10$ ppb and in healthy volunteers the average concentration of acetone is $(62 \pm 3) \times 10$ ppb.

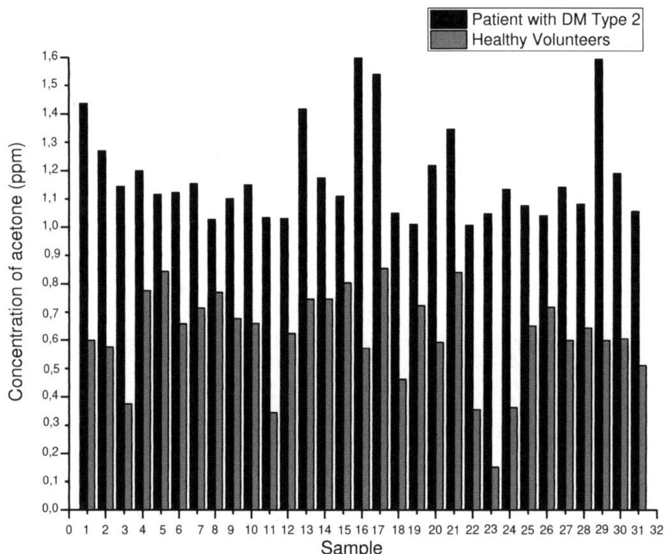

Figure 2. The acetone Gas Concentration in Healthy Volunteers' Breath and Patient with DM Type 2's Breath Measured by Photoacoustic

One limitation of some studies of measurement of acetone breath in DM patients is that the subjects always include controlled diabetes patients whereas some more information obtained when using uncontrolled diabetes patients [7]. The subjects of diabetes mellitus patients in this study are mostly also patients who fall into the category of diabetic patients controlled because they have received education from doctors and nurses at Dr. Sardjito to control blood sugar through several treatments/therapeutics so that certainly little effect on the results of this study.

4. Conclusion

SFA Optimization The CO_2 laser produces two groups of 10R and 10P with maximum intracavity $(49{,}96 \pm 0{,}02)$ W and a quality factor on the 10P20 was (28 ± 5). The SFA CO_2 laser demonstrates good performance in the measurement of voluntary volunteer gas DM type 2 patients and healthy person volunteers with the lowest detection limit on acetone line (10P20) was (30 ± 4) ppb. The highest acetone gas concentration for diabetes mellitus type 2 patients was $(162 \pm 3) \times 10$ ppb and the lowest one was $(101 \pm 3) \times 10$ ppb. Furthermore, for healthy volunteers, the highest acetone gas concentration was $(85 \pm 3) \times 10$ ppb and the lowest one was $(15 \pm 3) \times 10$ ppb. The average value of the concentration of acetone gas of type 2 DM patients $(118 \pm 3) \times 10$ ppb and in healthy volunteers the average concentration of acetone amounted to $(62 \pm 3) \times 10$ ppb).

References

[1] Wang C and Sahay P 2009 *Sensors* 8230-62

[2] Wojtas J, Bielecki Z, Stacewicz T, Mikołajczyk J, and Nowakowski M 2012 *Opto-Electronics Review* **20** 26-39

[3] Dumitras D C, Ana Maria B and Popa C 2012 *CO_2 Laser Photoacoustic Spectroscopy: I. Principles* (InTech Open Acces Publisher Croatia)

[4] Bijnen F G C, Harren F J M, Hanckstein JH P and Reuss 1996 *Applied Optics* 35 **27** 535-68.

[5] Trihono 2013 *Laporan Hasil Riset Kesehatan Dasar Riskesdas 2013* (Jakarta).

[6] Centers for Disease Control and Prevention 2014 *National Diabetes Statistics Report: Estimates of Diabetes and Its Burden in the United States* (Atlanta GA: U.S. Department of Health and Human Services)

[7] Wang Z and Wang C 2013 *Journal of Breath Res*earch **7** 18

Preliminary panoramic study of river calm muscle using neodymium:yttrium-aluminum-garnet (Nd: YAG) laser-induced breakdown spectroscopy (LIBS)

N Idris[1,*], K Lahna[1], T N Usmawanda[1], Herman[1], M Ramli[2], R Hedwig[3], A M Marpaung[4], K H Kurniawan[5]

[1]Department of Physics, Faculty of Mathematics and Natural Sciences, Syiah Kuala University, Jl. Syech Abdurrauf No. 3 Darussalam, 23111 Banda Aceh, Aceh, Indonesia
[2]Department of Chemistry, Faculty of Mathematics and Natural Sciences, Syiah Kuala University, Jl. Syech Abdurrauf No. 3 Darussalam, 23111 Banda Aceh, Aceh, Indonesia
[3]Department of Computer Engineering, Bina Nusantara University, 9 K.H. Syahdan, Jakarta 14810, Indonesia
[4]Department of Physics, Faculty of Mathematics and Natural Sciences, Jakarta State University, 10 Rawamangun, Jakarta, Indonesia
[5]Research Center of Maju Makmur Mandiri Foundation, 40/80 Srengseng Raya, Jakarta 11630, Indonesia
* Email: nasrullah.idris@unsyiah.ac.id

Abstract. A wide coverage spectral investigation on the muscle of river calm sample has been carried out using laser-induced breakdown spectroscopy for examining the overall profile of the emission spectra from the produced plasma. The basic apparatus of LIBS system used is a Nd-YAG laser and wide coverage optical multichannel analyzer (OMA) system. The river clam samples used is collected from Panga River in Aceh Jaya Regency, Aceh, Indonesia up streaming in a mountain of Gunong Ujeun, which is used as a location of the intensive traditional mining activity. Assuming that heavy metal accumulated in the clam muscle, LIBS experiments were carried out on the muscle of the calm. The sample used was fresh muscle sliced and attached to a copper plate. Plasma was generated by focusing the laser beam on the sample surface under air surrounding gas at 1 atmosphere. It is found that there are only major elements of host organic, namely C, H, O, N and the minor element of salts can be detected from fresh the clam sample when using a high pulse laser energy under air surrounding at high pressure of 1 atmosphere. There is no emission lines from any metal can be detected. Several experimental parameters were explored to study the panoramic dynamic of the emission spectra. It is found that the lower energy and the lower pressure is better for obtaining better emission spectra showing the possibility for determination of the analyte.

1. Introduction

Traditional mining activities exploiting gold, gemstones and other minerals were suspected to cause heavy metal pollution in the environment around the mining location since the heavy metals including mercury used the mining activities. It is blamed that the traditional miners have no required skills for handling the heavy metals and the post use management. If the mining tailing is flown directly into the environment, it possibly encounters into food chain resulting in a lethal effect on human beings. The locations of the mining activity are typically in the mountain region. Thus rivers up streaming to the mining location are potentially polluted by the heavy metals from the mining tailing, polluting the biotas in the river. As river calm lives on the bottom of the river and obtaining food from surrounding organic matter supposed to contain heavy metal, it causes an accumulation of the heavy metals in the clam organs. Thus the river calm is considered as a good precursor of heavy metal pollution due to the mining tailing and other anthropogenic sources [1]. There are several laboratory methods for detecting and measuring the heavy metal pollution in biological samples. However, the methods mostly use a wet chemical with tiring sample preparation and environmentally unfriendly residue. In this study, due to its excellence over the conventional analytical tools, LIBS technique as emerging the analytical tool is used to analyze river clam sample, especially clam muscle [2-3]. However, calm muscle is complex sample due to the soft matrix and a large content of water. Therefore, the preliminary panoramic study of LIBS over wide wavelength range was made on the calm muscle sample. The panoramic LIBS study is carried out to the obtained general profile of the plasma emission and the possibility of conducting a qualitative and quantitative analysis of major, minor and trace elements including heavy metals in the sample [4].

2. Experimental Procedure

LIBS apparatus used in this experiment is a Nd-YAG laser (Quanta-Ray; LAB SERIES; λ=1.064 nm; 500 mJ, 8 ns, 10 Hz) as plasma excitation source, a focusing lens (f= +150 mm), and an optical multichannel analyzer (OMA) system. The OMA system consisted of an Echelle spectrograph and intensified charged couple device (ICCD) detector (Andor Mechelle ME5000). The OMA system covers simultaneously wide wavelength in one acquisition ranging from 200 nm to 975 nm with the focal length of 195 mm, the slit of F/7, and the spectral resolution ($\lambda/\Delta\lambda$) of 1 (corresponding to 3 pixels at full width at half maximum, FWHM). The OMA system accuracy is better than ± 0.05 nm. The pulse energy of the laser can be varied by controlling the input voltage of the pumping laser diode ranging from 0 till 500 mJ. The laser pulse energy can further be reduced using a set of transmission filters.

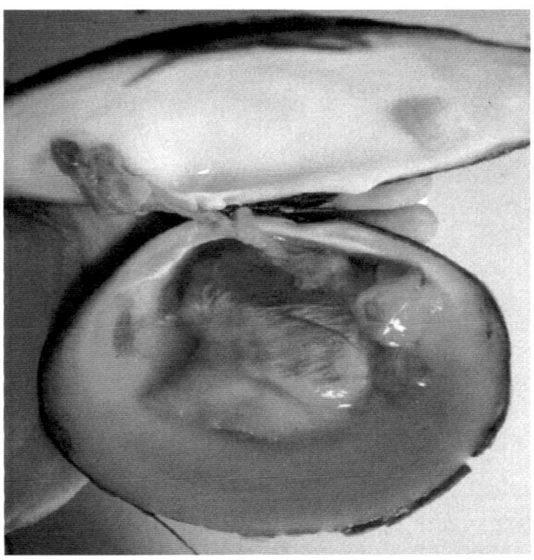

Figure 1. Photograph of the fresh muscle sample of the river clam

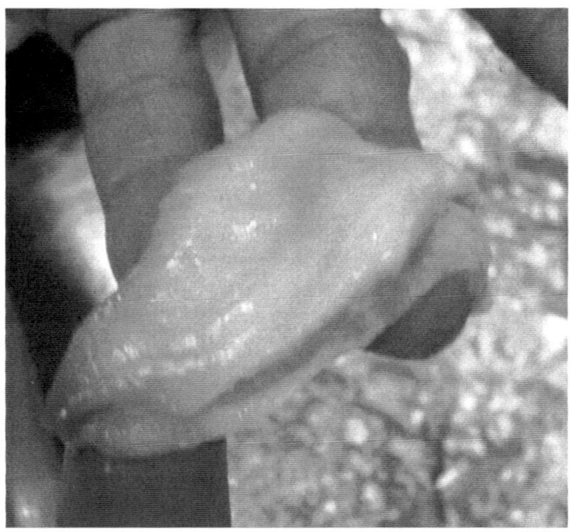

Figure 2. Photograph of a slice the fresh muscle of the river clam sample

The sample used in this study was collected from Panga River which up streams in Gunong Ujeun, a location used to intensive traditional gold mining activities. LIBS experiment was carried out on the fresh muscle as shown in Figure 2(b). The fresh muscle was cut into a slice with a dimension of 2 cm x 2 cm and a thickness of about 2 mm. The fresh muscle slice was then attached to a copper plate, as shown in Figure 2(b), and placed in the circular metal chamber. The metal chamber equipped with several windows can be filled with different gas kind and evacuated to keep the sample surrounding gas at the desired pressure. In this preliminary experiment, LIBS experiment was conducted under air surrounding gas.

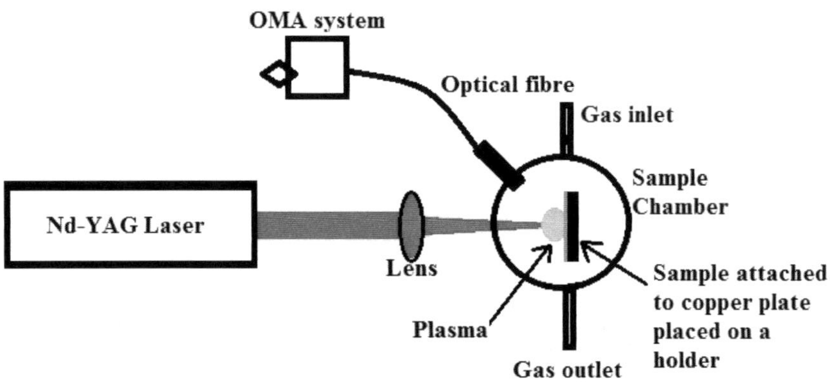

Figure 3. The experimental setup used this study.

For generating plasma from the calm sample, the Nd-YAG laser beam was focused onto the sample surface in the chamber. The emission of the produced plasma was collected using an optical fiber probe. One end of the optical fiber probe was set at an inclined position of about 45° to the laser beam path through a window of the metal chamber. The optical fiber probe is about 5 cm apart from the center of the plasma. The optical signal collected by the optical fiber probe was then fed into the entrance slit of the spectrograph of the OMA system equipped with the ICCD camera. The OMA system operation is controlled by a personal computer using AndorSOLIS software. The detected emission spectrum was displayed on the monitor and was recorded in the computer memory. The OMA system was operated under gating mode by setting the delay and width of the gate function. The acquisition of the emission spectrum was carried out by 5 accumulations and was repeated 2 times.

3. Results and Discussion

Figure 4 exhibits the emission spectrum detected from plasma produced after focusing Nd-YAG laser beam onto the fresh calm sample of air surrounding gas at an atmospheric pressure as typically adopted in LIBS technique. The laser pulse energy was 93 mJ. The use of a wide range coverage OMA system allows investigating the spectral emission lines in a wide wavelength region, starting from 200 nm till 925 nm. It can be seen in the figure that the emission spectra are dominated by many atomic emission lines due to host element of the organic sample namely carbon (C), hydrogen (H), oxygen (O) and nitrogen (N) appear with strong emission intensity. The emission lines are C I 247.9 nm, H I 656.32 nm, N I 742.36 nm, N I 744.22 nm, N I 746.83 nm and N I 818.80 nm, N I 821.63 nm, N I 822.31 nm, N I 824.23 nm as well as N I 868.02 nm, and N I 868.23 nm. The two nitrogen lines of N I 868.02 nm, and N I 868.23 nm are overlapping. Not only atomic emission lines but the ionic emission line due to nitrogen also occur namely N II 500.59 nm along with many molecular bands due to surrounding gas breakdown because high laser pulse energy was employed at high pressures of the surrounding gas. Similarly, there are many atomic emission lines due to oxygen arise including the typical three atomic lines of O I 777.19 nm, O I 777.41 nm and O I 777.54 nm. In the emission spectra, the three lines are overlapped. We also can observe other emission lines of oxygen, O I 844.62 nm and O I 844.63 nm, which are also overlap. This is because the resolution of the OMA system is not sufficient for resolving the emission lines. Together with the atomic and ionic emission lines and molecular emission band due to the host element of the organic sample and the air surrounding gas, the emission spectral lines from several salts can be seen with very high emission intensity. The emission lines due to salts are magnesium (Mg II 279.55 nm and Mg II 280.27 nm) and calcium (Ca II 393.39 nm and Ca II 396.88 nm). This spectrum implies that all host elements of the organic sample can be detected with strong intensity including atomic lines, ionic lines and molecular bands from the plasma produced on the fresh calm muscle.

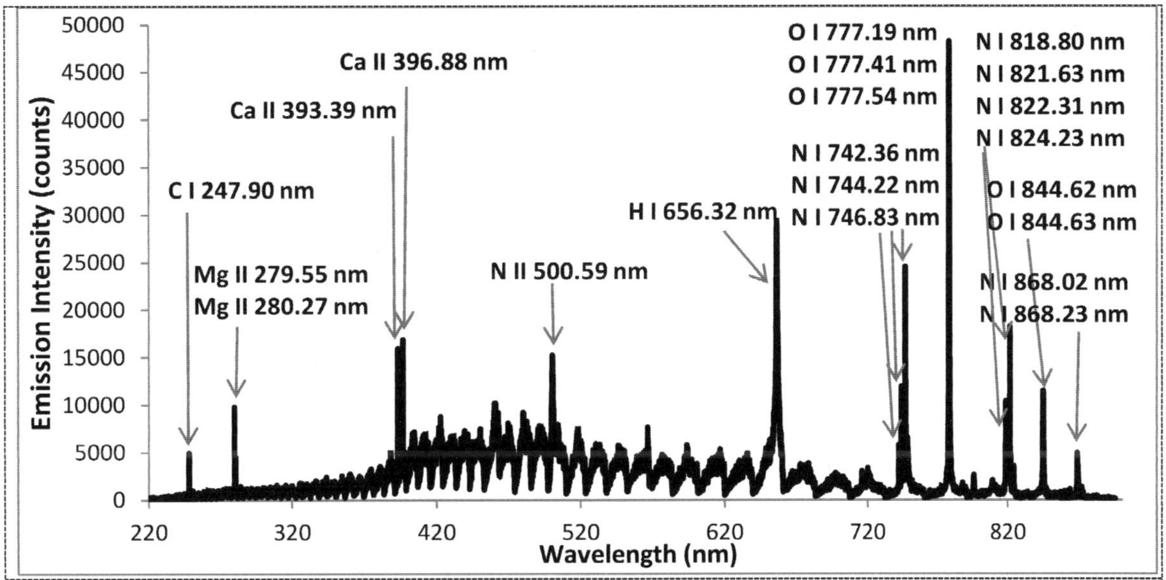

Figure 4. The emission spectrum is taken from plasma produced on the fresh clam muscle by focusing Nd-YAG laser beam of 93 mJ under air surrounding gas at 1 atmosphere. The clam sample was collected from Panga river in Sineubok Padang village, Panga District, Aceh Jaya Regency, Aceh, Indonesia.

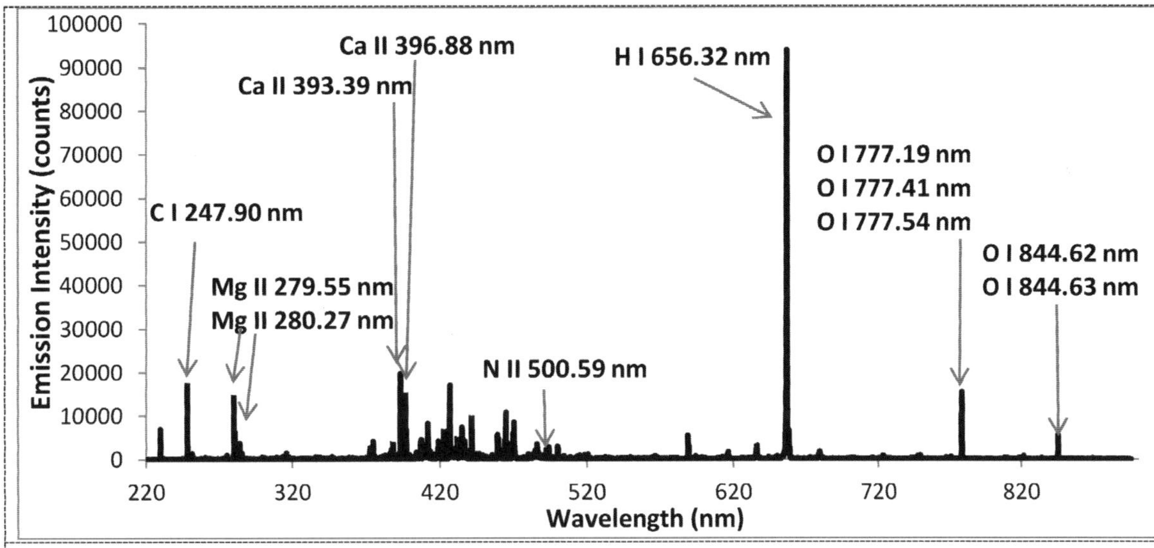

Figure 5. The emission spectrum is taken from plasma produced on the fresh clam muscle by focusing Nd-YAG laser beam of 60 mJ under air surrounding gas at a pressure of about 5 Torr.

In general as mentioned above, the emission spectra detected from plasma produced on the fresh calm muscle sample using the Nd-YAG laser beam with a pulse of energy of 93 mJ under air surrounding gas at atmospheric pressure is dominated mainly by the emission lines and the molecular band due to the host constituents of the sample and the surrounding gas. This is probably due to the high energy of the laser beam induces high pressure and high-density plasma not only from the sample but also from strong gas breakdown of the air surrounding gas. This results in strong emission of atomic and ionic emission lines as well as the molecular band. This is indicated by the appearance of strong ionic emission lines and molecular bands due to nitrogen since the air consists of more 78% nitrogen. In this

experimental condition of high laser pulse energy and high pressure of surrounding gas, there are no any emission lines from metals especially heavy metal can be detected. There are two possible explanations; the calm muscle sample contains no metals including heavy metal, or it might contain metal at very low concentration. Probably, the low concentration of metal results in weak emission intensity but overwhelmed by the numerous strong emission spectral lines. Strong emission spectral lines from the host constituent of the organic sample are induced by the high pulse energy of Nd-YAG laser at high pressure of surrounding gas. Thus, in next experiment, the plasma was produced using lower laser pulse energy of 60 mJ under air surrounding gas at 1 atmospheric pressure. However, the resulted emission spectrum, in this case, displayed very similar features to that of using the high laser pulse energy of 93 mJ. Therefore, the experiment then was continued using lower pulse energy of 60 mJ but reducing the surrounding gas pressure to around 5 Torr.
Figure 5 displays the emission spectrum detected from the plasma produced after focusing the Nd-YAG laser beam with a pulse energy of 60 mJ onto the fresh clam muscle under air surrounding gas at a low pressure of 5 Torr. The emission spectrum is largely different from the spectrum shown in Fig. 4. In contrary to the emission spectrum in Fig. 4 dominated by the molecular band, this emission spectrum is dominated mostly by atomic and ionic lines. The emission spectrum is free from the molecular band, and there are many emission lines from the host, minor and trace elements including heavy metals such as Fe observed in the wavelength region from 200 nm to 900 nm. The emission lines due to the host constituent of the calm sample appear clearly. This confirms the above interpretation that the molecular bands arise from the breakdown of the air surrounding gas. In this low-pressure case, 5 Torr, the amount of air in the surrounding reduces substantially. The low pressure spectrum is now under detailed investigation to confirm and identify the emission lines.

4. Conclusion

A preliminary wide spectral wavelength investigation of LIBS emission spectrum has been carried out on the fresh clam muscle sample. The emission spectrum detected from the plasma produced the calm muscle sample using the Nd-YAG laser energy of 93 mJ under air surrounding gas at high pressure is heavily dominated by the emission lines from the host constituent of the sample and the strong gas breakdown plasma of the surrounding gas. Reduction the laser pulse energy to 60 mJ together with lowering the surrounding gas pressure to 5 Torr yield to a much better profile of the emission spectrum with favorable features for spectrochemical analysis. Since the emission spectrum dominated by the emission lines due to host constituent of the clam sample and the surrounding, no emission lines from metal especially heavy metal can be detected at high laser pulse energy and high surrounding gas pressure. Further assessment of the emission lines for studying characteristics of plasma produced on the calm muscle in term of temperature and electron density and the possibility for making qualitative and quantitative analysis is being carried out and will be reported elsewhere shortly.

Acknowledgments

The authors thank to Research and Society Service Institute and Graduate School of Syiah Kuala University, Aceh, Indonesia for financial support to this research through the scheme of the postgraduate research incentive grant 2016 with a contract number: 1714/UN11/SP/PNBP/2016.

References

[1] Yuka S, Elvince R, Tomonori K, Noriyasu K, Yuji N, Takanobu I., Osamu, N., Takayuki, M., Syamsidik 2013 *Journal of Ecotechnology Research* **17**(1) 11- 15
[2] Miziolek A W, Palleschi V, and Schechter I (Eds) 2006 *Laser-Induced Breakdown Spectroscopy (LIBS), Fundamentals and Applications* (New York: Cambridge University Press)
[3] Winefordner J D, Gornushkin I B, Correll T, Gibb E, Smith B W, and Omenetto N 2004 *J. Anal. Atom. Spectrom.*, **19** 106-108
[4] Charfi B and Harith M A 2002 *Spectrochimica Acta Part B: Atom. Spectr.* **57** 1141–53.

IOP Conf. Series: Journal of Physics: Conf. Series **1011** (2018) 012058 doi:10.1088/1742-6596/1011/1/012058

Effect of milling time on microwave absorption ability on barium-hexaferrite nanoparticles

Y E Gunanto[1], M P Izaak[1], S S Silaban[1], W A Adi[2]
[1] Dept. of Physics Education, Faculty of Education, University of Pelita Harapan, Karawaci, Tangerang 15811, Indonesia
[2] Center for Science and Technology of Advanced Materials, BATAN, Tangerang Selatan 15314, Indonesia

E-mail: yohanes.gunanto@uph.edu, maya.izaak@uph.edu, sanny.silaban@uph.edu, dwisnuaa@batan.go.id

Abstract. In this study, $Ba_{0.6}Sr_{0.4}Fe_{10}MnTiO_{19}$ nanoparticles were prepared using solid state reaction method with three milling time variations, 25 hours, 50 hours and 75 hours. X-ray diffraction (XRD), scanning electron microscopy (SEM), transmission electron microscopy (TEM), vibration sample magnetometer (VSM) and vector network analyzer (VNA) were used to analyze the crystal phases, structure, morphology, static magnetic and microwave absorption properties respectively. The results obtained after three variations of milling time for each sample were a single phase with the hexagonal crystal structure. Grain size decreased with increasing milling time, about 100-150 nm for 25 hours milling time, 75-100 nm for 50 hours and 20-30 nm for milling time of 75 hours. The magnetization saturation Ms reached a maximum of 52.0 emu/g, and the minimum Hc coercivity was 0.32 T for 75 hours milling time. As for 50 hours milling time, Ms = 50.0 emu/g, Hc = 0.34 T and 25 hours, Ms = 44.0 emu/g, Hc = 0.37 T Maximum Reflection Loss (*RL*) was approximately -49 dB at a frequency 10.8 GHz with bandwidth 1.7 GHz for 75 hours milling time. For a 50 hours milling time the *RL* value was approximately -33 dB at a frequency of about 11.2 GHz with bandwidth 1.5 GHz and 25 hours milling time the *RL* value was approximately -20 dB at a frequency 10.9 GHz with bandwidth 1.3 GHz. The average particle size was about 200-350 nm for 20 hours milling time, and *RL* reflection loss was about - 29 dB at a frequency 12.2 GHz, while for 10 hours milling time, the particle size was about 460 nm, *RL* reflection loss was about 12 dB at a frequency 9.5 GHz [1].

1. Introduction

In recent years, the material in the hexaferrite family, especially the M-type barium hexaferrite is very attractive to many researchers [1-11]. This is because this material has unique magnetic properties such as high magnetization saturation, high Curie temperature, high resistivity, good chemical stability and high scientific resonance frequency. So it is very applicable to absorb radar, permanent magnet and recording medium [1-9]. There are several ways to further improve its performance. The usual ways used include substituting Ba atoms or Fe atoms or substituting Ba atoms and Fe atoms simultaneously [1-9] with solid state reaction [1,3,5,6,8], sol-gel [2,4], and chemical methods [7,9].
In addition to substituting, some researchers do variations in the manufacturing process, one with a variation of milling time that will result in smaller particle size as the length of milling increases. The average size of particle is about 200-350 nm with a milling time of 20 hours and the reflection loss RL is about -29 dB at 12.2 GHz, while for 10 hours milling time, the particle size is about 460 nm, RL

reflection loss is about -12 dB at a frequency of 9.5 GHz [1]. Thus by increasing the length of time, the milling will reduce the particle size and increase the ability to absorb microwaves.

In this research, $Ba_{0.6}Sr_{0.4}Fe_{10}MnTiO_{19}$ sample making is done by mechanical activation method using high energy milling (HEM) with three variation of milling time, 25 hours, 50 hours and 75 hours. X-ray diffractometer is used to analyze the phase and crystal structure. Scanning electron microscopy and transmission electron microscopy are used for morphological characterization and grain size of crystals. For magnetization, vibration sample magnetometer was used and characterization of microwave absorption capability used vector network analyzer.

2. Experimental

The raw materials were used in this work consist of $BaCO_3$, $SrCO_3$, Fe_2O_3, $MnCO_3$ and TiO_2 with the purity of each material not less than 99.5%. The sample $Ba_{0.6}Sr_{0.4}Fe_{10}MnTiO_{19}$ was prepared by stoichiometric calculation of the mixture of the five raw materials then put in a vial stainless steel with steel balls with an average diameter of 10 mm. The weight ratio of the spheres to the powder is 2: 1. High energy milling (HEM) with 750 rpm was used for milling with three-time variations, i.e. 25 hours, 50 hours and 75 hours. Each sample was calcined at a temperature of 1000 ^{0}C for 5 hours.

X-ray diffraction analysis was performed using X-ray diffraction device type Philip PW1710 with the Cu-K_α radiation (λ = 1.5406 Å) in the range of $2\theta = 10 - 80^0$. The morphology and grain size were performed by using scanning electron microscopy JEOL JED 2300 and transmission electron microscopy JEOL JEM 1400. The magnetic properties were performed using vibration sample magnetometer Oxford VSM1.2H Instrument 1.2 T and the reflection loss (RL) of samples were calculated by using vector network analyzer Advantest type R3770.

3. Results and discussion

3.1 Phase analysis

Figure 1 (a) - (b) shows the refinement results of the x-ray diffraction pattern of sample $Ba_{0.6}Sr_{0.4}Fe_{10}MnTiO_{19}$ with milling time (a) 50 hours and (b) 75 hours. This refinement result produces excellent fitting quality with very small R factor. R factor is criteria of fit and a factor of $\chi 2$ (chi-squared) which is of very small value, and according to Izumi, the χ^2 (chi-squared) value is allowed maximum 1,3 [12]. While the value of structural parameters, Criteria (factor R) and χ2 (chi-squared) refinement results are shown in table 1.

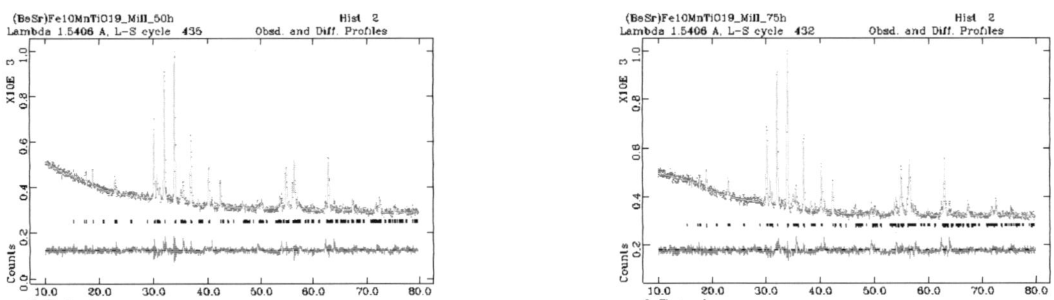

(a) (b)

Figure 1. The refinement X-ray diffraction patterns for $Ba_{0.6}Sr_{0.4}Fe_{10}MnTiO_{19}$ with milling time 50h (a) and 75h (b).

In Table 1, it is shown that the lattice parameters for phase $Ba_{0.6}Sr_{0.4}Fe_{10}MnTiO_{19}$ with variations of 25 hours, 50 hours and 75 hours milling times appear to be enlarged even slightly. This indicates that there has been an expansion of unit cell volume in which Ti^{4+} and Mn^{2+} ions have been successfully substituted into Fe^{3+} ions, where the radius of the ion Ti^{4+} = 0.68 Å, Mn^{2+} = 0.75 Å and Fe^{3+} = 0.67 Å.

Table 1. The lattice constants, chi-squared (χ^2) and R-factor (wRp)

t (hours)	Phase	*a* (Å)	*b* (Å)	*c* (Å)	V (Å)^3	χ^2	*w*Rp
25		5.8914(3)	5.8914(3)	23.143(1)	695.6(1)	1,041	3,40
50	$Ba_{0.6}Sr_{0.4}Fe_{10}MnTiO_{19}$	5.8986(2)	5.8986(2)	23.168(1)	698.13(9)	1,034	3,10
75		5.8984(3)	5.8984(3)	23.171(1)	698.1(1)	1,036	3,10

3.2 Microstructure analysis
Single phase $Ba_{0.6}Sr_{0.4}Fe_{10}MnTiO_{19}$ with this milling time variation has a homogeneous microstructure of surface morphological observations as shown in Figure 2. The surface morphological observation results show that the sample has a shape and particle size distribution of about 100-150 nm for 25 hours milling time (not shown here), 75-100 nm for 50 hours and 20-30 nm milling time for 75 hours milling time each of which has a single phase. And according to the results of this elementary analysis, it is shown that the phase composition of sample $Ba_{0.6}Sr_{0.4}Fe_{10}MnTiO_{19}$ with a milling time variation of 25 hours, 50 hours and 75 hours are all close to the nominal composition. Afghani *et al.*, obtained for a 20-hour milling time using a planetary ball mill obtained a particle size of about 200-350 nm [1, 6]. All samples are single domains because they have an average particle size smaller than the single domain's critical diameter of 460 nm [13].

3.3 Magnetic properties
The hysteresis loop in Fig. 4 for $Ba_{0.6}Sr_{0.4}Fe_{10}MnTiO_{19}$ with a milling time of (a) 25 hours, (b) 50 hours and (c) 75 hours displays the magnetic data and is summarized in table 2. It appears that when the milling time is longer, the coercivity of H_C will decrease; Magnetic remanence and magnetic saturation are increased. These results indicate that the sample is very promising to serve as a microwave absorber (Radar). With low coercivity, small remanence and high magnetic saturation.
The magnetic properties of the hexagonal ferrite are strongly dependent on several conditions on the synthesis and the selection of cations substituted between five different sublattices of Fe^{3+}, namely tetrahedral ($4f_1$), bipyramidal trigonal (2b) and octahedral (12k, 2a, And $4f_2$) on hexagonal structures. The replacement of Fe^{3+} ions by Mn^{2+} ions has no effect on the magnetization saturation for two ions having five unpaired electrons in the outer shell, wherein Mn ions will occupy 12k and 2a positions. The electrons at 12k, 2a, and 2b positions have spin-ups, whereas in positions $4f_1$ and $4f_2$ have spin-down. Therefore when some Fe^{3+} ions are replaced by more non-magnetic Ti^{4+} ions will cause exchange interactions between $4f_1$-12k and $4f_2$-2a so that it will decrease its magnetic saturation. Microwave absorption result of $Ba_{0.6}Sr_{0.4}Fe_{10}MnTiO_{19}$ synthesis with milling time 25 hours, 50 hours and 75 hours shown in figure 4.

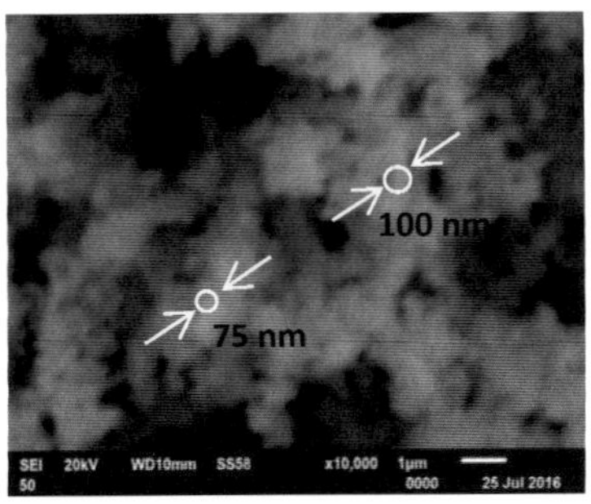

(a)

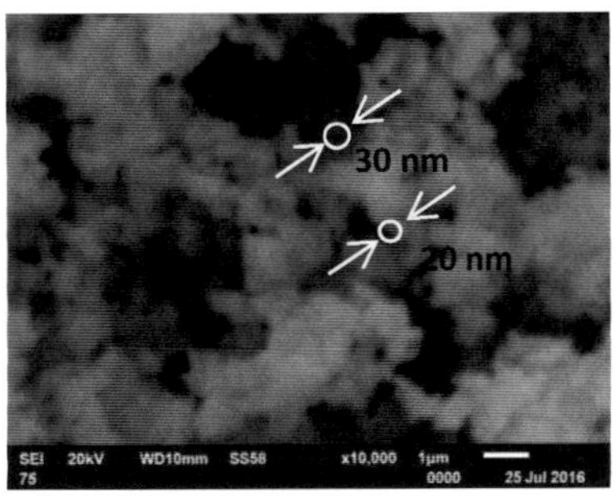

(b)

Figure 2. SEM micrograph of the samples $Ba_{0.6}Sr_{0.4}Fe_{10}MnTiO_{19}$ with milling time 50 h (a) and 75 h (b).

IOP Conf. Series: Journal of Physics: Conf. Series **1011** (2018) 012058 doi:10.1088/1742-6596/1011/1/012058

Figure 4 shows the relationship between $Ba_{0.6}Sr_{0.4}Fe_{10}MnTiO_{19}$ particle reflection loss to X-band microwave frequency in the range of 8-14 GHz with a variation of milling time 25 hours, 50 hours and 75 hours and a sample thickness of about 1.5 mm for each milling time. It appears that the longer milling time, the maximum peak absorption was increased strongly/highly. For 25-hour milling time, the absorption peak is about -20 dB at 10.9 GHz frequency. For a 50-minute milling time, the absorption peak is around -33 dB at a frequency of about 11.2 GHz. As for 75-hour milling time, the absorption peak is about -49 dB at 10.8 GHz frequency. This data shows that substitution of Mn and Ti at Fe position on Barium hexaferrite with $Ba_{0.6}Sr_{0.4}Fe_{10}MnTiO_{19}$ composition is very good as microwave absorber so it can be applied as radar absorber. The longer the milling time (75 hours), the absorption peak will be higher.

Tabel 2. Summary of $Ba_{0.6}Sr_{0.4}Fe_{10}MnTiO_{19}$ Magnetic Data

t (h)	Coercivity, *Hc* (T)	Saturation, *Ms* (emu/g)	Remanence, *Mr* (emu/g)
25	*0,37*	44,0	28,0
50	0,34	50,0	32,0
75	0,32	52,0	33,0

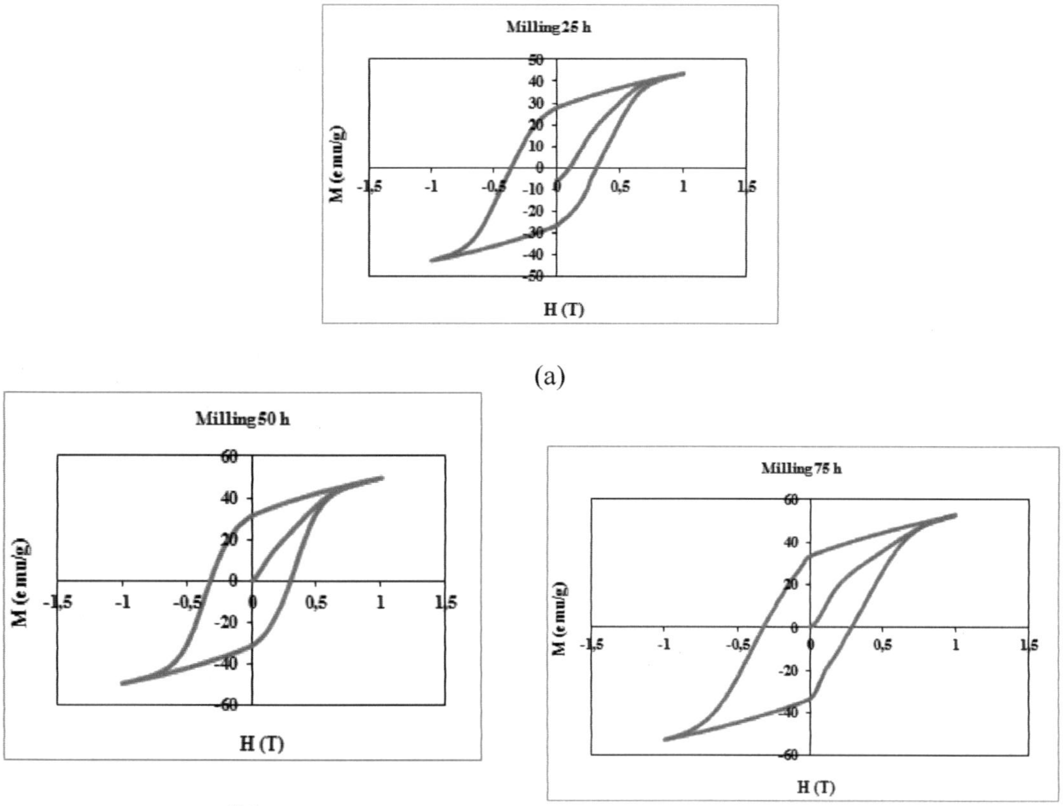

(b) (c)

Figure 3. Characterization of $Ba_{0.6}Sr_{0.4}Fe_{10}MnTiO_{19}$ particle with milling time variations (a) 25 hours, (b) 50 hours and (c) 75 hours using VSM

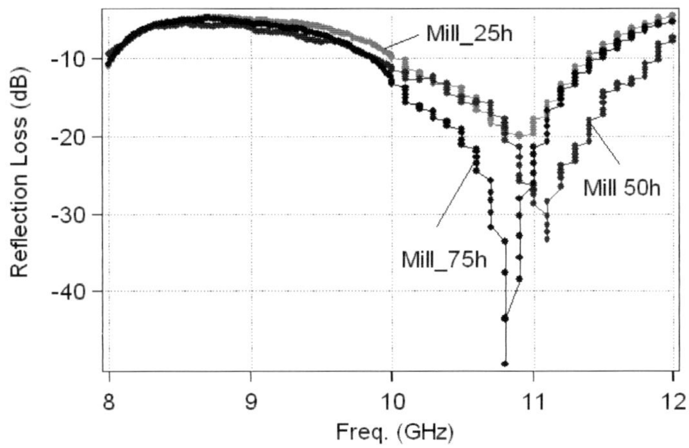

Figure 4. Reflection loss curve (*RL*) $Ba_{0.6}Sr_{0.4}Fe_{10}MnTiO_{19}$ with milling time 25 h, 50 h and 75 h

4. Conclusion

In this research, $Ba_{0.6}Sr_{0.4}Fe_{10}MnTiO_{19}$ sample is produced through mechanical activation process (high energy milling) with three variations of time. The XRD results indicate that the sample is in phase with the same hexagonal crystal structure. The average size of crystal grains ranged from 100 to150 nm for 25 hours milling time, 75-100 nm for 50 hours and 20-30 nm for milling time of 75 hours. The ability to absorb microwaves increases with increasing milling time. For 25-hours milling time, the absorption peak is about -20 dB at 10.9 GHz frequency, 50-hours milling time, the absorption peak is about -33 dB at 11.2 GHz frequency. As for 75-hours milling time, the absorption peak is about -49 dB at 10.8 GHz frequency.

Acknowledgments

This work was supported by University of Pelita Harapan Grants 2017 No. **P-005-FIP/I/2017**, Development of Anti-Radar Nanoparticle Material Based Barium-Strontium Hexaferrite

References

[1] Afghahi S S S, Jafarian M, Salehi M, and Atasi Y, 2017 *Journal of Magnetism and Magnetic Materials* **421** 340-343
[2] Chavan V C, Shirsath S E, Mane M L, Kadam R H, and More S S, 2016 *Journal of Magnetism and Magnetic Materials* **398** 32-37
[3] Gunanto Y E, Jobiliong E, and Wisnu A A, 2016 *J. Math. Fund. Sci* Vol. **48** No. 1, 55-65
[4] Mosleh Z, Kameli P, Poorbaferani A, Ranjbar M, and Salamati H, 2016 *Journal of Magnetism and Magnetic Materials* **397** 101-107
[5] Gunanto Y E, Cahyadi L and Ari Adi W, 2015 presented at the 9th Seminar on Magnetic Materials, University of Sriwijaya, Palembang, South Sumatera Indonesia, October 19 -21
[6] Afghahi S S S, Jafarian M, and Atasi Y, 2016 *Journal of Magnetism and Magnetic Materials* **406** 184-191
[7] Dawar N, Chitkara M, Sandhu I S, Jolly J S, and Malhotra S, 2016 *Cogent Physics* **3** 1208450
[8] Gunanto Y E, Cahyadi L, Ari Adi W, 2015 *AIP Conf. Proc.* **1725**, 020023-1–020023-6
[9] Durmus Z, Durmus A, and Kavas H, 2015 *J. Mater Sci* **50** 1201-1213
[10] Choi M, Cho S, Song Y, Baek S, Kim H, Jung J, Lee H, Park C, Park S, and Kim Y, 2014 *Curr. Appl. Physics* **14** 1208-1211
[11] Huang X, Zhang J, Liu Z, Sang T, Song B, Zhue H, and Wong C, 2015 *J. Alloy. Compd.* **648** 1072-1075

[12] Izumi F, A Rietveld-Refinement Program RIETAN-94 for Angle-Dispersive X-Ray and Neutron Powder Diffraction 1996 National Institute for Research in Inorganic Materials 1-1 Namiki, Tsukuba, Ibaraki 305, Japan, Revised on June 22

[13] Vinnik D A, Zherebtsov D A, Mashkovtseva L S, Nemrava S, Semisalova A S, Galimov D M, Gudkova S A, Chumanov I V, Isaenko L I, Niewa R, 2015 *J.Alloy.Compd.* **628** 480–484

The effect of synthesis parameter on crystal structure and magnetic properties of $Ni_{0.5}Zn_{0.5}Fe_2O_4$ magnetic nanoparticles

A Rifianto[1], J Widakdo[1], N Istikhomah[1], E Suharyadi[1]*, T Kato[2] and S Iwata[3]

[1] Department of Physics, Universitas Gadjah Mada, Yogyakarta, INDONESIA
[2] Department of Electronics, Nagoya University
[3] Institutes of Materials and Systems for Sustainability, Nagoya University, JAPAN

*Corresponding Author: esuharyadi@ugm.ac.id

Abstract. Nickel Zinc Ferrite ($Ni_{0.5}Zn_{0.5}Fe_2O_4$) magnetic nanoparticles have been successfully synthesized using the co-precipitation method with variation of synthesis temperature and concentration of NaOH. X-ray diffraction (XRD) analysis confirmed that $Ni_{0.5}Zn_{0.5}Fe_2O_4$ nanoparticles have spinel cubic crystal structure. The XRD profile of sample showed that the crystallinity increases with the increase of synthesis temperature. Meanwhile the XRD profile of sample showed the crystallinity decreases with the increase of NaOH concentration. Crystallite size of samples were in the range of 11.1-16.0 nm. There is no significant change in the lattice parameter size for the samples synthesized with increase of temperatures and NaOH concentrations. Transmission electron microscope (TEM) image showed that the sample was agglomerated. The selected area electron diffraction (SAED) image showed the diffraction ring as representation of Miller plane and confirmed that sample was polycrystalline. The maximum magnetization (σ_S) at 15 kOe of the samples increases with the increase of crystallite size and crystallinity. The coercivity (H_c) of the samples increases with the increase of crystallite size.

1. Introduction

Ferrite nanoparticles (FNPs) are one of the type of magnetic nanoparticles (MNPs) that have great attention due to their wide application in various fields. They have used for large number of application such as in high density magnetic recording, microwive device, magnetic fluid etc [1]. FNPs are metal oxside with spinel structure, we know that the spinel ferrites have a general molecular formula $[M_{1-x}Fe_x]^A\ [M_xFe_{2-x}]^BO_4$ where 'A' represents tetrahedral site, 'B' represents octahedral site, and 'x' denotes the degree of inversion [2]. Each unit cell contains eight formula unit with 32 of O^{2-} anions packed in the lattice, and M^{3+} and M^{2+} cations occupaying the 16 (tetrahedral) and 8 (oktahedral) sites. The ocupations of tetraheral site entirely with a divalent transition metal (such as Zn) produces a normal spinel structure, wherease occupation of the octahedral site with the divalent transitions metal ions (such as Ni) yields an invers spinel structure. If divalent transition metal ion are present on both A and B sub-lattice, the structure is mixed spinel [3,4]. Because of them wide application, FNPs must have specific requirenment to be used in each application fields. Nickel-zinc ferrite have been extensively used in a number electronic devices due to their high electrical resistivity, large stability at high frequency, low coercivity, high magnetization, mechanical hardness, chemical stability and reasonable cost [5].

Selection of an apropriate process is an important factor to achived ferrites with high quality. Methode and synthesis techniques conditions affect the structural, electrical and magnetic properties of ferrites [6, 7]. Kurian and Nair studied the effect of different preparation conditions on structural changes of nickel-zinc ferrite [8]. Deraz and Abd-Elkader studied the structural and magnetic properties of nickel-zinc ferrite using spinel solid solution method [9]. Both of research reported the three aspect (preparation condition, structural, and magnetic properties) separately. The purpose of this work is to investigate the effect of synthesis temperature and NaOH concentration on the crystal structure. The relation between crystal structure and magnetic properties of $Ni_{0.5}Zn_{0.5}Fe_2O_4$ nanoparticles were be also investigated in detail in order to give more comprehensive information.

2. Experimental detail

$Ni_{0.5}Zn_{0.5}Fe_2O_4$ nanoparticles were synthesized using coprecipitation method with variation of synthesis parameters (temperatures and concentrations of NaOH). Nickel cloride hexahydrate ($NiCl_2.6H_2O$), zinc sulphat heptahyrate ($ZnSO_4.7H_2O$), ferric chloride hexahydrate ($FeCl_3.6H_2O$), cloride acid (HCl) and sodium hydroxide (NaOH) were used as starting materials.

Principally, the stage of coprecipitation was initialized by the formation of hydroxide compound via coprecipitation reaction of a metal salt precursor with NaOH. The HCl solution used to break the ionic bond between metal cation and chloride anions or sulfate group. Those metal cations were highly attracted by hydroxide anion OH- to form hydroxide compound. Each hydroxide compound of $Ni(OH)_2$, $Zn(OH)_2$ and $Fe(OH)_3$ has different reaction rate to form ferrite nanoparticles. According to the chemical reaction of $Ni_{0.5}Zn_{0.5}Fe_2O_4$ nanoparticles, the molar ratio of Ni, Zn and Fe is 1 : 1: 4, respectively. Variation of synthesis temperature used in this work were 30, 60, 90, and 120°C (with constant of NaOH concentration at 1.5 M) and variation of NaOH concentration used in this work were 1.5, 2.5, 3.5, and 4.5 M (with constant of synthesis temperature at 60°C).

The crystal structure and phase identification were confirmed by the X-ray diffractometer (XRD). The morpologhy was studied by Transmission electron microscope (TEM). The magnetic measurements of $Ni_{0.5}Zn_{0.5}Fe_2O_4$ nanoparticles were carried out by vibrating sample magnetometer (VSM) at room temperature.

3. Result and discussion

3.1. Crystal structure from X-ray diffraction

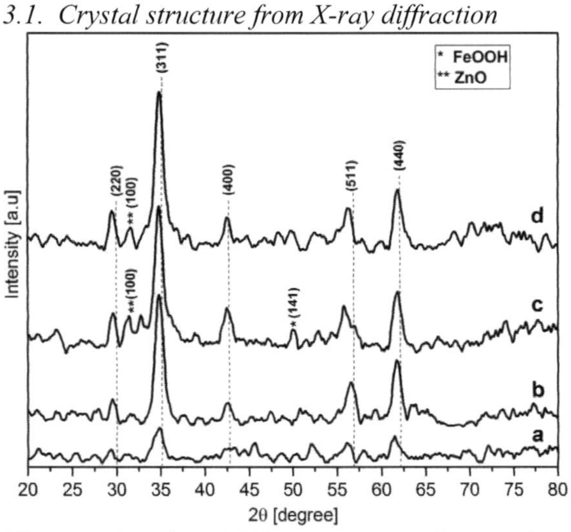

Figure 1 The XRD spectra of samples synthesized at (a) 30°, (b) 60°, (c) 90° and (d) 120°C of temperatures (with constant of NaOH concentration at 1.5 M)

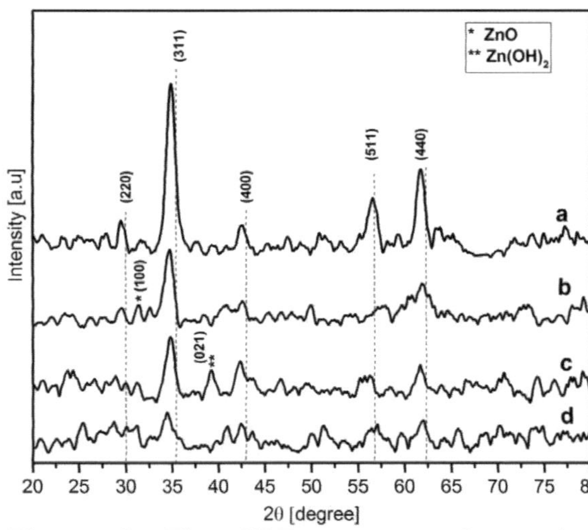

Figure 2 The XRD spectra of samples synthesized at (a) 1.5 M, (b) 2.5 M, (c) 3.5 M and (d) 4.5 M of NaOH concentrations (with constant of synthesis temperature at 60°C)

IOP Conf. Series: Journal of Physics: Conf. Series **1011** (2018) 012059 doi:10.1088/1742-6596/1011/1/012059

The XRD spectra of $Ni_{0.5}Zn_{0.5}Fe_2O_4$ nanoparticles that were obtained from variation of synthesis temperatures and NaOH concentrations are shown in figure 1 and 2. Analysis of XRD spectra showed that ferrite phases were obtained. The peaks identified the main diffraction plane indexed to (220), (311), (400), (511) and (440) planes, which indicated the presence of cubic spinel structure. Those diffraction planes are in good agreement with those of Ni-Zn ferrite JCPDS no 08-0234 [10]. The XRD spectra typically broad shape and low intensity reflections indicate the small of ferrite crystals [11].

Figure 1 showed that all the XRD peaks become narower and sharper with increase of synthesis temperatures. It indicated that there is enhancement of crystallite size and crystallinity [12]. Additional peaks observed at figure 1 (c and d) indicates the impurity phase (non spinel $Ni_{0.5}Zn_{0.5}Fe_2O_4$). Figure 2 showed that all the XRD peaks become broader and the intensity decreases with the increase of NaOH concentrations. It indicated that crystallite size and crystallinity decreases. Additional peaks observed at (Figure 2 b and c) indicates the impuryti phase (non spinel $Ni_{0.5}Zn_{0.5}Fe_2O_4$).

Table 1. Lattice parameter (a), crystallite size (t), and lattice strain (ε) of $Ni_{0.5}Zn_{0.5}Fe_2O_4$ nanoparticles.

Synthesis parameter		a (Å)	t (nm)	ε	Synthesis parameter		a (Å)	t (nm)	ε
Temperature	30	8,56	12,3	0,04	NaOH	1,5	8,53	14,4	0,03
(°C)	60	8,53	14,4	0,03	Concentration	2,5	8,59	14,4	0,03
	90	8,54	15,3	0,03	(M)	3,5	8,56	13,8	0,03
	120	8,53	16,0	0,03		4,5	8,63	11,1	0,04

Table 1 shows that there is no significant change in the lattice parameter for the samples which is synthesized with increase of temperatures and NaOH concentrations. it is attributed to homogenity of Fe^{3+}, Ni^{2+} and Zn^{2+} in spinel structure and there is a strong electrostatic force between ions [13]. The crystallite size of sample sythesized at temperature of 30°C was 12.3 nm and then increased with the increase of synthesis temperatures. The increased in crystallite size is attributed to rapid diffusion of cations at high temperature. Higher temperature might be benefical to accelerate the rate of the diffusion and integration process, and thus, the crystallite size became larger. Meanwhile the crystallite size of sample synthesized with 1.5 M of NaOH was 14.4 nm and then it decreased with the increase of NaOH concentrations. The decrease in crystallite size is attributed to high concentration of hydroxide anions results in an extremly fast nucleation process that generate tyni nuclei [14]. Due to high concentration and small hydration sphere in solution, the sodium ions can probably be absorbed significantly onto all the facets of the nuclei, wich hinder incorporation and subsequent growth of neighbouring nuclei [15].

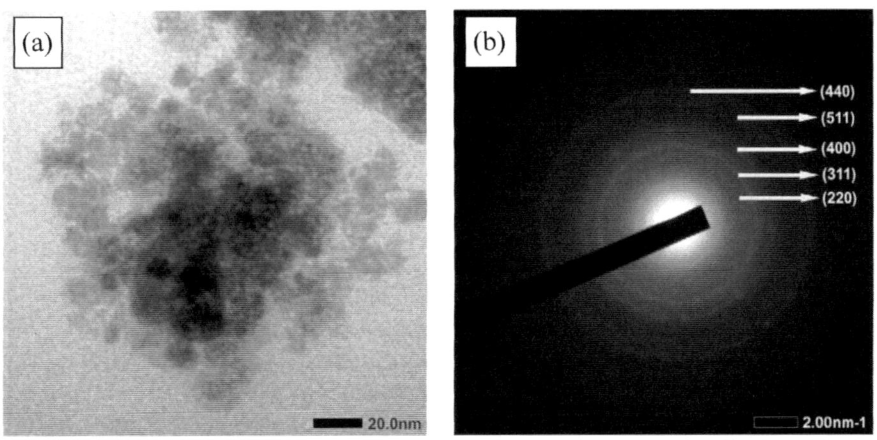

Figure 3 (a) TEM image, and (b) SAED image of $Ni_{0.5}Zn_{0.5}Fe_2O_4$ nanoparticles

IOP Conf. Series: Journal of Physics: Conf. Series **1011** (2018) 012059 doi:10.1088/1742-6596/1011/1/012059

3.2. Morphological studies

Morphologi of nanoparticles $Ni_{0.5}Zn_{0.5}Fe_2O_4$ that is synthesized at 60°C with 1.5 M of NaOH was shown in Figure 3. Figure 3 (a) shows that nanoparticles are nearly spherical and aglomeration occured. The powders of nanoparticles $Ni_{0.5}Zn_{0.5}Fe_2O_4$ are easy to form aglomerated due to nanosize effect and intrinsic magnetism. Because of very small in size, thermal energy of nanoparticles is large enough to overcome the energy of the magnetic interactions among the nanoparticles [16]. The selected area electron diffraction (SAED) image (Figure 3 (b)) showed the diffraction ring as representation of Miller plane and confirmed that sample was polycrystalline. The corresponding SAED pattern of the selected region indicates the sample is well crystallized.

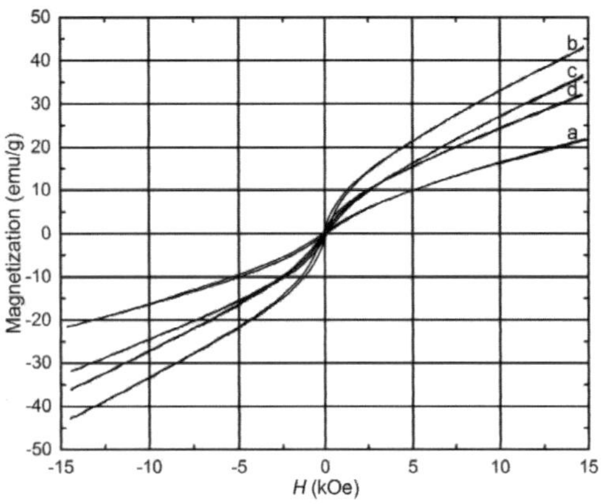

Figure 4 Hysteresis curve of samples synthesized at (a) 30°, (b) 60°, (c) 90° and (d) 120°C of temperatures (with constant of NaOH concentration at 1.5 M)

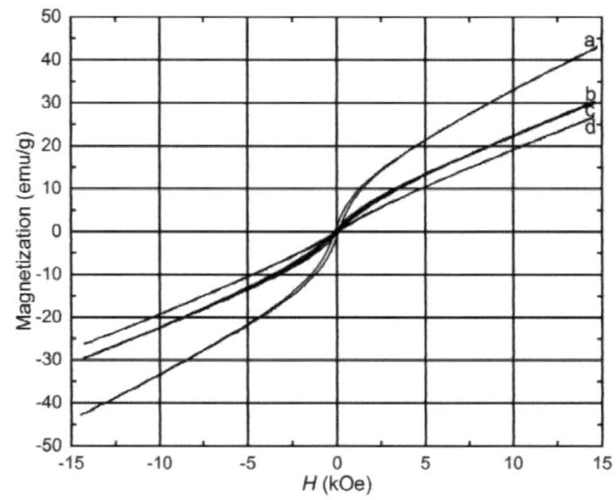

Figure 5 Hysteresis curve of samples synthesized at (a) 1.5, (b) 2.5, (c) 3.5 and (d) 4.5 M of NaOH concentrations (with constant of synthesis temperature at 60°C)

Table 2. Crystallite size (*t*), maximum magnetization (σ_S), and coercivity (H_c) of $Ni_{0.5}Zn_{0.5}Fe_2O_4$ nanoparticles

Variable of Synthesis	t (nm)	σ_S (emu/g)	H_c (Oe)	Variable of Synthesis	t (nm)	σ_S (emu/g)	H_c (Oe)
temperature	12,3	21,9	99,6	NaOH Concentration	14,4	43,3	112,5
	14,4	43,3	112,5		14,4	30,3	84,6
	15,3	36,8	86,9		13,8	29,9	91,7
	16,0	32,2	108,3		11,1	26,9	53,3

3.3. Magnetic properties

The magnetic properties of all samples were investigated with a vibrating sample magnetometer (VSM) at room temperature. Figure 4 and figure 5 shows hysteresis curve of $Ni_{0.5}Zn_{0.5}Fe_2O_4$ nanoparticles synthesized with variation of synthesis temperatures and NaOH concentrations.

All samples were not fully saturate at maximum external field of 15kOe. Lack of saturation magnetization can be related to single domain nature of the particles and magnetocrystalline anisotropy energy of the nanoparticles. Maximum magnetization (σ_S) increase with the crystallite size mainly due to the spin non-colinearity at the surface [17]. Lower σ_S that is assosiated with the smaller crystallite size could be attributed to magnetocrystalline anisotropy of the particles that depends on the crystallinity. Coercivity (H_c) shows the fluctuate value with the increase of crystallite size, for the samples with variation of temperature. It occure due to the different of magnetocrystalline anisotropy energy. The difference of magnetocrystalline anisotropy energy may due to the irregular shape of the particles. Meanwhile coercivity increased with the increase of crystallite size for the sample synthesized with variations of NaOH concentrations. This phenomena indicated that the samples have single domain structure. The H_c is in direct proportion to the volume singgle-domain grains. Therefore, H_c become gradually larger as the single-domain particle size increase. Particles which has single domain, have the magnetostatic energy dominant at the surface of particles, so the coercivity increases with the increase of crystallite size.

4. Conclusion

Nickel zinc ferrite ($Ni_{0.5}Zn_{0.5}Fe_2O_4$) magnetic nanoparticles have been successfully synthesized using the co-precipitation method with variation of synthesis temperatures and NaOH concentrations. Crystallite size of samples were in the range of 11.1-16.0 nm. The XRD studies confirm the presence of cubic spinel structure. The SAED image showed the diffraction ring as representation of Miller plane and confirmed that sample was polycrystalline. The VSM analysis showed that all samples were not fully saturate at maximum external field of 15kOe. The maximum magnetization increases with the increasing of crystallite size. The coercivity increases with the increasing of crystallite size.

Acknowledgments

Nanofabrication Platform Consortium Project of Nagoya University, Ministry of Culture, Sports, Science and Technologi (MEXT) Nano-Project Platform, Japan and Grant of “Penelitian Berbasis Kompetensi”, Ministry of Research, Technology and Higher Education of the Republic Indonesia (Kementerian Ristek Dikti) for providing financial support for this work.

References

[1] Kumar S, Singh V, Aggarwal S, Mandal U K and Kotnala R K 2010 *Mater. Sci. Eng. B* **166** 76-78
[2] Chen, D G, Tang, X G, Wu J B, Zhang W, Liu Q X and Jiang Y P 2011 *J. Magn. Magn. Mater.* **323** 1717-1721
[3] Shokrollahi H and Avazpour L 2016 *Particuology* **26** 32-39
[4] Indrayana I P T, Siregar N, Suharyadi E, Kato T and Iwata S 2016 *J. Phys.: Conf. Ser.* **776** 012021-1 – 12021-7
[5] Rezlescu N, Sachelarie L, Rezlescu E, Sava C-L and Popa D 2003 *Ceram. Int.* **29** 107-111
[6] Rezlescu N, Rezlescu E, Paniscus C and Carus M L 1994 *J. Phys.: Condens. Matter* **6** 5707-5716
[7] Wang J M, Tsuo W J and Yen F S 1999 *Solid State Chem.* **145** 50-57
[8] Kurian M and Nair D S 2016 *J. Saudi Chem. Soc.* **20** S517-S522
[9] Deraz N M and Abd-Elkader O H 2014 *J. Anal. App. Pyrolysis* **106** 171-176
[10] Rashad M M, Elsayed E M, Moharam M M, Abou-Shahba R M and Saba A E 2009 *J. Alloy. Compd.* **486** 759-767
[11] Kumar S, Kumar P, Singh V, Mandal U K and Kotnala R K 2015 *J. Magn. Magn. Mater.* **379** 50-57
[12] Azhagushanmugam S J, Suriyanarayanan N and Jayaprakash R 2013 *Physics Procedia* **49** 44-48
[13] Nabiyouni G, Fesharaki M J, Mozafari M and Amighian J 2010 *Chin. Phys. Lett.* **27** No 12 12640
[14] Lv J, Qiu L and Qu B 2004 *J. Cryst. Growth* **267** 676-684
[15] Ghasemi A, Ekhlasi S and Mousavina M 2014 *J. Magn. Magn. Mater.* **354** 136-145
[16] Mascolo M C, Pei Y and Ring T A 2013 *Materials* **6** 5549-5567
[17] Gharagozlou M and Bayati R 2015 *Super. Microstruct.* **78** 190-200

IOP Conf. Series: Journal of Physics: Conf. Series **1011** (2018) 012060 doi:10.1088/1742-6596/1011/1/012060

Biomolecule detection using wheatstone bridge giant magnetoresistance (GMR) sensors based on CoFeB spin-valve thin film

P Elda Swastika[1], G Antarnusa[1], E Suharyadi[1*], T Kato[2] and S Iwata[3]
[1] Department of Physics, Universitas Gadjah Mada, Yogyakarta, INDONESIA
[2] Department of Electronics, Nagoya University, JAPAN
[3] Institutes of Materials and Systems for Sustainability, Nagoya University, JAPAN

*Corresponding Author: esuharyadi@ugm.ac.id

Abstract. A potential wheatstone bridge giant magnetoresistance (GMR) biosensor have been successfully developed for biomolecule detection. [IrMn(10 nm)/CoFe(3 nm)/Cu(2.2 nm)/CoFeB(10 nm)] spin-valve structure has been chosen as the magnetic sensing surface, showing a magnetoresistance (MR) of 6% fabricated by DC magnetron sputtering method. The Fe_3O_4 magnetic nanoparticles used as biomolecular labels (nanotags) was synthesized by co-precipitation method, exhibiting soft magnetic behavior with saturation magnetization (*Ms*), remanent magnetization (*Mr*) and coercivity (*Hc*) is 77.2 emu/g, 7.8 emu/g and 51 Oe, respectively. The X-ray diffraction (XRD) patterns and transmission electron microscopy (TEM) images showed that Fe_3O_4 was well crystallized and grew in their inverse spinel structure, highly uniform morphology with an average grain size was about 20 nm. Fe_3O_4 was coated with polyethylene-glycol (PEG)-4000 for surface functionalization. Detection of biomolecule such as formalin, gelatin from bovine-skin and porcine-skin were dispersed in ethanol at room temperature. Induction would cause a shift in output voltage with a minimum delta output voltage (ΔV) 4.937 mV (10%) for formalin detection, 2.268 mV (7%) for bovine-skin gelatin and 2.943 mV (7%) for porcine-skin gelatin detection. The ΔV of the wheatstone bridge in real-time measurement decrease by increase in biomolecules concentration. The change of ΔV with various concentration of biomolecule indicates that the spin-valve thin film with wheatstone-bridge circuit is potential as a biosensor.

1. Introduction

Nowadays, technologies for biomolecular diagnostics have broad applications in biology and medicine. It is necessary to establish a reliable, sensitive and simple quantitative method to detect biomolecule. Magnetism provides great opportunities for researchers to remotely control and detect small biological samples. Magnetic immunoassay is one type of immunoassays that can quantitatively detect biomolecules.

One of the most attracting bio-detection sensors to be studied due to its wide application is GMR. GMR is the change in electrical resistance of some materials in response to an external magnetic field. Among the numerous magnetic sensors, giant magnetoresistance (GMR) sensor is one of magnetic sensor that can be the powerful tools for bio-detection due to their merits of low cost, portability, high sensitivity, and real-time electronic readout [1].

The mechanism of GMR biosensors is detecting the magnetic fringing fields of the particle labels after capture by target-probe biomolecular recognition. Wheatstone-bridge can be used in sensor system because its high sensitivity that can read signal output in millivolt, so if there is small change in sensor resistance will change signal output voltage. Magnetic fringing field will be induced by the magnetic nanoparticles which in turn exerts on the free layer of the GMR sensor along the opposite direction of the externals field. Thus, the resistance of the detecting spin-valve in GMR sensors will be reduced, which results in an unbalance of Wheatstone bridge and gets an output voltage.

To achieve the desired detection sensitivity and quantification, nanometer-sized magnetic particles should be used as the biomolecular labels because of their comparable size to biomolecules and monodispersity of dimensional and magnetic properties. One of magnetic nanoparticles (MNPs) which gained considerable attention as labels to the GMR sensors is Fe_3O_4 that have superparamagnetic characteristics and high magnetic saturation (*Ms*) [2]. Thus make the Fe_3O_4 MNPs responsive to the external magnetic field [3]. To capture biomolecules and achieved better dispersibility, the Fe_3O_4 MNPs are surface-modified or coated by biocompatible polymers. Polyethylene glycol (PEG) is a popular one since their long polymeric chains are highly soluble in water and nontoxic [4], biocompatible, and low-costed [5].

In recent work, Suharyadi et al. [6] had investigated the detection of Fe_3O_4 MNPs and formaldehyde using GMR sensor by four point probe system. However, that experiment shows that signal output of GMR sensor was resistance. Even though a unique resistance can be used as sensing element, a Wheatstone bridge setup is always a good recommendation as the starting step in the design of resistive sensors [7]. Hence, Nurpriyanti et al. [8] investigated the detection of Fe_3O_4 MNPs using wheatstone bridge GMR sensors. However, the sensors was not portable and real-time.

Therefore, the aim of the present work is to develop more portable wheatstone-bridge GMR biosensor and able to read signal output in real time for various concentration of formaldehyde, bovine and porcine skin gelatin with Fe_3O_4 nanoparticles as nanotags.

2. Experimental Detail

The spin valve thin film as the magnetic sensing surface was fabricated by DC magnetron sputtering method. Spin valve structure consists of [Ta (2nm)/IrMn (10nm)/CoFe (3nm)/Cu (2,2nm)/CoFeB (10nm)/Ta (5nm)]. The magnetic properties were characterized by Vibrating sample magnetometer (VSM). Spin Valve structure was chosen as the sense element because of its high field sensitivity at room temperature [9].

The Fe_3O_4 magnetic nanoparticles were synthesized by chemical co-precipitation method. For pure Fe_3O_4 nanoparticles, $FeSO_4.7H_2O$ and $FeCl_3.6H_2O$ (Merck Germany) as a source of Fe^{3+} and Fe^{2+} were dissolved in deionized water. The obtained dark orange solution was stirred at 60°C in order to obtain homogenously solution. An aqueous NH_4OH solution 10% then added dropwise to that mixture under constant magnetic stirring 450rpm for 90 min. After that, the Fe_3O_4 nanoparticles were separated and purified by magnetic separation and washed with deionized water to make them free of any residual salts, and then dried in furnace for 2 hours at 80°C [6,8]. Finally, the black solid powder was obtained and characterized. The process was continued by synthesis Fe_3O_4 nanoparticles functionalized with polyethylene-glycol 4000 (PEG-4000). This functionalization was intended to capture biomolecules. The crystal structure and phase identification were confirmed by the X-ray diffractometer Shimadzu XD with Cu-K radiation (λ=1.5406 Å). The vibrational spectra of nanoparticles were recorded by using Fourier transform infrared spectroscope Prestige Shimadzu-21. The morphology and size distribution were studied by using transmission electron microscope (TEM) Jeol JEM-1400. The magnetic measurements of nanoparticles were carried out by using vibrating sample magnetometer (VSM) Riken Denshi Co Ltd at room temperature.

For biomaterial detection, 1mg of PEG-coated Fe_3O_4 nanoparticles was added with various biomolecules concentration. The concentration of formalin were 2%, 4%, 6%, 8% and 10% and for bovine-skin and porcine-skin gelatin were 1%, 3%, 5% and 7%. Furthermore, the mixture dissolved in 1 ml of ethanol by means of sonicated for 30 min at room temperature. 10μl of the solution was dispersed on the surface of the GMR sensor under a bias magnetic field along the in-plane direction. The GMR sensor was connected to a printed circuit board (PCB) with two part probe and was wire-bounded directly to the pads on the PCB. The devices are driven by a control signal which depends on the system status that makes the devices portable and real-time.Various applied magnetic fields of 0-650 Gauss have been performed using electromagnetic. 10 mA constant DC current was applied to the wheatstone bridge for measure signal output voltage (V_{out}). The current was provided by Keithley-2401 which also functions as a voltmeter. For a small variation of resistance in sensors, the constant-current mode can offer more linear response and higher sensitivity than the constant-voltage mode [10]. Moreover, it can avoid the heating effect in sensors as well [9].

3. Result and Discussion

The Spin valve [Ta(2nm)/IrMn(10nm)/CoFe(3nm)/Cu(2,2nm)/CoFeB(10nm)/Ta(5nm)] as magnetic sensing surface were deposited on silicon substrate by DC magnetron sputtering under pressure was about 3×10^{-5} Pa. 8 Oe exchange bias field (*Hexc bias)* was obtained due to the exchange coupling between ferromagnetic and antiferromagnetic layers. Moreover, the saturation magnetization (*Ms*) was about 1162.5 emu/cc [8]. The Magnetoresistance was measured as a function of the external magnetic field (*H*) showing magnetoresistance 6% [8]. MR curve has shown the GMR phenomenon as the change in resistance that influenced by magnetic field. It can be explained with origin of GMR by spin-dependent scattering. In the parallel magnetization case, scattering is strong for the electrons with spin anti-parallel (minority carries) to the magnetization of the ferromagnetic layers and weak for electrons with spin parallel (majority carries). The anti-parallel of alignment of the magnetic layers result in appreciable scattering from electrons of both spin directions. The lower amount of scattering events in the parallel case leads to a lower resistance state than in the anti-parallel case.

The Fe_3O_4 nanoparticles exhibit superparamagnetic behaviour. *Ms* decrease from 77.2 emu/g to to 49.6 emu/g and *Mr* decrease from 7.8 emu/gram to 6.4 emu/gram after Coated with PEG [6,8]. *Hc* increase from 51.2 Oe for Fe_3O_4 nanoparticles to 61.5 Oe for PEG-Coated Fe_3O_4 nanoparticles [6,8]. The X-ray diffraction peaks corresponding to the (111), (220), (311), (400), (511) and (440) planes, which indicates the cubic spinel crystal structure of pure Fe_3O_4. The estimated average crystallite size of Fe_3O_4 was about 20.1 nm and increase to 22.3 nm after coated with PEG. TEM images in figure 1(a) and (b) shows that Fe_3O_4 nanoparticles and PEG-Coated Fe_3O_4 nanoparticles exhibit a spherical shapes with uniform grain size. Moreover, the addition of PEG can decrease the agglomeration. The SAED patterns for both of them, is closely related to XRD analysis [6,8].

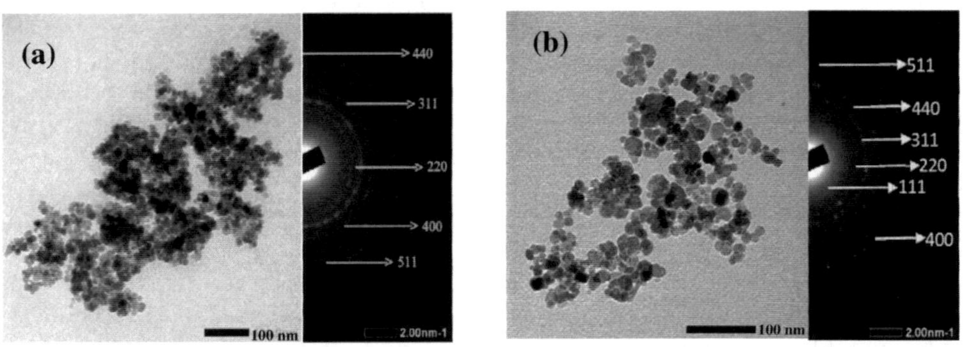

Figure 1. TEM images and selected area electron diffraction (SAED) patterns of (a) Fe_3O_4 nanoparticles and (b) PEG-Coated Fe_3O_4 nanoparticles.

For biomolecule detection, 1µl solution of PEG-coated Fe_3O_4 nanoparticles and biomolecules were dropped on the surface of the GMR sensor to detect the output voltage (V_{out}). When the sensors are at the high resistance state, the V_{out} of Wheatstone bridge will decrease because the magnetic fringing field of the magnetic particles exerting magnetic moments in the free layer of the spin valve is opposite to the external magnetic field, and vice versa. The different value of V_{out} between without with and without 650 Gauss magnetic field is defined as $\Delta V=V(H)-V(0)$, where V(H) and V(0) refer to the V_{out} of the

GMR sensor with and without 650 Gauss magnetic field, respectively. ΔV is proportional to the MR value where the greater the ΔV, the magnetic moments on the free layer are easier to magnetize by the external magnetic field.

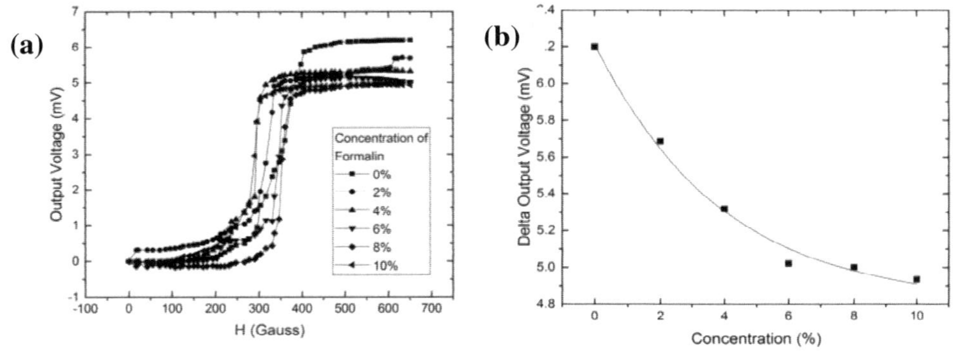

Figure 2. (a) The output voltage of wheatstone-bridge and (b) delta output voltage as function of formalin concentration.

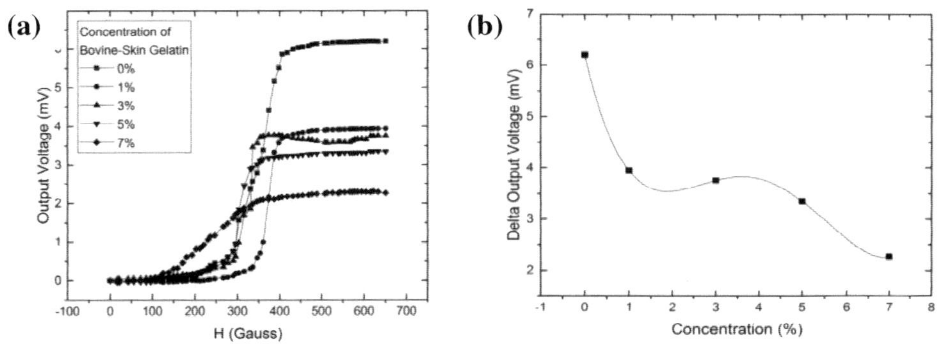

Figure 3. (a) The output voltage of wheatstone-bridge and (b) delta output voltage as function of bovine-skin gelatin concentration.

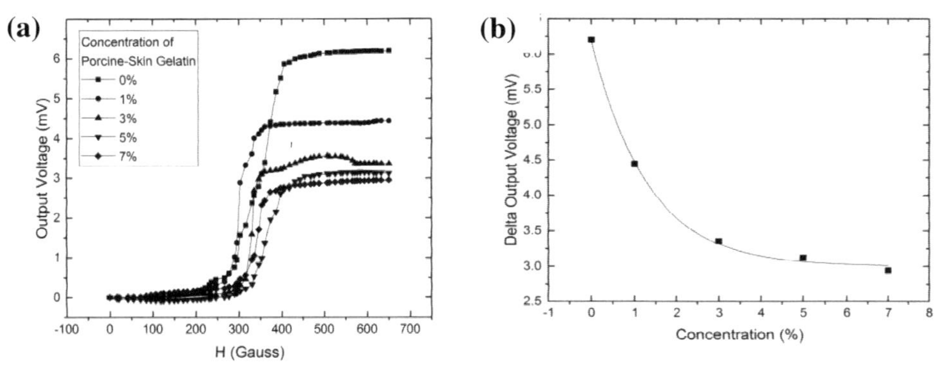

Figure 4. (a) The output voltage of wheatstone-bridge and (b) delta output voltage as function of porcine-skin gelatin concentration.

Figure 2(a), 3(a) and 4(a) shows that sensor output voltage rising from H=0 for various biomolecule concentration and then tend to saturated at H=650 Gauss, implying that magnetic particles binding has reached equilibrium on sensor surface. Lower saturated signal level is detected for higher concentration of biomolecule. The changes of the output voltage are a result of changes in resistivity of the CoFeB spin-valve thin film after coated with various biomolecules concentration. The different value of V_{out} between without with and without 650 Gauss magnetic field are plotted in figure 2(b), 3(b) and 4(b).

The delta output also decrease with the increase of biomolecule concentration. When the biomolecule concentration increase, PEG-coated Fe_3O_4 nanoparticles bond more biomolecules. Paramagnetic properties of the biomolecule made decrease in MNPs saturation magnetization. It caused a decrease in the effect of magnetic induction and made interaction between magnetic moments of nanoparticles and ferromagnetic layer weakened. This resulted magnetic moments of ferromagnetic layer was easier to magnetize by external magnetic field. All of this made magnetoresistance of thin film decrease with the increase of biomolecule concentration. The ΔV is proportional to the MR value. The schematic of biomolecule detection was shown in Figure 5.

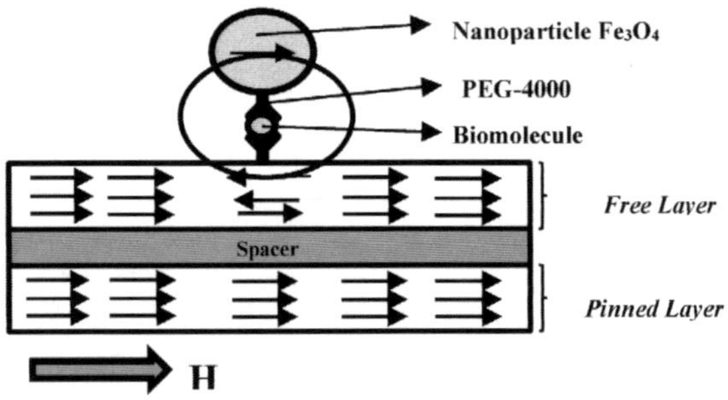

Figure 5. Schematic of biomolecule detection and interaction between ferromagnetic layer and nanoparticles Fe_3O_4.

4. Conclusion

A wheatstone bridge giant magnetoresistance (GMR) biosensor was successfully developed for biomolecule detection. The delta output voltage decrease with the increase of biomolecule concentration. This shift in delta output voltage is a direct proof that detection of formaldehyde and gelatin in various concentrations has been achieved.

Acknowledgement

This work was financially supported by Nanofabrication Platform Consortium Project of Nagoya University, Ministry of Culture, Sport, Science and Technology (MEXT) Nano-Project Platform, Japan and Grant of "Penelitian Berbasis Kompetensi", Ministry of Research, Technology and Higher Education of the Republic Indonesia.

References

[1] Li, G., Sun S., Wilson, R.J., White, R.L., Pourmand, N., and Wang, S.X 2006 *Sensors Actuators A: Phys.* **126** p. 98–106.
[2] Rife, J., Miller, M., Sheenan, P., Tamanaha, C., Tondra, M., and Whitman, L 2003 *Sensors Actuators A: Phys* **107** p. 209-18.
[3] Wang, S.X.. and Li, G. 2008, *IEEE Transaction on Magnetics* **44** p. 1687-1691.
[4] Yang, J., Zou, P., Yang, L., Ciao, J., Sun, Y., and Han, D. 2014 *Journal of Applied Surface Science* **303** p. 425-432.
[5] Basel, B 2009 *PEGylated Protein Drugs:Basic Science and Clinical Applications*, ISBN 978-3-7643-8678-8 147.
[6] Suharyadi, E., Pardede, I., Hasibuan, F.A 2016 *Proceeding Progress In Electromagnetic Research Symposium (PIERS)* p. 566-571.
[7] Reig, C., Cubells-Beltran, M.D., and Mu˜noz, D.R 2009 *Sensors* **9** 7919–7942.
[8] Nurpriyanti, I., Pardede, I., and Suharyadi, E. 2016 *Proceeding International Seminar on Sensors, Instrumentation, Measurement and Metrology (ISSIMM)* p. 32-36.
[9] Xu, J., Li, Q., Zhong, W., Zhang, Y., and Li, S 2016 *Journal of Magnetism and Magnetic Materials* **417** p. 25-29.
[10] Mathivanan 2007 N PC-based Instrumentation: Concepts and Practice *Prentice-Hall Of India.*

Wheatstone bridge-giant magnetoresistance (GMR) sensors based on Co/Cu multilayers for bio-detection applications

G Antarnusa, P Elda Swastika and E Suharyadi*
Department of Physics, Universitas Gadjah Mada, Yogyakarta, Indonesia

*Corresponding author: esuharyadi@ugm.ac.id

Abstract. A Wheatstone bridge-giant magnetoresistance (GMR) sensor was successfully developed for a potential biomaterial detection. In order to achieve this, a giant magnetoresistive $[Co(1.5nm/Cu(1.0nm)]_{20}$ multilayer structures have been fabricated by DC magnetron sputtering method, showing a magnetoresistance (MR) of 2.7%. The X-Ray diffraction (XRD) patterns showed that Co/Cu film multilayer has a high degree of crystallinity with a single peak corresponding to face-centered cubic (111) structure at $2\theta = 44.1°$. Co/Cu multilayers exhibit a soft magnetic behavior with the saturation magnetization (*Ms*) of 1489 emu/cc and the coercivity (*Hc*) of 11.2 Oe. The magnetite Fe_3O_4 nanoparticles used as a bimolecular labels (nanotags) were synthesized via co-precipitation method, exhibiting a soft magnetic behavior with *Ms* of 77.16 emu/g and *Hc* of 49 Oe. XRD patterns and transmission electron microscopy (TEM) images showed that Fe_3O_4 was well crystallized and it grew in their inverse spinel structure with an average size of around 10 nm. The GMR sensor design was used to detect a biomolecules of streptavidin magnetic particles with concentration 10, 20, 30, and 40 μl/ml and α-amylase enzyme with consentration 10, 20, 30, and 40 μl/ml captured using polyethylene glycol (PEG)/ Fe_3O_4 nanoparticles. Various applied magnetic fields of 0-650 Gauss have been performed using electromagnetic with the various currents of 0-5 A. Here, the final value of the output voltage signals for the streptavidin magnetic particles concentration is 1.2 mV (10 μl/ml). The output voltage changes with the increase of concentration. It was reported that the output voltage signal of the Wheatstone bridge exhibits log-linear function in real time measurement of the concentration of streptavidin magnetic particles and α-amylase enzyme respectively, making the sensor suitable for use as a biomolecule concentration detector. Thus, the combination of Co/Cu multilayer, Wheatstone bridge, magnetite and PEG polymer has potential application to be used in bio-detection applications where ultra-small bio-labels are needed.

1. Introduction

Magnetic bio-detection based on magnetic particles has been extensively studied nowadays. The basic principle is first labeling the targeting biomolecules with magnetic particles, and then these attached magnetic particles are captured by target-probe bio-molecular recognition and measured by magnetic sensors. There are remarkable advantages to use magnetic particles in the detection of biomolecules [1]. The detection of biological agents using magnetic sensors and labels, such as giant magnetoresistance (GMR) biosensor, Hall phenomena, magnetic tunnel junction (MTJ) sensor, and superconducting quantum interference device (SQUID) [2], etc., draws an increasing attention.

One of the most attracting bio-detection sensors to be studied due to its wide application is GMR. GMR is the change in electrical resistance of some materials in response to an external magnetic field. Among the numerous magnetic sensors, the GMR sensors are the most powerful tools for bio-detection due to their advantages, for instance their low-cost, portability, high sensitivity, real-time electronics readout, low-power consumption, and less complex instruments [3]. GMR effects can be observed by using thin film multilayers. Some thin film multilayers which can be used as GMR sensors are Co/Cu multilayer [4], Ag/Co multilayer [5], Fe/Cr multilayer [6], and NiFe/Ag multilayer [7].

GMR sensors can be used in detecting the presence of biomolecule which is called biosensor. The working principles of biosensor based on GMR sensor are: the labeling of biomolecule with nano-tags since not all biomolecules employ magnetic characteristic and the detection of GMR sensor which is indicated by the change of magnetoresistance (MR). Magnetic nanoparticles (MNPs), especially the iron oxide magnetite (Fe_3O_4) have gained considerable attention because of its characteristics, such as high magnetic saturation (Ms), biomolecule-disperse, biomolecule-responsive, and superparamagnetic. However, the use of MNPs may accompany risks and deleterious effects associated with their increased usage, particularly when they are used without appropriate characterization as biological agents. For these reasons, MNPs are typically surface-modified or coated with biocompatible polymer molecules. One of the most commonly used polymers is polyethylene glycols (PEGs) which can improve their dispersibility, and is biocompatible, nontoxic, and low-costed.

In the recent work, Suharyadi et al. had investigated the detection of Fe_3O_4 magnetic nanoparticles using GMR sensors based on Co/Cu multilayer by four-point probe system (FPPS). However, that experiment shows that signal output of GMR sensor was resistance [8]. Even though a unique resistance can be used as sensing element, a Wheatstone bridge setup is always a good recommendation as the starting step in the design of resistive sensors [9]. Thus, Nurpriyanti et al. investigated the detection of Fe_3O_4 magnetic nanoparticles using GMR sensors based on Co/Cu multilayer by Wheatstone bridge circuit to provide output signal of voltage [10]. However, this sensor was not portable and real-time. As a solution to these problems, auto-balancing bridge-based topologies were set up in order to improve the Wheatstone bridge features. Hence, the devices are driven by a control signal which depends on the system status that makes the devices portable and real-time.

In this paper, the possibility of detection streptavidin magnetic particles and α-amylase enzyme utilizing GMR sensors in Wheatstone bridge is investigated using Co/Cu multilayer films. Furthermore, Fe_3O_4 magnetic nanoparticles are used to be more comparable for conjugating biomolecules. In addition, developing Wheatstone bridge-GMR biosensor more portable and able to read signal output in real-time for various biomolecules is also studied.

2. Experiment details

A multilayer GMR film structure was designed as $[Co(1.5nm)/Cu(1.0nm)]_{20}$ fabricated by DC magnetron sputtering under pressure was about 3×10^{-5} Pa on silicon substrates. The magnetic property measurement of the Co/Cu multilayer was carried out at room temperature by using vibrating sample magnetometer (Riken Denshi Co Ltd, VSM). The GMR sensor was connected to a printed circuit board (PCB) with two part probe and was wire-bounded directly to the pads on the PCB. The Keithley 2401 source meter provides constant current for the GMR sensor and measures the V_{out} of the Wheatstone bridge in magnetic fields using electromagnetics between 0 and 650 Gauss in the field in-plane geometry at room temperature. In all measurement, a DC current of 10 mA was applied to the Wheatstone bridge for the measurement of V_{out}. For a small variation of resistance in sensors, the constant-current mode can offer more linear response and higher sensitivity than the constant-voltage mode [11]. Moreover, it can avoid the heating effect in the sensor as well.

The Fe_3O_4 magnetic nanoparticles were synthesized by chemical co-precipitation method [8,10]. Room temperature VSM measurements have been conducted by using Riken Denshi Co Ltd vibrating sample magnetometer (VSM). Transmission electron microscopy (TEM) analysis was performed using a JEOL JEM-1400. The streptavidin magnetic particles (Sigma-Aldrich) were dispersed in ethanol by sonication for 5 min and vortexed for 15 s. Four different concentrations (10, 20, 30, and 40 μl/ml) were prepared. Fe_3O_4/PEG was synthesized following a simple two steps co-precipitation approach with polyetthylene glycole 4000 (PEG-4000) with concentration ratio of 0.5 mg: 0.5 mg respectively for Fe_3O_4/PEG. For labeling process, 0.1 gram of Fe_3O_4/PEG was stirred and added with enzyme α-amylase (Sigma-Aldrich) with concentrations 10, 20, 30, and 40 μl/ml. Furthermore, the mixture dissolved in 1 ml of distilled water by means of sonicated for 30 min at room temperature.

Fourier Transform Infrared (FTIR) spectra of the Fe_3O_4 and Fe_3O_4/PEG were measured with FTIR spectroscopy. Transmission Electro Microscope (TEM) JEOL JEM-1400 was used for characterizing the microstructure of the Fe_3O_4 and Fe_3O_4/PEG. The characterization of Fe_3O_4/PEG and the addition of α-amylase enzyme were only done for observing the functional groups by using FTIR spectroscopy.

3. Result and discussion

The Fe_3O_4 have exhibited good magnetic response and are easily attracted to a magnet placed beside with superparamagnetic behavior. The saturation magnetic moments, coercivity field and remanent magnetization obtained are 77.16 emu/gram, 49 Oe and 7.68 emu/gram respectively. TEM performed for both Fe_3O_4 and Fe_3O_4/PEG. The low-resolution image shows the existence of spherical Fe_3O_4 of about 10 nm with uniform grain size and good dispersibility although in some place still agglomerate. The addition of PEG can decrease the agglomeration and reduce the particle size. In addition, ring diffraction patterns for both of them, are closely related to XRD analysis, which corresponding to the crystal plane (220), (311), (400), (440), and (511) [8,10].

A structure of the Co/Cu multilayer thin films for the GMR sensor was applied for the detection of biomolecules for streptavidin magnetic particles and α-amylase enzyme. The magnetoresistance (MR) of the sensor changes linearly with the logarithm of streptavidin magnetic particles and α-amylase enzyme concentration as shown in Fig. 1(a) and (b), respectively.

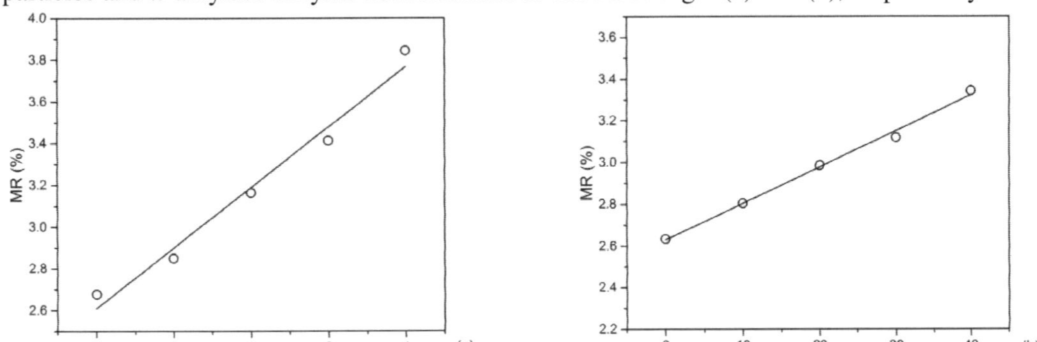

Figure 1. MR ratio as a function of (a) streptavidin magnetic particles concentration and (b) α-amylase enzyme concentration.

The direction of Co/Cu multilayer is in-plane to electromagnetics strips. In the absence of an external magnetic field, GMR sensors have median magnetoresistance (MR) and they work at the most sensitive region. When magnetic nanoparticle labels are bound to the sensor surface, they can be magnetized as dipoles, and they introduced stray fields may reduce the magnetic field. The change in magnetic field leads to a change in MR of the sensor, which can be measured and read out primary detection signal using the Wheatstone bridge.

The typical real-time binding curves (Change of output voltage vs. external magnetic field) for magnetic streptavidin with different concentrations (10, 20, 30, and 40 μl/ml) on the GMR sensors in Wheatstone bridge are shown in Fig. 2a. The signal from the GMR sensors were treated with the same volume of ethanol without streptavidin magnetic particles served as the background reference. Beginning at H=0, the signal for various streptavidin magnetic particles concentration showed a rise, which reflects real-time binding of MNPs to sensor surface. These rising signals saturate within 650 Gauss, implying that magnetic particles binding has reached equilibrium on sensor surface. Higher saturated signal level is detected for higher concentration of streptavidin magnetic particles. Their final average signals for logarithm of streptavidin magnetic particles concentrations are plotted in Fig. 2b.

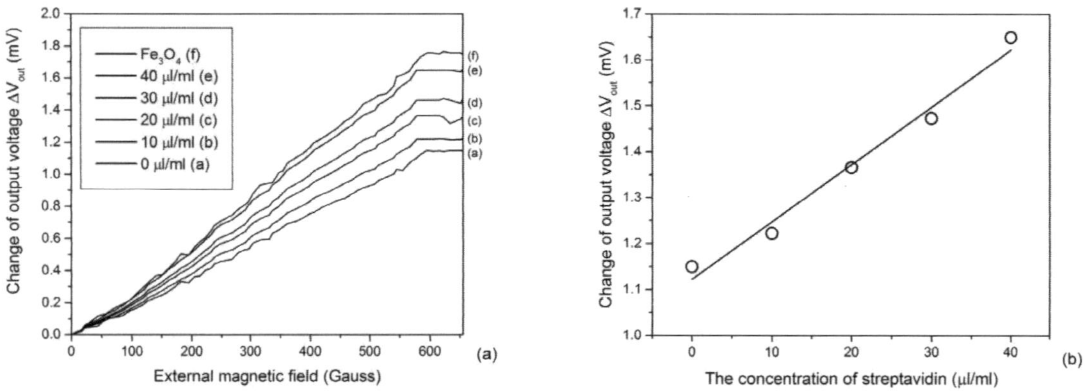

Figure 2. (a) Output voltage versus External magnetic field for the GMR sensor bio-detection of the magnetic streptavidin with different concentrations and (b) Plot of the sensor output signal versus the concentration of streptavidin magnetic particles.

Furthermore, we also demonstrate detection of α-amylase enzyme with concentration 10, 20, 30, and 40 μl/ml. Similar binding curves are also observed for streptavidin magnetic particles. The reported signal values are shown in Fig 3. For GMR bio-detection technology; the final output signal has a relationship with the bound number of MNPs [12]. Consequently, for detecting a biological or chemical agent using GMR biosensor, it is critical to build a model that the number of bound MNPs is dependent on the added amount of the biological target.

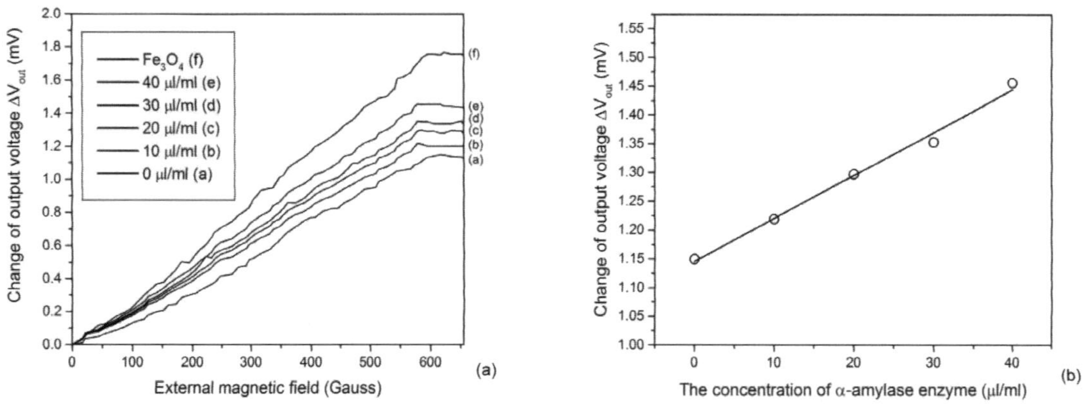

Figure 3. (a) Output voltage versus external magnetic field for the GMR sensor bio-detection of the α-amylase enzyme with different concentrations and (b) Plot of the sensor output signal versus the concentration of α-amylase enzyme.

The increase of output voltage is caused by interaction between magnetic moments of MNPs and ferromagnetic strength. The magnetic moments of ferromagnetic layer were harder to magnetize by external magnetic fields. Thus, these cause an increase of the output voltage. The addition of streptavidin magnetic particles and α-amylase enzyme respectively caused an increase in magnetic induction so the voltage increased with the increase of streptavidin magnetic particles and α-amylase enzyme concentrations respectively.

4. Conclusion

The output voltage changes with the increase of concentration. It was reported that the output voltage signal of the Wheatstone bridge exhibits log-linear function in real-time measurement of the concentration of streptavidin magnetic particles and α-amylase enzyme respectively, making the sensor suitable for use as a biomolecule concentration detector since it can detect the quantity of biomolecules. Thus, the combination of Co/Cu multilayer, Wheatstone bridge, magnetite and PEG polymer has potential application to be used in bio-detection applications where ultra-small bio-labels are needed.

Acknowledgement

This work was financially supported by Nanofabrication Platform Consortium Project of Nagoya University, Ministry of Culture, Sports, Science and Technology (MEXT) Nano-Project Platform, Japan and Grant of "Penelitian Berbasis Kompetensi", Ministry of Research, Technology and Higher Education of the Republic Indonesia.

References

[1] Rife, J., Miller, M., Sheenan, P., Tamanaha, C., Tondra, M., and Whitman, L 2003 *Sensors Actuators A: Physics* **107** 209-218
[2] Kotitz, R., Matz, H., Trahms, L., Koch, H., Weitschies, W., Rheinlander, T., Semmler, W., and Bunte, T 1997 *IEEE Trans. Appl. Supercond* **7** 3678–3681
[3] Li, G., Sun, S., Wilson, R.J., White, R.L., Pourmand, N., and Wang, S.X 2006 *Sensors Actuators A: Physics* **126** 98–106
[4] Rajasekaran, N., Mohan, N., Chelvane, J.A., and Jagannathan, R 2012 *Journal of Magnetism and Magnetic Materials* **324** 2983-2988
[5] Angelakeris, M., Papaioannou, E.T., Poulopoulos, P., Valassiades, O., and Flevaris, N.K 2003 *Sensors and Actuators A* **106** 91-95
[6] Baibich, M.N., Broto, J.M., Fert, A., Nguyen Van Dau, F., Petroff, F., Etienne, P., Creuzet, G., Friederich, A., and Chazelas, J 1988 *Physical Review Letters* **61** 2472–2475
[7] Rochaz, L.V., Cuchet, R., and Vaudaine, M.H 2000 *Sensors and Actuators* **81** 53-56
[8] Suharyadi, E., Pardede, I., and Hasibuan, F.A 2016 *Proceeding Progress In Electromagnetic Research Symposium (PIERS)* 566-571
[9] Reig, C., Cubells-Beltran, M.D., and Mu˜noz, D.R 2009 *Sensors* **9** 7919–7942
[10] Nurpriyanti, I., Pardede, I., and Suharyadi, E 2016 *Proceeding International Seminar on Sensors, Instrumentation, Measurement and Metrology (ISSIMM)* 32-36
[11] Mathivanan 2007 N PC-based Instrumentation: Concepts and Practice *Prentice-Hall Of India.*
[12] Schotter, J., Kamp, P.B., Becker, A., Pühler, A., Reiss, G., and Brückl, H 2004 *Biosensors Bioelectron* **19** 1149–1156

UV Light-Driven Photodegradation of Methylene Blue by Using $Mn_{0.5}Zn_{0.5}Fe_2O_4/SiO_2$ Nanocomposites

I P T Indrayana[1], T Julian[2], E Suharyadi[2]*

[1] Department of Physics, Universitas Halmahera, Maluku Utara, INDONESIA
[2] Department of Physics, Universitas Gadjah Mada, Yogyakarta, INDONESIA

*email : esuharyadi@ugm.ac.id

Abstract. The photodegradation activity of nanocomposites for 20 ppm methylene blue solution has been investigated in this work. Nanocomposites $Mn_{0.5}Zn_{0.5}Fe_2O_4/SiO_2$ have been synthesized using coprecipitation method. The X-ray diffraction (XRD) pattern confirmed the formation of three phases in sample $Mn_{0.5}Zn_{0.5}Fe_2O_4/SiO_2$ i.e., $Mn_{0.5}Zn_{0.5}Fe_2O_4$, $Zn(OH)_2$, and SiO_2. The appearance of SiO_2 phase showed that the encapsulation process has been carried out. The calculated particles size of $Mn_{0.5}Zn_{0.5}Fe_2O_4/SiO_2$ is greater than $Mn_{0.5}Zn_{0.5}Fe_2O_4$. Bonding analysis via vibrational spectra for $Mn_{0.5}Zn_{0.5}Fe_2O_4/SiO_2$ confirmed the formation of bonds Me-O-Si stretching (2854.65 cm^{-1}) and Si-O-Si asymmetric stretching (1026.13 cm^{-1}). The optical gap energy of $Mn_{0.5}Zn_{0.5}Fe_2O_4/SiO_2$ was smaller (2.70 eV) than $Mn_{0.5}Zn_{0.5}Fe_2O_4$ (3.04 eV) due to smaller lattice dislocation and microstrain that affect their electronic structure. The $Mn_{0.5}Zn_{0.5}Fe_2O_4/SiO_2$ showed high photodegradation ability due to smaller optical gap energy and the appearance of SiO_2 ligand that can easily attract dye molecules. The $Mn_{0.5}Zn_{0.5}Fe_2O_4/SiO_2$ also showed high degradation activity even without UV light radiation. The result showed that photodegradation reaction doesn't follow pseudo-first order kinetics.

1. Introduction

Heterogeneous photocatalyst has remarkable properties, both photocatalytic and magnetic properties. Among all alternative heterogeneous photocatalysts, ferrite nanoparticles have gained a great interest of researchers. Ferrite nanoparticles offer many advantages of having narrow band gap which capable of absorbing visible light radiation. Ferrites posses' spinel crystal structure which is able to improve the efficiency because of the availability of extra catalytic sites [1]. Some types of ferrite materials that have been used for photocatalyst are Fe_3O_4, Fe_2O_3, $CuFe_2O_4$, $NiFe_2O_4$, $ZnFe_2O_4$, $CoFe_2O_4$, $MnFe_2O_4$, $CaFe_2O_4$, $MgFe_2O_4$, $ZnFe_2O_4/SnO_2$, $BiFeO_3$, etc [2]. Another type of ferrite nanoparticles is Zn-doped $MnFe_2O_4$. The $Mn_{1-x}Zn_xFe_2O_4$ is a mixed spinel ferrite which has high magnetic properties and narrow band gap (E_g < 3.0 eV) [3-5]. Polymer coated $Mn_{1-x}Zn_xFe_2O_4$ nanoparticles also showed interesting properties for their biocompatibility. For instance, a remarkable microstructure, morphology, and magnetic properties of silica-coated nanoparticles $Mn_{0.4}Zn_{0.6}Fe_2O_4$ had reported by [5].

To the best of our knowledge, only small number of researchers reported the photocatalytic activity of $Mn_{0.5}Zn_{0.5}Fe_2O_4/SiO_2$ nanocomposites. The magnetic property of $Mn_{0.5}Zn_{0.5}Fe_2O_4$ is useful for the separation between nanocomposites and the solution only by using external magnetic field after the photodegradation process was conducted. Herein, we predominantly focus on reporting the photodegradation activity of $Mn_{0.5}Zn_{0.5}Fe_2O_4/SiO_2$ for 20 ppm methylene blue dye with a driven of UV light radiation. Further, microstructural and optical properties of nanocomposites were also investigated for main consideration in explaining the photodegradation phenomena. The reaction mechanism during photodegradation was also studied. It was found that the photodegradation activity of $Mn_{0.5}Zn_{0.5}Fe_2O_4/SiO_2$ is much higher than bare $Mn_{0.5}Zn_{0.5}Fe_2O_4$. The degradation activity of $Mn_{0.5}Zn_{0.5}Fe_2O_4/SiO_2$ for methylene blue without UV light radiation was also investigated.

2. Experimental Method

2.1 Preparation and Characterization of $Mn_{0.5}Zn_{0.5}Fe_2O_4/SiO_2$

There are many researchers have reported the procedure of preparing ferrite nanoparticle such as $Mn_{0.5}Zn_{0.5}Fe_2O_4$. In this research, the nanoparticles were prepared by using coprecipitation method such as reported by [4,6]. For an encapsulation, the Si_2O_3 solution was used as the starting material for the formation of SiO_2 layer. The encapsulation process was conducted by using Ströber method as reported by [7]. The nanocomposites were characterized by their microstructural parameters by using XRD technique. The FTIR measurement was performed to study the vibrational mode of functional group bonds within samples. Particles' morphology and electron diffraction image of samples were investigated by using Transmission electron microscope (TEM). The optical properties of samples were investigated by using diffuse reflectance UV-visible spectrophotometer.

2.2 Photodegradation Testing of $Mn_{0.5}Zn_{0.5}Fe_2O_4/SiO_2$

The photodegradation testing was carried out for sample $Mn_{0.5}Zn_{0.5}Fe_2O_4/SiO_2$ nanocomposites and bare $Mn_{0.5}Zn_{0.5}Fe_2O_4$. Each sample was tested for their photodegradation activity for 120 minutes. The measurement of absorbance spectra was performed in every 30 minutes using UV-visible light spectrophotometer with a wavelength of 664.095 nm. The $Mn_{0.5}Zn_{0.5}Fe_2O_4/SiO_2$ degradation activity was also investigated in condition without UV light radiation. The 10W UV source of 275 nm in wavelength was used in this testing. Powder $Mn_{0.5}Zn_{0.5}Fe_2O_4/SiO_2$ nanocomposites and bare $Mn_{0.5}Zn_{0.5}Fe_2O_4$ of each mass 20 mg were used to degrade 30 ml of 20 ppm methylene blue dye. Each prepared powder was added into methylene blue solution in Erlenmeyer flask. The mixer was stirred at room temperature with and without UV light radiation.

3. Result and Discussion

3.1 XRD Analysis

The X-ray diffraction pattern of samples nanocomposite $Mn_{0.5}Zn_{0.5}Fe_2O_4/SiO_2$ and nanoparticles $Mn_{0.5}Zn_{0.5}Fe_2O_4$ can be shown in Figure 1a. These patterns showed that samples contain two different phases, such as $Mn_{0.5}Zn_{0.5}Fe_2O_4$ (JCPDS card no. 10-0467) and $Zn(OH)_2$ (card no. 41-1359). A phase of SiO_2 (card no. 45-0131) was investigated in samples of $Mn_{0.5}Zn_{0.5}Fe_2O_4/SiO_2$. The appearance of $Zn(OH)_2$ might be caused by an incomplete ferritization reaction of zinc hydroxide compound into ferrite $Mn_{0.5}Zn_{0.5}Fe_2O_4$. Figure 1a shows that both diffraction patterns have four characteristic peaks of $Mn_{0.5}Zn_{0.5}Fe_2O_4$ which represent the mixed spinel cubic FCC structure and space group of Fd3m. Peak of Miller plane (311) has the highest intensity. The formation of sharp peaks indicates that samples have high crystallinity. The calculated particles size by using Scherrer's method [5,6] of $Mn_{0.5}Zn_{0.5}Fe_2O_4/SiO_2$ is 20.73 ± 0.24 nm greater than $Mn_{0.5}Zn_{0.5}Fe_2O_4$ of 18.71 ± 0.24 nm. The lattice parameter of $Mn_{0.5}Zn_{0.5}Fe_2O_4/SiO_2$ is approximately close to $Mn_{0.5}Zn_{0.5}Fe_2O_4$.

3.2 Structural Morphology and Electron Diffraction

Structural morphology and electron diffraction of $Mn_{0.5}Zn_{0.5}Fe_2O_4/SiO_2$ was performed by using TEM image and SAED (Selected area electron diffraction) such Figure 1b. The structural morphology of $Mn_{0.5}Zn_{0.5}Fe_2O_4/SiO_2$ couldn't be determined. Although the encapsulation with SiO_2 layer has been carried out, the sample of nanocomposites still exhibited agglomeration. Hence, the specific layer of SiO_2 on the surface of bare nanoparticles definitely difficult to be investigated of its size. The SAED image shows that the formation of polycrystalline materials. These are indicated by the dotted rings appear in electron diffraction images. The SAED image also confirmed the appearance Miller planes of (220), (311) and (511) that could be diffracted by the electron in the crystal lattice of the sample. Somehow, the intensity of electron diffraction of $Mn_{0.5}Zn_{0.5}Fe_2O_4/SiO_2$ is lower than $Mn_{0.5}Zn_{0.5}Fe_2O_4$ nanoparticles. The diffraction rings of (220), (311) and (511) were consistent with result presented by X-ray diffraction pattern (Figure 1). Among these rings, the diffraction ring of Miller plane (311) has the sharpest intensity in agreement with the highest intensity shown by X-ray diffraction pattern.

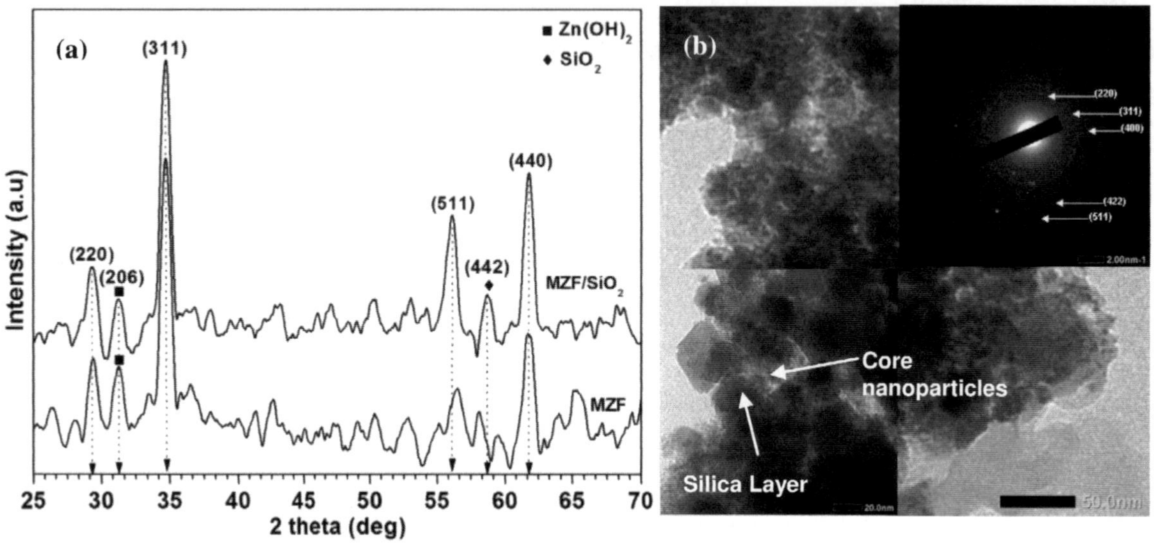

Figure 1. (a) X-ray diffraction patterns of sample nanocomposites and (b) TEM images of $Mn_{0.5}Zn_{0.5}Fe_2O_4/SiO_2$ nanocomposites with its SAED image.

3.3 Vibrational Spectra

Bonding analysis of samples can be performed base on spectra provided by FTIR in Figure 2. The bands at 570.93 cm^{-1} and 447.49 cm^{-1} of $Mn_{0.5}Zn_{0.5}Fe_2O_4$ spectra are assigned as Me-O tetrahedral stretching and Me-O octahedral stretching vibration consecutively. These two characteristic bands show the formation of a bond between metal salt and oxygen in entire spinel ferrite structure. Sample $Mn_{0.5}Zn_{0.5}Fe_2O_4/SiO_2$ nanocomposites show some absorption bands. Peaks at 1026.13 cm^{-1} are assigned as Si-O-Si asymmetric stretching vibration. This vibrational frequency shifted toward lower value due to a higher bond length between Si-O after interacting with the metal ion of bare nanoparticles. The appearance band at 2854.65 cm^{-1} is assigned as Me-O-Si stretching which definitely shows the formation of a bond between bare nanoparticles and the silica layer [5]. It implicates that the encapsulation of silica layer on the surface of bare nanoparticles was successfully carried out. Further, bands at 1627.92 cm^{-1}, 2337.72 cm^{-1} and 3394.72 cm^{-1} also confirmed the existence of hydroxyl group bonds within sample $Mn_{0.5}Zn_{0.5}Fe_2O_4/SiO_2$ nanocomposites.

3.4 Optical Properties

The optical properties of samples were studied in accordance to the diffuse reflectance UV-visible spectra. The absorbance spectra vs photon wavelength can be shown in Figure 3. The calculated optical band gap energy of sample $Mn_{0.5}Zn_{0.5}Fe_2O_4$ and $Mn_{0.5}Zn_{0.5}Fe_2O_4/SiO_2$ in this result are bigger than reported by [5]. The optical band gap energy shows the energy that required by an electron to excite from valence band to conduction band. Nevertheless, in this case, we specifically can't determine the type of inter-band excitation whether direct or indirect optical band gap energy. The optical gap energy of $Mn_{0.5}Zn_{0.5}Fe_2O_4/SiO_2$ nanocomposite is smaller than bare $Mn_{0.5}Zn_{0.5}Fe_2O_4$ due to two reasons. First, surface encapsulation of bare nanoparticle with SiO_2 layer might cause a change in microstrain and lattice density of the sample. These changes will affect the electronic band structure of the material that makes smaller optical gap energy. Second, the decrease of the optical band gap is caused by the increase of particle size as an effect of quantum confinement [8].

3.5 Photodegradation Activity

Photodegradation activity of samples was carried out with variation of irradiation time such Figure 4a. For a pure methylene blue solution, the dye degradation actually run very slow due to the absent of photocatalyst which drives the reduction-oxidation reaction. The existence of $Mn_{0.5}Zn_{0.5}Fe_2O_4$ can drive gradually photodegradation reaction of methylene blue dye. Photodegradation rate of dye with only $Mn_{0.5}Zn_{0.5}Fe_2O_4$ nanoparticle was still lower than $Mn_{0.5}Zn_{0.5}Fe_2O_4/SiO_2$ nanocomposites. A sample of $Mn_{0.5}Zn_{0.5}Fe_2O_4/SiO_2$ nanocomposites has the highest photodegradation rate up to 97.72%.

First, the existence of SiO_2 layer on the surface of bare nanoparticles increases the surface reactivity of samples. The ligand of vicinal silanols group that is resulted by SiO_2 layer increases the adsorption ability of nanocomposites to attract dye molecules to the surface of particles. Further, the SiO_2 layer is a good adsorbent to address several organic molecules such methylene blue dye. This fact enhances the adsorption ability of nanocomposites toward methylene blue molecules on the surface of nanocomposites [9]. Second, the optical band gap energy of sample $Mn_{0.5}Zn_{0.5}Fe_2O_4/SiO_2$ nanocomposites is smaller than $Mn_{0.5}Zn_{0.5}Fe_2O_4$ nanoparticles. Therefore, the generation of electron-hole pairs wills more effective than larger optical band gap energy materials for the same photon energy radiation. Studying the photodegradation rate of each sample can give chance to understand the photodegradation kinetic. The photodegradation kinetic was displayed in Figure 4b. All samples do not follow the pseudo-first order kinetic reaction. Somehow, it is difficult to decide what model of kinetic reaction the samples follow for this case. Furthermore, it needs to be studied in more detail about the mechanism of reaction and possibly to carry out an experiment by using appropriate probes such as GC-MS and photoluminescence spectrophotometer. Moreover, $Mn_{0.5}Zn_{0.5}Fe_2O_4/SiO_2$ can be considered as a candidate for nanophotocatalyst or nano-adsorbent for organic pollutant because of their potential that has been proven even though without UV-light radiation.

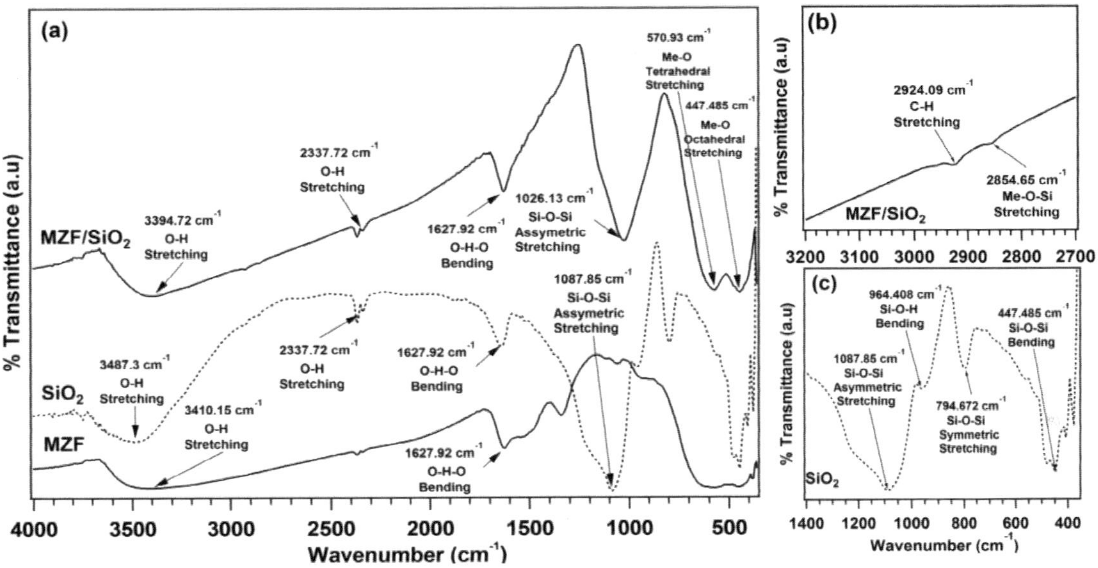

Figure 2. (a) Full vibrational spectra of samples nanocomposites; (b) zoom spectra of nanocomposites $Mn_{0.5}Zn_{0.5}Fe_2O_4/SiO_2$; (c) zoom spectra of silica layer.

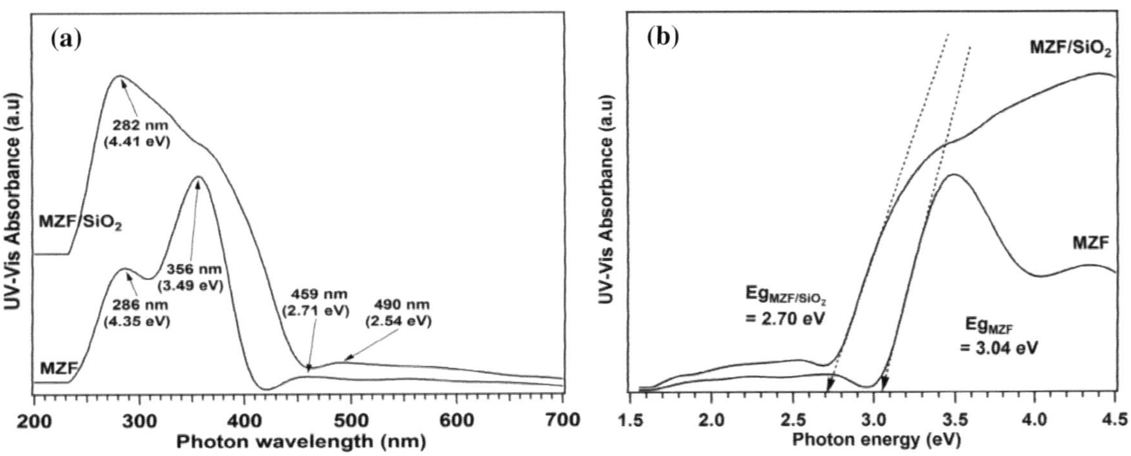

Figure 3. (a) Absorbance spectra vs photon wavelength and (b) the Tauch's plot.

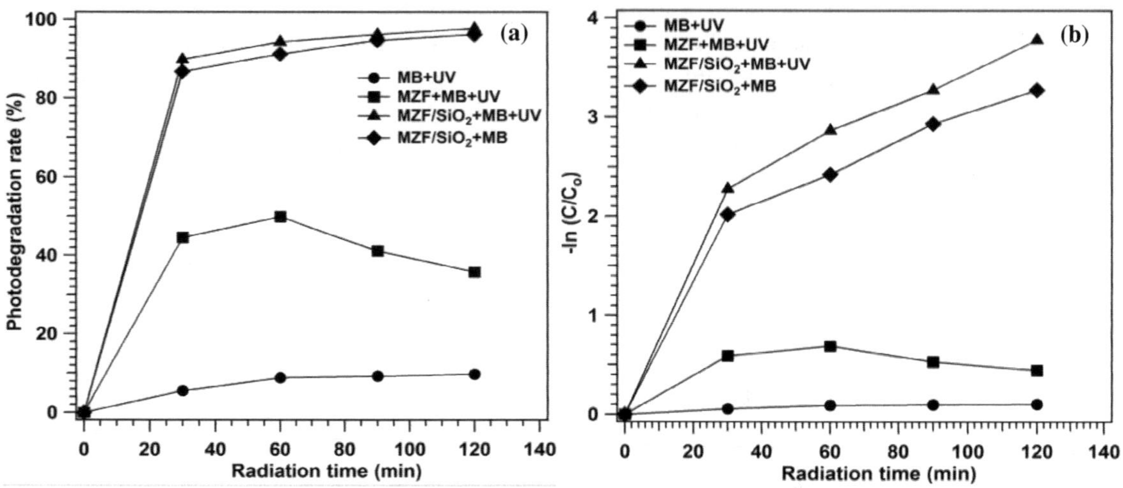

Figure 4. (a) Photodegradation rate of $Mn_{0.5}Zn_{0.5}Fe_2O_4$ and $Mn_{0.5}Zn_{0.5}Fe_2O_4/SiO_2$ nanocomposites under and without UV-light radiation; (b) Photodegradation kinetic of samples.

4. Conclusion

The $Mn_{0.5}Zn_{0.5}Fe_2O_4$ and $Mn_{0.5}Zn_{0.5}Fe_2O_4/SiO_2$ nanocomposite have been synthesized using coprecipitation method. The bare nanoparticle $Mn_{0.5}Zn_{0.5}Fe_2O_4$ has a particles size smaller than $Mn_{0.5}Zn_{0.5}Fe_2O_4/SiO_2$. The FTIR spectra confirmed the formation of the nanoferrites structure and the appearance of vibration band Si-O-Si stretching and Me-O-Si stretching. Nanocomposites $Mn_{0.5}Zn_{0.5}Fe_2O_4/SiO_2$ have smaller optical band gap than bare $Mn_{0.5}Zn_{0.5}Fe_2O_4$ due to smaller surface defect and larger particles size. The encapsulation of SiO_2 layer on the surface on bare nanoparticles enhances photodegradation activity of $Mn_{0.5}Zn_{0.5}Fe_2O_4/SiO_2$. Nanocomposites $Mn_{0.5}Zn_{0.5}Fe_2O_4/SiO_2$ also showed a high degradation activity for methylene blue dye even though without UV-light radiation. Nanocomposites $Mn_{0.5}Zn_{0.5}Fe_2O_4/SiO_2$ can be a promising nano-photocatalyst and nano-adsorbent for organic pollutants.

5. Acknowledgments

The author is grateful to the laboratory staffs of Material and Instrumentation Laboratory, Physics Department, Universitas Gadjah Mada for the facilities. This reported work was financially supported by LPDP Thesis Scholarship 2016 of contract ID: PRJ-681 /LPDP.3/2016 and Nanofabrication Platform Consortium Project of Nagoya University, Minister of Culture, Sports, Science and Technology (MEXT) Nano-Project Platform, Japan of Period 2012 - 2016.

6. References

[1] Oliveira L C A, Fabris J D, Rios R R V A, Mussel W N and Lago R M 2004 *Appl. Catal.* A **259** 253–259
[2] Kharisov B I, Dias H V R and Kharissova O V K 2014 *Arab. J. Chem.* **10** 1–13
[3] Laohhasurayotin K, Pookboonmee S, Viboonratanasri D and Kangwansupamonkon W 2012 *Mater. Res. Bull.* **47** 1500–07
[4] Indrayana IP T, Siregar N, Suharyadi E, Kato T and Iwata S 2016 *J. Phys.: Conf. Ser.***776** 012021
[5] Baykal A, Güner A, Demir A, Esir S, and Genç F 2014 *Ceramics Int.* **40** 13401–13408
[6] Siregar N, Indrayana IP T, Suharyadi E, Kato T and Iwata S 2017 *J. Phys.: Conf. Ser.* **202** 012048
[7] Liberman A, Mendez N, Trogler W C and Kummel A C 2014 *Surf. Sci. Rep.* **69** 132 - 158
[8] Mote V D, Dargad J S, urushotham Y and Dole B N 2015 *Ceramics Int.* **41** 15153–15161
[9] Ali M B, Bouaifel F B, Elhouichet H, Sieber B, Addad A, Boussekey L, Férid M and Boukherroub R 2015 *J. Colloid Interf. Sci.* **457** 360–369

IOP Conf. Series: Journal of Physics: Conf. Series **1011** (2018) 012063 doi:10.1088/1742-6596/1011/1/012063

Crystallite Size-Lattice Strain Estimation and Optical Properties of $Mn_{0.5}Zn_{0.5}Fe_2O_4$ Nanoparticles

I P T Indrayana[1] and E Suharyadi[2*]

[1] Department of Physics, Universitas Halmahera, Maluku Utara, INDONESIA
[2] Department of Physics, Universitas Gadjah Mada, Yogyakarta, INDONESIA

*email : esuharyadi@ugm.ac.id

Abstract. In the present work, we performed William-Hall plot using uniform deformation model (UDM) to estimate the crystallite size and lattice strain of $Mn_{0.5}Zn_{0.5}Fe_2O_4$ with various calcination temperature. The calculated crystallite sizes are 25.86 nm, 29.55 nm and 24.97 nm for nanoparticles which were calcined at a temperature of 600°C, 800°C and 1000°C, consecutively. The strain of nanoparticles has value in the order of 10^{-3}. Controlling the calcination temperature will facilitate a change in crystallinity of nanoparticles and influence their crystallite size and strain of the crystal lattice. The optical band gap energy of samples nanoparticles is in a range of 1.09 eV – 3.30 eV. Increasing calcination temperature increased the direct and indirect band gap energy. The Urbach energy was found to increase with increased of gap energy. These results demonstrated that higher structural and optical properties of $Mn_{0.5}Zn_{0.5}Fe_2O_4$ can be obtained from a higher calcination temperature.

1. Introduction

Ferrite nanocrystalline semiconductor such as $Mn_{1-x}Zn_xFe_2O_4$ has been widely studied because the need for materials which have excellent magnetic and optical properties for data storage media [1]. For this reason, since a decade $Mn_{1-x}Zn_xFe_2O_4$ has gained a great interest of many researchers. The $Mn_{1-x}Zn_xFe_2O_4$ ferrite nanoparticle possesses a cubic close-packed crystal structure with FCC bravais lattice. These cations both divalent and trivalent are distributed in two crystal sites such as tetrahedral and octahedral sites. Specifically, the preferred site of those cations in an entire crystal structure of $Mn_{1-x}Zn_xFe_2O_4$ is in accordance with the cation distribution rule of mixed-spinel structure [2]. The cation distribution is practically affected by the preparation method, calcination temperature and addition of proper cations into the spinel structure [3,4,5,6]. Those factors generate a migration of cations between the two sites and possibly cause defect and deformation of a lattice. Lattice defect and deformation can be in form of lattice strain and lattice dislocation. Physically, lattice strain is in connection with the optical properties of nanoparticles as a study reported by Sheena et al [5]. For this reason, studying the lattice strain of $Mn_{1-x}Zn_xFe_2O_4$ nanoparticles and the method for its estimation is very important for understanding the electronic structure as well as its optical and magnetic properties.
In the previous work, we had performed a study about the effect of calcination temperature on vibrational spectra and magnetic properties of $Mn_{0.5}Zn_{0.5}Fe_2O_4$ nanoparticles [7]. The particle size of nanoparticles was calculated by using Scherrer's equation. The Scherrer's equation gives rougher crystallite size estimation because it doesn't take attention to the contribution of strain and instrumental correction on diffraction peak broadening [8]. In the purpose for addressing the effect of lattice strain on the microstructure of $Mn_{0.5}Zn_{0.5}Fe_2O_4$ nanoparticles, herein we perform a study of crystallite size and lattice strain estimation using William-Hall (W-H) plotting method with uniform

deformation model (UDM). In addition, the crystallite size of nanoparticle obtained by UDM was compared with the Scherrer's equation. Furthermore, the optical properties of $Mn_{0.5}Zn_{0.5}Fe_2O_4$ nanoparticles were also studied in relation to the lattice strain which was tuned by calcination temperature.

2. Experimental Method

2.1 Preparation and Characterization of $Mn_{0.5}Zn_{0.5}Fe_2O_4$

Samples of $Mn_{0.5}Zn_{0.5}Fe_2O_4$ nanoparticles with difference calcination temperature were synthesized by using coprecipitation method such as reported by [6,7]. The calcination temperatures used in this study were 600°C, 800°C and 1000°C. The samples were characterized by their microstructural parameters using XRD technique. The optical properties of samples were characterized by using diffuse reflectance UV-visible spectrophotometer.

3. Result and Discussion

3.1 XRD Analysis

In a previous study [7], we had performed the diffraction pattern of $Mn_{0.5}Zn_{0.5}Fe_2O_4$ nanoparticles with various of calcination temperature. In accordance with those patterns, we can estimate the crystallinity and structural characteristics of $Mn_{0.5}Zn_{0.5}Fe_2O_4$ matrix along with the texture coefficient (TC). A study of TC was carried out by considering the grains of nanoparticle which have certain preferred orientations and are diffracted by X-ray as diffraction planes. A single crystalline material possesses a preferred orientation while a polycrystalline material possesses some preferred grain orientations. Physically, TC value informs the distribution of crystallographic orientations. The TC value also informs the degree of crystallinity of samples. Sample with high TC values (TC > 1) has the high degree of crystallinity. Practically, TC determines the comparison between relative intensities of the observed peak with those of the standard powder diffraction intensities of the same compound [9]. The calculation of TC can be carried out by using equation 1 [9],

$$TC(hkl) = \frac{I(hkl)/I_o(hkl)}{(1/N)\sum_{n=1}^{N} I_n(hkl)/I_{on}(hkl)}. \quad (1)$$

The *Io* (*hkl*) determines the intensities of standard powder diffraction for the (*hkl*) diffraction plane. The *I* (*hkl*) is the relative intensity of the observed peak, while *N* determines the number of the observed diffraction peak in X-ray diffraction pattern. The TC (*hkl*) values of sample $Mn_{0.5}Zn_{0.5}Fe_2O_4$ with various calcination temperatures can be displayed in Figure 1.

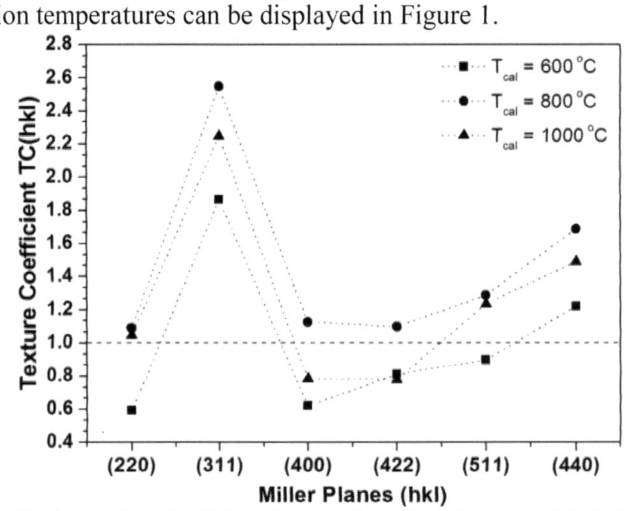

Figure 1. Texture coefficient of each diffraction peaks related to specific Miller plane within crystal $Mn_{0.5}Zn_{0.5}Fe_2O_4$ nanoparticles.

Figure 1 shows the variation of texture coefficient for three system of the planes of the matrix. If the value of TC (*hkl*) ≈ 1 for all the diffraction planes (*hkl*), then the sample nanoparticles are randomly oriented crystalline such as the standard powder (JCPDS). The results are shown in Figure 1 confirmed that the diffraction plane of (311) has the highest value of TC (*hkl*) ascribed that this direction is the highest preference orientation of grains. The peaks of (220), (400), (422) and (511) for samples calcined with 600°C and 800°C have TC (*hkl*) in range of 0 to 1. This indicates the lack of grains to orient in those directions. As increasing calcination temperature, the TC (311) increases, the preferential growth of crystallites in this direction which is perpendicular to (*hkl*) plane is greater [9]. Consequently, increasing the calcination temperature will facilitate tuning the crystallinity of nanoparticles.

The estimation of crystallite size and lattice strain have been done by using W-H plot with uniform deformation model (UDM). Mathematically, the UDM can be represented by equation 2 [10],

$$\beta_{hkl}\, cos\,\theta = \left(\frac{k\lambda}{d}\right) + 4\varepsilon\, sin\,\theta \tag{2}$$

where d is the crystallite size of nanoparticles, k is the Scherrer's constant (≈ 0.89); ε is the lattice strain; β_{hkl} is the broadening of the diffraction peak measured at FWHM (in radians); λ is the wavelength of X-Ray (1.5406 Å) and θ is the diffraction angle of peak. Plots are drawn between term 4sinθ (X-axis) and term β_{hkl} cosθ (Y-axis). The results can be displayed in Figure 2 and Table 1.

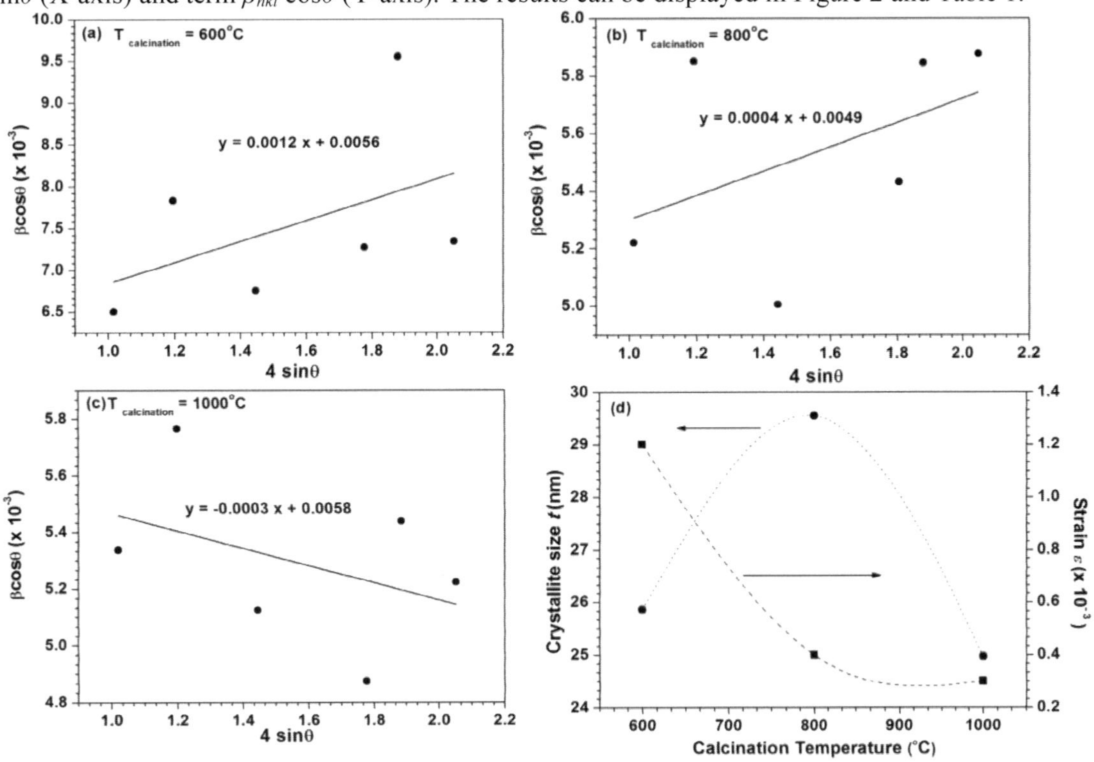

Figure 2. W-H plot of $Mn_{0.5}Zn_{0.5}Fe_2O_4$ calcined at temperature 600°C (a), 800°C (b), 1000°C (c) and the relation between crystallite size-strain nanoparticle and calcination temperature (d).

Table 1. Crystallite size and strain of $Mn_{0.5}Zn_{0.5}Fe_2O_4$ with Schrerr and W-H UDM method.

No.	Calcination Temperature (°C)	Scherrer' crystallite size (nm) [6]	W-H UDM	
			Crystallite size (nm)	Strain (x10^{-3})
1	600	18.30	25.86	1.2
2	800	24.80	29.55	0.4
3	1000	22.30	24.97	0.3

Figure 2a – Figure 2c show that all data are poorly linear. These linear trendlines are really poor compared than result reported by [11]. The results reported by [11,12] also show that UDM method give worse linear trendline of the data than Size-Strain Plot (SSP) method. This suggests the use of SSP method to obtain better result. The crystallite sizes of nanoparticle which have been calculated by using the UDM were greater than Scherrer method. These results agree with the calculation reported by [7,9,10]. Nanoparticles size estimated using Scherrer's equation is quite different from that obtained by WH plotting. This difference due to the strain that can induce a greater broadening in the diffraction peak [11]. Figure 2 shows that there are positive and negative slopes in the plots. The positive and negative slopes indicate the existence of compressive and tensile strain on the lattice of nanoparticles [10]. The crystallite size of nanoparticles decreased with an increase of calcination temperature due to the densification process exhibited by nanoparticles.

3.2 Optical Properties of $Mn_{0.5}Zn_{0.5}Fe_2O_4$

The optical properties of $Mn_{0.5}Zn_{0.5}Fe_2O_4$ nanoparticles can be represented by the UV-visible absorbance spectra, absorption index and the Urbach energy such as Figure 3. There are two absorbance peaks found in each absorbance spectra. The maximum electronic excitation occurs at a wavelength of 341 nm for sample calcined at a temperature of 600°C. Hence the wavelengths of 348 nm and 327 nm were found to be the maximum electronic excitation for sample calcined at a temperature of 800°C and 1000°C, consecutively. Small absorbance peaks at 493 nm, 446 nm, and 436 nm were found due to scattering effect of photon and surface defect of crystal nanoparticles [9].

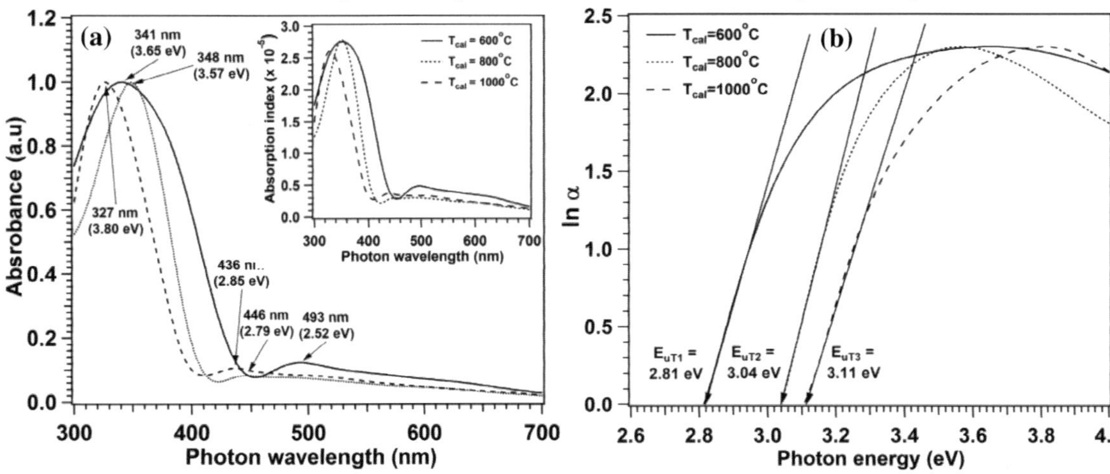

Figure 3. The Absorbance spectra with absorption index (a) and the Urbach energy (b) of $Mn_{0.5}Zn_{0.5}Fe_2O_4$ nanoparticles for each calcination temperature of 600°C, 800°C and 1000°C.

By plotting of the absorption index against wavelength as shown in Figure 3(b) indicating that the absorption index of $Mn_{0.5}Zn_{0.5}Fe_2O_4$ is very small (in order of 10^{-5}). These small values of the absorption index indicate the low and weakly absorption were possessed by the samples [9].

The optical gap energy of nanoparticles is determined by using Tauch's plot which represented by the equation 3 [9] and displayed in Figure 4,

$$\alpha h\nu = \left(\frac{A}{t} h\nu\right) = B\left(h\nu - E_g\right)^n \quad (3)$$

where α represents the absorption coefficient, $h\nu$ is the photon energy, A is the normalized absorbance, t the sample's thickness, E_g is the optical gap energy, B is an energy independent constant and n is a constant which represents the optical transitions of electrons in electronic band edge of nanoparticles. The $n = 2$ for the indirect allowed transition and $n = ½$ for the direct allowed transition.

The direct optical band gap energy of $Mn_{0.5}Zn_{0.5}Fe_2O_4$ nanoparticles increased with an increase of calcination temperature. This direct gap energy facilitates electron to excite from valence band to conduction band with the same wavenumber k. The $Mn_{0.5}Zn_{0.5}Fe_2O_4$ nanoparticles show two indirect optical band gap energies. Their values are smaller than direct optical gap energy. These energies are needed by the electron to excite from valence band to near conduction band edge via phonon with different wavenumber *k*. Increasing calcination temperature is followed by increase indirect optical gap energy due to change in the electronic band structure of nanoparticles during calcination. In fact, the E_g value changed as the E_u value changes. The E_g value increased when the E_u value increased. Therefore, the excited electron preferred to transit from the valence band to the band tailing.

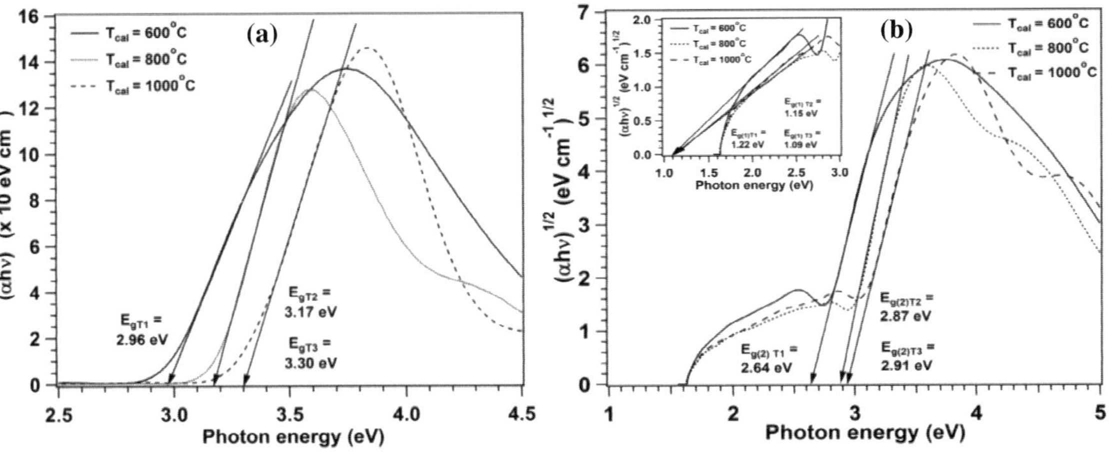

Figure 4. Direct optical band gap energy (a) and indirect optical band gap energy (b) of $Mn_{0.5}Zn_{0.5}Fe_2O_4$ nanoparticles for each calcination temperature of 600°C, 800°C and 1000°C.

4. Conclusion

The crystallite size and lattice strain of $Mn_{0.5}Zn_{0.5}Fe_2O_4$ nanoparticles with different calcination temperature have been successfully estimated by using the W-H plotting with the UDM method. The calculated crystallite size is greater than that obtained by using Scherrer's formula due to the contribution of strain which facilitates defect and lattice imperfection of nanoparticles. The prepared $Mn_{0.5}Zn_{0.5}Fe_2O_4$ nanoparticles have weak absorption ability to the UV-visible radiation and low optical band gap energy which depend practically on the calcination temperature.

5. Acknowledgments

The authors thank the laboratory staffs of Material and Instrumentation Laboratory, Physics Department, Universitas Gadjah Mada for the facilities. This reported work was financially supported by LPDP Thesis Scholarship 2016 of contract number PRJ-681 /LPDP.3/2016 and Nanofabrication Platform Consortium Project of Nagoya University, Japan of Period 2012 - 2016.

6. References

[1] Demir A, Guner S, Bakis Y, Esir S and Baykal A 2014 *J. Inorg. Organomet. Polym.* **24** 729–736
[2] Mathew D S and Juang R S 2007 *Chem. Eng. J.* **129** 51–65
[3] Mozaffari M, Arani M E and Amighian J 2010 *J. Magn. Magn. Mater.* **322** 3240–3244
[4] Klencsár Z, Tolnai G, Korecz L, Sajó I, Németh P, Osán J, Mészáros S and Kuzmann S 2013 *Solid State Sci.* **24** 90-100
[5] Sheena P A, Priyanka K P, Sabu N A, Sabu B and Varghese T 2014 *Nanosystem: Phys., Chem., Math.* **5** 441 – 449
[6] Siregar N, Indrayana IP T, Suharyadi E, Kato T and Iwata S 2017 *J. Phys.: Conf. Ser.* **202** 012048
[7] Indrayana IP T, Siregar N, Suharyadi E, Kato T and Iwata S 2016 *J. Phys.: Conf. Ser.* **776** 012021
[8] Gonçalves N S, Carvalho J A, Lima Z M and Sasaki J M 2012 *Mater. Lett.* **72** 36–38
[9] Hafez M, Yahia I S and Taha S 2014 *Spectro. Acta. A: Molec. Biomolec. Spect.* **127** 521–529
[10] Sivakami R, Dhanuskodi S and Karvembu R 2016 *Spectro. Acta. A: Molec. Biomolec. Spect.* **152** 43–50
[11] Zak K A, Majid W H A, Abrishami M E and Yousefi R 2011 *Solid State Sci.* **13** 251-256
[12] Tatarchuk T, Bououdina M, Macyk W, Shyichuk O, Paliychuk N, Yaremiy I, Al-Najar B and Pacia M 2017 *Nanoscale Res. Lett.* **12** 141

IOP Conf. Series: Journal of Physics: Conf. Series **1011** (2018) 012064 doi:10.1088/1742-6596/1011/1/012064

Reflectance spectra characteristics from an SPR grating fabricated by nano-imprint lithography technique for biochemical nanosensor applications

Jalu Setiya Pradana and Rahmat Hidayat

Physics of Magnetism and Photonics Research Division, Faculty of Mathematics and Natural Sciences, Bandung Institute of Technology, Bandung-Indonesia

E-mail: jspradana@gmail.com, rahmat@fi.itb.ac.id

Abstract. In this paper, we report our research work on developing a Surface Plasmon Resonance (SPR) element with sub-micron (hundreds of nanometers) periodicity grating structure. This grating structure was fabricated by using a simple nano-imprint lithography technique from an organically siloxane polymers, which was then covered by nanometer thin gold layer. The formed grating structure was a very well defined square-shaped periodic structure. The measured reflectance spectra indicate the SPR wave excitation on this grating structure. For comparison, the simulations of reflectance spectra have been also carried out by using Rigorous Coupled-Wave Analysis (RCWA) method. The experimental results are in very good agreement with the simulation results.

1. Introduction

A sensor is a device that can convert a measurable physical quantity into an electrical signal, which often requires a transducer for that purpose [1]. One example of transducers is an optical transducer by utilizing the Surface Plasmon Resonance (SPR) phenomenon. SPR is a plasmon resonance phenomenon with electromagnetic waves causing quantized plasmon oscillation on the surface of metal-dielectric interface. SPR is very sensitive to the dielectric constant change of the dielectric medium so that it is widely applied for biochemical nanosensors applications [2]. The refractive index of may be strongly influenced by the type of molecule and intermolecular interactions. SPR based sensors are very suitable for label-free sensing and real-time sensing, which can measure the concentration and kinetic of biomolecules as well as distinguish the specific interaction between two molecule pairs, such as an antigen-antibody pair.

The requirement for excitation of SPR wave is that the vector component of light waves propagating at the boundary must be equal to the surface plasmon vector (SP) [3]. SPR wave can be created if the incident light can be coupled to the metal layer by using a prism, grating, or waveguide structure. This is related to the intersection between the dispersion curve of free-space light propagation and the dispersion curve of the SP wave on the metal/dielectric interface. SPR excitation by using grating has many advantages such as simpler instrumentation set-up and more practical. However, the measured SPR spectra are more complicated to be analyzed, particularly for a grating with non-sinusoidal structure, such as a square-wave periodic structure. In this case, the incident light will be also subjected to diffraction, where the difraction pattern is more complex while some portions of the diffracted beams may be not coupled to the SPR waves [4].

In order to estimate the reflectance spectra of periodic structures, simulation based on the rigorous coupled wave analysis (RCWA) method may be very useful for that purpose. RCWA calculation is different from the coupled wave analysis (CWA) method in which RCWA is a full-vectorial calculation derived from the Maxwell equation without using an approximation as in CWA. [6-9] In this RCWA formulation, the dielectric constant of the structure and the light waves are transformed into their Fourier series representation. Therefore, this method is very relevant for periodic structure case. The calculation results are then a representation of electric fields in each mode. Another benefit from RCWA is that the differential equations of Maxwell equation are transformed into a matrix form and solved as an eigen value problem. The wave equation (for two-dimensional structure) is then given by

$$\frac{\partial^2}{\partial z^2}\begin{bmatrix} s_x \\ s_y \end{bmatrix} - \boldsymbol{\Omega}^2 \begin{bmatrix} s_x \\ s_y \end{bmatrix} = 0 \tag{1}$$

where $\mathbf{s}_x$ and $\mathbf{s}_y$ are the electric fields, which can be found as the eigenvector solution, while $\boldsymbol{\Omega}$ stands for wave propagation constants. [6] All $\mathbf{s}_x$, $\mathbf{s}_y$ and $\boldsymbol{\Omega}$ are wrtitten in the matrix form.

In the present work, the grating structures with sub-micron (hundreds of nanometers) periodicity were made from an organically modified siloxane hybrid polymers by employing nano-imprint lithography. The advantages of this polymer are optically transparent and easy to be fabricated from its gel phase without requiring high temperature. The cross-linking formed in this polymer gives some advantages leading to good mechanical strength and thermal stability as well as chemical resistance so that the fabricated device can be more durable.

2. Experiments

2.1. Precursor preparations
The preparation of the polymer precursor was similar as in our previous work, where we have demonstrated the fabrication of sinusoidal corrugated grating structure with hundreds of nanometers periodicity by laser interferometry from this polymer precursor. [10,11] Cross-lingking of siloxane groups of the 3-(Trimethoxysilyl) propyl methacrylate 98% (TMSPMA, Aldrich) monomer was formed during the sol-gel process. For that purpose, TMSPMA was dissolved in ethanol and added by HCl 1 N as the catalyst. The solution was continuously stirred for 72 hours at 50°C until forming a gel phase. This gel was added by photo-initiator Irgacure-819 (Ciba Speciality Chemical Inc.), which was then used as the polymer precursor for the nano-imprint lithography process. This process was carried out in dark room to avoid undesired photo-polymerization.

2.2. SPR grating fabrication
The sub-wavelength grating structure was made by employing the nanoimprint lithography technique. The polymer precursor was poured over a silicon stamp and then followed by a vacuum process for 30 minutes. The next step was UV-curing, where the polymer precursor layer on the top of the silicon stamp was irradiated by the 405 nm laser for 15 minutes. The resulted grating structure was peeled off from the silicon stamp and then evaporated with very thin layer of gold with the thickness of about 50 nm.

2.3. SPR Characterization
The fabricated grating SPR structure was characterized by measuring its reflectance spectrum by using an experimental setup as shown in Figure 1. A tungsten lamp was used as the polychromatic light source. Slit and pin holes were used to keep the light beam from spreading out beyond the grating size. The polarization was adjusted by using a polarisator. The reflectance light beam was measured by a CCD based spectrophotometer.

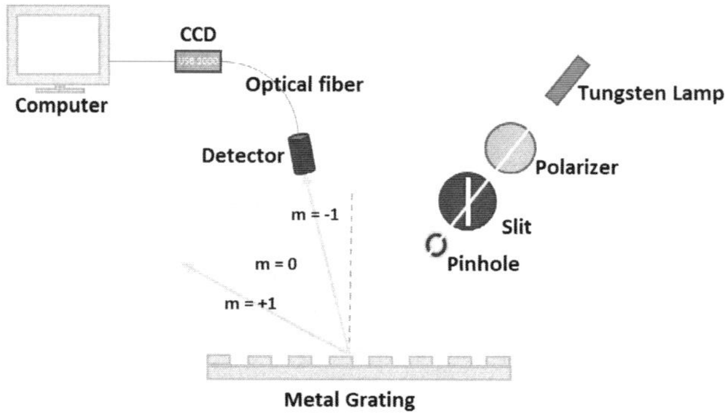

Figure 1. Schematic diagram of the SPR reflectance measurement setup.

3. Results and Discussions

Figure 2 shows the scanning electron microscopy (SEM) images of the fabricated grating structure with a periodicity of about 500 nm. These images show that the grating has a surface corrugated structure of a periodic square wave shape with corrugation depth of around 200 nm.

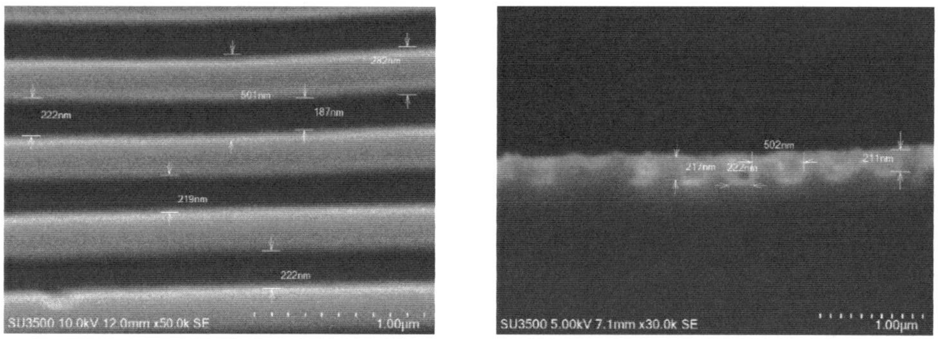

(a) (b)

Figure 2. The SEM images of (a) the surface and (b) cross-section area measured from the fabricated grating structure with a periodicity of about 500 nm.

In this paper, we shall just focus on the reflection spectrum from the diffracted beam at the zero order (i.e. $m = 0$). For this small periodicity, the appearance of high order diffraction beams ($m > 0$) is not easy because it can appear at a relatively small incident angle or large diffraction angle. This is related to the relationship given by

$$n_{diff} \sin\theta_{diff} = n_{inc} \sin\theta_{inc} + m\frac{\lambda}{\Lambda} \quad (2)$$

where m is the diffraction order, λ is the light wavelength and Λ is the grating periodicity. For $m = 0$, the diffraction angle is same as the reflectance angle. For small diffraction angle, however, high diffraction order (larger than $m = 0$) can be observed only when $\lambda \leq \Lambda$, but this wavelength is in the region with strong light absorption from the gold layer.

Figure 3 clearly shows the presence of SPR phenomenon in this fabricated grating, which is indicated by the appearance of SPR dip in the reflection spectrum for the *p*-polarization or TM mode. The dip is not observed for the *s*-polarization. For both s- and p- polarization, there is a strong light absorption by gold at wavelengths shorter than 550 nm. The s-polarization shows even stronger absorption because the direction of the electric field in the s-polarized is parallel to the grating structure repetition.

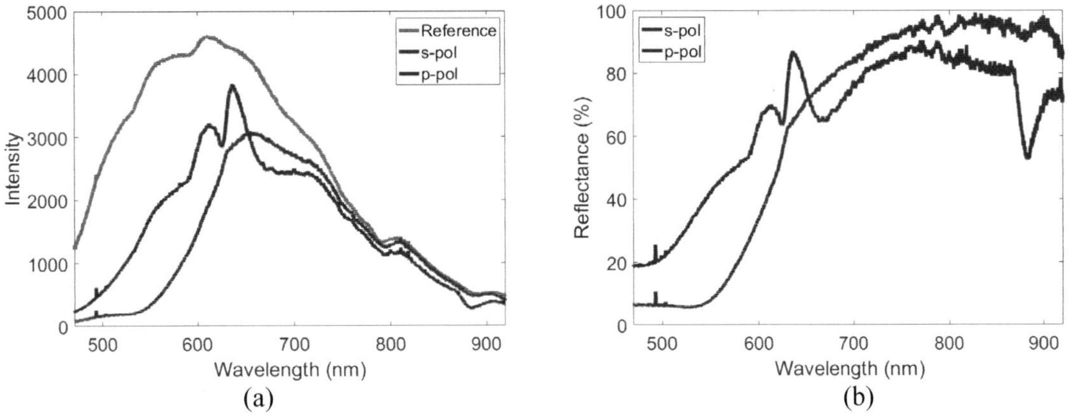

Figure 3. (a) The reflection intensity spectra and (b) the reflectance spectra of the zero order diffraction beam ($m = 0$) measured from the fabricated metal grating for the s- and p-polarization at an incident angle of 10°.

Figure 4 shows the reflectance spectra taken for several incident angles. For s-polarization, there is no dip and no change of spectral shape with increasing incident angle. For the p-polarization, however, the SPR dip is clearly red-shifted with increasing incident angle. The third dip (3) shifts to a longer wavelength, while the first and second dips (1 and 2) seem to shift and merge into one dip. The appearance of multi dips here is slightly different from the SPR spectrum characteristics in perfect sinusoidal SPR grating. [10]

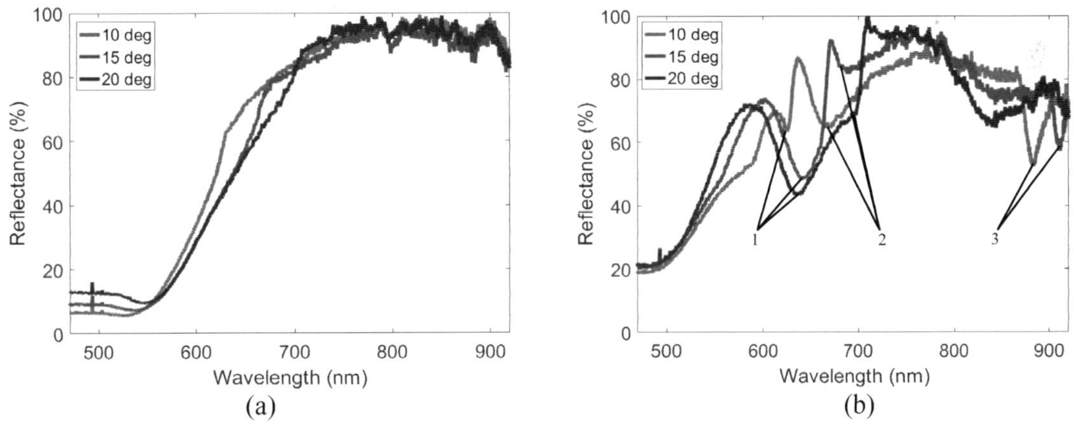

Figure 4. The reflectance spectrum for (a) the s-polarization and (b) the p-polarization of the grating for diffraction order $m = 0$ taken at several incident angles.

Comparing Figure 4 and Figure 5, we can clearly see that the reflectance spectra calculated from the RCWA method are similar to the experimental reflectance spectra. The simulation was done by using a modified computation program initially developed by the Zhang group of the Georgia Institute of Technology. [12] The spectral shape and the trend of dip shifting dip are almost similar, where the maximum difference between the dip wavelengths obtained from the simulation results and from the experimental measurements, for the same incident angle, is just about 3%.

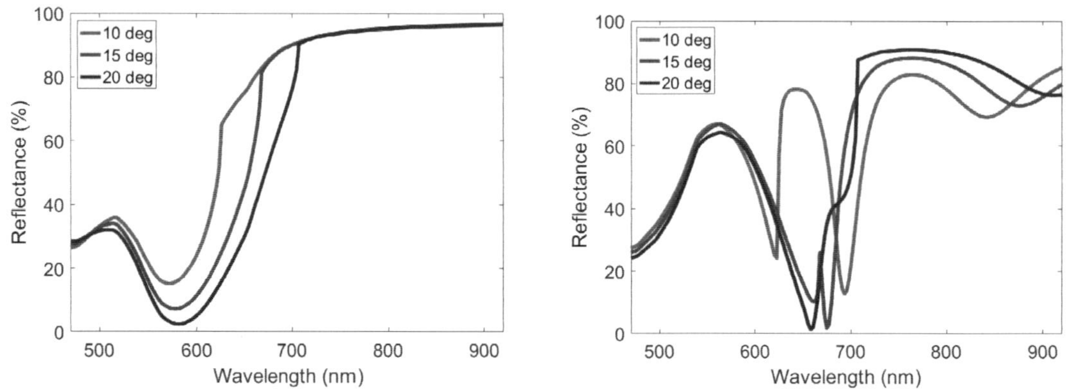

Figure 5. The simulation results of reflectance spectra for (a) the s-polarization and (b) the p-polarization of a grating with the same structure as the fabricated grating in this work, which are showed for zero diffraction order (m = 0) at various incident angles.

4. Conclusion

The nano-imprint lithography technique and hybrid polymer precursor used in this work is very effective for fabricating submicron (hundreds nanometer periodicity) grating structure. The SPR wave can be effectively excited in the fabricated grating structure, which is indicated by red-shifting of SPR dip with increasing incident angle. This systematic SPR dip shifting will be useful as a sensing tool in a biochemical nanosensors. The experimental results are in good agreement with the simulation results calculated by using the RCWA method.

5. References

[1] Wilson J 2005 *Sensor Technology Handbook* (Oxford: Elsevier Inc)
[2] Homola J *et al* 2006 Surface Plasmon Resonance (SPR) Sensors *Surface Plasmon Resonance Based Sensors* (*Springer Series on Chemical Sensors and Biosensors* vol 4) ed J Homola (Berlin: Springer) chapter 2 pp 69 - 91
[3] Maier S A 2007 *Plasmonics: Fundamentals and Applications* (UK: Springer)
[4] Liang G *et al* 2016 Fiber Optic Surface Plasmon Resonance–Based Technique: Fabrication, Advancement, and Application *Crit. Rev. Anal. Chem.* **46** 213-223
[5] Wismanto W Y *et al* 2012 Emission enhancement characteristics of oxazine in pmma matrix influenced by surface plasmon polariton induced on sinusoidal silver grating *J. Nonlinear Opt. Phys. Mater.* **21** 1250013
[6] Merhari L 2009 *Hybrid Nanocomposites For Nanotechnology* (US: Springer)
[7] Pathak A *et al* 2016 SPR Based Cone Tapered Fiber Optic Chemical Sensor for the Detection of Low Water in Ethanol *AIP Conference Proceedings* vol 1728 (US: AIP Publishing) pp 020017
[8] Han K S *et al* 2013 Fabrication of 3D nano-structures using reverse imprint lithography *Nanotechnology* **24** 045304
[9] Peng S and Morris G M 1995 Efficient implementation of rigorous coupled-wave analysis for surface-relief gratings *J. Opt. Soc. Am. A.* **12** 1087-1096
[10] Hidayat R *et al* 2012 *Journal of Engineering and Technological Sciences* **44** 207-219
[11] Hidayat R *et al* 2012 *Polymer. Adv. Tech.* **23** 1264-1270
[12] http://zhang-nano.gatech.edu/

Acknowledgement
The authors would like to appreciate the financial support from Program Penelitian, Pengabdian kepada Masyarakat, dan Inovasi (P3MI) KK ITB 2017.

Effect of TiO_2 Nanoparticles on Conductivity and Thermal Stability of PANI-TiO_2/Glass Composite Film

M Diantoro[1,2], **M Z Masrul**[1], **and A Taufiq**[1,2]

[1]Department of Physics, Faculty of Mathematics and Natural Sciences, Universitas Negeri Malang, Jl. Semarang 5, Malang 65145, Indonesia
[2]Minerals and Advanced Materials Laboratory, Universitas Negeri Malang Malang, Jl. Semarang 5, Malang 65145, Indonesia

Email: markus.diantoro.fmipa@um.ac.id

Abstract. Polyaniline (PANI), one of the conductive polymers has been intensively studied. The PANI has practically used in electronic device, and gas sensor. However, the PANI based on the electronic device has weaknesses such as fragile, and unstable of conductivity at high temperature and in an acidic environment. This obstacle can be solved by inducing other better thermal properties materials. One kind of the materials that meet the criteria is TiO_2. However, TiO_2 has limitation due to the low conductivity compared to the conductivity of PANI. Theoretically, this problem can be solved by reducing the particles size of TiO_2 down to nanometric size. In this study, we prepared a composite film with a various amount of nano TiO_2 to obtain information related the microstructure, electrical conductivity, and their thermal stability. The results of FTIR spectra analysis showed that the PANI-TiO_2 composites films were successfully synthesized which is indicated by the appearance of PANI and TiO_2 characteristic bonds associated with the references. XRD spectra showed the increased crystallinity of PANI-TiO_2 composite films along the addition of TiO_2. The SEM images presented the decreasing porosity of the samples by increasing the TiO_2 contents. Further, the conductivity of PANI-TiO_2 composite films decreased by increasing the TiO_2 nanoparticles. On the other hands, the thermal stability of the PANI composite films increased by increasing the TiO_2 nanoparticles.

Keywords: PANI, TiO_2, Nanoparticles, Film, Composites, Conductivity, Thermal Stability

1. Introduction

Polyaniline (PANI) is one of the conductive polymers that are intensively studied. The PANI has some advantages like easy synthesis and inexpensive compared to other conductive polymers such as polypyrrole, polythiophene and inorganic semiconductors [1]. Also, its conductivity can be easily controlled via doping processes. These advantages made PANI potentially applied in various fields especially gas sensor.

The change of conductivity in a PANI Emeraldine Salt (ES) can be easily measured associated with its sensitivity when being exposed to various gases. So that, it was suitable to use in gas sensor applications. The PANI based gas sensor shows high sensitivity when exposed to the gas at room temperature. However, The PANI-based gas sensor has disadvantages such as fragile, and unstable conductivity at high temperature and in an acidic environment [3]. This condition will decrease the level of its sensitivity. Thermal stability of the PANI ES become lower caused by chemical reaction of PANI conjugate system with oxygen or loss of dopant as chlorine [14] on a polymer chain that can limit its application. It was reported that the conductivity of the PANI-HCl decreased about 0.4 ± 0.03% S/cm [15] or 0.08% [12] by heating at 145 or 150 °C respectively from 25 S/cm. The decrease in conductivity is presumably caused by the formation of a PANI Emeraldine Base (EB).

The thermal stability of the PANI can be enhanced by introducing other materials such us a Titanium Dioxide (TiO_2) [4]. On the other hands, The TiO_2 has limitation due to the low conductivity compared to the PANI. Introducing TiO_2 reduced conductivity from 1.91 S/cm to 1.06 S/cm and from 4.65 S/cm to 0.054 S/cm [20]. The nano TiO_2 films synthesized using sol-gel with a *Tetrabutyl orthotitanate* and *Titanium Isopropoxide* (TTIP) precursor have conductivity 7×10^{-2} S/cm [5] and 1×10^{-2} S/cm [6] respectively. In this work, we report the influence of TiO_2 nanoparticles addition on microstructures, conductivity, and thermal stability related temperature change of the PANI- nano-TiO_2 film using a PVA solvents.

2. Experimental Methods

2.1 Synthesis Pani-TiO_2/PVA Composite Film

Anilin ($C_6H_5NH_2$) Merck (99%, PA), Distilled Water, Ammonium Peroxidisulfat $(NH_4)_2S_2O_8$ Merck (99%, PA), (Hydrochloric Acid) HCl (32%), $TiCl_3$ Merck (99%, PA), Glass Substrate, NH_3 Merck (99%, PA), *Polyvinyl Alcohol* (PVA) Merck (99%, PA). Pani was synthesized by using chemical polymerization methods. Meanwhile, TiO_2 was synthesized by using Co-Precipitation methods.

The PVA solvent was prepared by adding 0.05 g of PVA in 1.5 mL of distilled water, stirred at 400 rpm for two hours. A 0.2 g of PANI and 0, 0.02, 0.04, 0.06, 0.08 and 0.1 grams of TiO_2 powders dissolved in a solvent stirred at 750 rpm for 2 hours. The composite films were prepared by using spin coating method in 1×1 cm^2 glass substrate and spun it for 30 seconds at the speed of 500 rpm. The produced film was then dried at room temperature for three days.

2.2 Characterization

The functional chemical bonds that were appeared in the sample was characterized using spectrophotometer ZHIMADU IR Prestige-21. The crystallinity, grain size, and phase that was formed in the samples Characterized using diffractometer x-ray PHILIPS XPERT-type PRO with CuKα radiation sources ($\lambda 1 = 1.54060$). The morphological of the sample characterized using SEM FEI type INSPECT S50. Morphological that's mean grain from and porosity of the samples. Electrical conductivity measurement was obtained from 4-probe methods, with the distance between probes is 0.2 cm and 50 at room temperature, 60, 80, 100, and 120 °C using a constant current of 0.05 A.

3 Results and Discussion

3.1 Microstructure Data Analysis

The PANI Emeraldine Salt (ES) has functional bonding characteristics on the range wavenumber 400-2000 cm^{-1} as shown in Figure 1a. Bonding characteristic formed on the wavenumber in the range 1440-1600 cm^{-1} indicated the presence of the C= C bond of benzenoid rings and quinoid ring. The peak at a wavenumber of 1307 cm^{-1} indicates the presence of C-N bond or quinoid ring. This quinoid ring shows the existence π electron delocalization demonstrate through protonation process of polyaniline. The alternating bonding characteristic C = C and C-C in the PANI chain made the PANI potentially has conductive properties [7].

Figure 1b showed the bond nature that was formed on the PANI-ES film with PVA solvent. The form of films and powders and also the addition of the PVA solvent affected the strength properties of bonding force. The decrease of intensity occurred at C-Cl, C-H, C-N, and C = C bond [8], but the increase of intensity also happened on the C-O bond. It can be affected by remaining the PVA due to the temperature of the synthesis of the PVA solution just at 80 °C. The formation of hydrogen bonds occurred between hydroxyl PVA group with immine and the amino group of PANI chains that have a positive charge. IR spectrum of the PANI-TiO_2 film in Figure 1c was not much different with the IR spectrum of the PANI ES film, but it appeared that the peak intensity became weaker than the beginning. The new peak showed the appearance bonding of Ti-O-Ti at a wavenumber of 418.55 cm^{-1}

[4]. The appearance of this bonding characteristics without omitted PANI bonding characteristic was evidently explained the formation of the PANI-TiO_2/PVA composites film [11].

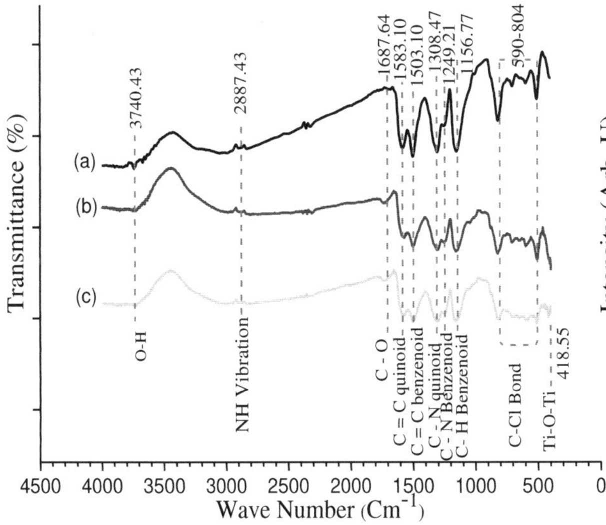

Figure 1. FTIR Spectrum of (a). PANI ES Powder, (b). PANI ES Film, and (c). PANI ES-Nano-TiO_2 Film with 0.08 g of TiO_2.

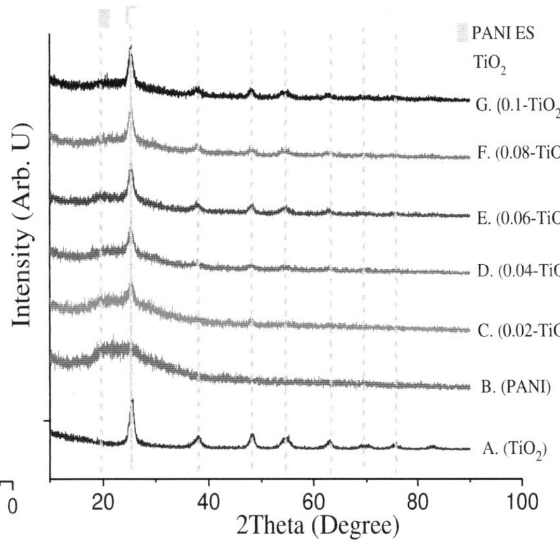

Figure 2. X-Ray Diffraction Patterns of Nano-TiO_2 Powder and PANI-Nano-TiO_2 Film with various TiO_2.

Figure 2 showed a powder diffraction pattern of the TiO_2 and the PANI ES-Nano TiO_2 film with a different mass of TiO_2. Figure 2a confirmed the formation of the TiO_2 anatase phase according to the match analyses the experimental data using High Score Plus software with reference of 00-021-1272. The data model used for the Cellref analysis was a tetragonal crystal structure with lattice parameters of a = 3.780 Å, b = 3.780 Å, c = 9.510 Å. The peaks analysis using Cellref indicates that the experimental data fit correctly with the reference of the TiO_2 anatase phase.

The grain size of the TiO_2 was 8.14 nm from the calculation using Scherrer equation. The diffraction patterns of PANI ES without TiO_2 (Fig. 3b) showing an amorphous phase with the bump positions around 2θ of 22° and 25° [8]. The use of PVA solvent does not give impact on the peak position of the amorphous phase at 2θ of 11, 20, and 23° [8]. The increase of TiO_2 nanoparticles in Figure 2 c-g, showed the increase of peaks intensity match with the reference [11], showing the increase of crystallinity from 0, 3.83, 8.25, 15.17, 15.47, 17.37% as function as the TiO_2 addition.

SEM image(doesnot included) analyses of the PANI film and PANI-TiO_2 Film showed that the PANI ES has a spherical grain-shaped morphology with an inhomogeneous distribution and some porosity or cavity. On the other hand, the PANI ES-TiO_2 film has granular shaped morphology with the homogenous distribution and little cavity or porosity. The spherical shape of TiO_2 particles prominent visible out around the surface of PANI ES. The spherical shape of TiO_2 seems fused with a surface of the PANI through adsorption mechanism when the process of synthesis. This process will lead agglomeration on the structure of the PANI ES-TiO_2[12]. The porosity of PANI-TiO_2 film with mass variations of 0 and 0.08 grams analyzed by ImageJ software analysis, indicates that the porosity of the film was 4.6% and 1.33% respectively. This porosity showed that the addition TiO_2 mass on the PANI ES enhanced the density of the PANI. The presence of the TiO_2 with 13.6 Wt % in the PANI ES film also confirmed by EDX data which was supported by XRD data. The addition of TiO_2 into the PANI ES film relatively caused reducing C, N, O elements. This occurred because these elements have been replaced by a Titanium (Ti) that from the TiO_2 compound.

3.2 Electrical Conductivity Measurement Against to Temperature

The room temperature electrical conductivity of PANI ES decreased proportionally with the increase of TiO_2 of 286.5, 226.3, 223.4, 205.5, 145.2 and 135.5 S/cm respectively for 0.0, 0.02, 0.04, 0.06, 0.08, and 1.0 fractions. The decrease of the film conductivity attributed to TiO_2 conductivity which is lower than the conductivity of PANI ES. Several factors may also contribute such as the approximate ratio of oxidant/aniline 1.25, protonation process involving the HCl and the drying process was done at 40 °C during the synthesis made its conductivity can be improved reached to 265.8 S/cm [13]. Another factor could also be induced by intrinsic defect or purity of nano TiO_2.

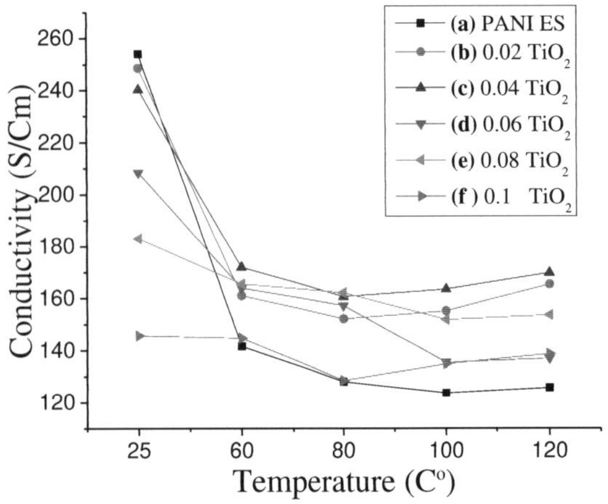

Figure 3. Conductivity of PANI and PANI-TiO_2 Film against Temperature.

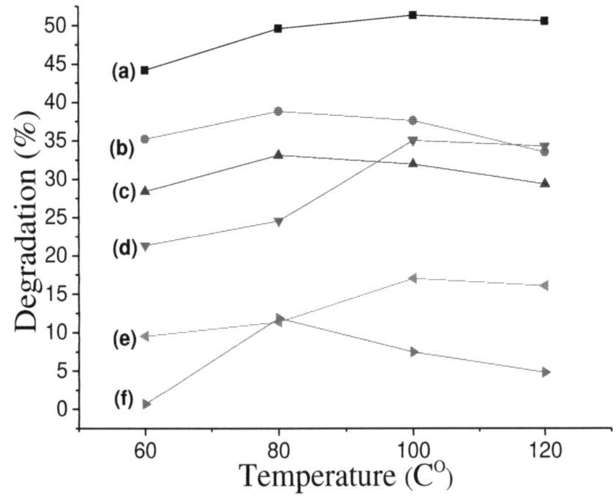

Figure 4. Conductivity Degradation of PANI and PANI-TiO_2 Film against Temperature.

In general, the electrical conductivity of the composite film decrease drastically from room temperature to about 60 °C, then somewhat insignificant change as indicated by Figure 3. A detail observation, the conductivity above 60 °C is slightly increased as the increase in temperature. These features suggest that the conductivity becomes stable due to incorporating TiO_2 nanoparticles in the PANI ES. A relative decreasing of electrical conductivity from its room temperature is displayed in Figure 4. It is seen that the strongest reduce cnductivity is showing by the pure PANI ES, followed by the increase of TiO_2 incorporation. This feature is also shown by PANI ES-TiO_2 with NMP solvent [14].

The PANI has covalent bond types, while TiO_2 has ionic bond types. These bonding types appropriated with the binding energy (E_0). The binding energy of PANI ES is 400 eV [15]. On the other hand, the binding energy of TiO_2 is 459.2 eV [16]. The presence of TiO_2 in PANI ES-TiO_2 composite film may increase its binding energy. This condition caused the higher thermal energy in degradation process of PANI ES-TiO_2 composite film. The presence of nano- TiO_2 in PANI ES-TiO_2 composite film may increase the crystallinity of the film PANI ES from 0; 3.83; 8.25; 15.17; 15.48 and 17.38%. In addition, the density of the PANI ES-TiO_2 films is higher compared to the PANI ES film. When PANI composed with 0.08 of TiO_2, demonstrated the reduced porosity surface PANI from 4.6 to 1.33%. The increased of crystallinity and density of PANI composite films caused the higher thermal energy for degradation process of the PANI ES-TiO_2 composite film. According to this theory, we can explain the thermal stability of the PANI-TiO_2 film increased as function as the TiO_2 addition.

4. Conclusion

Thermal stability of the PANI-TiO_2 film increased as function as the TiO_2 addition. This feature suggests that conductivity of the PANI becomes more stable at high temperature due to incorporating of the TiO_2 nanoparticles. There is many factors influence this condition, such as different the binding energy, phase, and density of PANI compare to PANI-TiO_2. Further study is required to examine thermal stability of the PANI-TiO_2 in more various range temperature.

Acknowledgments

This work was supported by Ministry of Research Technology and Higher Education through the grant of PUPT, and INSINAS.

References

[1] Saeed M, Shakoor A and Ahmad E 2013 Temperature dependent electrical conductivity of polyaniline/Y2O3 composites *Polym. Sci. Ser. A* **55** 549–555

[2] Sutar D S, Padma N, Aswal D K, Deshpande S K, Gupta S K and Yakhmi J V 2007 Preparation of nanofibrous polyaniline films and their application as ammonia gas sensor *Sens. Actuators B Chem.* **128** 286–292

[3] Song E and Choi J-W 2013 Conducting polyaniline nanowire and its applications in chemiresistive sensing *Nanomaterials* **3** 498–523

[4] Khan A F, Mehmood M, Durrani S K, Ali M L and Rahim N A 2015 Structural and optoelectronic properties of nanostructured TiO 2 thin films with annealing *Mater. Sci. Semicond. Process.* **29** 161–169

[5] Pomoni K, Vomvas A and Trapalis C 2008 Dark conductivity and transient photoconductivity of nanocrystalline undoped and N-doped TiO 2 sol–gel thin films *Thin Solid Films* **516** 1271–1278

[6] Amirtharajan S, Jeyaprakash P, Natarajan J and Natarajan P 2016 Electrical investigation of TiO2 thin films coated on glass and silicon substrates—effect of UV and visible light illumination *Appl. Nanosci.* **6** 591–598

[7] Dai L 2004 *Intelligent macromolecules for smart devices: from materials synthesis to device applications* (Springer Science & Business Media)

[8] Bhadra J and Sarkar D 2010 Electrical and optical properties of polyaniline polyvinyl alcohol composite films *Ind J Pure Appl Phys* **48** 425–428

[9] Zhu J, Liu X, Wang X, Huo X and Yan R 2015 Preparation of polyaniline–TiO 2 nanotube composite for the development of electrochemical biosensors *Sens. Actuators B Chem.* **221** 450–457

[10] Gu L, Wang J, Qi R, Wang X, Xu P and Han X 2012 A novel incorporating style of polyaniline/TiO 2 composites as effective visible photocatalysts *J. Mol. Catal. Chem.* **357** 19–25

[11] Asha A, Goyal S L, Kumar D, Kumar S, Kishore N and others 2015 Synthesis and characterization of polyaniline/TiO2 composites *Indian J. Pure Appl. Phys. IJPAP* **52** 341–347

[12] Geethalakshmi D, Muthukumarasamy N and Balasundaraprabhu R 2014 Effect of dopant concentration on the properties of HCl-doped PANI thin films prepared at different temperatures *Opt.-Int. J. Light Electron Opt.* **125** 1307–1310

[13] Chuanyu S and Yu W 2014 Factors influencing preparation of polyaniline doped with hydrochloric acid *Mater. Sci.-Pol.* **32** 521–525

[14] Ansari M O and Mohammad F 2012 Thermal stability of HCl-doped-polyaniline and TiO2 nanoparticles-based nanocomposites *J. Appl. Polym. Sci.* **124** 4433–4442

[15] Rodrigue D, Domingue M, Riga J and Verbist J J 1993 Electronic structure of polyaniline: experimental study of oligomers *Synth. Met.* **57** 4802–4806

[16] Linh N T T, Tuan P D and Dzung N V 2014 The Shifts of Band Gap and Binding Energies of Titania/Hydroxyapatite Material *J. Compos.* **2014**

IOP Conf. Series: Journal of Physics: Conf. Series **1011** (2018) 012066 doi:10.1088/1742-6596/1011/1/012066

Nanostructural Characters of β-SiC Nanoparticles Prepared from Indonesian Natural Resource using Sonochemical Method

A Fuad[1,2], U Kultsum[1], A Taufiq[1,2], Hartatiek[1,2], E Latifah[1]

[1]Department of Physics, Faculty of Mathematics and Natural Sciences, Universitas Negeri Malang (State University of Malang), Jl. Semarang No. 5, Malang 65145, Indonesia
[2]Central Laboratory for Mineral and Advanced Materials, Faculty of Mathematics and Natural Sciences, Universitas Negeri Malang (State University of Malang), Jl. Semarang No. 5, Malang 65145, Indonesia
*Corresponding Author: abdulloh.fuad.fmipa@um.ac.id

Abstract. Silicon carbide (SiC) nanoparticles become one of the interesting non-oxide ceramics due to their physical and chemical properties. For an extended period, SiC nanoparticles have been prepared by several methods that usually performed at high temperatures ranging from 1200 - 2000 °C from inexpensive commercial precursors. In this work, we prepared SiC nanoparticles from the low priced precursor of Indonesia natural resource using the sonochemical method at a temperature that is lower than 1000 °C. To produce samples with particular characters, we varied the sintering holding time (1, 10, and 20 hours) and the sintering temperatures (850, 950, and 1050 °C) during the synthesis. The samples were then characterized using XRD, SEM-EDX, TEM, and FTIR. The XRD data analysis showed that the samples have a dominant phase of SiC in the form of β-SiC with a 3C-SiC structure and SiO_2 phase in a low composition within a good agreement with the EDX characterization. Interestingly, the sample prepared at the sintering temperature of 850 °C for 1 hour showed a non-crystallite phase. Using a Scherer's equation, the particles of the samples sized from 13 to18 nm, which were validated by SEM and TEM images. Furthermore, the FT-IR spectra presented several peaks, i.e., at wavenumbers of 482.2 and 1150 cm^{-1} representing Si-O-Si bonding and also at 798.5 cm^{-1}regarding with Si-C bonding.

1. Introduction

Silicon Carbide (SiC) is an important non-oxide ceramic with unique properties such as mechanical toughness and chemical resistance so that it is utilized for numerous applications [1]. SiC is used intensively in military, aerospace, automotive, nuclear industry, electronic and optoelectronic devices such as solar cell, detector, modulator, and a semiconductor laser, specifically for high frequency, intense radiation, or high temperature [2,3].

Two most familiar SiC crystal structures are β-SiC crystal structure with 3C-SiC phase and α-SiC crystal structure with 2H-, 4H-, and 6H-SiC phase [4]. The α-SiC components start to form at approximately 2000 °C or above, and the β-SiC components start to form under 2000 °C. The SiC synthesis could be performed by different methods, for example, a carbothermal method with 1450 °C sintering temperature [5], a sol-gel method using *Tetraethyl orthosilicate* (TEOS) [6], and solar furnace method with 1500 °C sintering temperature [2].

The temperature and materials required for SiC synthesis are still relatively expensive. Hence, it is necessary to conduct research and synthesis of SiC at low temperature, i.e., under 1000 °C, by using

IOP Conf. Series: Journal of Physics: Conf. Series **1011** (2018) 012066 doi:10.1088/1742-6596/1011/1/012066

affordable local materials. One method that can be a candidate for the SiC synthesis is a sonochemical method [7]. The sonochemical method is an efficient and interesting method to form a mini-emulsion since it could be broken down into a nanometer-sized material of around 20-200 nm by ultrasonic waves [8]. This method is useful for reducing the particle size, controlling morphology, and producing crystal with excellent phase purity and crystallinity.

This research utilizes local materials for the SiC synthesis. The SiC synthesis utilized pyrophyllite that has been synthesized into SiO_2 with a purity of up to 98.2 % as the primary material; the produced silica has a nanometer size ranging from 40 until 90 nm [9].

2. Experimental Method

2. 1 Synthesis of SiC Nanoparticle

The SiC synthesis was performed via a sonochemical method utilizing SiO_2 and sucrose ($C_{12}H_{22}O_{11}$). In the synthesis, 3 g $C_{12}H_{22}O_{11}$, 1 g SiO_2, ethanol, and distilled water were mixed for 45 minutes until it became homogeneous at a speed of 350 rpm. The process was continued by a sonication process applied to the solution for 4 hours. Next, the sample was drained at 180 °C for 1 hour. Next, the solution was sintered at 850, 950, and 1050 °C in Ar gas atmosphere for 1, 10, and 20 hours.

2. 2 Characterization

X-Ray Diffraction (XRD, Philips in X'Pert PRO type), Fourier Transform Infrared Spectroscopy (FT-IR, Shimadzu in Prestige-21 type), Scanning Electron Microscopy (FEI in INSPECT-S50 type), and Transmission Electron Microscopy (FEI in Talos F200X TEM type) characterized the SiC nanoparticle synthesis.

3. Results and Discussion

3. 1 Diffraction Pattern and Crystal Structure

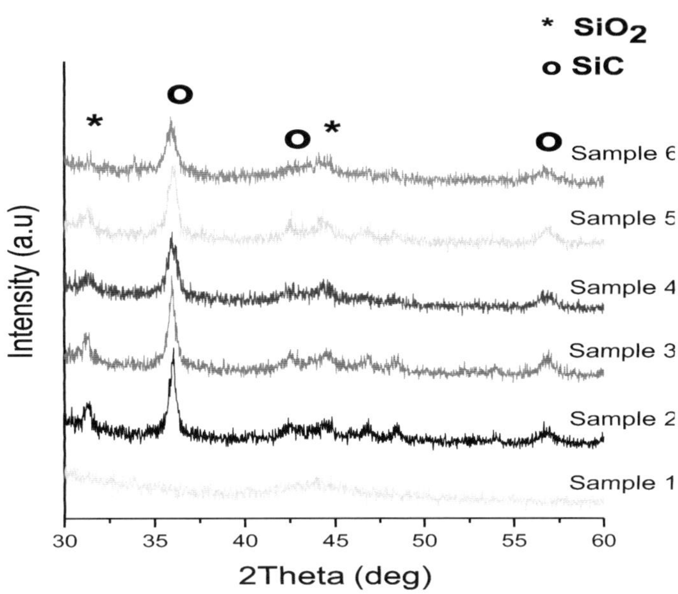

Figure 1. X-Ray diffraction pattern of synthesized SiC nanoparticles.

IOP Conf. Series: Journal of Physics: Conf. Series **1011** (2018) 012066 doi:10.1088/1742-6596/1011/1/012066

Figure 1 presents the sample 1 sintered for 1 hour and at 850 °C, sample 2 with a 1-hour sintering time at 950 °C, sample 3 that is sintered for 1 hour at 1050 °C, sample 4 sintered for 10 hours at 850 °C, sample 5 sintered for 10 hours at 950 °C, and sample 6 sintered for 20 hours at 850 °C. The XRD result shows no formation of SiC peaks in sample 1 with a consistent amorphous structure. However, multiphase was found in other samples with SiC and SiO_2 peaks forming the diffraction pattern. Sample 5 showed an optimum condition of SiC synthesis via a Sonochemical method while sample 6 showed a decline in peak intensity due to the over prolonged sintering time. Sintering temperature and time affect the intensity of material and tend to reduce the material intensity if its melting point is passed [10]. A sintering process is usually performed in 1 to 1000 minutes [11].

Two phases were formed during the analysis, i.e., Low-cristoballite phase (SiO_2) and Moissanite-3C (3C-SiC) with PDF numbers of 01-076-0939 [12] and 01-075-0254 [13], respectively. The sample showed several SiO_2 crystal peaks at angle 2θ of approximately 31.27° with (101) hkl field and 44.56° with (212) hkl field, while SiC peaks were displayed at angle 2θ of around 35.96° with (111) hkl field, 42.7° with (020) hkl field, and 56.62° with (202) hkl field.

3. 2 Particle Size

A Scherrer equation measured the average crystal size on the SiC nanoparticle synthesis. Based on the XRD results, the smallest angle of diffraction formed was $\theta = 18.001°$ ($2\theta = 36.003°$) with the $\beta = 0.0096$ (radian). Such calculation presented the average crystal size on a nanometer scale of approximately 13 up to 18 nm.

Figure 2. SEM image of SiC nanoparticles

Figure 3. TEM image of SiC nanoparticles

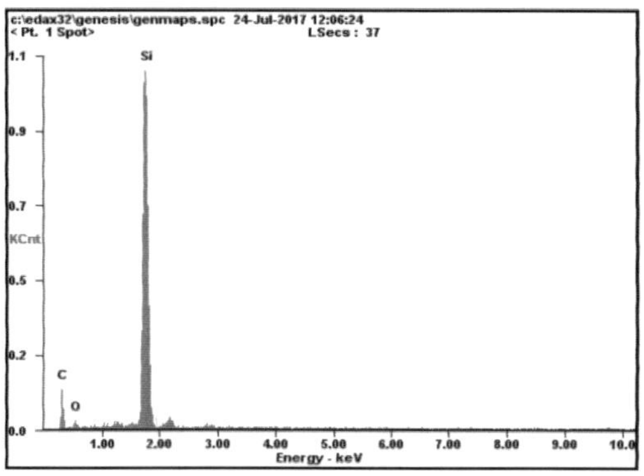

Figure 4. EDX spectrum of SiC nanoparticles

The SiC crystal size has been inevitably confirmed by SEM, TEM, and EDX characterization presented in Figure 2-4. The Energy Dispersive X-Ray (EDX) test was conducted to analyze the constituent elements of the material. Figure 2 and 3 displays an agglomeration in the sample due to its nano size. It affected the SEM and TEM morphology test results that showed non-uniform sizes of the sample, i.e., approximately 10 nm up to 18 nm. Aside from that, it is already certain that the particle already had a nano size. In other words, the SiC nanoparticle was already formed. Figure 4 is a spectrum of the EDX test performed on the SiC sample showing a smaller Si element concentration than that of the C element of 44.05 % and 50.94%, respectively. Also, the EDX identified another element of O in a concentration of 5.01 so that it the SiC synthesis could be regarded as successful, which is proven by the formation of two phases, namely SiC and SiO_2 phases.

3. 3 FT-IR Characterization Results

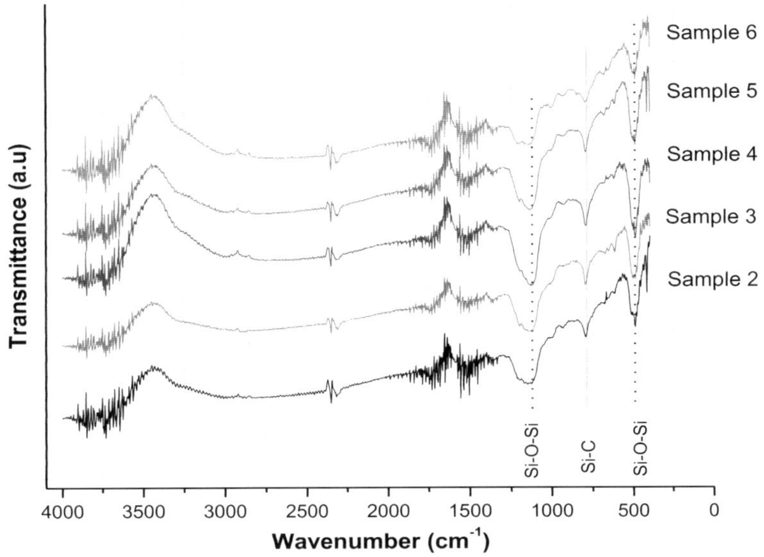

Figure 5. IR spectra of SiC nanoparticles

Figure 5 displays the FT-IR test results on the SiC sample. All samples showed the existence of a Si-C functional group at the wavenumber of 798.5 cm^{-1} with different transmissions, and a Si-O-Si functional group was found at the wavenumbers of 482.2 and 1150 cm^{-1}. Such findings are in good agreement with the database of experimental results of the researchers, in which the Si-C functional group formed at a wave number of around 754 until 840 cm^{-1} [14–17].

4. Conclusions

To conclude, in this research, the β-SiC has been successfully synthesized from SiO_2 and $C_{12}H_{22}O_{11}$ via a sonochemical method at low temperatures with nanoscale particles of around 10-18 nm. The XRD test results showed a multiphase formation of SiC and SiO_2. According to the FT-IR test results, the sample displayed a formation of a Si-C functional group. Hence, the SiC was well-synthesized by a sonochemical method.

References

[1] Vennekamp M, Bauer I, Groh M, Sperling E, Ueberlein S, Myndyk M, Mäder G, Kaskel S 2011 Formation of SiC Nanoparticles in An Atmospheric Microwave Plasma *Beilstein J. Nanotechnol.* **2** 665

[2] Nishioka K, Komori J, Maeda K, Ota Y, Kaneko H, Sato K 2015 Formation of Silicon Carbide Using Volcanic Ash as Starting Material And Concentrated Sunlight as Energy Resource *Int. J. Photoenergy* **2015** 394605 (*Preprint* http://dx.doi.org/10.1155/2015/394605)

[3] Legba E T 2007 *Synthesis and Characterization of α-Silicon Carbide Nanostructures* (University of Kentucky Master's These)

[4] Luo X, Goel S, Reuben R L 2012 A Quantitative Assessment of Nanometric Machinability of Major Polytypes of Single Crystal Silicon Carbide *J. Eur. Ceram. Soc.* **32** 3423

[5] Moshtaghioun B M , Poyato R, Cumbrera F L, de Bernardi-Martin S, Monshi A, Abbasi M H, Karimzadeh F, Dominguez-Rodriguez A 2012 Rapid Carbothermic Synthesis of Silicon Carbide Nano Powders by Using Microwave Heating, *J. Eur. Ceram. Soc.* **32** 1787

[6] Najafi A, Golestani-Fard F, Rezaie H R, Ehsani N 2011 Synthesis And Characterization of Silicon Carbide Nano Powder By Sol-Gel Processing *Iran. J. Mater. Sci. Eng.* **8** 41

[7] Sivasankaran S, K K MJ 2015 A Novel Sonochemical Synthesis of Nano-Crystalline Silicon Carbide Ceramic Powder and Its Characterization *Int. J. Recent Sci. Res.* **6** 2630

[8] Badnore A U, Pandit A B 2015 Synthesis of Nanosized Calcium Carbonate Using Reverse Miniemulsion Technique: Comparison between Sonochemical and Conventional Method *Chem. Eng. Process. Process Intensif.* **98** 13

[9] Fuad A, Mufti N, Diantoro M, Subakti 2016 Synthesis and Characterization of Highly Purified Nanosilica from Pyrophyllite Ores *AIP Conf. Proc., AIP Publishing* p. 030020.

[10] Kim J, Creasy T S 2004 Selective Laser Sintering Characteristics of Nylon 6/Clay-Reinforced Nanocomposite *Polym. Test.* **23** 629

[11] Shimosaka A, Ueda Y, Shirakawa Y, Hidaka J 2003 Sintering Mechanism of Two Spheres Forming A Homogeneous Solid Solubility Neck [Translated] *KONA Powder Part. J.* **21** 219

[12] Peacor D B 1973 High-Temperature Single-Crystal Study of The Cristobalite Inversion *Z. Für Krist.-Cryst. Mater.* **138** 274

[13] Burdick C L, Owen E A 1918 The Atomic Structure of Carborundum Determined By X-Rays *J. Am. Chem. Soc.* **40** 1749

[14] Zhao B, Zhang H, Tao H, Tan Z, Jiao Z, Wu M 2011 Low-Temperature Synthesis of Mesoporous Silicon Carbide via Magnesiothermic Reduction *Mater. Lett.* **65** 1552

[15] Raman V, Bahl O P, Dhawan U 1995 Synthesis of Silicon Carbide Through The Sol-Gel Process from Different Precursors *J. Mater. Sci.* **30** 2686

[16] Abro D M, Dablé P, Cortés-Salazar F, Amstutz V, Girault H 2016 Characterization of Surface State of Inert Particles: Case of Si and SiC *J. Miner. Mater. Charact. Eng.* **4** 62

[17] Liu Z, Shen W, Bu W, Chen H, Hua Z, Zhang L, Li L, Shi J, Tan S 2005 Low-Temperature Formation of Nanocrystalline β-SiC With High Surface Area And Mesoporosity via Reaction of Mesoporous Carbon And Silicon Powder *Microporous Mesoporous Mater.* **82** 137

Polyacrylonitrile nanofiber as polar solvent N,N-dimethyl formamide sensor based on quartz crystal microbalance technique

A Rianjanu[1], T Julian[1], S N Hidayat[1], E A Suyono[3], A Kusumaatmaja[1,2], and K Triyana[1,2]

[1] Department of Physics, Universitas Gadjah Mada Sekip Utara Yogyakarta, 55281, Indonesia

[2] Nanomaterial Research Group, Universitas Gadjah Mada Sekip Utara Yogyakarta, 55281, Indonesia

[3] Department of Biology, Universitas Gadjah Mada Sekip Utara Yogyakarta, 55281, Indonesia

aditya.rianjanu@mail.ugm.ac.id

Abstract. Here, we describe an N,N-dimethyl formamide (DMF) vapour sensor fabricated by coating polyacrylonitrile (PAN) nanofiber structured on quartz crystal microbalance (QCM). The PAN nanofiber sensors with an average diameter of 225 nm to 310 nm were fabricated via electrospinning process with different mass deposition on QCM substrate. The nanostructured of PAN nanofiber offers a high specific surface area that improved the sensing performance of nanofiber sensors. Benefiting from that fine structure, and high polymer-solvent affinity between PAN and DMF, the development of DMF sensors presented good response at ambient temperature. Since there is no chemical reaction between PAN nanofiber and DMF vapour, weak physical interaction such absorption and swelling were responsible for the sensing behavior. The results are indicating that the response of PAN nanofiber sensors has more dependency on the nanofiber structure (specific surface area) rather than its mass deposition. The sensor also showed good stability after a few days sensing. These findings have significant implications for developing DMF vapour sensor based on QCM coated polymer nanofibers.

Keywords: polyacrylonitrile (PAN), nanofiber, quartz crystal microbalance (QCM), N,N-dimethyl formamide (DMF), vapour sensing

1. Introduction

Polar organic solvent such N,N-dimethyl formamide (DMF) are widely used the chemical in various industries, such as the production of synthetic fibers, films, and coatings, and leather tanning [1]. Recent studies have suggested its serious health concerns that associated with exposure to DMF vapor through skin contact and inhalation. Many studies show the effect of DMF vapor both on exposed worker [2] and animal experiments [3] have confirmed the hepatotoxicity of DMF. The other experiment suggested embryotoxicity [4] and carcinogenesis [5] effect due to exposure to DMF vapors. Because of the health-related issues and the widespread use in industry, DMF has been

IOP Conf. Series: Journal of Physics: Conf. Series **1011** (2018) 012067 doi:10.1088/1742-6596/1011/1/012067

prioritized for field study. The detection and standard sampling for DMF vapor in the air require lengthy and complicated procedures and the use of gas chromatography (GC) [6]. Therefore an efficient, selective, and convenient sensor for DMF vapor is desirable.

Recently, various vapor sensors have been produced by using different sensing materials, such as inorganic thin films [7,8], semiconductor metal oxides [9,10], and conducting polymers [11] with excellent sensitivity results. Various vapor sensor was also have been developed based on different sensing mechanisms such as photoelectric sensors [12], resistance sensor [13], optical sensor [14], amperometric sensor [15] and acoustic wave sensor [15]. Quartz crystal microbalance (QCM) is one example of acoustic wave (AW) sensor platform based on acoustic-electric effect and mass deposition. A QCM sensor offers an accurate, real-time detection, and suitable sensors with high sensitivity response that is widely used as vapor/gas sensors [16]. Nanostructure material especially nanofibers have received increasing attention for various applications including gas sensor application due to their large specific surface area, high porosity and interconnected porous structures [17-19]. Many researchers have been used electrospun nanofiber [20,21] and polymer modified nanofiber [16,22] as sensing layer for various vapor/gas sensing.

The sensing mechanism of vapor sensor mainly dominated by chemical interaction between the active layer and the analytes. However, many volatile organic compounds like methanol, benzene, and toluene are not reactive at room temperature and mild conditions; fortunately, they may have weak physical interactions with the sensing layer, involving absorbing or swelling the polymer matrices. These interactions do not change the oxidation levels of conducting polymers, although, can also affect the properties of the sensing layer and make these vapor/gases detectable [23]. Several studies have been done investigating the swelling behavior of various polar solvent at electrospun nanofiber membrane to improve their electrical or mechanical properties [24–27]. Huang et al. (2013) investigated the swelling effect of polyacrylonitrile (PAN) and polysulfone (PSU) nanofiber after solvent vapor treatment with N,N-dimethyl formamide. The results confirmed that swelling behavior occurred when PAN nanofiber was exposure with DMF. The swelling behavior of PAN nanofiber and DMF vapor is expected due to their high polymer-solvent affinity. Based on their results, we are developing an N,N-dimethyl formamide (DMF) sensor based on polyacrylonitrile (PAN) nanofiber as sensing layer. We suspected that the weak physical interaction including absorption and swelling phenomena between PAN nanofiber and DMF vapor is responsible for DMF sensing.

2. Experimental

Typical Polyacrylonitrile (PAN) with a molecular weight of 150,000 from Sigma–Aldrich was used for the preparation of nanofiber. Non-ionic surfactant Triton x-100 was purchased from Merck, Germany. The solvents used for dissolving PAN and vapor sensing N, N-dimethyl formamide (DMF) was purchased from Merck, Germany. AT-Cut QCM sensor with the gold electrode and 10 MHz resonant frequency were purchased from Novaetech, Italy. All above materials used as received without any further purification.

Polyacrylonitrile (PAN) nanofiber was prepared by dissolving 0.5 g and 0.6 g polymer PAN into 10 mL DMF followed by mechanical stirring at 1000 rpm for 24 hours at ambient temperature until a homogeneous solution was achieved. After PAN fully dissolved in DMF, a small amount (1%) of non-ionic surfactant Triton X-100 was added to the solution. The non-ionic surfactant is added to lower the surface tension of the solution. Followed by mechanical stirring at 900 rpm for 30 min to fully dissolve the surfactant.

The solution then transferred into 10 mL plastic syringes for electrospinning process. When the electrospinning process was performed, a DC voltage of 5 kV and 10 kV was applied, and the tip to collector distance was set as 10 cm. The collector used in the electrospinning process was metal plate covered with aluminum foil with crystal QCM put in the collector. The electrospinning process was carried out for 30 s to get a suitable thickness nanofiber film. Finally, the nanofiber film was placed in the desiccator for 12 hours to evaporate the residual solvent before further treatment. To vary the nanofiber morphology, we used a different parameter in electrospinning process. Detailed listed of electrospinning parameter used in this study is listed in Table. 1. Scanning electron microscope (SEM) JEOL-JSM-6510 with auto fine coater JEOL JEC-3000FC was used for investigating the morphologies of electrospun nanofiber PAN.

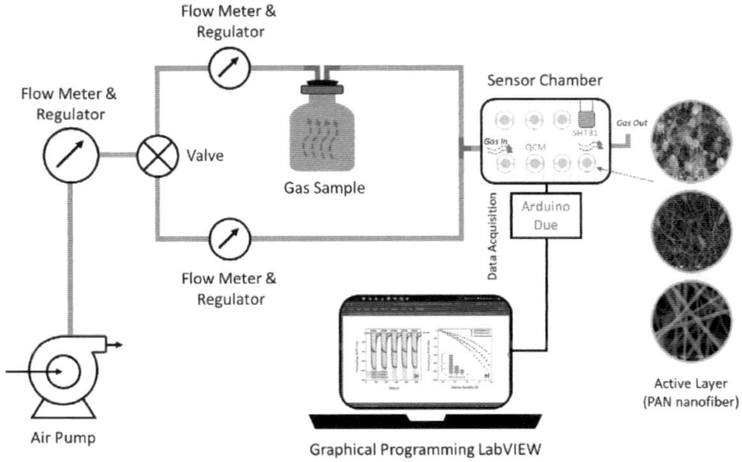

Figure 1. Schematic illustration of PAN nanofiber QCM vapor sensing process

A schematic diagram of the N,N-dimethyl formamide vapor sensing system is presented in Fig. 1. The QCM sensor was installed in the testing chamber with Sensirion SHT 31 (temperature and humidity sensor) included to monitoring the humidity and the temperature of the chamber. During the experiments process, several of gas flow rate was used. The sensing of the QCM sensor was measured by a shift in the resonance frequency. The shift in the resonance frequency is directly related to mass deposition on sensing layer according to Sauerbrey equations [28]. The resonance frequency sent to PC accompanies RS232 or serial protocol with graphical programing language LabVIEW.

3. Results and Discussion

The representative SEM image of the PAN nanofiber membrane used in this study is shown in Fig. 2. The as-spun nanofibers were randomly oriented three-dimensional structure (3D) porous structure with relatively random diameter varies from 225 nm to 310 nm. The electrospinning parameter was varying to obtain three different nanofiber morphologies. Several studies have been revealed that the sensitivity of gas sensors can be efficiently improved by increasing the specific surface area (SSA) per unit mass of the sensing layer. The specific surface area of smooth nanofiber is higher compared than the beaded fiber. However, the smooth nanofiber with higher diameter has lower specific surface area [29]. So, the optimum morphologies are a crucial aspect to improve the sensor sensitivity.

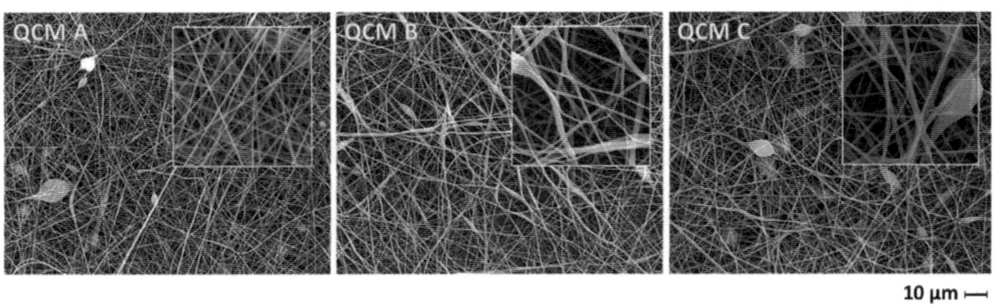

Figure 2. SEM image of PAN nanofiber with different electrospinning parameters

The parameter used and detailed nanofiber form with frequency shift after the deposition was shown in Table. 1. It shows that the frequency shift after coating increase from QCM A to QCM C. It means that the mass deposition of nanofiber PAN in QCM substrate was also increasing. Fig. 3 shows the mass deposition and nanofiber diameter of PAN nanofiber sensors. The nanofiber diameter was found about 225 nm, 242 nm, and 310 nm for QCM A–C, respectively. The increase in the nanofiber diameter is due to the increased viscosity of the solution. The viscosity of the polymer solution directly depends on their solution concentration. In the electrospinning process, high viscosity solution means that the polymer chain bonding is high enough to maintain its shape thus make nanofiber with higher diameter.

Table 1. The parameter used in the electrospinning process with frequency shift after coating and nanofiber formation.

Sample	Electrospinning parameter (t =20 s)			Frequency Shift
	d (cm)	c (wt%)	V (kV)	after Coating (Hz)
QCM A		5	5	1344
QCM B	10		10	2163
QCM C		6	5	4130

The correlation of frequency shift with mass deposition as mention before is described by Sauerbrey equations. We found that the mass deposition is about 296 ng, 476 ng, and 909 ng for QCM A–C, respectively. The difference in the mass deposition is due to the different deposition rate when the electrospinning process. The deposition rate of electrospinning process depends on the velocity of the polymer molecule under the influence of processing voltage. The increasing of processing voltage would increase the velocity of polymer molecules and make the deposition rate increased. Both nanofiber mass deposition and nanofiber diameter increase from QCM A to QCM C. The response of the QCM sensor is highly dependent on the change in humidity and usually utilized as humidity sensors [30]. Since the response of QCM sensor is highly dependence to the alteration of relative humidity; we also investigated the response of our sensor in the influence of relative humidity change. The linear correlation between humidity shift and frequency response of our sensor is shown in inset Fig. 3 (a). It shows that the correlation is about 2.17, 0.68, 1.96, and 2.72 Hz/%RH for uncoated (blank), QCM A, QCM B, and QCM C, respectively. The value is insignificant compared to the response of PAN nanofiber sensors, and being neglectable to the response measurement.

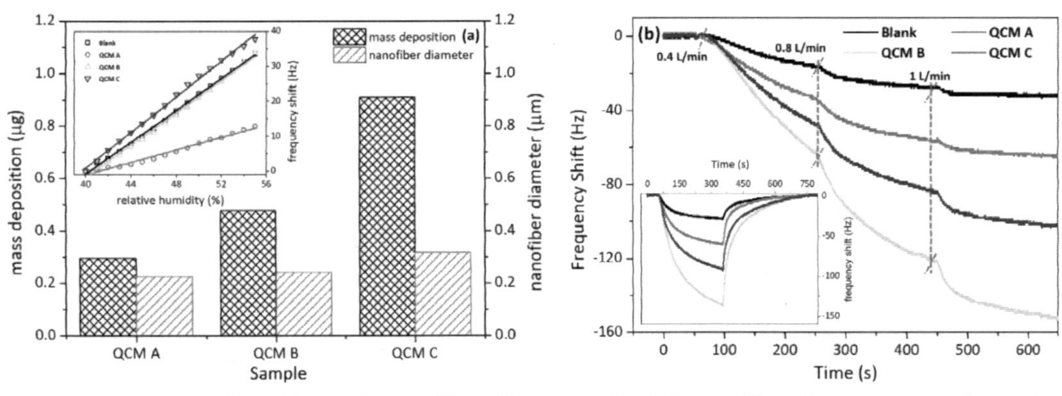

Figure 3. (a) Mass deposition and nanofiber diameter of PAN nanofiber QCM sensors (inset: humidity dependence of PAN nanofiber QCM sensors), (b) Dynamic response of QCM sensors coated with PAN nanofiber upon exposure to increasing the gas flow of DMF vapor, (inset: one cycle of QCM response against DMF vapor 1 L/min).

Fig. 3 (b) shows the dynamic response of PAN nanofiber QCM sensors with several of gas flow rate. It shows that the response is increasing consecutively from blank to QCM A, QCM C, and the highest response is achieved with QCM B. For commonly QCM sensors increasing mass deposition would increasing its sensor sensitivity [22]. However, in our results, the highest response was achieved by QCM B which has lower mass deposition than QCM C. These results were occurred due to the higher specific surface area (SSA) that QCM B possessed. Jia et al. (2017) found that smooth nanofiber formation with larger diameter has lower specific surface area [29]. This particular reason is the causes of the higher response of QCM B. In our cases, the specific surface area of nanofiber sensor has more influence rather than its mass depositions. Inset Fig. 3 (b) shows the dynamic response (adsorption/desorption) of the sensor of PAN nanofiber sensors with 1 L/min gas flow rate. The results show that the response time is faster than its recovery time. The slow response time (over 300 s) is suspected due to the low evaporation rate of DMF solvent. The DMF solvent has a boiling point of 153 °C which makes the DMF solvent hardly to evaporate at ambient temperature.

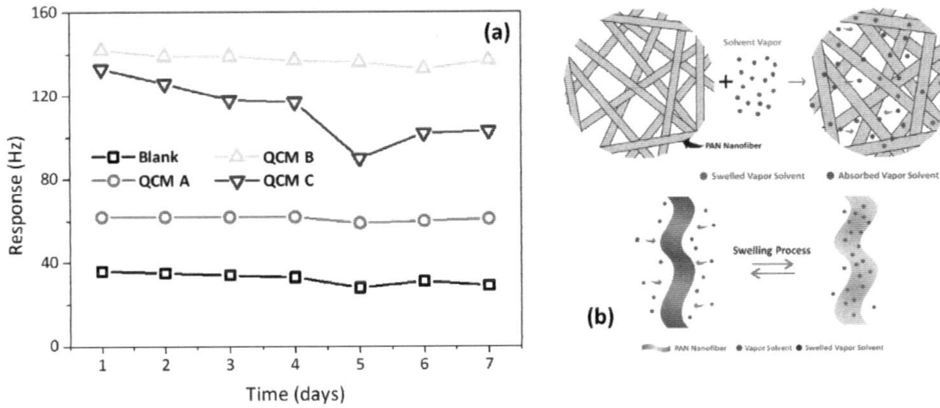

Figure 4. (a) Respons stability of PAN nanofiber QCM sensors with 1 L/min gas flow of DMF vapor, (b) Schematic illustration of sensing mechanism

We also investigated the stability of the response frequency of PAN nanofiber sensors under the influence of DMF vapour. Fig. 4 (a) shows the response of sensor after sensing DMF vapor with vapor flowrate 1 L/min. It shows that QCM C is quite unstable for one week used, while the other sample is relatively stable. This is probably due to beaded morphologies that QCM C possessed. The sensing mechanism of PAN nanofiber gas sensor under the influence of DMF vapour shows in Fig. 4 (b). We believe that the weak physical interaction such as absorption and swelling is responsible for the change in the resonance frequency. From Fig. 4 (b) we illustrated the phenomena that occurred after the PAN nanofiber exposure with DMF vapour. The DMF molecules are partly absorbed into the PAN nanofiber utilizing the high affinity between DMF molecules and PAN polymers via the swelling process. However, this mechanism alone couldn't deliver such high response. In addition to swelling behavior, we believe that the porous structure in PAN nanofiber film may contribute as a temporary shelter for DMF molecules and increase the mass loading of the active layer. Both phenomena in our belief are responsible for the change in the resonance frequency of the PAN nanofiber sensor under the influence of DMF vapour.

4. Conclusion

N,N-dimethyl formamide vapour sensor based on polyacrylonitrile nanofiber deposited on QCM substrate has been successfully developed. The increase in nanofiber specific surface area and its mass deposition increasees the sensor sensitivity. The change in the resonance frequency of the sensing element upon exposure to DMF vapour is caused by weak physical interaction such as absorption and swelling of DMF molecules and into polyacrylonitrile nanofibers. The PAN nanofiber sensors show excellent stability in sensing DMF after one week.

Acknowledgment

This work was supported by the "Penelitian Pendidikan Magister Menuju Doktor Sarjana Unggulan (PMDSU), Contract No. 1991/UN1-P.III/DIT-LIT/LT/2017" from the Ministry of Research Technology and Higher Education, the Republic of Indonesia.

References

[1] Li Y, Zhang S and Song D 2013 *Angewandte Chemie - International Edition* **52** 710–3
[2] Long G and Meek M E 2001 *Journal of Environmental Science and Health, Part C* **19** 161–87
[3] Rui D, Daojun C and Yongjian Y 2011 *Environmental Toxicology and Pharmacology* **31** 357–63
[4] Saillenfait A M 1997 *Fundamental and Applied Toxicology* **39** 33–43
[5] Senoh H, Aiso S, Arito H, Nishizawa T, Nagano K, Yamamoto S and Matsushima T 2004 *Journal of Occupational Health* **46** 429–39
[6] Rimatori V and Carelli G 1982 *Scandinavian Journal of Work, Environment & Health* **8** 20–3
[7] Kaushik A, Kumar R, Arya S K, Nair M, Malhotra B D and Bhansali S 2015 *Chemical Reviews* **115** 4571–606
[8] Evans S D, Johnson S R, Cheng Y L and Shen T 2000 *Journal of Materials Chemistry* **10** 183–8
[9] Dhahri R, Leonardi S G, Hjiri M, Mir L El, Bonavita A, Donato N, Iannazzo D and Neri G 2017 *Sensors and Actuators, B: Chemical* **239** 36–44
[10] Hidayat S N and Triyana K 2016 *AIP Conference Proceedings* vol 1755p 20001
[11] Bora A, Mohan K, Pegu D, Gohain C B and Dolui S K 2017 *Sensors and Actuators B: Chemical* **253** 977–86
[12] Cui J, Wang D, Xie T and Lin Y 2013 *Sensors and Actuators B: Chemical* **186** 165–71
[13] Wang J, He E, Liu X, Yu L, Wang H, Zhang R and Zhang H 2017 *Sensors and Actuators B: Chemical* **239** 898–905
[14] Evyapan M, Kadem B, Basova T V., Yushina I V. and Hassan A K 2016 *Sensors and Actuators B: Chemical* **236** 605–13
[15] He K, Wang X, Meng X, Zheng H and Suye S I 2014 *Sensors and Actuators, B: Chemical* **193** 212–9
[16] Wang N, Wang X, Jia Y, Li X, Yu J and Ding B 2014 *Carbohydrate Polymers* **108** 192–9
[17] Ding B, Wang M, Yu J and Sun G 2009 *Sensors* **9** 1609–24
[18] Triyana K, Mu'min M S, Faizah K, Yusuf Y, Kusumaatmaja A and Harsojo S 2015 *Materials Science Forum* **827** 321–5
[19] Kusumaatmaja A, Sukandaru B, Chotimah and Triyana K 2016 *AIP Conf. Proc. 1755*
[20] Jia Y, Yu H, Zhang Y, Chen L and Dong F 2015 *Sensors and Actuators B: Chemical* **212** 273–7
[21] Jia Y, Chen L, Yu H, Zhang Y and Dong F 2015 *RSC Adv.* **5** 40620–7
[22] Huang W, Wang X X, Jia Y, Li X, Zhu Z, Li Y, Si Y, Ding B, Wang X X and Yu J 2013 *RSC Advances* **3** 22994–3000
[23] Bai H and Shi G 2007 *Sensors* **7** 267–307
[24] Huang L, Manickam S S and McCutcheon J R 2013 *Journal of Membrane Science* **436** 213–20
[25] Chotimah, Rianjanu A, Winardianto B, Munir M, Kartini I and Triyana K 2016 *International Journal on Advanced Science, Engineering, and Information Technology* **6** 675–81
[26] Liu C, Li X, Liu T, Liu Z, Li N, Zhang Y, Xiao C and Feng X 2016 *Journal of Membrane Science* **512** 1–12
[27] Kusumaatmaja A, Fauji N, and Triyana K 2017 Mater. Sci. Forum 886 145–9
[28] Sauerbrey G 1959 *Zeitschrift für Physik* **155** 206–22
[29] Jia Y, Yu H, Cai J, Li Z and Dong F 2017 *Sensors and Actuators B: Chemical* **243** 1042–5
[30] Hidayat SN, Julian T, Rianjanu A, Kusumaatmadja A, Triyana K and Roto 2017 *International Seminar on Sensors, Instrumentation, Measurement, and Metrology (ISSIMM) (IEEE)* 119–23.

Crystal Structures and Magnetic Properties of Polyethylene Glycol (PEG-4000) Encapsulated $Zn_{0.5}Ni_{0.5}Fe_2O_4$ Magnetic Nanoparticles

J. Widakdo[1], N. Istikhomah[1], A. Rifianto[1], E. Suharyadi[1*], T. Kato[2] and S. Iwata[3]

[1] Department of Physics, Universitas Gadjah Mada, Yogyakarta, Indonesia
[2] Department of Electronics, Nagoya University, Japan
[3] Institutes of Materials and Systems for Sustainability, Nagoya University, Japan

*Corresponding Author: esuharyadi@ugm.ac.id

Abstract. Nanocrystalline mixed spinel ferrite of $Zn_{0.5}Ni_{0.5}Fe_2O_4$ magnetic nanoparticles (MNPs) has been successfully synthesized by coprecipitation method and encapsulated and PEG-4000 with various concentrations. X-Ray Diffraction (XRD) patterns showed that nanoparticles contained $Zn_{0.5}Ni_{0.5}Fe_2O_4$ spinel ferrite with the particles size of 15.2 nm. After PEG-4000 encapsulation, particles size decrease became 13.3 nm. Transmission Electron Microscopy (TEM) image showed that the nanoparticles still aglomerate after PEG-4000 encapsulation. The coercivity (H_c) of $Zn_{0.5}Ni_{0.5}Fe_2O_4$ was 50.7 Oe. The H_c decrease after PEG-4000 encapsulation became 47.9 Oe, respectively. It is due to the decrease of particle size. The saturation magnetization (M_s) of $Zn_{0.5}Ni_{0.5}Fe_2O_4$ was 18.4 emu/g, and decrease to 17.1 emu/g after PEG-4000 encapsulation. This is because PEG-4000 has paramagnetic properties.

1. Introduction

Ferrite nanoparticles (MFe_2O_4, M = Zn, Ni, Mn, Co, Mg) have been attracted interest to the researchers in the field of functional materials due to their high electrical and chemical properties, and thermal stabilities [1]. Ferrite nanoparticles are widely used in technologies like drug delivery, magnetic recording media, Magnetic Resonance Imaging (MRI), contrast agent, and tomography imaging [1,2].

Zn–based ferries is one of the spinel ferrites that attractive to researchers due to its low cost, high electromagnetic performance, and high permeability [3,4,5]. Many researchers have studied the synthesis of Ni–Zn ferrites [3,4,5]. However, for the applications, pure Ni–Zn ferrite nanoparticles usually lead to problems associated with the formation of large aggregates and the alteration of magnetic properties [6]. An encapsulation process is performed to improve the quality of Zn-Ni ferrite. This process will make Zn-Ni ferrite nanoparticles more reduce aggregates and control grain size.

Different techniques to prepare ferrite nano-particles have been reported such as low temperature solid state reaction [7], co-precipitation, sol-gel, hydrotermal, gel combustion, etc. [8], have advantages over physical methods because of low cost, the reaction taking place at room temperature and large scale possibility [9]. Among the methods, co-precipitation method is the most effective methods and relative simple.

Sertkol *et al.*, [10] for Zn-Ni ferrite nanoparticles encapsulated with PEG have some drawbacks such as PEG used is PEG-400, it so expensive, have short chain polymer. Then we use PEG-4000 because have long chain polymer and to reduce the cost incurred. Therefore, In this research, work we

IOP Conf. Series: Journal of Physics: Conf. Series **1011** (2018) 012068 doi:10.1088/1742-6596/1011/1/012068

have used co-precipitation method for making $Zn_{0.5}Ni_{0.5}Fe_2O_4$ nanoparticles. The aim of our investigation was to understand the effect of variation of concentration of PEG-4000 as encapsulate substrate. To study the influences of encapsulating substrate on crystal structures, morphological, and magnetic properties of $Zn_{0.5}Ni_{0.5}Fe_2O_4$ nanoparticles. Further, the information mechanism is discussed in detail.

2. Experimental Details

The precursor such as $FeCl_3.6H_2O$ (Merck, Germany), $ZnSO_4.7H_2O$ (Merck, Germany), $NiCl_2.6H_2O$ (Merck, Germany) and NaOH (Merck, Germany) was weighed carefully by the required stoichiometric proportion of salt. Mix $ZnSO_4.7H_2O$ salt and $NiCl_2.6H_2O$ salt and dissolved it in 20 ml aquadest. Dissolved each $FeCl_3.6H_2O$ salt and NaOH salt in 50 ml of aquadest. Then, mix the metal solution and add 3,37 ml HCl 37% (Merck, Germany) into it. The precipitation was carried out by dropping the alloys solution into 1.5 M NaOH solution called sample respectively, at temperature 90°C respectively, for 1 hour and stirred at 1000 rpm. The precipitate then washed six times for 30 minutes and dried at 90°C for 5 hours.

The various PEG solution was dropped into 20 ml NaOH when co-precipitation process. The concentration of PEG-4000 is used as in Table 1.

TABLE 1. Variation of the concentration of PEG-4000

Sample	Comparison of PEG-4000 in molarity	
	Proportion NaOH : PEG 4000 at 50 ml aquades	PEG Molarity mixed in NaOH (M)
P1	2.99 gr : 0.5 gr	0.0026 molar
P2	2.99 gr : 1 gr	0.0052 molar
P3	2.99 gr : 1.5 gr	0.0079 molar
P4	2.99 gr : 2 gr	0.0105 molar
P5	2.99 gr : 2.5 gr	0.0130 molar
P6	2.99 gr: 3 gr	0.0157 molar

The crystal structures are analyzed by X-ray difractometer (XRD) Shimadzu XD with CuKα1 radiation (λ = 1.5406 Å). The magnetic properties are analyzed by Vibrating Sample Magnetometer (VSM) Riken Denshi Co. Ltd. with H_{max} = 15 KOe at room temperature. The morphology was analyzed by Transmission Electron Microscopy (TEM) JEOL Jem-4000. The bonding analysis are characterized by IR spectrometer Shimadzu Prestige-21.

3. Results and Discussion

Structural Studies

Structural studies were done with XRD measurement and then compared to standard data (JCPDS no 45-0131) in 2θ range of 20° – 80° to $Zn_{0.5}Ni_{0.5}Fe_2O_4$ nanoparticles are shown in Fig. 1. XRD patterns showed that nanoparticles contained $Zn_{0.5}Ni_{0.5}Fe_2O_4$ spinel ferrite because all sample show peak 311. At Table 2 is the value of analysis crystallite size, X-Ray density and lattice parameter. For $Zn_{0.5}Ni_{0.5}Fe_2O_4$ before encapsulated crystallite size of 15.2 nm. After PEG-4000 encapsulation crystallite size decreased became 12.9 nm and 13.3 nm. This is because interaction $Zn_{0.5}Ni_{0.5}Fe_2O_4$ nanoparticles with long chain PEG-4000 that is causing the crystal growth will trap. Lattice parameter and X-Ray density constant after encapsulated PEG-4000.

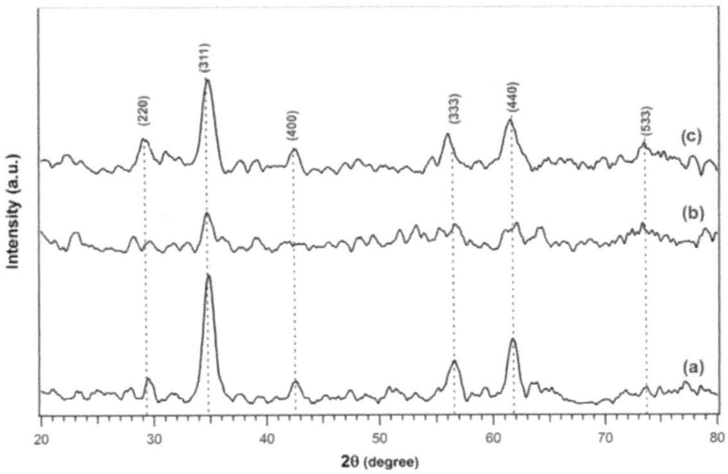

Figure. 1. XRD patterns of $Zn_{0.5}Ni_{0.5}Fe_2O_4$ nanoparticles, (a) before, (b) after encapsulated 1 g PEG , and (c) 3 g PEG.

TABLE 2. Analysis XRD of $Zn_{0.5}Ni_{0.5}Fe_2O_4$ before and after encapsulated with PEG

MATERIAL	Crystallite size (nm)	X-Ray density (g/cm^3)	Lattice parameter (nm)
$Zn_{0.5}Ni_{0.5}Fe_2O_4$	15.2	5.09	0.85 ± 0.05
$Zn_{0.5}Ni_{0.5}Fe_2O_4$ + 1 g PEG	12.9	5.07	0.85 ± 0.03
$Zn_{0.5}Ni_{0.5}Fe_2O_4$ + 3 g PEG	13.3	5.08	0.85 ± 0.02

Morphological Studies

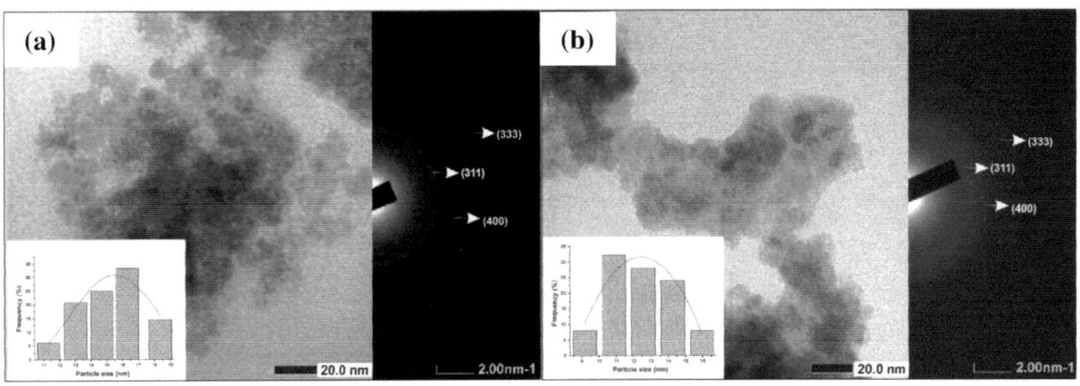

Figure 2. Morphology, SAED pattern and histogram of the grain size from $Zn_{0.5}Ni_{0.5}Fe_2O_4$ nanoparticles, (a) before, and (b) after encapsulated 3 g PEG-4000.

The TEM images in Figure 2 show the morphology, shape and size distribution of the $Zn_{0.5}Ni_{0.5}Fe_2O_4$ nanoparticles before and after encapsulated with PEG-4000. The morphology of all samples before and after encapsulated is still agglomerated. This is due to coprecipitation method resulting small particle size and encapsulated method not enough to properly encapsulates nanoparticles, and then the concentration of PEG to much large cause nanoparticles still agglomerate because PEG-4000 have long chain polymer.

Figure 2 shows an image of a diffraction ring and a grain size distribution of $Zn_{0.5}Ni_{0.5}Fe_2O_4$ nanoparticles (a) before, and (b) after encapsulated 3 grams of PEG-4000. For the diffraction rings seen at $Zn_{0.5}Ni_{0.5}Fe_2O_4$ have peak diffraction (311), (400) and (333), but when encapsulated with PEG-4000 3 g still there. The obtained diffraction ring confirmed the results from the previous XRD data. The resulting grain size distribution was obtained on the $Zn_{0.5}Ni_{0.5}Fe_2O_4$ nanoparticles of 15.5 nm, which was encapsulated with PEG-4000 to 13.3 nm. The size of the grains and the crystallite size already obtained in the XRD data are almost identical.

Magnetic behavior

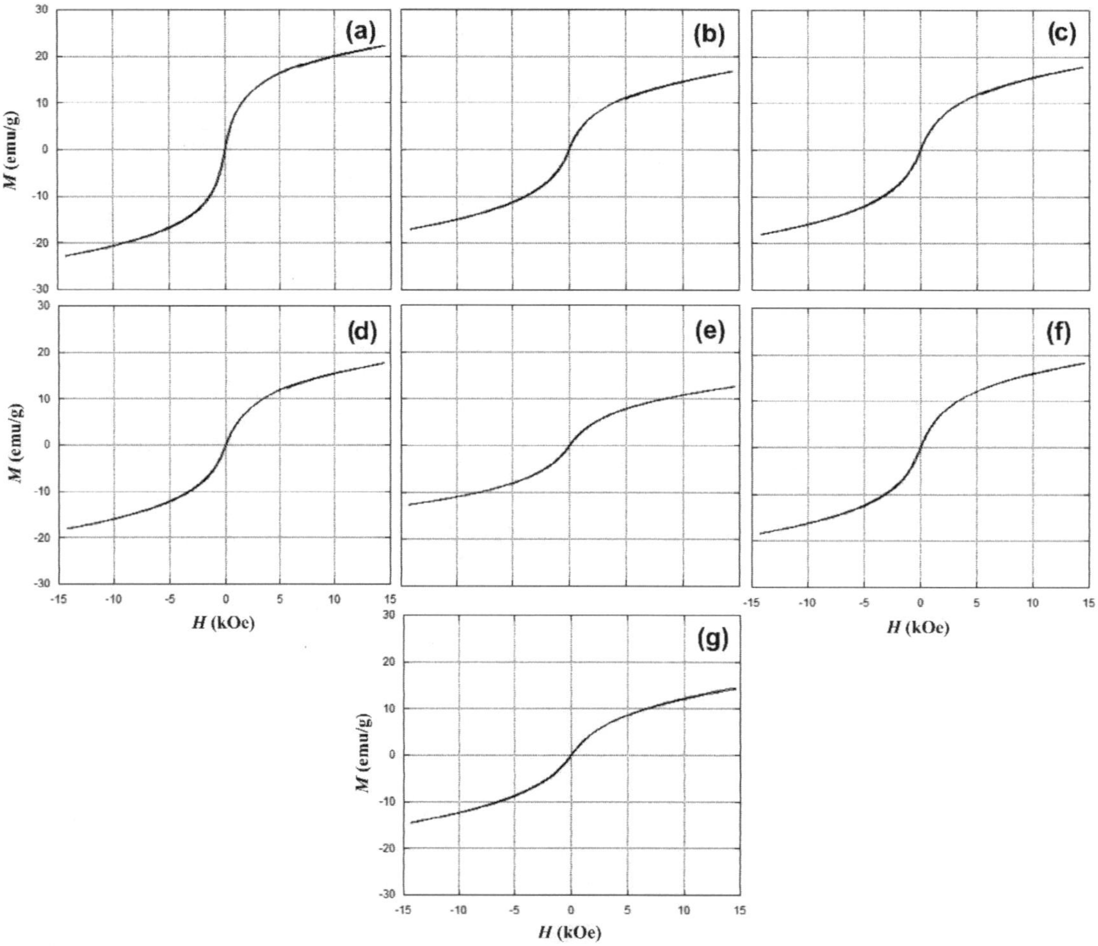

Figure 3. Hysteresis curves of $Zn_{0.5}Ni_{0.5}Fe_2O_4$ encapsulated PEG-4000 (a) before, and after encapsulated (b) 0.5 g, (c) 1 g, (d) 1.5 g, (e) 2 g, (f) 2.5 g, and (g) 3 g.

The magnetic properties for all samples were studied using VSM. The magnetic hysteresis loop of $Zn_{0.5}Ni_{0.5}Fe_2O_4$ before and after encapsulated with PEG-4000 as shown in Figure 3. In Table 3 as we can see the maximum magnetization (M_{max}) of $Zn_{0.5}Ni_{0.5}Fe_2O_4$ was 18.4 emu/g, and decrease after PEG-4000 encapsulation (see Table 3), its due to PEG-4000 is paramagnetic [11] which weaken the magnetic properties of $Zn_{0.5}Ni_{0.5}Fe_2O_4$. After the concentration of PEG-4000 increased, then the amount of paramagnetic material increase which lead the value of saturation magnetization decrease by the increasing of PEG-4000 concentration. Magnetic anisotropy constants is not linear with the coercivity. This due to the shape of particles that has an irregular shape.

The coercivity (H_c) of $Zn_{0.5}Ni_{0.5}Fe_2O_4$ was 50.7 Oe. The H_c decrease after encapsulated with variation concentration of PEG-4000. H_c decrease because of the particle size decreases. The result of this study is in the multidomain region. The multidomain particles and the energy anisotropy will be smaller to demagnetization require a smaller external field. Thus the value of coercivity decrease.

TABLE 3. Maximum magnetization (M_{max}), remanen magnetization (M_r), coercivity (H_c) and anisotropy constant (K) of $Zn_{0.5}Ni_{0.5}Fe_2O_4$ before and after encapsulated PEG-4000.

No	PEG-4000 concentration (gram)	M_{max} (15 kOe) (emu/g)	M_r (emu/g)	H_c (Oe)	K (erg/g)
1	0	22.7	0.6	45.8	0.5
2	0.5	17.2	0.3	47.9	0.4
3	1	18.2	0.3	48.9	0.4
4	1.5	17.9	0.3	49.2	0.4
5	2	18.4	0.3	50.7	0.5
6	2.5	14.6	0.1	43.4	0.3
7	3	12.8	0.2	44.9	0.3

4. Conclusion

Nanocrystalline mixed spinel ferrite of $Zn_{0.5}Ni_{0.5}Fe_2O_4$ nanoparticles have been successfully synthesized by coprecipitation method. The crystallite size before encapsulation was 15.23 nm, After encapsulated PEG-4000 decreased to 13.26 nm. After the encapsulated PEG-4000 nanoparticles $Zn_{0,5}Ni_{0,5}Fe_2O_4$ still agglomerated. Coercivity value before encapsulated was 50.7 Oe, After encapsulated PEG-4000 decreased to 47.9 Oe. The maximum magnetization before encapsulation of 18.4 emu/g After encapsulated PEG-4000 decreased to 17.1 emu/g.

Acknowledgment

This work was financially supported by Nanofabrication Platform Consortium Project of Nagoya University, Ministry of Culture, Sports, Science and Technology (MEXT) Nano-Project Platform, Japan and Grant of "Penelitian Berbasis Kompetensi", The Ministry of Research, Technology and Higher Education of the Republic of Indonesia

References

[1] Goldman, A, Modern Ferrite Technology Second Edition, Springer, USA, 2006.
[2] Schmid, G, Nanoparticles: From Theory to Application, Weinheim : WILEY-VCH Verlag GmbH & Co. KGaA, USA, 2010.
[3] Zhou, B., Liu, Z., Wang, X., Siu, Y., Huang, X., Lu, Z., and Su, W 2010 *Physica B* **405** 374
[4] Godbole, B., Badera, N., Shrivastava, S.B, Jain, D., Chandra, L.S.S and Ganesan, V 2013 *Physics Procedia* **49** 58
[5] Indrayana, I.P.T., Siregar, N., Suharyadi, E., Kato, T., and Iwata, S *J. Phys : Conf. Ser* **776** 2016 012021
[6] Liu, R., Shen, X., Jing, C., Song, F., dan Li, H 2012 *Journal of Alloys and Compounds* **511** 163
[7] Deraz, N.M and Abd-Elkader, O.H 2014 *Journal of Analytical and Applied Pyrolisis* **106** 171
[8] Vivekanadhan, S., Venkateswarlu, M., and Satyanarayana, N 2004 *Mater. Lett* **58** 2717
[9] Sivakumar, P., Ramesh., Ramanand, A., Ponnusany, S., and Muthamizhcelvan, C 2013 *J. Alloys Compd* **563** 6
[10] Serkol, M., Köseoğlu, Y., Baykal, A., Kavas, H., and Basaran, A.C 2009 *J. Magnetism and Magnetic Materials* **321** 157
[11] Puspitarum, D. L., Hermawan, A., Suharyadi, E 2016 *AIP conference proceedings* **1725** 20079

Study on Photocatalytic Properties of TiO_2 Nanoparticle in various pH condition

Nasikhudin[1,2], M Diantoro[2], A Kusumaatmaja[1], K Triyana[1]

[1] Departement of Physics, Faculty of Mathematics and Natural Sciences Universitas Gadjah Mada

[2] Departement of Physics, Faculty of Mathematics and Natural Sciences, Universitas Negeri Malang

Email: triyana@ugm.ac.id

Abstract: Titanium dioxide has been widely studied for its ability to photocatalytic and applications have high performance for photovoltaic applications. In this paper TiO_2 nanoparticle was investigated for the degradation of methylene blue under UV light in various pH condition. The TiO_2 nanoparticle was characterized by SEM and XRD. The results showed that TiO_2 nanoparticle has the structure of anatase and have a particle size of 27 nm. The photocatalytic activity of TiO_2 nanoparticle show that the degradation of methylene blue under UV light have dye removal of 97% dye was degraded in 3 h, but the degradation of methylene blue without UV light have dye removal of 15% dye was degraded in 3 h. It indicated that The photocatalytic activity of TiO_2 nanoparticle could occur if there the UV light. If not UV light the photocatalytic activity cannot occurs, the degradation of Methylene Blue 15% is not a photocatalytic activity but it is adsorption of Methylene Blue by TiO_2 nanoparticle. The photocatalytic activity of TiO_2 nanoparticle has pH-sensitive. The photocatalytic activity of TiO_2 nanoparticle in acid condition (pH 4.1) is 40%, in neutral condition (pH 7.0) is 90%, and in base condition (pH 9.7) is 97%. The highest photocatalytic activity occurs in base condition, it causes in base condition OH- can be direct reaction with a hole to produce hydroxyl radical (OH*).

Keywords: TiO_2, pH, Methylene Blue, photocatalytic.

1 Introduction

Titanium dioxide has been widely studied for its ability to photocatalytic and applications have high performance for photovoltaic applications [1]. Recently, environmental purification using TiO2 as a photocatalytic has attracted a great interest due to its high Activity [2]. It also shows chemical stability, low toxicity, low pollutant load, and availability at low-cost [3] [4]. TiO_2 size has been reduced to increase the reactive surface area to improve efficiency [5]. Wastewater treatment for environmental preservation-based TiO_2 photocatalytic has been proven to degrade many organic pollutants effectively [6]. TiO_2 has been widely used as photocatalytic due to its chemical stability and availability in the commercial. Many forms of TiO_2 nanomaterial have been prepared including nanoparticle, nanotube, nanofiber, and nanosheet [7]. Many efforts are focused on the application of TiO_2 in the wastewater treatment system [8]. An important parameter in the photocatalytic reactions taking place on the particulate surfaces is the pH of the solution, since it dictates the surface charge properties of the photocatalyst and size of aggregates it forms [9] Water and oxygen molecules dissociation is a phenomenon commonly observed as a result of the achieved extremely conditions. [10] The pH might play an important role for the elimination mechanism of the pharmaceuticals in the environment. About photolysis, the different species might have various photolytic degradation pathways, transformation products, and kinetics of mechanism-based degradation [11]. The better degradation rate in acidic or basic pH may also be attributed by the fact that the structural orientation of the molecule is favoured for the attack of the reactive species under that condition. The degradation rate for the decomposition and mineralization of acid red was found to increase with the increase pH solution, while the degradation rate for the decomposition and mineralization of the pesticide was found to decrease with the increase pH solution [9]. The interpretation of pH effects on the photocatalytic process is very difficult task because of its multiple roles such as electrostatic interactions between the semiconductor surface, solvent molecules, substrate and charged radicals formed during the reaction process. Many efforts has been done, but no clear explanation. In this study we will discuss the effect of pH of Methylene Blue solution as a pollutant on photocatalytic activity of TiO_2 nanoparticle.

2 Experiment Methods

The material used in the research were: Titanium (IV) oxide, anatase, nanopowder < 25 nm particle size, 99.7% trace metal basis and methylene blue powder, all of them purchased from Sigma-Aldrich. Characterizations of TiO_2 nanoparticle using XRD and SEM. The nanofibers surface morphology was studied using SEM (FEI, Type: Inspect-S50). Image processing software was used to measuring the grain size of the nanofibers captured in the SEM micrograph. XRD (PanAnalytical, Type: Expert Pro) were used to determine the structure of TiO_2 nanoparticle. Retica program was used to analysis XRD pattern. Photocatalytic activities of TiO_2 nanoparticle were investigated by observing degradation of methylene blue (MB) under UV light. The experiment was set up with conditions 4 x 10 W lamp as the UV light source (365 nm). The reaction was prepared by using 0.1 g of TiO_2 nanoparticle into the beaker containing 10 ppm of Methylene blue solution (10mL). Degradation of methylene blue was observed by UV-vis spectroscopy each 1 hours during 5 hours which suggested TiO_2 nanoparticle could degrade the methylene blue solution. Photocatalytic activities of TiO_2 nanoparticle were also investigated in various condition (without UV and with UV light) and various pH (acid, neutral, and base).

3 Results and Discussion

The SEM image of TiO_2 nanoparticle is shown in Figure 1. The morphology of TiO_2 nanoparticle shows that TiO_2 nanoparticles have agglomeration [12]. The nature of the agglomeration of TiO_2 nanoparticles in aqueous suspension [13]. The agglomeration occurs because the TiO_2 is unstable in the nanoparticles form so that they will tend to join each other until they are relatively stable. The result of SEM analysis in Figure 1.(a) with image j program shows the agglomeration size of TiO_2 is about 975 nm. To know the size of these tio2 particles, SEM performed with 50,000 times magnification. SEM results with 50000 magnification this time can be seen the arrangement of TiO_2 nanoparticles that form agglomeration. SEM analysis results with image j program in Figure 1.(c) shows the particles size of TiO_2 is about 25 nm.

(a) (b) (c)

Figure 1. SEM image of TiO_2 nanoparticle with magnification of (a) 5000x, (b) 10.000x, and (c) 50.000x.

The XRD used to investigate the crystal structure of TiO_2 nanoparticle. The XRD patterns of TiO_2 nanoparticle have some diffraction peaks shown in Fig. 2. The peaks occur at 2θ of 25.3° for crystal plane 101; 37.90 for crystal plane (004); 48.2 for crystal plane (200), and other peaks show in figure 2.
The XRD result was analysis with Retica program to identify cited the structure and parameter of the crystal. The result show that the crystal structure of the TiO_2 is anatase (tetragonal) with parameter a = b = 3.789603 Å, and c = 9.516015 Å, cell volume = 136.660339 Å^3, molecular weight = 319.600, density = 3.882, particle size of 25.25 nm..

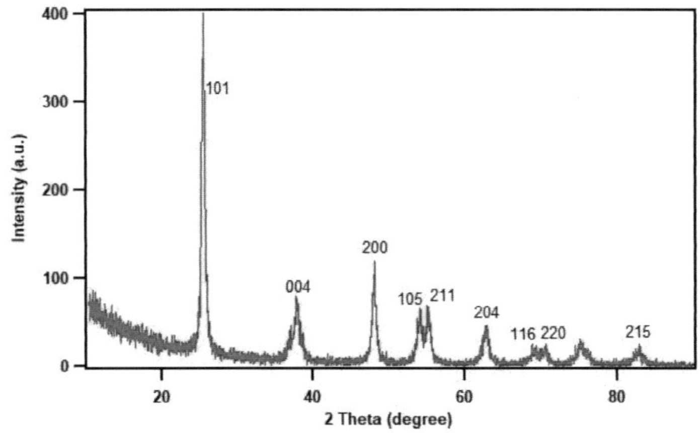

Figure 2. XRD pattern of TiO_2 nanoparticle

The UV-Vis spectroscopy used to observe the concentration of Methylene Blue solution after photocatalytic treatment. The UV-Vis spectroscopy observes every 1 hours during 5 hours. Figure 3 shows that the Methylene Blue (MB) solution is degraded by TiO_2 nanopowder. Degradation of Methylene Blue solution by TiO_2 without UV light very less than TiO_2 with UV light. Comparison of photocatalytic activity to degrade Methylene Blue solution between TiO_2 without UV light and TiO_2 with UV light is shown in Figure 3.

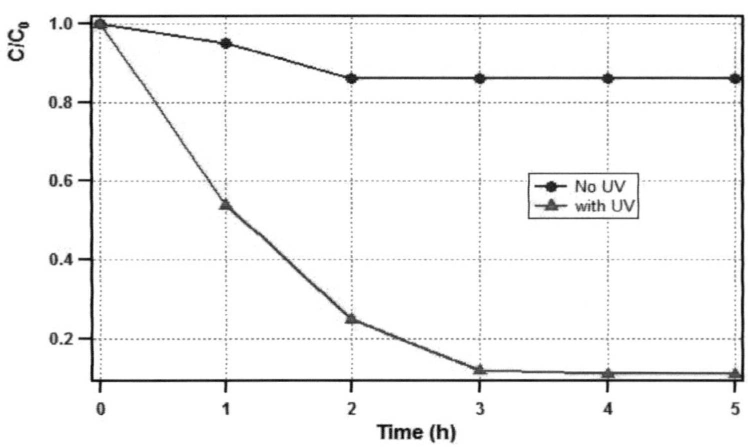

Figure 3. Photocatalytic activity of TiO_2 nanoparticle in dark (No UV) and bright (with UV) condition.

Figure 3 shows that in the absence of UV rays there will be no photocatalytic process, MB degradation of about 15% occurring is not a photocatalytic process, but this is due to the adsorption of MB by the TiO_2 particles. On the other hand, the process using UV mc degradation rays that occurred showed significant results. TiO_2 can degrade MB to 95% for 3 hours. This suggests that the photocatalytic process occurs only when there is light (UV light), this can be explained by the photocatalytic mechanism.

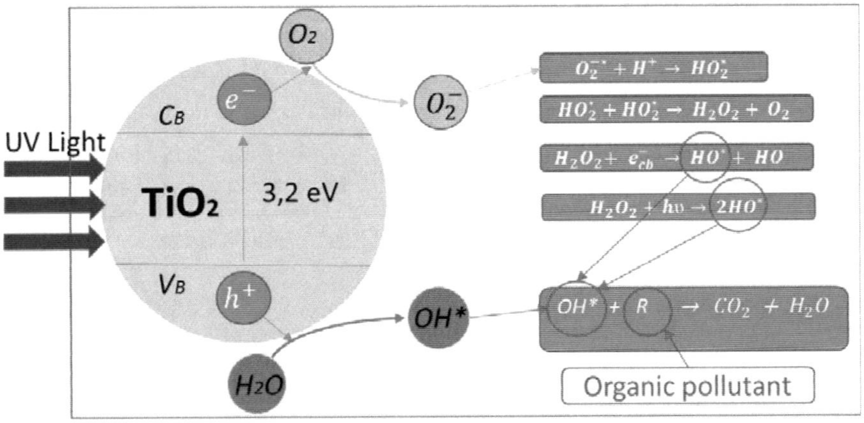

Figure 4. Photocatalytic mechanism of TiO_2.

The photocatalytic mechanism of TiO_2 can be explained as follows.

$$TiO_2 + h\upsilon \rightarrow e_{cb}^- + h_{vb}^+$$

$$h_{vb}^+ + H_2O \rightarrow H^+ + HO^*$$

$$e_{cb}^- + O_2 \rightarrow O_2^{-*}$$

$$O_2^{-*} + H^+ \rightarrow HO_2^*$$

$$HO_2^* + HO_2^* \rightarrow H_2O_2 + O_2$$

$$H_2O_2 + e_{cb}^- \rightarrow HO^* + HO^-$$

$$H_2O_2 + h\upsilon \rightarrow 2HO^*$$

$$R + HO^* \rightarrow CO_2 + H_2O$$

The photocatalytic mechanism of TiO_2 can be explained as follows in figure 4. When TiO_2 semiconductor is illuminated with greater energy than a band gap, an electron will move from the valence band to the conduction band to produce holes (h_{vb}^+) in the valence band and the electrons (e_{cb}^-) in the conduction band. The electrons will interact with the surrounding oxygen and produce superoxide radicals (O_2^{-*}), while the holes will interact with the surrounding water to produce hydroxyl radicals (HO^*). Both of these free radicals will interact with water-soluble compounds in the vicinity, especially organic compounds or polymers. This hydroxyl radical will decompose the organic compound or polymer into water (H_2O) and carbon dioxide (CO_2). The superoxide radicals (O_2^{-*}) will interact with water (H_2O) to produce hydrogen peroxide (H_2O_2), then this hydrogen peroxide will react with electrons that will produce hydroxyl radicals, Peroxide (H_2O_2) reacts with the light energy ($h\upsilon$) which will also produce a hydroxyl radical (HO^*). This hydroxyl radical (HO^*) will decompose the organic polymer (R) into water (H_2O) and carbon dioxide (CO_2).

Photocatalytic activities of TiO_2 nanoparticle were also investigated in various condition (without UV and with UV light) and various pH; acidic pH conditions (pH = 4.1), neutral (pH = 7), and base (pH = 9,7). The results of the photocatalytic investigation are shown in Figure 5.

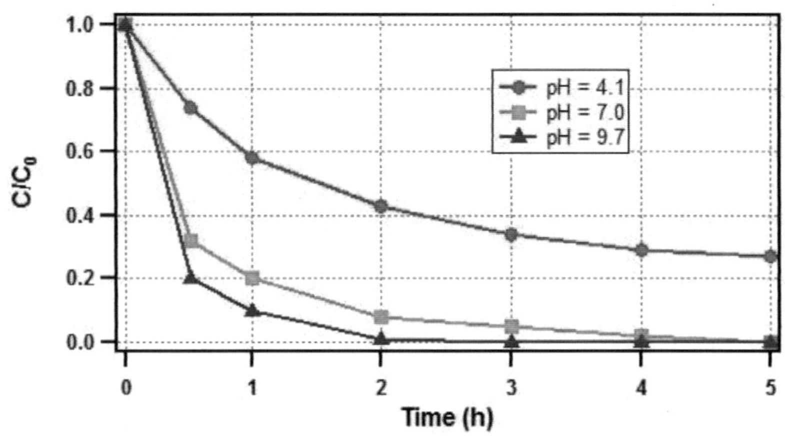

Figure 5. Photocatalytic activity of TiO_2 nanoparticle in acid, neutral, and alkaline condition.

Figure 5 shows that a photocatalytic process that takes place under alkaline conditions will be faster than in neutral conditions. Otherwise the photocatalytic process that takes place in acidic conditions will be slower than in neutral conditions, this can be explained as follows.

In the neutral condition the hydroxyl radical (HO^*) is produced from the reaction

$$h_{vb}^{+} + H_2O \rightarrow H^+ + HO^*$$

Under alkaline conditions, the hydroxyl radical may also be produced directly from the (OH^-) ion which interacts with the hole according to the following reaction

$$h_{vb}^{+} + OH^- \rightarrow HO^*$$

Therefore under alkaline conditions, the resulting hydroxyl radical (HO^*) will be more numerous so that its photocatalytic ability will be better.

In acidic conditions there exists an excess of H^+, so that this excess H^+ will interact with the free electrons (e^-) to form radicals (H^*). The formation of this radical (H^*) will cause a backlash with HO^* and H^* return to H_2O.

$$H^+ + e^- \rightarrow H^*$$

$$H^* + HO^* \rightarrow H_2O$$

This reduces the amount of the resulting hydroxyl radical resulting in a photocatalytic activity under acidic conditions the ability of the photocatalytic activity to decrease. It indicate that in the alkaline pH conditions the ability of photocatalytic activity is better than in neutral conditions, otherwise the acid pH conditions of the photocatalytic activity are slower than in neutral conditions.

4 Conclusions

The photocatalytic activity of TiO_2 nanoparticle for degradation of methylene blue under UV light have dye removal of 97% dye was degraded in 3 h, but the degradation of methylene blue without UV light have dye removal of 15% dye was degraded in 3 h. It indicated that The photocatalytic activity of TiO_2 nanoparticle could occur if there the UV light. The photocatalytic activity of TiO_2 will increase if the photocatalytic activity is performed under alkaline conditions, otherwise it will decrease if done under acidic conditions.

5 Acknowledgement

This research was supported by the Directorate General of Higher Education (DIKTI), the ministry of research and higher education, the Republic of Indonesia Contract No PUPT Lanjutan Fiscal Year 2017: 2383/UN1.PIII/DIT-LIT/LT/2017. The authors would also thank Mr. Kunto Wandono for maintaining the electrospinning machine.

6 References

[1] Gupta S M and Tripathi M 2011 A review of TiO2 nanoparticles *Chinese Sci. Bull.* **56** 1639–57
[2] Nasikhudin, Diantoro M, Kusumaatmaja A, Triyana K, Diantoro M, Kusumaatmaja A and Triyana K 2016 Preparation of PVA / Chitosan / TiO2 nanofibers using electrospinning method Preparation of PVA / Chitosan / TiO 2 Nanofibers using Electrospinning Method **150002** 2–8
[3] Mau D, Risti A, Dra G and Pintar A 2017 Titania versus zinc oxide nanoparticles on mesoporous silica supports as photocatalysts for removal of dyes from wastewater at neutral pH
[4] Wu Y, Zhao J, Li Y and Lu K 2016 Preparation and freezing behavior of TiO 2 nanoparticle suspensions *Ceram. Int.* **42** 15597–602
[5] Yun J, Jin D, Lee Y and Kim H 2010 Photocatalytic treatment of acidic waste water by electrospun composite nano fi bers of pH-sensitive hydrogel and TiO 2 *Mater. Lett.* **64** 2431–4
[6] Im J S, Kim M Il and Lee Y-S 2008 Preparation of PAN-based electrospun nanofiber webs containing TiO2 for photocatalytic degradation *Mater. Lett.* **62** 3652–5
[7] Bo Y I N, Ji-tong W, Wei X U, Dong-hui L and Wen-ming Q 2013 Preparation of TiO 2 / mesoporous carbon composites and their photocatalytic performance for methyl orange degradation *New Carbon Mater.* **28** 47–54
[8] Qu X, Alvarez P J J and Li Q 2013 Applications of nanotechnology in water and wastewater treatment *Water Res.* **47** 3931–46
[9] Haque M M, Bahnemann D and Muneer M 2012 Photocatalytic Degradation of Organic Pollutants : Mechanisms and Kinetics *Photocatalytic Degrad. Org. Pollut. Mech. Kinet.* **3** 294–326
[10] Hassani A, Khataee A, Karaca S, Karaca C and Gholami P 2016 Ultrasonics Sonochemistry Sonocatalytic degradation of ciprofloxacin using synthesized TiO 2 nanoparticles on montmorillonite *Ultrason. - Sonochemistry* 1–12
[11] Salma A, Thoröe-boveleth S, Schmidt T C and Tuerk J 2016 Dependence of transformation product formation on pH during photolytic and photocatalytic degradation of ciprofloxacin *J.*

Hazard. Mater. **313** 49–59
[12] Nasikhudin, E P Ismaya, M Diantoro, A Kusumaatmaja and K T Preparation of PVA / TiO 2 Composites Nanofibers by using Electrospinning Method for Photocatalytic Degradation **012011**
[13] Li G, Lv L, Fan H, Ma J, Li Y, Wan Y and Zhao X S 2010 Journal of Colloid and Interface Science Effect of the agglomeration of TiO 2 nanoparticles on their photocatalytic performance in the aqueous phase *J. Colloid Interface Sci.* **348** 342–7

IOP Conf. Series: Journal of Physics: Conf. Series **1011** (2018) 012070 doi:10.1088/1742-6596/1011/1/012070

Effect of external electric field on spin-orbit splitting of the two-dimensional tungsten dichalcogenides WX_2 (X = S, Se)

Y Affandi[1], M A U Absor[1]* and K Abraha[1]

[1]Department of Physics, Universitas Gadjah Mada, BLS 12 Yogyakarta, Indonesia

*Email : adib@ugm.ac.id

Abstract. Tungsten dichalcogenides WX_2 (X=S, Se) monolayer (ML) attracted much attention due their large spin splitting, which is promising for spintronics applications. However, manipulation of the spin splitting using an external electric field plays a crucial role in the spintronic device operation, such as the spin-field effect transistor. By using first-principles calculations based on density functional theory (DFT), we investigate the impact of external electric field on the spin splitting properties of the WX_2 ML. We find that large spin-splitting up to 441 meV and 493 meV is observed on the K point of the valence band maximum, for the case of the WS_2 and WSe_2 ML, respectively. Moreover, we also find that the large spin-orbit splitting is also identified in the conduction band minimum around Q points with energy splitting of 285 meV and 270 meV, respectively. Our calculation also show that existence of the direct semiconducting – indirect semiconducting – metallic transition by applying the external electric field. Our study clarify that the electric field plays a significant role in spin-orbit interaction of the WX_2 ML, which has very important implications in designing future spintronic devices.

1. Introduction

Recent discovery of graphene as a pioneer of a two-dimensional material with an extraordinary electronic, optical and magnetic properties motivates researchers to find new materials of two-dimensional structures. Especially in spintronics applications, these new materials are including topological insulators, Dirac materials and majonara fermions [1–3]. In addition to these materials, the two-dimensional material of transition metal dichalcogenides (TMDs) has also attracted much attention recently due to its unique properties such as large spin-splitting, spin-valley coupling and the sensitivity of its electronic structure [4–6]. Bulk TMDs have a MX_2 stoichiometry with M is a transition metal atoms and X is a chalcogen atoms. In contrast to bulk TMDs which have an indirect band gap with valence band maximum (VBM) at the Γ point and conduction band minimum (CBM) at the Q point, monolayer (ML) TMDs has a direct band gap with VBM and CBM are centered at the K point. This transition from indirect to direct band gap is due to missing of van der Waals interactions [4,7,8], which has important implications in photonics, optoelectronics, and sensing devices [7].

Another unique phenomenon in ML TMDs is emergence of spin-splitting in the VBM and CBM. Especially the WS_2 and WSe_2 ML which attracted much attention since their spin-splitting are predicted to be the largest among the ML families of TMDs [4]. In the WX_2 (X = S, Se) ML, the maximum spin-splitting is observed at K point of the VBM, which varies from 426 to 463 meV [4,6]. This large spin-splitting is believed to be responsible for inducing some interesting phenomena such as spin Hall effect, spin-dependent selection rule for optical transition, and magnetoelectric effect in

TMDs [9]. It is also reported that the large spin-splitting is also observed at the *Q* point of the CBM, which is from ranging 200 to 330 meV. This splitting is also plays important role in the properties of spintronics such as spin-conserving scattering [5]. However, the electronic structure of WX_2 ML is sensitive to some external treatments such as external electric field [10]. Therefore it is important to clarify the electric field effect on the spin-splitting properties of the WX_2 ML material, which is expected to be useful for designing spintronic devices.

In this paper, we report fully relativistic calculations within density functional theory on electronic properties of the WS_2 and WSe_2 ML. For all such systems, we clarified the emergence of spin splitting in the VBM and CBM. We have also examined the electronic structure and the spin-splitting properties of the WX_2 ML by applying electric field. Finally, the possibility of implementing such systems for spintronics will be discussed.

2. Computational Method

Crystal structure of the bulk WX_2 (X = S, Se) has a hexagonal structure (2H) with P63/mmc (D_{6h}^4) point group symmetry. It consists of several layers *X*-W-*X* characterized by a weak van der Waals interaction. Each layer consists of W atom layer that is sandwiched by two *X* atoms with a strong covalent bonding. Due to the weak van der Waals interaction, it is possible to create WX_2 ML by using micromechanical cleavage and liquid exfoliation [11].

In contrast to the bulk WX_2, inversion symmetry is broken in the WX_2 ML (figure 1. (*a*)), thus reduces its symmetry becomes P6m2 (D_{3h}^1). The D_{3h}^1 symmetry itself consists of C_{3v} and mirror symmetry M_{xy} (figure 1. (*a*)). The *k* space in first Brillouin zone characterizing the crystal structures of the WX_2 ML is shown in figure 1. (*b*). Electronic structure calculations are carried-out using density functional theory within the generalized gradient approximation (GGA) [12] as implemented in the OpenMX code [13]. We use norm-conserving pseudo-potentials, and the wave functions are expanded by the linear combination of multiple pseudo-atomic orbitals (LCPAOs) generated using a confinement scheme [14,15]. The orbitals are specified by W7.0-$s^2p^2d^1$, S9.0-$s^1p^1d^1$ and Se9.0-$s^1p^1d^1$, which means that the cut off radii are 7.0, 9.0, and 9.0 bohr for the W, S and Se atoms, respectively. A 12 × 12 × 1 *k*-point grid and energy cutoff 200 Ry are used. Spin orbit coupling (SOC) was also included in these fully relativistic calculations. The WX_2 ML are modelled as a periodic slab with a sufficiently large vacuum layer (24 Å) in order to avoid interaction between another layers. The structures are relaxed until the force acting on each atom was less than 0.5×10^{-3} eV/Å. The external electric field are applied in the direction of the positive z axis.

3. Result and discussion

3.1. Electronics structures of the WX_2 (X = S, Se) ML

First, we optimize lattice parameter, with the result shown in table 1. Consistent with previous calculation results [4], we find that optimized in-plane lattice constants are 3.194 Å and 3.4 Å for the case of WS_2 and WSe_2 ML, respectively. However, these values are slightly larger than that of the experiment [16,17].

Table 1. Calculated lattice constants of WX_2 (X = S, Se) ML using GGA-PBE. Experimental values and recently report calculated for WX_2 ML are also given.

System	Calculated Lattice constant (Å)	Experiment Lattice constant (Å)	Previous Calculation (Å)
WS_2 ML	3.194	3.154[a]	3.197[c]
WSe_2 ML	3.401	3.280[b]	3.310[c]

[a] Experimental value for WS_2 ML [16]
[b] Experimental value for WSe_2 ML [17]
[c] Previous calculation result [4]

Next, we investigate the electronic properties of WX_2 ML. As shown in figure 2, both WS_2 and WSe_2 ML show a direct-gap where the VBM and CBM are centered at the *K* point. We find that the energy gap obtained from our calculations are 1.53 eV and 1.00 eV for WS_2 and WSe_2 ML, respectively. Our partial density of state (PDOS) analysis show that at the *K* point of the VBM is dominated by $Wd_{x^2-y^2}+d_{xy}$ orbitals, while at the *K* point of CBM mainly originated from $Wd_{3z^2-r^2}$ orbitals (figure 1. (*c*) and (*d*)).

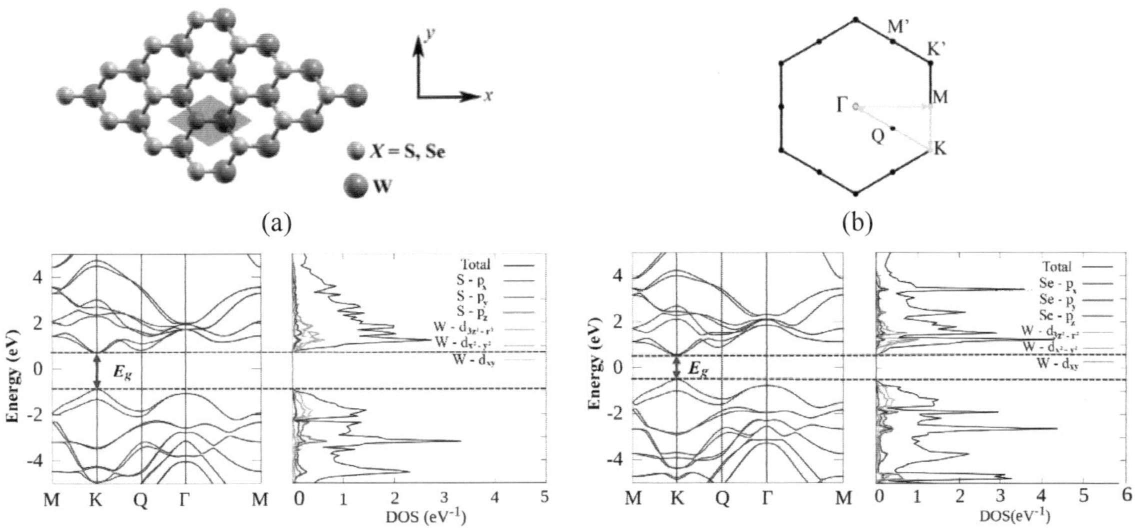

Figure 1. (*a*) Top view of crystal structure of tungsten dichalcogenides. (*b*) Brillouin zone of tungsten dichalcogenides with high symmetry point**.** The band structure corresponding to the density of state, projected to the atomic orbitals for (*c*) WS_2 and (*d*) WSe_2.

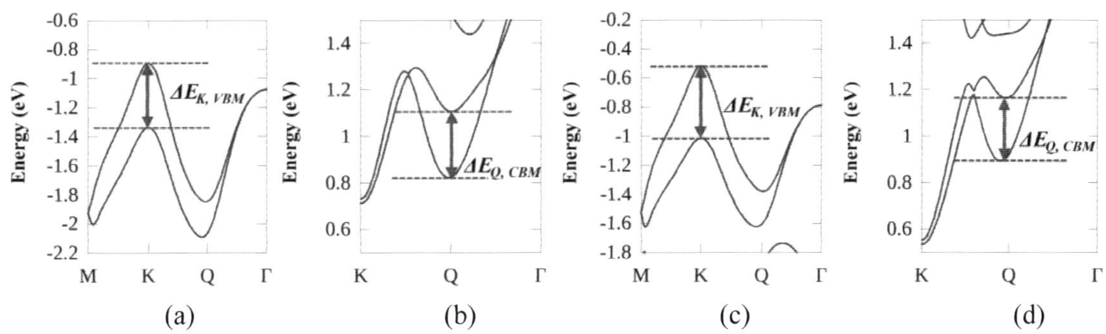

Figure 2. Spin-splitting in (*a*) the VBM and (*b*) CBM for WS_2 ML and in (*c*) the VBM and (*d*) the CBM for WSe_2 ML.

By concerning to the spin-splitting, we find that a substantial spin-splitting is visible along *M-K-Γ* as show in figure 2. The energy of spin-splitting at VBM (*K* point) are identified to be 441 meV and 492 meV for WS_2 and WSe_2 ML, respectively (figure 2 (*a*) and (*c*)). These results are in a good agreement with previous experimental result [18,19]. In addition, the spin-splitting is also observed at *Q* point in the CBM for both WS_2 and WSe_2 ML, which is found to be 285 meV and 270 meV, respectively (figure 2 (*b*) and (*d*)). The origin of the spin-splitting is due to the presence of the broken inversion symmetry. In bulk WX_2 Kramer's degeneracy $[E^{\uparrow}(\vec{k}) = E^{\downarrow}(\vec{k})]$ is suppressed by the existence of the time reversal symmetry $[E^{\uparrow}(\vec{k}) = E^{\downarrow}(-\vec{k})]$ and inversion symmetry $[E^{\uparrow}(\vec{k}) = E^{\uparrow}(-\vec{k})]$.

However, the inversion symmetry is broken in the monolayer, thus spin degeneracy is lifted, inducing the spin-splitting. Because the monolayer has in-plane dipole moment, the spin-splitting is established on the band except for the *Γ-M* direction (figure 2(*a*) and (*b*)). Moreover, at *Γ* point, the spin degeneracy is due to time reversal symmetry alone, whereas at the *M* point, the spin degeneracy suppressed by the combination of time reversal symmetry and translation symmetry [4].

3.2. The electric field effect on the electronic structure of the WX_2 ML

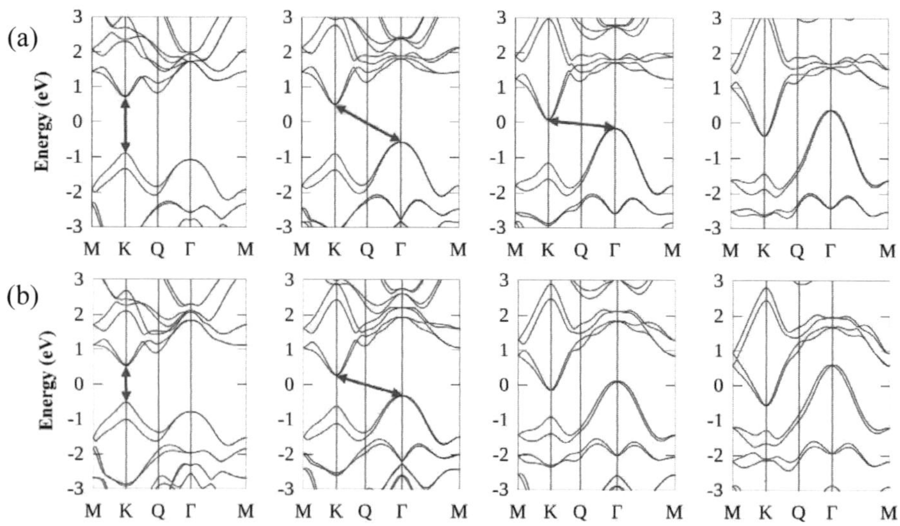

Figure 3. Electronic structure of (*a*) WS_2 and (*b*) WSe_2 ML with external electric field 0, 0.2, 0.4 and 0.6 V/Å, from left to right, respectively.

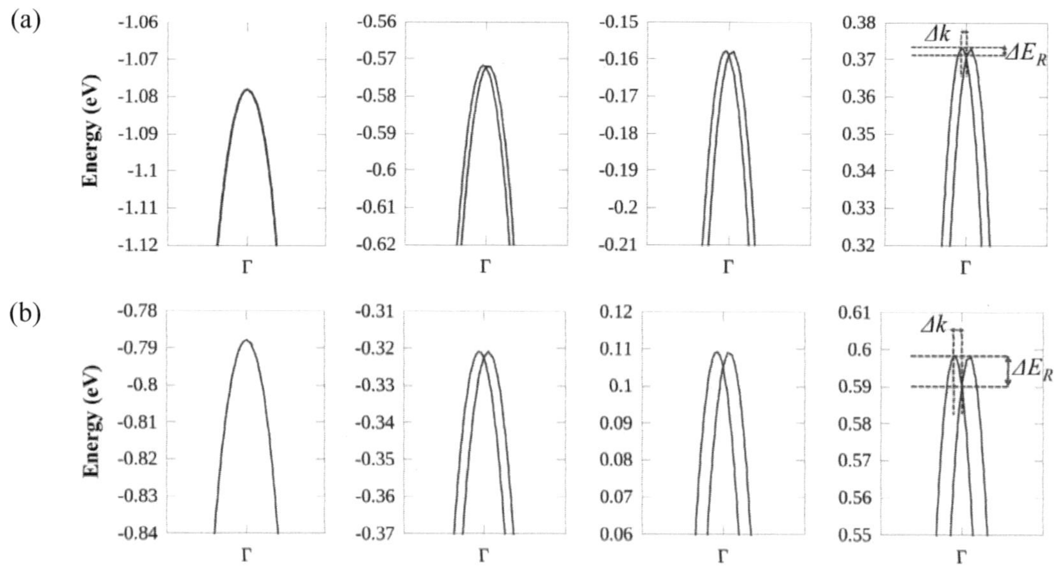

Figure 4. Zoom in rashba spin-splitting of (*a*) WS_2 and (*b*) WSe_2 ML by applied electric field 0, 0.2, 0.4 and 0.6 V/Å, from left to right, respectively.

Next, we study the effect of the electric field on the electronic properties of the WX_2 ML. We introduce the electric field ranging from 0.2 to 0.6 V/Å. As show in figure 3, we find direct semiconducting - indirect semiconducting - metallic transition, when the electric field increases. Consequently, the shifting of the VBM and CBM appears, which effects to the spin-splitting energy.

Interestingly, we find new spin-splitting along Γ-M direction, which is not found in the system without the electric field. As a result, the Rashba-type spin-splitting is induced around the Γ point as show in figure 4.

4. Conclusion

In summary, we report fully relativistic calculations within density functional theory to study the spin-splitting band in VBM and CBM of WS_2 and WSe_2 ML. We found that substantial spin-splitting bands are identified in VBM and CBM of WS_2 and WSe_2 ML. In VBM around K point, we obtained the energy of spin-splitting are 441 meV and 492 meV for WS_2 and WSe_2 ML. Meanwhile, in CBM around Q point, we found the large spin-splitting energy are 285 meV and 270 meV for WS_2 and WSe_2 ML, respectively. The spin-splitting bands is due to a loss of the inversion symmetry in the ML case. Those results make the WS_2 and WSe_2 ML exhibits great potential for application, e.g. spin field-effect transistor. Applied electric field also appears direct semiconducting – indirect semiconducting – metallic transition of electronics structure. Controlling electronics structures by external electric field, makes WX_2 (X = S, Se) ML potential candidates for future spintronic devices.

Acknowledgments

This work was supported by the Fundamental Research Grant (No.2237/UN1.PIII-DITLIT-LT/2017) funded by ministry of research and technology and higher education, Republic of Indonesia. The computations in this research were performed using high computer (laboratory of material and instrumentation physics) at Universitas Gadjah Mada, Indonesia.

References

[1] Alicea J 2010 *Phys. Rev. B* **81** 1–10
[2] Hasan M Z and Kane C L 2010 *Rev. Mod. Phys.* **82** 3045–67
[3] Wehling T O, Black-Schaffer A M and Balatsky A V 2014 *Adv. Phys.* **63** 1–76
[4] Zhu Z Y, Cheng Y C and Schwingenschlögl U 2011 *Phys. Rev. B* **84** 1–5
[5] Absor M A U, Kotaka H, Ishii F and Saito M 2016 *Phys. Rev. B* **94** 115131
[6] Komider K, González J W and Fernández-Rossier J 2013 *Phys. Rev. B* **88** 1–7
[7] Wang Q H, Kalantar-Zadeh K, Kis A, Coleman J N and Strano M S 2012 *Nat. Nanotechnol.* **7** 699–712
[8] Lv R, Terrones H, Laura A, Perea-lópez N, Gutiérrez H R, Cruz-silva E, Pulickal L, Dresselhaus M S and Terrones M 2015 *Nano Today* **10** 559–92
[9] Gong Z, Liu G, Yu H, Xiao D, Cui X, Xu X and Yao W 2013 *Nat. Commun.* **4** 1–6
[10] Yuan H, Bahramy M S, Morimoto K, Wu S, Nomura K, Yang B, Shimotani H, Suzuki R, Toh M, Kloc C, Xu X, Arita R, Nagaosa N and Iwasa Y 2013 *Nat. Phys.* **9** 563–9
[11] Novoselov K S, Jiang D, Schedin F, Booth T J, Khotkevich V V, Morozov S V and Geim A K 2005 *Proc. Natl. Acad. Sci. U. S. A.* **102** 10451–3
[12] Perdew J P, Burke K and Ernzerhof M 1996 *Phys. Rev. Lett.* **77** 3865
[13] Ozaki T, Kino H, Yu J, Han M J, Ohfuchi M, Ishii F, Sawada K, Kubota Y, Ohwaki T, Weng H, Toyoda M, Okuno Y, Perez R, Bell P P, Duy T V T, Xiao Y, Ito A M and Terakura K 2013 *User's manual of OpenMX Ver. 3.7* (http://www.openmx-square.org/)
[14] Ozaki T 2003 *Phys. Rev. B* **67** 1–5
[15] Ozaki T and Kino H 2004 *Phys. Rev. B* **69** 1–19
[16] Wilson J A and Yoffe A D 1969 *Adv. Phys.* **18** 193–335
[17] Benhida S, Bernède J C, Pouzet J and Barreau A 1993 *Thin Solid Films* **224** 39–45
[18] Zhang Y, Ugeda M M, Jin C, Shi S F, Bradley A J, Martín-Recio A, Ryu H, Kim J, Tang S, Kim Y, Zhou B, Hwang C, Chen Y, Wang F, Crommie M F, Hussain Z, Shen Z X and Mo S K 2016 *Nano Lett.* **16** 2485–91
[19] Dendzik M, Michiardi M, Sanders C, Bianchi M, Miwa J A, Gronborg S S, Lauritsen J V., Bruix A, Hammer B and Hofmann P 2015 *Phys. Rev. B* **92** 1–7

IOP Conf. Series: Journal of Physics: Conf. Series **1011** (2018) 012071 doi:10.1088/1742-6596/1011/1/012071

The Role of the Oxygen Impurity on the Electronic Properties of Monolayer Graphene: A Density-Functional Study

Muh. Yusrul Hanna[1], **Iman Santoso**[1], **Moh. Adhib Ulil Absor**[1,a]

[1]Departement of Physics, Universitas Gadjah Mada, Bulaksumur, Yogyakarta 55281, Indonesia

Abstract. Oxidation during the fabrication of a two-dimensional carbon-based system (graphene) can not be avoided so that will affect the electronic properties of the system. In this study, we investigate the effect of the oxygen impurity on the electronic properties of monolayer graphene using density functional theory calculations. The calculations were performed using the (3×3) supercell model and were validated by the (1×1) supercell calculations. We find that by increasing the oxygen concentration from $6\%, 11\%, 17\%, 33\%$ and 50%, the band gap enhanced from 0.1 to 2.48 eV. An analysis using the partial density-of-states projected to the atoms confirmed that strong hybridizations between $O - P_y$ and $C - P_z$ orbitals are responsible for inducing the enhancement of the bandgap. Therefore, we conclude that oxygen atoms are highly sensitive to changes in the electronic properties of monolayer graphene.

1. INTRODUCTION

During the last year, technological development is very advanced in terms of searching for new material such as carbon-based materials known as graphene. Graphene is two-dimensional of carbon atoms arranged in a hexagonal crystal and that has a single-atoms thickness[1,2]. Monolayer graphene has extraordinary properties such as high electron mobility in the ambient condition about $20\ m^2/Vs$[3], incredible mechanical strength with strain strength up to 130 Gpa[4] and thermal conductivity above 2000 $Wm^{-1}K^{-1}$[5]. In the pristine graphene, a linear dispersion appears around Fermi level known as the Dirac cone[2]. Moreover, the Fermi velocity up to $v_F \approx 1.0 \times 10^8$ m/s depending on the Fermi momentum as confirmed by data analysis of Shubnikov-de Haas oscillations (SdHO) on magnetoresistances[6].

Due to zero band-gap in the pristine graphene, it is important to induce opening band gap, which is expected to be useful for the electronic device. There are many methods to provide opening band gap in monolayer graphene, one of them is using atomic covalent functionalization such as hydrogen[7], flourine[8] and oxygen[9,10] or molecule[11] either above or below the graphene surface. However, the used of oxygen for graphene functionalization is more promising. On the other hand, introducing oxygen is expected to cause symmetry breaking on the electronic properties[9].

In present work, we used a first-principle density functional theory (DFT) calculation to clarify the electronic properties of graphene with oxygen-functionalization.

[a] Corresponding author :adhib@ugm.ac.id

IOP Conf. Series: Journal of Physics: Conf. Series **1011** (2018) 012071 doi:10.1088/1742-6596/1011/1/012071

2. COMPUTATIONAL METHODS

We performed first-principle electronic structure calculation on the monolayer graphene based on the density functional theory (DFT) within generalized gradient approximation (GGA)[12] using the OpenMX code[13]. Norm-conserving pseudopotentials[14] is used and the wave function is expanded by the linear combination of pseudo-atomic orbitals (LCPOs) generated using a confinement scheme[15,16]. The orbitals are specified by $C6.0 - s^2p^2d^1$ and $O7.0 - s^2p^2d^1$ which means that the cutoff radii are 6.0 and 7.0 bohr for the C and O atoms, respectively, in the confinement scheme[15,16]. For the C dan O atoms, two primitive orbitals expand the s and p orbitals, and one primitive orbital expands the d orbital.

The monolayer graphene is modelled as a periodic slab with a large vacuum layer (25Å) in order to avoid interaction between adjacent layers. We used 3×3 supercell consisting of 18 C atoms. We introduced the oxygen atoms with concentrations are varied for $0\% - 50\%$. A $12\times12\times1$ k-point grid and energy cutoff 250 Ry are used. The geometries were fully relaxed until the force acting each atom was less than 2 meV/Å. We find that the optimized lattice constant of pristine graphene is 2.47 Å, which is good agreement with recently calculated reported data[10].

We calculated the adsorption energy defined as follows[10]:

$$E_{ads} = E_{tot}(\text{graphene+O}) - E_G - NE_O \quad (1)$$

Where E_{tot} is total energy of graphene with various oxygen concentration, E_G is the energy of pristine graphene, N is the number of O atoms and E_O is the energy of an isolated O atom.

3. RESULT AND DISCUSSION

First, we examine the structural and energetic stability of graphene-oxygen functionalization. Table 1 shows calculated the result of structural optimisation and adsorption energy. In the pristine graphene, we find that dc-c = 1.42 Å which is consistent with the previous result[10]. Furthermore, oxygen-functionalization graphene will cause an increase of total energy.

Table 1. Summary of the structural optimisation. $d_{c\text{-}c}$ (Å) denotes to the distance between of two C atoms which are bonded. $d_{c\text{-}o}$ (Å) denotes to the distance between of C atom and O atom which are bonded. E_{ads} (eV/O) is oxygen adsorption energy for the most stable configuration, and E_{tot} (eV) is total energy a system.

O/C(%)	0%	6%	11%	17%	33%	50%
$d_{c\text{-}c}$ (Å)	1.42	1.42	1.39	1.37	1.35	1.41
$d_{c\text{-}o}$ (Å)	-	1.42	1.45	1.45	1.48	1.44
E_{ads} (eV/O)	-	-2.19	-2.02	-1.94	-2.36	-1.90
E_{tot} (eV)	-286154.91	-330182.93	-374156.53	-418157.34	-550377.45	-682053.34

Figure 1 shows the relaxed geometry of the oxygen-functionalization graphene, we find buckling structures at 11%, 17% and 33% concentrations. These buckling structures occur due to the symmetry breaking induced by the presence of the O atoms.

Next, we investigate the effect of oxygen-functionalization on the electronic properties graphene. As shown in table 2, the opening band gap is visible by introducing oxygen atoms, which is in a good agreement with previous results[10].

Table 2. Summary of band gaps energy of graphene with the different oxygen concentration.

O/C(%)	0%	6%	11%	17%	33%	50%
E_g(eV)	0	0.1	0.19	0.23	0.73	2.48

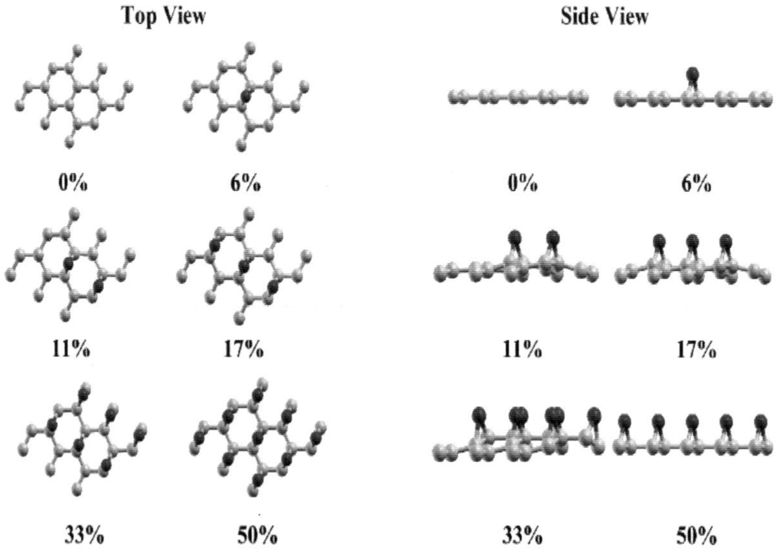

Figure 1. Structure optimization result for monolayer graphene systems with 3×3 supercell with various oxygen concentration (O/C) is 0%, 6%, 11%, 17%, 33% and 50%.

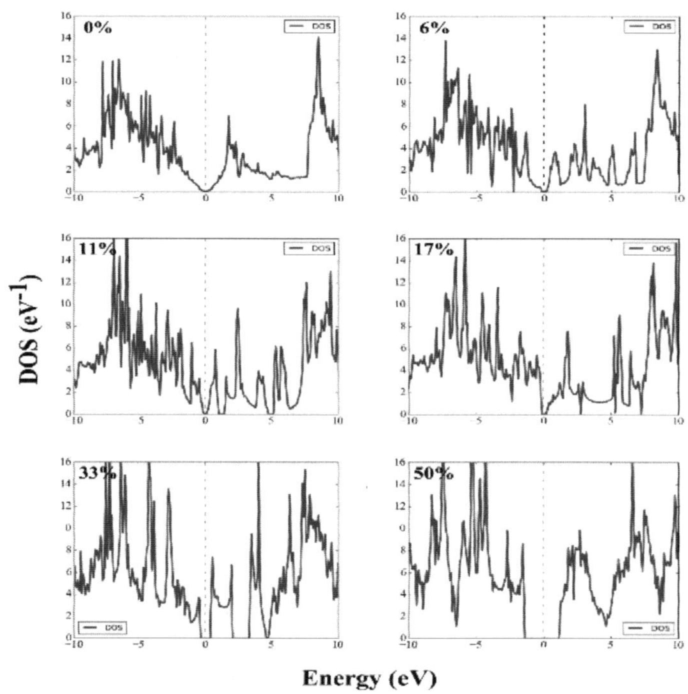

Figure 2. A Density of states (DOS) of monolayer graphene with the different oxygen concentration.

To clarify the origin of the opening the band gap, we consider the partial density of states (PDOS) projected the atomic orbitals as shown in figure 3.

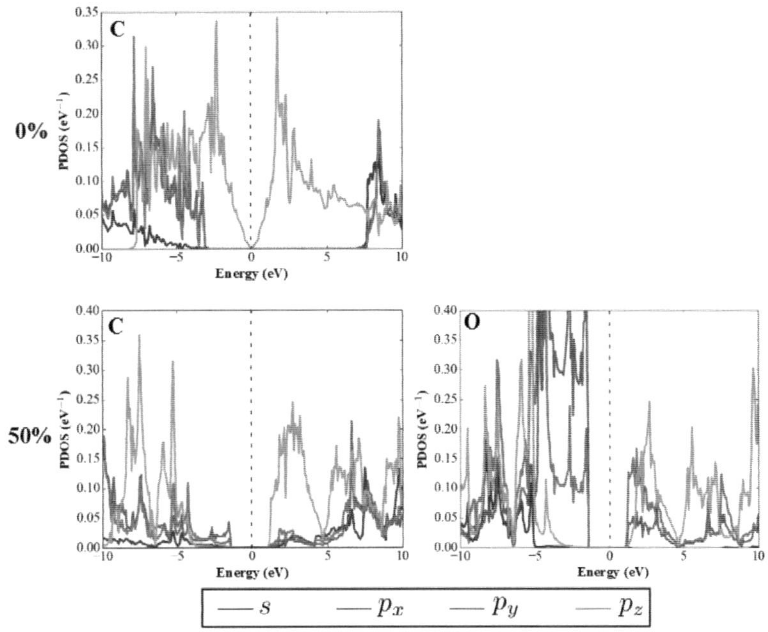

Figure 3. PDOS for C and O atoms with 0% and 50% oxygen concentration.

Figure 3 shows the PDOS of monolayer graphene with 0% oxygen concentration, The p_z orbital of the C atom dominates the band energy in the area around the Fermi level. Taking the system with 50% concentration, we find that strong hybridization between the $C - p$ and $O - p$ orbitals play an important role in the opening band gap. Here, the valence band maximum (VBM) is originated from the contribution of the $O - p_z$ and $O - p_y$ orbitals, while the conduction band minimum (CBM) predominantly by $C - p_z$ and $O - p_y$ so that the energy level in the valence band and the conduction shifting from the Fermi level and causing the bandgap in the systems.

4. CONCLUSION

In conclusion, we have successfully calculated the structure electronic properties of oxygen-functionalization of graphene using first-principle DFT calculation. We found that enhancement of the band gap is visible by increasing the oxygen concentrations. An analysis using the partial density-of-states projected to the atoms confirmed that strong hybridizations between $O - P_y$ and $C - P_z$ orbitals are responsible for inducing the enhancement of the bandgap.

Acknowledgment

This work was supported by the Fundamental Research Grant (2017) funded by Ministry of Research and Technology and the Higher Education Republic of Indonesia. The computations in this research were performed using a high computer (Laboratory of material and instrumentation physics) at Universitas Gadjah Mada.

References

[1] Wallace P R 1947 The band theory of graphite *Phys. Rev.* **71** 622

[2] Neto A H C, Guinea F, Peres N M R, Novoselov K S and Geim A K 2009 The electronic properties of graphene *Rev. Mod. Phys.* **81** 109

[3] Morozov S V, Novoselov K S, Katsnelson M I, Schedin F, Elias D C, Jaszczak J A and Geim A K 2008 Giant Intrinsic Carrier Mobilities in Graphene and Its Bilayer *Phys. Rev. Lett.* **100** 16602

[4] Lee C, Wei X, Kysar J W and Hone J 2008 Measurement of the elastic properties and intrinsic strength of monolayer graphene *Science (80-.).* **321** 385–8

[5] Balandin A A, Ghosh S, Bao W, Calizo I, Teweldebrhan D, Miao F and Lau C N 2008 Superior thermal conductivity of single-layer graphene *Nano Lett.* **8** 902–7

[6] Zhang Y, Small J P, Amori M E S and Kim P 2005 Electric field modulation of galvanomagnetic properties of mesoscopic graphite *Phys. Rev. Lett.* **94** 176803

[7] Putz S, Gmitra M and Fabian J 2014 Optical conductivity of hydrogenated graphene from first principles *Phys. Rev. B* **89** 35437

[8] Yuan S, Rösner M, Schulz A, Wehling T O and Katsnelson M I 2015 Electronic structures and optical properties of partially and fully fluorinated graphene *Phys. Rev. Lett.* **114** 47403

[9] Santoso I, Singh R S, Gogoi P K, Asmara T C, Wei D, Chen W, Wee A T S, Pereira V M and Rusydi A 2014 Tunable optical absorption and interactions in graphene via oxygen plasma *Phys. Rev. B* **89** 75134

[10] Nasehnia F, Lima S M, Seifi M and Mehran E 2016 First principles study on optical response of graphene oxides: From reduced graphene oxide to the fully oxidized surface *Comput. Mater. Sci.* **114** 112–20

[11] Wehling T O, Novoselov K S, Morozov S V, Vdovin E E, Katsnelson M I, Geim A K and Lichtenstein A I 2008 Molecular doping of graphene *Nano Lett.* **8** 173–7

[12] Perdew J P, Burke K and Ernzerhof M 1996 Generalized gradient approximation made simple *Phys. Rev. Lett.* **77** 3865

[13] Ozaki T, Kino H, Yu J, Han M J, Kobayashi N, Ohfuti M, Ishii F, Ohwaki T, Weng H and Terakura K http://www.openmx-square.org/

[14] Troullier N and Martins J L 1991 Efficient pseudopotentials for plane-wave calculations *Phys. Rev. B* **43** 1993

[15] Ozaki T 2003 Variationally optimized atomic orbitals for large-scale electronic structures *Phys. Rev. B* **67** 155108

[16] Ozaki T and Kino H 2004 Numerical atomic basis orbitals from H to Kr *Phys. Rev. B* **69** 195113

IOP Conf. Series: Journal of Physics: Conf. Series **1011** (2018) 012072 doi:10.1088/1742-6596/1011/1/012072

Theoretical study of the dependence of single impurity Anderson model on various parameters within distributional exact diagonalization method

L P Syaina, M A Majidi†

Department of Physics, Faculty of Mathematics and Natural Sciences, Universitas Indonesia, Kampus UI Depok, Depok 16424, Indonesia

†Corresponding author's email: aziz.majidi@sci.ui.ac.id

Abstract. Single impurity Anderson model describes a system consisting of non-interacting conduction electrons coupled with a localized orbital having strongly interacting electrons at a particular site. This model has been proven successful to explain the phenomenon of metal-insulator transition through Anderson localization. Despite the well-understood behaviors of the model, little has been explored theoretically on how the model properties gradually evolve as functions of hybridization parameter, interaction energy, impurity concentration, and temperature. Here, we propose to do a theoretical study on those aspects of a single impurity Anderson model using the distributional exact diagonalization method. We solve the model Hamiltonian by randomly generating sampling distribution of some conducting electron energy levels with various number of occupying electrons. The resulting eigenvalues and eigenstates are then used to define the local single-particle Green function for each sampled electron energy distribution using Lehmann representation. Later, we extract the corresponding self-energy of each distribution, then average over all the distributions and construct the local Green function of the system to calculate the density of states. We repeat this procedure for various values of those controllable parameters, and discuss our results in connection with the criteria of the occurrence of metal-insulator transition in this system.

1. Introduction

Systems of non-interacting electrons have been pretty well understood in how the electronic, magnetic, and optical properties arise. Further, systems with weak electron-electron interactions are usually still understandable as they may still behave qualitatively similar to the non-interacting ones. However, the theories established for non- or weakly-interacting electrons fail to describe the properties of the electron systems in which the strength of the electron-electron interactions is comparable to or larger than the kinetic energy. These systems are now referred to as strongly-correlated electron systems. Discoveries on these systems include, for instance, heavy-fermion compounds[1], superconductors[2], and Mott-Hubbard metal-insulator transition (MIT)[3]. Nowadays many new concepts have been constructed to explain these phenomena, but the understanding of these systems are not complete until now. A Mott insulator, as an example, can undergo phase change from metal to insulator and vice versa at certain condition. Its physical properties change dramatically with variation of control parameters such as carrier concentration, temperature, the interaction energy, or external magnetic field[4].

One of basic models that can theoretically describe strongly correlated electron systems is Anderson impurity model, or more specifically, the single impurity Anderson model (SIAM). This model describes the influence of the presence of transition element impurity in non-interacting host metal. Rigorous theoretical solutions (both analytic and numeric) for such an impurity problem are still quite complicated or demanding high computational cost. Hence, alternative methods which are simple, effective, and computationally inexpensive are still needed to be developed. Despite the existance of many impurity-problem solvers that have been tested, the distributional exact diagonalization (DED) method proposed by Granath *et al* [5] seems promising for the possibility of its application in small computer cluster. In this research, we develop computational algorithm of the DED method for solving our SIAM problem with various values of impurity concentration. The behavior of each impurity system is observed by varying temperature, interaction energy and hybridization parameters.

2. Model

SIAM which is proposed by P. W. Anderson[6] is a simple model to describe the interplay of charge and spin fluctuations of a localized interacting impurity in a metallic host environment. This magnetic impurity could be for example the d- or f-level of a transition metal atom or rare-earth atom embedded in a non-magnetic metal. In such systems one observes an anomalous minimum in the electrical resistivity at very low temperatures, which is caused by the interaction of the conduction bath electrons with the electron of impurities.

The Hamiltonian of the SIAM consists of three parts

$$\hat{H}_{SIAM} = \hat{H}_{bath} + \hat{H}_{hyb} + H_{imp}. \tag{1}$$

$\hat{H}_{bath}$ serves as the unperturbed Hamiltonian term describing the non-interacting conduction electrons. It is given by $\hat{H}_{bath} = \sum_{\mathbf{k},\sigma} \epsilon_{d\mathbf{k}} \hat{d}^{\dagger}_{\mathbf{k}\sigma} \hat{d}_{\mathbf{k}\sigma}$; where $\epsilon_{d\mathbf{k}}$ is the kinetic energy of the bath state d with momentum $\mathbf{k}$. The expression $\hat{d}^{\dagger}_{\mathbf{k}\sigma} \hat{d}_{\mathbf{k}\sigma}$ acts as the number operator describing the occupation number of electrons with spin σ in the band state with wave vector $\mathbf{k}$.

The coupling of the impurity site and the bath levels due to hybridization is described by the Hamiltonian term $\hat{H}_{hyb} = \sum_{\mathbf{k},\sigma} V_{\mathbf{k}} \left(\hat{f}^{\dagger}_{\sigma} \hat{d}_{\mathbf{k}\sigma} + \hat{d}^{\dagger}_{\mathbf{k}\sigma} \hat{f}_{\sigma} \right)$. This type of d-f coupling is characterized by the hybridization parameter $V_{\mathbf{k}}$. The bath electron with with momentum $\mathbf{k}$ and spin σ is annihilated by operator $\hat{d}_{\mathbf{k}\sigma}$ and then created in the impurity site with spin σ by operator $\hat{f}^{\dagger}_{\sigma}$. Meanwhile, electron which formerly occupies the impurity site is annihilated by operator $\hat{f}_{\sigma}$ and then created in the bath with momentum $\mathbf{k}$ by operator $\hat{d}^{\dagger}_{\mathbf{k}\sigma}$. In this model, for simplicity, we assume that the hybridization strength is constant in momentum space, giving us $V_{\mathbf{k}} = V = const$. Hence, the hybridization Hamiltonian term can be simplified to be $\hat{H}_{hyb} = V \sum_{\mathbf{k},\sigma} \left(\hat{f}^{\dagger}_{\sigma} \hat{d}_{\mathbf{k}\sigma} + \hat{d}^{\dagger}_{\mathbf{k}\sigma} \hat{f}_{\sigma} \right)$.

The last term of SIAM Hamiltonian, $\hat{H}_{imp}$, describes the impurity site that is defined as $\hat{H}_{imp} = \sum_{\sigma} \epsilon_f \hat{f}^{\dagger}_{\sigma} \hat{f}_{\sigma} + U \hat{f}^{\dagger}_{\uparrow} \hat{f}_{\uparrow} \hat{f}^{\dagger}_{\downarrow} \hat{f}_{\downarrow}$. It consists of two parts: the first part describes the on-site energy of the localized f-state on the impurity atom, while the second term describes the local electron-electron Coulomb interaction for two electrons occupying the impurity state.

3. Computational method

We aim to address problem of impurity systems with various impurity concentrations. For this, we choose to use the symmetric SIAM and solve it using a stochastic ED method on several small systems containing one f-state and N_d d-states, where N_d is to be varied. To mimic different impurity concentrations, we define the impurity concentration x, such that $x = 1/N_d$ is the ratio of the number of impurity sites to the total number of sites in the entire crystal sample. For every system with a fixed x or N_d, the energies of d-states (ϵ_d) are generated randomly with a constraint that they are distributed symmetrically with respect to zero, while the f-on-site energy (ϵ_f) is set fixed at $\epsilon_f = -U/2$. The symmetry of our model implies that the chemical potential (μ) is 0 for all temperatures. Accordingly, we sample the corresponding representative **k** points of $\epsilon_{d_\mathbf{k}}$ with only N_d base states denoted by an index α running from 1 to N_d. Thus, the simplified model Hamiltonian for each υth- sampling reads as

$$\mathcal{H} = \sum_\sigma \sum_{\alpha=1}^{N_d} \epsilon_{d_\alpha} \hat{d}^\dagger_{\alpha\sigma}\hat{d}_{\alpha\sigma} + \sum_\sigma \epsilon_f \hat{f}^\dagger_\sigma \hat{f}_\sigma + V\sum_\sigma \sum_{\alpha=1}^{N_d}\left(\hat{d}^\dagger_{\alpha\sigma}\hat{f}_\sigma + \hat{f}^\dagger_\sigma \hat{d}_{\alpha\sigma}\right) + U\hat{f}^\dagger_\uparrow \hat{f}_\uparrow \hat{f}^\dagger_\downarrow \hat{f}_\downarrow. \quad (2)$$

Our particular goal is to compute the retarded Green's function (GF) of SIAM from ED results for zero and finite temperatures. For this purpose, we need to obtain the eigenstates of $\mathcal{H}$ for N, $N+1$, and $N-1$ particle systems, respectively. Here, for the symmetric SIAM, N is always set to be half of the total number of orbitals considered in the system, i.e. $N = N_d + 1$. The calculated eigenvalues and eigenstates are then used to compute the retarded GF, G_{ff}, for the f-state, through Lehmann representation[7]. The need for obtaining G_{ff} is particularly for the purpose of extracting the self-energy of system[8]. In Lehmann representation, the corresponding retarded GF for the f electrons can be defined as:
For zero temperature ($T = 0$)

$$G^R_{ff_{\sigma\upsilon}}(\omega + i0^+) = \frac{1}{Z}\sum_{N=1}^{N_{max}}\left\{\sum_{l=1}^{\#N+1}\frac{|\langle l|\hat{f}^\dagger_\sigma|GS^{(N)}\rangle|^2}{\omega + i0^+ - (E_l^{(N+1)} - E_{GS}^{(N)} - \mu)} + \sum_{n=1}^{\#N-1}\frac{|\langle n|\hat{f}_\sigma|GS^{(N)}\rangle|^2}{\omega + i0^+ + (E_n^{(N-1)} - E_{GS}^{(N)} + \mu)}\right\}. \quad (3)$$

For finite temperature ($T > 0$)

$$G^R_{ff_{\sigma\upsilon}}(\omega + i0^+) = \frac{1}{Z}\sum_{N=1}^{N_{max}}\left\{\sum_{l=1}^{\#N+1}\sum_{m=1}^{\#N}\frac{\left(e^{-\beta(E_l^{(N+1)}-\mu_{(N+1)})} + e^{-\beta(E_m^{(N)}-\mu_{(N)})}\right)|\langle l|\hat{f}^\dagger_\sigma|m\rangle|^2}{\omega + i0^+ - (E_l^{(N+1)} - E_m^{(N)} - \mu)} + \sum_{n=1}^{\#N-1}\sum_{m=1}^{\#N}\frac{\left(e^{-\beta(E_n^{(N-1)}-\mu_{(N-1)})} + e^{-\beta(E_m^{(N)}-\mu_{(N)})}\right)|\langle n|\hat{f}_\sigma|m\rangle|^2}{\omega + i0^+ + (E_n^{(N-1)} - E_m^{(N)} + \mu)}\right\} \quad (4)$$

In SIAM, we believe that under no external magnetic field spontaneous magnetization would not occur. Hence, we would always have $G_\uparrow = G_\downarrow$. Thus, after obtaining GF of the system at both zero and finite temperature from Eq. (3) and (4), we can then drop the spin index using either $\sigma =\uparrow$ or $\sigma =\downarrow$. The self-energy of the SIAM problem for every υ^{th} sampling can be extracted through $\Sigma_\upsilon(\omega) = \left(G^0_{ff_\upsilon}(\omega + i0^+)\right)^{-1} - \left(G_{ff_\upsilon}(\omega + i0^+)\right)^{-1}$; where $G^0_{\upsilon ff}(\omega + i0^+)$ is bare GF for the f-state which is defined as $G^0_{ff_\upsilon}(\omega + i0^+) = \frac{1}{\omega + i0^+ - \epsilon_f - \frac{1}{x}\sum_\alpha \rho(\epsilon_{d_\alpha})\frac{V^2}{\omega + i0^+ - \epsilon_{d_\alpha}}}$. Here, as mentioned before, $x = 1/N_d$ is the concentration of impurity in the system. Finally, after a sufficient number of samplings ($N_{sampling}$), the complete self-energy is the average over all self-energy samples, that is $\Sigma(\omega) = \frac{1}{N_{sampling}}\sum_{\upsilon=1}^{N_{sampling}}\Sigma_\upsilon(\omega)$. After obtaining the averaged self-energy, we can construct the full GF of d- electrons ($G_{dd}(\omega + i0^+)$) and the corresponding DOS as follows

$$G_{dd}(\omega + i0^+) = \int d\varepsilon \frac{\rho(\varepsilon)}{\omega + i0^+ - \varepsilon - \frac{V^2}{\omega + i0^+ - \epsilon_f - \Sigma(\omega)}} \text{ and } DOS_d(\omega) = -\frac{1}{\pi}\mathrm{Im}G_{dd}(\omega + i0^+). \quad (5)$$

In this calculation, we simplify the bare density of states of the metallic host (i.e. the d electron density of states) to be semicircular, that is $\rho(\varepsilon) = \frac{2}{\pi}\sqrt{(1-\varepsilon^2)}$.

4. Results and discussion

As our specific research goal is to understand the influence of the parameter values in the SIAM problem, we vary temperature (T), hybridization parameter (V), interaction energy (U), and impurity concentration (x). We start with the most concentrated impurity system by taking $x = \frac{1}{2}$. For this x value, we search for the right combination of critical U and V values that lead to metal-insulator phase transition. Then, subsequently we do the same for more dilute systems with $x = \frac{1}{3}, \frac{1}{4}$ and $\frac{1}{5}$. However, for brevity, we only show the results of variation of U, V, and T for the most concentrated system ($x = \frac{1}{2}$) and the most dilute system ($x = \frac{1}{5}$) that we have modeled.

When the $d - f$ hybridization parameter is set to 0 ($V = 0$, the electrons in d-orbitals do not feel any effect of the presence of impurity. Therefore, the d-orbital band shows bare density of states that represents half-filled metallic band with Fermi energy lying at the middle of the band. If we turn on the hybridization parameter ($V > 0$), the existence of impurity will influence the movement of the bath electrons. As we increase the hybridization parameter, the charge fluctuations become larger, electrons can hop from bath to the impurity sites. This causes the$d$-band to redistribute into wider range of energy. Instead of gathering around the Fermi level, the electrons spread out and occupy the lower energy states. At some critical values of hybridization parameter V_c and interaction strength U_c, the coupling between d and f orbitals becomes large enough to allow the electrons to more easily go back and forth from d to f states, however the strong interaction at the f site tends to prevent the electrons to do. As a result, this situation leads the d band to split into two parts separated by an energy gap, while the Fermi energy falls inside the energy gap. Hence the metallic band undergoes a transition to become insulating band. Note that the formation of energy gap depends on all the parameters x, V, U, and T. The smaller x, the larger V and U, and the lower T we need, in order for the gap to form. The transitions of the d electron density of states from having no gap until the gap forms upon variation of the parameters x, V, U, and T are illustrated in Figures 1, 2, 3.

To vary the impurity concentration, we change the number of bath (d) states while keeping one impurity (f) state for every system. Therefore, the effect of impurity scattering will be different although we do not change the number of impurity state for our models. As we keep our systems to be always at half filling, the larger number of bath states implies the larger number of non-interacting electrons. It means that the possibility of impurity to scatter the conduction electron becomes more diminished since the entire electrons experience less disorder potential. From Figure 4, we can see that at those specific values of U and V, the MIT occurs for system $x = \frac{1}{2}$.

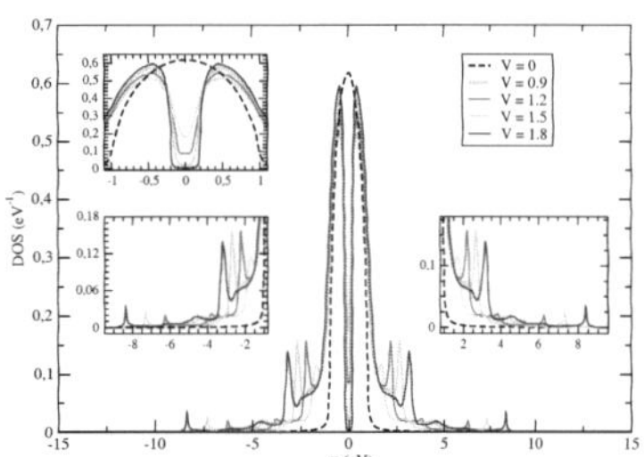

Figure 1: d-band DOS with various V values for system with $x = \frac{1}{2}$ at $T = 0$ K and $U = 4$ eV

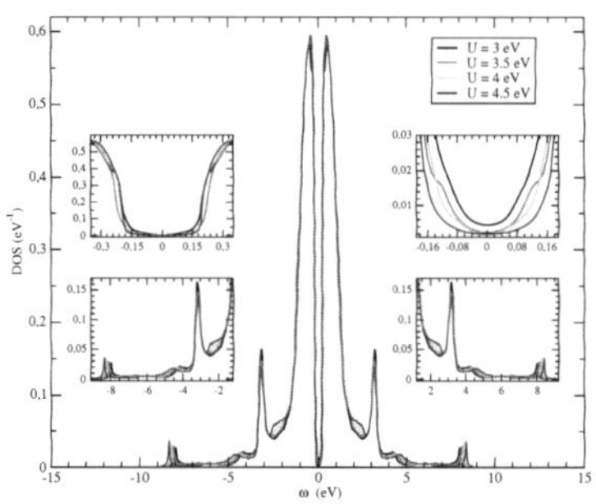

Figure 2: d-band DOS with various U values for system with $x = \frac{1}{2}$ at $T = 0$ K ($U_c = 4$ eV, $V_c = 1.8$ eV, $E_g = 0.32$ eV)

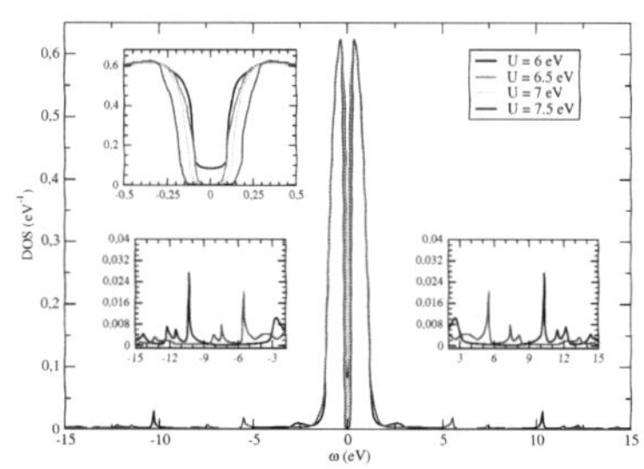

Figure 3: d-band DOS with various U values for system with $x = \frac{1}{5}$ at $T = 0$ K ($U_c = 6.5$ eV, $V_c = 3.6$ eV, $E_g = 0.28$ eV)

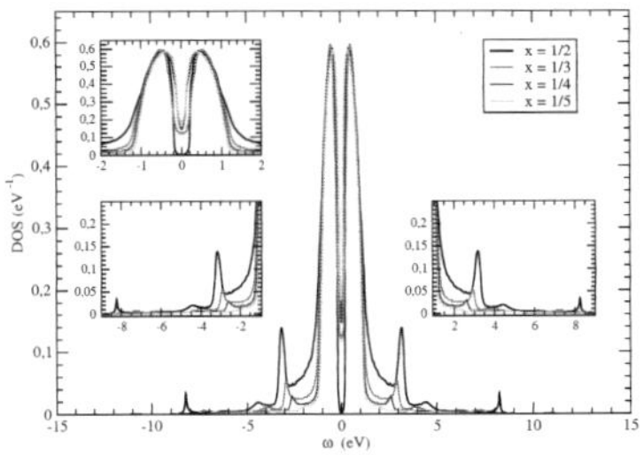

Figure 4: d-band DOS for various impurity concentration at $T = 0$ K ($U_c = 4$ eV, $V_c = 1.8$ eV, $E_g = 0.32$ eV)

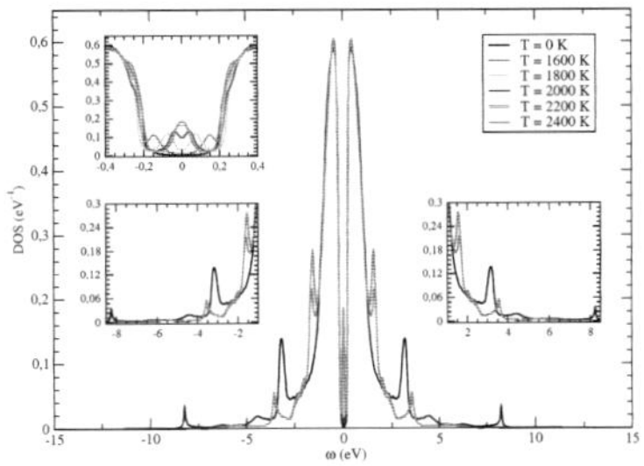

Figure 5: d-band DOS at some temperatures for system with $x = \frac{1}{2}$ having $T_c = 1600$ K

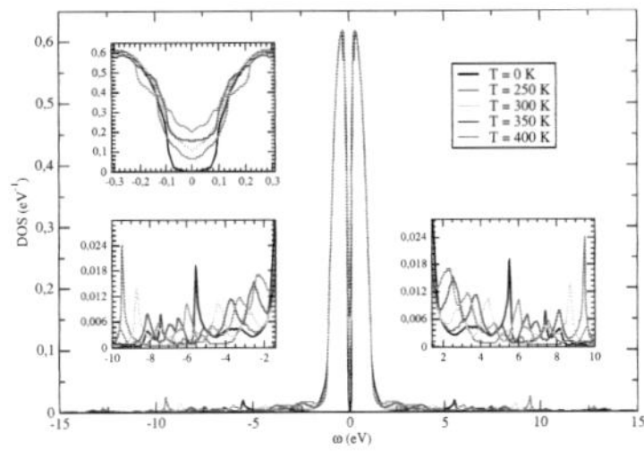

Figure 6: d-band DOS at some temperatures for system with $x = \frac{1}{5}$ having $T_c = 250$ K

When we apply those values of U and V to more dilute systems ($x = \frac{1}{3}$ and $\frac{1}{4}$), the gap starts to dissapear and insulating-like band will become 'bad' metallic band that has poor conducting ability. Then, for the last system, ($x = \frac{1}{5}$), the metallic properties get more obvious since the dip of band around Fermi energy become shallow. Thus, the higher impurity concentration, the more probable MIT to occur. In real materials pure Anderson transition often appeas to be temperature-independent. That is because such disorder-induced insulating phase usually has robust gap that remains existing upto very high temperatures[9]. In our model, however, we also vary the interaction energy, so that our systems can undergo Mott-Anderson transition which is temperature-dependent. This hypothesis is verified by our model and calculation, which is shown in the Figure 5 and 6 respectively for system with $x = \frac{1}{2}$ and $x = \frac{1}{5}$.

5. Conclusions

We have developed detailed computational algorithm to implement the DED method on SIAM for various parameter values. Our calculation has proven that DED is a reliable method to solve SIAM problem. From the results, our modeled systems capture the occurrence of Mott-Anderson metal to insulator transition (MIT) at some critical values of impurity concentration (x), temperature (T), interaction energy (U), and hybridization parameter (V). Overall, our study has demonstrated that doping a metallic system with a strongly-correlated element can transform the system into an insulator, depending on the values of controllable parameters, which is consistent with the picture of Anderson localization.

Acknowledgments

We are very grateful to Universitas Indonesia for supporting us with full funding for this project through PITTA Research Grant No. 666/UN2.R3.1/HKP.05.00/2017.

References

[1] Hewson A C 1993 *The Kondo Problem to Heavy Fermions* (Cambridge: Cambridge University Press)
[2] Yanase Y, Jujo T, Nomura T, Ikeda H, Hotta T and Yamada K 2003 *Phys. Rep.* **387** 1-149
[3] Mott N F 1990 *Metal-Insulator Transitions 2nd ed* (London: Taylor & Francis Ltd)
[4] Dobrosavljević V, Trivedi N and Valles J M 2012 *Introduction to Metal-Insulator Transitions* (Oxford: Oxford University Press)
[5] Granath M and Strand H U R 2012 *Phys. Rev.* B **86** 115111
[6] Anderson P W 1961 *Phys. Rev.* **124** 1
[7] Fetter A L and Walecka J D 2003 *Quantum Theory of Many-Particle Systems* (San Fransisco: McGraw-Hill)
[8] Mahan G D 2000 *Many-Particle Physics 3rd ed* (New York: Springer)
[9] Gebhard F 1997 *The Mott Metal-Insulator Transition: Models and Methods* (Germany: Springer-Verlag Berlin Heidelberg)

IOP Conf. Series: Journal of Physics: Conf. Series **1011** (2018) 012073 doi:10.1088/1742-6596/1011/1/012073

Exploring excitonic signal in optical conductivity of ZnO through first-order electron-hole vertex correction

Humaira Khoirunnisa[1] **and Muhammad Aziz Majidi**[1†]

[1]Department of Physics, Faculty of Mathematics and Natural Sciences, Universitas Indonesia, Depok 16424, Indonesia

[†]Corresponding author's email: `aziz.majidi@sci.ui.ac.id`

Abstract. The emergence of exitonic signal in the optical response of a wide band-gap semiconductor has been a common knowledge in physics. There have been numerous experimental studies exploring the important role of excitons on influencing both the transport and optical properties of the materials. Despite the existence of much information on excitonic effects, there has not been much literature that explores detailed theoretical explanation on how the exitonic signal appears and how it evolves with temperature. Here, we propose a theoretical study on the optical conductivity of ZnO, a well-known wide band-gap semiconductor that we choose as a case study. ZnO has been known to exhibit excitonic states in its optical spectra in the energy range of ∼3.13-3.41 eV, with a high exciton binding energy of ∼60 meV. An experimental study on ZnO in 2014 revealed such a signal in its optical conductivity spectrum. We present a theoretical investigation on the appearance of excitonic signal in optical conductivity of ZnO. We model the wurtzite ZnO within an 8-band *k.p* approximation. We calculate the optical conductivity by incorporating the first-order vertex correction derived from the Feynman diagrams. Our calculation up to the first-order correction spectrum qualitatively confirms the existence of excitons in wurtzite ZnO.

1. Introduction

ZnO is a wide band gap semiconductor (∼3.37eV) [1] that has been interesting for decades for which a lot of research have been done considering its potential for optoelectronic application [2–5]. ZnO has been known to exhibit excitonic signals in its absorption and reflection spectra [6]. A pretty recent experimental study [7] shows further that ZnO posses a quite distinct exitonic level below the band gap. The combination of measurements of dielectric functions, photoluminescence, and optical conductivity unveil the excitonic signals at the energy range of ∼3.16-3.41 eV. Here we present a theoretical study on the appearance of such an excitonic signal in optical conductivity of wurtzite ZnO. We model the ZnO band structure within an 8-band *k.p* approximation, which has proved to be succesful for studying bulk and quantum-well semiconductors [4, 8, 9]. We calculate the optical conductivity using Feynman diagrams up to the first-order electron-hole vertex correction of the current-current correlation function defined in Matsubara frequency domain. Later we do analytic continuation to transform the optical conductivity into real frequency domain. The electron-hole interaction is incorporated in the calculation to reveal the excitonic signal in the optical conductivity spectrum.

IOP Conf. Series: Journal of Physics: Conf. Series **1011** (2018) 012073 doi:10.1088/1742-6596/1011/1/012073

2. Model

Our 8-band $k.p$ model Hamiltonian is constructed as an extended version of the 6×6 $\mathbf{k} \cdot \mathbf{p}$ Hamiltonian for a strained wurtzite crystal derived by S. L. Chuang and C. S. Chang [10]. With the addition of the conduction s-like band and leaving out the strain elements, the Hamiltonian matrix we use to calculate the physical properties of our ZnO system is

$$H_{tot} = \begin{pmatrix} U_c & 0 & 0 & 0 & 0 & 0 & 0 & 0 \\ 0 & F & -K^* & -H^* & 0 & 0 & 0 & 0 \\ 0 & -K & G & H & 0 & 0 & 0 & \Delta \\ 0 & -H & H^* & \lambda & 0 & 0 & \Delta & 0 \\ 0 & 0 & 0 & 0 & U_c & 0 & 0 & 0 \\ 0 & 0 & 0 & 0 & 0 & F & -K & H \\ 0 & 0 & 0 & \Delta & 0 & -K^* & G & -H^* \\ 0 & 0 & \Delta & 0 & 0 & H^* & -H & \lambda \end{pmatrix}, \tag{1}$$

with matrix elements

$$\begin{aligned} U_c &= E_c + A_2'(k_x^2 + k_y^2) + A_1' k_z^2 \\ F &= \Delta_1 + \Delta_2 + \lambda + \theta \\ G &= \Delta_1 + \Delta_2 + \lambda + \theta \\ \lambda &= \frac{\bar{h}}{2m_o}\left[A_1' k_z^2 + A_2'(k_x^2 + k_y^2)\right] \\ \theta &= \frac{\bar{h}}{2m_o}\left[A_3' k_z^2 + A_4'(k_x^2 + k_y^2)\right] \end{aligned}
\qquad
\begin{aligned} K &= \frac{\bar{h}}{2m_o} A_5'(k_x^2 + ik_y)^2 \\ H &= \frac{\bar{h}}{2m_o} A_6'(k_x^2 + ik_y)k_z \\ \Delta &= \sqrt{2}\Delta_3 \\ A_1' &= \frac{\bar{h}}{2m_e^{\parallel}} - \frac{(P)_1^2}{E_g} \\ A_2' &= \frac{\bar{h}}{2m_e^{\perp}} - \frac{(P)_2^2}{E_g} \end{aligned}$$

$$\begin{aligned} (P)_1^2 &= \frac{\bar{h}}{2m_o}\left(\frac{m_o}{m_e^z} - 1\right) \frac{(E_g + \Delta_1 + \Delta_2)(E_g + 2\Delta_2) - 2\Delta_{\frac{2}{3}}}{(E_g + 2\Delta_2)} \\ (P)_2^2 &= \frac{\bar{h}}{2m_o}\left(\frac{m_o}{m_e^z} - 1\right) \frac{E_g[(E_g + \Delta_1 + \Delta_2)(E_g + 2\Delta_2) - 2\Delta_3^2]}{(E_g + \Delta_1 + \Delta_2)(E_g + \Delta_2) - \Delta_3^2}. \end{aligned} \tag{2}$$

Δ_1 is the crystal-field splitting parameter, Δ_2 and Δ_3 are parameters related to the spin-orbit splitting energy Δ_{so}, $\Delta_2 = \Delta_3 = \frac{\Delta_{so}}{3}$.

3. Calculation

The calculation of optical conductivity is performed in two steps according the Feynman diagrams of the current-current correlation function up to first-order in electron-hole vertex correction as depicted in Fig.1. First, we calculate the "bare" optical conductivity which does not incorporate the electron-hole interaction. The formula can be derived from the simple "bubble" diagram of the current-current correlation function (see Fig.1a) as

$$\Pi_{\alpha\beta}(i\omega_n) = e^2 \frac{1}{\beta \bar{h}} \sum_{l=-\infty}^{\infty} \sum_{\mathbf{k}} \mathrm{Tr}\left[v^\alpha(\mathbf{k}) G_0(\mathbf{k}, i\omega_l) v^\beta(\mathbf{k}) G_0(\mathbf{k}, i\omega_l + i\omega_n)\right], \tag{3}$$

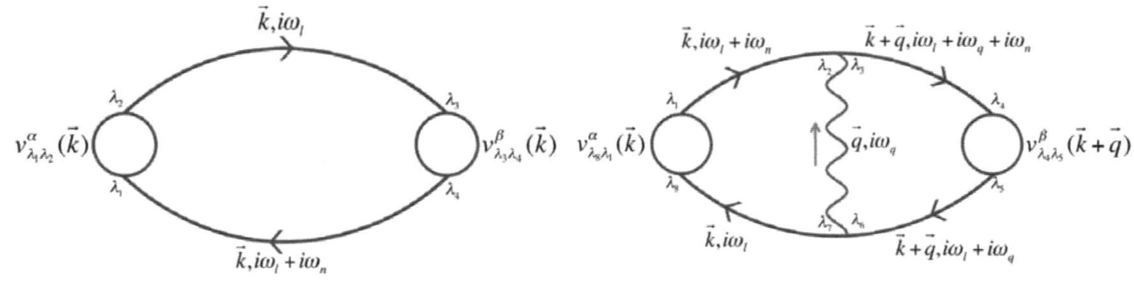

(a) (b)

Figure 1: Feynman diagrams showing the creation of an electron above the chemical potential and a hole below the chemical potential, as the result of adding an extra photon with frequency ω_n into the system, and the recombination of the excited electron back to the hole(a). Feynman diagrams with first-order electron-hole vertex correction(b).

which is known as the Kubo formula. Second, we calculate the first-order electron-hole vertex correction to the "bubble" diagram (see Fig.1b). The attractive Coulomb interaction between the excited electron in the conduction band and the hole created in the valence band is represented by the wiggly line. The second diagram can be translated into a mathematical form as

$$\begin{aligned}\Pi^{1}_{i\omega_n} =& T_2 \Sigma^{\infty}_{\omega_l=-\infty} \Sigma^{\infty}_{\omega_q=-\infty} \Sigma_{\mathbf{k}} \Sigma_{\lambda_7} \Sigma_{\lambda_2} \Sigma_{\lambda_3} \Sigma_{\lambda_6} \\ & \left(\Sigma_{\lambda_8} \Sigma_{\lambda_1} G_{\lambda_7\lambda_8}(\mathbf{k}, i\omega_l) v^{\alpha}_{\lambda_8\lambda_1}(\mathbf{k}) G_{\lambda_1\lambda_2}(\mathbf{k}, i\omega_l + i\omega_n)\right) U(\mathbf{q}) \\ & \left(\Sigma_{\lambda_4} \Sigma_{\lambda_5} G_{\lambda_3\lambda_4}(\mathbf{k}+\mathbf{q}, i\omega_l + i\omega_q + i\omega_n) v^{\beta}_{\lambda_4\lambda_5}(\mathbf{k}+\mathbf{q}) G_{\lambda_5\lambda_6}(\mathbf{k}+\mathbf{q}, i\omega_l + i\omega_q)\right).\end{aligned} \tag{4}$$

To transform the formula from Matsubara- to real-frequency domain we have to do analytic continuation. For the "bare" part of the optical conductivity, the analytic continuation from Matsubara- to real-frequency domain can be carried out analytically through Hilbert transformation, resulting the expression of optical conductivity as used in Ref. [11]. For the first-order diagram, however, Hilbert transformation is not practically implementable since we deal with the vertex function having two Matsubara frequency variables. To get around of this problem, we turn into an approximate method by using Padé approximant to transform our optical conductivity calculation data defined in imaginary (Matsubara) frequency domain into the physical optical conductivity defined in real-frequency domain. Note that the use of Padé approximant may affect the accuration of our calculation results as we will discuss in Section 4.

4. Results and Discussion

For our overall calculation we use the following parameters: lattice constants $a = 3.25$Å and $c = 5.21$Å for the wurtzite ZnO; $k.p$ parameters as used in Ref. [9] to calculate the band strucutre of strained wurtzite ZnO; and the empirical value of ~ 3.3 eV for the energy gap taken at room temperature. Figure 2 displays the calculated density of states (DOS) of our $k.p$ model for the wurtzite ZnO, where the red vertical line marks the position of the chemical potential.

The calculated "bare" optical conductivity ($\sigma_1(\omega)$) is shown in Fig. 3. The onset of the uprising spectrum reflects the energy gap of ~3.3 eV, which is the minimum energy to excite an electron from the valence band to the conduction band. The spectrum is getting higher as the given photon energy is increased, showing that more electrons contribute to the conductivity. The anisotropy along the three different axes are expected as the system posses a hexagonal symmetry.

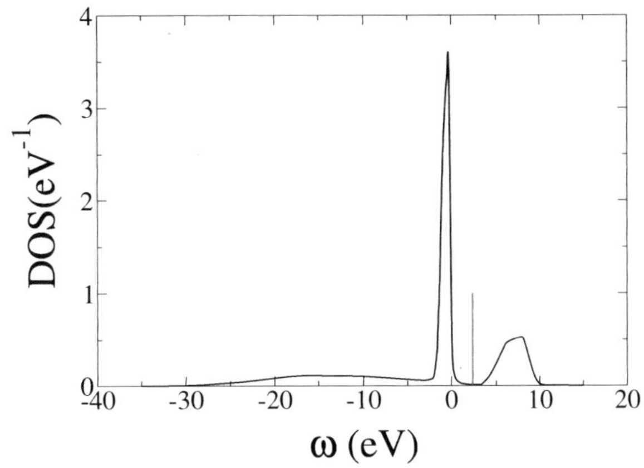

Figure 2: DOS

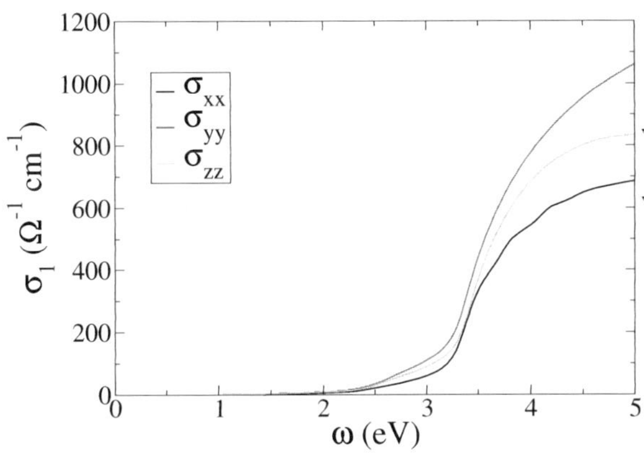

Figure 3: ZnO optical conductivity without vertex correction

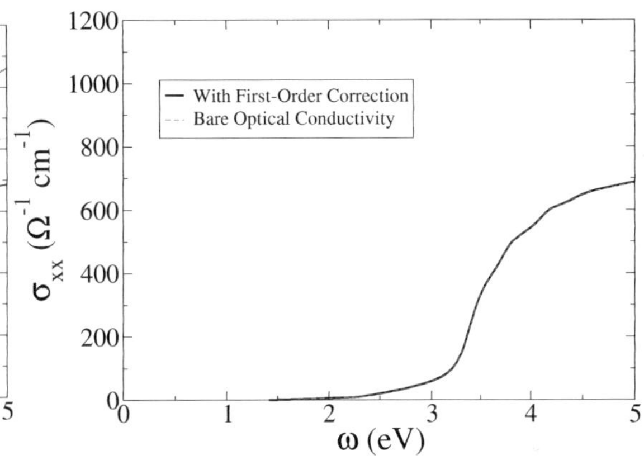

Figure 4: ZnO optical conductivity with vertex correction in x direction

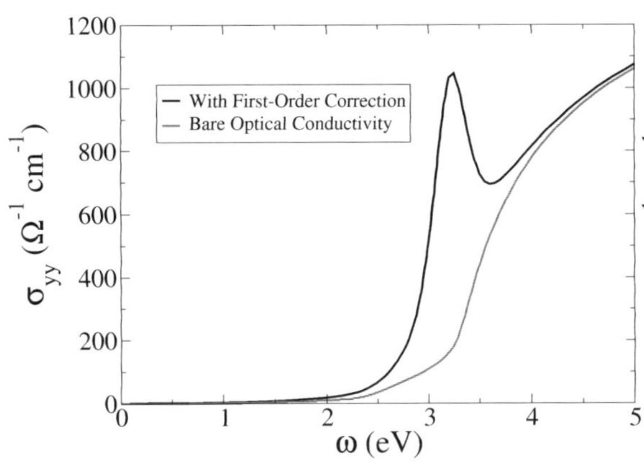

Figure 5: ZnO optical conductivity with vertex correction in y direction

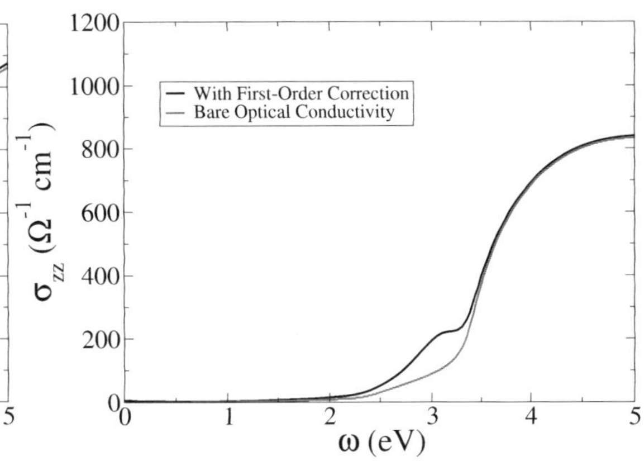

Figure 6: ZnO optical conductivity with correction in z direction

The anisotropy behavior that appears in the "bare" $\sigma_1(\omega)$ also shows up in the first-order correction results proportionally. $\sigma_{yy}(\omega)$ and $\sigma_{zz}(\omega)$ spectra are both showing a rising peak below the band gap energy, which we identify as the excitonic signal formed due to the electron-hole interaction. The peak in $\sigma_{yy}(\omega)$ arises at ~3.2 eV and the peak in $\sigma_{zz}(\omega)$ arises at ~3.1 eV, which are just around the value obtained from the experimental measurement. Being the smallest in value, $\sigma_{xx}(\omega)$ does not show any excitonic signal. We believe that this is not physically correct, and it appears so because of the uncontrollable error in the Padé approximant we use.

5. Conclusion

We have presented our 8-band $k.p$ model and calculation for the optical conductivity of wurtzite ZnO. The calculated DOS reveals the correct value of the band gap energy, which also manifests as the onset of the optical conductivity spectrum. Further, our calculation for the optical conductivity incorporating the electron-hole interaction through the first-order electron-hole vertex correction qualitatively confirms the existence of excitons in ZnO. We acknowledge some lack of accuracy in our calculation that may have resulted in the absence excitonic signal in z direction. We suspect that one possible source of lack of accuracy is the sensitivity of the application of Padé approximation against any small error in the Matsubara frequency data of the quantity being analytically continued.

Acknowledgement

We are grateful to Universitas Indonesia for providing us a full funding through PITTA Research Grant No. 666/UN2.R3.1/HKP.05.00/2017. Numerical calculations of this work were carried out on the parallel computing cluster of Theoretical Condensed Matter Physics (TCMP) research group at Universitas Indonesia.

References

[1] Chowdhury R, Adhikari S and Rees P 2010 *Physica B.* **405** 4763-67.
[2] Kabongo G L, Mbule P S, Mhlongo G H, Mothudi M B, Hillie K T and Dhlamini M S 2016 *Nanoscale Research Letters.* **11** 418.
[3] Djurišić A B, Ng A M C and Chen X Y 2010 *Progress in Quantum Electronics.* **34** 191-259.
[4] Stroud A and You J H 2012 *J. Cryst. Growth.* **340** 92-97
[5] Vettumperumal R, Kalyanaraman S and Thangavel R 2013). Enhancement of optical conductivity in the ultra-violet region of Cs doped ZnO sol gel thin films. *J Sol-Gel Sci Technol.* **66** 206-11.
[6] Thomas D G 1960 *J. Phys. Chem. Solids.* **15** 86-96
[7] Darma Y, Seng H T, Marlina R, Fauziah R, Ding J and Rusydi A 2014 *Appl. Phys. Let.* **104** 081922
[8] Valence hole subbands and optical gain spectra of GaN/Ga1xAlxNstrained quantum wells Fan W J, Li M F, Chong T C and Xia J B 1996 *J. Appl. Phys.* **80** 3471-78
[9] Fan W J, Abiyasa A P, Tan S T, Yu S F, Sun X W, Xia J B, Yeo Y C, Li M F and Chong T C 2006 *J. Cryst. Growth.* **287** 28-33
[10] Chuang S L and Chang C S 1996 *Phys. Rev. B.* **54** 2491-2504
[11] Majidi M A, Su H, Feng Y P, Rübhausen M and Rusydi A 2011 *Phys. Rev. B.* **84** 075136

Biaxial strain effects on the electronic properties of silicene: the density-functional-theory-based calculations

Khaerul Umam, Sholihun, Pekik Nurwantoro, Moh Adhib Ulil Absor, Ari Dwi Nugraheni, and Romy H. S. Budhi

Department of Physics, Faculty of Mathematics and Natural Sciences, Universitas Gadjah Mada, Sekip Utara BLS 21, Yogyakarta, Indonesia 55281

E-mail: sholihun@ugm.ac.id

Abstract. The effects of biaxial strain on the electronic properties of silicene are studied by carrying out the density-functional-theory (DFT) calculations. We simulate planar and buckled structures and find that the buckled structure is more stable than the planar one. We next apply biaxial strain up to 12% and find that the applied tensile and compressive strains below 8% do not change the electronic structure. At tensile strains of 8% or more, the Dirac point at K-point shifts up from the Fermi level, indicating that such tensile strain behaves as p-type doping. Meanwhile at the compressive strains of 8% or more, the Dirac point at K-point shifts down from the Fermi level, indicating that the such compressive strain behaves as n-type doping. We find that silicene remains stable for the applied strain up to 12%. We also calculate the Fermi velocity around K-point, which is found to be 9.1×10^5 m/s.

Keywords: Silicene, buckled structure, biaxial strain

1. Introduction

Since graphene was successfully fabricated [1], theoretical and experimental studies of two dimensional materials (2D- materials) having graphene-like structure such as 2D-silicone (silicene) and 2D-germanium (germanene) have attracted much attention [2–7]. It was reported that π-band contributes to the electron conduction band [8, 9]. An experimental observation of the the electronic properties of graphene shows that the Dirac cone appears on the edge of the Brillouin zone [10]. Graphene was grown on various substrates such as sapphire [11], h-BN [12], and Pt (111) [13]. As for the 2D-silicone, silicon sheets was synthesized [5]; however, the formed silicon sheet was immediately capped by foreign atoms. The difficulty of fabrication of silicene is that, in nature, there is no silicon whose structure resembles like graphite so that silicene cannot be fabricated by using mechanical stripping [10]. Silicene could be synthesized by using chemical reduction method ; however, the produced material behaves as insulator [7].

One of efforts to have understanding of silicene is by performing quantum-based DFT simulations [14–18] since the DFT calculational results [19–22] frequently agree with those of the experimental measurements. The DFT studies predicted that silicene has stable buckled structure instead of planar structure [14, 15, 17], and a linear dispersion characteristic appears at the K-point with the Fermi velocity of 5.3×10^5 m/s [23]. Several studies were conducted

IOP Conf. Series: Journal of Physics: Conf. Series **1011** (2018) 012074 doi:10.1088/1742-6596/1011/1/012074

to study the effect of strain on the electronic properties of silicene [24–26]. It was shown as the strain increases, the band gap at K-point decrease. The structure of silicene breaks at strain 14% and 18% for zigzag and armchair geometry, respectively [18]. As for the influence to the Dirac cone, at strain more than 5%, the Dirac cone is 0.06 eV shifted above the Fermi level [27]. The Dirac cone is also shifted above the Fermi level when the tensile strain of +10% is applied, and it is shifted below the Fermi level when the compressive strain of 8% is applied [28]. The stability of silicene was also studied by using a molecular dynamics simulation that silicene is stable at 0 K but it breaks at room temperature [29]. Therefore, the study of the electronic structure of silicene is essential to explore the behavior of such materials to be plausible candidates for the electronic devices such as a tunneling-field-effect transistor. In this paper, we study the electronic properties of silicene by performing the DFT calculations. We first check the stability of silicene by carrying out calculation on planar and buckled structures. We apply tensile and compressive strain up to 12% and analyze its influence to the electronic structures.

2. Computational Methods

We carry out the DFT calculations on the strain application on silicene. We use code PHASE [30], which is based on the plane wave basis set. We use the norm-conserving pseudopotential and the local density approximation (LDA) [31] as the exchange-correlation functional. A cutoff energy of 40 Ry and a 6×6×1 Monkhorst-Pack mesh grid [32] are used in the calculation. We relax atoms so that the atomic force is less than 5.0×10^{-3} eV/Å. To neglect the interlayer interaction, we apply a 15 Å vacuum parameter perpendicular to the silicene sheet. The unit cell consists of two atoms whose lattice vectors are $\vec{a}[1/2, \sqrt{3}/2, 0]$ and $\vec{b}[-1/2, \sqrt{3}/2, 0]$.

We first calculate the total energy with respect to the cell volume to find the optimized lattice constant. The calculated data is then fitted by using the Birch-Murnaghan equation of sates (BM-EOS) in Eq. 1 [33, 34].

$$E(V) = E_0 + \frac{9}{16} V_0 B_0 \left[\left(\left(\frac{V_0}{V}\right)^{\frac{2}{3}} - 1 \right)^3 B'_0 + \left(\left(\frac{V_0}{V}\right)^{\frac{2}{3}} - 1 \right)^2 \left(6 - 4 \left(\frac{V_0}{V}\right)^{\frac{2}{3}} \right) \right]. \qquad (1)$$

where E_0, V_0, B_0, and B'_0 are parameters obtained from the fitting, which are the minimum energy, optimized cell volume, bulk modulus, and first derivative of the bulk modulus at $V = V_0$, respectively. To investigate the stable configuration of silicene, we simulate planar and buckled structures by introducing a buckling parameter and calculating cohesive energy by using $E_{coh} = E_{silicene} - 2E_{Si}$, where E_{coh}, $E_{silicene}$, and E_{Si} are the cohesive energy, total energy of silicene in two-atoms unit cell, and energy of an individual silicon atom calculated from the bulk structure, respectively.

3. Results and Discussion

We calculate the total energies with respect to the cell volumes for planar and buckled structures and fit the calculated results to the BM-EOS. As shown in Fig. (Fig. 1), the buckled structure has lower energies than the planar one, indicating that the buckled structure is more stable than the planar one. We also calculate the cohesive energies and find that the buckled structure has 0.02 eV lower cohesive energy than the planar one. Comparing with graphene, the calculated cohesive energy of graphene is 0.19 eV, which is 1.37 eV lower than that of silicene (1.56 eV).

We apply the biaxial strain and calculate the buckling parameters and Si-Si bondlengths (Fig. 2). As the strain increases the buckling parameter decreases while the Si-Si bondlength increases. The increase of strain enlarges the lattice parameter; then the distance between silicon atoms increases, and vice versa, to maintain the stability of the silicene structure. The optimized buckling parameter and Si-Si bondlength of the zero-strain are 0.40 Å and 2.24 Å, respectively. The largest buckling parameter calculated at strain -10% is found to be 112.9% higher while

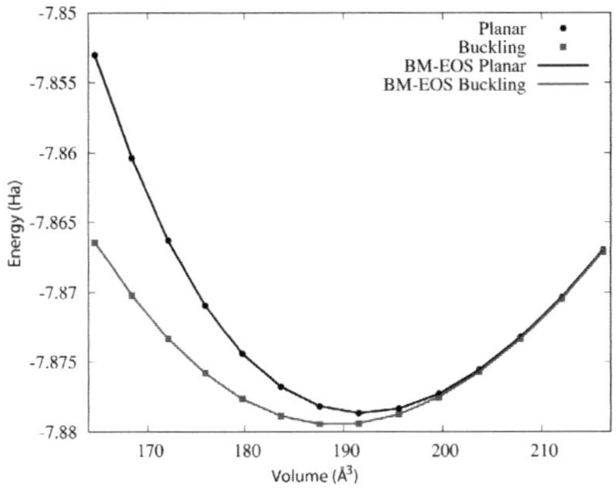

Figure 1. The calculated data (circle and square) and the BM-EOS fitting result (solid line) of total energies versus cell volumes for planar (black) and buckled (blue) structures.

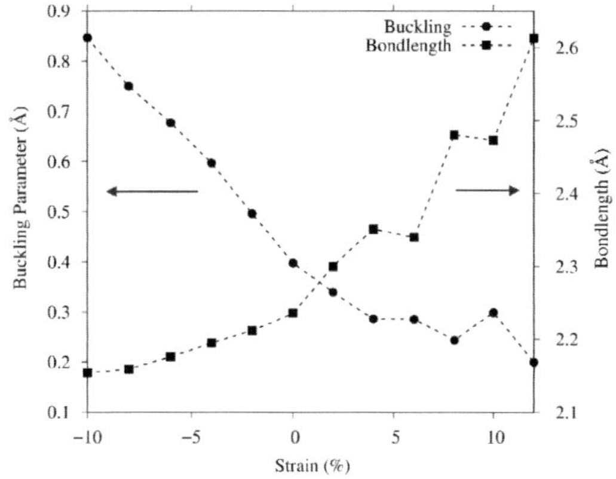

Figure 2. Strain effect to the buckling parameter (left axis, circle points) and Si-Si bondlength in silicene (right axis, square points).

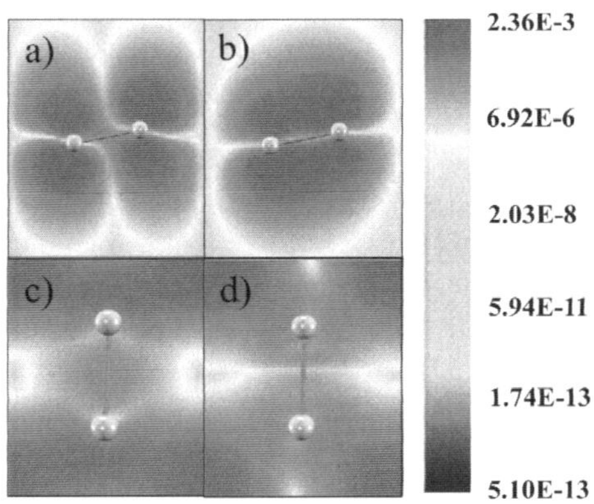

Figure 3. Charge density distribution near the (a) π-bond valence, (b) π-bond conduction, (c) σ-bond valence and (d) σ-bond conduction band in unit $e/(a.u)^3$, calculated at ± 0.05 eV from the Fermi level.

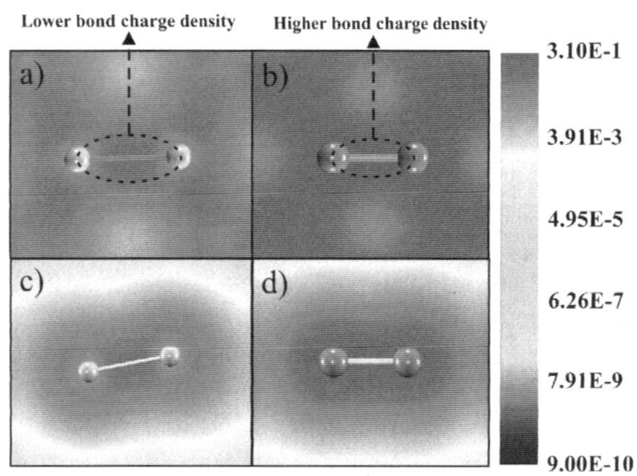

Figure 4. Full charge density distribution near σ-bond for (a) silicene and (b) graphene. The π-bonds are shown in (c) and (d) for silicene and graphene, respectively. The unit is in $e/(a.u)^3$.

the smallest one calculated at strain 12% is 49.9% lower than that of the zero-strain calculation. As for the bondlength, the largest and the smallest ones are 16.9% higher and 3.7% lower than that of the zero-strain calculation. We next calculate the partial charge density distribution of π-bonds and σ-bonds in the valence and conduction bands calculated at ± 0.05 eV from the Fermi level, as shown in Fig. 3. The π-bond charge in the valence band (Fig. 3(a)) is lower than that in the conduction band (Fig. 3(b)). This lower charge originates from the asymmetric wave function, which does not obey the inversion symmetry while the higher π-bond charge in the conduction band originates from the symmetric wave function. Conversely, the higher σ-bonds occur in the valence band (Fig. 3(c)) instead of the conduction band (Fig. 3(d)). We also calculate the total charge density for silicene (Fig. 4(a)) and graphene (Fig. 4(b)) and we show that silicene has lower bond charge density than graphene, indicating that the bond in silicene is weaker than that in graphene. The larger bondlength of Si-Si bond in silicene, which is found to be 2.24 Å comparing to that in graphene of 1.44 Å, contributes to the lower bond in silicene. Meanwhile, the lower π-bond causes silicene has a buckled structure, as we mentioned above that the most stable structure of silicene is the buckled structure.

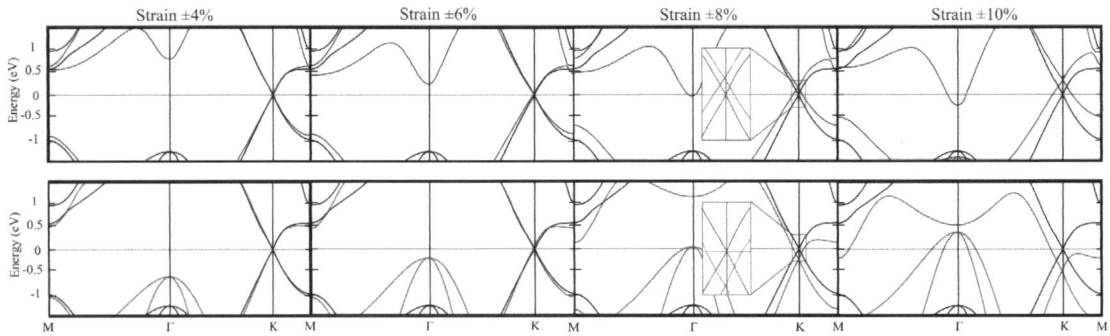

Figure 5. Band structures of the strained silicene (a) ±4%, (b) ±6%, (c) ±8% and (d) ±10%. The band structures of the tensile, compressive, and zero-strain are shown in blue, red and black lines, respectively.

To know the effects of strain to the electronic structures, we calculate band structures (Fig. 5). The calculated band structure of the zero-strain is shown by the black line in Fig.5. The presence of cross linear at the K-points shows the characteristic of silicene, the Dirac cone, as in graphene. The center point of the Dirac cone lays at the Fermi level, which is in agreement with results of the previous studies [14,35]. The Dirac cone is due to electrons following the relativistic Dirac fermion [10]. Our calculated Fermi velocity is 9.1×10^5 m/s, which is in agreement with that of the previous studies ($\approx 10^5$ m/s) [23, 36, 37].

We apply compressive and tensile strains up to 10%. Our calculational results show that the applied strains equal or less than 6% do not change the electronic structure, for instance, the Dirac point remains at the Fermi level. The applied strain of 8% slightly shifts the position of the Dirac cone at the K-point, the tensile strain shifts up while the compressive strain shifts down. The applied strains of ±10% result larger shift than those of ±8% but have similar behavior that the tensile strain shifts the Dirac cone up and the compressive strain shifts it down from the Fermi level. This shift indicates that the applied tensile strain behaves as the p-type doping and the compressive strain behaves as the n-type doping. We also calculate the tensile strain of 12% and clarify the similar behavior with the results of the strain 8% and 10%. From our calculation, silicene remains stable by the application of strain 12%.

4. Conclusion

We have carried out calculations on the electronic structures of silicene. As for the silicene structure, we found that the buckled structure is stable while the planar one is metastable. We applied biaxial strain and found that the applied strain of 8% – 12% shifts the Dirac point up from the Fermi level while the compressive strain shifts it down. Thus, the tensile strain behaves as p-type doping while the compressive strain behaves as n-type doping.

Acknowledgments
This work was supported by a research grant of Penelitian Unggulan Perguruan Tinggi (PUPT) No. 2451/UN1.P.III/DIT-LIT/LT/2017, from the General Directorate of Higher Education (DIKTI), Ministry of Research, Technology and Higher Education of the Republic of Indonesia. The computations in this research were performed using the High-Performance Computing at DSSDI Universitas Gadjah Mada and LIPI (Lembaga Ilmu Pengetahuan Indonesia), Indonesia.

References

[1] Novoselov K S, Geim A K, Morozov S V, Dubonos S V, Zhang Y and Jiang D 2004 *Science* **306** 666
[2] Takeda K and Shiraishi K 1994 *Phys. Rev.* B **50** 14916
[3] Voon L W, Zhu J and Schwingenschlögl U 2016 *Appl. Phys. Rev.* **3** 040802
[4] Chowdhury S and Jana D 2016 *Rep. Prog. Phys.* **79** 126501
[5] Nakano H, Mitsuoka T, Harada M, Horibuchi K, Nozaki H and Takahashi N 2006 *Angew. Chem. Int. Ed.* **45** 6303
[6] Vogt P, Padova P D, Quaresima C, Avila J, Frantzeskakis E, Asensio M C , Resta A, Ealet B and Lay G L 2012 *Phys. Rev. Lett.* **108** 155501
[7] Varykhalov A, Sánchez-Barriga J, Shikin A, Biswas C, Vescovo E and Rybkin A 2008 *Phys. Rev. Lett.* **101** 157601
[8] Johnston D F 1955 *Proc. R. Soc. Lond* A Vol 227 (London : Mathematical and Physical Sciences) pp 349-358
[9] Lomer W M 1955 *Proc. R. Soc. Lond* A Vol 227 (London : Mathematical and Physical Sciences) pp 330-349
[10] Geim A K and Novoselov K S 2007 *Nature Materials* **6** 18
[11] Wang S, dos Santos L F, Wurstbauer U, Wang L, Pfeiffer L N, Hone J, Garcia J M and Pinczuk A 2014 *Solid State Commun.* **189** 15
[12] Garcia J M *et al* 2012 *Solid State Commun.* **152** 975
[13] Hernandez-Rodriguez I, Garcia J M, Martin-Gago J A, de Andres P L and Mendez J 2015 *Diam. Relat. Mater.* **57** 58
[14] Cahangirov S, Topsakal M, Aktürk E, Sahin H and Ciraci S 2009 *Phys. Rev. Lett.* **102** 236804
[15] Mortazavi B, Rahaman O, Makaremi M, Dianat A, Cuniberti G and Rabczuk T 2017 *Physica* E **87** 228
[16] Sholihun, Saito M, Ohno T and Yamasaki T 2015 *Jpn. J. Appl. Phys.* **54** 041301
[17] Roome N J and Carey J D 2014 *ACS Appl. Mater Interfaces* **6** 7743
[18] Zhao H 2012 *Phys. Lett.* A **376** 3546
[19] Makhov D V and Lewis L J 2004 *Phys. Rev. Lett.* **92** 255504
[20] Sholihun, Ishii F and Saito M 2016 *Jpn. J. Appl. Phys.* **55** 011301
[21] Fitriana A S, Pranowo H D and Purwono B 2016 *Indones. J. Chem.* **16** 80
[22] Spiewak J P, Vanhellemont and Kurzydlowski K J 2011 *Jpn. J. Appl. Phys.* **110** 063534
[23] Matthes L, Gori P, Pulci O and Bechstedt F 2013 *Phys. Rev.* B **87** 035438
[24] Yan J, Gao S, Stein R and Coard G 2015 *Phys. Rev.* B **91** 245403
[25] Mohan J, Kumar A and Ahluwalia P 2014 *Physica* E **61** 40
[26] Qin R, Zhu W, Zhang Y and Deng X 2014 *Nanoscale. Res. Lett.* **9** 521
[27] Kaloni T P, Cheng Y C and Schwingenschlögl U 2013 *J. Appl. Phys.* **113** 104305
[28] Wang Y and Ding Y 2012 *Solid State Commun.* **155** 6
[29] Pun G P and Mishin Y 2017 *Phys. Rev.* B **95** 224103
[30] PHASE, http://www.ciss.iis.u-tokyo.ac.jp/dl/index.php
[31] Hohenberg P and Kohn W 1964 *Phys. Rev.* B **136** 864
[32] Monkhorst H J and Pack J D 1976 *Phys. Rev.* B **13** 12
[33] Birch F 1947 *Phys. Rev.* **71** 809
[34] Zhuravlev K 2007 *Physica* B **394** 1
[35] Miro P, Audiffred M and Heine T 2014 *Chem. Soc. Rev.* **43** 6537
[36] Guzmán-Verri G and Voon L C 2007 *Phys. Rev.* B **76** 075131
[37] Huang S, Kang W and Yang L 2013 *Appl. Phys. Lett.* **102** 133106

IOP Conf. Series: Journal of Physics: Conf. Series **1011** (2018) 012075 doi:10.1088/1742-6596/1011/1/012075

A toy model to investigate the existence of excitons in the ground state of strongly-correlated semiconductor

H R Karima, M A Majidi[†]
Department of Physics, Faculty of Mathematics and Natural Sciences, Universitas Indonesia, Kampus UI Depok, Depok 16424, Indonesia
[†]Corresponding author's e-mail: aziz.majidi@sci.ui.ac.id

Abstract. Excitons, quasiparticles associated with bound states between an electron and a hole and are typically created when photons with a suitable energy are absorbed in a solid-state material. We propose to study a possible emergence of excitons, created not by photon absorption but the effect of strong electronic correlations. This study is motivated by a recent experimental study of a substrate material $SrTiO_3$ (STO) that reveals strong exitonic signals in its optical conductivity. Here we conjecture that some excitons may already exist in the ground state as a result of the electronic correlations before the additional excitons being created later by photon absorption. To investigate the existence of excitons in the ground state, we propose to study a simple 4-energy-level model that mimics a situation in strongly-correlated semiconductors. The four levels are divided into two groups, lower and upper groups separated by an energy gap, E_g, mimicking the valence and the conduction bands, respectively. Further, we incorporate repulsive Coulomb interactions between the electrons. The model is then solved by exact diagonalization method. Our result shows that the toy model can demonstrate band gap widening or narrowing and the existence of exciton in the ground state depending on interaction parameter values.

1. Introduction

Strongly-correlated systems have attracted attention of researchers for decades because of their exotic phenomena. Just to mention some examples, colossal magnetoresistance effect (CMR) in several manganites that describes a large change of the electrical resistance in an applied magnetic field [1], high temperature superconductivity in several cuprates [2], metal-insulator-transitions as observed e.g. in magnetite [3] or certain vanadites, are due to strong electronic correlations.

A number of investigations over the past decades have demonstrated that Coulomb repulsive interactions among electrons dominate the nonlinear optical response of semiconductors [4]. Arita *et al* [5] have done experimental study using angle-resolved photoemission spectroscopy (ARPES) that reveals the transition from semiconductor into metal in FeSi as temperature increases, which has been attributed to strong correlations among thermally induced carriers near the Fermi energy.

In semiconductors, there are conditions where an electron is excited from the valence band to the conduction band, leaving a hole in the valence band, but the electron and the hole undergo Coulomb attractive interaction. If screening effect of the environment is weak, the electron and the hole may form a bound state acting as a neutral quasiparticle called exciton [6]. This quasiparticle is known to have an important role in determining the properties of semiconductors, especially energy transport. Excitonic signals usually appear on optical absorption and emission of direct-gap semiconductors [7]. A recent experimental study [8] of manganite thin film deposited on a substrate material $SrTiO_3$ (STO) has revealed anomalous excitonic signal in its optical conductivity. Some of these excitonic signals are

IOP Conf. Series: Journal of Physics: Conf. Series **1011** (2018) 012075 doi:10.1088/1742-6596/1011/1/012075

found to have energy above the band-gap edge. Based on the study, we conjecture that some excitons may already exist in the ground state as a result of the electronic correlations before the additional excitons being created later by photon absorption. Here, we propose to do a theoretical study to prove, at least in principle, the possibility of existence of excitons in the ground state of strongly correlated semiconductors through a simple model we refer to as a toy model.

In this study we hypothesize that, due to the presence of repulsive Coulomb interactions between electrons, especially the intra-band interactions, electrons would not feel the most comfortable configuration by fully occupying the valence band all the time, while the conduction band is left empty. As a result, the most stable electron configuration, i.e. corresponding to the ground state energy of the system, would be that some electrons are excited to the conduction band leaving some holes in the valence band. We also hypothesize that each electron-hole pair is bound to form an exciton. Our results show that the toy model can demonstrate the band gap widening and narrowing depending on the criteria of parameter values of interaction among electrons. We also find the criteria in which excitonic state occurs in the ground state of our model.

2. Model and computational method

The actual problem we would like to solve may be represented by the following Hamiltonian:

$$H = \sum_{n\in\{c,v\}} \sum_{\boldsymbol{k}} \sum_{\sigma} \epsilon_{n,\boldsymbol{k}} \hat{a}^{\dagger}_{n,\boldsymbol{k},\sigma} \hat{a}_{n,\boldsymbol{k},\sigma} + \frac{1}{2} \sum_{n,m,n',m'} \sum_{\boldsymbol{k}\boldsymbol{k}',\boldsymbol{q}} U_{n,m,n',m'}(\boldsymbol{q}) \sum_{\sigma,\sigma'} \hat{a}^{\dagger}_{m',\boldsymbol{k}'-\boldsymbol{q},\sigma'} \hat{a}^{\dagger}_{n',\boldsymbol{k}+\boldsymbol{q},\sigma} \hat{a}_{n,\boldsymbol{k},\sigma} \hat{a}_{m,\boldsymbol{k}',\sigma'} \quad (1)$$

Here n, m, n', m' are indices denoting the valence (v) and the conduction (c) bands. $\boldsymbol{k}$ and $\boldsymbol{k}'$ represent momenta of the electrons, while $\boldsymbol{q}$ is the transfer momentum between two interacting electrons. σ is the spin index denoting the up (↑) and the down (↓) spin components. $\epsilon_{n,\boldsymbol{k}}$ is the energy of an electron in band n with momentum $\boldsymbol{k}$. $U_{n,m,n',m'}(\boldsymbol{q})$ is the Coulomb interaction defined in momentum space between two electrons initially in band n' and m' that are scattered off each other into band n and m, with transfer momentum $\boldsymbol{q}$. The Coulomb interaction couplings $U_{n,m,n',m'}(\boldsymbol{q})$ are predicted to be much larger in magnitude for the intra-band ($n = m = n' = m'$) scatterings compared to those for inter-band ($n \neq m, n' \neq m'$) scatterings.

In the real problem of semiconductor we have to deal with a large number of electrons. It is then practically impossible to solve the actual problem exactly. We consider to restrict our goal to demonstrate the effect of exciton creation due to strong electron correlations only qualitatively. For this purpose, we propose to replace the actual problem with a "toy" model in which we drastically reduce the number of single-electron states to a minimum number that still accommodates possible transitions of electron states, both intra and inter-bands. Here, we define four $\boldsymbol{k}$ states named 1, 2, 3, and 4, with the corresponding energies ϵ_1, ϵ_2, ϵ_3, and ϵ_4, respectively. We group these four energies into two bands (lower and upper) mimicking the valence and the conduction bands as shown in Fig. 1.

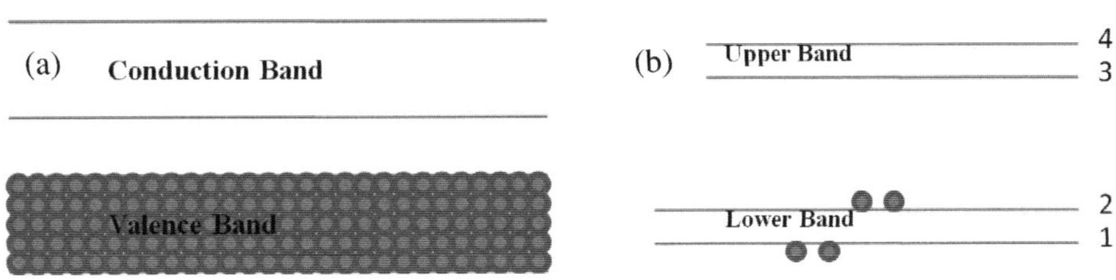

Figure 1. Electronic configuration of actual semiconductor (a) and electronic configuration of the toy model (b) in $\boldsymbol{k}$-space.

Each two states in the same band may be considered to be nearly degenerate, that is, $\epsilon_1 \cong \epsilon_2$, $\epsilon_3 \cong \epsilon_4$. Incorporating the spin degree of freedom, we now have $2 \times 4 = 8$ single-electron states.

Since we are interested in the half-filling case, we set the number of electrons to be $\frac{1}{2}\times 8 = 4$ electrons. With these four electrons distributed among the eight single-electron states, the dimensionality of our 4-electron Hilbert space becomes: $\#_4 = \frac{8!}{4!(8-4)!} = \frac{5\times6\times7\times8}{1\times2\times3\times4} = 70$.

Thus, the corresponding 4-electron Hamiltonian matrix is of size 70×70, which can be easily constructed and diagonalized numerically using standard exact diagonalization subroutine. Now that we only deal with 4 energy levels in our toy model, equation (1) can be reduced into

$$H = \sum_{r,\sigma} \epsilon_r \hat{a}^\dagger_{r,\sigma}\hat{a}_{r,\sigma} + \sum_{lmno}\sum_q \frac{1}{2} U_{lmno}\ (\boldsymbol{q}) \sum_{\sigma\sigma'} \hat{a}^\dagger_{l\sigma}\hat{a}^\dagger_{m\sigma'}\hat{a}_{n\sigma'}\hat{a}_{o\sigma}. \quad (2)$$

In our model, we assume that the interaction between electrons in the lower band and that between electrons in the upper band has the same strength, we have

$$U_{lower-lower}(\boldsymbol{q}=0) = U_{upper-upper}(\boldsymbol{q}=0) = U_{intra-direct}$$
$$U_{lower-lower}(\boldsymbol{q}\neq 0) = U_{upper-upper}(\boldsymbol{q}\neq 0) = U_{intra-exchange}.$$

We also make abbreviations

$$U_{lower-upper}(\boldsymbol{q}=0) \equiv U_{inter-direct} \text{ and } U_{lower-upper}(\boldsymbol{q}\neq 0) \equiv U_{inter-exchange}.$$

To examine how parameter values affect the band gap in our model, we use mean field approximation (MFA) where we assume that an electron 'feels' the mean field caused by the presence of other electrons in certain states. The mean field value which is felt by an electron depends on the mean occupation number of other electrons lying in a particular state. Using the MFA we have

$$n_i n_j = \langle n_i\rangle n_j + n_i\langle n_j\rangle - \langle n_i\rangle\langle n_j\rangle \quad (3)$$

Assuming that mean occupation number of each orbital $\langle n_1\rangle = 2$, $\langle n_2\rangle = 2, \langle n_3\rangle = 0$, and $\langle n_4\rangle = 0$, we can derive the criterion for the widening of band gap due to the presence electron-electron interactions

$$4U_{inter-direct} + U_{intra-exchange} > 3U_{intra-direct} + 2U_{inter-exchange}. \quad (4)$$

Before we solve the Hamiltonian in Eq. (2), we want to find an approximate criterion such that the ground-state electron configuration has the existence of one exciton. We do this by applying MFA again and comparing the total energy of state where the valence band is fully occupied by four electrons (i.e. no exciton), E_0, and the total energy of the excitonic state in which there are three electrons in the lower band and an electron in the upper band (i.e. one exciton), E_1. To have an exciton in the ground state we need $E_1 < E_0$. Application of MFA gives us the criterion we need

$$6U_{intra-direct} - 2U_{intra-exchange} >$$
$$E_g + 3U_{intra-direct} + 3U_{inter-direct} - U_{intra-exchange} - 2U_{inter-exchange}. \quad (5)$$

We can apply this criterion to tune our interaction parameters in the Hamiltonian Eq. (2) so that the ground-state electron configuration will have an exciton, then obtain the ground-state energy and the corresponding eigen state by means of numerical exact diagonalization.

After we have obtained the ground-state configuration with one exciton, we may calculate its binding energy. We propose an approximate way to calculate the exciton binding energy, that is

$$E_{binding} = \langle\Psi|H_{MF} - H|\Psi\rangle \quad (6)$$

With $|\Psi\rangle$ being the eigen vector of the excitonic ground state of the system, H_{MF} is the MFA version of the Hamiltonian, and H is the Hamiltonian stated in Eq. (2).

To display the exact diagonalization results in terms of the density of states (DOS), we need to construct the corresponding retarded Green function through Lehmann representation [9]. For zero temperature the expression is

$$G_{i\sigma}(\omega + i0^+) = \frac{1}{Z}\left\{\sum_{l=1}^{\#_{N+1}} \frac{\left|\langle l|\hat{a}^\dagger_{i,\sigma}|GS^{(N)}\rangle\right|^2}{\omega + i0^+ - \left(E_l^{(N+1)} - E_{GS}^{(N)}\right)} + \sum_{n=1}^{\#_{N-1}} \frac{|\langle n|\hat{a}_{i,\sigma}|GS^{(N)}\rangle|^2}{\omega + i0^+ + \left(E_n^{(N-1)} - E_{GS}^{(N)}\right)}\right\}. \quad (7)$$

The DOS can then be obtained through the relation

$$\mathrm{DOS}(\omega) = -\frac{1}{\pi}\sum_{i=1}^{4} \mathrm{Im}\, G_{i\sigma}\ (\omega + i0^+). \quad (8)$$

Further, the corresponding chemical potential or the Fermi energy can be extracted by solving

$$\int d\omega\ \mathrm{DOS}(\omega)\, f(\mu, T, \omega) = N, \quad (9)$$

where $f(\mu, T, \omega) = \frac{1}{e^{\frac{\omega-\mu}{k_B T}} + 1}$ is the Fermi distribution function, and $N = 4$.

3. Results and discussion

3.1. Band gap

In this section we want to examine our model on how the resulting band gap depends on the value of interaction parameters. The table below lists the parameters we use in the calculation of DOS to see the band gap narrowing and widening based on the criterion we have presented before.

Table 1. Parameters used to calculate DOS in band gap widening and narrowing

Physical Parameters	Value (eV)	Physical Parameters	Value (eV)	
			Narrowing Gap	Widening Gap
ϵ_1	0	$U_{intra-direct}$	1.5	1.5
ϵ_2	0	$U_{inter-direct}$	1	1.25
ϵ_3	3	$U_{intra-exchange}$	0.003	0.003
ϵ_4	3	$U_{inter-exchange}$	0.002	0.0021

Using the interaction parameters listed in Table 1 we obtain the results as shown in Fig. 2. The figure clearly shows an energy shift due to the presence of interactions. The DOS in Fig. 1a gives chemical potential about 6.26 eV, while in Fig. 1b around 5.76 eV, which are inside the band gap as expected. By comparing the band gap of the two conditions we see that Fig. 1a exhibits the band gap narrowing while Fig. 1b shows the band gap widening.

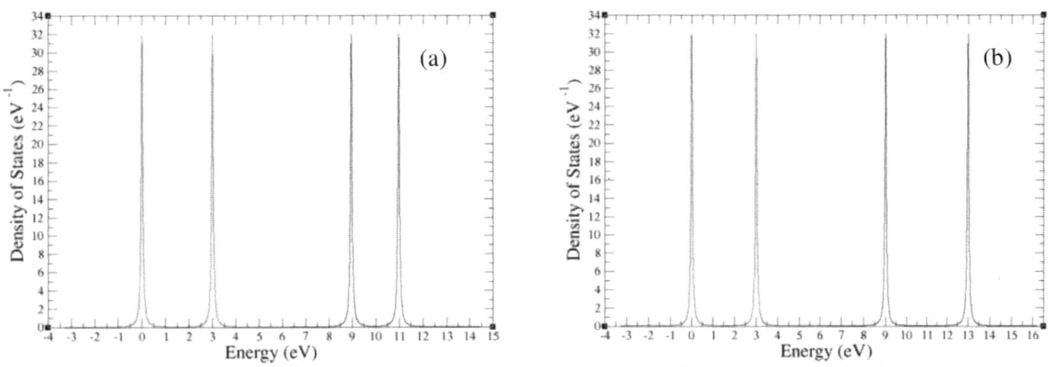

Figure 2. DOS of interacting system (black) and non-interacting system (red) showing the narrowing (a) and the widening (b) of the band gap.

3.2. Ground state of strongly-correlated system

Earlier we have proposed the condition for obtaining the ground state with the presence of of one exciton. Here we vary the interaction parameters to examine the criterion we presented before.

Table 2. Parameters used to investigate the existence of exciton in the ground state with band gap narrowing and widening.

Physical Parameters	Value (eV)	Physical Parameters	Value (eV)	
			Narrowing Gap	Widening Gap
ϵ_1	0	$U_{intra-direct}$	3.5	4.6
ϵ_2	0	$U_{inter-direct}$	2.5	3.5
ϵ_3	3	$U_{intra-exchange}$	0.0055	0.008
ϵ_4	3	$U_{inter-exchange}$	0.004	0.007

Figure 3 shows the calculated DOS using interaction parameters in Table 2. We can see the difference of DOS using MFA and the exact solution. Figure 3a shows that there is a gap narrowing, but the DOS shows splitting of the energy level in both bands. The exact solution gives the chemical

IOP Conf. Series: Journal of Physics: Conf. Series **1011** (2018) 012075 doi:10.1088/1742-6596/1011/1/012075

potential about 11.67 eV with an electron in the upper band and three electrons in the lower band. Using Eq. (6) we find the exciton binding energy of 2.49 eV. Although the parameter values we use exhibit excitonic state, we see that there is a gap narrowing in the energy band spectrum. This is because the parameter values satisfy the condition needed for excitonic state, but do not satisfy the condition for the gap widening. Figure 3b shows the situation when the parameter values meet both criteria for the emergence of an exciton in the ground state and the widening of the band gap. In this case the exciton binding energy is 3.41 eV and the chemical potential of the system is at 15.46 eV.

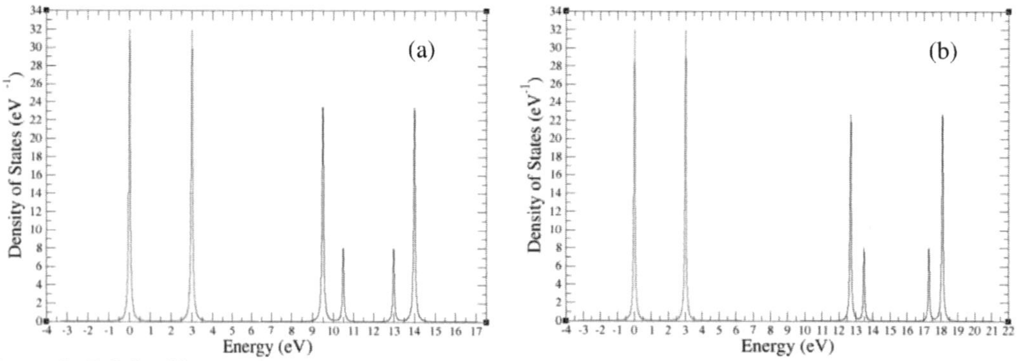

Figure 3. DOS of interacting system (black) with an exciton in the ground state and non-interacting system (red) showing narrowing (a) and widening (b) of the band gap.

4. Conclusion

We have presented our theoretical study to investigate the existence of excitons in the ground state of a toy model mimicking strongly-correlated semiconductor. The toy model can predict both band gap widening and narrowing due to the interaction of electrons in the system based on the criterion we obtain using MFA. In addition, we have also presented the criterion for the parameter values that yields an exciton in the ground state. We have proposed a way to calculate the exciton binding energy and implemented the formula for certain parameter values. The exciton binding energies we obtained in these calculations happen to be much larger than typical exciton binding energy. However, this should not be too surprising, as we did the calculations on the toy model, not the actual semiconductor. Overall, we believe that our toy model and method of calculations can give us some physical insight about the mechanism of exciton formation in the ground state of a strongly-correlated semiconductor.

Acknowledgments

This work was funded in full through PITTA research Grant from Universitas Indonesia, Indonesia, No. 665/UN2. R3.1/HKP.05.00/2017.

References

[1] Haghiri-Gosnet A M and Renard J 2003 *Journal of Physics D: Applied Physics* ***36*** R127
[2] Dagotto E 1994 *Reviews of Modern Physics* ***66*** 763
[3] Verwey E J W 1939 *Nature* ***144*** 327-328
[4] Chemla D S, and Shah J 2001 *Nature* ***411*** 549-557
[5] Arita M, Shimada K, Takeda Y, Nakatake M, Namatame H, Taniguchi M, and Kanomata T 2008 *Physical Review B* **77** 205117
[6] Scholes G D and Rumbles G 2006 *Nature materials* **5** 683-696
[7] Koch S W, Kira M, Khitrova G, and Gibbs H M 2006 *Nature materials* **5** 523-531
[8] Gogoi P K, Sponza L, Schmidt D, Asmara T C, Diao C, Lim J C, and Rusydi A2015 *Physical Review B* ***92*** 035119
[9] Fetter A L and Walecka J D 2003 *Quantum Theory of Many-Particle Systems* (San Fransisco: McGraw-Hill)

IOP Conf. Series: Journal of Physics: Conf. Series **1011** (2018) 012076 doi:10.1088/1742-6596/1011/1/012076

Optical conductivity calculation of a $k.p$ model semiconductor GaAs incorporating first-order electron-hole vertex correction

Maryam Nurhuda and Muhammad Aziz Majidi[†]

Department of Physics, Faculty of Mathematics and Natural Sciences, Universitas Indonesia, Kampus UI Depok, Depok 16424, Indonesia

[†]Corresponding author's e-mail: aziz.majidi@sci.ui.ac.id

Abstract. The role of excitons in semiconducting materials carries potential applications. Experimental results show that excitonic signals also appear in optical absorption spectra of semiconductor system with narrow gap, such as Gallium Arsenide (GaAs). While on the theoretical side, calculation of optical spectra based purely on Density Functional Theory (DFT) without taking electron-hole (e-h) interactions into account does not lead to the appearance of any excitonic signal. Meanwhile, existing DFT-based algorithms that include a full vertex correction through Bethe-Salpeter equation may reveal an excitonic signal, but the algorithm has not provided a way to analyze the excitonic signal further. Motivated to provide a way to isolate the excitonic effect in the optical response theoretically, we develop a method of calculation for the optical conductivity of a narrow band-gap semiconductor GaAs within the 8-band $k.p$ model that includes electron-hole interactions through first-order electron-hole vertex correction. Our calculation confirms that the first-order e-h vertex correction reveals excitonic signal around 1.5 eV (the band gap edge), consistent with the experimental data.

1. Introduction

Semiconductor responds to light in a wide variety of ways, including the formation of short-lived quasiparticles which can affect the optical response of the material. Excitonic signals may appear in optical absorption spectra of semiconductors not only those with wide band gap, but also in narrow gap semiconductors, such as Gallium Arsenide (GaAs) [1]. GaAs is a promising material due to its superior properties, including high electron mobility and good opto-electrical transfer effeciency [2]. The formation of excitons inside GaAs also has an important role in increasing solar cell efeciency and light formation in LED and lasers.

There is much interest in the characterization of GaAs crystal over the past decades, mostly in its optical properties. Despite that excitonic signals have been observed in the GaAs optical spectra experimentally, the optical properties of GaAs have not yet been much explored theoretically. Calculation based purely on Density Functional Theory (DFT) without taking electron-hole interactions into account does not lead to the appearance of any excitonic signal [3]. Meanwhile, existing DFT-based algorithms that include a full vertex correction through Bethe-Salpeter equation may reveal an excitonic signal, but the algorithm has not provided a way to analyze the excitonic signal further [4]. In this study, we develop a method of calculation for the optical conductivity of a narrow band gap semiconductor GaAs within the 8-band $k.p$ model, which includes electron-hole (e-h) interactions through first-order electron-hole vertex correction. By this method we could isolate the excitonic effect in the optical response theoretically. Our calculation confirms that the first order e-h vertex correction reveals excitonic signal around the band gap edge.

2. Model

We construct our model Hamiltonian using the following basis states: $|s\rangle|\uparrow\rangle$ and $|s\rangle|\downarrow\rangle$ (forming the conduction band), and $|p_x\rangle|\uparrow\rangle$, $|p_x\rangle|\downarrow\rangle$, $|p_y\rangle|\uparrow\rangle$, $|p_y\rangle|\downarrow\rangle$, $|p_z\rangle|\uparrow\rangle$, and $|p_z\rangle|\downarrow\rangle$ (forming the three degenerated valence band, known as the heavy hole, light hole, and split off bands [5]. By operating the Hamiltonian $H = \frac{p^2}{2m_0} + V + \frac{\hbar}{4m_0^2c^2}(\nabla V \times p)\cdot\sigma$ on the above basis states, following Bahder [6], we obtain our 8-band $k.p$ model Hamiltonian matrix as

$$H_0(k) = \begin{bmatrix} A & 0 & T^*+V^* & 0 & -\sqrt{3}(T-U) & \sqrt{2}(W-U) & W-U & \sqrt{2}(T^*+V^*) \\ 0 & A & \sqrt{2}(W-U) & -\sqrt{3}(T^*+V^*) & 0 & T-V & -\sqrt{2}(T-V) & W^*+U \\ T+V & \sqrt{2}(W^*-U) & -P+Q & -S^* & R & 0 & \sqrt{1.5}\,S & -\sqrt{2}\,Q \\ 0 & -\sqrt{3}(T+V) & -S & -P-Q & 0 & R & -\sqrt{2}\,R & \sqrt{1.5}\,S \\ -\sqrt{3}(T^*-V^*) & 0 & R^* & 0 & -P-Q & S^* & \sqrt{0.5}\,S^* & \sqrt{2}\,R^* \\ \sqrt{2}(W^*-U) & T^*+V^* & 0 & R^* & S & -P+Q & \sqrt{2}\,Q & \sqrt{1.5}\,S^* \\ W^*-U & -2(T^*-V^*) & \sqrt{1.5}\,S^* & -\sqrt{2}\,R^* & \sqrt{0.5}\,S & \sqrt{2}\,Q & Z & 0 \\ \sqrt{2}(T+V) & W+U & -\sqrt{2}\,Q & \sqrt{0.5}\,S^* & \sqrt{2}\,R & \sqrt{1.5}\,S & 0 & Z \end{bmatrix} \quad (1)$$

for which

$$A = E_c + \left[A' + \frac{\hbar^2}{2m_0}\right](k_x^2 + k_y^2 + k_z^2)$$

$$U = \frac{1}{\sqrt{3}} P_0 k_z$$

$$V = \frac{1}{\sqrt{6}} P_0 (k_x - ik_y)$$

$$W = i\frac{1}{\sqrt{3}} B k_x k_y$$

$$T = \frac{1}{\sqrt{6}} B k_z (k_x + ik_y)$$

$$P = -E_v + \frac{1}{2}\gamma_1 \frac{\hbar^2}{m_0}(k_x^2 + k_y^2 + k_z^2)$$

$$Q = \frac{1}{2}\gamma_2 \frac{\hbar^2}{m_0}(k_x^2 + k_y^2 + k_z^2)$$

$$R = -\frac{\sqrt{3}}{2}\left[\gamma_2 \frac{\hbar^2}{m_0}(k_x^2 - k_y^2) - 2i\gamma_3 k_z k_y\right]$$

$$S = \sqrt{3}\gamma_3 \frac{\hbar^2}{m_0} k_z (k_x - ik_y)$$

$$Z = E_v - \Delta - \frac{1}{2}\gamma_1 \frac{\hbar^2}{m_0}(k_x^2 + k_y^2 + k_z^2)$$

$$E_p = \frac{2m_0}{\hbar^2} P_0^2$$

$$E_g = E_c - E_v$$

with the empirical parameter values taken from Ref. [7] as: A=5.65 Å, Δ =0.314 eV, $\boldsymbol{E_c}$=1.519 eV, A'=0 eVÅ^2, B=4.19 eVÅ^2, E_p=22.827 eV, Γ_{1l}=6.672, Γ_{2l}=1.866, and Γ_{3l}=2.669.

3. Calculation Method

To perform 0-5 eV optical conductivity which reveals excitonic signal, we calculate bare optical conductivity and add it up with the corresponding first order electron-hole vertex correction. We first calculate the retarded Green functions, density of states (DOS), and chemical potential. Then by using the obtained Green functions and chemical potential, we calculate the corresponding optical conductivity. The bare optical conductivity formula is derived using the Kubo formula [8]. As $\sigma_{\alpha\beta}$ is a retarded quantity then it is more convinient to first defined it in Matsubara frequency $\sigma_{\alpha\beta}(i\omega_n)$, where $\omega_n = 2\pi n k_B T/\hbar$ is the "bosonic" Matsubara frequencies, with T being the temperature and n being an integer running from $-\infty$ to ∞ [9].

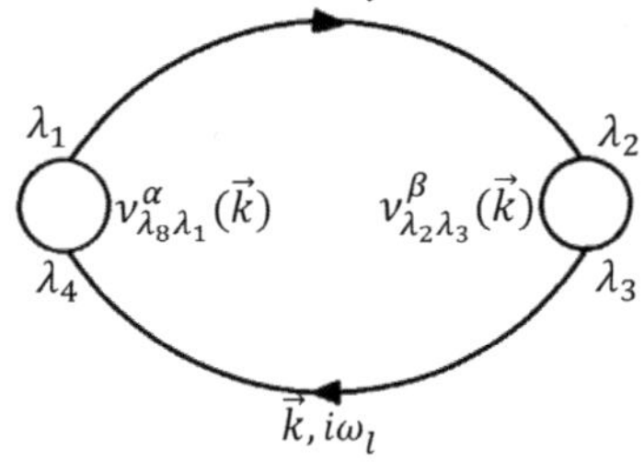

Figure 1. Bubble diagram of bare optical conductivity.

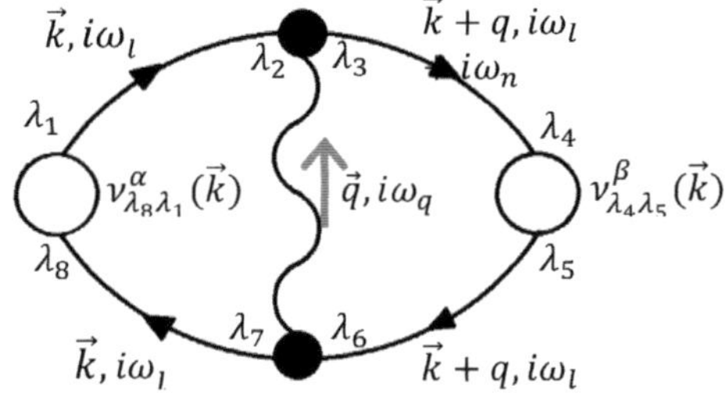

Figure 2. Bubble diagram of first order vertex correction.

Figure 1 and 2 display the Feynman diagrams of the current-current correlation function. The diagram in Fig. 2 can be translated into the bare optical conductivity formula in Matsubara frequency domain as

$$\sigma_{\alpha\beta}(i\omega_n) = e^2 \frac{1}{\beta\omega\hbar^2} \sum_{m=-\infty}^{\infty} \sum_{\boldsymbol{k}} \mathrm{Tr}\, [v^\alpha(\boldsymbol{k})][\mathcal{G}_0(\boldsymbol{k}, i\omega_m)]\left[v^\beta(\boldsymbol{k})\right][\mathcal{G}_0(\boldsymbol{k}, i\omega_m + i\omega_n)]\,. \quad (2)$$

Because physical properties need to be extracted from retarded optical conductivity in real frequencies, we do analytical continuation through the Hilbert transformation to take optical conductivity from Matsubara to real frequency domain. The real part of bare optical conductivity could then be expressed by

$$\mathrm{Re}\,\sigma_{\alpha\beta}(\omega) = \frac{\pi e^2}{\hbar\omega} \int_{-\infty}^{\infty} d\nu \,\{f(\nu,\mu,T) - f(\nu+\omega,\mu,T)\} \sum_{\boldsymbol{k}} \mathrm{Tr}\, [v^\alpha(\boldsymbol{k})][A_0(\boldsymbol{k},\nu)]\left[v^\beta(\boldsymbol{k})\right][A_0(\boldsymbol{k},\nu+\omega)] \quad (3)$$

with

$$[A_0(\boldsymbol{k},\nu)] = \frac{1}{2\pi i}\{[G_0(\boldsymbol{k},\nu + i0^+)] - [G_0(\boldsymbol{k},\nu - i0^+)]\}, \quad (4)$$

and the velocity matrix

$$v^\alpha_{\lambda_1\lambda_2}(\boldsymbol{k}) = \frac{1}{\hbar}\frac{\partial}{\partial k_\alpha} H_{\lambda_1\lambda_2}(\boldsymbol{k}). \quad (5)$$

As the attractive Coulomb interaction between electron in the conduction band and hole created in the valence band is incorporated, we then deal the diagram in Fig. 2, where the electron-hole interaction is represented by a wavy line connecting the electron and the hole propagators. This diagram translates mathematically as

$$\sigma^{(1)}_{\alpha\beta}(i\omega_n) = e^2 \frac{1}{\hbar\omega} \frac{1}{(\beta_T\hbar)^2} \sum_{\mathrm{l}=-\infty}^{\infty} \sum_{\mathrm{q}=-\infty}^{\infty} \sum_{k} \sum_{\lambda_7} \sum_{\lambda_2} \sum_{\lambda_3} \sum_{\lambda_6} \Big(\sum_{\lambda_8} \sum_{\lambda_1} \mathcal{G}_0(\boldsymbol{k}, i\omega_l) v^\alpha(\boldsymbol{k}) \mathcal{G}_0(\boldsymbol{k}, i\omega_l + i\omega_n)\Big)\, \mathrm{U}(\boldsymbol{q}) \Big(\sum_{\lambda_4} \sum_{\lambda_5} \mathcal{G}_0\big(\boldsymbol{k}+\boldsymbol{q}, i\omega_l + i\omega_n + i\omega_q\big) v^\beta(\boldsymbol{k}) \mathcal{G}_0(\boldsymbol{k}+\boldsymbol{q}, i\omega_l + i\omega_n)\Big). \quad (6)$$

For this formula, it is rather tedious to implement the Hilbert transformation, as it will generate many-many integrals over real frequency variables, resulting in expensive computation. To overcome this problem, we choose to use Padé approximant technique [10] to do the analytic continuation to transform the optical conductivity from Matsubara- to real-frequency domain.

4. Results and Discussion

Figure 3 shows the bare optical conductivity. Here we don't have anisotropy as all the three curves corresponding to diagonal components of optical conductivity tensors overlay each other. The optical conductivity remains zero up to 1.4eV. At the frequency 1.4eV, it starts to rise and undergoes a significant change at 3eV until it reaches its highest peak at 4.4eV. This means that, below 1.4eV all electrons are confined in valence band and none in the conduction band, thus no carrier contibutes to conduct the material. As the photon frequency exceeds the band gap value, photons start to excite electrons to the conduction band. The higher the photon energy, more free electrons are produced, and more conducting the material becomes.

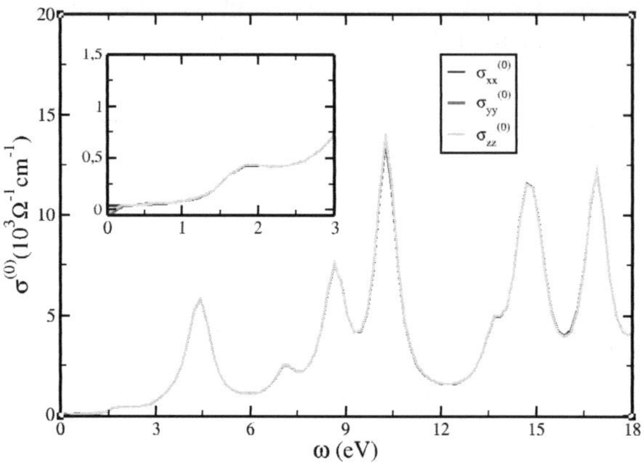

Figure 3. Calculation result of bare optical conductivity.

To see how electron – hole interaction effects the optical response of GaAs, we add up the two corresponding curves, the bare optical conductivity and the corresponding first order correction. The results are displayed results in Fig. 4.

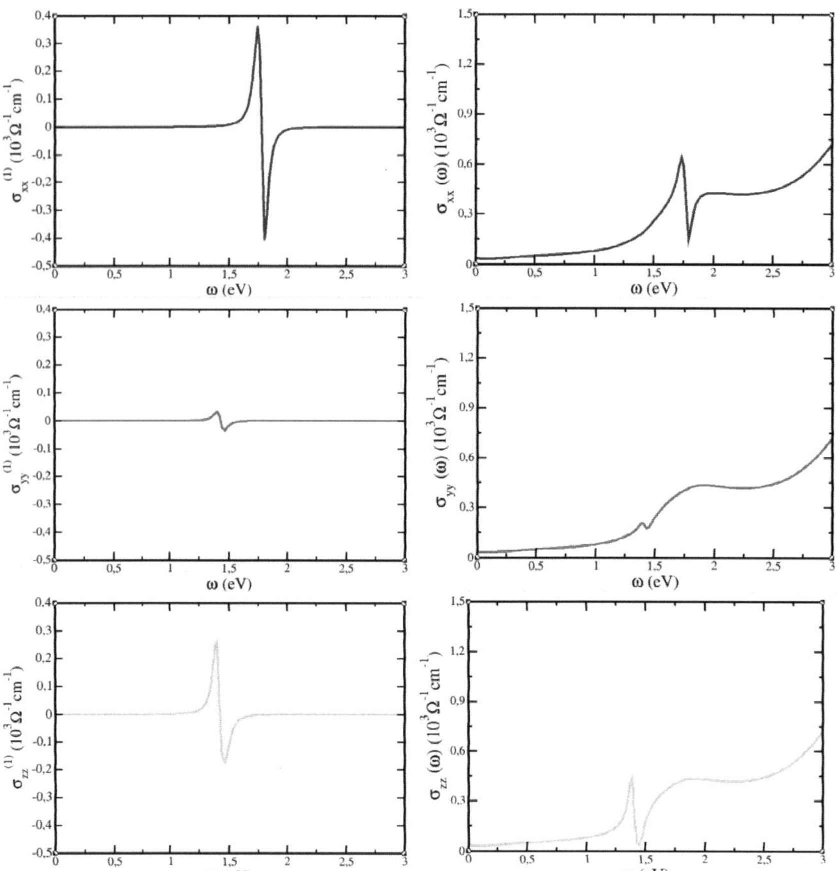

Figure 4. Optical conductivity caluculation including first-order e-h vertex correction. Left figures: first order vertex correction; right figures: bare optical conductivity added with its first order vertex correction.

The excitonic signal appears in all directions of optical conductivity σ_{xx}, σ_{yy} and σ_{zz}. It appears slightly below the band gap energy. Our calculated first-order correction optical conductivity reveals excitonic signals starting at 1.27 eV for σ_{xx}, σ_{yy} and 1.36 eV for σ_{zz}. The first order highest peak strikes at 1.4 eV for both σ_{yy} and σ_{zz}, while for σ_{xx}, the peak strikes at frequency 1.75 eV. The excitonic signal from measurements also appear at about these frequencies. As to compare with the bare optical conductivity curve, a little discrepancy occurs at the first-order electron-hole vertex correction calculation which does not preserve isotropic behavior as the bare one does. We conjecture that such a discrepancy results from the uncontrollable error occurring in the use of Padé approximant technique.

5. Conclusion

In conclusion, we have developed a $k.p$ based model for GaAs and the method to calculate optical conductivity up to first order e-h vertex correction and obtained optical conductivity spectrum up to 18 eV. Our results for the first order e-h vertex correction reveals excitonic signal around 1.5 eV (the band gap edge), consistent with the experimental data. Overall, we asses that our model and method works pretty well. Despite some lack of accuracy in the Padé approximation that needs to be improved.

Acknowledgement

We are very grateful to Universitas Indonesia for supporting us with full funding for this project through PITTA Research Grant No. 666/UN2.R3.1/HKP.05.00/2017.

References

[1] Sturge M 1962, Physical Review **127** 768
[2] Lee M M Teuscher J Miyasaka T Murakami T N and Snaith H J 2012, Science **338** 643
[3] Wang C and Klein B M 1981, Physical Review B **24** 3417
[4] Benedict L X, Shirley E L and Bohn R B 1998, Physical Review B **57** R9385
[5] Yu P Y and Cordona M 2010 *Fundamentals of Semiconductors Physics and Materials Properties* (London: Springer)
[6] Bahder T 1990, Physical Review B **41** 11992
[7] Ostromek TE 1996, Physical Review B **24** 60
[8] Mahan G 1990 *Many-Particle Physics* (New York: Plenum Publishers)
[9] Altland A and Simons B D 2010 *Condensed Matter Field Theory* (USA: Cambridge University Press)
[10] Baker G.A Jr, Graves-Morris P 1996, *Padé Approximants*. Cambridge U.P.

Investigation of the charge-orbital ordering mechanism in single-layered $Pr_{0.5}Ca_{1.5}MnO_4$

C N Rangkuti, M A Majidi[†]

Department of Physics, Faculty of Mathematics and Natural Sciences Universitas Indonesia, Kampus UI Depok, Depok 16424, Indonesia

[†]Corresponding author's e-mail: aziz.majidi@sci.ui.ac.id

Abstract. Motivated by the experimental study of half-doped single-layered $Pr_{0.5}Ca_{1.5}MnO_4$ showing charge, orbital, and spin orderings [1], we propose a model to theoretically study the system to explain such ordering phenomena. The ground state electron configuration reveals that the charges form a checkerboard pattern with alternating Mn^{3+}/Mn^{4+} sites, while the orbitals are aligned in zigzag chains [1, 2]. We calculate the ground state energy of this system to find the most preferable configuration by comparing three types of configurations (charge-unordered, charge-ordered, and charge-orbital-ordered states). The calculations are based on a tight-binding model representing effective electron hoppings among Mn ions in MnO_2-plane. We take into account the horizontally- and vertically-oriented orbital and spin degrees of freedom at Mn sites. We assume that the hopping integral values depend on the relative orientation between the corresponding orbitals of adjacent Mn ions. The interaction terms we incorporate into our effective Hamiltonian include inter-orbital, intra-orbital Hubbard repulsions, and Jahn-Teller distortion [2]. We absorb the exchange interaction between spins into local self-energy that we calculate within dynamical mean field algorithm [2]. Within our model we show a circumstance in which the charge-orbital ordered configuration has the lowest energy, consistent with the ground state ordering revealed by the experimental data.

1. Introduction

The emergence of various properties in manganites has become hot research topics in the area of condensed-matter physics since the last few decades [3, 4]. The manganites present a very rich phase diagram involving charge, orbital and spin orderings. Compounds of Pr-Ca-Mn-O are among those of manganites that can exhibit charge and orbital ordering phenomena [5]. In 2D system, single-layered $Pr_{1-x}Ca_{1+x}MnO_4$ shows the domination of insulator anti-ferromagnetic phase along the hole doping range [5, 6]. The most interesting phenomenon in this system is the formation of charge-orbital ordering (CO/OO). The system below charge-orbital ordering temperature $T_{CO/OO}$ shows that Mn^{3+}/Mn^{4+} are arranged alternately as a checkerboard pattern and e_g orbitals are arranged in zigzag chain. Meanwhile, the spins of Mn ions, formed by their t_{2g} electrons, are ferromagnetically coupled along zigzag chain and anti-ferromagnetically coupled between adjacent zigzag chains [1]. Such an ordering pattern is referred to as the CE type order as depicted in Figure 1. The appearance of such an ordering contributes to transport, optical and magnetic properties of the system. Therefore, in hope of being able to exploit the system for applications, we need more understanding about the mechanism of this fascinating phenomenon.

In this paper, we present our theoretical study on single-layered half-doped $Pr_{0.5}Ca_{1.5}MnO_4$ to investigate the mechanism governing its ground-state charge and orbital configurations. Recent experimental study has revealed that this system shows charge-orbital ordering at temperature below $T_{CO/OO} \sim 325$ K and forms insulating phase [1]. We propose a model incorporating relevant interactions and seek the circumstances leading to the lowest energy with such configuration. We hypothesize that the charge-orbital ordering arises as a result of the interplay between various interactions involving hopping integral values that depend on relative orientations between e_g orbitals of adjacent Mn ions.

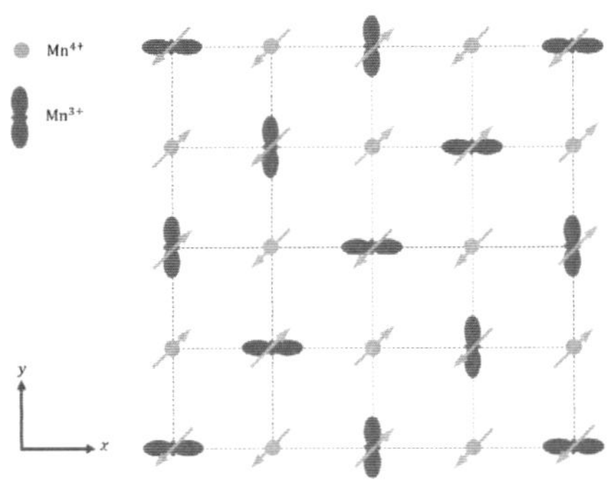

Figure 1. CE type and AF ordering scheme on MnO_2-plane.

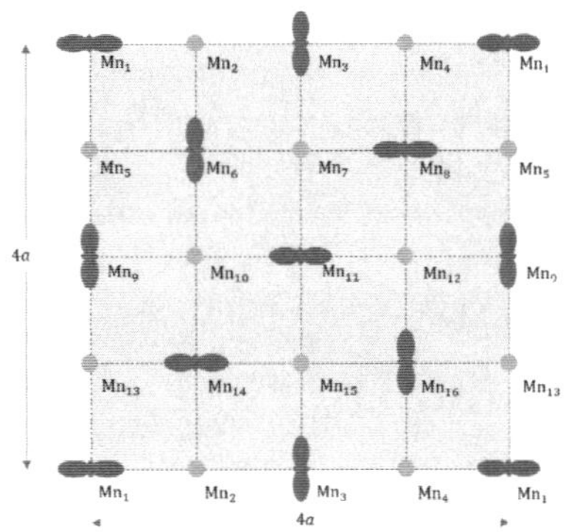

Figure 2. Unit super-cell of our model for MnO_2-plane.

2. Model

We construct our model to accommodate the charge-orbital ordering phenomena as follows. As we would like to consider only the dynamics of electrons in Mn orbital, we do not incorporate the presence of Pr, Ca and O ions explicitly in our model. In addition, we also neglect coupling between layers assuming that it is very weak compared to the hopping electrons on the same layer. Furthermore, to accommodate the formation of charge ordering and orbital ordering, we construct a unit super-cell with an area of $4a \times 4a$ as shown in Fig 2. Each unit super-cell consists of 16 atomic sites (s) where each atomic site has 2 e_g orbitals (A, B) with 2 possible space orientations (x,y) and 2 spin orientations ($\uparrow, \downarrow$). Hence, we use $16 \times 2 \times 2 \times 2 = 128$ basis states out of Mn e_g orbitals in the unit super-cell to construct our Hilbert space, with each basis state denoted as $|\text{Mn } d^{\alpha}_{s\gamma\sigma}\rangle$. This basis set is used to form our model Hamiltonian as below

$$\begin{aligned} H &= \frac{1}{N}\sum_{\mathbf{k}} \eta^{\dagger}_{\mathbf{k}}[H_0(\mathbf{k})]\eta_{\mathbf{k}} + \sum_{i,s,\gamma,\sigma,\sigma'} U n^{A}_{is\gamma\sigma} n^{B}_{is\gamma\sigma'} + \sum_{i,s,\gamma,\alpha} U_{\alpha} n^{\alpha}_{is\gamma\uparrow} n^{\alpha}_{is\gamma\downarrow} + \sum_{i,<s,s'>,\gamma,\alpha,\sigma} g u_i (n^{\alpha}_{is\gamma\sigma} - n^{\alpha}_{is'\gamma\sigma}) \\ &+ \sum_{i,s,\gamma,\sigma} g v_i (d^{A\dagger}_{is\gamma\sigma} d^{B}_{is\gamma\sigma} + d^{B\dagger}_{is\gamma\sigma} d^{A}_{is\gamma\sigma}) - \sum_{i,s,\gamma,\alpha} J\mathbf{S}_{is}.\mathbf{s}^{\alpha}_{is\gamma}. \end{aligned} \quad (1)$$

The first term of Eq. 1 is the kinetic part, where the matrix $[H_0(\mathbf{k})]$ is a 128×128 matrix containing the on-site and hopping energies as functions of the wavevector $\mathbf{k}$ in the Brillouin zone (BZ), and $\eta^{\dagger}_{\mathbf{k}}$ ($\eta_{\mathbf{k}}$) is a row (column) vector containing the corresponding creation (annihilation) operators. The second term represents the inter-orbital Coulomb repulsion that we shall treat at mean-field level. The third term represents intra-orbital Coulomb repulsion between spin up and down electrons in the same orbital. In this case, we set U_{α} to be infinitely large to avoid double occupancy on each Mn orbitals. The fourth and fifth terms represent two modes of vibrations related to electron-phonon Jahn-Teller interaction, with u_i and v_i being the corresponding local lattice displacements, and g the electron-phonon coupling. In our approximation, we ignore the dynamics of the Jahn-Teller displacement, rather we take the equilibrium values of u_i and v_i defined as u_0 and v_0 [2]. We can derive u_0 and v_0 from the equation: $\frac{\partial}{\partial u}[\frac{k}{2}u^2+gu\langle n^{\alpha}_{is\gamma\sigma}\rangle]_{u=u_0} = 0$. Then, with u_0 and v_0 as new parameters, the constants in the fourth and fifth terms can be reexpressed, respectively, as $gu_0 \equiv U_{JT}\langle n^{s}_{i\alpha\sigma}\rangle$ and $gv_0 \equiv V_{JT}$. The last term represents the magnetic interaction between spin of itinerant electron and total spin of the localized (t_{2g}) electrons of Mn, which we shall treat within dynamical mean field algorithm.

Our specific goal is to compare the total energy of three different ordering configurations: charge-unordered, charge-ordered, and charge-orbital-ordered. The description of the three ordering configurations is the following. Consider that there are 8 itinerant electrons per unit super-cell such that the electron filling satisfies $\sum_{s\alpha\gamma\sigma}\langle n^{\alpha}_{s\gamma\sigma}\rangle = 8$. In charge-unordered state, electrons have same probability to occupy each site, so that each occupation takes a value $\langle n^{\alpha}_{s\gamma\sigma}\rangle = \frac{8}{128} = \frac{1}{16}$. In charge-ordered state, electrons choose to occupy one of two neighboring sites. So if site s is occupied by an electron, the nearest site s' will be empty. There are 64 combinations of s, α, γ and σ, so that $\langle n^{\alpha}_{s\gamma\sigma}\rangle = \frac{8}{64} = \frac{1}{8}$. The occupation of charge-orbital-ordered state is similar to charge-ordered state but the electron only occupies one of two e_g orbitals at a site. There are 32 combinations of s, γ and σ so that $\langle n^{A}_{s\gamma\sigma}\rangle = \frac{8}{32} = \frac{1}{4}$.

3. Result and Discussion

At the same atomic site, there are three Coulomb repulsive interactions we may consider to be distinct: 1) Coulomb repulsion between electrons in different e_g orbitals A and B, regardless of space (x,y) and spin orientation; 2) Coulomb repulsion between electrons in the same e_g orbital A or B, but at different space-orientation states (x and y), regardless of the spin orientations; and 3) Coulomb repulsion between electrons in the same e_g orbital A or B at the same space-orientation state (x or y), with different spin orientations. We want to calculate the total energy of each of the above three ordering configurations at $T = 0$ for two different possible scenarios. Since we are interested only in the ground-state properties, i.e. $T = 0$, for all calculations we freeze the Mn-spin configuration according the zigzag CE-type configuration as depicted in Fig. 1. Note that, unless we make a specific statement about the values of certain parameters we use in this paper, we use all the parameters given in Ref. [2].

3.1. Scenario 1

We assume the strength of the Coulomb repulsion 1) and 2) above to be of finite value U, while 3) is taken to be infinity. Figures 3 and 4 show how the total energy of each of the three ordering configurations vary with U, assuming the hopping parameters being the same and different for different space-configuration states, respectively. Regardless of whether the hopping parameters are the same or different, we see that the energy of charge-unordered configuration increases and becomes the highest among the three configurations. Meanwhile, the energy of charge-ordered and charge-orbital-ordered configurations decrease as U increases. But the charge-orbital-ordered energy is found to be much higher than that of charge-ordered. This is unexpected result as the experimental study at low temperatures [1] indicates that the ground-state phase of the system is charge-orbital ordered.

3.2. Scenario 2

Now, we assume the strength of the Coulomb repulsion 1) to be of finite value U, while both 2) and 3) are taken to be infinity. In Figure 5 and 6, we can see the total energy of the different ordering configurations. Again, regardless of whether the hopping parameters are the same or different, the energy of all the three ordering configurations increase as U increases. More importantly, as $U > 1$ eV charge-unordered phase has the highest energy indicative of being unfavorable phase at $T = 0$. The formation of charge-ordered or charge-orbital-ordered phase is also followed with the transition from metallic to insulating-like phase (not shown)

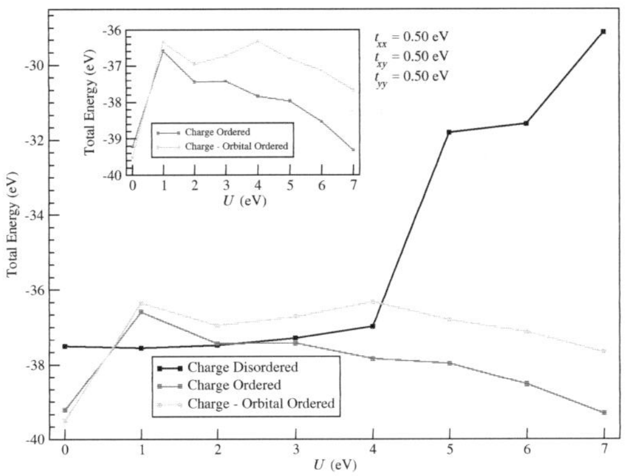

Figure 3. Total energy for the same hopping parameters.

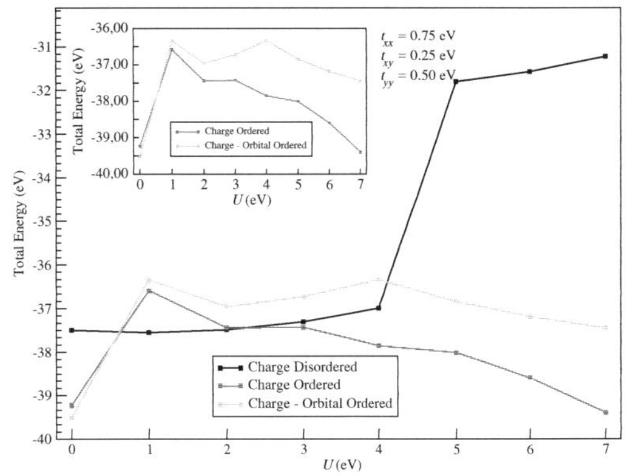

Figure 4. Total energy for different hopping parameters.

[8]. Meanwhile, the charge-orbital-ordered phase has lower in energy as compared to the charge-ordered phase with the difference around $\sim$ 18 meV. This signifies that the charge-orbital-ordered phase is the true ground state for this system.

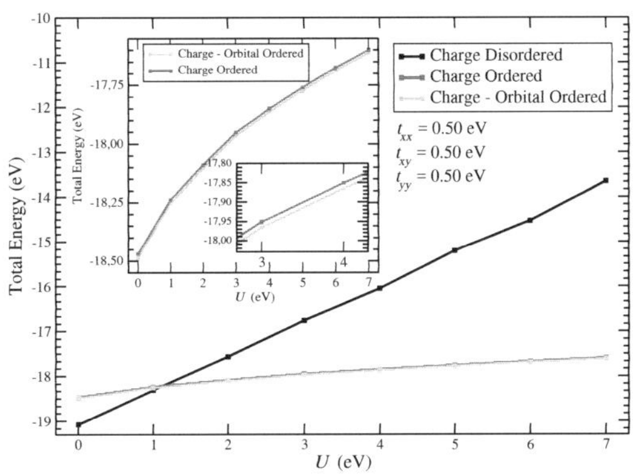

Figure 5. Total energy for the same hopping parameters.

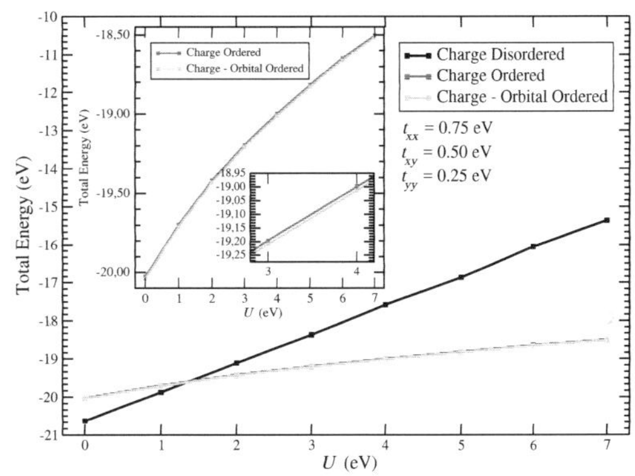

Figure 6. Total energy for different hopping parameters.

Still with scenario 2, Figs. 7 and 8 show us the effect of Jahn-Teller parameter (U_{JT}) on the splitting energy between charge-ordered and charge-orbital-ordered states as U_{JT} increases, while $U = 4$ eV is set fixed.

If we inspect further through Fig. 9, there is a significant difference in energy of the charge-orbital-ordered phase between the case with the same hopping parameters and that with different parameters. The dependence of the hopping parameter values on the relative orbital orientation decreases further the charge-orbital-ordered energy of the system.

4. Conclusion

We have developed a model to describe the mechanism of charge-orbital ordering in the ground state of single-layered $Pr_{0.5}Ca_{1.5}MnO_4$ by comparing the total electronic energy among three different possible charge configurations. We test two scenarios to see which Coulomb repulsion should be considered much larger than others. We find that the scenario 2 (as detailed above) is more suitable for the system being studied, for which it can show that the charge-orbital-ordered phase is lowest in energy compared to the charge-unordered and the charge-ordered phases. In

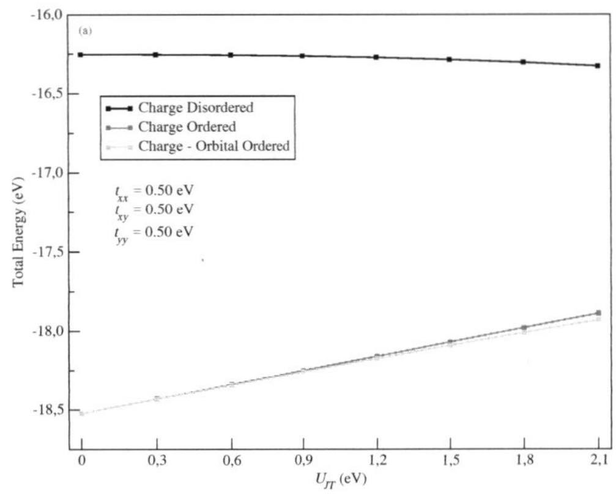

Figure 7. Total energy for the same hopping parameters.

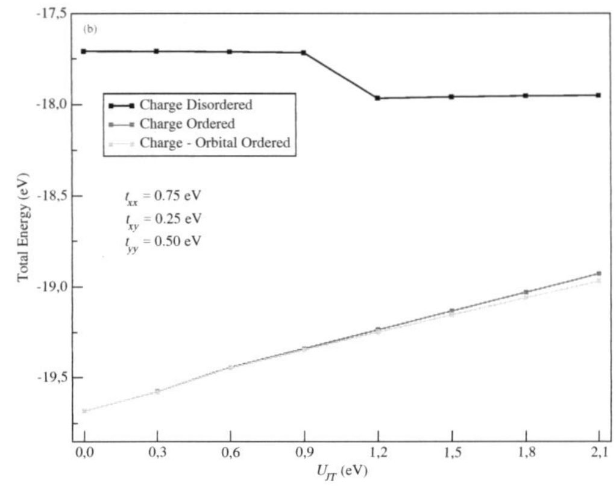

Figure 8. Total energy for different hopping parameters.

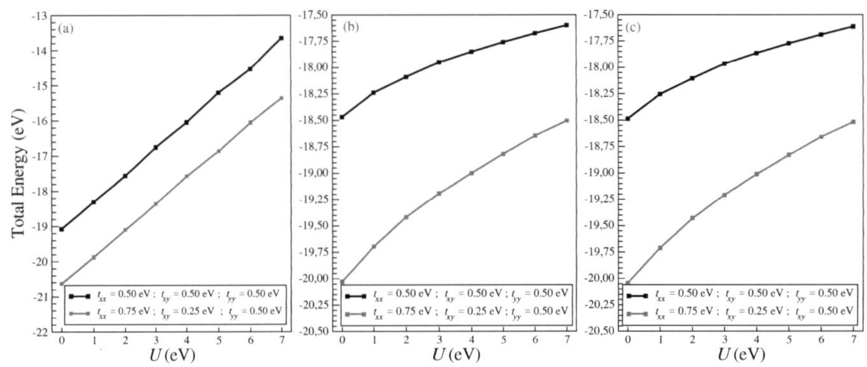

Figure 9. Total energy vs U showing the importance of different hopping parameters along different relative orbital orientation.

addition, the model with the choice of scenario 2, demonstrates that the charge-ordered and the charge-orbital-ordered phases can be lower in energy compared to the charge-unordered phase only for U above some threshold value, confirming the important role of the U parameter for the formation of charge-ordered and charge-orbital-ordered phases [2, 7]. Moreover, we have also demonstrated that at some finite U value (e.g. $U = 4$ eV) the Jahn-Teller effect and the dependence of the hopping parameters on the relative orbital orientation increase the energy difference between charge-ordered state and charge-orbital-ordered phases.

5. Acknowledgement

We are very grateful to Universitas Indonesia for supporting this project through full-funding by PITTA Research Grant No.666/UN2.R3.1/HKP.05.00/2017.

References

[1] Chi S *et al* 2007 *Proc. Natl. Acad. Sci. USA.* **104** 10796.
[2] Majidi M A *et al* 2013 *Phys. Rev. B.* **87** 235135.
[3] J. B. Goodenough 1955 *Phys. Rev.* **100** 564
[4] E. O. Wollan and W. C. Koehler 1955 *Phys. Rev.* **100** 545.
[5] Ibarra M 2003 *J. Solid State Chem.* 170, 361.
[6] Yu X Z *et al* 2007 *Phys. Rev. B* **75**, 174441.
[7] Van den Brink J *et al* 1999 *Phys. Rev. Lett.* **83**, 5118.
[8] Verwey 1939 *Nature* **144**, 327328.

Numerical Study of Fermi Surface on Cuprate Using One-Band Hubbard Model: Role of Charge Density Wave in The Antiferromagnetic Mott-Insulator and Pseudogap Region

M E I Akbar and I Santoso*
Departemen Fisika, Universitas Gadjah Mada, Sekip Utara BLS 21 Yogyakarta, INDONESIA

*Corresponding Author: iman.santoso@ugm.ac.id

Abstract - Numerical study of Fermi surface has been done using one-band Hubbard model with exact diagonalization method to study the effect of doping factor on Charge Density Wave (CDW) development in *Anti ferromagnetic Mott Insulator* (AFM) and pseudogap region in cuprate material. In this study the effect of CDW is characterized by Fermi Surface Nesting (FSN) vector in which it might be formed in pseudogap region. The wavelength of CDW increases when the value of disorder, which is representing the doping level in cuprate material, is increased. However, the CDW pattern will dissapear as the doping reaches the value which represented by Hubbard parameters: $t = 1$ *eV* , $U = 4$ *eV*, $\mu = 2$ *eV*, and $v_i = -6$ *eV*, marked by the decreasing of charge density wavelength.

1. Introduction

Cuprate is a High Temperature Superconductor (HTSC) material which has superconductivity phase with T_c above 30 K as first studied by Berdnoz and Muller on the LBCO system [1]. Besides superconductivity phase, there are several other phases in cuprate material such as Anti Ferromagnetic Mott-Insulator (AFM), pseudogap, strange metal and Fermi liquid. The phase change is affected by the temperature and the amount of doping of charge carriers applied to the system.

One of interesting problem in cuprate is that how the competing order mentioned above evolved or played some role on mechanism of electron pairing [2] that drives the superconductivity,for example is pseudogap becomes the necessary precursor for superconductivity? Meanwhile, the interaction between electrons can also produce other phenomena such as Charge Density Wave (CDW). It has been known that CDW is broken symmetry states of metals, due to electron-lattice instability and found in some materials such as cuprate material [3-5]. It is argued that CDW is a driven force for in pseudogap phase in cuprate [6]. Thus it is important to study the role of CDW in pseudogap phase since it is believed to be the key for explaining the mechanism of HTSC in cuprates.

The behavior of electrons in a material can be characterized by Density of State (DOS), the magnitude of interaction between electrons and energy state around Fermi surface. CDW can be characterized by vector nesting on the Fermi surface.

In this work, we report the numerical study band structures and Fermi surfaces from cuprate materials in the AFM and pseudogap regions to investigate the pattern of CDW formation associated with the given disorder factor. The disorder factor in this study represents the potential energy of the given doping. The Hamiltonian used in this study is Hubbard model which can describe the phase transition from conductor to insulator or vice versa. Chemical potential (μ) and disorder (v_i) factor were added to this Hamiltonian.

2. Method

In the Hubbard model the atomic state in the material is illustrated by 2 parameters: t which describes the energy hopping and U which describes the potential energy of the interaction between

IOP Conf. Series: Journal of Physics: Conf. Series **1011** (2018) 012078 doi:10.1088/1742-6596/1011/1/012078

electrons in the same state. The formulation of Hubbard's Hamiltonian model can be written as follows:

$$H = -t \sum_{<i,j>\sigma} c^{\dagger}{}_{i\sigma}\, c_{j\sigma} + U \sum_j n_{i\uparrow} n_{i\downarrow} + h.c \ , \tag{1}$$

with $c^{\dagger}{}_{i\sigma}$ is creation operator of electron with spin σ on site i, $c_{j\sigma}$ is annhilation operator of electron with spin σ on site j, $n_{i\sigma} = c^{\dagger}{}_{i\sigma} c_{i\sigma}$ is the number operator and h.c is *Hermitian conjugate*.
Hubbard's model in real space can be transformed into reciprocal space by using Fourier transform for the second quantization as follows:

$$c^{\dagger}{}_{k\sigma} = \frac{1}{\sqrt{N}} \sum_l e^{ik.l}\, c^{\dagger}{}_{l\sigma}\ , \tag{2}$$

$$c_{k\sigma} = \frac{1}{\sqrt{N}} \sum_l e^{-ik.l}\, c_{l\sigma}\ , \tag{3}$$

The energy value is obtained when the Hamiltonian is converted into a diagonal matrix. The matrix is obtained by transform the Hamiltonian by the following methods:

$$\begin{pmatrix} c_{i\sigma} & c_{j\sigma} \end{pmatrix} \begin{pmatrix} H_{11} & H_{12} \\ H_{21} & H_{22} \end{pmatrix} \begin{pmatrix} c^{\dagger}{}_{i\sigma} \\ c^{\dagger}{}_{j\sigma} \end{pmatrix} . \tag{4}$$

The component values of a matrix a which are complex values can be converted into real values using the following methods:

$$\begin{pmatrix} \boldsymbol{b} & -\boldsymbol{c} \\ \boldsymbol{c} & \boldsymbol{b} \end{pmatrix} \begin{pmatrix} x \\ y \end{pmatrix} = \lambda \begin{pmatrix} x \\ y \end{pmatrix} . \tag{5}$$

where b and c are respectively the real and imaginary parts of a. λ is the eigenvalue related to the energy eigen of particular eigen state $\begin{pmatrix} x \\ y \end{pmatrix}$.

The Householder transformation is used to convert symmetry matrices into tridiagonal matrices, then the Jacobi transformation is used to transform the tridiagonal matrix into a diagonal matrix. The value of each element of the transformed matrix represents the value of Eigen energy in each of the energy bands of the material.

The Hamiltonian model used in this study is one-band Hubbard model as presented in the following equations:

$$H = \sum_{<i,j>\sigma} t_{ij}\, c^{\dagger}{}_{i\sigma}\, c_{j\sigma} - (\mu + U) \sum_{i\sigma} n_{i\sigma}\ , \tag{6}$$

with μ denotes the chemical potential.

In reciprocal space equation (6) change into:

$$H = \sum_k \begin{pmatrix} -v_i & \xi_k \\ \xi_k & -v_i \end{pmatrix}, \tag{7}$$

with,

$$\xi_k = \varepsilon_k - \mu - \mu^H\ , \tag{8}$$

and

$$\varepsilon_k = -t \sum_k e^{ikr}. \tag{9}$$

In this study, we used a Plaquetee model of cuprate material in which one copper atom (Cu) is surrounded by four oxygen (O) atoms with distance r, then equation (9) becomes:

$$\varepsilon_k = -t\,(e^{ik_x r} + e^{-ik_x r} + e^{iyr} + e^{-iyr}) \tag{10}$$

with,

$$e^{i\theta} = \cos\theta + i \sin\theta. \tag{11}$$

Equation (12) shows the matrix model (with complex elements) that is to be used in this study. Since the matrix in equation (7) is the Hermitian matrix, the transformation of the $n \times n$ Hermitian matrix to the real symmetry matrix $2n \times 2n$ is required, using the transformation as shown in (5), the Hamiltonian model becomes:

IOP Conf. Series: Journal of Physics: Conf. Series **1011** (2018) 012078 doi:10.1088/1742-6596/1011/1/012078

$$\sum_{k_x k_y} \begin{pmatrix} -v_{i_{real}} & \xi_{k_{real}} & v_{i_{imajiner}} & -\xi_{k_{imajiner}} \\ \xi_{k_{real}} & -v_{i_{real}} & -\xi_{k_{imajiner}} & v_{i_{imajiner}} \\ -v_{i_{imajiner}} & \xi_{k_{imajiner}} & -v_{i_{real}} & \xi_{k_{real}} \\ \xi_{k_{imajiner}} & -v_{i_{imajiner}} & \xi_{k_{real}} & -v_{i_{real}} \end{pmatrix} \tag{12}$$

The Hamiltonian model of equation (12) was used as a model for cuprate material with 6×6, 12×12 and 18×18 cluster size variations. The exact diagonalization then has been used for each k-point in first Brillouin Zone of Cuprate. The CDW wavelength was calculated based on the value of vector nesting in the antinodal region.

3. Result

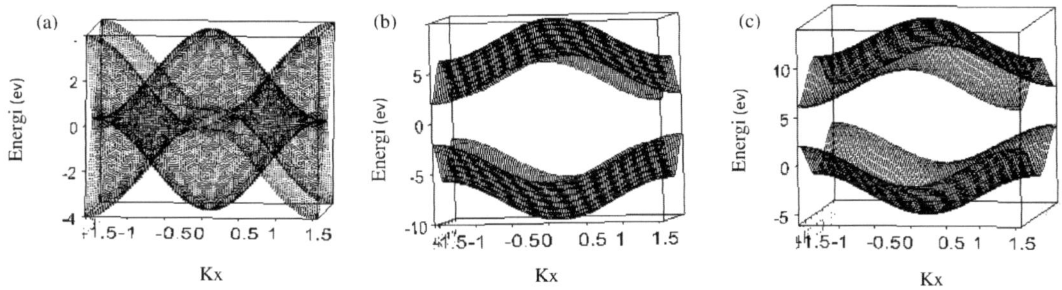

Figure 1. Band structure of cuprate with Hubbard parameters (a) $t = 1\ eV; U = 0\ eV;\ \mu = 0\ eV;\ v_i = 0\ eV$ (b) $t = 1\ eV; U = 4\ eV;\ 2eV;\ v_i 0\ eV$ (c) $t = 1\ eV; U = 4\ eV;\ \mu = 2eV;\ v_i = -4\ eV$

The results of this study show that Hubbard model can describe the phase transition from a metallic phase (fig. 1(a)) to an insulator phase (fig. 1(b)) or vice versa. 3-dimensional of energy bands of the first Brilouin zone of cuprate material show in fig.1. When the parameter U and μ were added, the band structure were divided into two parts as shown in fig 1(b) namely the upper Hubbard band and the lower Hubbard band, and form a gap between energy bands. This means that the phase of the material changes from a metallic state to an insulator. This phase indicates the AFM phase. The parameter disorder shifts the two energy bands, so that the $E = 0$ plane can cut one of the two bands as shown in fig 1(c). This phase may indicate the pseudogap phase.

IOP Conf. Series: Journal of Physics: Conf. Series **1011** (2018) 012078 doi:10.1088/1742-6596/1011/1/012078

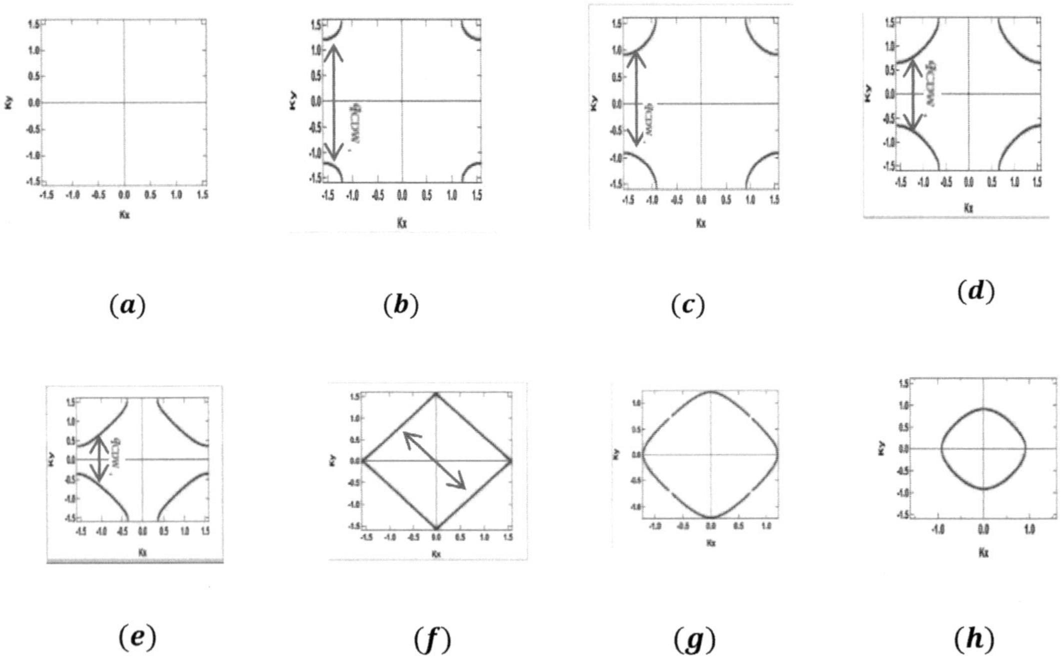

Figure 2. CDW pattern formation based on the development of Nesting vector on the Fermi surface with Hubbard parameter values : $t = 1\ eV\ ; U = 4\ eV\ ;\ \mu = 2\ eV\ (a)\ v_i = -1.5\ eV\ (b)\ v_i = -2.5\ eV (c)\ v_i = -3.5\ eV (d)\ v_i = -4.5\ eV (e)\ v_i = -5.5\ eV (f)\ v_i = -6.0\ eV (g)\ v_i = -6.5\ eV (h)\ v_i = -7.5\ eV$

The formation of a nesting vector based on a variation of the given disorder value. In the Fig. 2(b), 2(c), 2(d), and 2(e) the formation of the antinodal region is seen. The length of the nesting vector in this area is used to calculate the CDW wavelength. Fig. 2(f) shows the Fermi surface of ordinary metal where gap in the antinodal regions were not formed.

Table 1 CDW wavelength

No	t (ev)	U(ev)	μ(ev)	v_i(ev)	K_f	λ(Å)
1.	1	4	2	-1,5	-	-
2.	1	4	2	-2,5	1,2	2,62
3.	1	4	2	-3,5	1,1	2,85
4.	1	4	2	-4,5	0,6	5,23
5.	1	4	2	-5,5	0,4	7,85
6.	1	4	2	-6,0	2,1	1,48
7.	1	4	2	-6,5	-	-
8.	1	4	2	-7,5	-	-

The Fermi surface nesting shows that CDW patterns can be formed in antinodal areas. The greater the given factor of the disorder, the greater the wavelength value of the CDW is. However when the doping reaches a certain value which is represented by Hubbard parameters with the value: t = 1 eV, U = 4 eV, μ = 2 eV and v_i = -6 eV The CDW pattern disappears which is marked by the returning of the wavelength value as shown in table 1.

4. Conclusion

IOP Conf. Series: Journal of Physics: Conf. Series **1011** (2018) 012078 doi:10.1088/1742-6596/1011/1/012078

The CDW phase might be manifested in the Pseudogap region when the material is doped¬ with a certain concentration. When the doping factor is added the CDW wavelength also increases, up to a certain doping level (v_i = -6 eV) the wavelength of the material returns to normal and the CDW phase is no longer found.

5. Refererences

[1] Bednorz J G, Muller K A 1986 *Z. Phys. B.* **64** (1) pp 189–193.
[2] Bardeen J, Cooper L N, Schriffer J R 1957 *Phys. Rev.* **108** (5) 1175.
[3] Pate M, Mottershead J D F, Elfimov I S, Peets D C, Liang R, Boon D A, Hardy W N, Chiuzbaian S, Falub M, Shi M, Patthey L, Damascelli A 2005 *Phys. Rev. Lett.* **95** 077001.
[4] Robertson J A, Kivelson S A, Fradkin E, Fang A C, Kapitulnik A. 2006 *Phys. Rev.* B **74** 134507.
[5] Shen K M, Ronning F, Lu D H, Baumberger F, Ingle N J C, Lee W S, Me*ev*asana W, Kohsaka, Y, Azuma M, Takano,M, Takagi H, Shen Z X 2005 Science **307** pp 901-904.
[6] Timusk T, Statt B W 1999 *Rep. Prog. Phys.* **62** pp 61-122.
[7] Loh Y L, Trivedi N 2012 *Theoretical Studies of Superconductor-Insulator Transitions* chapter 17, pp. 492-548 in *Conductor-Insulator Quantum Phase Transitions* (ed. V Dobrosavljevic, N Trivedi, and J M Valles Jr Oxford University Press)

IOP Conf. Series: Journal of Physics: Conf. Series **1011** (2018) 012079 doi:10.1088/1742-6596/1011/1/012079

Critical properties of the antiferromagnetic Ising model on rewired square lattices

Tasrief Surungan[1], Bansawang BJ[1] and Muhammad Yusuf[2]
[1]Department of Physics, Hasanuddin University, Makassar, South Sulawesi, 90245, Indonesia
[2]Department of Physics, Gorontalo State University, Gorontalo, Gorontalo 96128, Indonesia

E-mail: tasrief@unhas.ac.id

Abstract. The effect of randomness on critical behavior is a crucial subject in condensed matter physics due to the the presence of impurity in any real material. We presently probe the critical behaviour of the antiferromagnetic (AF) Ising model on rewired square lattices with random connectivity. An extra link is randomly added to each site of the square lattice to connect the site to one of its next-nearest neighbours, thus having different number of connections (links). Average number of links (ANOL) κ is fractional, varied from 2 to 3, where $\kappa = 2$ associated with the native square lattice. The rewired lattices possess abundance of triangular units in which spins are frustrated due to AF interaction. The system is studied by using Monte Carlo method with Replica Exchange algorithm. Some physical quantities of interests were calculated, such as the specific heat, the staggered magnetization and the spin glass order parameter (Edward-Anderson parameter). We investigate the role played by the randomness in affecting the exisiting phase transition and its interplay with frustration to possibly bring any spin glass (SG) properties. We observed the low temperature magnetic ordered phase (Néel phase) preserved up to certain value of κ and no indication of SG phase for any value of κ.

1. Introduction

The cooperative phenomena are ubiquitous in nature, driven by the presence of coupling interaction between each individual constituent of materials [1, 2]. This is exemplified by the spontaneous magnetization by which certain magnetic material is entering a ferromagnetic phase below its transition temperature T_c (Curie temperature) and able to attract the nearby metals surrounding it. The high temperature disorded phase and the cooperative ordered phase at low temperature are separated by T_c. A phase transition is essentially rendered by the competition between the external fields, such as temperature, which tend to destroy the order and the coupling interaction which tends to create order.

Based on the nature of free energy function, a phase transition is grouped into the first and the second order phase transition[3]. The transition is called as first order if the first order derivative of the free energy of the system is discontinuous. If the derivative is continuous, the transition is second order, also called continuous phase transition, at which no latent heat involved in the course of transition and no co-existing phases[2, 4]. A firm example of these is the phase transition experienced by PVT systems above the critical point[5]. The phenomena nearby critical point are known as critical phenomena, and affected by limited number of controlling variables such as the type of coupling interaction, symmetry of the microscopic constituents

IOP Conf. Series: Journal of Physics: Conf. Series **1011** (2018) 012079 doi:10.1088/1742-6596/1011/1/012079

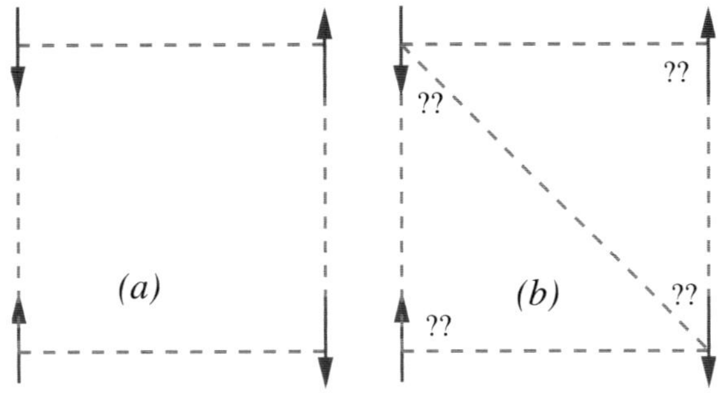

Figure 1. (a) Unfrustrated spins and (b) geometrical frustrated spins on rewired plaquettee. The AF couplings are represented by the dashed lines.

(e.g., spins) and spatial dimensionality. Critical phenomena are charaterized by a set of critical exponents which are in general universal, where different systems may be grouped in the same universality class, i.e., having similar value of exponents.

The type of phase transitions and universality classes are subject to the controlling variables and affected by the presence of randomness. The presence of randomness is a crucial aspect in the study of phase transition; and has long been intensively investigated. A randomness can be realized in a variety of forms such as defect (site vacancies) and diluted bonds. The effect of bond dilution for system undergoing Kosterlitz-Thouless (KT) transition has been investigated and found that the KT phase preserved so long as the bonds of the lattice are percolated[6]. According to Harris criterion[7], randomness is relevant (able to change the type or destroy the order or alter the universality classes) for the pure system experiencing second order phase transition with specific heat exponent $\alpha > 0$, and irrelevant for $\alpha < 0$; otherwise it is marginal.

In this study we investigated the role played by the connectivity randomness in affecting critical properties of the Ising model on rewired square lattice with AF couplings. The pure system of this model is well known to experience second order phase transition with $\alpha > 0$[8]. The crucial point is that the AF couplings bring geometrical frustration due to the presence of diagonal neighbours randomly added to the square lattice. A spin is frustrated if it can not find a unique orientation in interacting with its neighbors to stay in a minimum energy state[9]. There will be an interplay between randomness and the geometrical frustration, which may affect the existing low temperature ordered phase (the Néel phase) and could possibly lead to SG ordering[10]. In fact, various spin models on rewired lattices with AF interaction have been probed in conection to SG problem[11, 12, 13]. Previously it was found that no finite temperature Ising SG phase observed on rewired square lattice for $\kappa = 3$[14]. The study did not pay attention to the role played by the randomness in affecting the existing order phase. To systematically study this topic, here we consider fractional values of κ, ranging from 2 to 3. We probed the existing second order phase transition and the Néel phase at low temperature due to the presence of randomness and frustration. In particular we searched for the disappearance of AF orderings due to the increasing value of κ.

The subsequent parts of the paper is organized as follows: Section II explained the model and method. Results and discussion are presented in Section III. Section IV is devoted to the summary and concluding remarks.

2. Model and Simulation

The model is written in the following Hamiltonian

$$H = -\sum_{\langle ij \rangle} J_{ij} s_i s_j \tag{1}$$

where J_{ij} is the coupling interaction between the Ising spin s occupying respectively the site i-th and j-th of the lattice. As we probed a purely AF system, $J_{ij} = J < 0$. The summation is carried out over all pairs of spins directly connecting. For regular lattices such as the square and the cubic lattice, all spins possess similar number of neighbours, associated with a fixed coordination number. Currently, we consider rewired square lattices with fractional connectivity, i.e., the average number of links (ANOL), κ, varies from 2 to 3. A rewired lattice is considerd as a quasi-regular structure as the translational symmetry partially preserved. Unlike a random structure or a scale free network[15], where the notion of spatial dimension is obscure, for a quasi-regular lattice dimensionality remains. Our procedure of constructing the lattice is by randomly connecting each site to one of the diagonal neighbors of the site. There are 4 native nearest neighbors of the original square lattice; after being rewired, where one link is randomly added to each site, we obtain a lattice with fractional ANOL. Each individual site may have in minimum of 4 and maximum of 7 neighbors.

The randomly rewired lattices with AF couplings inherits the main ingredients of SG system and is considered as a new type of SG model, called non-canonical SG [11, 13, 16, 17]. It is distinguished from the canonical model as the type of randomness and frustration are not due to the presence of both FM and AF couplings. In a canonical model, both FM and AF coupling exist, while for non-canonical model only AF couplings exist, symbolically indicated by the dashed lines in Figure 1. The larger the value of κ the larger the number of triangular units. As shown in Fig. 1, spins in AF triangular unit are frustrated. If we further add some extra links the degree of frustration of the system increases. The number of triangular units determines the degree of frustration. This is analogous to the canonical SG system with various distribution FM and AF couplings, resulting SG phase diagram with Nishimori line[18].

As a standard method in statistical mechanics, we used Monte Carlo (MC) simulation to compute the physical quantities. Due to the complexity of the energy landscape, which may be difficult to tackle using the the conventional Metropolis algorithm, here we used the Replica Exchange (RE) algorithm[19]. It is a powerful MC algorithm introduced for studying complex problems such as randomly frustrated system, in particular to overcome the problem of slow dynamics. This problem is commonly found in complex system because of the presence of local minima in the energy landscape, where a random walker can easily get trapped at certain local minimum. The basic idea of this algorithm is to extend the conventional Metropolis algorithm by duplicating the original systems into several replicas. Each replica is then independently simulated using a standard Metropolis algorithm. As each replica belongs to a heat bath with certain temperature, then the whole result of the simulation is obtained by combining each individual result. During the computation, two replicas associated with to neighboring temperatures are exchanged. This is the essential trick for the RE algorithm, by which a random walker is able to escape from any local minimum.

If we begin the simulation with a number K of replicas; where every replica is in equilibrium with a heat bath of a corresponding temperature $T = 1/\beta$, then the probability distribution of observing the whole system in a state $\{X\} = \{X_1, X_2, \ldots, X_M\}$ is given by,

$$P(\{X, \beta\}) = \prod_{m=1}^{M} \tilde{P}(X_m, \beta_m), \tag{2}$$

with

$$\tilde{P}(X_m, \beta_m) = Z(\beta_m)^{-1} \exp(-\beta_m H(X_m)), \tag{3}$$

Here, the partition function of the m-th replica is represented by $Z(\beta_m)$. Next, we can define an exchange matrix between two replicas, $W(X_m, \beta_m | X_n, \beta_n)$. This matrix is the probability to swap the configuration X_m of β_m with the configuration X_n of β_n. In order to maintain the entire system at equilibrium, the detailed balance condition is imposed on the transition matrix

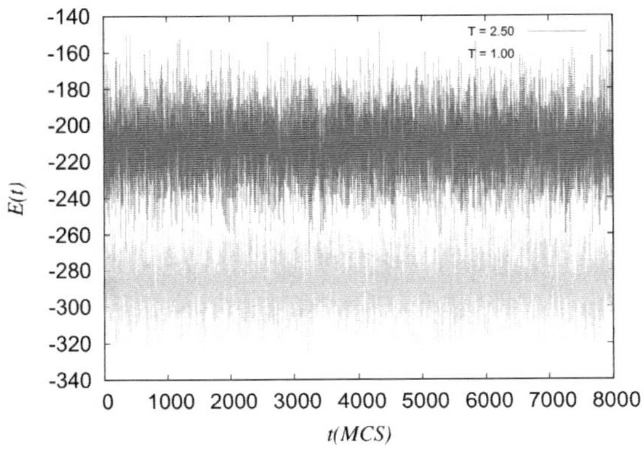

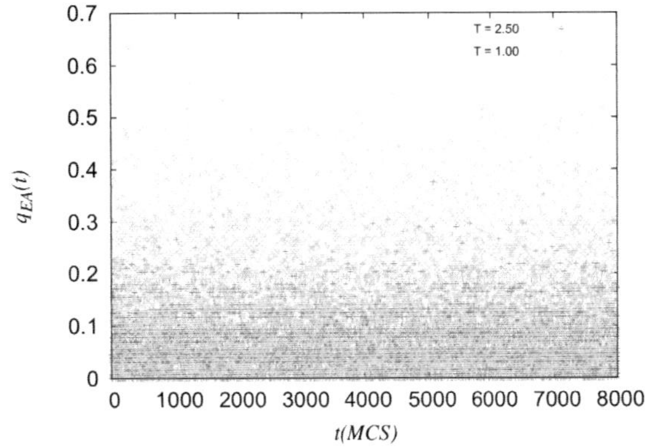

Figure 2. (Color online) Time series of enegy (left); and Edward-Anderson order parameter (right) at $T = 2.50$ and 1.00 for system size $L = 16$ with κ=3.0.

$$P(\{X_m, \beta_m\}, \ldots, \{X_n, \beta_n\}, \ldots) \cdot W(X_m, \beta_m | X_n, \beta_n)$$
$$= P(\{X_n, \beta_m\}, \ldots, \{X_m, \beta_n\}, \ldots) \cdot W(X_n, \beta_m | X_m, \beta_n), \quad (4)$$

along with Eq. (3), so that we obtain

$$\frac{W(X_m, \beta_m | X_n, \beta_n)}{W(X_n, \beta_m | X_m, \beta_n)} = \exp(-\Delta), \quad (5)$$

where $\Delta = (\beta_n - \beta_m)(H(X_m) - H(X_n))$. Based on this constraint, we can choose the coefficient of matrix according to the standard Metropolis algorithm which gives the following

$$W(X_m, \beta_m | X_n, \beta_n) = \begin{cases} 1 & \text{if} \quad \Delta < 0, \\ \exp(-\Delta) & \text{if} \quad \Delta > 0. \end{cases} \quad (6)$$

Since the ratio of acceptance ratio exponentially decays with $(\beta_n - \beta_m)$, the swap is carried out only between teo adjacent temperatures, i.e., the terms $W(X_m, \beta_m | X_{m+1}, \beta_{m+1})$.

In carrying out the simulation, we define a step of MC (MCS) as updating each spin once for each replica, either randomly or consecutively. After enough MCSs, we swapped each pairs of configurations belonging to neighboring temperatures according to the probability according to Eq. 6. For example, we swap replicas X_1 and and X_2, based on the probability defined Eq. 6. Then, two consecutive swaps are assigned as even and odd and. For the even swap, we exchange the pairs of replicas X_2 and X_3; X_4 and X_5; $\cdots$; X_{M-1} and X_M; while for an odd one we exchange X_1 and X_2; $\cdots X_{M-1}$ and X_{M-3}.

For a certain realization of connectivity, we started from a random spin configuration. Then, we equilibrated the system with 10^4 MCSs before doing calculation with total of 5×10^4 samples for each temperature. This data make up time series of corresponding physical quantities, from which we extract the thermal averages. Several additional MCSs between two samples were taken in order to avoid high time-correlation. We simulated the system with reasonable number of temperatures, so that covering the high and low temperatures. The temperature dependence of a corresponding physical quantities, apart from being averaged over samples, it is also averaged over many lattice connectivities for each system size. This is a standard procedure in MC simulation in dealing with random systems. In the next section, we present and discuss our results.

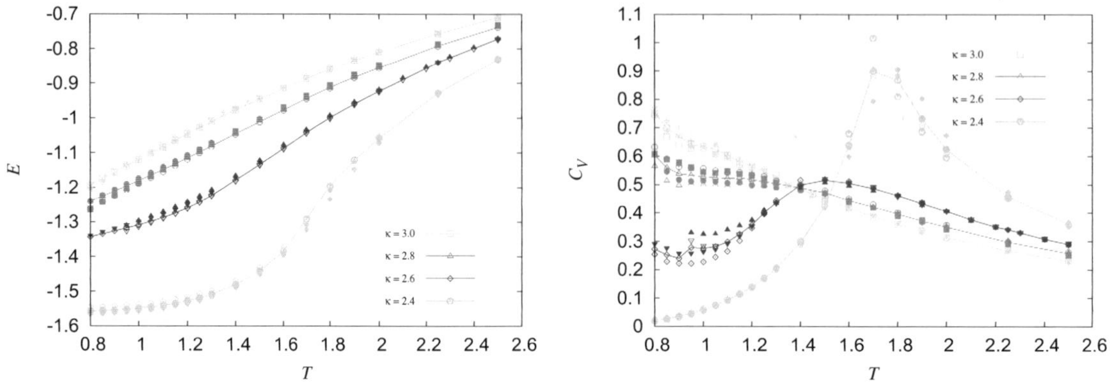

Figure 3. Temperature dependence of energy (left) and specific heat (right) of lattices with different values of κ. As shown, for $\kappa = 2.4$, the specific heat is diverging, signifying a presence of second order phase transition.

3. Results and Discussion

3.1. Time series of energy and specific heat

We have studied AF Ising model on rewired square lattices with various values of κ. Several linear sizes, $L = 16, 24, 32$ and 48, were simulated, with $N = L^2$ is the total number of sites (spins). Since the periodic boundary condition is imposed, each site of the native square lattice has the same number of nearest neighbors. For each system size we took many realizations of connectivities, then average the results over the number of realizations. Each realization corresponds to a particular connectivity distribution which is randomly generated. As a random system, results of a particular realization tend to be different from that of the other. Therefore, for a reliable result (better statistics), we took reasonable number of realizations. In previous studies on various randomly frustrated systems[16, 11, 12, 13], 1000 realizations were taken, here we took 500 realizations.

In MC simulation, time corresponds to a series of MCSs. We perform M MCSs for each temperature and take a number (N_s) samples out of M_s. To check whether the system is well equilibrated, we evaluated the energy time series of the system, from which the average energy and specific heat can be extracted. The presence of any existing phase transition could be signified in the temperature dependence of specific heat. To make sure the system is well equilibrated we perform enough initial MCSs, in the order of 10^4 MCSs, before doing measurement.

The energy time series (ETS) of two different temperatures were taken, i.e., T = 2.50 and T = 1.00, for linear size $L = 16$ of rewired square lattice with $\kappa = 3.0$. As shown in Fig. 2, both plots indicate that system has achieved equilibrium state after a number of preliminary MCSs. Fluctuation at smaller system is more pronounced compared to larger system size. We extracted two quantities from ETS, namely the ensemble average of energy, $\langle E \rangle = \frac{1}{N}\sum_N E_i$, and the specific heat defined as follows

$$C_v = \frac{N}{kT^2}\left(\langle E^2 \rangle - \langle E \rangle^2\right) \tag{7}$$

where N and k are respectively the number of spins and Boltzmann constant. These quantities are shown in Fig. 3(a) and Fig. 3(b). Each curve in the plot of average energy $\langle E \rangle$ corresponds to a system with certain κ.

The specific heat plot shown in Fig. 3 possesses a clear diverging peak for lattice with κ=2.4.

This is a strong indication of the existence of phase transition. Whether this transition is related to SG phase or not can be resolved from the data of order parameter, the EA parameter and the staggered magnetization M_s which is presented in the next sub-section. The reason of calculating M_s is due to the fact that that the original square lattice with AF coupling, before being rewired, does experience conventional magnetic phase transition with characterized by the finite value of staggered magnetization at low temperature.

3.2. Order Parameters

To probe the interplay between randomness and geometrical frustration in affecting the existing AF magnetic order phase, we calculated two corresponding order parameters, the staggered magnetization and the overlapping parameter, also known as Edward-Anderson (EA) order parameter[21]. The former is the quantity to characterize the low temperature AF order phase, known as Néel phase; while the latter is to search for the existence SG phase. At very low temperature, the AF system on pure square lattice will have perfect Néel phase which is characterized as follows

$$M_s = \sum_i (-1)^i M_i; \qquad M_i = \sum_{j\epsilon\{i\}} s_j \tag{8}$$

where M_i is the magnetization of the sub-lattice$i$-th, j is the spin index, depending on the sub-lattice. As here are two sub-lattices; j is running from 1 to N/2; where N is the number of sites.

The EA parameter is defined as the following

$$q = q_{EA} \equiv \left[\langle | \frac{1}{N} \sum_i s_i^{(\alpha)} s_i^{(\beta)} | \rangle \right]_{\text{av}}, \tag{9}$$

where s_i^α and s_i^β are two spin configurations with similar connectivity structure. This parameter is more like a trick in order to capture the condition of frozen configuration. Certainly, it is only relevant for numerical simulation and not relevant for experiment. It is not found in real material as almost impossible to obtain two duplicated systems with exactly the same realization of randomness. In experiment, the frozen state is measured directly by observing how the configuration change with time, in other words, one measures the time correlation of the spin configuration. The EA parameter for the Ising case is much simpler compared to the Heisenberg model which involves tensor product [11, 12]. This parameter is finite if the system is in the phase of frozen random spin configuration and vanishes otherwise.

The temperature dependence of q_{EA} for various system sizes, $L = 16, 24, 32$ and 48 is shown in Fig. 4(a). This parameter is to search for the existence of SG phase transition in randomly frustrated system.. We also calculated and the order parameter for AF system, the staggered magnetization M_s, shown in Fig. 4(b). The characteristics of the order parameter which increases with the decrease in temperature is the indication of phase transition. This trend is indicated by Fig. 4, where for lattice with $\kappa = 2.40$, both parameters, q_{EA} and M_s are increasing. However, it is to be noticed that the increasing q_{EA} can not be considered as an indication of presence of SG phase in the case M_s also increases. SG phase which is frozen random spin orientation can not co-exist with Néel phase. Therefore, the incerasing values of both parameters indicates the existence of the same ordered phase, namely the Néel phase. The plot of q_{EA} for larger value of κ, i.e, $\kappa > 2.40$, indicates that this parameter increases with the decrease in temperature. However, it tends to decrease as system sizes increase. This exhibits that there is no SG phase at thermodynamic limit. The role played by the randomess, induced by random connectivity, although results in frustrated state, it fail to bring low temperature SG phase.

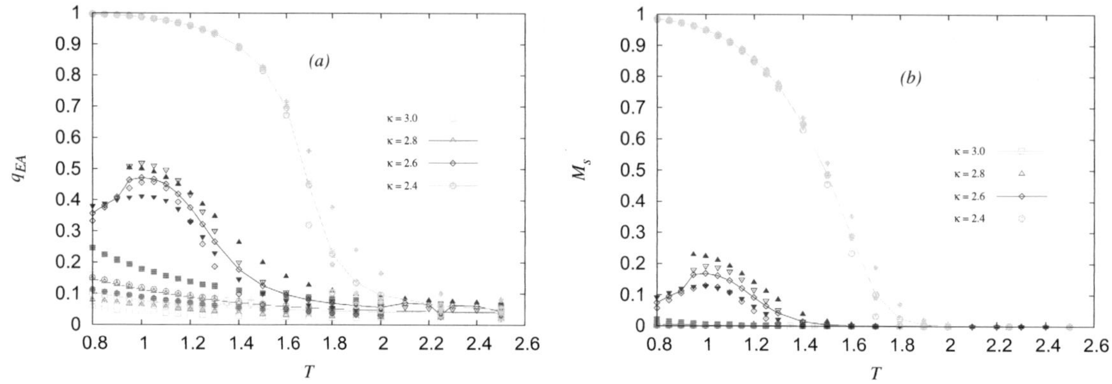

Figure 4. (a) Temperature dependence of Edward-Anderson parameter q_{EA} and (b) Staggered magnetization M_s of lattices with different values of κ. As shown, for $\kappa = 2.4$, both quantities significantly increasing as temperature decreasing. The finite value of M_s indicates a spatially ordered phase, signifies the absence of SG phase (frozen random orientation).

The absence of SG phase and the presence of AF phase at low temperature can be explained from the perspective of Harris criterion[7], which describes the role played by randomness on second order phase transition. This is exactly what observed here. The pure AF Ising model on square lattice does experience second order phase transition, with diverging specific heat. As the number of links increase, more precisely, the degree of randomness, the existing phase transition disappear at certain value of κ, i.e., around 2.7. The precise value of κ which destroy the AF order phase at low temperature will be the subject of our next study and will be published elsewhere. With this paper, we report that non-canonical SG phase transition is not found for Ising model in the rewired square lattices with $2 \leq \kappa \leq 3$.

4. Summary and Conclusion

This research studied the AF Ising model on rewired square lattices with various average number of links (ANOL), associated with coordination number for regular lattices. The lattices are obtained by randomly adding one extra link to each site of the original square lattice, lead the rewired latticess to have fractional ANOL, symbolized as κ, varied from 2 to 3. The added link is constrained to connect a site to one of its diagonal neighbors. Because of rewiring, there exist abundance of triangular units in the lattice, in which spins are frustrated due to AF couplings. Accordingly, the system under investigated is in the class of random system with geometrical frustration. We probed the role played by the randomness and frustration in affecting the existing second order phase transition of the native square lattice with AF interaction. We used Replica Exchange Monte Carlo method which is considered to be very powerful in dealing with randomly frustrated systems such as SGs. From the time series of energy, we extracted energy and calculated the specific heat. We also calculated the SG order parameter (Edward-Anderson parameter) as well as the staggered magnetization, which is the order parameter for AF ordering.

We observed a clear indication of the effect of randomness on the existing AF phase is clear and found no SG phase transition. The remnant of AF phase remains upto certain value of ANOL, i.e., $\kappa \sim 2.7$. The pure AF square lattice is associated with $\kappa = 2.0$ (no extra links added) while for $\kappa = 3.0$, each site has obtained one extra link. The current topic has opened another direction of research, i.e., to study the interplay between randomness rendered by rewiring procedure and the partially frustrated state. In addition, to find the lower dimension for the existence of Ising SG phase on rewired lattices is still desirable. It will be our next topic to probe, for example searching for Ising SG phase on rewired cubic lattices.

Acknowledgment

The author wishes to thank A. G. Williams, D. Tahir and M. Troyer for stimulating discussions. The computation of this work was carried out using parallel computing facility in the Physics Department, Hasanuddin University and the HPC facility of The Indonesian Institute of Science. The work is financially supported by PUPT Research Grant, FY 2017 from The Indonesian Ministry of Research, Technology and Higher Education.

References

[1] E. Ising, Z. Phys., 31, 253 (1925).
[2] S. Sachdev, *Quantum Phase Transition*, 2nd. ed., Cambridge Univ. Press, (2011).
[3] P. Ehrenfest and T. Ehrenfest, Phys. Z. **8**, 311 (1907).
[4] H. Nishimori and G. Ortiz, *Elements of Phase Transition and Critical Phenomena*, Oxford Univ. Press, (2011).
[5] L. E. Reichl, *A Modern Course in Statistical Physics*, 2nd. John Wiley and Sons, (2009).
[6] T. Surungan, Y. Okabe, Phys. Rev B71, 184438, (2005).
[7] A. B. Harris, J. Phys. C 7, 1671, (1974).
[8] C. N. Yang, Phys. Rev. 85, 808, (1952).
[9] H. T. Diep (Ed.), *Frustrated Spin Systems*, 2nd. Ed., World Scientific, (2013).
[10] V. Cannella and J. A. Mydosh, Phys. Rev. B6, 4220, (1972).
[11] T. Surungan, F. P. Zen, and A.G. Williams, JPCS 640, 012005 (2015).
[12] T. Surungan, Bansawang BJ, D. Tahir, AIP Conf. Proc. 1719, 030006 (2016).
[13] T. Surungan, JPCS 759, 012012, (2016).
[14] T. Surungan, (*unpublished*).
[15] A.-L. Barabási and R. Albert, Science **286**, 509, (1999).
[16] M. Bartolozzi, T. Surungan, D.B. Leinweber and A.G. Williams, Phys. Rev. B73, (2006) 224419.
[17] C. P. Herrero, Phys. Rev. E **77**, (2008) 041102.
[18] H. Nishimori, Prog. Theor. Phys. 66, 1169 (1981).
[19] K. Hukushima and K. Nemoto, J. Phys. Soc. Japan 65, 1863, (1996).
[20] T. Surungan, Y. Okabe, and Y. Tomita, J. Phys. A37, 4219, (2004).
[21] S. F. Edwards and P. W. Anderson, J. Phys. F: Metal Phys. **5**, 965, (1975).

IOP Conf. Series: Journal of Physics: Conf. Series **1011** (2018) 012080 doi:10.1088/1742-6596/1011/1/012080

Electron hopping integral renormalization due to anharmonic phonons

Muhammad Aziz Majidi[†]

Department of Physics, Faculty of Mathematics and Natural Sciences, Universitas Indonesia, Kampus UI Depok, Depok 16424, Indonesia

[†]Corresponding author's e-mail: aziz.majidi@sci.ui.ac.id

Abstract. Interpretation of some experimental data of temperature-dependent transport properties of a solid material has led to a hypothesis that some renormalization has occurred to the electron hopping integral, making it decrease as temperature increases up to some limit. Such a renormalization may be attributed to the effect of coupling of electrons with anharmonic phonons. Namely, the presence of anharmonic vibration of two adjacent atoms yields an increase of their equilibrium distance as temperature increases, this in turn weakens the electron hopping integral connecting the two atomic sites. Despite that the above qualitative picture is pretty clear, little has been explored on the detail of how such an electron hopping integral evolves as a function of temperature. Motivated by this issue, here, we propose a tight-binding model incorporating the presence of anharmonic phonons and electron-phonon coupling, and solve it using Green function technique. We define the renormalized electron hopping integral from the band width of the interacting system, and perform self-consistent calculations to demonstrate the evolution of the electron hopping integral as a function of equilibrium distance between adjacent atoms and temperature.

1. Introduction

Previous experimental and theoretical studies [1, 2] have led us interpret that electron hopping integrals in the materials may evolve with temperature. This has motivated us to do a theoretical study to investigate how this phenomenon actually occurs. For this purpose, below we provide some theoretical background of what electron hopping integral is, how it is affected by temperature, and how we develop the algorithm to compute it.

In studying physical properties of a material theoretically, it has been common, whenever still adequate, that one approximates the relevant part of the band structure of the system through a tight-binding model [3]. This is usually aimed to avoid complexities in the calculations resulting from the presumably-irrelevant details of the system in connection to specific material properties being studied. A tight-binding model relies on an assumption that the electronic degrees of freedom in the system can be effectively reduced by considering only a few selected atomic orbitals such that the dynamics of the electrons occurs as a result of the movement, or popularly termed as hopping, of electrons from one orbital to another in the same atom or between nearby atoms. The parameters for the tight-binding model may simply be guessed based on some intuitive reasons, or may be derived by mapping the results of an *ab intio* calculation [4]. In the second-quantized formalism a tight-binding Hamiltonian can be expressed as

$$\mathcal{H} = \sum_{i,j} \sum_{\alpha,\beta} \sum_{\sigma} H_{i,j}^{\alpha,\beta} c_{i,\sigma}^{\alpha\dagger} c_{j,\sigma}^{\beta}, \qquad (1)$$

where i and j denote the atomic sites, while α and β are indices representing the atomic orbitals in the basis set, σ is the spin index, and $c^{\alpha\dagger}_{i,\sigma}$ and $c^{\beta}_{j,\sigma}$ are the corresponding creation and annihilation operators, respectively. Here, $H^{\alpha,\beta}_{i,j,\sigma}$ are the tight-binding parameters formally derived from

$$H^{\alpha,\beta}_{i,j} = \int dr\ \phi^*_\alpha(\mathbf{r}-\mathbf{R}_i)\left\{-\frac{\hbar}{2m}\nabla^2 + \sum_n V(|\mathbf{r}-\mathbf{R}_n|)\right\}\phi^*_\beta(\mathbf{r}-\mathbf{R}_j). \tag{2}$$

The simplest version of a tight-binding model assumes that the unit cell of the system contains only one atom and there is only one orbital per atom taken in the basis set. Further, if the hopping of electrons is assumed to be only between nearest-neighbor atoms, the Hamiltonian form reduces to

$$\mathcal{H} = \epsilon_0 \sum_{i,\sigma} c^\dagger_{i,\sigma}c_{j,\sigma} - t\sum_{<i,j>}\sum_{\sigma} c^\dagger_{i,\sigma}c_{j,\sigma}. \tag{3}$$

In Eq. (3) $\epsilon_0 \equiv H_{i,i,\sigma}$ is referred to as the on-site energy which is assumed equal for all sites; while $t \equiv -H_{i,j,\sigma}$ with i and j being indices of nearest-neighbor sites, is referred to as the nearest-neighbor hopping integral; and $\sum_{<i,j>}$ denones the summation over only nearest-neighbor pairs of i, j sites. This Hamiltonian is easily solved by performing Fourier transformation to **k**-space, leading to

$$\mathcal{H} = \sum_{\mathbf{k},\sigma} \epsilon(\mathbf{k}) c^\dagger_{\mathbf{k},\sigma}c_{\mathbf{k},\sigma}, \tag{4}$$

where $\epsilon(\mathbf{k}) = \epsilon_0 - t\sum_n e^{-i\mathbf{k}.\delta_n}$ is the energy dispersion of the model, with δ_n being a translation vector connecting a site with its nth nearest neighbor. If we assume the system to be a simple cubic lattice with lattice constant a, and set $\epsilon_0 = 0$, then we have

$$\epsilon(\mathbf{k}) = -2t\left(\cos k_x a + \cos k_y a + \cos k_z a\right). \tag{5}$$

The energy dispersion of Eq. (20) spans from $-6t$ to $6t$, with the band width being $W = 12t$. From this simple relation, we get the idea that the band width of some energy dispersion is, to some extent, proportional to the value of the hopping integral of the system. Or, conversely, we may say that the hopping integral t is proportional to the band width W of the energy dispersion of the system.

Our focus of study is the hopping integral t, which according to Eq.(2) may change if the equilibrium distance between the nearest neighbor atoms changes. Here, we are interested in exploring how t changes if we vary the temperature of the material. The most plausible way to manifest the effect of temperature on t is through atomic vibrations, namely phonons. From elementary solid-state physics we know that equilibrium distance between atoms does not change with purely harmonic atomic vibrations, instead it requires some anharmonicity in addition to the harmonic part [5]. Hence, we hypothesize that the presence of anharmonic phonons would also affect the effective value of electron hopping integral t. In this study, we aim to confirm through explicit calculations how t changes with temperature as a result of the presence of anharmonic phonons.

2. Model

Generically an electron-phonon system can be described by the following Hamiltonian terms

$$\mathcal{H} = \mathcal{H}_{\text{el}} + \mathcal{H}_{\text{ph}} + \mathcal{H}_{\text{el-ph}}, \tag{6}$$

where $\mathcal{H}_{\text{el}}$ is the purely electronic part of the Hamiltonian that may be constructed from a tight-binding model, e.g. Eq. (3). Here, $\mathcal{H}_{\text{ph}}$ is the purely phononic part that we define in the simplest way to include anharmonic effects such that

$$\mathcal{H}_{\text{ph}} = \sum_i \frac{P_i^2}{2M} + \frac{1}{2}M\Omega^2 x_i^2 - \lambda_3 x_i^3 - \lambda_4 x_i^4, \tag{7}$$

where P_i is the momentum operator of the ith atom; M is the mass of each atom which is assumed the same for all atoms; Ω is the frequency of lattice vibration, which for simplicity is assumed to be non-dispersive, i.e. only a single value like in the Einstein model. x_i is the displacement of the ith atom from its equilibrium position along any direction in three-dimensional space; and λ_3 and λ_4 are positive coefficients measuring the strength of the third- and fourth-order anharmonic terms, respectively. Lastly, $\mathcal{H}_{\text{el-ph}}$ represents the electron-phonon coupling that we construct within Holstein model [6, 7] as

$$\mathcal{H}_{\text{el-ph}} = g \sum_i c_{i,\sigma}^\dagger c_{i,\sigma} x_i, \tag{8}$$

with g being the electron-phonon coupling strength.

3. Method

Our technical goal is to compute the equilibrium change of interatomic distance, $\langle x \rangle$, and the renormalized electron hopping integral, $\tilde{t}$, as functions of temperature. In the subsequent discussion below, we present how we fomulate the method and algorithm to eventually compute $\langle x \rangle$ and $\tilde{t}$.

We start by constructing the electron Green function in $\mathbf{k}$ space, $G(\mathbf{k}, z)$, through the Dyson equation

$$G_\sigma(\mathbf{k}, z) = \frac{1}{z - \epsilon(\mathbf{k}) - \Sigma_\sigma(\mathbf{k}, z)}, \tag{9}$$

where σ is the spin index and z is complex frequency variable. Note that in our formulation, frequency and energy are set in the same unit, i.e. eV. We shall work in both Matsubara and real-frequency domains, where in the Matsubara formalism, z is to be substitued with $z = i\omega_n + \mu$, with $\omega_n = (2n+1)\pi k_B T$ being the fermionic Matsubara frequency, k_B the Boltzmann constant, T temperature, μ the chemical potential of the system, and n integer numbers ranging from $-\infty$ to ∞; while in the real-frequency domain, $z = \omega + i0^+$. Here, $\Sigma_\sigma(\mathbf{k}, z)$ is the self energy formally manifesting all interactions occuring in the system experienced by the electrons. In the computation process, we initialize the self energy value with an initial guess, then we recompute it later on through some approximate procedure until the guessed and the calculated values come into convergence. We detail the self-consistency algorithm below.

To tackle this problem, we choose to treat the phonon field semi-classically, and implement the dynamical mean-field theory (DMFT) algorithm [8] suitable for this choice [9, 10]. Within DMFT spatial correlations are suppressed yielding the self-energy become $\mathbf{k}$ independent, namely $\Sigma_\sigma(\mathbf{k}, z) \to \Sigma_\sigma(z)$. Further, since we do not consider any spin-dependent interactions, we can keep $\Sigma_\uparrow(z) = \Sigma_\downarrow(z)$, such that $G_\uparrow(\mathbf{k}, z) = G_\downarrow(\mathbf{k}, z)$ as well, accordingly. Upon this choice of treatment, the next step is to coarse-grain the Green function over the Brillouin zone to make it $\mathbf{k}$ independent,

$$\overline{G_\sigma(z)} = \frac{1}{N} \sum_{\mathbf{k}} \frac{1}{z - \epsilon(\mathbf{k}) - \Sigma_\sigma(z)}, \tag{10}$$

and then extact the mean-field Green function through

$$G_\sigma(z)_{\text{MF}} = \left(\overline{G_\sigma(z)}^{-1} + \Sigma_\sigma(z) \right)^{-1}. \tag{11}$$

IOP Conf. Series: Journal of Physics: Conf. Series **1011** (2018) 012080 doi:10.1088/1742-6596/1011/1/012080

From here, we compute the local interacting Green function by adding the electron-phonon interaction to the mean-field Green function as

$$G_\sigma(z,x)_{\text{loc}} = \left(G_\sigma(z)_{\text{MF}}^{-1} - gx\right)^{-1}. \tag{12}$$

$G_\sigma(z,x)_{\text{loc}}$ has a physical meaning as the interacting Green function defined at a particular site of atom undergoing a displacement of x from its equilibrium position. Since atoms at other sites are undergoing displacements of different values of x, then we must average $G_\sigma(z,x)_{\text{loc}}$ over x to obtain the x-independent interacting Green function representing the whole system. To do this, we first need to define the statistical weighting factor $\mathcal{P}(x,T)$ that depends on temperature T and the x value. That is,

$$\mathcal{P}(x,T) = e^{-S_{\text{eff}}(x,T)}, \tag{13}$$

where $S_{\text{eff}}(x,T)$ is the effective action defined for this particular model as

$$S_{\text{eff}}(x,T) = -\sum_n\sum_\sigma \ln\det\left\{G_\sigma(z)_{\text{MF}}G_\sigma(i\omega_n+\mu,x)_{\text{loc}}^{-1}\right\}e^{i\omega_n 0^+} + \frac{\frac{1}{2}M\Omega^2x^2 - \lambda_3x^3 - \lambda_4x^4}{k_BT}. \tag{14}$$

Once we have obtained $\mathcal{P}(x,T)$ we can do the averaging process over x variable for all the relevant quantities, such as the Green function itself

$$G_\sigma(z)_{\text{ave}} = \int_{-\infty}^{\infty} dx\ \mathcal{P}(x,T)G_\sigma(z,x)_{\text{loc}}, \tag{15}$$

and the equilibrium change of interatomic distance

$$x_{\text{ave}} = \int_{-\infty}^{\infty} dx\ \mathcal{P}(x,T)x. \tag{16}$$

Along the self-consistent process we need to update the chemical potential μ. We may do so by first obtaining the density of states (DOS), that is

$$\text{DOS}(\omega) = -\frac{1}{\pi}\ \text{Im}\ \sum_\sigma G_\sigma(\omega+i0^+)_{\text{ave}}, \tag{17}$$

and then apply the constraint of the fixed electron filling $\langle n\rangle$ such that

$$\int d\omega\ \text{DOS}(\omega)f(\omega,T,\mu) = \langle n\rangle, \tag{18}$$

where $f(\omega,T,\mu)$ is the Fermi-Dirac distribution function.

The most essential part of the self-consistency process is to recompute the self energy. We do this by applying the Dyson equation again, that is

$$\Sigma_\sigma(z) = G_\sigma(z)_{\text{MF}}^{-1} - G_\sigma(z)_{\text{ave}}^{-1}. \tag{19}$$

The computed self energy will then be compared with the initial guessed value. If the difference is bigger than some tolerance value, the new guessed value may be defined by taking the computed result mixed with previous guessed value. The self-consistency process repeats with a new guessed self energy and the updated chemical potential until the convergence condition is achieved.

As we aim to compute the renormalized electron hopping integral $\tilde{t}$, we may use the computed DOS profile to extract the information of the interacting band width $\tilde{W}$. But note that this procedure is good only as long as the profile of the interacting energy dispersion, assuming simple cubic lattice, is roughly still of the form

$$\tilde{\epsilon}(\mathbf{k}) \approx -2\tilde{t}\left(\cos k_xa + \cos k_ya + \cos k_za\right). \tag{20}$$

With this condition, we can then take $\tilde{t} \approx \tilde{W}/12$.

4. Discussion

Now that we have set up the algorithm we need to explore the model by fixing or varying the necessary parameters. For making the calculation easier, we choose the system to be half filled by taking one electron per site, i.e. $\langle n \rangle = 1$. We may choose the lattice constant value a to be of a typical value of a few angstroms. The the value of product of $M\Omega^2$ may be set somewhat arbitrarily, but it is desirable to make the $(1/2)M\Omega^2 x^2$ value falls in the range comparable to the band width W. The g parammeter in the electron-phonon coupling is desired to set such that it delivers the effect of the presence of phonons to the electron DOS, but not strong enough to form a pseudo gap or gap in the DOS when the anharmonic terms are switched-off. Next, we need to turn on and carefully vary the anharmonic parameters λ_3 and λ_4, such that we can observe the evolution of the DOS profile due to the anharmonic effects. Once we have obtained the best combination of parameters that can give us meaningful changes in the DOS as compared to states with no phonons or no anharmonic effects, we may fix that combination of parameters and then vary the temperature to see how $\tilde{t}$ and x_{ave} evolve as functions of temperature.

5. Summary

We have introduced the real physical problems from some previous experimental studies, giving us motivation to theoretically study the temperature dependence of the electron hopping integral in the system. We hypothesize that the change in the electron hopping integral is due to anharmonic effects. We formulate our Hamiltonian of the electron-phonon system within Holstein model with additional anharmonic terms. We then develop the dynamical mean-field theory algorithm to compute the renormalized electron hopping integral $\tilde{t}$ and the equilibrium change in the interatomic distance x_{ave} as functions of temperature. Finally, we discuss how to explore our model with the so-constructed algorthm. The computational step of this project is being in progress.

6. Acknowledgement

We gratefully acknowledge the funding support from the Directorate of Research and Community Services of Universitas Indonesia to this project through PITTA Research Grant No. 666/UN2.R3.1/HKP.05.00/2017.

References

[1] Xinmao Y *et al* 2015 *NPG Asia Materials* **7** e196.
[2] Fauzi AD *et al* 2017 *J. Phys.: Condens. Matter* **29** 135802 (8pp).
[3] Slater JC, Koster GF 1954 *Physical Review* **94** (6) 14981524.
[4] Marzari N and Vanderbilt D 1997, *Phys. Rev.* B **56**, 12847.
[5] Kittel C 2005 *Introduction to Solid State Physics - 8th Ed.* John-Wiley and Sons, Inc.
[6] Holstein T 1959 *Ann. Phys.* **8** 325.
[7] Alexandrov AS and Mott NF 1995, *Polarons and Bipolarons*, World Scientific, Singapore.
[8] Georges A *et al* 1996 *Rev. Mod. Phys.* **68** 13.
[9] Furukawa N 1994 *J. Phys. Soc. Jpn.* **63** 3214.
[10] Millis AJ, Mueller R, and Shraiman BI 1996, *Phys. Rev.* B **54**, 5389.

IOP Conf. Series: Journal of Physics: Conf. Series **1011** (2018) 012081 doi:10.1088/1742-6596/1011/1/012081

Stability enhancement of high Prandtl number chaotic convection in an anisotropic porous layer with feedback control

M. N. Mahmud

Universiti Kuala Lumpur Malaysian Institute of Chemical & Bioengineering Technology, 78000 Alor Gajah, Melaka. Malaysia.

E-mail: nasir@unikl.edu.my

Abstract. The chaotic dynamical behaviour of thermal convection in an anisotropic porous layer subject to gravity, heated from below and cooled from above, is studied based on theory of dynamical system in the presence of feedback control. The extended Darcy model, which includes the time derivative has been employed in the momentum equation to derive a low dimensional Lorenz-like equation by using Galerkin-truncated approximation. The classical fourth-order Runge-Kutta method is used to obtain the numerical solution in order to exemplify the dynamics of the nonlinear autonomous system. The results show that stability enhancement of chaotic convection is feasible via feedback control.

1. Introduction

Chaotic convection of fluid in a saturated porous medium has been studied intensively for its significance in industrial applications. Most of the systems in engineering industry exhibit some kind of chaotic behaviour such as stabilization in semiconductor lasers, turbulent flows of fluids in chemical reactors and mechanical vibrations on aircraft structure. The effect of anisotropy in permeability and thermal conductivity on convective flow in a rectangular porous layer is studied and resulted in Nusselt number decreases continuously as permeability increases [1]. Next, the natural convection with opposing horizontal heat and solute gradients in an anisotropic porous cavity has been explored using the Darcy model [2]. The linear and nonlinear thermal instability in an anisotropic saturated porous layer with internal heat source is explored and found that Rayleigh number increases as the anisotropic parameters increase thus stabilize the system [3]. Moreover, it was reported that the mechanical anisotropy parameter actually advanced the convection which contradict to the thermal anisotropy parameter as it delayed the convection [4]. Later, the proportional relationships between scaled Rayleigh number and scaled mechanical and thermal anisotropic parameters have been presented [5].

Several researchers have realized that chaos can literally be advantageous in many situations and as it inevitably presents, it can frequently be managed to acquire desired results. The first model-free chaos control, known as OGY (Ott-Grebogi-Yorke) method, forces some trajectory to reach a particular unstable periodic orbit via the linearization of a Poincaré map [6]. Subsequently, a time-delayed feedback control has been introduced and it could be applied in order to stabilize the chaotic system [7],[8]. Also, it was found that the Lorenz chaos could be

IOP Conf. Series: Journal of Physics: Conf. Series **1011** (2018) 012081 doi:10.1088/1742-6596/1011/1/012081

stabilized through the use of a controller with input saturation, even when the system external distraction exists [9]. Later, a robust controller has been proposed to stabilize dynamical system of unstable periodic orbits and chaos control can be performed even in conditions where high uncertainties are implicated [10]. Besides, it was discovered that amount of controller gain is proportional to scaled Rayleigh number after applying the feedback control strategy [11]–[13]. Recently, the effects of feedback control on chaotic and periodic dynamics for small and moderate Prandtl number has been studied in a porous layer subject to vibration [14].

In the present study, the work of [5] on the chaotic nature of anisotropic parameters is developed further to include consideration of the influence of feedback control. Truncated Galerkin approximation was applied to the governing equations for thermal convection in an anisotropic porous cavity subject to gravity and heated from below for high Prandtl number in order to arrive at a low dimensional model.

2. Mathematical structure of the problem

We consider an anisotropic square cavity in horizontal porous-saturated layer of depth H_* and width L_* with stress free boundaries, which is heated from below and cooled from above as presented in Fig. 1. A Cartesian co-ordinate system is used such that the vertical axis z is collinear with gravity, i.e. $\hat{e_g} = \hat{e_z}$. The actuators are placed at the bottom heated surface and the sensors placed at the top plate of the fluid-saturated porous layer are used to detect the departure of the surface temperature from its conductive state as inspired from the proportional feedback control [15]. The determination of a control, $\widehat{q}(t)$, can be performed using the proportional-integral-differential (PID) controller [16].

$$\widehat{q}(t) = r + K[e(t)], \qquad e(t) = \widehat{m}(t) - m(t), \tag{1}$$

where r is the calibration of the control, $e(t)$ an error or deviation from the state measurement, $\widehat{m}(t)$ from some desired or reference value, $m(t)$, $K = C_p + C_D \dot{/} + C_I \int_0^t$ with C_p as the proportional gain, C_D differential gain, and C_I integral gain. Based on (1), for one sensor plane and proportional feedback control, the actuator rectifies the heated surface temperature using a proportional relationship between the upper ($z = z_u$) and lower ($z = z_l$) thermal boundaries for the perturbation field, thus:

$$\theta(x, z_l, t) - \theta(x, z_l) = 0.5 + K[\theta(x, z_u, t) - \theta(x, z_u)], \tag{2}$$

or, equivalently,

$$\theta'(x, z_l, t) = 0.5 + K\theta'(x, z_u, t), \tag{3}$$

where θ' denotes the deviation of the fluid's temperature from its conductive value and K is the scalar gain controller.

The extended-Darcy equations that comprise the time derivative term is engaged for the momentum equation. The governing equations are given by

$$\nabla \cdot \vec{q} = 0, \tag{4}$$

$$\frac{1}{\varepsilon}\frac{\partial \vec{q}}{\partial t} = -\frac{1}{\rho_0}\nabla p + \frac{\rho}{\rho_0}\vec{g} - \frac{\nu}{P_p}\vec{q}, \tag{5}$$

$$\frac{\partial T}{\partial t} + (\vec{q} \cdot \nabla) T = \nabla \cdot (\kappa \nabla \cdot T) \tag{6}$$

By applying the Boussinesq approximation to the equation of state, we obtain

$$\rho = \rho_0 \left[1 - \alpha_T (T - T_0)\right] \tag{7}$$

where $\vec{q}$ is the velocity of fluid in porous medium, ε the porosity, P_p the permeability of the porous domain, p the fluid pressure, κ the thermal diffusivity, T is the temperature, $\vec{g}$ the gravitational acceleration, ν the kinematic viscosity. The height of the layer H_* was used for scaling the variables x_*, y_*, z_* and H_*^2/κ for scaling the time t_*. Accordingly, $x = x_*/H_*$, $y = y_*/H_*$, $z = z_*/H_*$ and $t = t_*\kappa/H_*^2$. The time derivative term was included in Darcy's equation (5).

The governing equations can be presented in terms of a stream function defined by $u = \partial\psi/\partial z$ and $w = -\partial\psi/\partial x$, as for convective rolls having axes parallel to the shorter dimension (i.e. y) when $v = 0$. Applying the curl $(\nabla\times)$ operator on Eq. (5) yields the following PDEs:

$$\left[\frac{1}{Va}\frac{\partial}{\partial t}\left(\frac{\partial^2}{\partial x^2}+\frac{\partial^2}{\partial z^2}\right)+\left(\frac{\partial^2}{\partial x^2}+\frac{1}{\xi}\frac{\partial^2}{\partial z^2}\right)\right]\psi + Ra_T\frac{\partial T}{\partial x} = 0, \tag{8}$$

$$\left(\frac{\partial}{\partial t}-\eta\frac{\partial^2}{\partial x^2}-\frac{\partial^2}{\partial z^2}\right)T = -\frac{\partial\psi}{\partial x}+\frac{\partial(\psi,T)}{\partial(x,z)}, \tag{9}$$

The set of partial differential Eqs. (8) and (9) form a nonlinear coupled system with the boundary conditions: $T(1) = -0.5, T(0) = 0.5 + KT(1)$ and $\partial T/\partial x = 0$ at $x = 0$ and $x = Ar$ will accept a basic motionless conduction solution and has been solved by using Galerkin method.

3. Method of solution

In order to obtain the solution of the nonlinear coupled system of partial differential Eqs. (8) and (9), we represent the stream function and temperature in the form

$$\psi = A_{11}\sin\left(\frac{\pi x}{Ar}\right)\sin(\pi z), \tag{10}$$

$$T = \frac{1}{2}\left[1-K-(2-K)z\right] + B_{11}\cos\left(\frac{\pi x}{Ar}\right)\sin(\pi z) + B_{02}\sin(2\pi z). \tag{11}$$

Substituting Eqs. (10) and (11) into Eqs. (8) and (9), multiplying the equations by the orthogonal eigenfunctions corresponding to Eqs. (10) and (11) and integrating them yields a set of ODEs for the time evolution of the amplitudes:

$$\frac{\partial A_{11}}{\partial\tau} = -\frac{Va\gamma}{\pi^2}\left[A_{11}+\frac{Ra}{\pi\theta}B_{11}\right], \tag{12}$$

$$\frac{\partial B_{11}}{\partial\tau} = -B_{11}-\frac{(2-K)}{2\pi\theta}A_{11}-\frac{1}{\theta}A_{11}B_{02}, \tag{13}$$

$$\frac{\partial B_{02}}{\partial\tau} = \frac{1}{2\theta}A_{11}B_{11}-4\gamma B_{02}, \tag{14}$$

Rescaling the amplitudes with respect to their convective fixed points provided the following set of scaled equations which are equivalent to Eqs. (12)–(14):

$$\dot{X} = \alpha\zeta\,(Y-X), \tag{15}$$

$$\dot{Y} = \frac{1}{\zeta}\left[\frac{1}{2}(2-K)RX-\zeta\chi Y-\left(\frac{1}{2}(2-K)R-\zeta\chi\right)XZ\right], \tag{16}$$

$$\dot{Z} = 4\gamma(XY-Z), \tag{17}$$

where the dots $(\cdot)$ denote the time derivative $\mathrm{d}()/\mathrm{d}\tau$. Eqs. (15)–(17) are equivalent to the Lorenz equations (see [17] and [18]), although with different coefficients. Setting $K = 0$, we recover the uncontrolled Lorenz-like system [5].

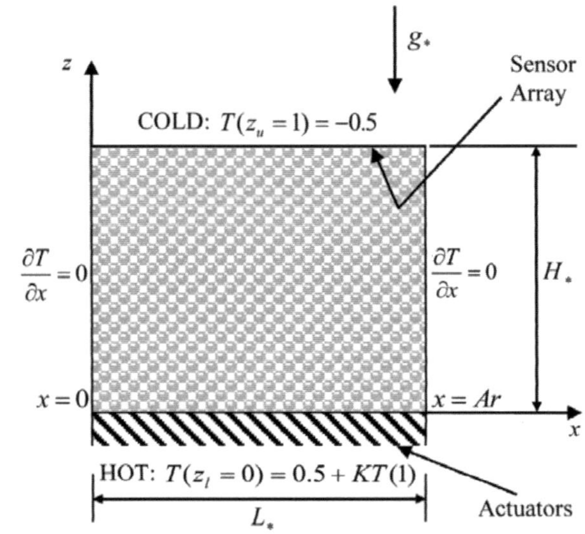

Figure 1. Physical configuration of an anisotropic porous layer heated from below. K is the proportional controller's gain.

4. Results and discussion

Maple's built-in 4th-order Runge-Kutta method on double precision with stepsize 0.001 has been used for the numerical simulations of the system (15)–(17) for the time domain $0 < t < 210$, unless otherwise specified. In all solutions, we considered the initial conditions $X(0) = Y(0) = 0.8$ and $Z(0) = 0.9$ and fixed the parameters $\gamma = 0.5$, $\alpha = 500$ ($Va = 9870$) and taking $R = 750, 1250$. For $R = 750$, a low mechanical and thermal anisotropic parameters have been considered, i.e. $\zeta = 0.8$ and $\chi = 0.9$, correspondingly. As the Rayleigh number increased i.e. $R = 1250$, the mechanical and thermal anisotropic parameters are also increased to $\zeta = \chi = 1.5$.

The phase portrait diagrams depict how feedback control affect the dynamics of the thermal convection for different kinds of scaled Rayleigh number and anisotropy parameters. The sequence of occurrences leading to the stabilization of the homoclinic convective regime is presented in Figs. 2–5 in terms of projections of trajectories corresponding to $R = 1250$. Fig. 2 shows a homoclinic chaotic regime corresponding to the uncontrolled case, $K = 0$. It is delayed and a solitary limit cycle (period-1) occured as depicted in Fig. 3 at $K = 0.2125$. As soon as the value of K is slightly higher, $K = 0.2147$, we observe that a concave periodic explosion takes over as illustrated in Fig. 4. We rendered our computations up to $K = 0.25$ and found that a steady bifurcation converging toward the fixed point occurs as shown in Fig. 5.

Periodic results are presented in Figs. 6(a)–(h) subject to time-domain $200 < t < 400$ as the outcome of the solution obtained at different proportional controller gain values for $R = 750$. A period-2 periodic solution appears as observed in Figs. 6(a)–(d) corresponding to $K = 0$. Increasing the proportional controller gain to $K = 0.3$ yields the results presented in Figs. 6(e)–(h) identifying a period-1 periodic solution. This period-halving bifurcation sequence leading to the stabilization of the periodic bifurcations as presented in Figs. 6(i)–(l) when $K = 1.155$.

IOP Conf. Series: Journal of Physics: Conf. Series **1011** (2018) 012081 doi:10.1088/1742-6596/1011/1/012081

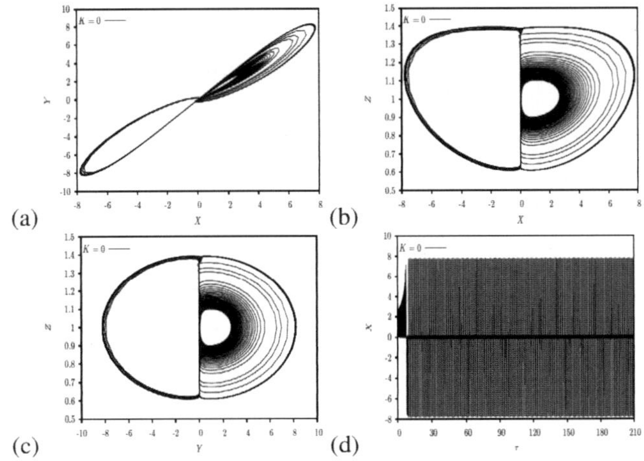

Figure 2. Phase trajectories and time domain diagram show a homoclinic bifurcation at $K = 0$ with parameters $R = 1250$, $\alpha = 500, \gamma = 0.5, \zeta = \chi = 1.5$, with $0 < t < 210$.

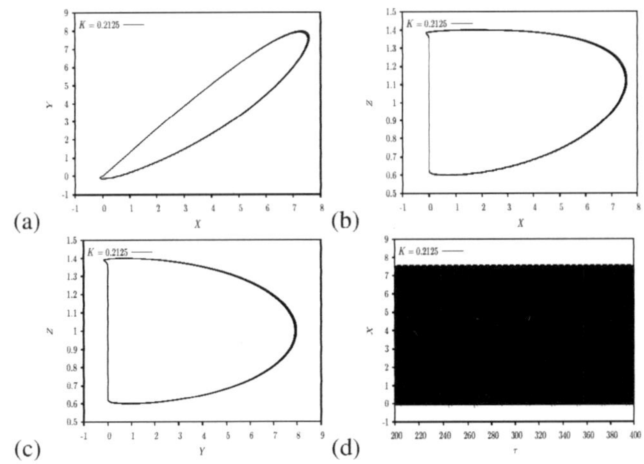

Figure 3. Phase trajectories and time domain diagram show a limit cycle at $K = 0.2125$ with parameters $R = 1250$, $\alpha = 500, \gamma = 0.5, \zeta = \chi = 1.5$, with $200 < t < 400$.

IOP Conf. Series: Journal of Physics: Conf. Series **1011** (2018) 012081 doi:10.1088/1742-6596/1011/1/012081

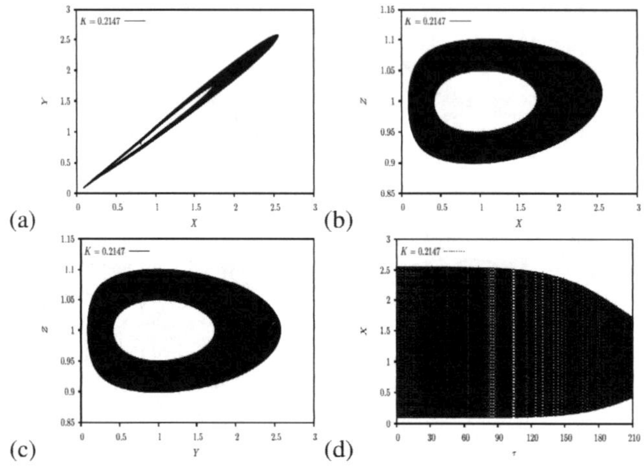

Figure 4. Phase trajectories and time domain diagram show a periodic bifurcation at $K = 0.2147$ with parameters $R = 1250$, $\alpha = 500, \gamma = 0.5, \zeta = \chi = 1.5$, with $0 < t < 210$.

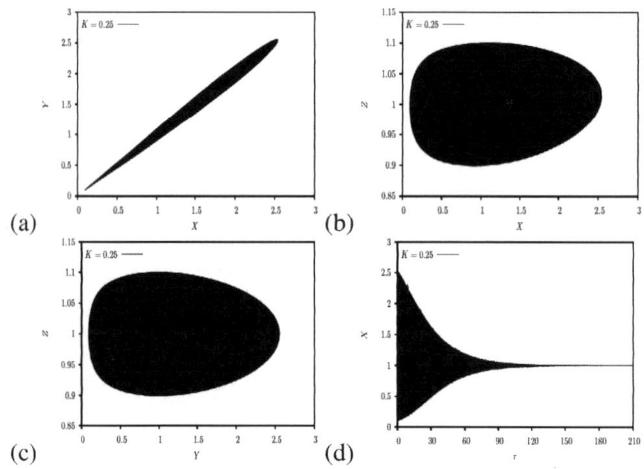

Figure 5. Phase trajectories and time domain diagram show a steady bifurcation at $K = 0.25$ with parameters $R = 1250$, $\alpha = 500, \gamma = 0.5, \zeta = \chi = 1.5$, with $0 < t < 210$.

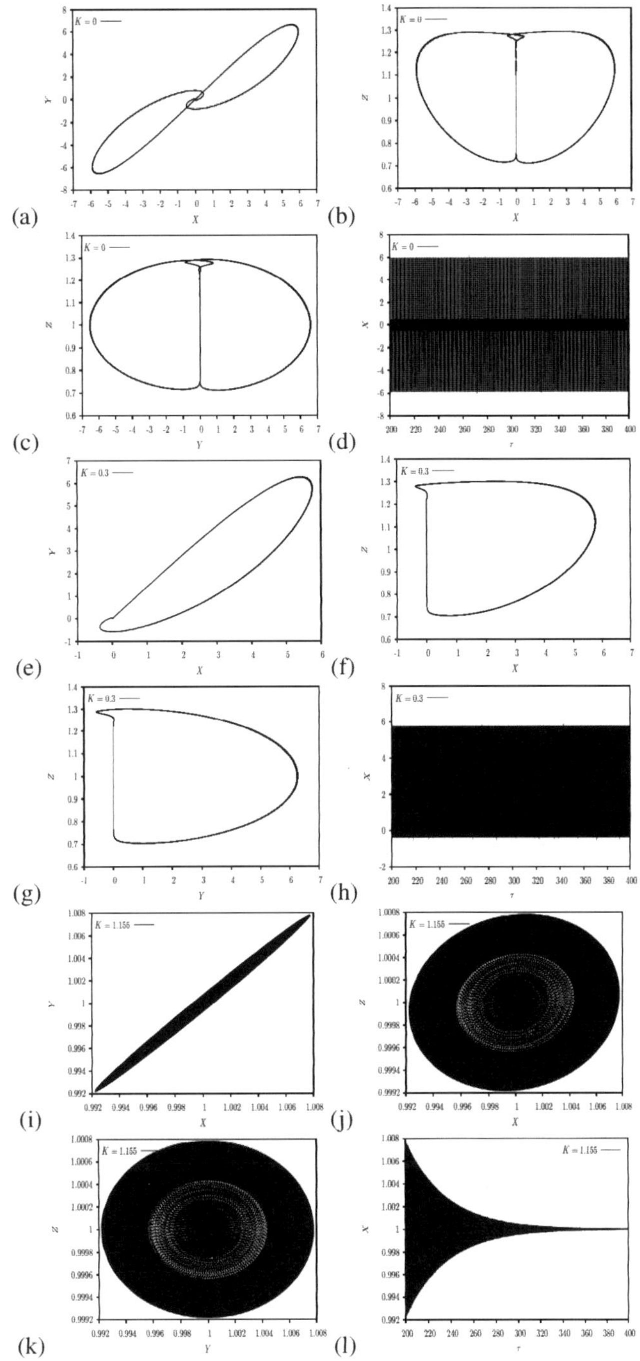

Figure 6. Projections of phase portraits and time domain diagram for parameters $R = 750$, $\alpha = 500, \gamma = 0.5, \zeta = 0.8, \chi = 0.9$, with $200 < t < 400$. Phase trajectories show Figs. 6(a)–(d) a period-2 at $K = 0$; Figs. 6(e)–(h) a period-1 at $K = 0.3$; Figs. 6(i)–(l) a steady solution at $K = 1.155$.

IOP Conf. Series: Journal of Physics: Conf. Series **1011** (2018) 012081 doi:10.1088/1742-6596/1011/1/012081

5. Conclusions

In this work, by using phase portrait and time domain diagrams the following findings are achieved.

(i) The proportional controller gain K has tendency to inhibit the homoclinic chaos and periodic behaviour in the system.

(ii) The structural variation in transition from homoclinic chaotic convection to steady state via a solitary limit cycle is observed under the influence of various values of the controller gain K when $R = 1250$, $\alpha = 500$, $\gamma = 0.5$, $\zeta = 1.5$, $\chi = 1.5$.

(iii) A sequence of period halving followed by the convergence to the final steady-state have been discovered as the proportional controller gain K increased in the range $0 \leq K \leq 1.155$, at the fixed values of $R = 750$, $\alpha = 500$, and $\gamma = 0.5$ for low anisotropic parameters.

(iv) There is proportional relation between the amount of feedback control and scaled Rayleigh number.

(v) The results indicate that feedback control has stabilizing character according to the significantly slightly higher controller gain K than in an uncontrolled case and will confer a tremendous advantage for controlling homoclinic chaos and periodic dynamics in many industrial applications.

6. References

[1] Chang W J and Hsiao H F 1993 *International Journal of Heat and Mass Transfer* **36** pp 3361–3367
[2] Bera P and Khalili A 2002 *International Journal of Heat and Mass Transfer* **45(15)** pp 3205–3222
[3] Bhadauria B S, Kumar A, Kumar J, Sacheti N C and Chandran P 2011 *Transport in Porous Media* **90** pp 687–705
[4] Bhadauria B S 2012 *Transport in Porous Media* **92** pp 299–320
[5] Gupta V K and Singh A K *International Journal of Energy Technology* **5(27)** pp 1–12
[6] Ott E, Grebogi C and Yorke J A 1990 *Physical Review Letters* **64** pp 1196–1199
[7] Pyragas K 1992 *Physics Letters A* **170** pp 421–428
[8] Pyragas K 2006 *Philosophical Transactions of the Royal Society A* **364** pp 2309–2334
[9] Yau H T and Chen C L 2007 *Chaos, Solitons and Fractals* **34** pp 1567–1574
[10] Bessa W M, de Paula A S and Savi M A 2009 *Chaos, Solitons Fractals* **42** pp 784–791
[11] Mahmud M N and Hashim I 2010 *Transport in Porous Media* **84(2)** pp 421–440
[12] Mahmud M N and Hashim I 2011 *Physics Letters A* **375** pp 2382–2393
[13] Roslan R, Mahmud M N and Hashim I 2011 *International Journal of Heat and Mass Transfer* **54(1-3)** pp 404–412
[14] Mahmud M N 2017 *International Journal of Advanced Science, Engineering and Technology* **5(1)** pp 71–76
[15] Tang T and Bau H H 1994 *Proceedings of the Royal Society of London A* **447** pp 587–607
[16] Bau H H 1999 *International Journal of Heat and Mass Transfer* **42** pp 1327–1341
[17] Lorenz E N 1963 *Journal of the Atmospheric Sciences* **20** pp 130–142
[18] Sparrow C 1982 *The Lorenz Equations: Bifurcations, Chaos, and Strange Attractors* (Springer-Verlag, New York)

IOP Conf. Series: Journal of Physics: Conf. Series **1011** (2018) 012082 doi:10.1088/1742-6596/1011/1/012082

Correlation between diffuse interstellar bands (DIBs) and interstellar extinction using data from Bosscha Compact Spectrograph

L. Puspitarini[1], H. L. Malasan[1], Aprilia[1], M. I. Arifyanto[1], R. Lallement[3], M. Irfan[1], E. Puspitaningrum[2]

[1]Department of Astronomy and Bosscha Observatory, FMIPA Institut Teknologi Bandung, Jalan Ganesha 10 Bandung 40132 Indonesia
[2] Master Program in Astronomy, Faculty of Mathematics and Natural Sciences, Bandung Institute of Technology, Jalan Ganesha 10, Bandung 40132, Indonesia
[3] GEPI, Observatoire de Paris, CNRS UMR8111, Université Paris Diderot, Place Jules Janssen, 92190 Meudon, France

E-mail: lucky.puspitarini@as.itb.ac.id

Abstract. A longstanding challenge in astronomical spectroscopy is to uncover the carriers of diffuse interstellar bands (DIBs). They are broad absorption features due to the interstellar matter (ISM). They are seen in stellar spectra of background stars or other astronomical objects. Although we do not know utterly the carriers of the DIBs, they can be a promising tracer of the ISM. One of the interesting properties is their correlations with the interstellar (IS) extinction. For each band, the correlation has considerable dispersion and differences that possibly due to the IS physical conditions. Some DIBs are sensitive to the stellar radiation field, and some are not. To study the effect, we measured the DIB observed in Be/B stars spectra. The stars were observed by using Bosscha Compact Spectrograph at the Bosscha Observatory, Lembang, Indonesia. We performed an automated fitting of a combination of a smooth stellar continuum, the DIB profile, and a synthetic telluric transmission to the spectrum. The DIB measurements were compared to the general DIBs-extinction relationship. The correlation is found to be in good agreement with previous determinations.

1. Introduction

Diffuse interstellar bands (DIBs) are mysterious absorption features seen in a spectrum of a star (or other astronomical object). DIBs behave like interstellar (IS) lines with regard to occurrence, strength, and Doppler shift. As a matter of fact, they are caused by the absorption of light by the interstellar matter (ISM). Instead of being narrow and sharp, as the well-known interstellar lines, DIBs profiles are widened (diffuse width) and irregular.

Despite the unknown origin/carrier, DIBs can be a promising tracer of the ISM. (1) DIBs are numerous. Around 500 DIBs are identified in optical, IR, and UV domain [3, 4, 7]. Identifying their carriers is a subject of active research for many years. (2) DIBs are not easily saturated. They can be used to trace distant Galactic IS structures (see[9]). (3) DIBs are correlated with interstellar extinction (AV) or reddening (E(B-V)) with some dispersion. Interstellar extinction or reddening occurs because interstellar dust absorbs and scatters the light. The relationship between DIBs and interstellar extinction can be used for extinction estimation of an unknown region in the sky; and to indicate interstellar physical conditions because some DIBs are sensitive to the stellar radiation field, and some are not.

Observation Program of Be Stars has been conducted in Bosscha Observatory, Lembang, Indonesia for many years [6]. Be stars are B-type stars surrounded by gaseous disks which exhibit (variable) Balmer line emission and an infrared excess. These stars may contribute to the interstellar environment through the stellar wind and the stellar radiation that may destroy or favor the DIB carrier(s). We are aiming at studying the effect of stellar radiation field through the correlation between DIBs and interstellar extinction.

2. Data

For the analysis, we used 34 target stars (Be and B-type star) that are part of Be Stars Observation Program ([6]; Aprilia, priv. comm.). The stellar spectra were observed by using Bosscha Compact Spectrograph [5] with $R \simeq 8000$ and spectral range of 6350-6800 Å. For the analysis we focus on the 6613.62 (6614) Å DIB that are strong enough to ensure a detection in most targets. Stellar parallaxes of the target stars were obtained using SIMBAD Astronomical Database, whereas interstellar extinction data were obtained using a computerized model by [2].

3. Method

Our goal is to extract and measure DIB equivalent width (EW) in series of stellar spectra in an automated way. The equivalent width is defined as the area between the line profile and the continuum. It corresponds to the measure of the strength of spectral features.

We fit the stellar spectra (data) to a triple combination of DIB Model (fit in strength and shift), telluric transmission model (also fit in strength and shift), and smooth continuum (with a polynomial function). The fitting uses the LevenbergMarquardt algorithm in the IGOR Pro software (WaveMetrics, Inc.).

The DIB Model is an empirical profile (see [8]), whereas a synthetic telluric transmission computed using the line-by-line radiative transfer model (LBLRTM) code and the HIgh-Resolution-TRANsmission molecular absorption (HITRAN) spectroscopic database [1, 10].

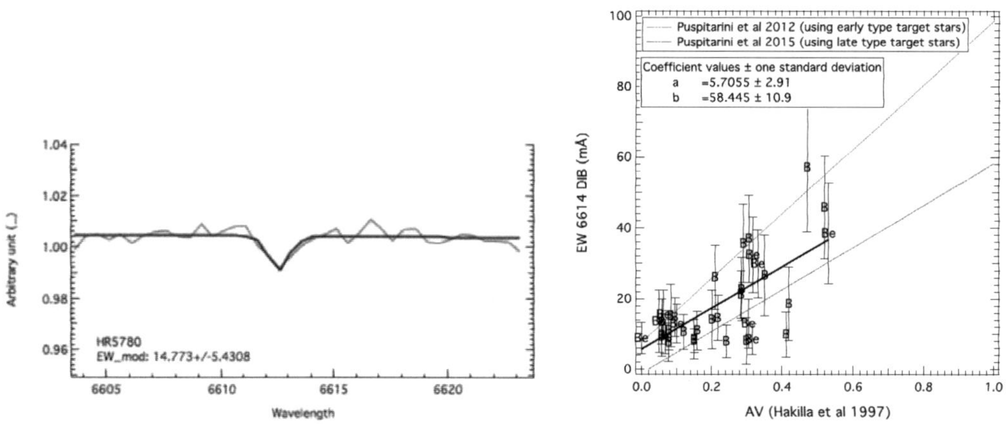

Figure 1. Left: An example of DIB fitting in the spectrum of HR5780 star. Red shows the stellar spectrum, whereas blue shows the fitting result. Right: The equivalent width measurements of the 6614 DIB vs. the interstellar extinction (AV) from [2]. The correlation is found to be in good agreement with previous determinations.

IOP Conf. Series: Journal of Physics: Conf. Series **1011** (2018) 012082 doi:10.1088/1742-6596/1011/1/012082

4. Results and Discussions

An example of the DIB fitting can be found in Fig. 1 (left). From the fitting, we can measure the equivalent width of 6614 DIB profile which corresponds to abundance of its carrier. We then plot the equivalent width measurements of the 6614 DIB vs. the interstellar extinction (AV) from [2] (see Fig. 1 right). We remark that the interstellar extinction in this case is totally independent from our data. It is mainly derived from photometric data and model [2].

The plot in Fig. 1 right shows that there is a positive correlation between the strength of the DIB and interstellar extinction. The Pearson correlation coefficient (V_{pr}) that we obtained is 0.7. The interstellar extinction itself is reddening and dimming that are mainly caused by dust. The positive correlation indicates that DIB's carrier and interstellar dust are quite well mixed in the interstellar cloud. However, one of the explanation we don't obtain perfect correlation (V_{pr}=1) is because of the stellar radiation field. The stellar radiation in this case tends to destroy the 6614 DIB carrier.

This correlation that we obtained is in good agreement with previous determinations, i.e., [8] that used hot O-type target stars and [9] that used cool or late-type target stars. The gradient of the correlation is found to be in between the ones derived from hot and cool target stars. This may indicate typical stellar radiation field from Be/B stars that is moderate (not too strong and not too weak). However, we notice some dispersions in the measurements. The dispersions could come from the measurement error (S/N of the low resolution data), blending with diffuse nearby DIB, i.e. 6607.07 Å, or the stellar radiation field itself.

5. Conclusions

We conclude from this study that 6614 DIB has been correctly extracted from our spectroscopic data of B/Be stars. The positive correlation between DIB's equivalent width (abundance) and interstellar extinction indicates that the mysterious DIB's carrier and interstellar dust are well-mixed in the interstellar cloud. The correlation parameter (V_{pr} = 0.7) is found to be in good agreement with previous determinations ([8] and [9]), within uncertainties. When compared with [8], our Pearson coefficient is higher. We believe that this is due to the smaller stellar radiation field influences, because our targets (B/Be) are cooler than those used by [8] (O-type stars). In reverse, our Pearson coefficient is smaller than the ones in [9] that used cooler stars (G, K, M type stars). Because these cooler stars produce smaller stellar radiation field influences that favor the carrier of 6614 DIB.

Acknowledgments

We thank the P3MI ITB for research fund and travel support.

References

[1] Clough S A, Shephard M W, Mlawer E J et al 2005 *Journal of Quantitative Spectroscopy & Radiative Transfer* **91** 233
[2] Hakkila J, Myers J M, Stidham B J, & Hartmann D H 1997 *ApJ* **114** 2043 D G, Snow T P et al 2008 *ApJ* **680** 1256
[3] Hobbs L M, York D G, Snow T P et al 2008 *ApJ* **680** 1256
[4] Hobbs L M, York D G, Thorburn J A et al 2009 *ApJ* **705** 32
[5] Malasan H L, Yamamuro T, Takeyama N et al 2001 *Proc. of the Indonesia-German Conf. on Instrumentation, Measurements and Communication for the Future* 159-165
[6] Malasan H L, Irfan M, Aprilia, Dawanas D 2014 *Tenth Pacific Rim Conference on Stellar Astrophysics* **482** 25
[7] McCall B J and Griffin R E 2013 *Proc. R. Soc. A* **469** 2151
[8] Puspitarini L, Lallement R, Chen H-C 2013 *A&A* **555** A25
[9] Puspitarini L, Lallement R, Babusiaux C et al 2015 *A&A* **573** AA35
[10] Rothman L S, Gordon I E, Barbe A et al 2009 *Journal of Quantitative Spectroscopy & Radiative Transfer* **110** 533

IOP Conf. Series: Journal of Physics: Conf. Series **1011** (2018) 012083 doi:10.1088/1742-6596/1011/1/012083

Solution of Einsteins Equation for Deformation of a Magnetized Neutron Star

R. Rizaldy, A. Sulaksono

Department of Physics, FMIPA, University of Indonesia, Kampus UI Depok, Indonesia

E-mail: ryan.rizaldy@ui.ac.id, anto.sulaksono@sci.ui.ac.id

Abstract. We studied the effect of very large and non-uniform magnetic field existed in the neutron star on the deformation of the neutron star. We used in our analytical calculation, multipole expansion of the tensor metric and the momentum-energy tensor in Legendre polynomial expansion up to the quadrupole order. In this way we obtain the solutions of Einstein's equation with the correction factors due to the magnetic field are taken into account. We obtain from our numerical calculation that the degree of deformation (ellipticity) is increased when the the mass is decreased.

1. Introduction

Observations of magnetic fields of pulsars are usually in the range of values from 10^8 to 10^{12} G. However, new classes pulsars such as Anomalous X-ray pulsars (AXPs) and Soft-Gamma Repeater pulsars (SGRs) have been identified produce very large magnetic fields. SGR is usually associated with remnants of a supernova, which is in fact a young neutron star [1, 2]. Furthermore, measurements of periods and derivatives of periods of the corresponding pulsar provide evidence that it neutron star with very large magnetic field with strength around $\sim 10^{15}$ G. Observation results of some AXPs indicate also that AXPs surface magnetic field around of $10^{14} - 10^{15}$ G. The relationship between SGR observation and AXPs are not yet clear enough up to now. However, it is confirmed that a class of neutron stars that have indeed a very large magnetic field and the corresponding neutron stars is called Magnetars [3–6] If we consider a neutron star matter that influenced by very strong magnetic field, we should take care two quantum effects, namely Pauli paramagnetism (i.e., the spin interaction of Fermions with a magnetic field) and Landau's magnetism. The other is the Lorentz force which causes deformation of stars due to magnetic field [7]. In this study, we focused on the deformation occurring in neutron stars caused by very large magnetic fields, in which the deformation of spherical stars are considered within the framework of general relativity using by Hartle and Thorne method [8, 9] and using also the assumptions that non-uniform pressure as a perturbation of the total (matter and magnetic) pressures as the one already done in Mallicks work [10].

IOP Conf. Series: Journal of Physics: Conf. Series **1011** (2018) 012083 doi:10.1088/1742-6596/1011/1/012083

2. Formalism

The total energy density and pressure neutron star with magnetic field can be written as follows[10]:

$$\varepsilon = \varepsilon_m + \frac{B^2}{8\pi} \tag{1}$$

$$P_\perp = P_m + \frac{B^2}{8\pi} \tag{2}$$

$$P_\parallel = P_m - \frac{B^2}{8\pi}, \tag{3}$$

where ε_m and P_m is the energy density and pressure of neutron star matter, $B^2/8\pi$ is the magnetic stress contribution. Note that for calculating ε_m and P_m we use effective field relativistic mean field (ERMF) model with BSP parameter set for nucleons coupling constants and standard SU(6) and hyperons potential energy depths to obtain hyperons (Λ, Σ and Ξ) coupling constants (see details of the corresponding model in Ref. [11]). For the pressure part, as it is done in Mallicks work [10], it can be written in isotropic form as

$$P = P_m \pm P_B \tag{4}$$

$$P = P_m + \frac{B^2}{8\pi}\left(1 - 2\cos^2\theta)\right), \tag{5}$$

P_B is the magnetic stress and θ is the polar angle with respect to the direction of the magnetic field. Total pressure can be written also Legendre polynomial expansion

$$P = P_m + \frac{B^2}{8\pi}\left[\frac{1}{3} - \frac{4}{3}P_2(\cos\theta)\right] \tag{6}$$

$$P = P_m + p_0 + p_2 P_2(\cos\theta), \tag{7}$$

$p_0 = B^2/24\pi$ is the monopole and $p_2 = -4B^2/24\pi$ is the quadrupole factors due to magnetic field. The most important thing in this study is that expansion of the magnetic field correction is only for the pressure part. Since the momentum-energy tensor is multipole expansion form, if we assume neutron star is initially has a sphere form then the Schwarzschild metric should also expand by using the Hartle and Thorne method [8, 9]

$$\begin{aligned} ds^2 =& - e^\nu \left[1 + 2\left(h_0 + h_2 P_2(\cos\theta)\right)\right] dt^2 \\ &+ e^\lambda \left[1 + \frac{e^\lambda}{r}\left(m_0 + m_2 P_2(\cos\theta)\right)\right] dr^2 \\ &+ r^2\left[1 + 2k_2 P_2(\cos\theta)\right]\left(d\theta^2 + \sin^2\theta d\phi^2\right). \end{aligned} \tag{8}$$

Where the metric ν or λ is a function of r and they can be expressed as

$$\nu' = -\frac{2}{(\varepsilon + P)}P' \tag{9}$$

$$e^\lambda = \left(1 - \frac{2Gm}{r}\right)^{-1}. \tag{10}$$

Here h_0, h_2, m_0, m_2, k_2 are also function of r, respectively and they can be considered as the correction factors due to magnetic field up to second order. By solving the Einsteins equation using above metric and energy momentum tensor and also by considering the conservation law of

IOP Conf. Series: Journal of Physics: Conf. Series **1011** (2018) 012083 doi:10.1088/1742-6596/1011/1/012083

total momentum-energy by employing the Bianchi identity we can obtain a usual TOV equations as

$$P' = -\frac{G}{r^2}\frac{(\varepsilon + P)(m + 4\pi r^3 P)}{1 - 2Gm/r} \tag{11}$$

$$m' = 4\pi r^2 \varepsilon, \tag{12}$$

the correction for monopole order as

$$m_0' = 12\pi r^2 p_0 \tag{13}$$

$$h_0' = 4\pi G r e^{\lambda} p_0 + \frac{1}{r} G\nu' e^{\lambda} m_0 + \frac{1}{r^2} G e^{\lambda} m_0, \tag{14}$$

as well as for quadrupole order as

$$h_2' + k_2' = h_2\left(\frac{1}{r} - \frac{\nu'}{2}\right) + \frac{e^{\lambda}}{r} G m_2 \left(\frac{1}{r} + \frac{\nu'}{2}\right) \tag{15}$$

$$h_2 + \frac{e^{\lambda}}{r} G m_2 = 0 \tag{16}$$

$$h_2' + k_2'\left(1 + \frac{r}{2}\nu'\right) = 4\pi G r e^{\lambda} p_2 + \frac{e^{\lambda}}{r^2} G m_2 + \frac{e^{\lambda}}{r} G\nu' m_2 + \frac{3}{r} e^{\lambda} h_2 + \frac{2}{r} e^{\lambda} k_2, \tag{17}$$

and from conservation law total momentum-energy we also obtained:

$$p_0' = -2\nu' p_0 - (\varepsilon + P)\, h_0' \tag{18}$$

$$p_2 = -(\varepsilon + P)\, h_2 \tag{19}$$

$$p_2' = -\frac{1}{2}\nu' p_2 - (\varepsilon + P)\, h_2'. \tag{20}$$

With some simple algebra and using equations (15), (19), and (20) we can also obtain h_2' and k_2' in the following form

$$k_2' = \frac{\frac{3}{2} p_2 \nu' + p_2'}{\varepsilon + P} \tag{21}$$

$$h_2' = \frac{-\frac{1}{2} p_2 \nu' - p_2'}{\varepsilon + P}. \tag{22}$$

Using Runge-Kutta methods, we can solve numerically Eqs. (11), (12), (13), (14), (21), and (22) with boundary condition for standard TOV equation is $P(0) = P_c, P(R) = 0$ and $m(0) = 0, m(R) = M$ where P_c is pressure in the neutron star center and R is radius of neutron stars while for monopole order $h_0(0) = 0$ and $m_0(0) = 0$, for quadrupole order $k_2(0) = h_2(0) = 0$ and $m_2(0) = 0$.

In this way wee can define total mass as [10],

$$M = M_0 + m_0, \tag{23}$$

where M_0 is the matter's mass and m_0 is mass correction due to the magnetic field. In the presence of mass correction of the neutron star, the anisotropic magnetic pressure on the neutron star is affected. Following Eq.(4) where the pressure of the matter and pressure due to magnetic

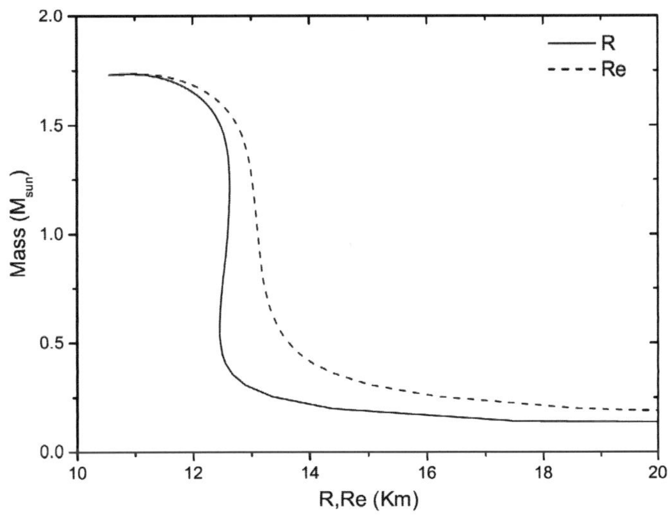

Figure 1. Straight line is the radius without magnetic field correction, dash line is radius with magnetic field correction.

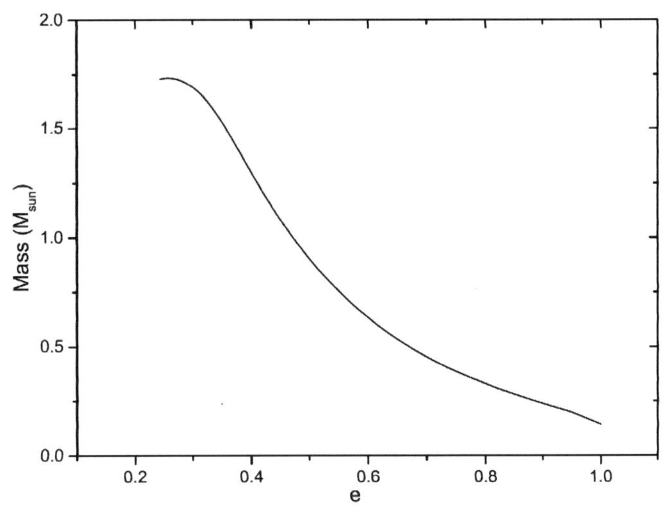

Figure 2. Relation between ellipticity of magnetized neutron star and the corresponding Mass.

field are summed along the equatorial direction and reduced the polar direction, we can write the radius equatorial and polar of the deformed neutron star as[10],

$$R_e = R + \xi_0(R) - \frac{1}{2}\left(\xi_2(R) + Rk_2(R)\right) \tag{24}$$

$$R_p = R + \xi_0(R) + \left(\xi_2(R) + Rk_2(R)\right), \tag{25}$$

where R radius of spherical neutron star (without magnetic field), ξ_0 and ξ_2 define by [8–10],

$$\xi_0(r) = -h_0 \frac{r(r-2Gm(r))}{G(4\pi r^3 P_m + m(r))} \tag{26}$$

$$\xi_2(r) = -h_2 \frac{r(r-2Gm(r))}{G(4\pi r^3 P_m + m(r))}. \tag{27}$$

Where the contribution ξ comes from the magnetic field on the magnetar surface and contribution k_2 comes from integration of magnetic pressure through inside of the star [10]. To know the degree of deformation in neutron stars, we can write the ratio of ellipticity as

$$e = \sqrt{1 - \left(\frac{R_p}{R_e}\right)^2} \tag{28}$$

3. Result and Discussion

By solving Eqs. (11), (12), (13), (14), (21), and (22), we have a relationship between the mass and the radius of the neutron stars as shown in Fig. 1. M is the total mass of stars as expressed in Eq. (23). We can see that in the magnetic field can increase the mass of the neutron star.

Fig. 2 shows the relation between the mass and the corresponding ellipticity of neutron star. If the polar radius of the neutron star increases then the mass decreases and if the equatorial radius of the neutron star increases the mass increases. the maximum deformation happen in the ellipticty ~ 0.257. In this case, $R_e > R_p$. Therefore, the neutron stars are maximally in oblate form.

4. Conclusion

Using this approach, we have obtained the solution of Einsteins equation in the form of standard TOV equations and added by some monopole and quadrupole order factor equations. The latters can also lead to first order differential equations Therefore, we can solve the corresponding differential equations by utilizing boundary conditions numerically. The solution informs us that in the presence of a very large magnetic field can change the form of the neutron star from sphere into the oblate shape if the corresponding mass is decreased.

ACKNOWLEDGMENT

We are partially supported by the UI's PITTA grant No. 622/UN2.R3.1/HKP.05.00/2017.

References

[1] Kulkarni S R and Frail D A 1993 Identification of a supernova remnant coincident with the soft gamma-ray repeater SGR1806-20 *Nature (London)* **365** 33-35
[2] Murakami T, Tanaka Y, Kulkarni S R, Ogasaka Y, Sonobe T, Ogawara Y, Aoki T,p and Yoshida A, 1994 X-ray identification of the soft -ray repeater 1806 20 *Nature* **368** 127-129
[3] Duncan R C and Thompson C 1992 Formation of very strongly magnetized neutron stars - Implications for gamma-ray bursts, *Astrophysical Journal*, **392** L9-L13
[4] Thompson C and Duncan R C 1993 Neutron star dynamos and the origins of pulsar magnetism *Astrophysical Journal* **408** 194-217
[5] Thompson C and Duncan R C 1995 The soft gamma repeaters as very strongly magnetized neutron stars - I. Radiative mechanism for outbursts *Monthly Notices of the Royal Astronomical Society* **275** 255-300
[6] Thompson C and Duncan R C 1996 The Soft Gamma Repeaters as Very Strongly Magnetized Neutron Stars. II. Quiescent Neutrino, X-Ray, and Alfven Wave Emission *Astrophysical Journal* **473** 322-342
[7] Diener J P W 2012 Ferromagnetic Phase Transitions in Neutron Stars Ph.D. Thesis University of Stellenbosch (*Preprint* nucl-th/1305.7346)
[8] Hartle J B 1967 Slowly Rotating Relativistic Stars *Astrophyical Journal* **150** 1005-1030
[9] Hartle J B and Thorne K S 1968 Slowly Rotating Relativistic Stars. II. Models for Neutron Stars and Supermassive Stars *Astrophysical Journal* **153** 807-834

[10] Mallick R and Schramm S 2014 Deformation of a magnetized neutron star *Physical Review C* **89** 045805-1 045805-7

[11] B. K. Agrawal, A. Sulaksono and P.-G. Reinhard 2012 Optimization of relativistic mean field model for finite nuclei to neutron star matter *Nuclear Physics A***882** 1-20;A. Sulaksono and B. K. Agrawal 2012 Existence of hyperons in the pulsar PSRJ1614-2230 *Nuclear Physics A***895** 44-58.

IOP Conf. Series: Journal of Physics: Conf. Series **1011** (2018) 012084 doi:10.1088/1742-6596/1011/1/012084

Investigating the effect of mutation on the thermo stability of GB1 protein

K N Sawitri[1, 3*], T Sumaryada[1], L Ambarsari[2] and S T Wahyudi[1]

[1]Computational Biophysics and Molecular Modelling Research Group (CBMoRG), Department of Physics, Bogor Agricultural University, Jalan Meranti Kampus IPB Dramaga Bogor 16680, Indonesia
[2]Department of Biochemistry, Bogor Agricultural University, Jalan Agatis Kampus IPB Dramaga Bogor 16680, Indonesia
[3]Department of Physics, University of Jambi, Jalan Lintas Jambi Kampus UNJA Mendalo 36122, Indonesia

E-mail : KaniaNurSawitri@unja.ac.id

Abstract. The thermo stability of wild-type and mutants of B1 domain of Protein G (GB1 protein) have been studied using molecular dynamics simulation and free energy perturbation simulation. This research is aimed to examine what residue or what interaction that has a major role in the thermo stability of GB1 protein thermo stability by using the point mutation method. Based on the analysis, the unfolding of wild-type protein occurred in 500 K simulation at 704 ps. The mutations were chosen based on the changes in some analysis parameters and the calculated net solvation free energy change. It was found that a simple replacement of a positively charged residue in the β-sheet (K4S) decreases the stability of GB1 protein (unfolding at 452 ps), while the replacement of a negatively charged residue in the α-helix (E27G) increases the stability (unfolding at 846 ps). It was also found that the K4A mutation will break the α-helix and all β-sheet into the coil and turn. All those results suggest that the non-bonded interaction has the major role in the thermo stability of GB1 protein with the β-sheets were identified as the most important structure in the thermo stability of GB1 protein..

1. Introduction

The B1 domain of protein G (PDB code 1GB1) is an immunoglobulin protein from G streptococcal group which helps an organism to evade the host defense via unique protein-binding properties [1]. This protein has played a central role in protein model studies to elucidate the folding and unfolding process due to several reasons. First, its unique structure: small size comprises only 56 amino acid residues, simple and highly symmetrical topology [2]. Second, its high thermo stability properties: denaturation points reach 87.5°C without disulfide bridge that cooperatively folds like the other globular proteins [3]. Third, its well characterized mechanical properties: the mechanical unfolding mechanism has been identified to be different from that of chemical denaturant unfolding at the atomic level [4].

IOP Conf. Series: Journal of Physics: Conf. Series **1011** (2018) 012084 doi:10.1088/1742-6596/1011/1/012084

Understanding how mechanically or non-mechanically the protein folding or unfolding is still unclear and a major challenge in life sciences. Hundreds of studies have been performed both experimentally using NMR and computationally using molecular dynamic simulation since the last two decades. With the increasing power of computers, simulations can generate highly detailed descriptions of the motions of proteins. As a model, many strategies have been applied on unfolding simulations of 1GB1 protein to expand the range of problems that may be addressed [5,6]. Some of them are applying the differences of pH using in lucem [2], steer forces using Gromacs 4.5.3 [7], pressure using X-PLOR [8], denature types [9] and temperature using SYMLIS [10].

One of the main targets in protein engineering is increasing the stability of protein by conducting experiment and simulation under various conditions, including mutation. The rational design of mutants is aimed to increase the stability of a protein by either improving the stabilizing interactions or reducing the factors which potentially destabilizing the protein [11]. In this paper we have conducted an investigation on the effect of mutation to the thermo stability of GB1 protein by using an in-silico approach via a molecular dynamics (MD) simulation and the free energy perturbation methods. Analyses of wild-type GB1 protein MD parameters were used as a basis for the screening and mutation selection. By conducting the point mutations on the non-bonded interactions of GB1 protein, we expect to see a thermo stability enhancement as compared to the wild-type protein.

2. Materials and Methods

This study was performed using two simulations. First is the molecular dynamics (MD) simulation for the wild type at various temperatures. The wild-type structure as reference structure was downloaded from http://www.rscb.org/pdb/ [Protein Data Bank (PDB) ID code 1GB1]. The second simulation is the free energy perturbation simulation to find the most stable mutants. Once the mutants were identified, then the mutants were simulated using MD simulation to be analyzed.

2.1. *Molecular dynamics simulation*

We used NAMD 2.9 [12] with CHARMM potential and VMD 1.9.1 [13], starting with protein placed in a TIP3P water box with dimension 65x54x50 Å. The simulation time step is 2 fs. Each simulation was started by minimization about 100 ps to place the protein on its lowest energy, and then continued by heating and equilibration. The wild-type was heated start from 0 K until the final point at 400 K, 425 K, 450 K, 475 K, and 500 K by increment 25 K. But the mutants were heated start from 0 K until 500 K. The equilibration was obtained by Langevin protocol about 60 ps. We eliminated the constraints in MD production run process so that the protein moves freely. The simulation time for final points 400 K - 475 K is 10 ns, but the simulation time for 500 K is 2 ns.

2.2. *Free energy perturbation simulation*

The Free Energy perturbation technique (FEP) is a common method to estimate the free energies of various physical processes including the free energy of a mutation of a side chain. The FEP simulation technique can be illustrated in figure 1 where the free energy (ΔG) was calculated via a thermodynamic cycle (figure 1) [14]. The vertical arrows correspond to the hydration of the wild-type and its mutant, while the horizontal arrows correspond to the point mutation. The difference between ΔG of mutant in vacuo ($\Delta G^1_{alch.}$) and aqueous($\Delta G^2_{alch.}$) condition will be compared with the one found in the wild-type protein (ΔΔG). If the ΔΔG in mutant is lower than the wild-type then the mutation has successfully produced a more stable structure. In this research, the FEP simulations were applied by two conditions: vacuo system and aqua system, using dual-topology approach. In vacuo simulation: the increment of λ is 0.05, equilibration time is 10 ps, and the free energy capture time is 100 ps for each λ. In Aqua simulation: the increment of λ is 0.05, equilibration time is 100 fs, and the free energy capture time is 500 fs for each λ. We didn't use soft-core potential in both conditions to remove the interaction between topology target and the system environment when the λ value near 0 or 1. Wild-type and mutant structure did not interact during FEP simulation. Solvation free energy change was obtained by subtraction of $\Delta G^1_{alch.}$ from $\Delta G^2_{alch.}$ [14].

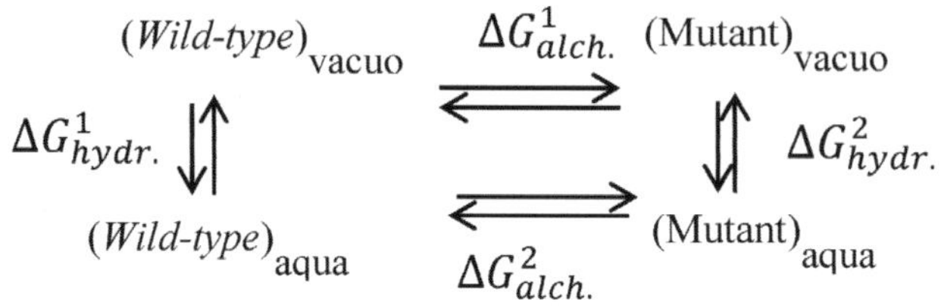

Figure 1. Thermodynamic cycle of solvation free energy

3. Result and Discussions

3.1. Thermo stability analysis of wild-type

Molecular dynamic simulations for wild-type structure were analyzed at temperature variance: 400 K, 425 K, 450 K, 475 K and 500 K. Simulation at high temperature is a common strategy to reduce the computing time [15]. The results show that 1GB1 protein unfolds at 500K on 704 ps as indicated by the breaking of α-helix structure which is transformed into coil and turn. In figure 2 we show the SASA (solvent accessible surface area) of various residues of the wild-type protein. The information from this figure is important for determining the residue candidate for mutation. As seen in figure 2, the apolar (non-polar) residue which implies hydrophobic interaction is dominant as compared to other type of residues, so the mutation on the apolar residues could potentially change the thermo stability of the GB1 protein. Earlier work shows that salt-bridge which implies non-bonded interaction play a role to stabilize C-termini of β-hairpin fragment [16] and second β-hairpin of domain B1 [17]. Those give insight to choose the mutants based on hydrophobic and non-bonded interactions.

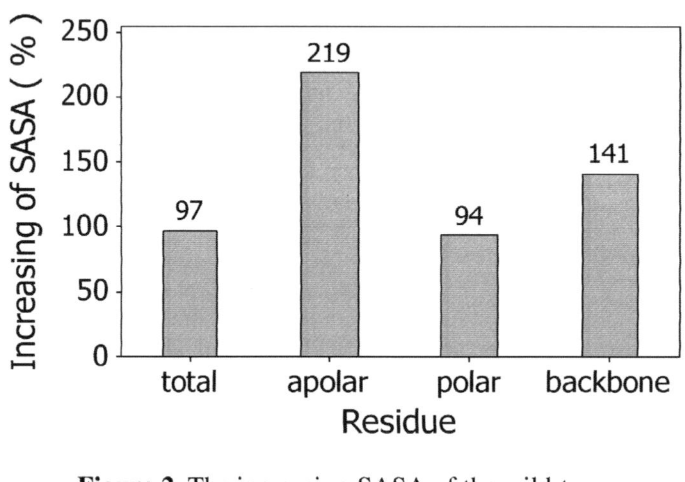

Figure 2. The increasing SASA of the wild-type.

3.2 Selection of the mutants

Based on the wild-type simulation analysis at 500K, there are 8 residues of the 1GB1 protein selected to be mutated: K4, K10, A23, A24, E27, K28, D40, and E56. The selection based on the scoring of several MD parameters such as root mean square deviation (RMSD), radius of gyration, Hydrogen bonding, SASA, RMSF, conformation energy. The stability of the protein was analyzed by inspecting six different properties: RMSF (root mean square fluctuation), salt-bridge occurrence, SASA, the stability of secondary structure elements, non-bonded energy, and hydrogen bond occurrence. For mutant selection, we consider the conformation preference of secondary structure. To select the best mutants, FEP simulations generate solvation free energy change (ΔΔG). The highest and lowest ΔΔG led to biggest changes in point mutation, so there is 16 mutants continue to molecular dynamic simulation (table 1). K10G means that lysine (residue number 10) was mutated into glycine, same with others.

Table 1. FEP simulation results.

No	Mutant	$\Delta\Delta G_{solv}$ (Kcal/mol)	No	Mutant	$\Delta\Delta G_{solv}$ (Kcal/mol)
1	K10G	10.5705	9	K10N	40.4396
2	D40T	33.0834	10	D40L	58.3459
3	E56V	32.5037	11	E56I	55.6443
4	A23S	0.29486	12	A23Q	6.1274
5	A24S	0.34247	13	A24T	8.46362
6	K4S	8.6961	14	K4A	23.6733
7	E27G	10.8084	15	E27S	53.3006
8	K28T	0.1974	16	K28C	26.6864

3.3 MD Simulation analysis of Mutants

Protein G B1 domain has 56 amino acid residues which are composed of secondary structures: a α-helix (residue 23-36) , 4-stranded β-sheet (residue 2-8, 13-19, 42-46, 51-55), 2 coils (residue 1, 20-22, 41, 51-55) and 2 turns (residue 9-12, 47-50) as shown in figure 3. They are linked by hydrogen bonds between oxygen atoms on carbonyl groups with hydrogen atoms on amide groups. The fluctuation on secondary structure represents that the residues are maintaining its existence. All mutants unfold at 500K as shown by the destruction of α-helix structures (figure 4). Three out of sixteen mutant simulations gave interesting results, and all will be discussed in this section. Wild-type 1GB1 unfolds on 704 ps (figure 4a), where the mutants K4S is the most unstable protein since it unfolds at 452 ps (figure 4b). Analysis of secondary structure shows that Lysine mutated into Serine cause most of the β-sheet at the N-terminal end was destroyed and transformed into coil and turn structures (figure 5b). This effect can be described by conformation preference that available in the web server located at http://bmrb.wisc.edu/referenc/choufas.shtml. Lys-4, positively charged residue, is located on first strand β-sheet, and lysine is a β-sheet breaker with p-value 0.74. But serine, neutrally charged residue, has higher conformation to break the β-sheet since its p-value is 0.72. Another impact of this mutation is the disappearance of salt-bridge pair D46-K4 and some hydrogen bonds K4 with residues on third and fourth strand β-sheet.

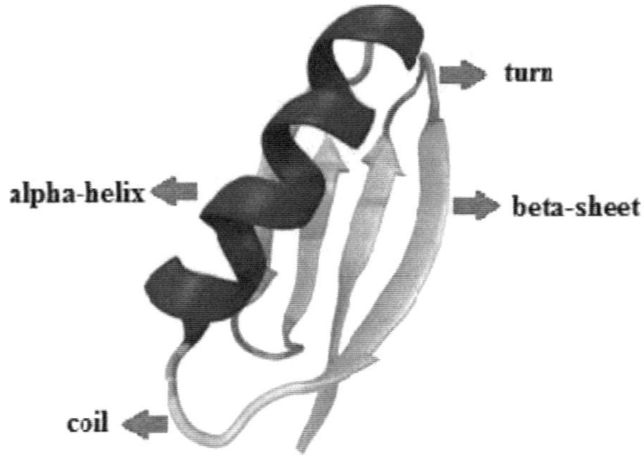

Figure 3. The cartoon representation of GB1 protein. The colors represent secondary structure: Purple (α-helix), Yellow(β-sheet), Blue (310 helix), Green (Turn) and White(Coil).

IOP Conf. Series: Journal of Physics: Conf. Series **1011** (2018) 012084 doi:10.1088/1742-6596/1011/1/012084

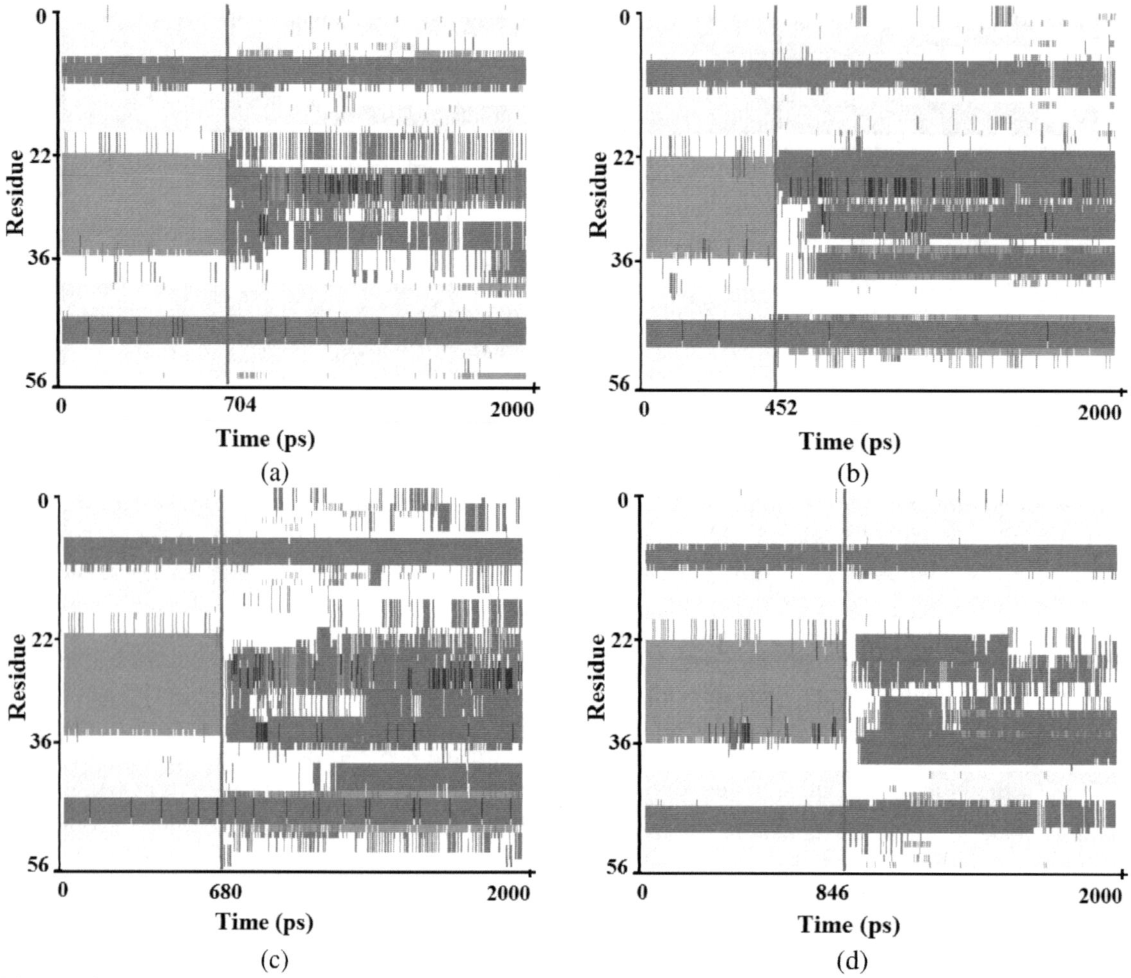

Figure 4. Analysis of secondary structure elements of protein G B1 domain on 500K simulation: (a) Wild-type, (b) mutant K4S, (c) mutant K4A, (d) mutant E27G. X-axis represents time (0-2000 ps), and Y-axis represents residue number (0-56).

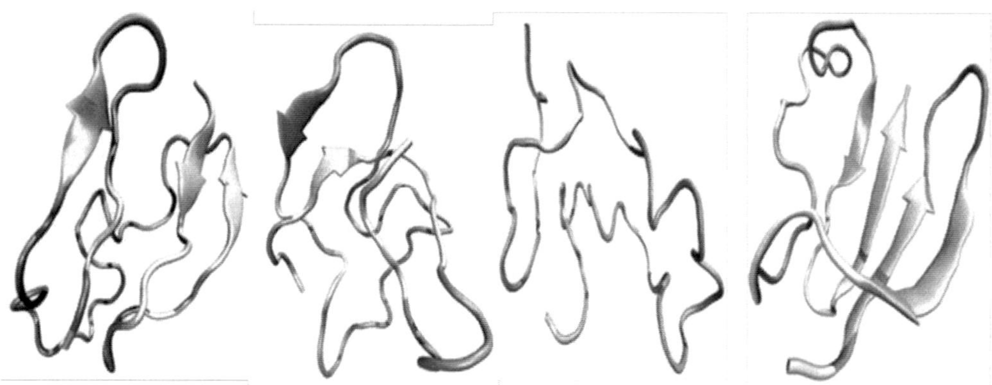

(a) (b) (c) (d)

Figure 5. Final conformation at the end of md 500K simulation: (a) Wild-type, (b) mutant K4S, (c) mutant K4A, (d) mutant E27G.

Second interesting mutant is K4A (Lysine-4 mutated into Alanine) that unfolds at 680 ps (figure 4c). K4A is the only one of sixteen mutants that all of its secondary structure (α-helix and β-sheet) were completely destroyed (figure 5c), and it will be elucidated through RMSF analysis. The K4A mutant unfolds faster than the wild-type since the mutation in principle totally eliminated the salt-bridge and hydrogen bonds contacts. The K4A mutant turned out to be the most flexible mutant, since its RMSF fluctuates within the range of 8-17 Å, while the wild-type only fluctuates in the range of 0-4 Å (figure 6). We suggest that this high flexibility of K4A mutant was caused by the point mutation on β-sheet which responsible for the protein stability. Moreover, hydrophobic contacts between protein and water become stronger since Alanine is non-polar residue. The third interesting mutant is E27G (Glu-27 mutated into Gly) that stabilize the GB1 protein as shown by a longer unfolding time at 846 ps, later than the wild-type structure (figure 4d). The final conformation of E27G has a complete three last strand of β-sheet as seen in figure 5d, while the wild-type has not.

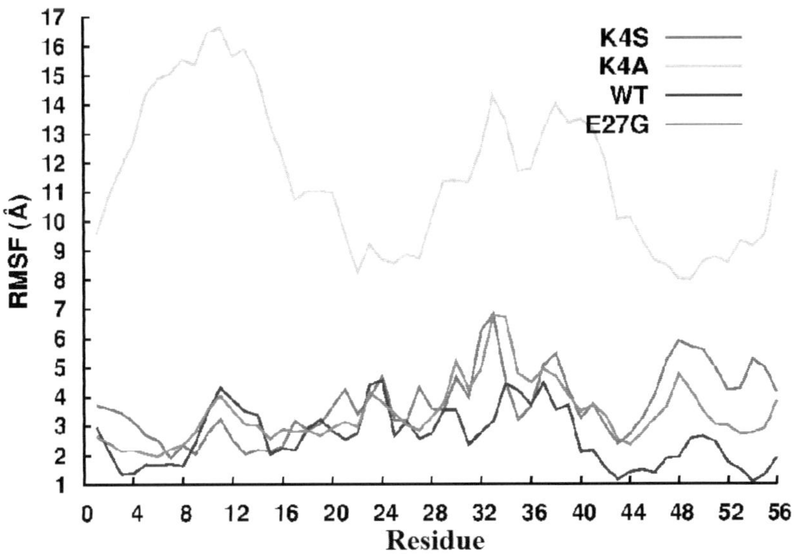

Figure 6. RMSF values over MD simulation.

4. Conclusion

Based on the MD and FEP simulation analysis, three mutants were selected to investigated the effect of point mutations to the thermo stability of GB1 protein. The point mutation simulations indicated that β-sheet is the important structure to maintain the thermo stability of GB1 protein. Mutation on this structure (K4S and K4A) suggests the protein unfolds faster than the wild-type (less-stable protein), while the mutation on α-helix (E27G) stabilize the protein as indicated by the longer time (more-stable protein) needed by the mutant to unfold as compared to the wild-type. The elimination of non-bonded interaction leads to the collapse of protein secondary structure especially in mutant K4A which has the highest RMSF. This result suggests that non-bonded interaction plays an important role in the thermo stability of GB1 protein.

5. References

[1] Gronenborn AM and Clore GM. 1993. *ImmunoMethods.* **2**(1): 3-8

[2] Morrone A, Giri R, Toofanny RD,Allocatelli CT, Brunori M, Dagget V and Gianni S. 2011. *Biophys J.* **101**(8): 2053-60

[3] Kobayashi N, Honda S, Yoshii H, Uedaira H and Munekata E. 1995. *FEBS Lett.* **366** (2-3): 99-103
[4] Shen T, Cao Y, Zuang S and Li H. 2012. *Biophys J.* **103** (11): 2361-68
[5] Granata D, Camilloni C, Vendruscolo M and Laio A. 2013. *PNAS.* **110**(17): 6817-22
[6] Sumaryada T, Hati J, Wahyudi ST, Malau ND, Sawitri KN, *IOP Conf. Series: Earth and Environmental Science* **31** (2016): 012008
[7] Kouza M and Hansmann UHE.2012. *J.Phys.Chem. B.* **116** (23): 6645-53
[8] Wilton DJ, Tunnicliffe RB, Kamatari YO, Akasaka K and Williamson MP. 2008. *Proteins.* **71** (3): 1432-40
[9] Wei H, Fan Y and Gao YQ. 2010. *J. Phys. Chem. B.* **114** (1): 557-68
[10] Lee J and Shin S. 2001. *Biophys J.* **81** (5): 2507-16
[11] Pikkemaat MG, Linssen AB, Berendsen HJ and Janssen DB. 2002. *Protein Eng.* **15** (3): 185-92
[12] Philips JC, Braun R, Wang W, Gumbart J, Tajkhorshid E, Villa E, Chipot C, Skeel RD, Kale L and Schulten K. 2005. *J Comput Chem.* **26** (16): 1781-802
[13] Humprey W, Dalke A and Schulten K. 1996. *J Mol Graph.* **14** (1): 33-8
[14] Henin J, Gumbart J, Harrison C and Chipot C. 2012. *In Silico Alchemy: A Tutorial for Alchemical Free-Energy Perturbation Calculations with NAMD* (University of Illinois at Urbana-Champaign)
[15] Pande VS and Rokhsar DS. 1999. *PNAS.* **96** (16): 9062-67
[16] Tsai J and Levitt M. 2002. *Biophys Chem.* **101-102**: 187-201.
[17] Huyghues-Despointes BM, Qu X, Tsai J and Scholtz JM. 2006. *Proteins.* **63** (4): 1005-17

The Modified Hartmann Potential Effects on γ -rigid Bohr Hamiltonian

A Suparmi, C Cari and Beta Nur Pratiwi

Physics Department, Graduate Program, Sebelas Maret University, INDONESIA

Email: soeparmi@staff.uns.ac.id

Abstract. In this paper, we present the solution of Bohr Hamiltonian in the case of γ -rigid for the modified Hartmann potential. The modified Hartmann potential was formed from the original Hartmann potential, consists of β function and θ function. By using the separation method, the three-dimensional Bohr Hamiltonian equation was reduced into three one-dimensional Schrodinger-like equation which was solved analytically. The results for the wavefunction were shown in mathematically, while for the binding energy was solved numerically. The numerical binding energy for the presence of the modified Hartmann potential is lower than the binding energy value in the absence of modified Hartmann potential effect.

1. Introduction

The collective model of atomic nucleus has been introduced by Bohr and Mottelson since more than 60 years ago, was well-known as Bohr-Mottelson hamiltonian or Bohr Hamiltonian. It is important and attractive topics in nuclear physics to explain the properties and behavior of the nucleus. The model of collection motion for atomic nucleus treats the nucleus as a vibrating and rotating liquid drop. In this model, the focus is on some collective parameters like γ, the angle measuring origin from axial symmetry, β ,which represents deformation coordinate from the spherical shape, $Q_k (k = 1,2,3)$, which are the components of angular momentum in the intrinsic frame [1-3]. The solution of Bohr Hamiltonian in various cases have been investigated by some researchers. The solution for Davidson potential was analyzed in 1974 [4] and others paper involving Kratzer potential [5] and Morse potential [6] has been solved. Iachello used in the critical point symmetries E(5) and X(5) [7,8]. In the framework of the Bohr Hamiltonian, applying minimal length effect of considering gravitational influences in the system and it was used to satisfy experimental data. Effect of the presence and the absence of minimal length has been investigated in Coulomb-like potential on γ -rigid by Alimohammadi et.al in 2016 [3].And then, in 2017, Alimohammadi et. al analyzed Bohr Hamiltonian in γ -rigid with position-dependent mass [9]. In this paper, we are interested in comparing these two situations: the collective motion of atomic nucleus in the absence of minimal length effects in the presence and the absence of modified Hartmann potential in γ -rigid Bohr Hamiltonian. The Hartmann potential is an exactly solvable ring-shaped potential was introduced by H. Hartmann [10-11], is given as:

$$V(r,\theta) = \eta\sigma^2\varepsilon_0\left(\frac{2a_0}{r} - \frac{1}{r^2}\frac{q\eta a_0^2}{\sin^2\theta}\right) \quad (1)$$

where $a_0 = \hbar^2/B_m e^2$ and $\varepsilon_0 = -B_m e^4/2\hbar^2$ represent the Bohr radius and the ground state energy of the hydrogen atom, respectively, B_m is rest mass electron, e is a charge of the electron and η, σ, q are three dimensionless parameters. When $q = 0, \eta\sigma^2 = Z$, the Hartmann potential was reduced to the Coulomb potential, which Z is protons numbers of an atom. This ring-shaped potential was introduced to describe ring-shaped molecules like cyclic polyenes and benzene[12]. For double-ring-shaped, this potential is given by

$$V(r,\theta) = \eta\sigma^2\varepsilon_0\left(\frac{2a_0}{r} - \frac{1}{r^2}\frac{q\eta a_0^2}{\sin^2\theta}\right) + \frac{\hbar^2}{2B_m}\frac{c}{r^2\cos^2\theta}, \quad c \geq 0 \tag{2}$$

where c is a dimensionless parameter, and $0 < r < \infty, 0 \leq \theta \leq \pi$. The Hartmann potential in equation (2) is formed to explain the behavior of electron [10], so we modified it becomes the suitable form which can be applied to explain the collective motion of nucleus using Bohr Hamiltonian by changing the variable r into deformation coordinate β, is given as

$$V(\beta,\theta) = \eta\sigma^2\varepsilon_1\left(\frac{2a_1}{\beta} - \frac{1}{3\beta^2}\frac{q\eta a_1^2}{\sin^2\theta}\right) + \frac{\hbar^2}{2B_1}\frac{1}{3\beta^2}\frac{c}{\cos^2\theta} \tag{3}$$

where $a_1 = r_0 A^{1/3}$ and $\varepsilon_1 = a_{vol}A + a_{surf}A^{2/3} + a_{coul}Z^2A^{-1/3} + a_{sym}\left((N-Z)^2/A\right)$, which B_l is nucleon mass, A is a number of neutrons plus protons, N is a number of neutrons, r_0 is a radius of the nucleon $a_{vol} = -16$ MeV, $a_{surf} = 20$ MeV, $a_{coul} = 0.751$ MeV, $a_{sym} = 21.4$ MeV and η, σ, q, c are three dimensionless parameters. a_1 represent the nucleus radius and ε_1 is binding energy of nucleus from the Bethe-Weizsacker formula [13]. The Hartman potential in equation (2) and the modified Hartmann potential in equation (3) are shapes invariant, so the modification which was applied can be used, mathematically.

2. The Bohr Hamiltonian in the case γ-rigid in the absence of minimal length

The collective Bohr Hamiltonian in the case γ-rigid operator is given as [3]:

$$\hat{H} = -\frac{\hbar^2}{2B_1}\Delta + \frac{\alpha\hbar^4}{B_1}\Delta^2 + V(\beta,\theta,\varphi) \tag{4}$$

with

$$\Delta = \left[\frac{1}{\beta^2}\frac{\partial}{\partial\beta}\beta^2\frac{\partial}{\partial\beta} + \frac{1}{3\beta^2}\Delta_\Omega\right] \tag{5}$$

and Δ_Ω is the angular part of Laplace operator, is given as

$$\Delta_\Omega = \left[\frac{1}{\sin\theta}\frac{\partial}{\partial\theta}\sin\theta\frac{\partial}{\partial\theta} + \frac{1}{\sin^2\theta}\frac{\partial^2}{\partial\phi^2}\right] \tag{6}$$

For the absence minimal length effect, where $\alpha = 0$, the Hamiltonian operator in equation (4) is reduced to,

$$\hat{H} = -\frac{\hbar^2}{2B_1}\Delta + V(\beta,\theta,\varphi) \tag{7}$$

By inserting equations (5-6) into equation (7), and by using $H\Psi(\beta,\theta,\varphi) = E\Psi(\beta,\theta,\varphi)$ we obtained

$$-\frac{\hbar^2}{2B_1}\left[\frac{1}{\beta^2}\frac{\partial}{\partial\beta}\beta^2\frac{\partial}{\partial\beta} + \frac{1}{3\beta^2}\left[\frac{1}{\sin\theta}\frac{\partial}{\partial\theta}\sin\theta\frac{\partial}{\partial\theta} + \frac{1}{\sin^2\theta}\frac{\partial^2}{\partial\phi^2}\right]\right]\Psi(\beta,\theta,\varphi) = (E - V(\beta,\theta,\varphi))\Psi(\beta,\theta,\varphi) \tag{8}$$

where

$$V(\beta,\theta,\varphi) = V(\beta) + \left(V(\theta)/3\beta^2\right) + \left(V(\varphi)/3\beta^2\sin^2\theta\right) \tag{9}$$

By letting $\Psi(\beta,\theta,\varphi) = \left(R(\beta)/\beta\right)\Theta(\theta)\Phi(\varphi)$ in equation (8), substituting equation (9) into equation (8) with applying equation (3) and by using separation variable method, we obtained the three one-dimensional differential equations, are given as

$$\frac{d^2R}{d\beta^2} + \left(\frac{2B_1}{\hbar^2}\left(E - \frac{2a_1\eta\sigma^2\varepsilon_1}{\beta}\right) - \frac{\lambda}{3\beta^2}\right)R = 0 \tag{10}$$

$$\frac{d^2\Theta}{d\theta^2} + \left(\cot\theta\frac{d}{d\theta} + \lambda - \frac{\Lambda^2}{\sin^2\theta} - \frac{2B_1}{\hbar^2}\left(\frac{q\eta^2a_1^2\sigma^2\varepsilon_1}{\sin^2\theta} + \frac{c}{\cos^2\theta}\right)\right)\Theta = 0 \tag{11}$$

$$\frac{d^2\Phi}{d\varphi^2}+\Lambda^2\Phi=0 \tag{12}$$

where we have introduced the separation constants $\lambda=L(L+1)$ and Λ^2. L denotes the orbital angular momentum quantum number [9].

3. Solution of the Bohr Hamiltonian equation for modified Hartmann potential

3.1. Solution for β-part

In this part, we used equation (10), which is like Coulomb potential. By setting $\rho=\tau\beta, \tau=\sqrt{-8B_1E/\hbar^2}, s=\left(-2B_1/\hbar^2\right)\left(2a_1\eta\sigma^2\varepsilon_1/\tau\right)$ in equation (10), we get

$$\frac{d^2R}{d\rho^2}+\left(\frac{s}{\rho}-\frac{1}{4}-\frac{L(L+1)/3}{\rho^2}\right)R=0 \tag{13}$$

By taking the physically acceptable solution $R(\rho)=\rho^{\sqrt{\frac{L(L+1)}{3}+\frac{1}{4}}+\frac{1}{2}}e^{\frac{-\rho}{2}}f(\rho)$ and substituting into equation (13), yield

$$\rho\frac{d^2f}{d\rho^2}+\left(\left(2\sqrt{\frac{L(L+1)}{3}+\frac{1}{4}}+1\right)-\rho\right)\frac{df}{d\rho}-\left(\left(\sqrt{\frac{L(L+1)}{3}+\frac{1}{4}}+\frac{1}{2}\right)-s\right)f=0 \tag{14}$$

Equation (14) is the confluent hypergeometric differential equation, so for the eigenvalue, we obtained from relation $\left(\sqrt{\frac{L(L+1)}{3}+\frac{1}{4}}+\frac{1}{2}\right)-s=-n, n=0,1,2,\ldots$, yield

$$E_{n,L}=-\frac{B_1}{2\hbar^2}\left(2a_1\eta\sigma^2\varepsilon_1\Big/\sqrt{\frac{L(L+1)}{3}+\frac{1}{4}}+\frac{1}{2}+n\right)^2 \tag{15}$$

which it is in agreement with the result of the energy equation in Ref. 3, but different in potential constants. The numerical energies are shown in Table 1 for isotope Palladium (Pd) with $n=0, n=1$ by using equation (18), where L is not dependent in solution for the angular part, so it is not influenced by the angular part of the modified Hartmann potential. The binding energy for isotope Pd increase for the heavier isotopes, while for the increasing value of L and n, energy decrease. The binding energy gives an explanation about the rigid properties of the nucleus. The heavy nucleus which is indicated by a number of protons plus neutrons A has the greater binding energy, it shows that the heavy nucleus more rigid than the small nucleus. While, in the greater value of n, which indicates the nucleus orbital, the binding energy decrease, because in the outer orbital, the nucleus force decrease.

Table 1. Numerical energy spectra (in TeV) for isotope Palladium ($_{46}$Pd) with $\eta=0.2, \sigma=0.2$

Nucleus	$E_{0,2}$	$E_{0,4}$	$E_{0,6}$	$E_{0,8}$	$E_{1,2}$	$E_{1,4}$	$E_{1,6}$	$E_{1,8}$
^{102}Pd	46.312	18.909	10.137	6.296	20.583	10.861	6.657	4.488
^{104}Pd	47.743	19.494	10.450	6.490	21.219	11.196	6.863	4.628
^{106}Pd	49.132	20.061	10.754	6.678	21.837	11.522	7.063	4.761
^{108}Pd	50.423	20.588	11.037	6.855	22.410	11.825	7.249	4.887
^{110}Pd	51.670	21.097	11.309	7.024	22.964	12.117	7.428	5.008

Then, for the wave function, we use equation (14). Equation (14) is confluent hypergeometric differential equation, so the generalized wave function was obtained in hypergeometric term, is given as

$$R(\rho)=\rho^{\sqrt{\frac{L(L+1)}{3}+\frac{1}{4}}+\frac{1}{2}}e^{\frac{-\rho}{2}}\,{}_1F_1\left(\left(\sqrt{\frac{L(L+1)}{3}+\frac{1}{4}}+\frac{1}{2}\right)+a_1\eta\sigma^2\varepsilon_1\sqrt{\frac{-2B_1}{\hbar^2E}};\ \left(2\sqrt{\frac{L(L+1)}{3}+\frac{1}{4}}+1\right);\ \rho\right) \tag{16}$$

The wavefunction in equation (16) can be transformed to β function to explain about the behavior nucleus in varying deformation coordinates. Every nucleus has specific deformation coordinate which is considered with the number of protons plus neutrons A.

3.2. Solution for θ-part

For the θ-part solution, we used the variable substitution $\cos\theta = x$ into equation (11), yield

$$\left(1-x^2\right)\frac{d^2\Theta}{d\theta^2}-2x\frac{d\Theta}{d\theta}+\left(L\left(L+1\right)-\frac{d^2}{1-x^2}-\frac{2B_1c}{\mathrm{h}^2}\frac{1}{x^2}\right)\Theta=0 \quad (17)$$

where $d^2=\Lambda^2+\left(2B_1/\mathrm{h}^2\right)q\eta^2\sigma^2{a_1}^2\varepsilon_1$. Equation (17) is universal associated-Legendre differential equation when $c=0, L=k+d, |\Lambda|, k=0,1,2,...$ which has the three regular singular points $x=0,\pm1$, so we take the form,

$$\Theta=\left(1-x^2\right)^{d/2}x^{\delta}y\left(x\right) \quad (18)$$

where δ is taken as $\delta\left(\delta-1\right)-\left(2B_1c/\mathrm{h}^2\right)=0$, yield

$$\delta=\left\{0 \text{ or } 1,\ c=0\ ;\ \left[1+\sqrt{1+\left(8B_1/\mathrm{h}^2\right)c}\right]/2,\ c>0\right. \quad (19)$$

By substituting equation (18) into equation (17), give the differential equation is given as

$$\left(1-x^2\right)\frac{d^2y}{d\theta^2}+\left(\frac{2\delta}{x}-2\left(d+\delta+1\right)x\right)\frac{dy}{d\theta}+\left(L\left(L+1\right)-\left(d+\delta\right)\left(d+\delta+1\right)\right)y=0 \quad (20)$$

To solve equation (20), we used the series method by choosing the form: $y\left(x\right)=\sum_{\kappa=0}^{\infty}y_{\kappa}x^{\kappa}$ (21)

and substituting into equation (20), yield the values of quantum number L are given as,

$$L=2k+d+\delta_1=\begin{cases}2k+\sqrt{\Lambda^2+\frac{2B_1}{\mathrm{h}^2}q\eta^2\sigma^2{a_1}^2\varepsilon_1}+\left[1+\sqrt{1+\frac{8B_1}{\mathrm{h}^2}c}\right]/2, & c>0\\ 2k+\sqrt{\Lambda^2+\frac{2B_1}{\mathrm{h}^2}q\eta^2\sigma^2{a_1}^2\varepsilon_1}\ ;\ 2k+\sqrt{\Lambda^2+\frac{2B_1}{\mathrm{h}^2}q\eta^2\sigma^2{a_1}^2\varepsilon_1}+1, & c=0\end{cases} \quad (22)$$

and we obtained,

$$y(x)=\sum_{\kappa=0}^{\infty}\frac{(-1)^{\kappa}}{2^L\kappa!(k-\kappa)!}\frac{\Gamma\left(k+\delta-\kappa+1\right)}{\Gamma\left(2k+2\delta-2\kappa+1\right)}\times\frac{\Gamma\left(2L-2\kappa+1\right)}{\Gamma\left(L-\kappa+1\right)}x^{2k-2\kappa} \quad (23)$$

For the wavefunction, we substituted equation (23) into equation (18), so we obtained

$$\Theta=\left(1-x^2\right)^{d/2}x^{\delta}\sum_{\kappa=0}^{\infty}\frac{(-1)^{\kappa}}{2^L\kappa!(k-\kappa)!}\frac{\Gamma\left(k+\delta-\kappa+1\right)}{\Gamma\left(2k+2\delta-2\kappa+1\right)}\times\frac{\Gamma\left(2L-2\kappa+1\right)}{\Gamma\left(L-\kappa+1\right)}x^{2k-2\kappa} \quad (24)$$

The solution in θ-part gives two solutions considering of δ value in equation (19).

3.3. Solution for φ-part

Equation (12) which is the ordinary differential equation, so we get the solution is given as

$$\Phi=\left(1/\sqrt{2\pi}\right)e^{i\Lambda\varphi}, \Lambda=0,\pm1,\pm2,\pm3,... \quad (25)$$

so, total wavefunction was obtained by using equations (16, 24-25),

$$\Psi\left(\beta,\theta,\varphi\right)=\frac{1}{\beta}\rho^{\sqrt{\frac{L(L+1)}{3}+\frac{1}{4}}+\frac{1}{2}}e^{\frac{-\rho}{2}}{}_1F_1\left(\left(\sqrt{\frac{L\left(L+1\right)}{3}+\frac{1}{4}}+\frac{1}{2}\right)+a_1\eta\sigma^2\varepsilon_1\sqrt{\frac{-2B_1}{\mathrm{h}^2E}};\ \left(2\sqrt{\frac{L\left(L+1\right)}{3}+\frac{1}{4}}+1\right);\ \rho\right) \quad (26)$$

$$\times\left(1-x^2\right)^{d/2}x^{\delta}\sum_{\kappa=0}^{\infty}\frac{(-1)^{\kappa}}{2^L\kappa!(k-\kappa)!}\frac{\Gamma\left(k+\delta-\kappa+1\right)}{\Gamma\left(2k+2\delta-2\kappa+1\right)}\times\frac{\Gamma\left(2L-2\kappa+1\right)}{\Gamma\left(L-\kappa+1\right)}x^{2k-2\kappa}\frac{1}{\sqrt{2\pi}}e^{i\Lambda\varphi}$$

The wavefunction in equation (26) is depending on β function, the polar in θ function and azimuth in φ function which has the related quantum number. In this paper, we only give the mathematical solution in the wavefunction. We solved numerically the binding energy from equation (15) which depend on the polar solution and azimuth solution to compare the result of energy between the presence and the absence of the modified Hartmann potential. The result is shown in Table 2 for isotope Pd with the value of $k=0,1;\Lambda=0;c=0$ which L is obtained from the θ -part solution in equation (22).

Table 2. Numerical energy spectra (in MeV) for isotope Palladium ($_{46}$Pd) in the presence the Hartmann potential effect

	$k=0$		$k=1$	
Nucleus	E_{0,L_1}	E_{0,L_2}	E_{1,L_1}	E_{1,L_2}
^{102}Pd	1.262	1.258	1.248	1.244
^{104}Pd	1.253	1.250	1.240	1.236
^{106}Pd	1.245	1.241	1.232	1.228
^{108}Pd	1.237	1.233	1.224	1.220
^{110}Pd	1.229	1.225	1.216	1.213

The binding energies in Table 2 show that in the presence of the Hartmann potential in θ -part give the smaller value of energy if it was compared with the energy result which is shown in Table1 for the absence of the Hartmann potential. It indicated that the effect of Hartmann potential in binding energy gives the decreasing value of energy. For the isotope Palladium, the increase of value A or for the heavier nucleus, the binding energy decrease, it is reverse with the results in the absence of the modified Hartmann potential. The result in Table 2, the value of L_1 is smaller than L_2 which give the effect in the value of energy, where for the greater value of L, the energy decrease. The φ -part solution does not give the influence in the value of binding energy, but in wavefunction only, because the modified Hartmann potential which was used in this research does not have the φ -part potential. The binding energy of nucleus in the presence of the modified Hartmann potential give the large different with the results in the absence the Hartmann potential effect.

4. Conclusion

This paper has developed to compare the presence and the absence of the modified Hartmann potential effect in Bohr Hamiltonian in the case of gamma rigid and without the minimal length effect. The solution for wavefunction was expressed in mathematically in equation (26). The numerically result for binding energy for some nucleus were solved and there are largely different between the value energy for the presence modified Hartmann potential and without this potential. The modified Hartmann potential give effect in decreasing value of binding energy.

Acknowledgement
This work is partly funded by Hibah PUT-MRG 2017, Sebelas Maret University.

References

[1] Bohr A 1952 *K. Dan. Vidensk. Selks. Mat-Fys. Medd* **26** 14
[2] Bohr A and Mottelson B 1953 *K. Dan. Vidensk. Selks. Mat-Fys. Medd* **27** 16
[3] Alimohammadi M, Hassanabadi H and Sobhani H 2016 *Modern Physics Letter A* **31** 1650193
[4] Rohozinski S G, Srebrny J and Horbaczewska K 1974 *Z. Phys.* **268** 401
[5] Bonatsos D, Georgoudis P E, Minkov N. Petrellis D and Quesne C 2013 *Phys. Rev. C* **88** 034316
[6] Boztosum I, Bonatsos D and Inci I 2008 *Phys. Rev. C* **77** 044302
[7] Iachello F 2000 *Phys. Rev. Lett.* **85** 3580
[8] Iachello F 2001 *Phys. Rev. Lett.* **87** 052502
[9] Alimohammadi M, Hassanabadi H and Zare S 2017 *Nuclear Physics A* **960** 78

IOP Conf. Series: Journal of Physics: Conf. Series **1011** (2018) 012085 doi:10.1088/1742-6596/1011/1/012085

[10] Hartmann H and Schuch D 1980 *Int. J. Quantum Chem* **18** 125
[11] Chang-Yuan C, Fa-Lin L, Dong-Sheng S and Shi-Hai D 2013 *Chin. Phys. B* **22** 100302
[12] Rahbar H nd Sadeghi J 2016 *Theoretical Phyics* **1** 7
[13] Greiner W and Maruhn J A 1996 Nu clear Models Springer-Verlag Berlin Heidelberg New York

IOP Conf. Series: Journal of Physics: Conf. Series **1011** (2018) 012086 doi:10.1088/1742-6596/1011/1/012086

Analysis of Eigenvalue and Eigenfunction of Klein Gordon Equation Using Asymptotic Iteration Method for Separable Non-central Cylindrical Potential

A. Suparmi, C. Cari, and Isnaini Lilis Elviyanti

Physics Department, Graduate Program, Sebelas Maret University

Email: soeparmi@staff.uns.ac.id

Abstract. Analysis of relativistic energy and wave function for zero spin particles using Klein Gordon equation was influenced by separable noncentral cylindrical potential was solved by asymptotic iteration method (AIM). By using cylindrical coordinates, the Klein Gordon equation for the case of symmetry spin was reduced to three one-dimensional Schrodinger like equations that were solvable using variable separation method. The relativistic energy was calculated numerically with Matlab software, and the general unnormalized wave function was expressed in hypergeometric terms.

1. Introduction

Klein Gordon equation is used to describe the characteristic of spin-zero microscopic particles in many physics fields, relativistic quantum mechanics, in particular[1,2]. Klein Gordon equation for various potentials has been solved using different methods such as Kratzer potential [3] and Trigonometric Poschl-Teller potential using Asymptotic Iteration Method (AIM) [4]. Also, Klein Gordon equation is solved by using Nikiforov Uvarof for Hulten Potential [5] and using factorization method for Makarov Potential [6].

In this paper, Klein Gordon equation for separable non-central cylindrical potential in the spin symmetry case reduces into three one-dimensional Schrodinger like equations with shape invariant potential in the centrifugal approximation scheme. The three-dimensional separable non-central cylindrical potential [7] consists of a radial part potential was using Hyperbolic Manning Rosen potential [8], a angular part potential is using Hyperbolic Scarf type II potential [7] and a axial part (z part) is using trigonometric Scarf potential [9], and each potential part is invariant shape potential.The Hyperbolic Manning Rosen potential has been used to study quark-gluon dynamic [5,10]. Scarf trigonometric and hyperbolic type II potentials have been used to study the electrodynamic [11] and crystal model in solid state [12].The Schrodinger like an equation with invariant shape potential is reduced to the generalized hypergeometric differential equation that can be solved by AIM [13-15]. The paper is organized as follows. Klein Gordon equation and the separable non-central cylindrical potential are discussed in section 2 and AIM is briefly discussed in section 3. Results and discussion are presented in section 4 and conclusion in section 5.

2. Klein Gordon Equation with Separable Non–central Cylindrical Potential

The separable non-central cylindrical potential is a potential that its variable can be separated completely and it is given as [7],

$$V(r,\theta,z)=\left(\frac{v(v-1)}{\sinh^2\omega_r r}-2q\coth\omega_r r\right)+\frac{1}{r^2}\left(\frac{b^2+a(a-1)}{\sin^2\omega_\theta\theta}-\frac{2b(a-\frac{1}{2})\cos\omega_\theta\theta}{\sin^2\omega_\theta\theta}\right)+\left(\frac{b^2-a(a+1)}{\cosh^2\omega_z z}-\frac{2b(a+\frac{1}{2})\sinh\omega_z z}{\cosh^2\omega_z z}\right) \quad (1)$$

where v,q,a,b and ω are positive constants. This potential is applied to cylindrical coordinates which its Laplacian is expressed,

$$\Delta=\frac{1}{r}\frac{\partial}{\partial r}\left(r\frac{\partial}{\partial r}\right)+\frac{1}{r^2}\frac{\partial^2}{\partial\theta^2}+\frac{\partial^2}{\partial z^2} \quad (2)$$

with $0<r\le\infty, 0\le\theta\le 2\pi$, and $0<z<\infty$. The Klein Gordon equation is given by [4,16],

$$\left(E-V(\hat{r})\right)^2\psi(r)=\left[P^2c^2+\left(M_o c^2+S(\hat{r})\right)^2\right] \quad (3)$$

where $V(\hat{r})$ and $S(\hat{r})$ are vector and scalar potentials, E and M_o are relativistic energy and rest mass, respectively. For spin symmetric case $\left(S(\hat{r})=V(\hat{r})\right)$, and by taking the vector potential is as half of separable non-central cylindrical potential, then by using equations (1-3), the associated Klein Gordon equation as follows,

$$\left(\frac{1}{r}\frac{\partial}{\partial r}\left(r\frac{\partial}{\partial r}\right)+\frac{1}{r^2}\frac{\partial^2}{\partial\theta^2}+\frac{\partial^2}{\partial z^2}\right)\psi(r,\theta,z)+\left[\left(E^2-M_o^2\right)-\left(E+M_o\right)\left(\begin{array}{l}\left(\frac{v(v-1)}{\sinh^2\omega_r r}-2q\coth\omega_r r\right)\\+\frac{1}{r^2}\left(\frac{b^2+a(a-1)}{\sin^2\omega_\theta\theta}-\frac{2b(a-\frac{1}{2})\cos\omega_\theta\theta}{\sin^2\omega_\theta\theta}\right)\\+\left(\frac{b^2-a(a+1)}{\cosh^2\omega_z z}-\frac{2b(a+\frac{1}{2})\sinh\omega_z z}{\cosh^2\omega_z z}\right)\end{array}\right)\right]\psi(r,\theta,z)=0 \quad (4)$$

By using variable separation method equation (4) reduces to three one-dimensional Schrodinger like equations that are solvable by AIM.

3. Asymptotic Iteration Method

Asymptotic Iteration Method is applied to solve the second order differential equation which is expressed as [13-15],

$$y_n''(x)=\lambda_0(x)y_n'(x)+s_0(x)y_n(x) \quad (5)$$

where $\lambda_0(x)\neq 0$ and $s_0(x)$ are coefficient of a differential equation. The parameter n is the quantum number. By writing the $(k+1)^{th}$ derivative of y which is obtained from equation (5) given as,

$$y_n^{k+1}(x)=\lambda_{k-1}(x)y_n'(x)+s_{k-1}(x)y_n(x) \quad (6)$$

where $\lambda_k(x)=\lambda'_{k-1}(x)+s_{k-1}(x)+\lambda_0(x)\lambda_{k-1}(x)$; $s_k(x)=s'_{k-1}(x)+s_0(x)\lambda_{k-1}(x)$; $k=1,2,3...$ The Eigenvalue is obtained from the quantization condition given by

$$\Delta_k(x)=\lambda_k(x)s_{k-1}(x)-\lambda_{k-1}(x)s_k(x)=0 \quad (7)$$

The one-dimensional Schrodinger like an equation is reduced to hypergeometric type differential equation which given as,

$$y_n''(x)=2\left(\frac{ax^{N+1}}{1-bx^{N+2}}-\frac{t+1}{x}\right)y_n'(x)-\frac{wx^N}{1-bx^{N+2}}y_n(x) \quad (8)$$

The Eigenfunction which is obtained from equation (8) is expressed,

$$y_n(x)=(-1)^n C'(N+2)^n(\sigma)_n\ {}_2F_1\left(-n,\rho+n;\sigma;bx^{N+2}\right) \quad (9)$$

where $(\sigma)_n=\frac{\Gamma(\sigma+n)}{\Gamma(\sigma)}$; $\sigma=\frac{2t+N+3}{N+2}$; $\rho=\frac{(2t+1)b+2a}{(N+2)b}$, C' is normalization constant and ${}_2F_1$ is a hypergeometric function. The wave function of Klein Gordon equation can be obtained by using equation (7-9) [13-15].

4. Result and Discussion

The equation (4) was solved by using variable separation method by setting wave function $\psi(r,\theta,z)=R(r)P(\theta)Z(z)$, so we had three part of second order differential equations, axial, angular, and radial parts as,

$$\left(\frac{\partial^2 Z(z)}{\partial z^2}\right)-\left[(E+M_o)\left(\frac{b^2-a(a+1)}{\cosh^2\omega_z z}-\frac{2b(a+\frac{1}{2})\sinh\omega_z z}{\cosh^2\omega_z z}\right)+\lambda_{11}\right]Z(z)=0 \tag{10}$$

$$\left(\frac{\partial^2 P(\theta)}{\partial\theta^2}\right)-\left[(E+M_o)\left(\frac{b^2+a(a-1)}{\sin^2\omega_\theta\theta}-\frac{2b(a-\frac{1}{2})\cos\omega_\theta\theta}{\sin^2\omega_\theta\theta}\right)+\lambda_{12}\right]P(\theta)=0 \tag{11}$$

$$\left(\frac{1}{r}\frac{\partial}{\partial r}\left(r\frac{\partial R(r)}{\partial r}\right)\right)-\left[(E+M_o)\left(\frac{v(v-1)}{\sinh^2\omega_r r}-2q\coth\omega_r r\right)-(E^2-M_o^2)-\lambda_{11}-\frac{\lambda_{12}}{r^2}\right]R(r)=0 \tag{12}$$

with λ_1 and λ_2 is variable separation constants. Equations (10-12) were solvable using AIM since all potentials in these three equations were shape invariant potentials.

4.1. *Solution of The Axial Part of Klein Gordon Equation*

By setting transformation variable of $\sinh\omega_z z=i(1-2u_z)$ and the new wave function of $Z=u_z^{\alpha_z}(1-u_z)^{\beta_z}f(u_z)$ in equation (10) which was reduced to AIM type equation by dividing it with $u_z(1-u_z)$, so we obtained,

$$f''(u_z)+\left[\frac{\left(2\alpha_z+\frac{1}{2}\right)-(2\alpha_z+2\beta_z+1)u_z}{u_z(1-u_z)}\right]f'(u_z)+\left[\frac{\lambda_z-(\alpha_z+\beta_z)^2}{u_z(1-u_z)}\right]f(u_z)=0 \tag{13}$$

where,

$$\eta_z=\frac{(E-M_o)}{\omega_z^2}\ ;\ \lambda_z=\frac{\lambda_{11}}{\omega_z^2}\ ;\ \alpha_z=\frac{\frac{1}{2}+\sqrt{-\eta_z\left[\left(b-i(a+\frac{1}{2})\right)^2+\frac{1}{4}\right]+\frac{1}{4}}}{2}\ ;\ \beta_z=\frac{\frac{1}{2}+\sqrt{-\eta_z\left[\left(b+i(a+\frac{1}{2})\right)^2+\frac{1}{4}\right]+\frac{1}{4}}}{2} \tag{14}$$

By comparing equation (5) and equation (13), yields,

$$\lambda_0=\frac{-(2\alpha_z+\frac{1}{2})}{u_z}+\frac{(2\beta_z+\frac{1}{2})}{1-u_z};\ s_o=\frac{(\alpha_z+\beta_z)^2-\lambda_z}{u_z}+\frac{(\alpha_z+\beta_z)^2-\lambda_z}{(1-u_z)} \tag{15}$$

Using equations (6-7) and (13), we obtained Eigenvalue function, was given as

$$\lambda_1=\omega_z^2(\alpha_z+\beta_z+n_z)^2 \tag{16}$$

The n_z is an axial quantum number. The axial parts of wave functions were obtained with equations (7-9), (13), $\sinh\omega_z z=i(1-2u_z)$ and $Z=u_z^{\alpha_z}(1-u_z)^{\beta_z}f(u_z)$, we got wave function for $n_z=0$ and $n_z=1$ as follows,

$$Z_0(z)=C'\left(\frac{1-i\sinh\omega_z z}{2}\right)^{\alpha_z}\left(\frac{1+i\sinh\omega_z z}{2}\right)^{\beta_z} \tag{17}$$

$$Z_1(z)=-C'\left(\frac{1-i\sinh\omega_z z}{2}\right)^{\alpha_z}\left(\frac{1+i\sinh\omega_z z}{2}\right)^{\beta_z}\left[\left(2\alpha_z+\frac{1}{2}\right)\left(1-\frac{(2\alpha_z-2\beta_z-1)\left(\frac{i\sinh\omega_z z-1}{2}\right)}{\left(2\alpha_z+\frac{1}{2}\right)}\right)\right] \tag{18}$$

4.2. *Solution of The Angular Part of Klein Gordon Equation*

By applying $\cos\omega_\theta\theta=(1-2u_\theta)$ and $P(\theta)=u_\theta^{\alpha_\theta}(1-u_\theta)^{\beta_\theta}f(u_\theta)$ in equation (11), we got an equation that had similar form with equations (5) and (8), was given as,

$$f''(u_\theta)+\left[\frac{\left(2\alpha_\theta+\frac{1}{2}\right)-(2\alpha_\theta+2\beta_\theta+1)u_\theta}{u_\theta(1-u_\theta)}\right]f'(u_\theta)+\left[\frac{\lambda_\theta-(\alpha_\theta+\beta_\theta)^2}{u_\theta(1-u_\theta)}\right]f(u_\theta)=0 \tag{19}$$

where,

$$\eta_\theta = \frac{(E-M_o)}{\omega_\theta^2};\ \lambda_\theta = \frac{\lambda_{12}}{\omega_\theta^2};\ \alpha_\theta = \frac{\frac{1}{2}+\sqrt{-\eta_\theta\left[\left(b-\left(a+\frac{1}{2}\right)\right)^2-\frac{1}{4}\right]+\frac{1}{4}}}{2};\ \beta_\theta = \frac{\frac{1}{2}+\sqrt{-\eta_\theta\left[\left(b+\left(a+\frac{1}{2}\right)\right)^2-\frac{1}{4}\right]+\frac{1}{4}}}{2} \tag{20}$$

By comparing equation (5) and equation (19), we obtained,

$$\lambda_0 = \frac{-\left(2\alpha_\theta+\frac{1}{2}\right)}{u_\theta}+\frac{\left(2\beta_\theta+\frac{1}{2}\right)}{1-u_\theta};\ s_o = \frac{(\alpha_\theta+\beta_\theta)^2-\lambda_\theta}{u_\theta}+\frac{(\alpha_\theta+\beta_\theta)^2-\lambda_\theta}{(1-u_\theta)} \tag{21}$$

Using equation (6-7) and (19), we obtained the Eigenvalue, as follows,

$$\lambda_{12} = \omega_\theta^2\,(\alpha_\theta + \beta_\theta + n_\theta)^2 \tag{22}$$

where, n_θ is an angular quantum number. By comparing equations (8) and (19), equations (9-10) and also it was substituted into $P(\theta) = u_\theta^{\alpha_\theta}(1-u_\theta)^{\beta_\theta} f(u_\theta)$ and $\cos\omega_\theta\theta = (1-2u_\theta)$, we had the wave functions of the angular part for $n_z=0$ and $n_z=1$ as,

$$P_0(\theta) = C'\left(\frac{1-\cos\omega_\theta\theta}{2}\right)^{\alpha_\theta}\left(\frac{1+\cos\omega_\theta\theta}{2}\right)^{\beta_\theta} \tag{23}$$

$$P_1(\theta) = -C'\left(\frac{1-\cos\omega_\theta\theta}{2}\right)^{\alpha_\theta}\left(\frac{1+\cos\omega_\theta\theta}{2}\right)^{\beta_\theta}\left[\left(2\alpha_\theta+\frac{1}{2}\right)\left(1-\frac{(2\alpha_\theta-2\beta_\theta-1)\left(\frac{\cos\omega_\theta\theta-1}{2}\right)}{\left(2\alpha_\theta+\frac{1}{2}\right)}\right)\right] \tag{24}$$

4.3. Solving The Radial Part of Klein Gordon Equation

By inserting $R(r) = Q(r)/\sqrt{r}$, $\frac{1}{r^2} = \frac{\omega_r^2}{\sinh^2(\omega_r r)}$ [14] and $\coth\omega_r r = 1-2u_r$ in equation (14), then it must be reduced to AIM type equation by setting $Q(r) = u_r^{\alpha_r}(1-u_r)^{\beta_r} f(u_r)$ and by dividing it with $u_r(1-u_r)$, then we obtained,

$$f''(u_r)+\left[\frac{(2\alpha_r+1)-(2\alpha_r+2\beta_r+2)y}{u_r(1-u_r)}\right]f'(u_r)+\left[\frac{E_r-(\alpha_r+\beta_r)(\alpha_r+\beta_r+1)}{u_r(1-u_r)}\right]f(u_r)=0 \tag{25}$$

where,

$$E_r = \left(\frac{1}{4}+\lambda_{12}-(E+M_o)v(v-1)\right);\ \eta_r = \frac{(E+M_o)}{\omega_r^2};\ \chi_r = \frac{(E^2-M_o^2)+\lambda_{11}}{\omega_r^2};\alpha_r = \sqrt{\frac{-2q(E+M_o)+(E^2-M_o^2)+\lambda_{11}}{4\omega_r^2}};\ \beta_r = \sqrt{\frac{2q(E+M_o)+(E^2-M_o^2)+\lambda_{11}}{4\omega_r^2}} \tag{26}$$

By comparing equation (5) and equation (19), we got

$$\lambda_0 = \frac{-(2\alpha_r+1)}{u_r}+\frac{(2\beta_r+1)}{1-u_r};\quad s_o = \frac{(\alpha_r+\beta_r)(\alpha_r+\beta_r+1)-E_r}{u_r}+\frac{(\alpha_r+\beta_r)(\alpha_r+\beta_r+1)-E_r}{(1-u_r)} \tag{27}$$

Using equation (6-7) and equation (19), we obtained the Eigenvalue, yields,

$$E_r = (\alpha_r+\beta_r+n_r)(\alpha_r+\beta_r+(n_z+1)) \tag{28}$$

From the equation (29), we obtained the relativistic energy equation by using equations (26) and (28). The relativistic energy equation of Klein Gordon equation for separable non-central cylindrical potential was,

$$(E^2-M_o^2) = \omega_r^2\left[\left(\sqrt{\left(\lambda_{12}+\frac{1}{4}-(E+M_o)v(v-1)\right)+\frac{1}{4}}-\left(n_r+\frac{1}{2}\right)\right)^2-\frac{\left(2q\left(\frac{(E+M_o)}{\omega_r^2}\right)\right)^2}{4\left(\sqrt{(\lambda_{12}+\delta_r)+\frac{1}{4}}-\left(n_r+\frac{1}{2}\right)\right)^2}\right]-\lambda_{11} \tag{29}$$

The n_r is an axial quantum number. The relativistic energy of equation (29) could not calculate analytically. Therefore, to get the equation (29), we had to apply Matlab software to calculate numerically. The result was shown in Table 1.

Table 1. The relativistic energy for n_r=n_z=1 and $n_\theta = 0$ with various parameters of a separable non-central cylindrical potential.

Value of constant *a, b, q,v*	*E (a)*	*E (b)*	*E (q)*	*E (v)*
0	-0.9898	-0.9896	-0.9981	-0.9895
1	-0.9895	-0.9895	-0.9895	-0.9895
2	-0.9887	-0.9889	-0.9984	-0.9887
3	-0.9870	-0.9872	-0.9993	-0.9869
4	-0.9864	-0.9868	-0.9996	-0.9600
5	-0.9852	-0.9861	-0.9997	-0.9812

Table 1 shows that the relativistic energy increased due to the increase a function of *a, b,* and *v*. The relativistic energy decreased by the increase a function of *q*. The radial wave functions were obtained with equation (8-10) and (19) , $Q(r)=u_r^{\alpha_r}(1-u_r)^{\beta_r} f(u_r)$ and together with $\coth\omega_r r=(1-2u_r)$, we had wave functions for $n_z=0$ and $n_z=1$, was given as

$$Q_0(r)=C'\left(\frac{1-\coth\omega_r r}{2}\right)^{\alpha_r}\left(\frac{1+\coth\omega_r r}{2}\right)^{\beta_r} \tag{30}$$

$$Q_1(r)=-C'\left(\frac{1-\coth\omega_r r}{2}\right)^{\alpha_r}\left(\frac{1+\coth\omega_r r}{2}\right)^{\beta_r}\left[(2\alpha_\theta+1)\left(1-\frac{(2\alpha_\theta-2\beta_\theta-2)\left(\frac{\coth\omega_r r-1}{2}\right)}{(2\alpha_\theta+1)}\right)\right] \tag{31}$$

The total wave function was obtained by multiplying equations for ground state wave function (18), (23), (30) and multiplication of equations (19), (24), (31) for the first excited state wave function.

5. Conclusion

Asymptotic iteration method (AIM) was used to analyze the solution Three-dimensional Klein Gordon equation with separable non-central cylindrical potential that was reduced to radial, angular, and axial of Klein Gordon equation by the use of cylindrical coordinates. The relativistic energy equation was obtained from the solution of radial Klein Gordon equation that was calculated numerically by using Matlab software. The relativistic energy increased with the increase of *a*, *b* and *v*, but it decreased by the increase of *q*. The wave function was expressed by hypergeometric terms.

Acknowledgement
This research was partly supported by Hibah PUT-MRG 2017 Sebelas Maret University.

References

[1] Ikhdair S M and Sever R 2008 *arXiv:0808.1002v1 [quant-ph]* 14.
[2] Momtazi E, Rajabi A A, and Yazarloo B H 2014 *Turkish J. Phys* 38 1 81–85.
[3] Nugraha D A, Suparmi A, Cari, C and Pratiwi B N 2017 *J. Phys. Conf. Ser* 795.
[4] Nugraha D A, Suparmi A, Cari, C and Pratiwi B N 2017 *J. Phys. Conf. Ser* 820.
[5] A N Ikot, L E Akpabio, and E J Uwah 2011 *Electron. J. Theor. Phys* 25 25 225-232.
[6] Z Min-cang and W Zhen-Bang 2007 *Chinese Phys* 16 7 1863–1867.
[7] Pratiwi B N, Suparmi A, Cari C, and Anwar F 2016 *J. Phys. Conf. Ser* 2016.
[8] Cari C, Suparmi A, Yunianto M, and Pratiwi B N 2016 *J. Phys. Conf. Ser.*
[9] Cari C, Suparmi A, Pratiwi B N, and Deta A U 2016 *AIP Conference proceeding.*
[10] D E Alvarez Castillo, C B Compean, and M Kirchbach 2011 arXiv:1105.1354v1[quant-ph.
[11] D E Alvarez Castillo and M Kirbach 2007 Rev. Mex. Fis 53.
[12] Negro J, Nieto L M, and Ortiz O R, 2004 *J. of Phy A: Mathematical and General* 37/4310079.
[13] H Ciftci, R L Hall, and N Saad 2003 *J. Phys. A. Math. Gen* 36 47 11807-11816.
[14] Pratiwi B N, Suparmi A, Cari C, and Husein S A 2017 *Pramana. J. Phys.*
[15] Pramono S, Suparmi A, and Cari C 2016 *Adv. High Energy Phys* 2016.
[16] Jana K T and Roy P 2009 *Physics Letters A* 373 1239–1241.

Transition from Markovian to Glassy Dynamics in Damped Kuramoto-Sivashinsky Turbulence

F Nugroho, H Hamadi, and P Nurwantoro

Department of Physics, Universitas Gadjah Mada, Sekip Utara BLS 21 Yogyakarta 55281, Indonesia

Email: fakhrud@ugm.ac.id

Abstract. In this research, we study Damped Kuramoto-Sivashinsky (DKS) equation, one of the famous nonlinear differential equation which exhibits turbulence. To solve DKS equation, we use Exponential Time Differencing (ETD) scheme which is combined with the Pseudospectral method. To know the effect of damping factor term, we analyze the resulting dynamics using Lyapunov exponent and total autocorrelation function. From that analysis, we elucidate the transition of Markovian to glassy dynamics in DKS equation. We conclude that there is a kind of mixing of several modes in the dynamics of the DKS equation which resulting the glassy dynamics in this system.

1. INTRODUCTION

One of the most famous and ubiquitous nonlinear phenomena is turbulence. It is observed in nature as well as in the experimental setup. The well-known experimental system showing turbulence are Rayleigh-Benard convection [1], electro convection in liquid crystal [2] and turbulence in chemical reaction [3]. The study of turbulence has also been conducted using mathematical modeling, i.e. by the nonlinear partial differential equations (PDEs). One of the fundamental models in the research of nonlinear phenomenon is Kuramoto-Sivashinsky (KS) equation.

In the present research, we study KS equation with damping factor as follows [4][5][6]

$$\frac{\partial u}{\partial t} = ru - u - 2\frac{\partial^2 u}{\partial x^2} - \frac{\partial^4 u}{\partial x^4} - u\frac{\partial u}{\partial t}, \tag{1}$$

where r is damping factor. Using a perturbation $u(x,t) \propto \exp[\omega t + ikx]$ in the linear term, one will obtain the dispersion relation of $\omega(k) = (r-1)k + 2k^2 - k^4$. From Fig.1, we see the shift of curve due to the damping factor r indicating that the wave number involves are decreased.

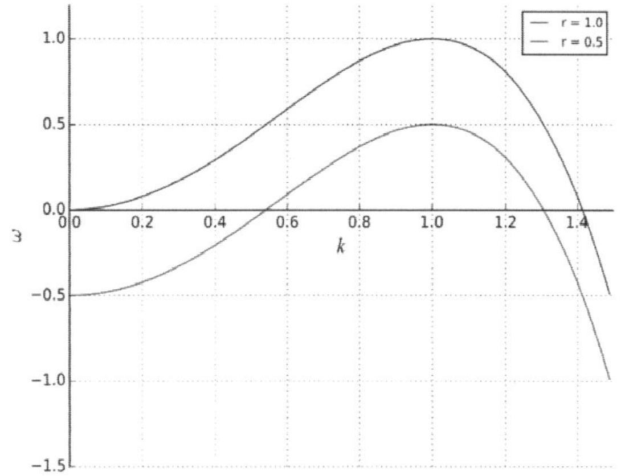

FIGURE 1. (color line) plot of dispersion relation $\omega(k) = (r-1)k + 2k^2 - k^4$ of DKS equation for two different values of *r*, namely *r=1* and *r=0.5*.

Several studies have been done to solve DKS equation using numerical treatment. Elder et al. [4] had examined the transition to spatiotemporal chaos in DKS equation. The transition to spatiotemporal chaos can be seen from the shift of energy spectrum. Brunet [7] using DKS equation as a model of secondary instabilities and dynamics in the cellular flow. Other research has completed in two-dimensional DKS equation, such as Gomes and Paris [6]. Here we report an evident of transition from Markovian to glassy dynamics at a certain value of r which hasn't been reported in the previous research.

2. Computational Methods

In this research, we solve the DKS equation using pseudo spectral method which combined with Exponential Time Differencing (ETD) scheme [8][9][10] and periodic boundary condition and using traveling wave as initial function. Note that the initial function in the present study arbitrary selected because it will not change the resulted solution. The ETD scheme is one of the numerical methods to solve differential equation especially stiff differential equation [8]. First, we perform the Fourier series of u as the following

$$u(x,t) = \sum_{n=1}^{N} \hat{u}_n \exp[iK_n x], \tag{2}$$

where $K_n = 2n\pi/L$. Using equation (2), the equation (1) can be rewritten as

$$\frac{d\hat{u}_n}{dt} = \left((r-1)K_n + 2K_n^2 - K_n^4\right)\hat{u} - \frac{iK_n}{2}\varpi_n, \tag{3}$$

Where ϖ_n is Fourier Transform of u^2.

To solve the first term of equation (1) the ETD2 written by

$$u_{n+1} = u_n e^{ch} + F_n \frac{\left((1+hc)e^{ch} - 1 - 2hc\right)}{hc^2} + F_{n-1}\frac{\left(-e^{ch} + 1 + hc\right)}{hc^2} \tag{4}$$

where $c = \left((r-1)K_n + 2K_n^2 - K_n^4\right)$ and $F_n = -\frac{iK_n}{2}\varpi_n$.

3. RESULT AND DISCUSSION

It is well known that the solution of KS equation depends on system size used (L) [11][12][13], in the present research *L*=256 and discretization number *N*=512. The maximum computational process is $t_{\max} = 1000$

with the time discretization of $h = 0.01$. The traveling wave equation $u(x, t = 0) = 0.5\sin(x)$ was used as an initial function. From such initial function, the solution the DKS equation shown by spatiotemporal plots in the Fig. 2.

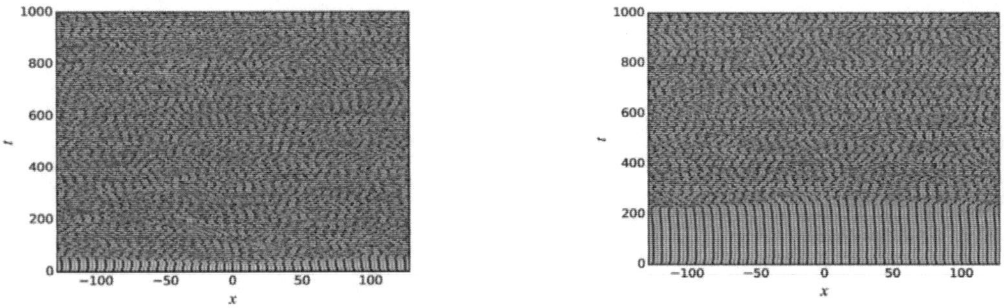

(a) (b)

FIGURE 2. Spatiotemporal plots of solution DKS equation using ETD2 from two damping value factor, $(a)\, r = 1.0$ and $(b)\, r = 0.9$.

Fig.2 shows that by setting the damping factor with higher value, the fluctuation in the DKS turbulence is suppressed. The smaller the damping factor, the longer the periodic structure in the spatiotemporal plot from $t \approx 25$ at (a) and $t \approx 200$ [arbitrary units] at (b). In Fig.2a the orderly area occurs when $t < 100$, while Fig.2b the order area increase for $t < 200$.

To characterize the transition from simple to the complex dynamic, we plot the orbit in the phase space for many values of r as shown in Fig.3. From Fig 3 we see that the attractor change from unstable periodic (see Fig.3a) to the stable fixed point (see Fig.3d), i.e., there is a transition from complex dynamics to a stable one.

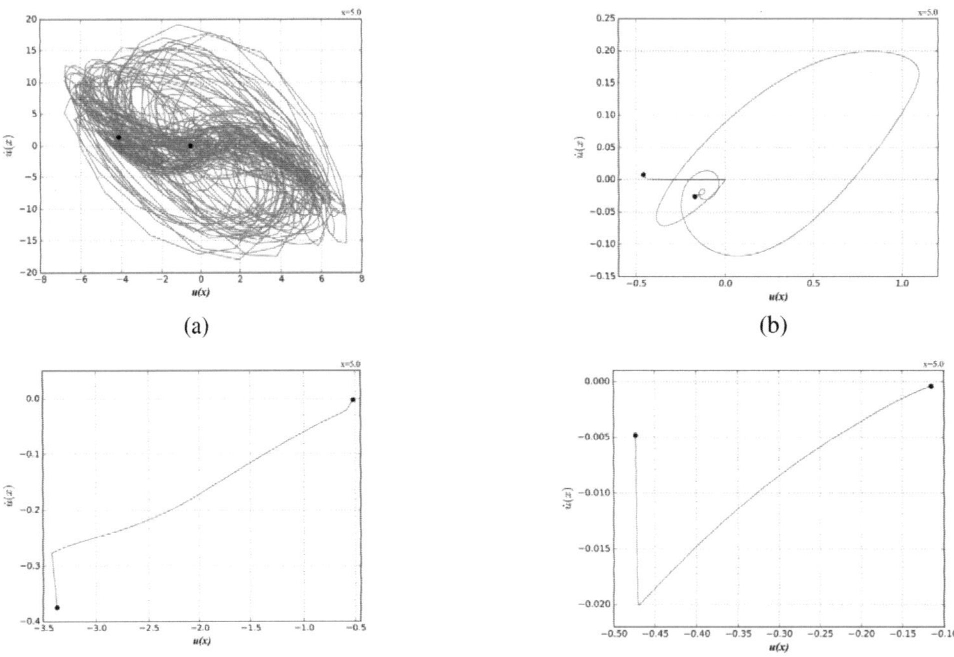

FIGURE 3. The orbit in phase space diagram of DKS equation, (a) r=0.9, (b) r=0.85, (c) r=0.5, and r=0.0.

IOP Conf. Series: Journal of Physics: Conf. Series **1011** (2018) 012087 doi:10.1088/1742-6596/1011/1/012087

For further consideration of transition of dynamics in DKS turbulence, we use dynamical analysis method using Lyapunov exponents (λ) [5][14]. In the present study, the Lyapunov exponent is calculated using an algorithm built by Rosenstein et al. [15].

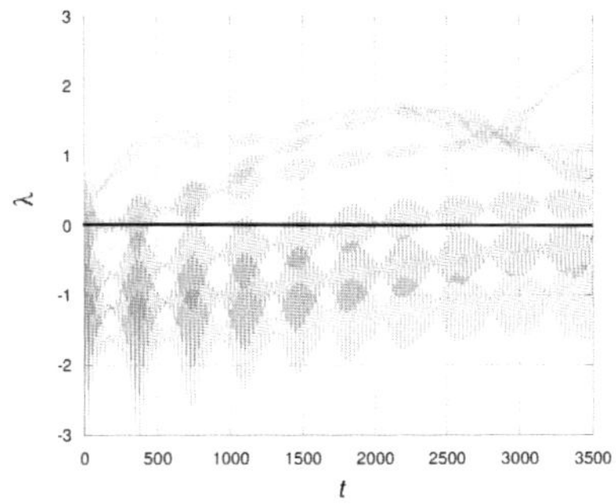

FIGURE 4. (a) Lyapunov Exponents (λ) of DKS equation for several values of damping factor *r* that are *r*=0.863(red), *r*=0.864(blue), *r*=0.865(dark-green), *r*=0.866(green), *r*=0.867(magenta), *r*=0.868(aqua), *r*=0.869(yellow), *r*=0.87(orange), and *r*=0.871(skyblue).

Figure 4 shows the Lyapunov exponents (λ) of DKS equation for several values of damping factor *r*. Fig.4 shows that there is a transition from ordinary to chaotic dynamics of fluctuations in DKS equation at a certain damping factor of $r \approx 0.87$.

Next, to study the statistical properties of DKS equation, we use the total autocorrelation function of the evolution of *u* as follows

$$Q(x,\tau) = \langle \Delta u(x,t+\tau)\Delta u(x,t)\rangle_t \langle \Delta u(x,t)^2 \rangle^{-1}, \tag{6}$$

To take into account all points in the simulation, we take the average of all 512 positional points. The autocorrelation function from various damping factor is shown in Fig.5.

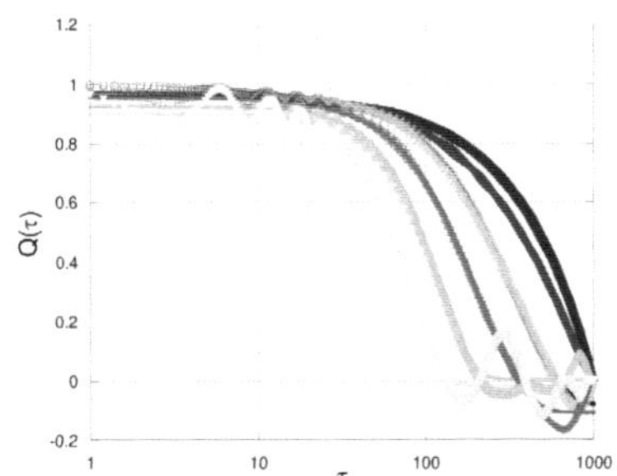

FIGURE 5. The autocorrelation function of DKS equation for *r*=0.865(red), *r*=0.861(blue), *r*=0.865(dark-green), *r*=0.868(green), *r*=0.87(magenta), *r*=0.872(aqua), *r*=0.875(yellow).

Figure 5 shows that the autocorrelation function is fitted by Kohlrausch–Williams–Watts (KWW) function [19] which fulfills

$$Q(\tau) = m_1 + m_2 \exp\left[-\left(\frac{t}{\tau}\right)^{\beta}\right], \tag{7}$$

where m_1, m_2, β, τ is a constant, normalization parameter, KWW exponent and correlation time, respectively. From parameter in KWW fitting function, we can get the relationship between β and τ with *r* as shown by Fig.6

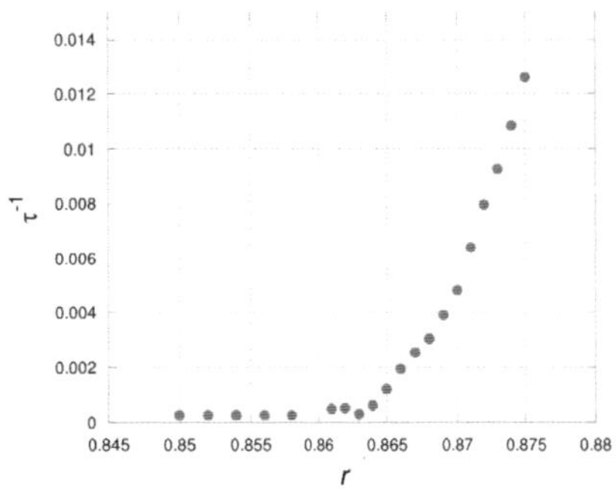

(a)

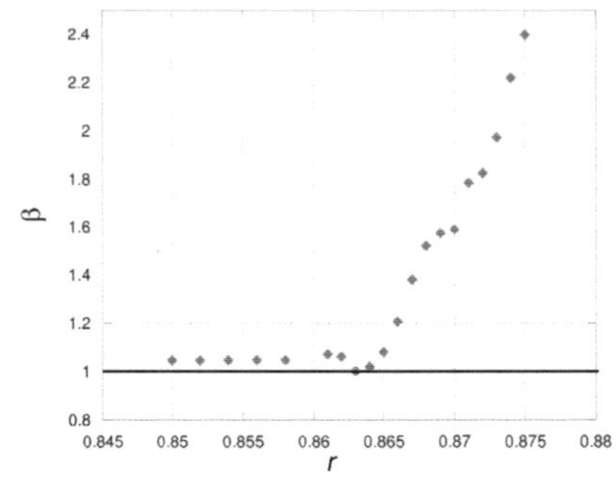

(b)

FIGURE 6. r dependence of τ^{-1} shown in (a) and β in (b).

Figure 6 (a) shows that at $r < 0.864$ the value of $\tau^{-1} \approx 0$ which corresponds to order dynamics. This behavior of the relaxation time, While at $r > 0.864$ the value of τ^{-1} increases indicating that the correlation time decreases meaning that complex dynamics occur. Fig 6 (b) shows the behavior of KWW exponent β, which is important to know what kind of dynamics produced in the system [16][17][18]. Note that $\beta = 1$ at $r < 0.864$, i.e simple exponential decay of autocorrelation indicates the Markovian dynamics. While the $\beta > 1$ at $r > 0.864$, i.e compressed exponential represents glassy dynamics [17][18]. Thus we conclude that there is a transition from Markovian to glassy dynamics in DKS by increasing damping factor.

4. CONCLUSION

In summary, we have solved DKS equation with damping factor is r with numerical treatment using ETD2 combined with the pseudo spectral method and analyzed the resulting dynamics using Lyapunov exponent and total autocorrelation function. In this research there are two kinds of attractors, the first attractor is an unstable periodic orbit that will generate chaotic dynamics for $r > 0.85$, and the second attractor type is a stable fixed point that produces order or ordinary dynamics for $r < 0.85$. The Lyapunov exponent and autocorrelation of the DKS equation show the transition from Markovian to glassy dynamics in DKS by increasing damping factor. Thus we report that a new kind of dynamics observed in the DKS equation.

ACKNOWLEDGMENT

The work has been financially supported by Indonesia Directorate General of Higher Education (DIKTI) with contract number 119/LPPM/2015 and 0204/J01.1.28/PL.06.02/2017. HH acknowledge the fresh graduate scholarship by DIKTI.

References

[1] Ahlers G, Walden RW 1980 *Phys. Rev. Lett.* **44**, 445.
[2] Hidaka Y, Huh J H, Hayashi KI, Kai S, Tribelsky MI 1997 *Phys. Rev. E.* **56**, R6256.
[3] Kuramoto Y, Tsuzuki T 1976 *Prog. Theor. Phys.* **55**, 356.
[4] Elder K, Gunton J, Goldenfeld N 1997 *Phys. Rev. E.* **56**, 1631.
[5] Schuster HG and Just W 2006 Deterministic chaos (Weinheim:WILEY-VCH Verlag GmbH & Co. KGaA,).
[6] Gomez H, Paris J 2011 Phys. Rev. E, **83**, 1.
[7] Brunet P 2007 *Phys. Rev. E*, **76**, 3.
[8] Cox SM, Matthews PC 2002 J. Comput. Phys. **176**, 430.
[9] Du Q, Zhu W 2004 J. Comput. Math. **22**, 1.

IOP Conf. Series: Journal of Physics: Conf. Series **1011** (2018) 012087 doi:10.1088/1742-6596/1011/1/012087

[10] Simbawa E, Matthews PC, Cox SM 2010 Phys. Rev. E **81**, 1.
[11] Shraiman B 1986 Phys. Rev. Lett. 27.
[12] Nicoli M, Vivo E, Cuerno R 2010 Phys. Rev. E **82**, 1.
[13] Hayot F, Jayaprakash C, Josserand C 1993 Phys. Rev. E. **47**.
[14] Geist K, Parlitz U, Lauterborn W 1990 Prog. Theor. Phys. **83**, 875.
[15] Rosenstein MT, Collins JJ, Luca CJ 1993 Phys. D. **65**, 117.
[16] Anderssen RS, Husain SA, Loy RJ 2004 ANZIAM Journal **45**, C800.
[17] Nugroho F, Narumi T, Hidaka Y, Yoshitani J, Suzuki M, Kai S 2012 Phys. Rev. E. 030701, 1.
[18] Bellour M, Knaebel A, Harden JL, Lequeux F, Munch J 2003 Phys. Rev. E. 67, 1.

IOP Conf. Series: Journal of Physics: Conf. Series **1011** (2018) 012088 doi:10.1088/1742-6596/1011/1/012088

Simulation of 2D Waves in Circular Membrane Using Excel Spreadsheet with Visual Basic for Teaching Activity

R Eso[1,a), L O Safiuddin[1], L Agusu[2], L M R F Arfa[1]

[1]Department of Physics Education, Faculty of Education, Universitas Halu Oleo, Indonesia
[2]Department of Physics, Faculty of Mathematics and Sciences, Universitas Halu Oleo, Indonesia

[a)]Corresponding author: ros_eso@yahoo.com

Abstract. We propose a teaching instrument demonstrating the circular membrane waves using the excel interactive spreadsheets with the Visual Basic for Application (VBA) programming. It is based on the analytic solution of circular membrane waves involving Bessel function. The vibration modes and frequencies are determined by using Bessel approximation and initial conditions. The 3D perspective based on the spreadsheets functions and facilities has been explored to show the 3D moving objects in transitional or rotational processes. This instrument is very useful both in teaching activity and learning process of wave physics. Visualizing of the vibration of waves in the circular membrane which is showing a very clear manner of m and n vibration modes of the wave in a certain frequency has been compared and matched to the experimental result using resonance method. The peak of deflection varies in time if the initial condition was working and have the same pattern with matlab simulation in zero initial velocity

Key words: *Visual Basic for Application, Vibration Modes, MS Excel*

1. Introduction

Teaching of physics sometimes difficult to be delivered without demonstration or visualization tools. Simulations can help teacher in creating or performing the real conditions in front of the class. Simulation can be expected to be an effective tool in helping students to explore their higher levels of learning capabilities such as analysis, synthesis and evaluation. Many authors plead for the use of spreadsheets demonstrating their multiple advantages, describing their facile use in physics lessons with examples of data processing according to tables and their graphic representation [1]. Moreover, an emphasis is placed upon their calculus capacity through the functions available for users, and the rapid feedback when modifying data [2]. Studies have shown that students have a better grasp of the definition of certain measures by dealing with basic calculi, as for instance the notion of velocity, while the instantaneous graphic response of the finite product represents an advantage in acquiring the concepts of physics [3]. In a series of recent papers, excel spreadsheets have helped to describe the simulation of motion in the case of a projectile motion in a gravitational field [4].

Successful in learning of physics is determined by the better understanding to the concept, because physics is generally a collection of concepts explaining the natural phenomena experienced by the students in their everyday life. The facts from natural phenomena are investigated and tested through repeated experiments, and then based on them, the theory is formulated.

In this report, we propose the simulation of vibration modes and its frequency on the circular membrane waves using the Excel interactive spreadsheets with the Visual Basic for Application (VBA) programming. The wave deflection on circular membran was analysed theoretically by using the separated variable method. Here, the Fourier series involving the Bessel function was selected as an initial condition. We have explored a series of facilities available on Excel spreadsheets, such as the perspective view of the 3D object in the further way as well as the 3D graphical scattering of in translational and rotational processes. The simulation results were compared to the experimental results from Radiansyah *et al.*, using the audio frequency generator as the wave source in resonance condition with vibration modes of mica membrane (diameter of 23.4 cm) poured by sand beach [5].

2. Methods

a. Two Dimensional Wave In Circular Membrane

The equation motion of 2D wave in circular membrane is described as

$$\frac{\partial^2 Z}{\partial t^2} = c^2 \; \frac{\partial^2 Z}{\partial r^2} + \frac{1}{r}\frac{\partial Z}{\partial r} + \frac{1}{r^2}\frac{\partial^2 Z}{\partial \theta^2} \tag{1}$$

with $0 \le r \le a$ and $t \ge 0$ [6]. Here, a is the radius of membrane. By performing the separation variable method, the deflection of wave in polar coordinate $Z(r,\theta,t)$ showed as

$$Z(r,\theta,t) = J_m(k_{mn}r)(A_{mn}\cos m\theta + B_{mn}\sin m\theta)\cos\omega_{mn}t + J_m(k_{mn}r)(A^*_{mn}\cos m\theta + B^*_{mn}\sin m\theta)\sin\omega_{mn}t$$

If the boundary condition at $r = a$ is $Z(a,\theta,t) = 0$, the wave number can be deduced from zeros of approximation of Bessel function [7],

$$J_m(ka) \approx \sqrt{\frac{2}{\pi ka}}\cos\;(k_{mn}a \;\; (2m+1)\frac{\pi}{4}\;) = 0; \qquad k_{mn} = (4n+2m-1)\frac{\pi}{4a}\;; k_{mn} = \frac{\omega_{mn}}{c}$$

For the case of wave moving stationary with the tip banded in circular membrane [8], the initial deflection and the initial velocity are respectively given by

$$Z(r,\theta,0) = f(r,\theta) \quad \text{and} \quad \frac{\partial}{\partial t}[Z(r,\theta,0)] = g(r,\theta) \tag{2}$$

Then, by performing the separation variable method and taking into account boundary condition, the deflection can be represented as

$$\begin{aligned} Z_{mn}(r,\theta,t) = &\sum_{n=1}^{\infty} A_{0n}J_0(\frac{u_{0n}}{a}r)\cos\omega_{mn}t + \sum_{m=1}^{\infty}\sum_{n=1}^{\infty} J_m(\frac{u_{mn}}{a}r) \\ &(A_{mn}\cos m\theta + B_{mn}\sin m\theta)\cos\omega_{mn}t \\ &+ \sum_{n=1}^{\infty} A^*_{0n}J_0(\frac{u_{0n}}{a}r)\sin\omega_{mn}t + \sum_{m=1}^{\infty}\sum_{n=1}^{\infty} J_m(\frac{u_{mn}}{a}r) \\ &(A^*_{mn}\cos m\theta + B^*_{mn}\sin m\theta)\sin\omega_{mn}t \end{aligned} \tag{3}$$

The initial deflection function of 2D waves in circular membrane, $f(r,\theta)$, is calculated using the initial conditions, $Z(r,\theta,0) = f(r,\theta)$ and $\frac{\partial}{\partial t}[Z(r,\theta,0)] = 0$,

$$f(r,\theta)=\sum_{m=0}^{\infty}\sum_{n=1}^{\infty}J_m(\frac{u_{mn}}{a}r)(A_{mn}\cos m\theta+B_{mn}\sin m\theta)$$
$$=\sum_{n=1}^{\infty}A_{0n}J_0(\frac{u_{0n}}{a}r)+\sum_{m=1}^{\infty}\sum_{n=1}^{\infty}A_{mn}J_m(\frac{u_{mn}}{a}r)\cos m\theta \quad (4)$$
$$+\sum_{m=1}^{\infty}\sum_{n=1}^{\infty}B_{mn}J_m(\frac{u_{mn}}{a}r)\sin m\theta$$

b. Shaped 2D Wave in Circular Membrane Using Spreadsheets Excel

On the process of making the wave shaped in circular membrane using spreadsheet excel, first we need to draw the circular membrane by combining two parameters, the radius ($0 \le r \le a$) and angle ($0 \le \theta \le 360$). Fig. 1 shows spreadsheet display of (a) circles drawing at interval of 0.05 m and (b) the circular membrane including angle interval of 18°. Here, the radius is chosen for 1 m. Other parameters, then, can be calculated by using several assumptions displayed in spreadsheet excel shown in Fig. 2.

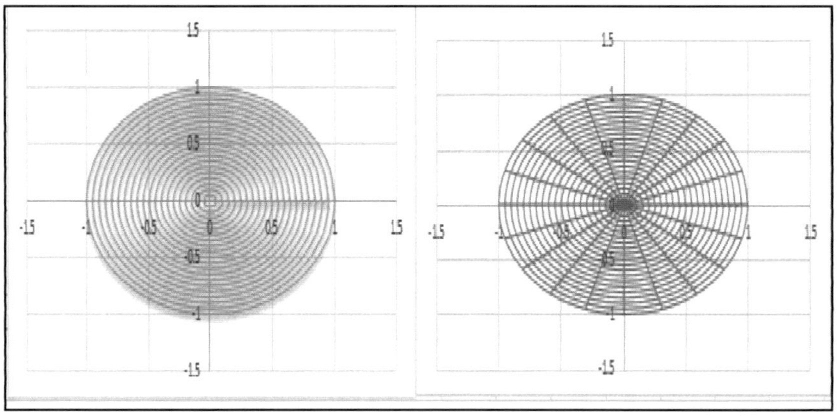

Figure 1. (a) The combination of circles with interval radius $0 \le r \le a$, a = *1m*, (b) The circular membrane in spreadsheet excel display.

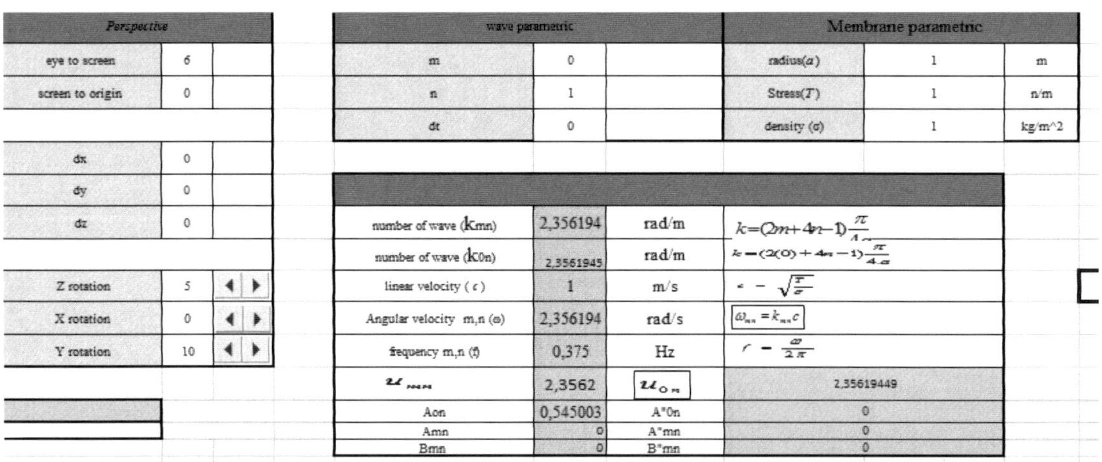

Perspective		
eye to screen	6	
screen to origin	0	
dx	0	
dy	0	
dz	0	
Z rotation	5	
X rotation	0	
Y rotation	10	

wave parametric			Membrane parametric		
m	0		radius(a)	1	m
n	1		Stress(T)	1	n/m
dt	0		density (σ)	1	kg/m^2

number of wave (kmn)	2,356194	rad/m	$k=(2m+4n-1)\frac{\pi}{4a}$
number of wave (k0n)	2,3561945	rad/m	$k=(2(0)+4n-1)\frac{\pi}{4a}$
linear velocity (c)	1	m/s	$c=\sqrt{\frac{T}{\sigma}}$
Angular velocity m,n (ω)	2,356194	rad/s	$\omega_{mn}=k_{mn}c$
frequency m,n (f)	0,375	Hz	$f=\frac{\omega}{2\pi}$
u_{mn}	2,3562	u_{0n}	2,35619449
Aon	0,545003	A"0n	0
Amn	0	A"mn	0
Bmn	0	B"mn	0

Figure 2. The proponent parametric of wave in circular membrane.

Furthermore, in order to show the coefficient of A_{0n}, A_{mn}, and B_{mn} in the spreadsheet excel cell, we must be consider the principle of Trapezoidal rule for performing the numerical integration

$$\int_a^b f(x)dx \approx \frac{\Delta x}{2}\{f(x_0)+2f(x_1)+2f(x_2)+...+2f(x_{n-1})+f(x_n)\} \quad (5)$$

This process are displayed in spreadsheet cells shown in Fig. 3.

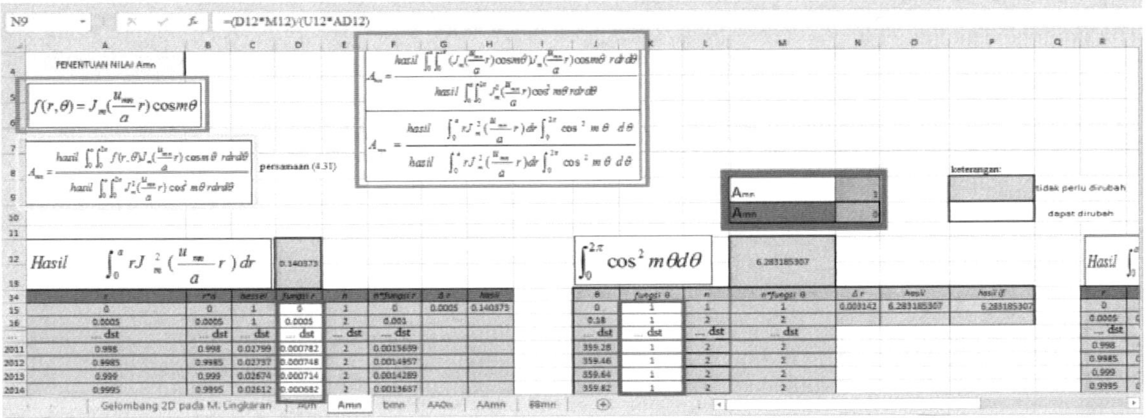

Figure 3. Calculation process of A_{0n} , A_{mn}, and B_{mn} coefficients shown in spreadsheet excel display.

c. *Application of Moving Shapes, Perspective and (XY) Scatter Chart*

Due to the limitation of *MS. Excel* in providing the 3D scattering of object, here the 2D or *XY Scatter Chart* function can be modified for moving shape and 3D perspective by applying the translation and rotation processes. The translation process is from (x,y,z) to $(x_{tr} = x + \Delta x,\ y_{Tr} = y + \Delta y, z_{Tr} = z + \Delta z)$. In cases rotation in z-axis, the vector components velocity becomes $\vec{v}_x = \hat{i} v_x \cos\theta + \hat{j} v_x \sin\theta$ $\vec{v}_y = \hat{j} v_y \cos\theta + \hat{i} v_y \sin\theta$ and $\vec{v}_z = \hat{k} v_z$ whereas rotation in y-axis is $X_{rot\_Y} = v_z \sin\theta + v_x \cos\theta$, $Y_{rot\_Y} = v_y$, $Z_{rot\_Y} = v_z \cos\theta + v_y \sin\theta$. The calculation process of moving shape in spreadsheet is shown in FIG. 4. The 3D object is look like 2D object illustrated using perspective application. An example of 2D circular wave plotted by using the *XY scatter chart* where m = 0 and n =1 is shown Fig. 5.

	L	M	N	O	P	Q	R	S	T	U	V	W
21	Tanslation			putar sumbu Z			putar sumbu X			Putar Sumbu Y		
22	x+Δx	y+Δy	z+Δz	X_ROT_Z	Y_ROT_Z	Z	X	Y_ROT_X	Z_ROT_X	X_ROT_Y	Y	Z_ROT_Y
23	1	0	0	1	0	0	1	0	0	1	0	0
24	0.951056516	0.309	0	0.951056516	0.309017	0	0.951056516	0.30901699	0	0.951056516	0.30901699	0
...	... dst	... dst	... dst	... dst	... dst	... dst	... dst	... dst	... dst	... dst	... dst	... dst
858	0.3	-7E-17	0.8789	0.3	-7.35E-17	0.87893486	0.3	-7.3509E-17	0.878934862	0.3	-7.3509E-17	0.878934862
859	0.25	-6E-17	0.9151	0.25	-6.13E-17	0.91511855	0.25	-6.1257E-17	0.915118551	0.25	-6.1257E-17	0.915118551
860	0.2	-5E-17	0.9452	0.2	-4.9E-17	0.94524926	0.2	-4.9006E-17	0.94524926	0.2	-4.9006E-17	0.94524926
861	0.15	-4E-17	0.969	0.15	-3.68E-17	0.96901491	0.15	-3.6754E-17	0.96901491	0.15	-3.6754E-17	0.96901491
862	0.1	-2E-17	0.9862	0.1	-2.45E-17	0.98616895	0.1	-2.4503E-17	0.986168954	0.1	-2.4503E-17	0.986168954
863	0.05	-1E-17	0.9965	0.05	-1.23E-17	0.99653323	0.05	-1.2251E-17	0.996533229	0.05	-1.2251E-17	0.996533229

Figure 4. The frame of moving shaped in spreadsheet.

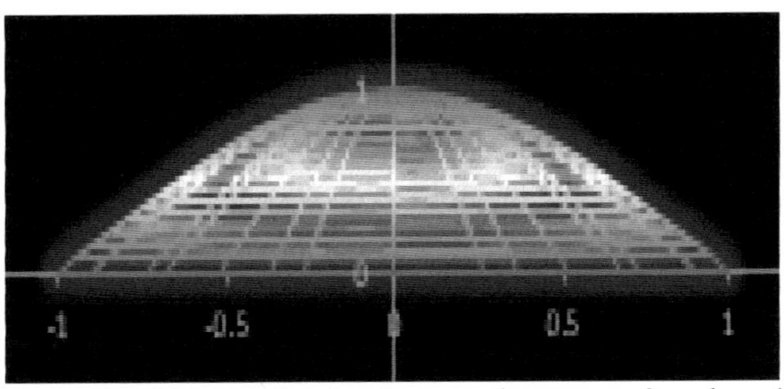

Figure 5. 2D wave in circular membrane m=0 and n=1.

d. Application of VBA Macro

VBA macro is a tool in MS. *Excel* used to move data with *action* button for *m, n, dt, Eye to screen (ES), Δx, Δy, Δz, rot_z, rot_x,* dan *rot_y*. The purpose of using this, is to move and to visualize those figures according to the real situation where they can be modified rapidly. The script program of 2D wave in circular membrane showed in following Table 1.

Table 1. VBA macro script of 2D circular membrane simulation.

Parameter	*Action Button*	*Action Script*
M	m_o	Private Sub m_o_Change() Range("m") = m_o.Value End Sub
N	n_o	Private Sub n_o_Change() Range("n") = n_o.Value End Sub
Dt	dt_o	Private Sub dt_o_Change() Range("dt") = dt_o.Value / 10 End Sub
rotation_z	z_rot	Private Sub z_rot _Change() If z_rot > 71 Then z_rot = 0 If z_rot < 0 Then z_rot = 71 Range("rot_z") = 5 * z_rot.Value End Sub
rotation_y	y_rot	Private Sub y_rot _Change() If y_rot > 71 Then y_rot = 0 If y_rot < 0 Then y_rot = 71 Range("rot_y") = 5 * y_rot.Value End Sub

3. Results and Discussions

The vibration mode is characterized by a modal frequency and shape. In general all modes have different frequencies and mode shapes. The low mode number have lower frequency. The node is the line on the structure that does not move (zero displacement) while the rest of the structure is vibrating. In a 2D system such as the circular membrane, each mode is entirely independent of all other modes. Some of the simulations of 2D wave in circular membrane using spreadsheet excel with several values of m (mode) and n (node) are represented in Table 2. The frequency of each mode is shown as a certain value multiplied to f_{01} (frequency of (0,1) vibration mode).

Table 2. Simulation results of circular membrane wave

m \ *n*	0	1	2	3
1	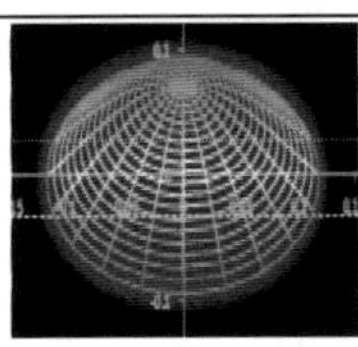 f_{01}	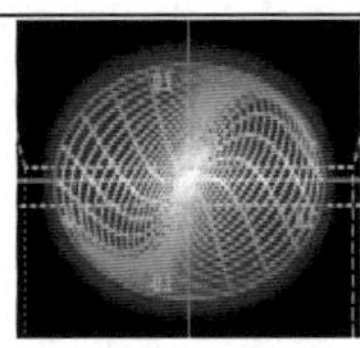 1.667 f_{01}	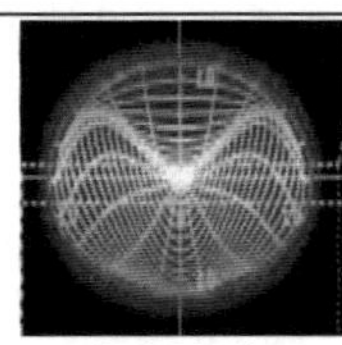 2.333 f_{01}	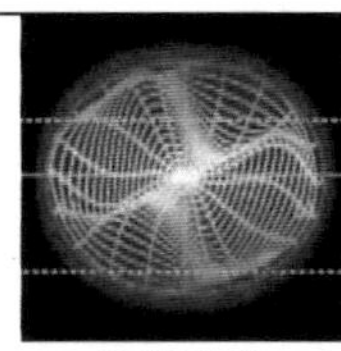 *3.000* f_{01}

2	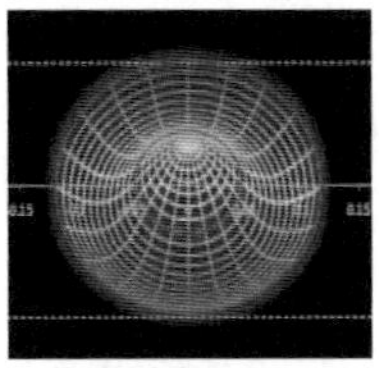 *2.333 f_{01}*	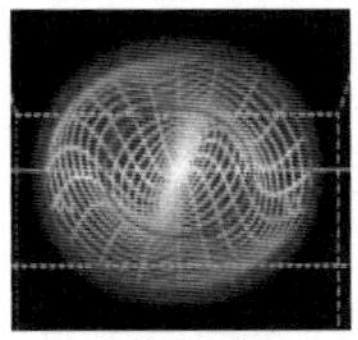 *3.000 f_{01}*	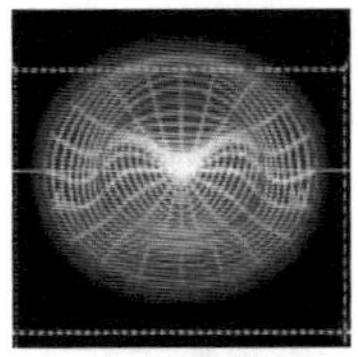 *3.667 f_{01}*	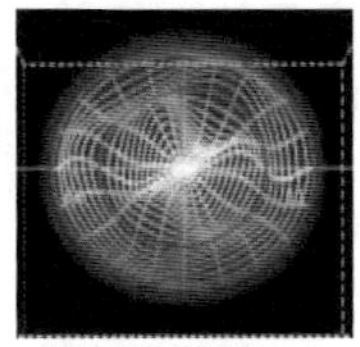 *4.333 f_{01}*
3	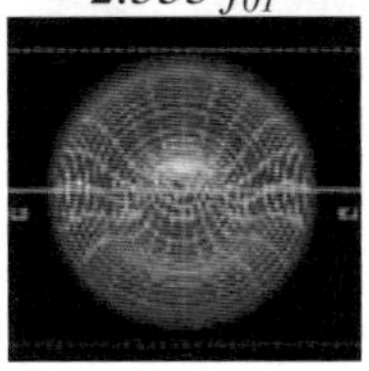 *3.667 f_{01}*	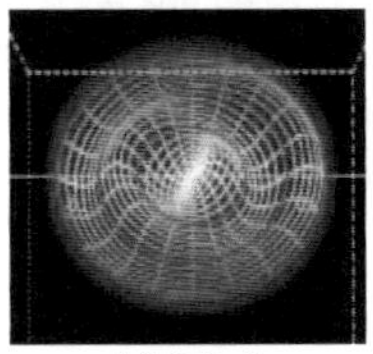 *4.333 f_{01}*	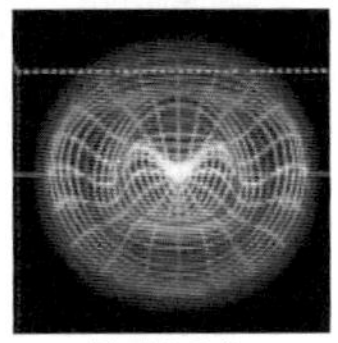 *5.000f_{01}*	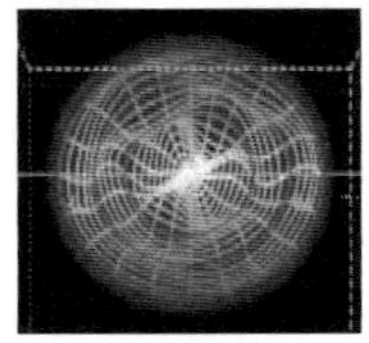 *5.667 f_{01}*

The frequency of vibration 2D wave in circular membrane in this research was calculated by using the Bessel approximation of the wave number. On the other side, the wave number is calculated from the roots of Bessel function [7]. Some conditions are given for verification of this simulation. In the case of m= 0 and n =1 or mode (0,1), the circular shape with no deflection occurs at the radius of $r_0 = 0.117$ m and frequency of 205 Hz. This result is similar to the experimental result observed from membrane diameter of 23.4 cm using sand beach under vibration at a specific frequency [5]. Using the same condition with experiment such as the radius and density of membrane, the simulation result can be verified and differentiation between them can be analyzed. In the simulation, the stress of 1000 N/m is introduced to the membrane. The membrane density is 0.2444 kg/m^2. For mode (0,2), the calculated frequency is 478 Hz where the circular shape formed at the radius of $r_0 = 0.117$ m and $r_1 = 0.0501$ m. The experimental result gave the circular shape at radius of r_0 0,117 m and $r_1 = 0.0525$ m (Fig. 6). The location of vibration mode is found a little bit different from resonance model in experiment but all of them have the same vibration pattern. The second way verify this simulation is to compare them with simulation results using MATLAB at the same conditions with zero initial velocity [9] (see in Fig. 7). A little bit difference result between them occur from different method in calculation of the wave number. In Matlab, the wave number were calculated from the root value of Bessel function.

	Mode (0,1)	Mode (0,2)	Mode (0,3)	Mode (2,2)	Mode (4,1)
Vibration mode by simulation MS. *Excel*	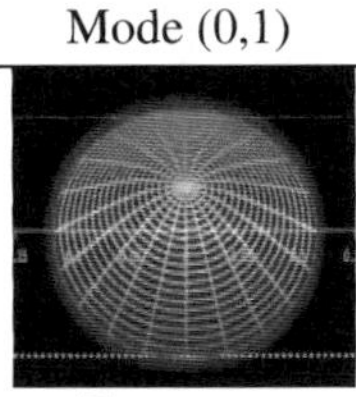	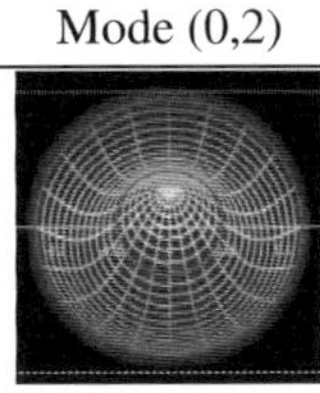	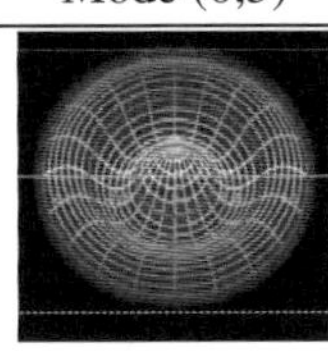	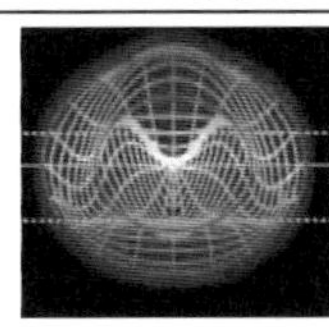	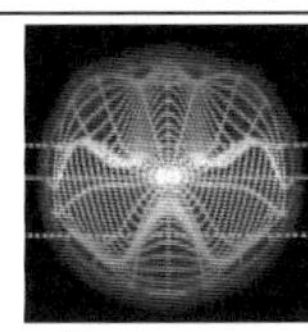
	Frequency result 205 Hz $r_0 = 0.117$ m	Frequency result 478 Hz $r_0 = 0.117$ m $r_1 = 0.0501$ m	Frequency result 752 Hz $r_0 = 0.117$ m $r_1 = 0.0744$ m $r_2 = 0.0319$ m	Frequency result 752 Hz $r_0 = 0.117$ m $r_1 = 0.0744$ m	Frequency result 752 Hz $r_0 = 0.117$ m

Figure 6. Vibration mode simulated by MS. Excel including frequency and circular node position.

IOP Conf. Series: Journal of Physics: Conf. Series **1011** (2018) 012088 doi:10.1088/1742-6596/1011/1/012088

	t = 0	*t=0.6 s*	*t=1.9 s*	*t= 2.5 s*	*t = 3.1 s*
Simulation with Matlab	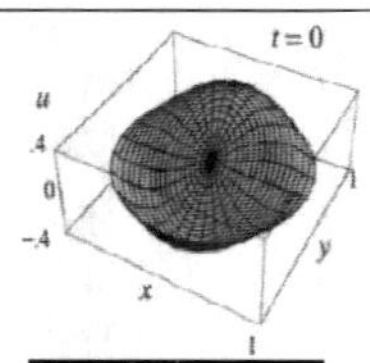	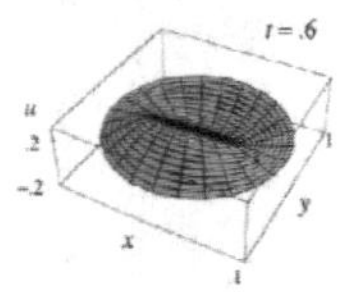	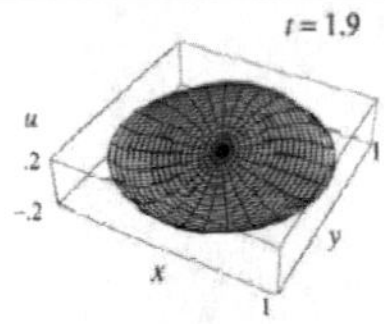	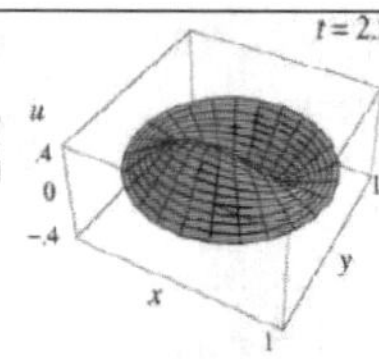	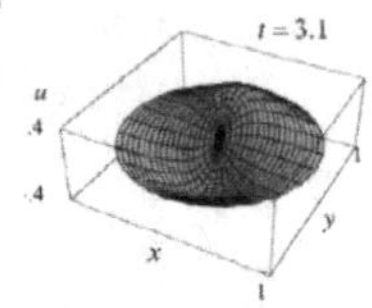
Simulation with spreadsheet *Excel*	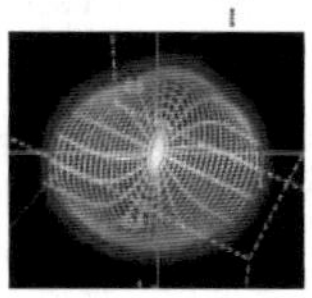	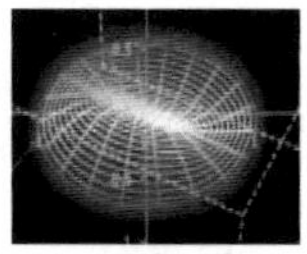	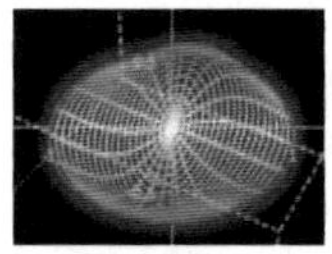	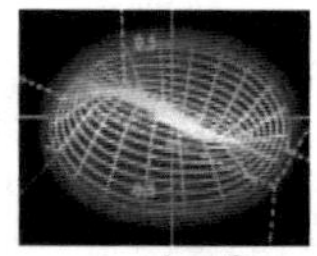	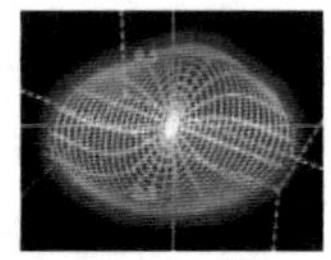
	peak of deflection 0.416044 m	peak of deflection 0.294188 m	peak of deflection 0.159213 m	peak of deflection 0.384375 m	peak of deflection 0.384375 m

Figure 7. Comparison of simulation result by using MS. Excel and Matlab.

4. Conclusion

Simulation of 2D wave in circular membrane has been created using the *spreadsheets* and *macro* VBA in *Ms Excel* with presence of initial deflection and initial zero velocity. The vibration modes have the same pattern compared to the resonance method in experiment with the membrane diameter of 23.4 cm. The peak of deflection in this simulation have also the same pattern compared to MATLAB simulation.

References

[1] Webb L 1993 J. *Physics Education 28* **2** p 77-82

[2] Cooke B A 1997 J. *Physics Education 32* **2** p 80-87

[3] Fowler M 2004 J. *Science & Education 13* (7-8) p 791-796.

[4] Benacka J 2010 J. Math. Ed. Sci. Technol. 41 **3** p 373-378

[5] Radiansyah H et al 2006 *The Analysis Of Circular Membrane Vibration Modes* (UNY: skripsi Jurusan Pendidikan Fisika)

[6] Ross S L 1984 *Differential Equations* (New York: John Wiley and Sons Inc)

[7] Boas L M 1983 *Mathematical Methods in the-Physical Sciences* (USA : John Wiley and Sons inc)

[8] Chaudhuri R N 2010 *Waves and Oscillations* (New Delhi : New Age International)

[9] Asmar N H 2005 *Partial Differential Equations with Fourier Series and Boundary Value Problems* (USA: *Pearson Prentice Hall*)

Institute of Physics
Dirac House, Temple Back
Bristol BS1 6BE UK

ISSN: 1742-6588
ISBN 978-1-5108-6288-3

CURRAN ASSOCIATES INC.
proceedings.com